CELL DEATH DURING
HIV INFECTION

CELL DEATH DURING
HIV INFECTION

edited by
Andrew D. Badley

Taylor & Francis
Taylor & Francis Group
Boca Raton   London   New York

A CRC title, part of the Taylor & Francis imprint, a member of the
Taylor & Francis Group, the academic division of T&F Informa plc.

Published in 2006 by
CRC Press
Taylor & Francis Group
6000 Broken Sound Parkway NW, Suite 300
Boca Raton, FL 33487-2742

International Standard Book Number-10: 0-8493-2827-6 (Hardcover)
International Standard Book Number-13: 978-0-8493-2827-5 (Hardcover)
Library of Congress Card Number 2005044006

Library of Congress Cataloging-in-Publication Data

Cell death during HIV infection / edited by Andrew D. Badley.
    p. cm.
  Includes bibliographical references and index.
  ISBN 0-8493-2827-6 (alk. paper)
   1. HIV infections. 2. Apoptosis. 3. Immunodeficiency. 4. T cells. I. Badley, Andrew D.

QR201.A37.C453 2005
616.97'9207--dc22
                                          2005044006

Taylor & Francis Group
is the Academic Division of Informa plc.

Visit the Taylor & Francis Web site at
http://www.taylorandfrancis.com

and the CRC Press Web site at
http://www.crcpress.com

Editor

Andrew D. Badley, M.D., is a consultant in the Division of Infectious Diseases, Department of Medicine, at the Mayo Clinic in Rochester, Minnesota. He is professor of medicine at the Mayo Clinic College of Medicine and associate director of the Mayo Clinic Translational Immunovirology and Biodefense Program, a National Institutes of Health Center of Excellence, as well. His duties include overseeing an active clinical practice, which focuses on the care of patients immunocompromised by human immunodeficiency virus infection, transplant recipients, and a research laboratory that currently is funded by the National Institutes of Health and private grants.

A Canadian by birth, Dr. Badley completed his undergraduate and medical studies at Dalhousie University in Halifax, Nova Scotia. After completing his internship in Halifax, he moved to the Mayo Clinic in Rochester, Minnesota, for his residency and clinical investigator fellowship. Following this fellowship, he returned to Canada for a position at Ottawa General Hospital that included clinical practice and basic research. In 2002, he returned to the Mayo Clinic as a consultant with his own laboratory. He actively mentors both clinical and research trainees who may be residents, fellows, or graduate students.

To date, he has published approximately 70 papers on human immunodeficiency virus, general infectious disease, infections in immunocompromised hosts, immunotherapy, and apoptosis dysregulation during infectious diseases. He has also been the principal or coinvestigator of numerous clinical trials.

Contributors

Andrew D. Badley, M.D.
Mayo Clinic College of Medicine
Rochester, Minnesota

Andreas Baur, M.D.
University of Miami School of Medicine
Department of Microbiology and Immunology
Miami, Florida

José A. M. Borghans, Ph.D.
Department of Immunology
University Medical Centre Utrecht
Utrecht, The Netherlands

Catherine Brenner, Ph.D.
Centre National de la Recherche Scientifique
Université de Versailles/St. Quentin
Versailles, France

David Camerini, Ph.D.
Department of Molecular Biology
  and Biochemistry
University of California
Irvine, California

Xian-Ming Chen, M.D.
Mayo Clinic College of Medicine
Rochester, Minnesota

Shailesh K. Choudhary, Ph.D.
Department of Molecular Biology
  and Biochemistry
University of California
Irvine, California

Luchino Y. Cohen, Ph.D.
Université de Montréal
Laboratoire d'Immunologie, CR-CHUM
Montréal, Quebec, Canada

Jacques Corbeil, Ph.D.
Université Laval
Quebec, Canada

Andrea Cossarizza, Ph.D.
Biomedical Sciences University
  of Modena
Modena, Italy

Demetre C. Daskalakis, M.D.
Massachusetts General Hospital
Division of Infectious Diseases
Boston, Massachusetts

Claude Desgranges, Ph.D.
Centre National de la Recherche Scientifique
UFR Biomédical
Paris, France

Marie-Lise Dion
Université de Montréal
Laboratoire d'Immunologie, CR-CHUM
Montréal, Quebec, Canada

Dara Ditsworth, M.S.
Abramson Family Cancer Research Institute
University of Pennsylvania
Philadelphia, Pennsylvania

David H. Dockrell, M.D.
Division of Genomic Medicine
University of Sheffield Medical School
Sheffield, United Kingdom

Gilad Doitsh, Ph.D.
Gladstone Institute of Virology
  and Immunology
University of California
San Francisco, California

Rebecca L. Elstrom
Abramson Family Cancer Research Institute
University of Pennsylvania
Philadelphia, Pennsylvania

Jérôme Estaquier, Ph.D
Unite de Physiopathologie des Infections
  Lentivirales
Institute Pasteur
Paris, France

Gregory J. Gores, M.D.
Mayo Clinic College of Medicine
Rochester, Minnesota

Daniel B. Graham, Ph.D.
Mayo Clinic College of Medicine
Rochester, Minnesota

Walter C. Greene, M.D., Ph.D.
Gladstone Institute of Virology
  and Immunology
University of California
San Francisco, California

Maria Eugenia Guicciardi, Ph.D.
Mayo Clinic College of Medicine
Rochester, Minnesota

Mette D. Hazenberg, M.D., Ph.D.
Gladstone Institute of Virology
  and Immunology
University of California
San Francisco, California

Georges Herbein, M.D.
Department of Virology
Université de Franche-Compte
Besançon, France

Gareth Jones, Ph.D.
Department of Microbiology
  and Infectious Diseases
University of Calgary
Calgary, Alberta, Canada

Scott H. Kaufmann, M.D., Ph.D.
Mayo Clinic College of Medicine
Rochester, Minnesota

Laurene M. Kelly
Research Institute for Genetic
  and Human Therapy
IRCCS Policlinic S. Matteo
Pavia, Italy

Jason F. Kreisberg, B.S.
Gladstone Institute of Virology
  and Immunology
University of California
San Francisco, California

Guido Kroemer, M.D., Ph.D.
Institut Gustave Roussy
Centre National de la Recherche
  Scientifique
Villejuif, France

Nicholas F. LaRusso, M.D.
Mayo Clinic College of Medicine
Rochester, Minnesota

Christophe Lemaire, Ph.D.
Centre National de la Recherche Scientifique
Université de Versailles/St. Quentin
Versailles, France

Michael J. Lenardo, M.D.
Laboratory of Immunology
National Institutes of Allergy
  and Infectious Diseases
National Institutes of Health
Bethesda, Maryland

Julianna Lisziewicz
Research Institute for Genetic
  and Human Therapy
IRCCS Policlinic S. Matteo
Pavia, Italy

Franco Lori, M.D.
Research Institute for Genetic
  and Human Therapy
IRCCS Policlinic S. Matteo
Pavia, Italy

Julain J. Lum, Ph.D.
Abramson Family Cancer Research Institute
University of Pennsylvania
Philadelphia, Pennsylvania

David J. McKean, Ph.D.
Mayo Clinic College of Medicine
Rochester, Minnesota

David R. McNamara, M.D.
Mayo Clinic College of Medicine
Rochester, Minnesota

Xue Wei Meng, M.D., Ph.D.
Mayo Clinic College of Medicine
Rochester, Minnesota

Frank Miedema, Ph.D.
Department of Immunology
University Medical Centre Utrecht
Utrecht, The Netherlands

Sylviane Muller, Ph.D.
Institut de Biologie Moleculaire et
  Cellulaire
Centre National de la Recherche
  Scientifique
Strasbourg, France

Zilin Nie, M.D.
Mayo Clinic College of Medicine
Rochester, Minnesota

David Nolan, M.D.
Center for Clinical Immunology
  and Biomedical Statistics
Royal Perth Hospital and Murdoch University
Perth, Western Australia

Giuseppe Nunnari, M.D.
Thomas Jefferson University
Institute for Human Virology and Biodefense
Philadelphia, Pennsylvania

Savita Pahwa, M.D.
University of Miami School of Medicine
Department of Microbiology and Immunology
Miami, Florida

Jean-Luc Perfettini, Ph.D.
Institut Gustave Roussy
Centre National de la Recherche
  Scientifique
Villejuif, France

Barbara N. Phenix, Ph.D.
North Shore LIJ Research Institute
Manhasset, New York

Michael J. Pinkoski, Ph.D.
Apoptosis Research Centre
Children's Hospital of Eastern Ontario
Ottawa, Ontario, Canada

Eric M. Poeschla, M.D.
Mayo Clinic College of Medicine
Rochester, Minnesota

Roger J. Pomerantz, M.D.
Thomas Jefferson University
Institute for Human Virology
  and Biodefense
Philadelphia, Pennsylvania

Christopher Power, M.D.
Department of Clinical Neurosciences
University of Calgary
Calgary, Alberta, Canada

Eric S. Rosenberg, M.D.
Massachusetts General Hospital
Division of Infectious Diseases
Boston, Massachusetts

Keiko Sakai, Ph.D.
National Institutes of Allergy and Infectious
  Diseases
National Institutes of Health
Bethesda, Maryland

David Schnepple, M.S.
Mayo Clinic College of Medicine
Rochester, Minnesota

Rafick-Pierre Sekaly, M.D.
Université de Montréal
Laboratoire d'Immunologie, CR-CHUM
Montréal, Quebec, Canada

Matthew Smith-Raska
Laboratory of Immunology
National Institutes of Allergy
  and Infectious Diseases
National Institutes of Health
Bethesda, Mary land

Craig B. Thompson, M.D.
Abramson Family Cancer Research
  Institute
University of Pennsylvania
Philadelphia, Pennsylvania

Anne J. Tunbridge, M.B.
Division of Genomic Medicine
University of Sheffield Medical School
Sheffield, United Kingdom

Stacey R. Vlahakis, M.D.
Mayo Clinic College of Medicine
Rochester, Minnesota

Nienke Vrisekoop
Department of Clinical Viro-Immunology
Laboratory of the Academic Medical Center
  Amsterdam
University of Amsterdam
Amsterdam, The Netherlands

Sara Warren, B.S.
Mayo Clinic College of Medicine
Rochester, Minnesota

John Zaunders, Ph.D.
Centre for Immunology
St. Vincent's Hospital
Darlington, New South Wales,
  Australia

Introduction

Establishing that the human immunodeficiency virus (HIV) is the etiologic agent of acquired immunodeficiency syndrome (AIDS) was the beginning of a revolution in science. Since then, the advances made in retrovirology have led to major advances in therapy for HIV, which, in turn, have allowed thousands worldwide to live longer lives. The advances in immunologic sciences have been rapid as well, and within a few short years, those advances will likely translate into immune-based therapies for use in clinical practice. In comparison, the study of apoptosis dysregulation during HIV-induced immunodeficiency has been less intense. Nevertheless, important lessons have been learned. Most scientists and clinicians now accept that cell death is not a passive process but, rather, a tightly coordinated and regulated one. As such, apoptotic cell death may be amenable to therapeutic intervention. In non-HIV disease states, clinical trials of apoptosis modifying agents used for HIV disease are already underway.

In HIV disease, enhanced understanding of the apoptotic pathways that are involved may lead to novel treatment strategies. It is apparent, however, that multiple, likely simultaneous, processes contribute to the enhanced CD4 T cell apoptosis, which together contribute to immune deficiency. In all likelihood, apoptotic dysregulations are cell-type dependent, because those cells that become latent reservoirs for HIV, by definition, do not succumb to the proapoptotic effects of infection. Such complexities offer multiple avenues for intervention: antiapoptotic approaches to limit HIV-induced T cell depletion and proapoptotic approaches designed to eradicate latent infections.

When I was approached by Taylor & Francis regarding this project, I envisioned soliciting coauthors who were conducting the most current and exciting bench science in apoptosis and HIV. I have taken their work and attempted to form a volume that may be used both as a reference source and as an incubator to stimulate future work. Thus, the book is divided into four sections: Basic Concepts, Mechanisms of HIV-Associated Cell Death, Clinical Consequences of HIV-Induced Cell Death, and Therapeutic Issues. Section 2, Mechanisms of HIV-Associated Cell Death, has three component parts: Viral Factors, Immune Mechanisms, and Infected vs. Uninfected Death.

Each of the authors in this book is a leader in his or her respective field. Their efforts ensure that each chapter meets the highest standards of scientific scholarship. I gratefully recognize those efforts and am sure that they will be appreciated by readers as well. I am indebted to them all.

I am equally indebted to the contributions of my many collaborators, colleagues, and past, present, and future members of my laboratory who were, over the years, a source of personal and professional inspiration and motivation. Early in my medical career, I was involved in the care of an HIV-positive patient at a time when no effective therapies existed and none were in development. The sense of hopelessness then and, again, more recently in the era of antiretroviral resistance was a recurring source of motivation to increase my understanding of the disease. To these patients and the numerous other patients who volunteered for clinical- and laboratory-based studies, I offer my sincere thanks and best wishes.

I also wish to thank Mark McClees and Carrie Rogness, from Mayo Clinic's Division of Infectious Diseases, for their efforts in assembling this edition and working with the contributing authors and publisher. Without their efforts, this volume would not be in print.

Finally, I acknowledge the support and love of my parents, my wife Nanci, and my children, Lauren, Andrew, Adrianne, and Caitlin. Their patience and understanding have allowed me to pursue my passions and obsessions; to each, I am eternally indebted.

Andrew D. Badley
Rochester, Minnesota

Table of Contents

Section 4
Therapeutic Issues

Section I

Basic Concepts

1 Lentiviral Biology and Cell Death

Eric M. Poeschla

CONTENTS

INTRODUCTION

The Lentivirus genus was named to connote the slow and inexorable progression of the degenerative diseases its members cause in a number of mammalian species. The median time to progression to acquired immunodeficiency syndrome (AIDS) after primary HIV (human immunodeficiency virus)-1 infection in humans, for example, is approximately 10 years. In other instances, however, a state of mutually benign accommodation seems to evolve, in which lentiviruses persist and replicate while causing no ill effects in host animals. The HIV-1 pandemic illustrates the pathogenicity that may ensue when a lentivirus has recently undergone cross-species transmission and remains unchecked by evolutionary coadaptation between parasite and host. Another biologically central issue, one that the prefix *lenti* belies, is the view we now have of HIV-1 replication and turnover in the body, where the process is anything but slow. Rather, a highly dynamic replication process is apparent, and the amount of HIV-1 circulating in plasma correlates very well with disease progression. This chapter will present an overview of these and other selected aspects of HIV-1 virology. The intent is to introduce the basic molecular biology of HIV-1 within a comparative lentiviral framework as a basis for considering the complex question that is the theme of the succeeding chapters in this volume: how do cells die from HIV-1 infection?

LENTIVIRAL BIOLOGY AND PATHOGENESIS

Lentiviruses were the first retroviruses associated with disease and were among the first filterable disease agents identified.[1] Three groups of lentiviruses infect primates, ungulates, and felines,

TABLE 1.1
Lentiviruses: Classification and Disease

Virus	Host	Cell Tropism	Disease Features
Primate			
HIV-1	Humans	Macrophages, CD4+ T cells	CD4+ T cell depletion, dementia, wasting
HIV-2	Humans	Macrophages, CD4+ T cells	CD4+ T cell depletion, dementia, wasting; less pathogenic and transmissible than HIV-1
SIVcpz	Chimpanzee; ancestral to HIV-1	Macrophages, CD4+ T cells	None
SIVmac	Macaques in captivity	Macrophages, CD4+ T cells	CD4+ T cell depletion, dementia, wasting
SIVsm	Sootey mangabey; ancestral to HIV-2	Macrophages, CD4+ T cells	None
Ungulate			
EIAV	Horses	Macrophages	Anemia, wasting
Maedi-visna	Sheep	Macrophages	Encephalitis, lung disease, mastitis, wasting
CAEV	Goats	Macrophages	Encephalitis, arthritis, mastitis, wasting
BIV, Jembrana	Cattle	Macrophages	Probably none in *Bos taurus* (BIV), anemia and wasting in *Bos javanicus* (Jembrana)
Feline			
FIV	Feline species	Macrophages, CD4+ T, CD8+ T, and B cells	CD4+ T cell depletion–AIDS in domestic cat; asymptomatic in numerous other Felidae

respectively (Table 1.1). The term "slow virus" was first applied by Bjorn Sigurdsson in the 1950s during his studies of a maedi-visna virus (MVV) epidemic that illustrates well the potentially severe effects of introducing lentiviruses into naive populations or species.[2] Importation of asymptomatic sheep from the European continent to Iceland in 1933 led to the subsequent death of more than 100,000 Icelandic sheep over several decades from MVV, which causes a pneumonic disease (*maedi* in Icelandic)[3] and chronic encephalitis (visna).[4,5] Both equine infectious anemia virus (EIAV), the disease agent described by Vallée and Carré in 1904,[1] and MVV were studied as model slow viruses before primate lentiviruses were discovered.[2–10] Feline immunodeficiency virus (FIV) and bovine lentiviruses (bovine immunodeficiency virus [BIV] and Jembrana disease virus) were isolated and characterized in the post-AIDS era.[11–13]

Lentiviruses differ in a number of respects from other retroviruses. Among the most striking difference is their ability to replicate in terminally differentiated, nondividing cells.[14,15] Recognition of this signature property inspired development of replication-defective vectors capable of permanent transgene integration in diverse nondividing cells of relevance to gene therapy.[16–18] All lentiviruses infect nondividing tissue macrophages, which are the principal reservoirs *in vivo* for the ungulate lentiviruses. For example, the primary pathology caused by EIAV is a cyclical hemolytic anemia caused by antigen–antibody complexes that bind to the surfaces of erythrocytes; however, the primary EIAV producer cell is the macrophage.[19,20] Lentiviral tropism is determined by receptor utilization. (See early events below.) FIV and the primate lentiviruses have evolved additional tropisms for lymphocytes. FIV has the broadest tropism, as it infects B cells and CD8+ T cells in addition to macrophages and CD4+ T cells.[21] HIV-1 infects CD4+ T cells and macrophages and glial cells, although a variety of other cell types have been reported to be infected to a lesser extent *in vivo*.

THE ORIGINS OF HIV-1 AND THE EMERGENCE OF LENTIVIRAL DISEASE IN PRIMATES

Each lentivirus has a rather narrow host range.[22] Infrequent interspecies transmissions give rise to new viruses and diseases. This potential is best illustrated by the origins of HIV-1 and HIV-2 in separate transmissions of ancestral nonhuman primate lentiviruses to humans.[23,24] HIV-1 and HIV-2 each seem to have arisen several times.[24–26] The three distinct genetic groups of HIV-1 (M, N, and O) resulted from independent cross-species transmission events.[27] The best estimates from maximum-likelihood phylogenetic methods place the last common ancestor of the M ("main") group before 1940.[25,26] Severe pathology (AIDS) was also observed when simian lentiviruses, which do not cause evident disease in their Old World host species, infected Asian macaques in captivity.[28] Similarly, FIV infects many large feline species throughout the world, but only domestic cats develop clearly recognizable disease, which is severe and closely mimics human AIDS clinically and virologically.[29,30]

REPLICATION AND PATHOGENESIS: A COMPARATIVE LENTIVIRAL PERSPECTIVE

Analyses of temporal and quantitative aspects of HIV-1 ribonucleic acid (RNA) in plasma before and after the institution of highly active antiretroviral therapy (HAART) have led to the realization that HIV-1 replication is highly dynamic *in vivo*, supplanting earlier notions that too little virus existed to account for the severity of the disease.[31–35] Plasma viral RNA will decrease by approximately 2 log units within 2 weeks of instituting HAART. Various estimates derived from such studies place the half-life of HIV-1 virions *in vivo* on the order of minutes to a few hours.[31–33] The half-life of productively infected cells is also very short, approximately a day or two. Most plasma virus is produced quite recently.[31 33] The relationship of viral load to HIV-1 to prognosis has been made clear in numerous studies.[36,37] Individual patients reliably establish a particular viral "set point," in which the plasma viral load remains fairly consistent over time within half a log unit after the resolution of the primary infection. Viral loads can exceed 10^7 RNA copies per ml of plasma during primary infection; set point values average between 10^4 and 10^5 but vary an order of magnitude or more in each direction. A single measurement of plasma RNA before treatment reliably reflects the steady state and also strongly predicts the rate of CD4+ lymphocyte depletion and progression to AIDS and death.[36,37] Regardless of the starting level, however, viremia can be reliably suppressed to undetectable (<50 copies per ml by standard assays) levels when effective HAART regimens are administered to individuals with drug-sensitive viral isolates.

The studies showing high viremia and cell turnover led to a dynamic model of HIV-1 pathogenesis in which CD4+ T cell destruction occurs at a rate that exceeds host compensatory mechanisms.[32] As such, replication may drive pathogenesis. A review of the many proposed mechanisms of immunopathology in HIV disease is beyond the scope of this chapter and will be the focus of others in this volume. However, in addition to direct and indirect means of CD4+ T cell destruction, there is evidence for a variety of other mechanisms. Decreased lymphocyte proliferation, probably from inhibition of thymic and bone marrow function, and destruction of lymphoid tissue architecture are likely to contribute to the progressive CD4+ T cell depletion characteristic of late-stage HIV disease.[38–44] The concept of a maladaptive host response has also drawn support from primate lentivirus models. A conceptual challenge to the "replication drives pathogenesis" rule is that it does not hold true for a number of other primate lentiviruses. In fact, no AIDS-like disease is caused by any nonhuman primate lentivirus in its natural host.[27] It is only with transmission of simian immunodeficiency viruses (SIVs) to nonnatural primate hosts, such as rhesus macaques, that CD4+ T cell depletion and immune deficiency occur.

By analogy, HIV-1 is a zoonosis comparable to SIV in rhesus macaques and to FIV in domestic cats.[27] SIVsm, which is prevalent in sooty mangabeys both in the wild and in captivity,[45] continues to intrigue and attract research interest because it strikingly illustrates a disconnect between persistent viral replication and disease. These animals have chronic high-level viremia—the levels are essentially the same as those observed in human HIV-1 disease, and the cellular substrates are the same

(CD4[+] T cells and macrophages)—yet lymphocyte depletion and AIDS are not observed.[46–49] A similar situation exists in African green monkeys that remain healthy despite chronic replication of SIVagm to more than 10^6 copies per ml in more than 50% of these animals in the wild.[50] Notably, SIVagm causes AIDS in Asian pig-tailed macaques when they are experimentally infected.[51]

Cross-sectional analyses of SIVsm-infected and uninfected sooty mangabeys showed that infected animals experience much less aberrant generalized immune activation and bystander T cell apoptosis than do HIV-1-infected humans or SIVmac-infected macaques.[49] The animals display preserved lymph node architecture as well as bone marrow and thymic function and mount comparatively little in the way of a proinflammatory cellular immune response or specific cytotoxic T lymphocyte (CTL) responses.[49,52] It has been argued that the attenuated cellular immune responses these animals display from primary infection onward suggest that their innate pathogen recognition mechanisms do not recognize SIVsm as a "dangerous" pathogen and support indirect mechanisms of lymphocyte depletion in primate AIDS.[49]

These observations have yet to be fit into a consensus pathogenesis model, but they suggest that the response of the host to primate lentiviral infection can be as problematic as direct viral destruction of T cells.

The nonprimate lentiviruses also illuminate the spectrum of chronic disease outcomes.[20] For example, most EIAV-infected animals survive and progress from a 1- to 12-month chronic phase characterized by recurring bouts of viremia, fever, and hemolytic anemia to an extended asymptomatic stage.[19] The chronic stage is characterized by very high titers of circulating cell-free virus. The virus also may impair the differentiation of erythroid precursors.[20,53] In the generally asymptomatic carrier state that follows the chronic phase, only quite low levels of EIAV replication can be detected.[54]

MVV and caprine arthritis-encephalitis virus (CAEV) replicate preferentially in macrophages rather than in circulating leukocytes, resulting in little cell-free viremia.[55–57] Their genetic diversity suggests that they undergo considerable replication chronically.[20] The arthritis seen in CAEV-infected goats resembles human rheumatoid arthritis in many respects.[58]

VIRAL LIFE CYCLE

Lentiviruses have a characteristic conical core structure that distinguishes them from type C retroviruses. In addition to Gag, Pol, and Env, three to six additional proteins with diverse regulatory roles contribute to their genetic complexity. HIV-1 encodes six such proteins: Tat, Rev, Nef, Vif, Vpr, and Vpu.

EARLY EVENTS

The main molecular events of HIV-1 (Figure 1.1) entry are now well understood, although the role of the viral envelope protein in cell death continues to be enigmatic.[59,60] The initial event in virus entry is the binding of the processed viral envelope glycoprotein (gp120) to the CD4 molecule on the cell surface. However, for years, it was known that introduction of human CD4 into nonhuman cells was insufficient to confer entry. The implicated second receptor was determined to be a group of chemokine co-receptors in 1996—CCR5 and CXCR4 predominated.[61] Subsequently, great structural detail has been revealed. After CD4 binding, a conformational shift seems to occur within gp120, which exposes a chemokine co-receptor binding site.[62] The dominant co-receptors used by HIV-1 are CXCR4 and CCR5. CXCR4-using (or X4) viruses efficiently infect T cells and are also referred to as T cell tropic. CCR5 (or R5) viruses efficiently infect cells of the monocyte/macrophage lineage and are referred to as macrophage tropic.

The limited exposure of the key surfaces within the critical intermediate structures that occur during gp120 binding and fusion is a main reason for the failure to generate neutralizing antibodies

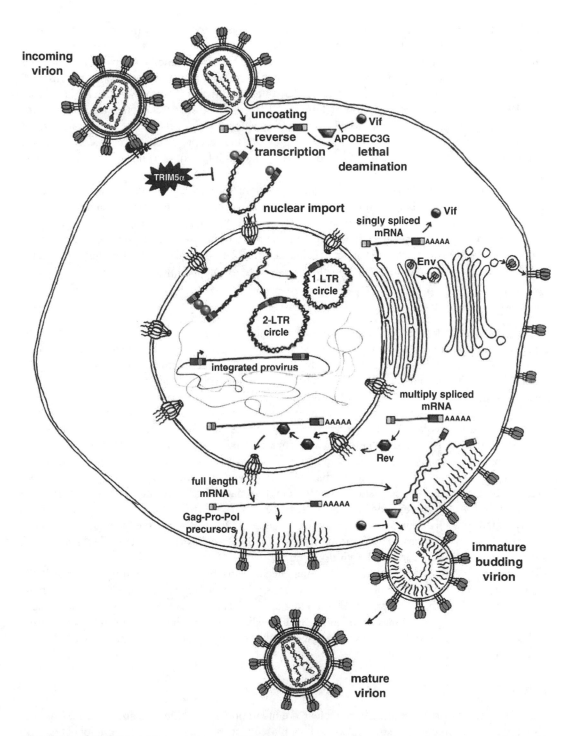

FIGURE 1.1 The HIV-1 life cycle. Entry is illustrated in a temporal sequence beginning at upper left, with assembly and budding at the lower right. The pre-integration complex (PIC) is depicted as a U-shaped intermediate with integrase protein molecules, presumably multimers, at the termini. TRIM5alpha[81] acts in the target cell via interaction with the capsid protein. In contrast, APOBEC3G is incorporated in the producer cell and subsequently appears to cause deamination of the minus strand DNA at C residues, leading to G to A hypermutation. Viral protease-mediated maturation of the core is synchronous with budding. (Drawing by Dyana T. Saenz.)

capable of controlling the infection.[63] Binding to the co-receptor induces further conformational changes within gp120 that dislocate it from gp41 and expose the gp41 fusogenic domain,[64] an amino-terminal hydrophobic peptide that inserts into the cell membrane. The gp41 then collapses into a trimer-of-hairpins structure that brings the two membranes close and mediates fusion.[65] Chemokine co-receptor usage seems to be fundamental to lentiviral biology, because FIV also uses CXCR4.[66,67] The recent identification of feline CD134[68] as the primary receptor rather than feline CD4 (or CD9[69]) explains the additional tropism of FIV for B cells and CD8 cells. Humans with a homozygous mutation of the CCR5 gene that prevents the receptor from being presented at the cell surface are less likely to be infected by HIV, and if infected, they progress to AIDS less rapidly.[70–72]

Cell–cell fusion mediated by gp120 (syncytium formation) and cell surface signal transduction triggered by gp120 can be shown to cause cell death *in vitro*; their roles *in vivo* remain unclear.[73] It is clear, however, that inhibition of entry can be used therapeutically. T-20 (enfurtivide) is an FDA-approved synthetic peptide that targets the amino-terminal region of gp41, which becomes exposed in the transient extended state.[74] This drug must be combined with other antiviral agents targeted at different stages of the life cycle or resistance develops rapidly.

Uncoating and Transport to the Nucleus

The uncoating, trafficking, and maturation steps that follow the entry of the viral core into the target cell cytoplasm constitute a poorly understood interval in the HIV-1 life cycle. HIV-1 seems to hijack a microtubule-dependent motor, dynein, to travel to the vicinity of the nucleus.[75] In this respect, it resembles other viruses.[76–79] However, we know little regarding the details of HIV-1 uncoating and the structure of the preintegration complex (PIC) as it evolves during trafficking. During this phase, there is a complex interplay between cellular restriction factors[80–82] and cellular factors that are exploited by the virus.[83] Recently, a viral defense mechanism that protects the integrity of the genome against a cellular antiretroviral activity was identified.[84] The virally encoded Vif (virion infectivity factor) protein, long known to enhance infectivity in a manner that was intriguingly cell-type dependent and to display a phenotype dependent on the producer cell rather than on the target cell, has been shown to promote degradation of the cellular protein APOBEC3G by targeting it to the proteasome.[85–91] In the absence of Vif, APOBEC3G, a cytidine deaminase, is incorporated into virion particles in the producer cell. APOBEC3G then mediates deamination of deoxycytidine to deoxyuridine within the minus-strand cDNA in the target cell, thus resulting in lethal G to A hypermutation of the viral genome.

Controversy persists about the underlying mechanisms by which lentiviral PICs transit the nucleopore complexes of mitotically inactive cells.[92] Proposed effectors have included peptide determinants in matrix,[93–100] Vpr,[96,101,102] and integrase.[98,103] The central DNA flap[104–106] has also been implicated.[104,105] Importin-7 has been suggested to mediate PIC translocation in semipermeabilized cell assays.[107] Countervailing views or reassessments exist for some of these proposals.[92,108–111,112] (See references[111,112] for reviews.)

Integration

The determinants of subsequent intranuclear preintegration PIC trafficking are not understood at all. However, its outcome is now known to be a biased distribution of HIV-1 integration sites, with actively transcribed genes being favored.[113,114] How cellular factors govern this important selectivity is unknown. In addition, little is known regarding how cellular proteins participate in the HIV-1 integration reaction as it occurs in cells.[115]

The linear viral cDNA that results from reverse transcription has four principal fates: bona fide integration; formation of 1-LTR circles, probably by homologous recombination[116]; conversion to 2-LTR circles by the host cell nonhomologous DNA end-joining (NHEJ) pathway[117]; and autointegration to form defective DNAs.[118] Only the linear form is a precursor to integration. Details of integrase catalysis have been reviewed elsewhere.[119]

There is some evidence that integration intermediates might trigger cell death under some circumstances (e.g., when DNA repair mechanisms are impaired). Early work by Temin and colleagues suggested that transient cytopathic effects correlated temporally with the transient accumulation of large numbers of unintegrated spleen necrosis virus DNA molecules.[120,121] More recently, the nonhomologous end joining (NHEJ) pathway has been implicated in completing the final stages of integration, and these studies have raised the possibility that integration intermediates can be toxic.[122] The NHEJ pathway is the principal repair system for double-stranded DNA breaks in eukaryotic cells. Such breaks occur during lymphocyte V(D)J recombination. They are also generated by ionizing radiation and are proapoptotic: just a few double-strand DNA breaks can induce p53-dependent G1 arrest.[123] The principal components of this pathway are DNA-dependent protein kinase (DNA-PK), Ku, XRCC4 ligase, and DNA ligase IV.[124,125] DNA-PK, which is activated by dsDNA ends, participates in this repair along with the Ku proteins, although the exact role played by DNA-PK has not been established. This pathway is currently understood to repair double-strand gaps and not single-strand gaps, which is of relevance to the following discussion.[125] Severe combined immunodeficient (SCID) mice have a truncation mutation in both alleles of DNA-PKCS, so they carry out the initial double-strand cleavage steps of V(D)J recombination but cannot repair the resulting double-strand breaks.[126] This defect results in an absence of mature B and T cells and greater susceptibility to spontaneous malignancy.

Retroviral reverse transcription also generates a double-stranded DNA with blunt ends. Before integration, two nucleotides are removed by integrase from each 3′ strand of the linear molecule. The recessed 3′ ends are then joined (the strand transfer reaction) to 5′ DNA ends in cellular DNA via a direct attack by the 3′ hydroxyl group.[119]

Daniel et al. found that retroviral vector integration was reduced in SCID cells and that high-titer infection induced apoptosis within approximately 12 h of infection.[122] In contrast, integration-defective vectors did not trigger cell death. They concluded that DNA-PK is required to complete integration, presumably by repairing the DNA gaps. Subsequent reports have led to refinement and some contradiction of this scenario.[117,127,128] In one study, increased cell death was also observed in SCID cells, but only after transduction at quite high multiplicity of infection (MOI) and subsequent cell division.[127] At an MOI of less than 1.0, transduction efficiency was actually higher in SCID cells than in control cells, and it proceeded normally *in vivo* in SCID mice brain without cell death, leading to the conclusion that DNA-PK and another DNA repair enzyme, PARP, are not essential for HIV-1 integration.[127] Similarly, Li et al. found that lower-titer HIV-1 vector infection proceeded normally in cells mutant for components of the NHEJ pathway.[117] Similar to other studies, high-titer infections were found to be toxic in NHEJ-mutant cells, apparently by an apoptotic mechanism. However, a key difference is that integrase mutant viruses were just as toxic, suggesting that it is unintegrated viral DNA ends that act as the apoptotic signal rather than unrepaired viral–cellular junctions. Unintegrated DNA was also found to be a substrate for the NHEJ pathway in that it was required for 2-LTR circle formation.[117]

The roles of DNA-PK and other cellular repair mechanisms in integration and cell death remain uncertain. Its relevance to the actual disease pathology of lentiviruses is not clear, because normal human CD4+ T cells are fully competent for the NHEJ pathway.

LATENCY

The factors that influence PIC trafficking to particular integration sites after nuclear import are unknown, but they are of clinical importance since recent data established that HIV-1 integration sites are not randomly distributed in the genome. This, in turn, may influence the ability of latent proviruses to serve as sources of recrudescence from resting memory T cells. Postintegration proviral latency is an area of intense interest at present. Viral rebound invariably occurs after the discontinuation of HAART, generally within 2 weeks, and virus can be isolated from T cells *in vitro* during effective HAART.[129–131] The capacity to remain transcriptionally silent for extended durations is one factor that permits the virus to evade immune surveillance.[132] The half-life of the latent reservoir in resting memory T cells is estimated to be approximately 4 years, indicating that eradication by decay on HAART is highly unlikely to be feasible.[133,134] Therefore, understanding the molecular mechanisms controlling integration site preference, silencing, and reactivation of the integrated HIV-1 provirus has important implications for understanding HIV disease pathogenesis and for therapy.[134,135] Chromatin possesses marked structural and functional heterogeneity, and extensive evidence demonstrates that retroviral transcription is strongly affected by the context of integration.[136–139] Epigenetic mechanisms such as histone deacetylation and DNA methylation play major roles in the silencing of viral promoters that occur progressively after integration.[140,141] Methylation-associated silencing is dependent on chromosomal position.[140]

Recent studies established definitively that several retroviruses do not integrate randomly with respect to genomic loci. MLV and HIV-1 both integrate preferentially into transcriptionally active regions.[113,114,142, 143] However, the placements of their proviruses seem to differ with respect to functional elements within the cellular transcription units. HIV-1 integration favors regions downstream of promoters.[113,114,142] In contrast, MoMLV integrations show a predilection for promoter regions, favoring the 2.5 kb surrounding the transcriptional start site.[114] In latently infected cells, a bias toward integration in heterochromatic regions, such as alphoid repeats, is also observed.[144] Integration site distributions will receive concentrated study with the newer methods available[113,114,142] because of the implications for HIV-1 latency[134,136,144–146] and to address the problem of insertional mutagenesis in human gene therapy.[147]

Cellular cofactors responsible for retroviral PIC trafficking or integration site specificity have not been identified.[148] Lentiviral integrase proteins localize to cell nuclei and associate with chromatin when expressed in the absence of other viral components.[98,149–153] These observations have led to efforts to identify a nuclear localization signal (NLS) and to implicate this protein in preintegration complex nuclear translocation. However, nuclear/chromatin targeting of lentiviral integrases now seems to be due not to any nuclear localization signal but rather to interaction with the transcriptional coactivator LEDGF/p75.[153,154] This interaction, which is lentivirus specific, is clearly not needed for nuclear import of the PIC,[153] but there is interest in its potential involvement in integration site choice. Interactions of integrase or PIC DNA with other cellular factors have been demonstrated as well.[148,155,156] Examples are BAF,[146,157,158] HMGa1 protein,[159–162] INI1,[155,163] Ku70 and Ku80,[117] the aforementioned DNA-PK,[122] and hRad18.[164] Some are known to enhance *in vitro* integration reactions,[155,157,165–169] as does the viral nucleocapsid (NC) protein,[159,169] although the significance in each case to the reaction in the natural life cycle of HIV-1 is not clear. INI1, an essential constituent of the human SWI/SNF chromatin remodeling machine, is incorporated into HIV-1 but not HIV-2, SIV, or other retroviral particles[170] and it seems to be required for efficient HIV-1 budding.[163] INI1 also stimulates the DNA-joining activity of HIV-1 integrase *in vitro*, and a role in targeting the viral DNA to active genes has been hypothesized.[155] Moreover, incoming HIV-1 and murine leukemia virus (MLV) PICs were found to trigger export of INI1 and the nuclear body constituent PML (promyelocytye leukemia protein) to the cytoplasm, which may be part of a cellular antiviral response.[168]

LATE EVENTS

Two small RNA-binding regulatory proteins, Tat and Rev, govern viral mRNA production. HIV-1 transcription is minimal in the absence of Tat, because transcriptional elongation does not occur efficiently. Tat binds to an RNA stem-loop (TAR) in the 5 end of the nascent viral transcript. The N-terminal domain of Tat interacts with CyclinT1, which, in turn, recruits the cyclin-dependent kinase 9 (CDK-9), leading to hyperphosphorylation of the large subunit of the RNA polymerase II.[171,172] The result is increased elongation efficiency of the Pol II complex.[172,173] Tat's role in modulating transcription seems to extend to the cellular transcriptome[174–176]; roles in regulating apoptosis,[177] and class I major histocompatibility complex (MHC) molecule expression have also been reported.[178] Rev binds to a secondary RNA structure in the *env* region, the Rev response element, and enables the virus to solve a central problem: export of its unspliced genomic RNA against the normal cellular checkpoint that prevents export of intron-containing messages.[179,180] The Nef and Vpr proteins have been reported to play a wide variety of roles in the viral life cycle and to be implicated in cell death; these proteins will be reviewed in Chapters 15 and 14, respectively.

Maturation of the virion via protease cleavage is synchronous with budding. Recently, viral budding has been shown to employ a complex cellular machinery involved in protein sorting and membrane budding at the multivesicular body[181]; this system is involved in the intracellular budding of HIV-1 virions that is observed in macrophages.[182,183] The HIV-1 protease cleaves the Gag/Pol precursor into virion proteins and was also implicated in cleaving cellular proteins such as caspase-8,[184] a cell death mechanism that will be reviewed in Chapter 17.

CONCLUSION

Lentiviruses cause severe disease in some host species. The present human pandemic is, unfortunately, only the first stage in the evolutionary coadaptation process that is likely to have resulted in the asymptomatic nature of lentiviral infection in Old World nonhuman primates and in some nonprimate lentiviral syndromes as well. Understanding at the cell biological level how lentiviral infections lead to such different outcomes in different hosts may yield new therapeutic approaches.

ACKNOWLEDGMENT

I thank graduate student Dyana T. Saenz for drawing Figure 1.1.

REFERENCES

1. Vallée, H. and H. Carré, Sur la nature infectieuse de l'anaemie du cheval, *CR Hebd. Seances Acad. Sci.,* 139, 331–333, 1904.
2. Sigurdsson, B., Observations on three slow infections of sheep: maedi, paratuberculosis, rida, a slow encephalitis of sheep, with general remarks on infections which develop slowly, and some of their special characteristics, *Br. Vet. J.,* 110, 255–270, 1954.
3. Sigurdsson, B., H. Grimsson, and P.A. Palsson, Macdi, a chronic, progressive infection of sheep's lungs, *J. Infect. Dis.,* 90, 233–241, 1952.
4. Sigurdsson, B., P.A. Palsson, and H. Grimsson, Visna: a demyelinating transmissible disease of sheep, *J. Neuropathol. Exp. Neurol.,* 16, 389–403, 1957.
5. Sigurdsson, B., H. Thormar, and P.A. Palsson, Cultivation of visna virus in tissue culture, *Arch Gesamte Virusforsch,* 10, 368, 1960.
6. Kobayashi, K., Studies on the cultivation of equine infectious anemia virus *in vitro*. I. Serial cultivation of the virus in the cuture of various horse tissues, *Virus,* 11, 177–189, 1961.
7. Haase, A.T., The slow infection caused by visna virus, *Curr. Top. Microbiol. Immunol.,* 72, 101–156, 1975.
8. Haase, A.T. and H.E. Varmus, Demonstration of a DNA provirus in the lytic growth of visna virus, *Nat. New Biol.,* 245, 237–239, 1973.
9. Haase, A.T., L. Stowring, P. Narayan, D. Griffin, and D. Price, Slow persistent infection caused by visna virus: role of host restriction, *Science,* 195, 175–177, 1977.
10. Brahic, M., L. Stowring, P. Ventura, and A.T. Haase, Gene expression in visna virus infection in sheep, *Nature,* 292, 240–242, 1981.
11. Pedersen, N.C., E.W. Ho, M.L. Brown, and J.K. Yamamoto, Isolation of a T-lymphotropic virus from domestic cats with an immunodeficiency-like syndrome, *Science,* 235, 790–793, 1987.

12. Gonda, M.A., M.J. Braun, S.G. Carter, T.A. Kost, J.W. Bess, Jr., L.O. Arthur, and M.J. Van der Maaten, Characterization and molecular cloning of a bovine lentivirus related to human immunodeficiency virus, *Nature,* 330, 388–391, 1987.

13. Chadwick, B.J., R.J. Coelen, L.M. Sammels, G. Kertayadnya, and G.E. Wilcox, Genomic sequence analysis identifies Jembrana disease virus as a new bovine lentivirus, *J. Gen. Virol.,* 76 (Pt 1), 189–192, 1995.

14. Roe, T., T.C. Reynolds, G. Yu, and P.O. Brown, Integration of murine leukemia virus DNA depends on mitosis, *EMBO J.,* 12, 2099–2108, 1993.

15. Lewis, P.F. and M. Emerman, Passage through mitosis is required for oncoretroviruses but not for the human immunodeficiency virus, *J. Virol.,* 68, 510–516, 1994.

16. Naldini, L., U. Bloemer, P. Gallay, D. Ory, R. Mulligan, F.H. Gage, I.M. Verma, and D. Trono, *In vivo* gene delivery and stable transduction of nondividing cells by a lentiviral vector, *Science,* 272, 263–267, 1996.

17. Poeschla, E., F. Wong-Staal, and D. Looney, Efficient transduction of nondividing cells by feline immunodeficiency virus lentiviral vectors, *Nat. Med.,* 4, 354–357, 1998.

18. Olsen, J.C., Gene transfer vectors derived from equine infectious anemia virus, *Gene Ther.,* 5, 1481–1487, 1998.

19. Leroux, C., J.L. Cadore, and R.C. Montelaro, Equine infectious anemia virus (EIAV): what has HIV's country cousin got to tell us? *Vet. Res.,* 35, 485–512, 2004.

20. Rosenberg, N. and P. Joliceur, Retroviral pathogenesis, in *Retroviruses,* Coffin, J.M., S.H. Hughes, and H.E. Varmus, Eds., Cold Spring Harbor Laboratory Press, Cold Spring Harbor, NY, 1999, pp. 161–203.

21. Elder, J.H. and T.R. Phillips, Molecular properties of feline immunodeficiency virus (FIV), *Infect. Agents and Dis.,* 2, 361–374, 1993.

22. Desrosiers, R.C., Nonhuman lentiviruses, in *Fields Virology,* Knipe, D.M. and P.M. Howley, Eds., Lippincott Williams and Wilkins, Philadelphia, 2001, pp. 2095–2121.

23. Hirsch, V.M., R.A. Olmsted, M. Murphey-Corb, R.H. Purcell, and P.R. Johnson, An African primate lentivirus (SIVsm) closely related to HIV-2, *Nature,* 339, 389–392, 1989.

24. Gao, F., E. Bailes, D.L. Robertson, Y. Chen, C.M. Rodenburg, S.F. Michael, L.B. Cummins, L.O. Arthur, M. Peeters, G.M. Shaw, P.M. Sharp, and B.H. Hahn, Origin of HIV-1 in the chimpanzee Pan troglodytes troglodytes [see comments], *Nature,* 397, 436–441, 1999.

25. Sharp, P.M., E. Bailes, R.R. Chaudhuri, C.M. Rodenburg, M.O. Santiago, and B.H. Hahn, The origins of acquired immune deficiency syndrome viruses: where and when? *Philos. Trans. R. Soc. London B Biol. Sci.,* 356, 867–876, 2001.

26. Korber, B., M. Muldoon, J. Theiler, F. Gao, R. Gupta, A. Lapedes, B.H. Hahn, S. Wolinsky, and T. Bhattacharya, Timing the ancestor of the HIV-1 pandemic strains, *Science,* 288, 1789–1796, 2000.

27. Hahn, B.H., G.M. Shaw, K.M. De Cock, and P.M. Sharp, AIDS as a zoonosis: scientific and public health implications, *Science,* 287, 607–614, 2000.

28. Letvin, N.L., M.D. Daniel, P.K. Sehgal, R.C. Desrosiers, R.D. Hunt, L.M. Waldron, J.J. MacKey, D.K. Schmidt, L.V. Chalifoux, and N.W. King, Induction of AIDS-like disease in macaque monkeys with T-cell tropic retrovirus STLV-III, *Science,* 230, 71–73, 1985.

29. Troyer, J.L., J. Pecon-Slattery, M.E. Roelke, L. Black, C. Packer, and S.J. O'Brien, Patterns of feline immunodeficiency virus multiple infection and genome divergence in a free-ranging population of African lions, *J. Virol.,* 78, 3777–3791, 2004.

30. Elder, J.H., G.A. Dean, E.A. Hoover, J.A. Hoxie, M.H. Malim, L. Mathes, J.C. Neil, T.W. North, E. Sparger, M.B. Tompkins, W.A. Tompkins, J. Yamamoto, N. Yuhki, N.C. Pedersen, and R.H. Miller, Lessons from the cat: feline immunodeficiency virus as a tool to develop intervention strategies against human immunodeficiency virus type 1, *AIDS Res. Hum. Retroviruses,* 14, 797–801, 1998.

31. Piatak, M., Jr., M.S. Saag, L.C. Yang, S.J. Clark, J.C. Kappes, K.C. Luk, B.H. Hahn, G.M. Shaw, and J.D. Lifson, High levels of HIV-1 in plasma during all stages of infection determined by competitive PCR, *Science,* 259, 1749–1754, 1993.

32. Ho, D.D., A.U. Neumann, A.S. Perelson, W. Chen, J.M. Leonard, and M. Markowitz, Rapid turnover of plasma virions and CD4 lymphocytes in HIV-1 infection [see comments], *Nature,* 373, 123–126, 1995.

33. Wei, X., S.K. Ghosh, M.E. Taylor, V.A. Johnson, E.A. Emini, P. Deutsch, J.D. Lifson, S. Bonhoeffer, M.A. Nowak, B.H. Hahn, M. Saag, and G.M. Shaw, Viral dynamics in human immunodeficiency virus type 1 infection, *Nature,* 373, 117–122, 1995.

34. Nowak, M.A., A.L. Lloyd, G.M. Vasquez, T.A. Wiltrout, L.M. Wahl, N. Bischofberger, J. Williams, A. Kinter, A.S. Fauci, V.M. Hirsch, and J.D. Lifson, Viral dynamics of primary viremia and antiretroviral therapy in simian immunodeficiency virus infection, *J. Virol.,* 71, 7518–7525, 1997.

35. Lifson, J.D., M.A. Nowak, S. Goldstein, J.L. Rossio, A. Kinter, G. Vasquez, T.A. Wiltrout, C. Brown, D. Schneider, L. Wahl, A.L. Lloyd, J. Williams, W.R. Elkins, A.S. Fauci, and V.M. Hirsch, The extent of early viral replication is a critical determinant of the natural history of simian immunodeficiency virus infection, *J. Virol.,* 71, 9508–9514, 1997.

36. Mellors, J.W., A. Munoz, J.V. Giorgi, J.B. Margolick, C.J. Tassoni, P. Gupta, L.A. Kingsley, J.A. Todd, A.J. Saah, R. Detels, J.P. Phair, and C.R. Rinaldo, Jr., Plasma viral load and CD4+ lymphocytes as prognostic markers of HIV-1 infection, *Ann. Intern. Med.,* 126, 946–954, 1997.

37. Mellors, J.W., C.R. Rinaldo, Jr., P. Gupta, R.M. White, J.A. Todd, and L.A. Kingsley, Prognosis in HIV-1 infection predicted by the quantity of virus in plasma, *Science,* 272, 1167–1170, 1996.

38. McCune, J.M., M.B. Hanley, D. Cesar, R. Halvorsen, R. Hoh, D. Schmidt, E. Wieder, S. Deeks, S. Siler, R. Neese, and M. Hellerstein, Factors influencing T-cell turnover in HIV-1-seropositive patients, *J. Clin. Invest.,* 105, R1–R8, 2000.

39. Hellerstein, M., M.B. Hanley, D. Cesar, S. Siler, C. Papageorgopoulos, E. Wieder, D. Schmidt, R. Hoh, R. Neese, D. Macallan, S. Deeks, and J.M. McCune, Directly measured kinetics of circulating T lymphocytes in normal and HIV-1-infected humans, *Nat. Med.,* 5, 83–89, 1999.

40. Hellerstein, M.K. and J.M. McCune, T cell turnover in HIV-1 disease, *Immunity,* 7, 583–589, 1997.

41. Douek, D.C., R.D. McFarland, P.H. Keiser, E.A. Gage, J.M. Massey, B.F. Haynes, M.A. Polis, A.T. Haase, M.B. Feinberg, J.L. Sullivan, B.D. Jamieson, J.A. Zack, L.J. Picker, and R.A. Koup, Changes in thymic function with age and during the treatment of HIV infection, *Nature,* 396, 690–695, 1998.

42. Haase, A.T., Population biology of HIV-1 infection: viral and CD4+ T cell demographics and dynamics in lymphatic tissues, *Annu. Rev. Immunol.,* 17, 625–656, 1999.

43. Moses, A., J. Nelson, and G.C. Bagby, Jr., The influence of human immunodeficiency virus-1 on hematopoiesis, *Blood,* 91, 1479–1495, 1998.

44. Zinkernagel, R.M. and H. Hengartner, T-cell-mediated immunopathology versus direct cytolysis by virus: implications for HIV and AIDS, *Immunol. Today,* 15, 262–268, 1994.

45. Fultz, P.N., D.C. Anderson, H.M. McClure, S. Dewhurst, and J.I. Mullins, SIVsmm infection of macaque and mangabey monkeys: correlation between *in vivo* and *in vitro* properties of different isolates, *Dev. Biol. Stand.,* 72, 253–258, 1990.

46. Rey-Cuille, M.A., J.L. Berthier, M.C. Bomsel-Demontoy, Y. Chaduc, L. Montagnier, A.G. Hovanessian, and L.A. Chakrabarti, Simian immunodeficiency virus replicates to high levels in sooty mangabeys without inducing disease, *J. Virol.,* 72, 3872–3886, 1998.

47. Chakrabarti, L.A., The paradox of simian immunodeficiency virus infection in sooty mangabeys: active viral replication without disease progression, *Front Biosci,* 9, 521–539, 2004.

48. Chakrabarti, L.A., S.R. Lewin, L. Zhang, A. Gettie, A. Luckay, L.N. Martin, E. Skulsky, D.D. Ho, C. Cheng-Mayer, and P.A. Marx, Normal T-cell turnover in sooty mangabeys harboring active simian immunodeficiency virus infection, *J. Virol.,* 74, 1209–1223, 2000.

49. Silvestri, G., D.L. Sodora, R.A. Koup, M. Paiardini, S.P. O'Neil, H.M. McClure, S.I. Staprans, and M.B. Feinberg, Nonpathogenic SIV infection of sooty mangabeys is characterized by limited bystander immunopathology despite chronic high-level viremia, *Immunity,* 18, 441–452, 2003.

50. Broussard, S.R., S.I. Staprans, R. White, E.M. Whitehead, M.B. Feinberg, and J.S. Allan, Simian immunodeficiency virus replicates to high levels in naturally infected African green monkeys without inducing immunologic or neurologic disease, *J. Virol.,* 75, 2262–2275, 2001.

51. Hirsch, V.M., G. Dapolito, P.R. Johnson, W.R. Elkins, W.T. London, R.J. Montali, S. Goldstein, and C. Brown, Induction of AIDS by simian immunodeficiency virus from an African green monkey: species-specific variation in pathogenicity correlates with the extent of *in vivo* replication, *J. Virol.,* 69, 955–967, 1995.

52. Kaur, A., R.M. Grant, R.E. Means, H. McClure, M. Feinberg, and R.P. Johnson, Diverse host responses and outcomes following simian immunodeficiency virus SIVmac239 infection in sooty mangabeys and rhesus macaques, *J. Virol.,* 72, 9597–9611, 1998.

53. Swardson, C.J., G.J. Kociba, and L.E. Perryman, Effects of equine infectious anemia virus on hematopoietic progenitors *in vitro, Am. J. Vet. Res.,* 53, 1176–1179, 1992.

54. Kim, C.H. and J.W. Casey, Genomic variation and segregation of equine infectious anemia virus during acute infection, *J. Virol.,* 66, 3879–3882, 1992.

55. Peluso, R., A. Haase, L. Stowring, M. Edwards, and P. Ventura, A Trojan Horse mechanism for the spread of visna virus in monocytes, *Virology,* 147, 231–236, 1985.

56. Narayan, O., S. Kennedy-Stoskopf, D. Sheffer, D.E. Griffin, and J.E. Clements, Activation of caprine arthritis-encephalitis virus expression during maturation of monocytes to macrophages, *Infect. Immunol.,* 41, 67–73, 1983.

57. Gendelman, H.E., O. Narayan, S. Kennedy-Stoskopf, P.G. Kennedy, Z. Ghotbi, J.E. Clements, J. Stanley, and G. Pezeshkpour, Tropism of sheep lentiviruses for monocytes: susceptibility to infection and virus gene expression increase during maturation of monocytes to macrophages, *J. Virol.,* 58, 67–74, 1986.

58. Crawford, T.B., D.S. Adams, W.P. Cheevers, and L.C. Cork, Chronic arthritis in goats caused by a retrovirus, *Science,* 207, 997–999, 1980.

59. Broder, C.C., R.G. Collman, C.C. Broder, and R.G. Collman, Chemokine receptors and HIV, *J. Leukocyte Biol.,* 62, 20–29, 1997.

60. Berger, E., HIV entry and tropism: the chemokine receptor connection, *AIDS,* 11 (Suppl A), S3–S16, 1997.

61. Feng, Y., C.C. Broder, P.E. Kennedy, and E.A. Berger, HIV-1 entry cofactor: functional cDNA cloning of a seven-transmembrane, G protein-coupled receptor [see comments], *Science,* 272, 872–877, 1996.

62. Wu, L., N.P. Gerard, R. Wyatt, H. Choe, C. Parolin, N. Ruffing, A. Borsetti, A.A. Cardoso, E. Desjardin, W. Newman, C. Gerard, and J. Sodroski, CD4-induced interaction of primary HIV-1 gp120 glycoproteins with the chemokine receptor CCR-5, *Nature,* 384, 179–183, 1996.

63. Rizzuto, C.D., R. Wyatt, N. Hernandez-Ramos, Y. Sun, P.D. Kwong, W.A. Hendrickson, and J. Sodroski, A conserved HIV gp120 glycoprotein structure involved in chemokine receptor binding [see comments], *Science,* 280, 1949–1953, 1998.

64. Weissenhorn, W., A. Dessen, S.C. Harrison, J.J. Skehel, and D.C. Wiley, Atomic structure of the ectodomain from HIV-1 gp41, *Nature,* 387, 426–430, 1997.

65. Chan, D.C. and P.S. Kim, HIV entry and its inhibition, *Cell,* 93, 681–684, 1998.

66. Willett, B.J., L. Picard, M.J. Hosie, J.D. Turner, K. Adema, and P.R. Clapham, Shared usage of the chemokine receptor CXCR4 by the feline and human immunodeficiency viruses, *J. Virol.,* 71, 6407–6415, 1997.

67. Poeschla, E. and D. Looney, CXCR4 is required by a non-primate lentivirus: heterologous expression of feline immunodeficiency virus in human, rodent and feline cells, *J. Virol.,* 72, 6858–6866, 1998.

68. Shimojima, M., T. Miyazawa, Y. Ikeda, E.L. McMonagle, H. Haining, H. Akashi, Y. Takeuchi, M.J. Hosie, and B.J. Willett, Use of CD134 as a primary receptor by the feline immunodeficiency virus, *Science,* 303, 1192–1195, 2004.

69. Willett, B.J., M.J. Hosie, O. Jarrett, and J.C. Neil, Identification of a putative cellular receptor for feline immunodeficiency virus as the feline homologue of CD9, *Immunology,* 81, 228–233, 1994.

70. Martinson, J.J., N.H. Chapman, D.C. Rees, Y.T. Liu, and J.B. Clegg, Global distribution of the CCR5 gene 32-base pair deletion, *Nat. Genet.,* 16, 100–103, 1997.

71. Dean, M., M. Carrington, C. Winkler, G.A. Huttley, M.W. Smith, R. Allikmets, J.J. Goedert, S.P. Buchbinder, E. Vittinghoff, E. Gomperts, S. Donfield, D. Vlahov, R. Kaslow, A. Saah, C. Rinaldo, R. Detels, and S.J. O'Brien, Genetic restriction of HIV-1 infection and progression to AIDS by a deletion allele of the CKR5 structural gene. Hemophilia Growth and Development Study, Multicenter AIDS Cohort Study, Multicenter Hemophilia Cohort Study, San Francisco City Cohort, ALIVE Study, *Science,* 273, 1856–1862, 1996.

72. de Roda Husman, A.M., M. Koot, M. Cornelissen, I.P. Keet, M. Brouwer, S.M. Broersen, M. Bakker, M.T. Roos, M. Prins, F. de Wolf, R.A. Coutinho, F. Miedema, J. Goudsmit, and H. Schuitemaker, Association between CCR5 genotype and the clinical course of HIV-1 infection, *Ann. Intern. Med.,* 127, 882–890, 1997.

73. Vlahakis, S.R., A. Algeciras-Schimnich, G. Bou, C.J. Heppelmann, A. Villasis-Keever, R.C. Collman, and C.V. Paya, Chemokine-receptor activation by env determines the mechanism of death in HIV-infected and uninfected T lymphocytes, *J. Clin. Invest.,* 107, 207–215, 2001.

74. Lazzarin, A., B. Clotet, D. Cooper, J. Reynes, K. Arasteh, M. Nelson, C. Katlama, H.J. Stellbrink, J.F. Delfraissy, J. Lange, L. Huson, R. DeMasi, C. Wat, J. Delehanty, C. Drobnes, and M. Salgo, Efficacy of enfuvirtide in patients infected with drug-resistant HIV-1 in Europe and Australia, *New Engl. J. Med.,* 348, 2186–2195, 2003.

75. McDonald, D., M.A. Vodicka, G. Lucero, T.M. Svitkina, G.G. Borisy, M. Emerman, and T.J. Hope, Visualization of the intracellular behavior of HIV in living cells, *J. Cell. Biol.,* 159, 441–452, 2002.

76. Suikkanen, S., T. Aaltonen, M. Nevalainen, O. Valilehto, L. Lindholm, M. Vuento, and M. Vihinen-Ranta, Exploitation of microtubule cytoskeleton and dynein during parvoviral traffic toward the nucleus, *J. Virol.,* 77, 10270–10279, 2003.

77. Dohner, K., A. Wolfstein, U. Prank, C. Echeverri, D. Dujardin, R. Vallee, and B. Sodeik, Function of dynein and dynactin in herpes simplex virus capsid transport, *Mol. Biol. Cell,* 13, 2795–2809, 2002.

78. Suomalainen, M., M.Y. Nakano, S. Keller, K. Boucke, R.P. Stidwill, and U.F. Greber, Microtubule-dependent plus and minus end-directed motilities are competing processes for nuclear targeting of adenovirus, *J. Cell Biol.,* 144, 657–672, 1999.

79. Petit, C., M.L. Giron, J. Tobaly-Tapiero, P. Bittoun, E. Real, Y. Jacob, N. Tordo, H. De The, and A. Saib, Targeting of incoming retroviral Gag to the centrosome involves a direct interaction with the dynein light chain 8, *J. Cell Sci.,* 116 (Pt 16), 3433–3442, 2003.

80. Hatziioannou, T., S. Cowan, S.P. Goff, P.D. Bieniasz, and G.J. Towers, Restriction of multiple divergent retroviruses by Lv1 and Ref1, *EMBO J.,* 22, 385–394, 2003.

81. Stremlau, M., C.M. Owens, M.J. Perron, M. Kiessling, P. Autissier, and J. Sodroski, The cytoplasmic body component TRIM5 restricts HIV-1 infection in Old World monkeys, *Nature,* 427, 848–853, 2004.

82. Bieniasz, P.D., Restriction factors: a defense against retroviral infection, *Trends Microbiol.,* 11, 286–291, 2003.

83. Towers, G.J., T. Hatziioannou, S. Cowan, S.P. Goff, J. Luban, and P.D. Bieniasz, Cyclophilin A modulates the sensitivity of HIV-1 to host restriction factors, *Nat. Med.,* 9, 1138–1143, 2003.

84. Sheehy, A.M., N.C. Gaddis, J.D. Choi, and M.H. Malim, Isolation of a human gene that inhibits HIV-1 infection and is suppressed by the viral Vif protein, *Nature,* 418, 646–650, 2002.

85. Zhang, H., B. Yang, R.J. Pomerantz, C. Zhang, S.C. Arunachalam, and L. Gao, The cytidine deaminase CEM15 induces hypermutation in newly synthesized HIV-1 DNA, *Nature,* 424, 94–98, 2003.

86. Mangeat, B., P. Turelli, G. Caron, M. Friedli, L. Perrin, and D. Trono, Broad antiretroviral defense by human APOBEC3G through lethal editing of nascent reverse transcripts, *Nature,* 424, 99–103, 2003.

87. Mariani, R., D. Chen, B. Schrofelbauer, F. Navarro, R. Konig, B. Bollman, C. Munk, H. Nymark-McMahon, and N.R. Landau, Species-specific exclusion of APOBEC3G from HIV-1 virions by Vif, *Cell,* 114, 21–31, 2003.

88. Sheehy, A.M., N.C. Gaddis, and M.H. Malim, The antiretroviral enzyme APOBEC3G is degraded by the proteasome in response to HIV-1 Vif, *Nat. Med.,* 9, 1404–1407, 2003.

89. Marin, M., K.M. Rose, S.L. Kozak, and D. Kabat, HIV-1 Vif protein binds the editing enzyme APOBEC3G and induces its degradation, *Nat. Med.,* 9, 1398–1403, 2003.

90. Yu, X., Y. Yu, B. Liu, K. Luo, W. Kong, P. Mao, and X.F. Yu, Induction of APOBEC3G ubiquitination and degradation by an HIV-1 Vif-Cul5-SCF complex, *Science,* 302, 1056–1060, 2003.

91. Harris, R.S., K.N. Bishop, A.M. Sheehy, H.M. Craig, S.K. Petersen-Mahrt, I.N. Watt, M.S. Neuberger, and M.H. Malim, DNA deamination mediates innate immunity to retroviral infection, *Cell,* 113, 803–809, 2003.

92. Goff, S.P., Intracellular trafficking of retroviral genomes during the early phase of infection: viral exploitation of cellular pathways, *J. Gene Med.,* 3, 517–528, 2001.

93. Bukrinsky, M.I., S. Haggerty, M.P. Dempsey, N. Sharova, A. Adzhubel, L. Spitz, P. Lewis, D. Goldfarb, M. Emerman, and M. Stevenson, A nuclear localization signal within HIV-1 matrix protein that governs infection of non-dividing cells, *Nature,* 365, 666–669, 1993.

94. Bukrinskaya, A.G., A. Ghorpade, N.K. Heinzinger, T.E. Smithgall, R.E. Lewis, and M. Stevenson, Phosphorylation-dependent human immunodeficiency virus type 1 infection and nuclear targeting of viral DNA, *Proc. Natl. Acad. Sci. U.S.A.,* 93, 367–371, 1996.

95. von Schwedler, U., R.S. Kornbluth, and D. Trono, The nuclear localization signal of the matrix protein of human immunodeficiency virus type 1 allows the establishment of infection in macrophages and quiescent T lymphocytes, *Proc. Natl. Acad. Sci. U.S.A.,* 91, 6992–6996, 1994.

96. Heinzinger, N.K., M.I. Bukinsky, S.A. Haggerty, A.M. Ragland, V. Kewalramani, M.A. Lee, H.E. Gendelman, L. Ratner, M. Stevenson, and M. Emerman, The Vpr protein of human immunodeficiency virus type 1 influences nuclear localization of viral nucleic acids in nondividing host cells, *Proc. Natl. Acad. Sci. U.S.A.*, 91, 7311–7315, 1994.

97. Gallay, P., S. Swingler, C. Aiken, and D. Trono, HIV-1 infection of nondividing cells: C-terminal tyrosine phosphorylation of the viral matrix protein is a key regulator, *Cell*, 80, 379–388, 1995.

98. Gallay, P., T. Hope, D. Chin, and D. Trono, HIV-1 infection of nondividing cells through the recognition of integrase by the importin/karyopherin pathway, *Proc. Natl. Acad. Sci. U.S.A.*, 94, 9825–9830, 1997.

99. Gallay, P., V. Stitt, C. Mundy, M. Oettinger, and D. Trono, Role of the karyopherin pathway in human immunodeficiency virus type 1 nuclear import, *J. Virol.*, 70, 1027–1032, 1996.

100. Haffar, O.K., S. Popov, L. Dubrovsky, I. Agostini, H. Tang, T. Pushkarsky, S.G. Nadler, and M. Bukrinsky, Two nuclear localization signals in the HIV-1 matrix protein regulate nuclear import of the HIV-1 pre-integration complex, *J. Mol. Biol.*, 299, 359–368, 2000.

101. Popov, S., M. Rexach, G. Zybarth, N. Reiling, M.A. Lee, L. Ratner, C.M. Lane, M.S. Moore, G. Blobel, and M. Bukrinsky, Viral protein R regulates nuclear import of the HIV-1 pre-integration complex, *EMBO J.*, 17, 909–917, 1998.

102. Fouchier, R.A., B.E. Meyer, J.H. Simon, U. Fischer, A.V. Albright, F. Gonzalez-Scarano, and M.H. Malim, Interaction of the human immunodeficiency virus type 1 Vpr protein with the nuclear pore complex, *J. Virol.*, 72, 6004–6013, 1998.

103. Bouyac-Bertoia, M., J. Dvorin, R. Fouchier, Y. Jenkins, B. Meyer, L. Wu, M. Emerman, and M.H. Malim, HIV-1 infection requires a functional integrase NLS, *Mol. Cell*, 7, 1025–1035, 2001.

104. Zennou, V., C. Petit, D. Guetard, U. Nerhbass, L. Montagnier, and P. Charneau, HIV-1 genome nuclear import is mediated by a central DNA flap, *Cell*, 101, 173–185, 2000.

105. Follenzi, A., L.E. Ailles, S. Bakovic, M. Geuna, and L. Naldini, Gene transfer by lentiviral vectors is limited by nuclear translocation and rescued by HIV-1 pol sequences, *Nat. Genet.*, 25, 217–222, 2000.

106. Whitwam, T., M. Peretz, and E.M. Poeschla, Identification of a central DNA flap in feline immunodeficiency virus, *J. Virol.*, 75, 9407–9414, 2001.

107. Fassati, A., D. Gorlich, I. Harrison, L. Zaytseva, and J.M. Mingot, Nuclear import of HIV-1 intracellular reverse transcription complexes is mediated by importin 7, *EMBO J.*, 22, 3675–3685, 2003.

108. Fouchier, R.A., B.E. Meyer, J.H. Simon, U. Fischer, and M.H. Malim, HIV-1 infection of non-dividing cells: evidence that the amino-terminal basic region of the viral matrix protein is important for Gag processing but not for post-entry nuclear import, *EMBO J.*, 16, 4531–4539, 1997.

109. Dvorin, J.D., P. Bell, G.G. Maul, M. Yamashita, M. Emerman, and M.H. Malim, Reassessment of the roles of integrase and the central DNA flap in human immunodeficiency virus type 1 nuclear import, *J. Virol.*, 76, 12087–12096, 2002.

110. Limon, A., N. Nakajima, R. Lu, H.Z. Ghory, and A. Engelman, Wild-type levels of nuclear localization and human immunodeficiency virus type 1 replication in the absence of the central DNA flap, *J. Virol.*, 76, 12078–12086, 2002.

111. Freed, E.O., G. Englund, F. Maldarelli, and M.A. Martin, Phosphorylation of residue 131 of HIV-1 matrix is not required for macrophage infection, *Cell*, 88, 171–174, 1997.

112. Vodicka, M.A., Determinants for lentiviral infection of non-dividing cells, *Somatic Cell Mol. Genet.*, 26, 35–49, 2001.

113. Schroder, A.R., P. Shinn, H. Chen, C. Berry, J.R. Ecker, and F. Bushman, HIV-1 integration in the human genome favors active genes and local hotspots, *Cell*, 110, 521–529, 2002.

114. Wu, X., Y. Li, B. Crise, and S.M. Burgess, Transcription start regions in the human genome are favored targets for MLV integration, *Science*, 300, 1749–1751, 2003.

115. Engelman, A., The roles of cellular factors in retroviral integration, *Curr. Top. Microbiol. Immunol.*, 281, 209–238, 2003.

116. Farnet, C.M. and W.A. Haseltine, Circularization of human immunodeficiency virus type 1 DNA *in vitro*, *J. Virol.*, 65, 6942–6952, 1991.

117. Li, L., J.M. Olvera, K.E. Yoder, R.S. Mitchell, S.L. Butler, M. Lieber, S.L. Martin, and F.D. Bushman, Role of the non-homologous DNA end joining pathway in the early steps of retroviral infection, *EMBO J.*, 20, 3272–3281, 2001.

118. Shoemaker, C., J. Hoffman, S.P. Goff, and D. Baltimore, Intramolecular integration within Moloney murine leukemia virus DNA, *J. Virol.*, 40, 164–172, 1981.

119. Brown, P.O., Integration, in *Retroviruses,* Coffin, J.M., S.H. Hughes, and H.E. Varmus, Eds., Cold Spring Harbor Laboratory Press, Cold Spring Harbor, NY, 1999, pp. 161–203.

120. Keshet, E. and H.M. Temin, Cell killing by spleen necrosis virus is correlated with a transient accumulation of spleen necrosis virus DNA, *J. Virol.,* 31, 376–388, 1979.

121. Temin, H.M., E. Keshet, and S.K. Weller, Correlation of transient accumulation of linear unintegrated viral DNA and transient cell killing by avian leukosis and reticuloendotheliosis viruses, *Cold Spring Harbor Symp. Quant. Biol.,* 44 (Pt 2), 773–778, 1980.

122. Daniel, R., R.A. Katz, and A.M. Skalka, A role for DNA-PK in retroviral DNA integration, *Science,* 284, 644–647, 1999.

123. Huang, L.C., K.C. Clarkin, and G.M. Wahl, Sensitivity and selectivity of the DNA damage sensor responsible for activating p53-dependent G1 arrest, *Proc. Natl. Acad. Sci. U.S.A.,* 93, 4827–4832, 1996.

124. Taccioli, G.E., T.M. Gottlieb, T. Blunt, A. Priestley, J. Demengeot, R. Mizuta, A.R. Lehmann, F.W. Alt, S.P. Jackson, and P.A. Jeggo, Ku80: product of the XRCC5 gene and its role in DNA repair and V(D)J recombination, *Science,* 265, 1442–1445, 1994.

125. Smith, G.C. and S.P. Jackson, The DNA-dependent protein kinase, *Genes Dev.,* 13, 916–934, 1999.

126. Blunt, T., N.J. Finnie, G.E. Taccioli, G.C. Smith, J. Demengeot, T.M. Gottlieb, R. Mizuta, A.J. Varghese, F.W. Alt, and P.A. Jeggo, Defective DNA-dependent protein kinase activity is linked to V(D)J recombination and DNA repair defects associated with the murine scid mutation, *Cell,* 80, 813–823, 1995.

127. Baekelandt, V., A. Claeys, P. Cherepanov, E. De Clercq, B. De Strooper, B. Nuttin, and Z. Debyser, DNA-dependent protein kinase is not required for efficient lentivirus integration, *J. Virol.,* 74, 11278–11285, 2000.

128. Coffin, J.M. and N. Rosenberg, Retroviruses. Closing the joint, *Nature,* 399, 413, 415–416, 1999.

129. Finzi, D., M. Hermankova, T. Pierson, L.M. Carruth, C. Buck, R.E. Chaisson, T.C. Quinn, K. Chadwick, J. Margolick, R. Brookmeyer, J. Gallant, M. Markowitz, D.D. Ho, D.D. Richman, and R.F. Siliciano, Identification of a reservoir for HIV-1 in patients on highly active antiretroviral therapy, *Science,* 278, 1295–1300, 1997.

130. Wong, J.K., M. Hezareh, H. Günthard, D. Havlir, C. Ignacio, C. Spina, and D. Richman, Recovery of replication-competent HIV despite prolonged suppression of plasma viremia, *Science,* 278, 1291–1295, 1997.

131. Chun, T.-W., L. Stuyver, S. Mizell, L. Ehler, J. Mican, M. Baseler, L. Lloyd, M. Nowak, and A.S. Fauci, Presence of an inducible HIV-1 latent reservoir during highly active antiretroviral therapy, *Proc. Natl. Acad. Sci. U.S.A.,* 94, 13193–13197, 1997.

132. Siliciano, J.D. and R.F. Siliciano, Latency and viral persistence in HIV-1 infection, *J. Clin. Invest.,* 106, 823–825, 2000.

133. Siliciano, J.D., J. Kajdas, D. Finzi, T.C. Quinn, K. Chadwick, J.B. Margolick, C. Kovacs, S.J. Gange, and R.F. Siliciano, Long-term follow-up studies confirm the stability of the latent reservoir for HIV-1 in resting CD4+ T cells, *Nat. Med.,* 9, 727–728, 2003.

134. Siliciano, J.D. and R.F. Siliciano, A long-term latent reservoir for HIV-1: discovery and clinical implications, *J. Antimicrob. Chemother.,* 54, 6–9, 2004.

135. Blankson, J.N., D. Persaud, and R.F. Siliciano, The challenge of viral reservoirs in HIV-1 infection, *Annu. Rev. Med.,* 53, 557–593, 2002.

136. Jordan, A., P. Defechereux, and E. Verdin, The site of HIV-1 integration in the human genome determines basal transcriptional activity and response to Tat transactivation, *EMBO J.,* 20, 1726–1738, 2001.

137. Rampalli, S., A. Kulkarni, P. Kumar, D. Mogare, S. Galande, D. Mitra, and S. Chattopadhyay, Stimulation of Tat-independent transcriptional processivity from the HIV-1 LTR promoter by matrix attachment regions, *Nucleic Acids Res.,* 31, 3248–3256, 2003.

138. Zentilin, L., G. Qin, S. Tafuro, M.C. Dinauer, C. Baum, and M. Giacca, Variegation of retroviral vector gene expression in myeloid cells, *Gene Ther.,* 7, 153–166, 2000.

139. Cherry, S.R., D. Biniszkiewicz, L. van Parijs, D. Baltimore, and R. Jaenisch, Retroviral expression in embryonic stem cells and hematopoietic stem cells, *Mol. Cell Biol.,* 20, 7419–7426, 2000.

140. Hoeben, R.C., A.A. Migchielsen, R.C. van der Jagt, H. van Ormondt, and A.J. van der Eb, Inactivation of the Moloney murine leukemia virus long terminal repeat in murine fibroblast cell lines is associated with methylation and dependent on its chromosomal position, *J. Virol.,* 65, 904–912, 1991.

141. Chen, W.Y., E.C. Bailey, S.L. McCune, J.Y. Dong, and T.M. Townes, Reactivation of silenced, virally transduced genes by inhibitors of histone deacetylase, *Proc. Natl. Acad. Sci. U.S.A.*, 94, 5798–5803, 1997.

142. Han, Y., K. Lassen, D. Monie, A.R. Sedaghat, S. Shimoji, X. Liu, T.C. Pierson, J.B. Margolick, R.F. Siliciano, and J.D. Siliciano, Resting CD4+ T cells from human immunodeficiency virus type 1 (HIV-1)-infected individuals carry integrated HIV-1 genomes within actively transcribed host genes, *J. Virol.*, 78, 6122–6133, 2004.

143. Trono, D., Virology. Picking the right spot, *Science*, 300, 1670–1671, 2003.

144. Jordan, A., D. Bisgrove, and E. Verdin, HIV reproducibly establishes a latent infection after acute infection of T cells *in vitro*, *EMBO J.*, 22, 1868–1877, 2003.

145. Pion, M., A. Jordan, A. Biancotto, F. Dequiedt, F. Gondois-Rey, S. Rondeau, R. Vigne, J. Hejnar, E. Verdin, and I. Hirsch, Transcriptional suppression of *in vitro*-integrated human immunodeficiency virus type 1 does not correlate with proviral DNA methylation, *J. Virol.*, 77, 4025–4032, 2003.

146. Mansharamani, M., D.R. Graham, D. Monie, K.K. Lee, J.E. Hildreth, R.F. Siliciano, and K.L. Wilson, Barrier-to-autointegration factor BAF binds p55 Gag and matrix and is a host component of human immunodeficiency virus type 1 virions, *J. Virol.*, 77, 13084–13092, 2003.

147. Hacein-Bey-Abina, S., C. Von Kalle, M. Schmidt, M.P. McCormack, N. Wulffraat, P. Leboulch, A. Lim, C.S. Osborne, R. Pawliuk, E. Morillon, R. Sorensen, A. Forster, P. Fraser, J.I. Cohen, G. de Saint Basile, I. Alexander, U. Wintergerst, T. Frebourg, A. Aurias, D. Stoppa-Lyonnet, S. Romana, I. Radford-Weiss, F. Gross, F. Valensi, E. Delabesse, E. Macintyre, F. Sigaux, J. Soulier, L.E. Leiva, M. Wissler, C. Prinz, T.H. Rabbitts, F. Le Deist, A. Fischer, and M. Cavazzana-Calvo, LMO2-associated clonal T cell proliferation in two patients after gene therapy for SCID-X1, *Science*, 302, 415–419, 2003.

148. Bushman, F.D., Host proteins in retroviral cDNA integration, *Adv. Virus Res.*, 52, 301–317, 1999.

149. Cherepanov, P., W. Pluymers, A. Claeys, P. Proost, E. De Clercq, and Z. Debyser, High-level expression of active HIV-1 integrase from a synthetic gene in human cells, *FASEB J.*, 14, 1389–1399, 2000.

150. Pluymers, W., P. Cherepanov, D. Schols, E. De Clercq, and Z. Debyser, Nuclear localization of human immunodeficiency virus type 1 integrase expressed as a fusion protein with green fluorescent protein, *Virology*, 258, 327–332, 1999.

151. Depienne, C., P. Roques, C. Creminon, L. Fritsch, R. Casseron, D. Dormont, C. Dargemont, and S. Benichou, Cellular distribution and karyophilic properties of matrix, integrase, and Vpr proteins from the human and simian immunodeficiency viruses, *Exp. Cell Res.*, 260, 387–395, 2000.

152. Petit, C., O. Schwartz, and F. Mammano, The karyophilic properties of human immunodeficiency virus type 1 integrase are not required for nuclear import of proviral DNA, *J. Virol.*, 74, 7119–7126, 2000.

153. Llano, M., M. Vanegas, O. Fregoso, D. Saenz, S. Chung, M. Peretz, and E.M. Poeschla, LEDGF/p75 determines cellular trafficking of diverse lentiviral but not murine oncoretroviral integrase proteins and is a component of functional lentiviral pre-integration complexes, *J. Virol.*, 78, 9524–9537, 2004.

154. Maertens, G., P. Cherepanov, W. Pluymers, K. Busschots, E. De Clercq, Z. Debyser, and Y. Engelborghs, LEDGF/p75 is essential for nuclear and chromosomal targeting of HIV-1 integrase in human cells, *J. Biol. Chem.*, 278, 33528–33539, 2003.

155. Kalpana, G.V., S. Marmon, W. Wang, G.R. Crabtree, and S.P. Goff, Binding and stimulation of HIV-1 integrase by a human homolog of yeast transcription factor SNF5 [see comments], *Science*, 266, 2002–2006, 1994.

156. Miller, M.D., C.M. Farnet, and F.D. Bushman, Human immunodeficiency virus type 1 preintegration complexes: studies of organization and composition, *J. Virol.*, 71, 5382–5390, 1997.

157. Lee, M.S. and R. Craigie, A previously unidentified host protein protects retroviral DNA from autointegration, *Proc. Natl. Acad. Sci. U.S.A.*, 95, 1528–1533, 1998.

158. Lin, C.W. and A. Engelman, The barrier-to-autointegration factor is a component of functional human immunodeficiency virus type 1 preintegration complexes, *J. Virol.*, 77, 5030–5036, 2003.

159. Gao, K., R.J. Gorelick, D.G. Johnson, and F. Bushman, Cofactors for human immunodeficiency virus type 1 cDNA integration *in vitro*, *J. Virol.*, 77, 1598–1603, 2003.

160. Brin, E. and J. Leis, HIV-1 integrase interaction with U3 and U5 terminal sequences *in vitro* defined using substrates with random sequences, *J. Biol. Chem.*, 277, 18357–18364, 2002.

161. Farnet, C.M. and F.D. Bushman, HIV-1 cDNA integration: requirement of HMG I(Y) protein for function of preintegration complexes *in vitro*, *Cell*, 88, 483–492, 1997.

162. Hindmarsh, P., T. Ridky, R. Reeves, M. Andrake, A.M. Skalka, and J. Leis, HMG protein family members stimulate human immunodeficiency virus type 1 and avian sarcoma virus concerted DNA integration *in vitro, J. Virol.,* 73, 2994–3003, 1999.

163. Yung, E., M. Sorin, A. Pal, E. Craig, A. Morozov, O. Delattre, J. Kappes, D. Ott, and G.V. Kalpana, Inhibition of HIV-1 virion production by a transdominant mutant of integrase interactor 1, *Nat. Med.,* 7, 920–926, 2001.

164. Mulder, L.C., L.A. Chakrabarti, and M.A. Muesing, Interaction of HIV-1 integrase with DNA repair protein hRad18, *J. Biol. Chem.,* 277, 27489–27493, 2002.

165. Chen, H. and A. Engelman, The barrier-to-autointegration protein is a host factor for HIV type 1 integration, *Proc. Natl. Acad. Sci. U.S.A.,* 95, 15270–15274, 1998.

166. Lee, M.S. and R. Craigie, Protection of retroviral DNA from autointegration: involvement of a cellular factor, *Proc. Natl. Acad. Sci. U.S.A.,* 91, 9823–9827, 1994.

167. Li, L., C.M. Farnet, W.F. Anderson, and F.D. Bushman, Modulation of activity of Moloney murine leukemia virus preintegration complexes by host factors *in vitro, J. Virol.,* 72, 2125–2131, 1998.

168. Turelli, P., V. Doucas, E. Craig, B. Mangeat, N. Klages, R. Evans, G. Kalpana, and D. Trono, Cytoplasmic recruitment of INI1 and PML on incoming HIV preintegration complexes: interference with early steps of viral replication, *Mol. Cell,* 7, 1245–1254, 2001.

169. Carteau, S., R.J. Gorelick, and F.D. Bushman, Coupled integration of human immunodeficiency virus type 1 cDNA ends by purified integrase *in vitro*: stimulation by the viral nucleocapsid protein, *J. Virol.,* 73, 6670–6679, 1999.

170. Yung, E., M. Sorin, E.J. Wang, S. Perumal, D. Ott, and G.V. Kalpana, Specificity of interaction of INI1/hSNF5 with retroviral integrases and its functional significance, *J. Virol.,* 78, 2222–2231, 2004.

171. Wei, P., M.E. Garber, S.M. Fang, W.H. Fischer, and K.A. Jones, A novel CDK9-associated C-type cyclin interacts directly with HIV-1 Tat and mediates its high-affinity, loop-specific binding to TAR RNA, *Cell,* 92, 451–462, 1998.

172. Garber, M.E. and K.A. Jones, HIV-1 Tat: coping with negative elongation factors, *Curr. Opinion Immunol.,* 11, 460–465, 1999.

173. Zhu, Y., T. Pe'ery, J. Peng, Y. Ramanathan, N. Marshall, T. Marshall, B. Amendt, M.B. Mathews, and D.H. Price, Transcription elongation factor P-TEFb is required for HIV-1 Tat transactivation *in vitro, Genes Dev.,* 11, 2622–2632, 1997.

174. Izmailova, E., F.M. Bertley, Q. Huang, N. Makori, C.J. Miller, R.A. Young, and A. Aldovini, HIV-1 Tat reprograms immature dendritic cells to express chemoattractants for activated T cells and macrophages, *Nat. Med.,* 9, 191–197, 2003.

175. Liou, L.Y., C.H. Herrmann, and A.P. Rice, Human immunodeficiency virus type 1 infection induces cyclin t1 expression in macrophages, *J. Virol.,* 78, 8114–8119, 2004.

176. Liou, L.Y., C.H. Herrmann, and A.P. Rice, HIV-1 infection and regulation of Tat function in macrophages, *Int. J. Biochem. Cell Biol.,* 36, 1767–1775, 2004.

177. McCloskey, T.W., M. Ott, E. Tribble, S.A. Khan, S. Teichberg, M.O. Paul, S. Pahwa, E. Verdin, and N. Chirmule, Dual role of HIV Tat in regulation of apoptosis in T cells, *J. Immunol.,* 158, 1014–1019, 1997.

178. Howcroft, T.K., K. Strebel, M.A. Martin, and D.S. Singer, Repression of MHC class I gene promoter activity by two-exon Tat of HIV, *Science,* 260, 1320–1322, 1993.

179. Malim, M.H., J. Hauber, R. Fenrick, and B.R. Cullen, Immunodeficiency virus Rev trans-activator modulates the expression of the viral regulatory genes, *Nature,* 335, 181–183, 1988.

180. Pollard, V.W. and M.H. Malim, The HIV-1 Rev protein, *Annu. Rev. Microbiol.,* 52, 491–532, 1998.

181. von Schwedler, U.K., M. Stuchell, B. Muller, D.M. Ward, H.Y. Chung, E. Morita, H.E. Wang, T. Davis, G.P. He, D.M. Cimbora, A. Scott, H.G. Krausslich, J. Kaplan, S.G. Morham, and W.I. Sundquist, The protein network of HIV budding, *Cell,* 114, 701–713, 2003.

182. Pelchen-Matthews, A., B. Kramer, and M. Marsh, Infectious HIV-1 assembles in late endosomes in primary macrophages, *J. Cell Biol,* 162, 443–455, 2003.

183. Nydegger, S., M. Foti, A. Derdowski, P. Spearman, and M. Thali, HIV-1 egress is gated through late endosomal membranes, *Traffic,* 4, 902–910, 2003.

184. Nie, Z., B.N. Phenix, J.J. Lum, A. Alam, D.H. Lynch, B. Beckett, P.H. Krammer, R.P. Sekaly, and A.D. Badley, HIV-1 protease processes procaspase 8 to cause mitochondrial release of cytochrome c, caspase cleavage and nuclear fragmentation, *Cell Death Differ.,* 9, 1172–1184, 2002.

2 Regulation of Apoptosis

Maria Eugenia Guicciardi
Gregory J. Gores

CONTENTS

INTRODUCTION

Multicellular organisms rely on a fine balance between cell proliferation and cell death for their health. Alterations in this balance inevitably lead to loss of tissue homeostasis and the onset of several diseases. More than 30 years ago, Kerr and coworkers described a novel form of cell death with peculiar morphological characteristics, and its crucial role in cellular biology became clear in the following years.[1] They named this type of cell death "apoptosis." Apoptosis is a highly organized and genetically controlled type of cell death, which has proven to be essential during embryonic development to ensure proper organogenesis.[2] It is characterized by a number of sequential morphological alterations, such as chromatin condensation, cell shrinkage, plasma membrane alterations with surface blebbing, and, in the latest stages, fragmentation of the cell into membrane-bound vesicles containing intact organelles called apoptotic bodies.[3] Removal of the apoptotic bodies is achieved via phagocytosis by neighboring cells and professional phagocytes such as macrophages. The prompt clearance of dying cells before the onset of secondary necrosis and lysis of the apoptotic bodies supposedly prevents the triggering of the immune response.

Apoptosis is also crucial in adult organisms, as it enables the elimination of unwanted aged, harmful, or infected cells to maintain normal cellular homeostasis and preserve tissue functions. Insufficient apoptosis can result in development of cancer or autoimmunity, whereas excessive cell death has been linked to acute injuries (such as stroke; myocardial infarct; fulminant hepatitis; reperfusion damage), chronic degenerative diseases (including multiple sclerosis, amyotropic lateral sclerosis, Alzheimer's disease, Parkinson's disease, and Huntington's disease), immunodeficiency (including acquired immunodeficiency syndrome [AIDS]); and infertility.

THE APOPTOTIC MACHINERY

Although several different stimuli can trigger apoptosis, the apoptotic signaling within the cell occurs mainly via two defined molecular pathways: the death receptor pathway (also called the extrinsic pathway) and the mitochondrial pathway (also called the intrinsic pathway). The extrinsic pathway is initiated by the engagement of a family of cell surface cytokine receptors named death receptors, and it generates an apoptotic signal that may or may not require the involvement of mitochondria for its execution. The intrinsic pathway is triggered by different external or internal signals that culminate in mitochondrial dysfunction, such as irradiations, oxidative stress, or some chemotherapeutic drugs. The mitochondrial dysfunction is characterized by disruption of mitochondrial architecture and function and release of proapoptogenic factors from the organelle (Figure 2.1). Both the intrinsic and the extrinsic pathways rely on a group of proteolytic enzymes called caspases (cysteinyl aspartate-specific proteases) as central executors of the cell death program. Indeed, activated caspases are responsible for the degradation of several cellular substrates that results in the morphological and biochemical alterations characteristic of the apoptotic phenotype. Endogenous inhibitors of caspases (i.e., inhibitors of apoptosis proteins [IAPs]) are present in the cell to prevent accidental caspase activation. Other proteolytic enzymes, such as calpain and the lysosomal cathepsins, can be activated during apoptosis, but their role in the apoptotic program has not been completely defined and seems to be dependent on cell type and nature of the apoptotic stimulus.[4]

Mitochondrial dysfunction in apoptosis is regulated by the Bcl-2 family of proteins, which includes both pro- and antiapoptotic members. These proteins are perhaps the most important regulators of the mitochondrial pathway, serving upstream and at the level of the mitochondria to integrate pro- and antiapoptotic signals. The balance between pro- and antiapoptotic members of the family, as well as their reciprocal interactions, determines whether the mitochondrial pathway of apoptosis is initiated or not. Although the extrinsic and the intrinsic pathways were initially thought to be independent, the BH3-only protein Bid has been found to provide a cross talk between the two pathways. Indeed, Bid is activated after death receptor engagement and translocates to the mitochondria, where it triggers the mitochondrial pathway of apoptosis. The above-mentioned components of the apoptotic machinery and their main endogenous inhibitors are here discussed in greater detail.

CASPASES

The signaling process by either the extrinsic or intrinsic pathway involves the activation of caspases. Caspases are cysteine proteases that cleave their substrates on the carboxyl side of aspartic acid residues. Their role as central executioner of cell death has been amply documented in the last decade[5] and is supported by the evidence that caspases seem to be evolutionarily conserved, being identified in *Drosophila melanogaster, Caenorhabditis elegans,* and *Xenopus laevis,* as well as in mammals. To date, 14 mammalian caspases have been identified, 12 of which have also been cloned in humans (Table 2.1). Caspases are constitutively and ubiquitously expressed as inactive proenzymes (zymogens) that require proteolytic processing for acquiring catalytic activity. The proenzyme possesses a variable-length N-terminal prodomain, a large subunit containing the active site cysteine in a conserved QACXG pentapeptide motif, and a C-terminal small subunit. By sequential processing of two Asp cleavage sites, the prodomain is removed, and the large and small subunits associate to provide the active form of the enzyme, a tetramer composed of two heterodimers.[6] Their specificity to cleave substrates on the carboxy-terminal side of an Asp residue, together with their need to be proteolytically activated by cleavage at Asp sites, renders the caspases capable of self-activation as well as of activating each other in a cascade-like process. The mammalian caspases can be divided into two families: those primarily involved in apoptosis (caspase-2, -3 -6, -7, -8, -9, -10, and -12) and those mainly involved in inflammatory processes and cytokine processing (caspase-1, -4, -5, and -11). Based on the length of their prodomains, the caspases involved in

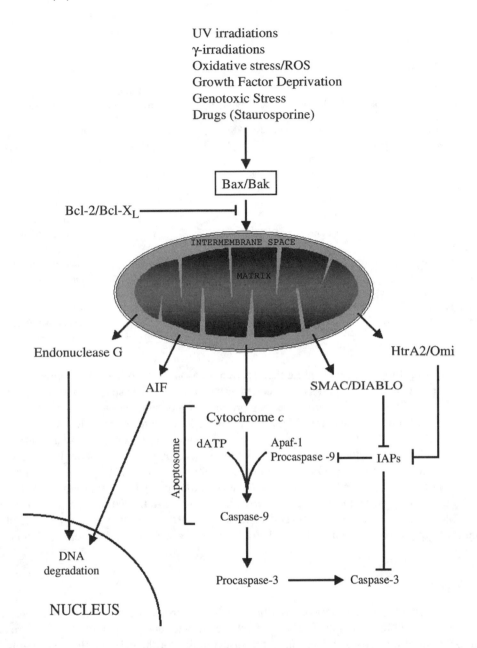

FIGURE 2.1 The mitochondrial/intrinsic pathway of apoptosis. Different stimuli, including UV and γ-irradiation, growth factor deprivation, oxidative stress with production of reactive oxygen species (ROS), genotoxic stress, and several drugs, trigger the intrinsic pathway via the activation of proapoptotic members of the Bcl-2 family of protein (i.e., Bax, Bak), which oligomerize on the outer mitochondrial membrane and cause mitochondrial dysfunction. The proapoptotic action of Bax and Bak can be antagonized by the antiapoptotic members of the same family Bcl-2 or Bcl-X_L. After mitochondrial dysfunction, several apoptogenic factors, including cytochrome c, Smac/DIABLO, Htr2A/Omi, AIF, and endonuclease G, are released from the mitochondrial intermembrane space. Cytochrome c binds to Apaf-1 and procaspase-9 to form a complex named apoptosome, which, in an ATP-requiring reaction, results in the activation of the initiator caspase-9. Caspase-9, in turn, activates the effector caspases (caspase-3, -6, and -7) responsible for the degradation of cellular substrates. Smac/DIABLO and Htr2A/Omi contribute to caspase activation by binding and inactivating the endogenous inhibitor of caspase IAPs. AIF and endonuclease G translocate to the nucleus, where they induce DNA degradation.

TABLE 2.1
Classification of Mammalian Caspases

Name	Human Homolog	Prodomain	Adaptor	Protein–Protein Interaction Motifs	Function
Caspase-1	Yes	Long	CARDIAK	CARD	Inflammation
Caspase-2	Yes	Long	RAIDD/CRADD	CARD	Apoptosis (initiator)
Caspase-3	Yes	Short			Apoptosis (effector)
Caspase-4	Yes	Long	Unknown	CARD	Inflammation
Caspase-5	Yes	Long	Unknown	CARD	Inflammation
Caspase-6	Yes	Short			Apoptosis (effector)
Caspase-7	Yes	Short			Apoptosis (effector)
Caspase-8	Yes	Long	FADD	DED	Apoptosis (initiator)
Caspase-9	Yes	Long	Apaf-1	CARD	Apoptosis (initiator)
Caspase-10	Yes	Long	FADD	DED	Apoptosis (initiator)
Caspase-11	Yes	Long	Unknown	CARD	Inflammation
Caspase-12	Yes	Long	Unknown	CARD	Apoptosis
Caspase-13	No	Long	Unknown	CARD	Apoptosis
Caspase-14	No	Short			Apoptosis (effector)

apoptosis can be further divided into either upstream caspases (also known as initiator or apical caspases) or downstream caspases (also called effector or distal caspases). Upstream caspases (caspase-2, -8, -9, and -10) generally have long prodomains containing caspase-recruitment domain (CARD) or death-effector domain (DED) motifs that mediate the recruitment to caspase-activating adaptors (such as Fas-associated protein with death domain [FADD] or apoptotic activating factor 1 [Apaf-1]). The long prodomain also promotes self-association of procaspase molecules, resulting in their autocatalytic activation. On the contrary, downstream caspases (caspase-3, -6, and -7) lack the ability to self-associate and do not possess CARD motifs due to the short prodomains; therefore, they require cleavage by initiator caspases for their activation. Activated downstream caspases are responsible for the degradation of several cellular substrates associated with the morphological changes of apoptosis, including DNA degradation, chromatin condensation, and membrane blebbing.

Bcl-2 Family

The Bcl-2 family of intracellular proteins arbitrates the life-or-death decision in response to diverse physiologic and cytotoxic stimuli. The balance between its anti- and proapoptotic members regulates the mitochondrial integrity and determines the susceptibility of the cell to mitochondria-mediated apoptosis. For this reason, the Bcl-2 family is considered a critical, if not the most important, regulator of the mitochondrial pathway of apoptosis (also referred to as the intrinsic pathway).

Bcl-2 (B-cell lymphoma-2), the first member discovered and the prototype of the family, was cloned from the t(14;18) break point in human follicular lymphoma as a proto-oncogene with antiapoptotic properties.[7] Since then, a number of proteins structurally similar to Bcl-2 have been discovered, and they are generally referred to as the Bcl-2 family. To date, the mammalian Bcl-2 family comprises at least 20 members sharing various degrees of homology within four conserved regions named Bcl-2 homology (BH) 1–4 domains.[8] The family can be further divided into three main subclasses, defined, in part, by this homology and, in part, by their functions (Table 2.2). A first subclass includes proteins with the highest homology to Bcl-2 in all the four conserved BH 1–4 domains. Members of this group are Bcl-2, Bcl-X$_L$, Mcl-1, A1, and Bcl-w, all of which strongly inhibit apoptosis in response to many cytotoxic insults. They are generally localized on the mitochondrial outer membrane, but some have

TABLE 2.2
The Bcl-2 Family

Class	Name	BH Domains	TM	Function in Apoptosis	Organism
Bcl-2 like	Bcl-2	BH 1–4	Yes	Inhibits	Mammalian
	Bcl-X$_L$	BH 1–4	Yes	Inhibits	Mammalian
	Bcl-X$_S$	BH 3–4	Yes	Promotes	Mammalian
	Bcl-W	BH 1–4	Yes	Inhibits	Mammalian
	Mcl-1	BH 1–4	Yes	Inhibits	Mammalian
	A1/Bfl-1	BH 1–4	No	Inhibits	Mammalian
	Boo/Diva	BH 4,1–2	Yes	Inhibits	Mammalian
	CED-9	BH 4,1–2	Yes	Inhibits	C. elegans
	E1B 19K	BH 1	No	Inhibits	Adenovirus
	BHRF-1	BH 1–2	Yes	Inhibits	Epstein–Barr virus
	KS-Bcl-2	BH 1–2	Yes	Inhibits	Human herpesvirus-8
	ORF-16	BH 1–2	Yes	Inhibits	Herpesvirus Saimiri
	LMW5-HL	BH 1–2	No	Inhibits	African swine fever virus
Bax-like	Bax	BH 1–3	Yes	Promotes	Mammalian
	Bak	BH 1–3	Yes	Promotes	Mammalian
	Bok	BH 1–3	Yes	Promotes	Mammalian
BH3-only	Bad	BH 3	No	Promotes	Mammalian
	Bid	BH 3	No	Promotes	Mammalian
	Bik	BH 3	Yes	Promotes	Mammalian
	Blk	BH 3	Yes	Promotes	Mammalian
	Hrk	BH 3	Yes	Promotes	Mammalian
	BNIP3/Nix	BH 3	Yes	Promotes	Mammalian
	Bim/Bod	BH 3	Yes	Promotes	Mammalian
	Bmf	BH 3	?	Promotes	Mammalian
	Puma	BH 3	?	Promotes	Mammalian
	Noxa	BH 3	No	Promotes	Mammalian
	EGL-1	BH 3	No	Promotes	C. elegans

been found also on the endoplasmic reticulum and the nuclear membrane. The mechanisms by which they prevent mitochondrial dysfunction are still largely unclear. Nonetheless, Bcl-2 has been found to inhibit activation and oligomerization of the proapoptotic member Bax.[9] A second subclass, the so-called Bax-like proteins, includes the multidomain proapoptotic members of the family that share homology in the BH 1–3 domains. Members of this group are Bax, Bak, and Bok. Whereas Bok expression seems to be limited to reproductive tissues, Bax and Bak are widely distributed. In healthy cells, Bax is found in the cytosol as a monomer, but after an apoptotic stimulus, it undergoes conformation changes, integrates into the outer mitochondrial membrane, and oligomerizes.[10–12] Bak is an oligomeric integral mitochondrial membrane protein, but it also undergoes conformation changes during apoptosis and forms larger aggregates.[12,13] The third subclass comprises the BH3-only members, proapoptotic proteins containing only the BH3 domain. Members of this group are Bid, Bik, Bad, Hrk, Bim, Bmf, Noxa, and Puma. They act as sensors to promote apoptosis in response to proximal death signals or intracellular damage, generally (but not exclusively) by binding to and neutralizing their prosurvival relatives. To prevent unnecessary triggering of apoptosis, in healthy cells, they are subjected to tight regulation either by transcriptional control or posttranslational modification or cellular compartmentalization. Bid requires proteolytic cleavage to be fully activated after death receptor engagement. Bad activation and inactivation in response to growth and survival factors are regulated by its phosphorylation status and by sequestration in the cytosol through binding to 14-3-3 proteins. Noxa and Puma are transcriptionally regulated by p53 in response to DNA damage. Bim and Bmf are sequestered on the cytoskeleton by binding to dynein light-chain complexes 1 and 2 in the microtubular dynein motor complex (Bim) and actin filamentous myosin V motor complex (Bmf), and they can be activated by multiple stimuli.

Both types of proapoptotic proteins are required to initiate apoptosis: specific death signals result in the activation of BH3-only proteins, which, in turn, activate the multidomain proapoptotic members Bax or Bak either by directly interacting with them[14] or by binding to the antiapoptotic Bcl-2-like proteins and antagonizing their inhibitory functions.[15] Activation of either Bax or Bak triggers mitochondrial dysfunction and is required for apoptosis.[16] Indeed, inactivation of either Bax or Bak does not result in any severe phenotypic alterations, but simultaneous inactivation of both genes dramatically impairs apoptosis in many tissues.[16,17]

INHIBITORS OF APOPTOSIS

Because all cells contain the apoptotic machinery required for cell death, it is clear that cells must also have inherent inhibitory mechanisms to control for inappropriate activation of the machinery and to ensure survival. Over the past several years, cellular mechanisms and proteins have been identified that function at various points in the cell death process to prevent or inhibit uncontrolled activation of the death program. A critical checkpoint of apoptosis exists at the level of the mitochondria, which is mainly regulated by the Bcl-2 family of proteins, as described earlier in this chapter. Here, we report on endogenous inhibition of apoptosis at another crucial point, the activation of caspases.

Inhibitors of Apoptosis Proteins (IAPs)

The inhibitors of apoptosis proteins (IAPs) constitute a family of proteins that function as intrinsic regulators of the caspase cascade. To date, IAPs are the only endogenous proteins identified that regulate the activity of both initiator (caspase-9) and effector caspases (caspase-3 and -7). IAPs are characterized by one or more baculoviral IAP repeat (BIR) domains, cysteine- and histidine-rich protein folding domains of approximately 70 amino acids that chelate zinc.[18] However, because not all BIR-containing proteins have been associated with cell death, IAPs have been defined as BIR-containing proteins (BIRPs) with antiapoptotic properties.[19] The other members of the BIRPs family generally contain only a single BIR domain and function mainly in cytokinesis and chromatin segregation. Members of the mammalian family of IAPs include, among others, neuronal apoptosis inhibitory protein (NAIP), cellular IAP-1 and IAP-2 (c-IAP1 and c-IAP2), X-linked IAP (XIAP), and survivin. The members c-IAP-1, c-IAP-2, and XIAP contain three BIR domains and a carboxy terminal RING finger[20,21]; c-IAP-1 and c-IAP-2 also contain a CARD domain near their C-termini, but this seems to not be essential for the functional activity of the proteins.[22] Several IAPs, including XIAP, c-IAP-1, and c-IAP-2, bind directly to caspase-3, -7, and -9, but not to other caspases, such as caspase-1, -6, and -8.[22] In particular, it seems that IAPs with multiple BIR motifs bind to procaspase-9 through the third BIR domain and to the active forms of caspase-3 and -7 through the second BIR domain.[22] By targeting procaspase-9 and inhibiting its activation, the IAPs block the mitochondrial pathway of apoptosis, whereas by targeting active caspase-3 and -7, they block those stimuli that can bypass the mitochondria and directly target the effector caspases. The RING-finger domain of XIAP and c-IAP-1 can trigger the ubiquitination and degradation of both caspase-3 and -7[23] and IAP proteins in response to apoptotic stimuli.[24] This apparent contradiction can be explained by different cellular responses to various degrees of apoptotic stress. The RING-finger may function to suppress apoptosis in conditions of low apoptotic stimulus (via ubiquitination and proteasome-mediated degradation of caspases), whereas higher levels of apoptotic stress may trigger self-degradation of the IAPs and cell death.

Whereas the inhibitory effect of IAPs is crucial in healthy cells to prevent aberrant activation of caspases, this effect must be relieved when the cell is signaled to undergo apoptosis. This process is mediated by a mitochondrial protein named second mitochondria-derived activator of caspases

(Smac)[25] or direct IAP-binding protein with low pI (DIABLO).[26] Smac usually resides in the mitochondrial intermembrane space and is released into the cytosol after an apoptotic stimulus, together with cytochrome c. Whereas cytochrome c contributes to the apoptotic cascade by directly activating Apaf-1 and caspase-9, Smac interacts with multiple IAPs and relieves their inhibitory effect on both initiator and effector caspases. Smac binds to both the BIR2 and BIR3 domains of XIAP and, possibly, of c-IAP-1 and c-IAP-2. The conserved IAP-binding motif in Smac competes with that in caspase-9 for the binding to IAPs and thus releases caspase-9 from inhibition by IAPs. A motif located between BIR1 and BIR2 on XIAP is responsible for binding and inhibiting caspase-3 or -7.

FLICE/Caspase-8-Inhibitory Proteins (FLIPs)

Death receptor-mediated caspase activation can be modulated by a family of proteins called the FLIPs. FLIPs were first identified in viruses (v-FLIPs) as proteins with antiapoptotic activity containing two tandem DEDs that enable them to bind to the Fas death-inducing signaling complex (DISC) and to other death receptor DISCs, and to block caspase-8 activation.[27] A human cellular homologue was later identified and called c-FLIP (also known as FLAME-1, I-FLICE, or Casper, or CASH, or MRIT, or CLARP, or Usurpin).[28] c-FLIP exists in short- and long-splicing variants. The short form, c-FLIP$_S$, consists only of two DEDs and is similar in structure to v-FLIP. The long form, c-FLIP$_L$ (also called I-FLICE), resembles caspase-8 and caspase-10 in its structure, and it consists of two DEDs along with an inactive caspase-like catalytic domain, where the active site cysteine residue embedded in the conserved pentapeptide QACRG or QACQG motif present in all caspases is replaced by a tyrosine. For this reason, c-FLIP$_L$ shows no cysteine protease activity. Both forms of c-FLIP are thought to be recruited to the DISC on stimulation through interaction with their death domain (DD).[29] FLIPs compete with the caspases for binding to the adaptor FADD. Whereas c-FLIP$_S$ may directly compete with procaspase-8 for binding to the DISC, c-FLIP$_L$ allows the recruitment and initial cleavage of procaspase-8 to the DISC but prevents its complete proteolytic processing to generate the active subunits.[30] Displacement or inhibition of caspases by FLIPs prevents the onset of the proteolytic caspase cascade and blocks death-receptor-mediated apoptosis.

MOLECULAR PATHWAYS OF APOPTOSIS

Several stimuli that trigger apoptosis seem to do so by inducing activation of proteolytic enzymes—in particular, but not exclusively, caspases. In order to activate initiator caspases, several molecules of caspase zymogens have to be brought to close proximity to allow self-processing thanks to the moderate catalytic activity of the zymogens. This is achieved through specific adaptor proteins that facilitate the recruitment of initiator caspases into large complexes. Activated upstream caspases then initiate a proteolytic cascade culminating in activation of effector caspases, degradation of cellular substrates, and cell death. Here, we describe the main signaling pathways that promote initiator caspase activation and apoptosis following a variety of stimuli.

THE MITOCHONDRIAL PATHWAY (THE INTRINSIC PATHWAY)

A large variety of different stimuli, including ultraviolet or gamma irradiation, oxidative stress, growth factor deprivation, and several cytotoxic compounds including kinase inhibitors such as staurosporine, have been reported to trigger caspase activation and apoptosis via a mitochondrial pathway (Figure 2.1).[31] Although very different in nature, all these stimuli result in activation of proapoptotic members of the Bcl-2 family, which tightly regulates the mitochondrial function. The multidomain proapoptotic proteins, such as Bax and Bak, can be activated directly, or indirectly, via interaction with BH3-only proteins of the same family. For example, Bid is cleaved by caspase-8 to generate a C-terminal fragment (t-Bid) that interacts with Bax and Bak, triggering activation of these proteins and facilitating their insertion into the mitochondrial membrane, possibly through destabilization of the membrane. When activated, Bax and Bak permeabilize the outer mitochondrial membrane, causing the release of proapoptotic proteins, such as cytochrome c, Smac/DIABLO, HtrA2/Omi, apoptosis-inducing factor (AIF), and endonuclease G, from the intermembrane space. The mechanism(s) by which Bax or Bak induce mitochondrial permeabilization are still debated. Studies on artificial liposomes suggested that Bax and Bak oligomers might form channels on the outer mitochondrial membrane.[32] Alternatively, it has been proposed that Bax might interact with components of the permeability transition (PT) pore, such as the voltage-dependent anion channel (VDAC), on the outer mitochondrial membrane to create larger channels or to induce pore opening with subsequent mitochondrial swelling and membrane rupture.[9,33,34] After they enter the cytosol, the proapoptogenic proteins released from the mitochondria promote either caspase activation or DNA degradation via multiple signaling pathways (Figure 2.1). Cytochrome c binds to the adaptor Apaf-1, an adenosine triphosphate (ATP)-binding protein. Complex formation with cytochrome c increases Apaf-1 affinity for deoxyadenosine triphosphate (dATP) and induces conformational changes in the protein leading to oligomerization and exposure of a CARD domain. Through the CARD domain, Apaf-1 recruits procaspase-9 to form a complex referred to as the apoptosome.[35] The proximity of several molecules of procaspase-9 results in its proteolytical autoactivation and in the subsequent activation of downstream effector caspases, such as caspase-3 and -7, via a cascade mechanism. The simultaneous release of Smac/DIABLO and HtrA2/Omi potentiates the mitochondria-mediated activation of caspases. Smac/DIABLO binds to the IAPs in the cytosol and antagonizes their inhibitory effect on both initiator caspases (i.e., caspase-9) and effector caspases (i.e., caspase-3 and -7). HtrA2/Omi, a serine protease, contains an N-terminal amino acid sequence almost identical to the N-terminal sequence of Smac/DIABLO and binds and inactivates IAPs in a way similar to Smac/DIABLO.[36] HrtA2/Omi serine protease activity also seems to be associated with its ability to activate caspases, but the mechanism is still largely unknown.[37] A second group of mitochondrial proapoptogenic factors contributes to the execution of the apoptotic program by promoting DNA degradation. AIF translocates from the cytosol to the nucleus after being released from the mitochondria.[38,39] Studies performed in cell-free systems demonstrated that, after entry into the nucleus, AIF induces chromatin condensation and large-scale DNA fragmentation, with no requirement for additional cytosolic factors.[38] Recombinant AIF, in the presence of cytosol, has also been reported to act on mitochondria with a feedback mechanism to induce disruption of the membrane potential and release cytochrome c. Both of these effects are caspase independent. Similar to AIF, endonuclease G also translocates to the nucleus after being released from the mitochondria, where it induces DNA fragmentation in a caspase-independent manner.[40,41]

THE DEATH RECEPTOR PATHWAY (THE EXTRINSIC PATHWAY)

The death receptors are a subset of the tumor necrosis/nerve growth factor (TNF/NGF) receptor superfamily. Members of the death receptor family include, among others, Fas (also called CD95 or APO-1), tumor necrosis factor-receptor 1 (TNF-R1, also called p55 or CD120a), and tumor necrosis factor-related apoptosis-inducing ligand-receptors 1 and 2 (TRAIL-R1 and TRAIL-R2, also called DR4 and DR5). When activated through interaction with their natural ligands, a group of complementary cytokines that belongs to the TNF family of proteins, these receptors are able to trigger an apoptotic signaling cascade.[42–44] The members of this family are all type-I transmembrane proteins with a cysteine-rich, extracellular N-terminal domain, a membrane-spanning region, and a C-terminal intracellular tail containing a sequence of approximately 80 amino acids (the DD). The DD allows for the recruitment of adaptor molecules, which function as "docking proteins" to bind the proform of the initiator caspase-8, -10, and, perhaps, -2.

Signal transduction by death receptors is initiated by the oligomerization of the receptor that follows the engagement of the ligand to the receptor's extracellular domain (Figure 2.2). As a

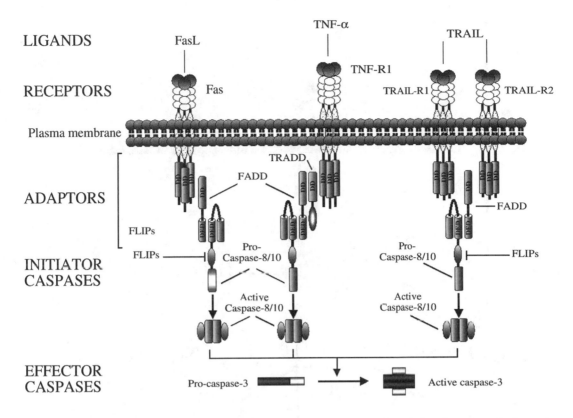

FIGURE 2.2 The death receptor/extrinsic pathway of apoptosis. Examples of signaling through the main death receptors (Fas/CD95/APO-1, TNF-R1, TRAIL-R1, and TRAIL-R2) are illustrated here. Engagement of death receptors by their cognate ligands results in oligomerization of the receptor and recruitment of adaptor proteins. The death domain (DD) on the adaptor protein interacts with the receptor's DD, whereas the death effector domain (DED) binds the correspondent DED in the prodomain of the inactive initiator caspase-8 or -10. The resulting complex is referred to as the death-inducing signaling complex (DISC). The proximity of several procaspase molecules results in the activation of the caspase by self-processing, which, in turn, activates downstream caspases such as caspase-3. The inhibitors of apoptosis cFLIPs prevent the recruitment or inhibit the processing of procaspase-8 to the DISC.

consequence, different DD-containing adaptor proteins, such as FADD or tumor necrosis factor receptor-associated protein with DD (TRADD), are recruited to the receptor through homophilic interaction of the receptor's and adaptor's DDs. Some adaptor proteins also contain a DED that mediates the recruitment of caspases through the association with correspondent DED or CARD motifs in the prodomain of the inactive initiator caspases. The resulting complex is referred to as the DISC. The proximity of several procaspase molecules recruited to the receptor results in self-processing and activation of the caspase, which subsequently starts a cascade of caspase activation by directly processing and activating the effector caspases (i.e., caspase-3, -6, and -7). With regard to Fas, this type of signaling is referred to as "type I" and is generally associated with high levels of caspase-8 recruited and activated at the DISC (Figure 2.3).[45] Alternatively, the activation of effector caspases may occur indirectly, via a mitochondria-mediated process, with a signal referred to as "type II." In type-II cells, the amount of caspase-8 activated at the DISC seems to be insufficient to directly cleave effector caspases, but it can efficiently cleave the BH3-only protein Bid to generate a fragment that translocates to the mitochondria and induces mitochondrial dysfunction.[45–47] After permeabilization of the outer mitochondrial membrane, several proapoptotic factors are released in the cytosol, where they contribute to caspase activation and apoptosis, as previously described for the mitochondrial pathway.

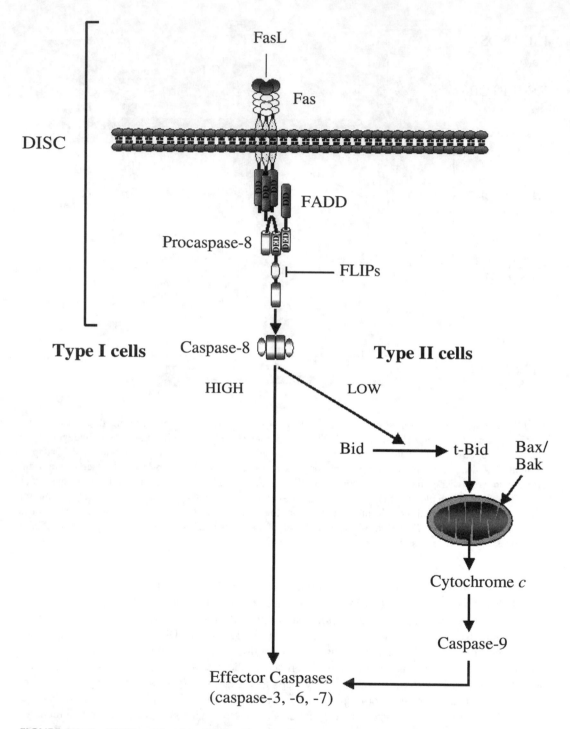

FIGURE 2.3 Fas/CD95-mediated apoptotic pathways in type I and type II cells. The amount of active caspase-8 generated at the DISC determines whether the cell activates a mitochondrial-independent (type I) or mitochondrial-dependent (type II) pathway of caspase activation and apoptosis.

SUMMARY AND CONCLUSION

From the initial description as a merely morphological phenomenon, the concept of apoptosis has evolved into an extremely complex, highly regulated biochemical and genetic process. Its crucial role in animal development and preservation of tissue homeostasis is documented by the presence of an evolutionarily conserved apoptotic machinery in virtually every organism. The capability of eliminating aged and potentially harmful cells in a controlled cellular environment is critical for maintaining the health of an organism, and it is not surprising that alterations in apoptosis, either leading to increased or reduced cell death, are generally associated with the development of pathologic conditions.

Although apoptosis can be triggered by many different extracellular or intracellular stimuli, the intracellular signaling proceeds generally through two main pathways: an intrinsic, mitochondria-mediated pathway and an extrinsic, death receptor–mediated pathway. Both pathways ultimately lead to the activation of effector caspases that are directly responsible for the degradation of cellular substrates. In a simplified version of the overall process of cell death, apoptotic stresses acting through the intrinsic pathway trigger the activation of proapoptotic members of the Bcl-2 family, the most important regulators of mitochondrial function. The balance of pro- and antiapoptotic Bcl-2 proteins determines the outcome of the process. Sufficient activation of the proapoptotic proteins results in permeabilization of the outer mitochondrial membrane and release of several proapoptogenic proteins, including cytochrome *c* and Smac/DIABLO. The release of cytochrome *c* triggers the formation of a complex (apoptosome) where initiator caspase-9 is activated. At the same time, Smac/DIABLO antagonizes the caspase-inhibiting effect of the endogenous IAPs. Apoptotic signals initiated at the level of plasma membrane after engagement of the death receptors result in activation of the initiator caspase-8, -10, and, possibly, -2. These caspases, then, can activate downstream caspases directly or indirectly by triggering the mitochondrial pathway via cleavage and activation of the BH3-only protein Bid. Ultimately, the two pathways converge when activated initiator caspases (caspase-8, -9, and -10) cleave and activate downstream caspases, such as caspase-3 and -7.

ACKNOWLEDGMENT

This work was supported by grants from the National Institute of Health DK 41876 (to G.J.G.) and the Mayo Foundation, Rochester, Minnesota.

REFERENCES

1. Kerr, J.F.R., Wyllie, A.H. and Currie, A.R., Apoptosis: a basic biological phenomenon with wide-ranging implications in tissue kinetics, *Br. J. Cancer,* 26, 239, 1972.
2. Jacobson, M.D., Weil, M. and Raff, M.C., Programmed cell death in animal development, *Cell,* 88, 347, 1997.
3. Wyllie, A.H., Kerr, J.F. and Currie, A.R., Cell death: the significance of apoptosis, *Int. Rev. Cytol.,* 68, 251, 1980.
4. Leist, M. and Jaattela, M., Four deaths and a funeral: from caspases to alternative mechanisms, *Nat. Rev. Mol. Cell Biol.,* 2, 589, 2001.
5. Nicholson, D.W. and Thornberry, N.A., Caspases: killer proteases, *Trends Biochem. Sci.,* 22, 299, 1997.
6. Earnshaw, W.C., Martins, L.M. and Kaufmann, S.H., Mammalian caspases: structure, activation, substrates, and functions during apoptosis, *Annu. Rev. Biochem.,* 68, 383, 1999.
7. Vaux, D.L., Cory, S. and Adams, J.M., Bcl-2 gene promotes haemopoietic cell survival and cooperates with c-myc to immortalize pre-B cells, *Nature,* 335, 440, 1988.
8. Cory, S. and Adams, J.M., The Bcl2 family: regulators of the cellular life-or-death switch, *Nat. Rev. Cancer,* 2, 647, 2002.

9. Antonsson, B., Conti, F., Ciavatta, A., Montessuit, S., Lewis, S., Martinou, I., Bernasconi, L., Bernard, A., Mermod, J.J., Mazzei, G., Maundrell, K., Gambale, F., Sadoul, R. and Martinou, J.C., Inhibition of Bax channel-forming activity by Bcl-2, *Science*, 277, 370, 1997.

10. Hsu, Y.T. and Youle, R.J., Bax in murine thymus is a soluble monomeric protein that displays differential detergent-induced conformations, *J. Biol. Chem.*, 272, 10777, 1998.

11. Antonsson, B., Montessuit, S., Sanchez, B. and Martinou, J.C., Bax is present as a high molecular weight oligomer/complex in the mitochondrial membrane of apoptotic cells, *J. Biol. Chem.*, 276, 11615, 2001.

12. Nechushtan, A., Smith, C.L., Lamensdorf, I., Yoon, S.H. and Youle, R.J., Bax and Bak coalesce into novel mitochondria-associated clusters during apoptosis, *J. Cell Biol.*, 153, 1265, 2001.

13. Wei, M.C., Lindsten, T., Mootha, V.K., Weiler, S., Gross, A., Ashiya, M., Thompson, C.B., and Korsmeyer, S.J., tBID, a membrane-targeted death ligand, oligomerizes BAK to release cytochrome c, *Genes Dev.*, 14, 2060, 2000.

14. Wang, K., Yin, X.M., Chao, D.T., Milliman, C.L. and Korsmeyer, S.J., BID: a novel BH3 domain-only death agonist, *Genes Dev.*, 10, 2859, 1996.

15. Cheng, E.H., Wei, M.C., Weiler, S., Flavell, R.A., Mak, T.W., Lindsten, T. and Korsmeyer, S.J., BCL-2, BCL-X(L) sequester BH3 domain-only molecules preventing BAX- and BAK-mediated mitochondrial apoptosis, *Mol. Cell*, 8, 705, 2001.

16. Wei, M.C., Zong, W.X., Cheng, E.H., Lindsen, T., Panoutsakopoulou, V., Ross, A.J., Roth, K.A., MacGregor, G.R., Thompson, C.B. and Korsmeyer, S.J., Proapoptotic BAX and BAK: a requisite gateway to mitochondrial dysfunction and death, *Science*, 292, 727, 2001.

17. Lindsten, T., Ross, A.J., King, A., Zong, W.X., Rathmell, J.C., Shiels, H.A., Ulrich, E., Waymire, K.G., Mahar, P., Frauwirth, K., Chen, Y., Wei, M., Eng, V.M., Adelman, D.M., Simon, M.C., Ma, A., Golden, J.A., Evan, G., Korsmeyer, S.J., MacGregor, G.R. and Thompson, C.B., The combined functions of proapoptotic Bcl-2 family members Bak and Bax are essential for normal development of multiple tissues, *Mol. Cell*, 6, 1389, 2000.

18. Deveraux, Q.L. and Reed, J.C., IAP family proteins — suppressors of apoptosis, *Genes Develop.*, 13, 239, 1999.

19. Miller, L.K., An exegesis of IAPs: salvation and surprises from BIR motifs, *Trends Cell. Biol.*, 9, 323, 1999.

20. Duckett, C.S., Nava, V.E., Gedrich, R.W., Clem, R.J., Van Dongen, J.L., Gilfilan, M.C. and al., e., A conserved family of cellular genes related to the baculovirus IAP gene and encoding apoptosis inhibitors, *EMBO J.*, 15, 2685, 1996.

21. Uren, A.G., Pakusch, M., Hawkins, C.J., Puls, K.L. and Vaux, D.L., Cloning and expression of apoptosis inhibitory protein homologs that function to inhibit apoptosis and/or bind tumor necrosis factor receptor-associated factors, *Proc. Natl. Acad. Sci. U.S.A.*, 93, 4974, 1996.

22. Roy, N., Deveraux, Q.L., Takahashi, R., Salvesen, G.S. and Reed, J.C., The c-IAP-1 and c-IAP-2 proteins are direct inhibitors of specific caspases, *EMBO J.*, 16, 6914, 1997.

23. Huang, H., Joazeiro, C.A., Bonfoco, E., Kamada, S., Leverson, J.D. and Hunter, T., The inhibitor of apoptosis, cIAP2, functions as a ubiquitin-protein ligase and promotes *in vitro* monoubiquitination of caspase-3 and -7, *J. Biol. Chem.*, 275, 26661, 2000.

24. Yang, Y., Fang, S., Jensen, J.P., Weissman, A.M. and Ashwell, J.D., Ubiquitin protein ligase activity of IAPs and their degradation in proteasomes in response to apoptotic stimuli, *Science*, 288, 874, 2000.

25. Du, C., Fang, M., Li, Y., Li, L. and Wang, X., Smac, a mitochondrial protein that promotes cytochrome c-dependent caspase activation by eliminating IAP inhibition, *Cell*, 102, 33, 2000.

26. Verhagen, A.M., Ekert, P.G., Pakusch, M., Solke, J., Connolly, L.M., Reid, G.E., Moritz, R.L., Simpson, R.J. and Vaux, D.L., Identification of DIABLO, a mammalian protein that promotes apoptosis by binding to and antagonizing IAP proteins, *Cell*, 102, 43, 2000.

27. Meinl, E., Fickenscher, H., Thome, M. and Tschopp, J., Anti-apoptotic strategies of lymphotropic viruses, *Immunol. Today*, 19, 474, 1998.

28. Tschopp, J., Irmler, M. and Thome, M., Inhibition of Fas death signals by FLIPs, *Curr. Opin. Immunol.*, 10, 552, 1998.

29. Scaffidi, C., Sanchez, I., Krammer, P.H. and Peter, M.E., The role of c-FLIP in modulation of CD95-induced apoptosis, *J. Biol. Chem.*, 274, 1541, 1999.

30. Krueger, A., Baumann, S., Krammer, P.H. and Kirchhoff, S., FLICE-inhibitory proteins: regulators of death receptor-mediated apoptosis, *Mol. Cell Biol.,* 21, 8247, 2001.

31. Green, D.R. and Reed, J.C., Mitochondria and apoptosis, *Science,* 281, 1309, 1998.

32. Antonsson, B., Montessuit, S., Lauper, S., Eskes, R. and Martinou, J.C., Bax oligomerization is required for channel-forming activity in liposomes and to trigger cytochrome c release from mitochondria, *Biochem. J.,* 345, 271, 2000.

33. Narita, M., Shimizu, S., Ito, T., Chittenden, T., Lutz, R.J., Matsuda, H. and Tsujimoto, Y., Bax interacts with the permeability transition pore to induce permeability transition and cytochrome c release in isolated mitochondria, *Proc. Natl. Acad. Sci. U.S.A.,* 95, 14681, 1998.

34. Vander Heiden, M.G., Chandel, N.S., Williamson, E.K., Schumacker, P.T. and Thompson, C.B., Bcl-xL regulates the membrane potential and volume homeostasis of mitochondria, *Cell,* 91, 627, 1997.

35. Li, P., Nijhawan, D., Budihardjo, I., Srinivasula, S.M., Ahmad, M., Alnemri, E.S. and Wang, X., Cytochrome c and dATP-dependent formation of Apaf-1/caspase-9 complex initiates an apoptotic protease cascade, *Cell,* 91, 479, 1997.

36. Suzuki, Y., Imai, Y., Nakayama, H., Takahashi, K., Takio, K. and Takahashi, R., A serine protease, HtrA2, is released from the mitochondria and interacts with XIAP, inducing cell death, *Mol. Cell,* 8, 613, 2001.

37. Suzuki, Y., Takahashi-Niki, K., Akagi, T., Hashikawa, T. and Takahashi, R., Mitochondrial protease Omi/HtrA2 enhances caspase activation through multiple pathways, *Cell Death Differ.,* 11, 208, 2004.

38. Susin, S.A., Zamzami, N., Castedo, M., Hirsch, T., Marchetti, P., Macho, A., Daugas, E., Geuskens, M. and Kroemer, G., Bcl-2 inhibits the mitochondrial release of an apoptogenic protease, *J. Exp. Med.,* 184, 1331, 1996.

39. Daugas, E., Susin, S.A., Zamzami, N., Ferri, K.F., Irinopoulou, T., Larochette, N., Prevost, M.C., Leber, B., Andrews, D., Penninger, J. and Kroemer, G., Mitochondrio-nuclear translocation of AIF in apoptosis and necrosis, *FASEB J.,* 14, 729, 2000.

40. Li, L.Y., Luo, X. and Wang, X., Endonuclease G is an apoptotic DNase when released from mitochondria, *Nature,* 412, 95, 2001.

41. van Loo, G., Schotte, P., van Gurp, M., Demol, H., Hoorelbeke, B., Gevaert, K., Rodriguez, I., Ruiz-Carrillo, A., Vandekerckhove, J., Declercq, W., Beyaert, R. and Vandenabeele, P., Endonuclease G: a mitochondrial protein released in apoptosis and involved in caspase-independent DNA degradation, *Cell Death Differ.,* 8, 1136, 2001.

42. Ashkenazi, A. and Dixit, V.M., Death receptors: signaling and modulation, *Science,* 281, 1305, 1998.

43. Locksley, R.M., Killeen, N. and Lenardo, M.J., The TNF and TNF receptor superfamily: integrating mammalian biology, *Cell,* 104, 487, 2001.

44. Nagata, S., Apoptosis by death factor, *Cell,* 88, 355, 1997.

45. Scaffidi, C., Fulda, S., Srinivasan, A., Friesen, C., Li, F., Tomaselli, K.J., Debatin, K.M., Krammer, P.H. and Peter, M.E., Two CD95 (APO-1/Fas) signaling pathways, *EMBO J.,* 17, 1675, 1998.

46. Li, H., Zhu, H., Xu, C.J. and Yuan, J., Cleavage of BID by caspase 8 mediates the mitochondrial damage in the Fas pathway of apoptosis, *Cell,* 94, 491, 1998.

47. Luo, X., Budihardjo, I., Zou, H., Slaughter, C. and Wang, X., Bid, a Bcl2 interacting protein, mediates cytochrome c release from mitochondria in response to activation of cell surface death receptors, *Cell,* 94, 481, 1998.

3 Approaches Used to Detect Apoptosis

Scott H. Kaufmann
Xue Wei Meng

CONTENTS

INTRODUCTION

During the last two decades, it has become evident that apoptosis plays an important role in a variety of physiological and pathological processes,[1] including acquired immunodeficiency syndrome (AIDS).[2] The growing awareness of the frequency and importance of apoptosis in these processes has resulted from an increased understanding of the biochemistry of apoptosis and concomitant development of methods for detecting this form of cell death. Moreover, a critical understanding of the relevant literature requires knowledge of the strengths and deficiencies of the assays involved. In the sections that follow, we outline the biochemical basis for many of the

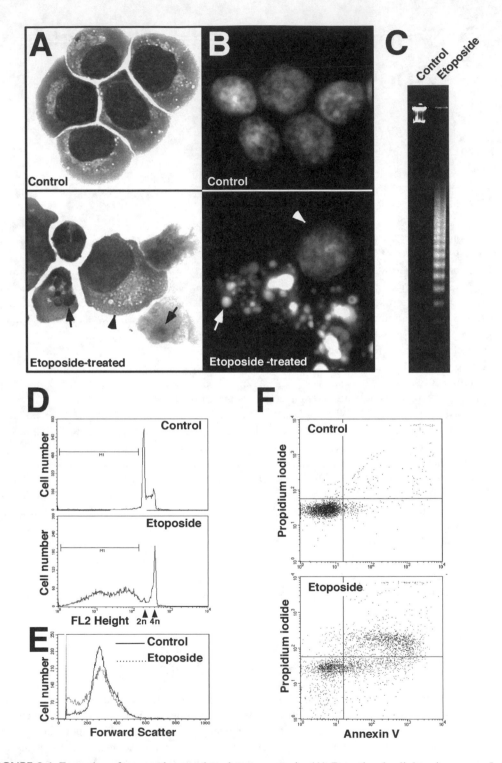

FIGURE 3.1 Examples of approaches used to detect apoptosis. (A) Detection by light microscopy. Jurkat human T cell leukemia cells were treated for 24 h with diluent (top) or 10 μM etoposide (bottom). Cytospins were prepared and stained with Wright's stain. Compared to nonapoptotic cells (top panel and arrowhead, bottom panel), apoptotic cells are smaller and contain one or more nuclear fragments (arrows, bottom panel).

techniques used to detect apoptotic cells and then summarize the strengths and weaknesses of each approach.

MORPHOLOGICAL ASSAYS OF APOPTOSIS

The original descriptions[3] characterized apoptosis as a form of cell death with distinct morphological features, which include shrinkage of the cell; detachment from its neighbors; condensation of the chromatin beginning in the areas adjacent to the inner nuclear membrane; fragmentation of the nucleus into multiple chromatin bodies, each surrounded by remnants of the nuclear envelope; packaging of the cell into multiple plasma membrane–enclosed cellular fragments; and phagocytosis of these fragments by neighboring cells.[4,5] Among these changes, chromatin condensation and nuclear fragmentation are the easiest to assess morphologically. The chromatin condensation reflects, at least to a first approximation, the action of nucleases that are activated during apoptosis.[6,7] Nuclear fragmentation requires cleavage of the lamins, major structural proteins of the nuclear envelope, by caspases.[8,9] Thus, the morphological changes described above reflect biochemical alterations that occur in apoptotic cells. These morphological changes can be assessed by electron microscopy, by conventional light microscopy, or by fluorescence microscopy, after staining of the DNA. Each of these approaches has advantages and disadvantages.

ELECTRON MICROSCOPY

The preceding description of apoptotic morphological changes was initially based on electron microscopy of cells after exposure to a variety of toxic stimuli.[4] Application of this technique to relatively uniform cell populations such as tissue culture cell lines or lymphoid cells undergoing apoptosis after treatment with proapoptotic stimuli permitted precise description and temporal

---

FIGURE 3.1 (*Continued*) (B) Detection by fluorescence microscopy after DNA staining. Jurkat cells were treated for 24 h with diluent (top) or 2.5 µM etoposide (bottom). After fixation in 3:1 (v:v) methanol:acetic acid, cells were dropped onto coverslips, treated with 1 µg/ml Hoechst 33258 in 50% glycerol containing 50 mM Tris-Hcl (pH 7.4), and examined by fluorescence microscopy. Compared to nonapoptotic cells (top panel and arrowhead, bottom panel), etoposide-treated cells show chromatin condensation (bright staining) and nuclear fragmentation (arrow). (C) Detection by agarose gel electrophoresis. HL-60 human acute myelomonocytic leukemia cells were treated for 6 h with diluent (left lane) or 68 µM etoposide (right lane). At the completion of the incubation, total cellular DNA was prepared by lysing cells in SDS containing 1 mg/ml proteinase K, extracting the resulting peptides in phenol/chloroform, and directly applying the DNA from 5 × 10^5 cells to nonadjacent lanes of a single 1.2% (w/v) agarose gel (see Kaufmann, S.H., *Cancer Res.*, 49, 5870, 1989.) Before drug treatment, the DNA runs at the exclusion limit of the gel. After induction of apoptosis, the DNA yields a typical nucleosomal ladder consisting of ~180 bp multiples beginning at 180 bp. (D) Detection of particles containing <2n DNA. Jurkat cells were treated for 24 h with diluent (top) or 4 µM etoposide (bottom) and washed once. After incubation at 4°C for 14 h in lysis buffer consisting of 0.1% (w/v) Triton X-100, 0.1% (w/v) sodium citrate, and 50 µg/ml prodium iodide (see Nicoletti, I., Migliorati, G., Pagliacci, M.C., Grignani, F., Riccardi, C. *J. Immunol. Methods*, 139, 271, 1991), cells were subjected to flow microfluorimetry. Arrowheads below lower panel indicate 2n and 4n DNA content. Note that the majority of nonapoptotic cells accumulate in the G2 phase of the cell cycle during etoposide exposure (see Cliby, W.A., Lewis, K.A., Lilly, K.K., Kaufmann, S.H. *J. Biol. Chem.*, 277, 1599, 2002). Particles containing a wide range of subdiploid DNA contents (denoted by M1 gates) contribute 6% of events in control cells and 77% of events in etoposide-treated cells. (E) Diminished forward scatter in etoposide-treated Jurkat cells. Jurkat cells treated for 24 h with diluent (control) or 10 µM etoposide were sedimented, resuspended in calcium-free and magnesium-free Dulbecco's phosphate-buffered saline, and immediately subjected to flow cytometry without fixation. Note the increased number of cells with low forward scatter after etoposide treatment. (F) Detection of annexin V binding. Jurkat cells treated with diluent (top) or 10 µM etoposide (bottom) for 24 h were stained with allophycocyanin-conjugated annexin V and 10 µg/ml propidium iodide before flow cytometry. Note the increase in annexin V-positive cells, some of which also stain with propidium iodide.

ordering of the apoptotic morphological changes. The relatively widespread availability of electron microscopy permitted the utilization of this approach in a variety of settings.

It is important to emphasize, however, that there are also several limitations in using electron microscopy to study apoptosis. When intact tissues are studied under physiological conditions or after exposure to minimally toxic stimuli, a "latent" phase that can vary dramatically from cell to cell is observed.[10] As a consequence, only a small fraction of cells might become apoptotic during any particular period of time. Because the number of cells examined in any thin section is also relatively small, electron microscopy is not appropriate for quantitation of the number of cells undergoing apoptosis. Finally, because the length of time between detachment of a particular cell from its neighbors and engulfment of the resulting apoptotic bodies by neighboring cells (the "active" phase of apoptosis) is typically only 30 to 60 minutes,[11] it is possible to underestimate the number of cells that have undergone apoptosis in a tissue if one looks only for cells that have the classical hallmarks of apoptosis.

LIGHT MICROSCOPY

Apoptotic cells can also be detected[4,11,12] in tissue sections after staining with standard histochemical dyes such as hematoxylin and eosin. Apoptotic chromatin condensation is evident in these cells as "pyknosis" (i.e., very dense staining of chromatin by hematoxylin). A similar approach can be applied to cytospin preparations of leukocytes or tissue culture cells, as illustrated in Figure 3.1A. This approach has several advantages. First, large numbers of cells can be examined. Second, this is a low-cost, low-technology technique that requires only a light microscope and, if sections are examined, a microtome. Third, stained sections from large numbers of patients with various diseases are available in pathology archives.

There are, however, several limitations to this approach. First, if a small fraction of the cells are undergoing apoptosis at the time the tissue is harvested, it can potentially be difficult to distinguish apoptotic cell fragments from cells that are grazed by the plane of sectioning when the slides are cut. Second, depending upon the degree of nuclear fragmentation, it can be difficult to distinguish apoptotic cells from normal cells with condensed chromatin (e.g., small lymphocytes). Nonetheless, in the hands of an experienced pathologist, this approach is quite useful.

FLUORESCENCE MICROSCOPY AFTER DNA STAINING

Apoptosis can also be detected in tissue culture by staining with a DNA binding dye, such as Hoechst 33258 or 4,6-diamidino-2-phenylindole (DAPI), and examining cells under a fluorescent microscope. Hallmark changes of chromatin condensation and nuclear fragmentation are readily visible by this technique (Figure 3.1B). For some dyes, such as Hoechst 33258, permeabilization with a detergent or organic solvent is required for the dye to enter the cell. For others (e.g., DAPI), dye uptake into nonpermeabilized cells is sufficient for the fluorescence of highly condensed nuclei to be readily visible.

Like light microscopy, this approach has the advantage that a large number of cells can be examined, potentially making quantitation of the number of apoptotic cells more accurate than electron microscopy. If this approach is used to examine slides prepared from tissue culture cells, however, it is important to assess whether apoptotic cells are uniformly distributed throughout the slide. In addition, when fluorescence microscopy is applied to tissue sections, nuclei of quiescent cells (e.g., small lymphocytes) must be carefully distinguished from apoptotic nuclei.

DETECTION OF DNA FRAGMENTATION

DNA from most mammalian cells undergoing apoptosis displays a characteristic series of bands (a so-called nucleosomal ladder, Figure 3.1C) after agarose gel electrophoresis.[13] This fragmentation pattern results from preferential cleavage of DNA in the linker regions between nucleosomes compared with the DNA that is wrapped around histone octamers.[14] Accordingly, this cleavage pattern is not specific for any particular endonuclease but, instead, reflects the structure of chromatin.

Because any nuclease that introduces DNA double-strand breaks will randomly yield this pattern when intact chromatin is digested,[15–17] there has been considerable debate about the nature and identity of the nuclease responsible for the apoptotic DNA fragmentation. Early indirect evidence implicated a variety of endonucleases, including DNase I,[18,19] DNase II,[20,21] and DNase γ.[22,23] Subsequent studies demonstrated that caspase-activated deoxyribonuclease (CAD) plays a critical, nonredundant role in apoptotic internucleosomal cleavage in a number of different cell types.[24–27] According to current understanding, this endonuclease is complexed with inhibitor of CAD (ICAD), which acts as both a chaperone (to permit folding of the nascent CAD polypeptide) and an inhibitor.[24] The constitutively expressed CAD/ICAD complex resides in nuclei,[28] where cleavage of ICAD by caspase-3, a protease activated during apoptosis (see below), frees CAD[29,30] and allows it to begin digesting chromatin.

Even though the discovery of CAD and ICAD provides a plausible link between other apoptotic events and endonuclease activation, other nucleases might also contribute to apoptotic internucleosomal DNA degradation in certain cell types. Targeted disruption of the ICAD[26] or CAD[27] genes abolishes apoptosis-associated internucleosomal cleavage in some cell types[26,27] but not others.[31] In the latter cells, endonuclease G released from mitochondria during apoptosis has been implicated in internucleosomal cleavage.[31]

Regardless of which endonuclease performs the cleavage, the demonstration that nucleases are activated during apoptosis has led to the development of a variety of techniques for detecting apoptotic cells.

CONVENTIONAL AGAROSE GEL ELECTROPHORESIS

The original description of internucleosomal DNA fragmentation during apoptosis relied on agarose gel electrophoresis.[13] For this technique, DNA was prepared from cells or tissues using any of a number of protocols. One commonly used approach involved solubilization of cellular contents in sodium dodecyl sulfate (SDS), digestion of polypeptides with protease K, and extraction with phenol to remove peptide fragments. Application of the resulting total cellular DNA to an agarose gel with suitable separation properties (e.g., 1 to 2% [w/v] agarose) readily demonstrated a ladder of oligonucleosomal fragments.

The advantages of detecting DNA fragmentation by this approach are its relatively low cost and simplicity. There are, however, several disadvantages as well. First, the technique is relatively insensitive. Because the nucleosomal fragments are spread throughout the length of the gel, a substantial fraction of the total DNA must be cleaved before fragments are detectable using ethidium bromide. Second, the technique remains inherently qualitative rather than quantitative. Although agarose gel electrophoresis distinguishes between internucleosomal DNA degradation and the random DNA fragmentation that accompanies necrosis,[11] it is difficult to determine whether 10% or 25% of the total DNA in a particular lane is present in the nucleosomal ladder. Merely photographing agarose gels and digitizing the images does not yield valid quantification unless appropriate controls to assure linearity are performed. Finally, even if the amount of cleaved DNA is precisely quantitated, the interpretation of the result is open to question. Conversion of 25% of the total DNA to nucleosomal fragments might indicate cleavage of a quarter of the DNA in 100% of the cells or cleavage of 100% of the DNA in a quarter of the cells. For these reasons, assays that assess apoptosis on a cell-by-cell basis (see the sections entitled "Flow Cytometry," "Terminal Deoxyribonucleotidyl Transferase-Mediated dUTP Nick-End Labeling (TUNEL)," and "Comet Assay") are preferred.

EXTRACTION OF SMALL FRAGMENTS

Alternative methods of quantitating DNA fragmentation have also been developed. The first of these is based on the observation that double-stranded DNA fragments below 10 to 20 kilobases are extracted from nuclei in buffers containing magnesium chelators.[32,33] To extract these fragments, cells undergoing apoptosis are lysed in buffer consisting of 20 mM Tris (pH 7 to 8), ethylenedi-aminetetraacetic acid (EDTA), and a neutral detergent. After a 10- to 30-minute incubation at 4°C, the intact chromatin is pelleted, which leaves the nucleosomal fragments in the supernatant. DNA in the two fractions can then be assayed using colorimetric or fluorimetric techniques. Alternatively, if DNA in the cells is uniformly radiolabeled (i.e., greater than one cell cycle) before application of the apoptotic stimulus, DNA in the various fractions can be quantified by scintillation counting.

This approach is simple and readily applicable to large numbers of samples. Moreover, it provides relatively precise quantitation of how much DNA has been cleaved to oligomers of nucleosome-sized fragments. This information can be very useful in time-course studies of nuclease activation during apoptosis. By using this technique alone, however, it is impossible to determine whether partial DNA fragmentation results from complete degradation in a subset of the cells or partial degradation of the DNA in all of the cells. Moreover, because small DNA fragments resulting from necrosis will also be extracted under the conditions described above, this technique cannot be used to distinguish apoptosis from necrosis.

FLOW CYTOMETRY

Solubilization of oligonucleosomal fragments in buffers lacking divalent cations also plays a role in the detection of apoptotic cells by flow cytometry after staining with propidium iodide.[34] When this approach is applied, cells treated with an apoptotic stimulus are fixed with ethanol, washed with an aqueous buffer lacking divalent cations (e.g., 0.1% sodium citrate), treated with RNase A, reacted with propidium iodide, and subjected to flow cytometry.[35] Alternatively, unfixed cells are lysed in buffer containing neutral detergent and EDTA, stained with propidium iodide, and subjected to flow cytometry (Figure 3.1D).[36] In either case, low-molecular-weight fragments are extracted from the cells, decreasing their apparent DNA content. In addition, cell remnants containing fragments of the nucleus ("apoptotic bodies") will contain less than the full cellular complement of DNA. Both of these effects presumably contribute to the detection of apoptotic cells as particles containing less than the diploid amount of DNA (so-called "subdiploid cells").

This technique is simple, relies on widely available equipment, and is amenable to high-throughput analysis. Moreover, because thousands of particles can be analyzed in each sample, a precise determination of the percentage of subdiploid events is possible. Nonetheless, this approach has two notable limitations. Because the low-molecular-weight DNA fragments generated during the course of necrosis should also be extractable in cation-free aqueous buffers, this technique does not distinguish between apoptotic and necrotic cells. In addition, because one cell can give rise to multiple apoptotic bodies, counting the number of particles that contain <2n DNA can lead to an overestimation of the number of cells that have undergone apoptosis. Conversely, gating only large particles and ignoring the cell fragments can potentially underestimate the number of apoptotic cells. Despite these concerns, however, we have often observed a close agreement between the results of morphological assays and this flow-cytometry-based assay in human leukemia cell lines.[37]

FILTRATION ASSAYS

An alternative approach is based on the separation of low- and high-molecular-weight DNA by filtration. It has been known for 30 years that the rate at which deproteinated DNA passes through the pores of a filter is related to the size of the pores and the length of the DNA fragments.[38] This observation forms the basis of the alkaline and neutral elution techniques, which were widely used in the past to study DNA breakage and repair.[39] A variation of this technique was also applied to the quantitation of oligonucleosomal DNA cleavage during apoptosis.[40] In brief, cells are incubated in radiolabeled nucleotide for at least one cell cycle. After incubation in radio-inert medium for a period of time, cells are exposed to an apoptotic stimulus, gently applied to filters, and lysed by addition of buffer containing a deproteinizing detergent (e.g., sodium sarkosyl). Low-molecular-weight DNA fragments rapidly flow through the filter with the lysing solution, whereas high-molecular-weight DNA is retained on the filter unless a large volume of neutral or alkaline buffer is pumped through. By determining the amount of radiolabel in the flowthrough and on the filter, the amount of DNA that has been degraded to low-molecular-weight fragments can be determined.

The simplicity and speed of this technique make it suitable for high-throughput screening. Several points, however, must be kept in mind when using this approach. First, it is important to radiolabel the DNA with ^{14}C rather than ^{3}H, because ^{3}H induces apoptosis in some cell types.[41] Second, it is important to "chase" all of the radiolabel into the mature DNA before applying the apoptotic stimulus, because newly synthesized DNA has a high avidity for some types of filters. In addition, this method suffers from the same limitations as the techniques based on differential extraction of fragmented DNA (see the above section "Extraction of Small Fragments"). These methods do not distinguish apoptotic DNA degradation (nucleosomal fragments) from necrotic DNA degradation (fragments of random size). Moreover, when 25% of the DNA passes through the filter, it is unclear whether this results from complete degradation of the DNA in 25% of the cells or partial degradation in all of the cells.

TERMINAL DEOXYRIBONUCLEOTIDYL TRANSFERASE–MEDIATED dUTP NICK-END LABELING (TUNEL)

Compared to extraction or filtration assays, the labeling of cleaved DNA within cells *in situ* allows for more accurate determination of the percentage of cells that contain damaged DNA. The principle of this technique is relatively simple: terminal deoxyribonucleotidyl transferase (TdT), an enzyme that adds nucleosides to a free 3′ OH terminus of DNA,[42] will transfer nucleosides to a blunt 3′ end, a protruding 3′ end, or a recessed 3′ end in the presence of Co^{2+}, albeit with differing efficiencies.[43] The observation that free 3′ OH groups are generated during apoptosis has permitted development of several TdT-based techniques that allow detection of cleaved DNA in individual cells.[44,45] In brief, cells undergoing apoptosis are fixed, permeabilized, and incubated with biotinylated deoxyuridine triphosphate in the presence of TdT and Co^{2+}. After several washes to remove unincorporated nucleotide, specimens are incubated with fluorochrome-coupled streptavidin, which avidly complexes with biotin, and are examined by fluorescence microscopy or flow cytometry. Alternatively, substitution of horseradish peroxidase-coupled streptavidin permits detection of biotinylated nucleotide by light microscopy after reaction with a chromogenic peroxidase substrate.

TUNEL staining has some advantages over agarose gel electrophoresis. First, because individual cells are examined, it is possible to determine the percentage of cells that contain fragmented DNA. Second, the ability of flow cytometry to detect low levels of fluorescence makes this approach extremely sensitive. Third, because cell morphology and staining intensity can be examined simultaneously, it is possible to compare the time course of DNA fragmentation with morphological changes in individual cells.

On the other hand, it is important to keep in mind that TUNEL staining is an assay for DNA cleavage, not for apoptosis. When the sensitivity of the TUNEL technique is pushed toward its limits, the technique will begin to pick up the nicks that occur under physiological conditions[46] (e.g., from topoisomerase action). In addition, the technique can stain DNA ends that result from the process of sectioning tissues.[46] Finally, several groups have reported that the TUNEL technique stains cells undergoing necrosis.[47–52] These observations serve as a reminder that the random DNA degradation occurring during necrosis also generates free 3′ OH ends.

COMET ASSAY

Single-cell agarose gel electrophoresis provides an alternative approach for detecting DNA fragmentation at the single-cell level.[53] After exposure to an apoptotic stimulus, cells are embedded in agarose, subjected to an electric field under neutral conditions, stained with a fluorescent DNA binding dye, and examined by fluorescence microscopy. After these manipulations, DNA that has not been cleaved remains in the nucleus, and fragmented DNA migrates toward the anode. The result is a structure resembling a comet.[54] With appropriate imaging, the amount of DNA in the head and tail can be quantified, and a weighted measure of DNA migration ("tail moment") can be calculated.[55] Because previous studies have demonstrated that small fragments migrate further than large fragments,[53] this tail moment will reflect both size and amount of DNA degraded.

In contrast to agarose gel electrophoresis, this technique allows assessment of DNA fragmentation at the single-cell level. Moreover, in contrast to TUNEL, the comet assay can be used to estimate the size of the fragments generated during the process of apoptosis.[53,55]

Despite these advantages, the comet assay has some potential limitations. First, this assay is relatively labor intensive compared with agarose gel electrophoresis or TUNEL staining. Second, application of the comet assay on a quantitative basis (e.g., to estimate the size distribution of DNA fragments within each cell) requires imaging and computational techniques that are not widely available. Finally, it is important to emphasize that the comet assay (like TUNEL) measures DNA strand breaks, not apoptosis. The primary strand breaks observed after γ-irradiation have been widely studied using the comet assay.[55,56] In addition, it has been reported that necrotic cells yield a comet when this approach is applied.[53] Nonetheless, when the occurrence of apoptosis has been documented by other techniques, the comet assay can be useful for studying the process on a cell-by-cell basis.

CASPASE ACTIVATION

Caspases comprise a family of cysteine proteases that cleave on the C-terminal side of aspartate residues.[57–59] Several observations have implicated these proteases in the apoptotic process. First, many of the proteolytic cleavages that occur during apoptosis[60] are catalyzed by caspases.[59,61] Second, caspase gene deletion delays or prevents apoptosis in certain cell types during development or after treatment with certain stimuli.[59,62] Finally, caspase inhibitors also prevent or alter the cell death phenotype after certain stimuli. It is important to emphasize, however, that other proteases, including cathepsins, calpains, and granzymes, were also implicated in proteolytic events during apoptosis in certain cell types.[63–66]

Not all caspases participate in apoptosis. Of the 12 human caspases, caspase-1, -4, -5, and -12 are thought to play a role in inflammatory cytokine maturation.[67,68] Whereas caspase-3, -6, -8, -9, and -10 have defined roles in apoptosis,[59,69] the exact functions of caspase-2, -7, and -14 remain to be defined.[70]

Based on their biochemistry and functions, the apoptotic caspases may be divided into two classes.[59,69] The initiator caspases (caspase-8, -9, and -10) transduce other biochemical signals into proteolytic activity. The effector caspases (caspase-3, -6, and possibly -7) are responsible for cleavages that help disassemble the cell. All of these proteases are synthesized as enzymatically intactive zymogens and then activated during apoptosis. The processes of activation, however, differ between initiator and effector caspases.

Initiator caspases are constitutively present as monomers and become activated as a consequence of multimerization.[71] In the case of caspase-8 (and presumably caspase-10), the inactive zymogen monomers[71] bind through their N-terminal protein–protein interaction domains to the adaptor molecule FADD (Fas-associated protein with death domain), which is assembled into multimers on the cytoplasmic surface of ligated death receptors (see Chapter 2). The resulting caspase-8 dimers undergo an apparent conformational change that results in acquisition of limited enzymatic activity.[72] These enzymatically active caspase-8 dimers then cleave neighboring caspase-8 dimers to yield the mature enzyme, which lacks the N-terminal protein interaction domains and has a different substrate specificity from the enzymatically active intermediates.[72]

Caspase-9 activation occurs by a related but distinct process. Treatment of cells with many apoptotic stimuli leads to release of cytochrome *c* and other intermembrane proteins from mitochondria.[73] After it is released, cytochrome *c* binds to a cytosolic scaffolding protein called apoptotic protease activating factor-1 (Apaf-1), which undergoes an ATP- or dATP-mediated conformational change and binds procaspase-9[74] to form a multimeric structure called an apoptosome.[75,76] In a manner that remains incompletely understood, binding to the apoptosome results in realignment of residues at the active site of caspase-9 into a conformation that allows catalytic activity.[77]

In contrast to initiator caspases, effector caspases are synthesized as enzymatically inactive homodimers.[71] Proteolytic cleavage at specific aspartate residues simultaneously produces two fragments (a small subunit and a large subunit that is linked to a short prodomain) and induces a conformational change that allows for realignment of the active site residues into a catalytically competent conformation.[78] The catalytically active effector caspase then cleaves the larger of the two fragments to remove the short prodomain from the large subunit.[79]

As should be evident from the preceding description, caspase activation can be detected by a variety of techniques, including assays for enzymatic activity (ability to cleave synthetic substrates *in vitro* or natural substrates *in situ*), changes in molecular weights of the caspase molecules as they are proteolytically activated, and appearance of new immunoreactive epitopes as the caspases undergo activating conformational changes.[80,81] Each of these approaches has strengths and limitations.

ACTIVITY ASSAYS

As indicated above, a subset of caspases acquires enzymatic activity during the course of apoptosis. This activity can be detected using suitable fluorogenic or chromogenic synthetic substrates. Because caspase-active sites usually accommodate only four amino acids to the N-terminal side of the scissile bond, most synthetic substrates consist of four amino acids followed by a detectable leaving group. The notable exception is caspase-2, which reportedly recognizes five amino acids.[82] The sequences of substrates that are widely used for these assays are indicated in Table 3.1.

In preparation for measurement of caspase activity, cells treated with an apoptotic stimulus are washed and lysed in a hypotonic buffer so that a desired subcellular fraction (typically cytosol) can be isolated. Alternatively, cells are lysed in buffer containing a nonionic or zwitterionic detergent. After removal of insoluble material, lysates are incubated at 30 to 37°C with saturating concentrations of substrate. Release of fluorescent or colored product can be monitored over time or at the end of the incubation using a fluorimeter or spectrophotometer, respectively.[80,81] If cells undergoing spontaneous apoptosis are removed from a population before addition of an apoptotic stimulus, a 20- to 100-fold increase in caspase activity can be detected in some model systems.[83–85] A similar approach can be used to study caspase activation under cell-free conditions. When the amount of product released is compared to a standard curve, the result can be expressed in units that can be compared from laboratory to laboratory (e.g., picomoles of product produced/minute/mg of protein in the extract).

Several points need to be kept in mind in order to perform valid assays. First, enzyme assays must be performed under conditions for which the amount of product is a linear function of enzyme activity. This condition is met only if extensive substrate depletion is avoided. Second, assays should be performed using saturating substrate concentrations. If a substrate concentration below the K_m for the enzyme is used, changes in the amount of product produced might reflect either changes in v_{max} (e.g., altered number of enzyme molecules) or K_m. The typical interpretation that the number of enzyme molecules has been altered is valid only if the activity has been assayed under conditions for which alterations in K_m would not substantially affect product release (i.e., under saturating conditions).

This approach has several advantages over alternative methods for detecting caspase activation described below. The present approach is relatively simple and relies on widely available equipment. It is readily adaptable to microtiter plate format, facilitating high-throughput screening. Finally, it is inherently more quantitative than immunoblotting methods.

TABLE 3.1

Cleavage of Various Synthetic Substrates by Purified Caspases

Sequence	Cleaved by[b]
YVAD-X[a]	Caspase **1**, 4, 5
DEVD-X	Caspase 1, **3**, 4, **7**, 8
VEID-X	Caspase 1, 3, **6**, 8, 10
IETD-X	Caspase 1, 3 6, **8**, **10**
LEHD-X	Caspase 1, 4, 5, 6, 8, **9**, 10
VDVAD-X	Caspase **2**

[a]X = a detectable leaving group such as *p*-nitroaniline, 7-amino-4-methylcoumarin, or 7-amino-4-trifluoromethyl-coumarin.

[b]Caspases reported to cleave the indicated substrate. Bold indicates substrate with the greatest activity.

Source: Adapted from Earnshaw, W.C., Martins, L.M., and Kaufmann, S.H., *Ann. Rev. Biochem.*, 68, 383, 1999; Talanian, R.V., Quinlan, C. Trautz, A., Hackett, M.C., mankovich, J.A., Banach, D., Ghayur, T. Brady, K.D., and Wong, W.W., *J. Biol. Chem.*, 272, 9677, 1997; Thornberry, N.A., Rano, T.A., Peterson, E.P., Rasper, D.M., Timkey, T., Garcia-Calvo, M., Houtzager, V.M., Nordstrom, P.A., Roy, S., Vaillancourt, J.P., Chapman, K.T., and Nicholson, D.W., *J. Biol. Chem.*, 272, 17907, 1997; Margolin, N., Raybuck, S.A., Wilson, K.P., Chen, W., Fox, T., Gu, Y., and Livingston, D.J., *J. Biol. Chem.*, 272, 7223, 1997. With permission.

Nonetheless, this approach also has several limitations. Because of overlap in caspase specificities,[86] the presence of an activity that cleaves one of the substrates listed in Table 3.1 does not necessarily identify the caspase that has been activated. For example, caspase-3, which cleaves DEVD-AFC, also cleaves LEHD-AFC, albeit less efficiently.[81] Thus, the detection of LEHD-AFC cleavage might reflect activation of caspase-9, caspase-3, or both. In order to specifically detect caspase-9 activity, it might be necessary to deplete the more abundant caspase-3. Although this could, in principle, be done using a caspase-3 inhibitor, currently available inhibitors suffer from the same lack of specificity. Accordingly, caspase-3 must be depleted using the caspase-3 binding domain of X-linked inhibitor of apoptosis (XIAP)[87] or immunological reagents. Alternatively, immunoblotting assays described below are required to identify the caspases that are proteolytically activated.

Two other limitations must also be kept in mind. First, in some cell types, activated caspase-3 and -9 are sequestered in cytoplasmic aggregates that lack enzymatic activity.[85,88] This aggregation can lead to underestimation of the amount of activated caspases unless other techniques are also employed. Finally, when activity increases, it is unclear whether this reflects caspase activation in all of the cells or only a small fraction. Thus, caspase assays are usually supplemented by some of the techniques described in the next three sections.

CLEAVAGE OF PROCASPASES AND SUBSTRATES

Caspase-mediated cleavage of more than 400 substrates has been described during the course of apoptosis.[59,61] Among the substrates are the procaspases, which undergo cleavage at specific aspartate residues during the course of activation. The precise cleavages sites have also been mapped for most other substrates. This cleavage of procaspases or other substrates into their signature fragments can be used to monitor caspase activation.

When this approach is employed, cells treated with an apoptotic stimulus are washed and solubilized directly in SDS sample buffer or another denaturing buffer. This is preferable to lysing the cells in buffer containing nonionic detergent (Nonidet P-40 or equivalent), as spurious caspase activation has been reported when T lymphocytes are solubilized under nondenaturing conditions.[89] Lysates containing equal amounts of protein are separated by electrophoresis in the presence of SDS, transferred to nitrocellulose, and probed with antibodies that recognize uncleaved or cleaved antigens. In addition to conventional antibodies that recognize both full-length antigens and their cleaved products, so-called "anti-neoepitope" antibodies are available for some antigens. These reagents recognize epitopes created when the polypeptide of interest has been cleaved (e.g., epitopes consisting of the free C-terminal aspartate of the polypeptide fragment generated by caspase cleavage and the preceding few amino acids). Antineoepitope antibodies are available for several of the caspases, for example.[90]

Strengths of this approach include its simplicity and the widespread availability of equipment. When reagents that recognize caspase substrates are employed, this technique will certainly indicate whether caspases have been activated *in situ*. In addition, for caspases that are cleaved during activation (e.g., caspase-3, -6, and -7), this method provides a potential way to assess activation status. In contrast, because caspase-9 can be activated without cleavage,[91,92] assessment of caspase-9 cleavage will underestimate the extent of caspase-9 activation. The same might also be true of caspase-8 and caspase-10.[72]

This approach has additional limitations. First, because immunoblotting is typically performed without controls to confirm its linearity, this approach is best viewed as qualitative rather than quantitative. Second, there is a frequent disparity between the amount of procaspase that has been cleaved and the levels of processed caspase species present in apoptotic cells. The ability of inhibitor of apoptosis proteins to ubiquitylate certain active caspase species,[93,94] thereby facilitating their proteasome-mediated degradation, presumably contributes to this disparity. As a consequence, the decrease of a particular procaspase species overestimates the amount of polypeptide present in the "active" form. Third, this technique is dependent on high-quality antisera. Most of the caspase neoepitope sera cross-react with multiple bands on SDS-polyacrylamide gels. Because of this cross-reactivity, these sera are not suitable for immunolocalization studies. In contrast, commercially available "cleaved PARP" (poly [ADP-ribose] polymerase) antiserum is highly specific for the N-terminus of the 89 kDa PARP fragment (T. Kottke and S. Kaufmann, unpublished observations). When this reagent is employed in indirect immunofluorescence, it is possible to determine whether caspases have been activated in all cells or only a fraction of the cells in a population.

Affinity Labeling

Affinity labeling was originally billed as an alternative technique that permitted determination of the percentage of cells with active caspases. As with other enzymes, treatment of caspases with substrate-like molecules fused to reactive groups can result in covalent modification (and inhibition) of the enzymes' active sites. A wide variety of sulfhydryl proteases can be covalently labeled with substrate-like peptides bound to chloromethyl ketone, fluoromethyl ketone, or acyloxymethyl ketone groups. The inclusion of a reporter moiety (e.g., biotin or a fluorescent group) allows for the detection of the covalently modified enzyme. In the case of caspases, the P2 amino acid side chain points away from the active site, permitting substitution of lysine coupled to biotin at this position in the affinity labeling reagent.[59,95] Alternatively, reporter groups can be coupled to the N-terminus of the substrate-like peptide.

The reaction between an affinity label and its target caspase has been used to probe the active site of caspase-1.[95] Likewise, when used in conjunction with two-dimensional electrophoresis[84] or affinity purification followed by mass spectrometry,[96] affinity labeling has helped identify the active caspase species that are present in lysates of apoptotic cells. Thus, this technique can be very useful under certain circumstances.

It was also suggested that this approach can be used to follow caspase activation on a cell-by-cell basis.[97,98] Treatment of cells with labeled aspartylglutamylvalinylaspartic acid fluoromethyl ketone (DEVD-fluoromethyl ketone), which contains the DEVD peptide recognized by caspase-3 and -7 fused to a reactive moiety, was proposed as a means to identify cells in which caspase-3 or -7 has been activated. In theory, if a population of cells is treated with the reagent and washed to remove unbound label, only cells that contain active caspase-3 or -7 will be labeled. Because the reagent also acts as a protease inhibitor, further apoptotic changes (including activation of caspases downstream of caspase-3)[99] will be inhibited.

Although the concept of using affinity labeling to quantitate the number of apoptotic cells is appealing, there are several potential problems with this approach. First, as indicated in Table 3.1, the cleavage specificities of the caspases are not absolute. Reagents designed for one caspase will label others. This lack of specificity becomes particularly problematic during prolonged incubation.[100] Second, because the reactive groups used in these affinity labels are designed to react with activated serine and sulfhydryl groups, cross-reaction with proteases other than caspases has been detected.[101] Because of these potential problems, the use of these reagents to detect caspase activation by immunofluorescence or flow cytometry might be less straightforward than originally envisioned.[102]

IMMUNODETECTION OF ACTIVATED CASPASES

Activation of caspase-3,[103] -7,[78] -9,[77] and presumably other caspases is accompanied by conformational changes. These observations have led to the development of conformation-sensitive antisera that can be used to probe for active caspase species.[103]

To apply this approach, cells undergoing apoptosis are fixed with a cross-linking fixative (e.g., formaldehyde or paraformaldehyde), permeabilized, and stained using the conformation-sensitive reagents as primary antibodies followed by peroxidase- or fluorochrome-labeled secondary antibodies as detection reagents. Samples can be examined by microscopy or, in the case of cell suspensions stained with fluorochrome-labeled secondary antibodies, flow cytometry.

In order for this approach to work, agents that denature polypeptides and destroy the conformational epitopes of interest (e.g., methanol and acetone) must be avoided during fixation and permeabilization. When applied successfully, this approach allows for the detection of caspase activation on a cell-by-cell basis. Because the immunological reagents employed can be highly specific, demonstration that a particular caspase has been activated can be relatively unequivocal. Moreover, this approach is applicable to tissue sections as well as single-cell suspensions.

There are, however, also limitations to this approach. First, conformation-sensitive antibodies to most of the caspases are not currently available. It is important to emphasize that these conformation-sensitive reagents are different from the "neoepitope antibodies" (as discussed in the section "Cleavage of Procaspases and Substrates") that recognize linear epitopes generated during caspase cleavage. Some of the neoepitope reagents currently available are relatively specific, but many exhibit cross-reactivity with multiple polypeptides on immunoblots and should not be used for staining. Second, the approach is inherently qualitative. Finally, because of subtle sequence and conformation differences, reagents that label cells from one species will not necessarily work in cells from another species.

ANNEXIN V STAINING

In healthy cells, certain phospholipids are asymmetrically arranged in the plasma membrane. Sphingomyelin, for example, is present predominantly in the outer layer, and phosphatidylserine is exclusively in the inner layer.[104] During the course of apoptosis, caspase-3-mediated cleavage of protein kinase Cδ separates the N-terminal inhibitory domain from the C-terminal catalytic domain.[105] The resulting constitutively active protein kinase Cδ catalytic domain then activates a lipid scramblase in the plasma membrane,[106] leading to exposure of phosphatidylserine on the outer face of the plasma membrane. Under physiological conditions, this cell surface phosphatidylserine serves as a recognition molecule for clearance of apoptotic cells by phagocytes.[107,108] The detection of this externalized phosphatidylserine by annexin V, a polypeptide that binds avidly and specifically to phosphatidylserine, forms the basis for widely used histochemical and flow cytometry–based methods of detecting apoptotic cells.[109,110]

To perform this assay, cells treated with an apoptotic stimulus are incubated with fluorochrome-coupled annexin V, often in the presence of a vital dye such as propidium iodide, and are examined by fluorescence microscopy or subjected to flow microfluorimetry (Figure 3.1(F)). Alternatively, cells can be fixed with a nonpermeabilizing fixative such as glutaraldehyde before annexin V staining. In either case, apoptotic cells will be labeled with annexin V, whereas intact cells will not.

The advantages of this technique include the ability to examine large numbers of cells and the widespread availability of the required equipment. In addition, if examined in conjunction with propidium iodide (PI) staining, annexin V binding can distinguish apoptotic cells (annexin V+, PI−) from necrotic (PI+) cells.[110] It is important to realize, however, that after plasma membrane integrity is lost, annexin V will have access to lipids of the inner face of the plasma membrane and will stain all cells. Accordingly, this technique is not suitable for examining tissue sections or cells that have been permeabilized. Moreover, although it was initially thought that annexin V staining detected an early apoptotic event,[110] it is now clear that phosphatidylserine becomes accessible to annexin V only after caspase activation.[106,111–113]

FLOW CYTOMETRY–BASED ASSAYS

Flow cytometers, which are widely used in the study of lymphocytes, offer several advantages in the study of apoptosis. These include the ability to analyze multiple parameters simultaneously and the possibility of analyzing up to 20,000 or more individual cells in each sample. Thus, it is not surprising that numerous techniques for the detection of apoptotic cells involve flow cytometry. In addition to the approaches described in the above sections ("Flow Cytometry," "Terminal Deoxyribonucleotidyl Transferase–Mediated dUTP Nick-End Labeling (TUNEL)," and "Annexin V Staining"), flow cytometry–based techniques used in apoptosis research include assessment of cell size, mitochondrial membrane potential, and vital dye uptake.

Altered Cell Size

One of the distinguishing features of apoptosis is cell shrinkage.[114] This change has been attributed to potassium ion loss[115] that occurs, at least in part, as a consequence of caspase-dependent activation of the Kv1.3 potassium channel[116] and degradation of the plasma membrane Na/K-ATPase.[115] Because larger cells scatter light more than smaller cells, the decrease in cell size can be monitored by changes in light scatter properties. These observations provide the basis for studies that use changes in light scatter to monitor apoptosis.

To apply this approach, cells treated with a proapoptotic stimulus for varying lengths of time are subjected to flow cytometry and are monitored for forward scatter, as illustrated in Figure 3.1E. This assay can be performed on either unfixed or fixed cells. In either case, apoptotic cells display less forward scatter than control cells.

Although this assay is rapid and simple, it has several limitations. Because each cell can give rise to multiple apoptotic bodies, merely quantifying the number of small particles can potentially result in overestimation of the number of cells that have undergone apoptosis, particularly at later time points when cells have fragmented. Even putting this problem aside, the method is most accurate when the cell population is extremely homogeneous in size (e.g., G0 lymphocytes). In proliferating cell populations, S and G2 phase cells are typically larger than G0/G1 phase cells. As the S and G2 phase cells become apoptotic, they will shrink; there might be considerable overlap between nonapoptotic G0/G1 cells and apoptotic S and G2 phase cells. Accordingly, cell size is one parameter that may be used in multiparameter flow cytometry, but it is problematic as a single discriminator of apoptosis in any particular cell.[117]

Loss of Mitochondrial Transmembrane Potential ($\Delta\Psi_M$)

During the last decade, there has been considerable interest in mitochondrial changes during apoptosis.[73] This interest was initially spurred by claims that loss of mitochondrial transmembrane potential was a particularly early event in apoptosis[118,119] and by demonstration that cytochrome c played a critical role in activation of the Apaf-1/caspase-9 pathway.[73,120] At one time, it was thought that loss of mitochondrial transmembrane potential might reflect inhibition of electron transport and concomitant diminution of proton translocation as a consequence of cytochrome c release. Subsequent experiments have shown that loss of $\Delta\Psi_m$ occurs after cytochrome c release and likely reflects caspase-mediated cleavage of another component of the electron transport chain.[121,122] Despite the fact that loss of $\Delta\Psi_m$ represents a late event in apoptosis, there remains considerable interest in its measurement.

The electrochemical potential difference across the inner mitochondrial membrane ordinarily drives mitochondrial ATP synthesis but can also be used to promote accumulation of membrane-permeant cationic dyes against a concentration gradient. The extent to which these dyes are concentrated in isolated mitochondria is directly related to the magnitude of the electrochemical gradient.[123] These observations form the basis for flow cytometry–based methods of assessing $\Delta\Psi_m$.

To apply this approach, cells treated with an apoptotic stimulus are incubated briefly with membrane-permeant cationic dyes (e.g., 5,5,6,6-tetrachloro-1,1,3,3-tetraethylbenzimidazolcarbocyanine iodide or 3,3-dihexyloxacarbocynine iodide) and are immediately subjected to flow cytometry. Cells containing a normal $\Delta\Psi_m$ accumulate larger amounts of these dyes than cells with diminished mitochondrial transmembrane potential.

The present approach can measure transmembrane potential relatively simply and easily in 20,000 or more cells/sample. It is important, however, to emphasize several limitations of this method. First, although it has been suggested that altered $\Delta\Psi_m$ might be a highly reliable early marker of apoptosis,[118,119,124] this change is not universally observed.[125] Second, this change is not specific for apoptosis. Instead, decreased $\Delta\Psi_m$ is also extensively documented in cells undergoing necrosis[126] and can result from any treatment that alters mitochondrial electron transport or proton translocation (e.g., metabolic poisons such as 2,4-dinitrophenol or sodium azide). Third, the quantum yield of many of the cationic dyes used to measure mitochondrial transmembrane potential is diminished by p-aminobenzoic acid derivatives that quench the fluorescent emission.[127] Thus, changes in fluorescent emission do not always reflect altered $\Delta\Psi_m$ and certainly do not reflect alterations specific to apoptosis.

Dye Uptake

A number of techniques for detecting apoptosis rely on changes in dye uptake that can be measured by flow cytometry. For example, cells can be incubated in Hoechst 33342 or 7-aminoactinomycin D (7-AAD) at 4°C and then subjected to flow microfluorimetry. Ordinarily, Hoechst 33342 and 7-AAD are excluded from cells at 4°C. Cells that have ruptured readily take up these agents and, therefore, fluoresce strongly. For reasons that are unclear, apoptotic cells take up intermediate amounts of these compounds at 4°C despite the presence of an intact plasma membrane. As a result, cells undergoing apoptosis can be quantitated as a population of weakly fluorescent cells.

Because Hoechst 33342 and 7-AAD freely penetrate the plasma membrane above its phase transition temperature, which is typically at 20°C, it is critical that cells be kept as close to 4°C as possible during exposure to the dye. When this precaution is achieved, the technique provides a straightforward means of assessing dye permeability in thousands of individual cells per sample. In addition, if the dyes listed above are combined with fluorochrome-labeled antibodies with suitable spectral properties, apoptosis can be readily quantified in a subset of cells (e.g., CD4-expressing lymphocytes) in a larger cell population.

Despite these advantages, two potential limitations of this approach should be kept in mind. First, when this approach is used to evaluate apoptosis in rare cell subsets, it can be difficult to use other techniques (morphology, agarose gel electrophoresis) to confirm that the cells taking up dye are truly undergoing apoptosis. In addition, it is not entirely clear that intermediate dye uptake is absolutely specific for apoptosis. It is conceivable that other changes within cells (e.g., ATP depletion or altered plasma membrane lipid composition) might also allow penetration of small amounts of these dyes into cells. As long as these potential limitations are kept in mind, dye uptake assays seem to offer a rapid and powerful technique for investigating apoptosis in mixed populations of cells.

ACKNOWLEDGMENTS

Research in the authors' laboratory is supported by NIH grant CA69008. We thank Son Le, William C. Earnshaw, Greg Gores, David Vaux, and Yuri Lazebnik for spirited discussions; David Locgering, Karen Flatten, and Julie Wahlstrand for technical assistance with the experiments illustrated in Figure 3.1; and Deb Strauss for editorial assistance.

REFERENCES

1. Thompson, C.B., Apoptosis in the pathogenesis and treatment of disease, *Science,* 267, 1456, 1995.
2. Badley, A.D., Dockrell, D., and Paya, C.V., Apoptosis in AIDS, *Adv. Pharmacol.,* 41, 325, 1997.
3. Kerr, J.F., Wyllie, A.H., and Currie, A.R., Apoptosis: a basic biological phenomenon with wide-ranging implications in tissue kinetics, *Brit. J. Cancer,* 26, 239, 1972.
4. Wyllie, A.H., Kerr, J.F.R., and Currie, A.R., Cell death: the significance of apoptosis, *Int. Rev. Cytol.,* 68, 251, 1980.
5. Duvall, E. and Wyllie, A.H., Death and the cell, *Immunol. Today,* 7, 115, 1986.
6. Wyllie, A.H., Morris, R.G., Smith, A.L., and Dunlop, D., Chromatin cleavage in apoptosis association with condensed chromatin morphology and dependence on macromolecular synthesis, *J. Pathol.,* 142, 67, 1984.
7. Samejima, K., Tone, S., Kottke, T., Enari, M., Sakahira, H., Cooke, C.A., Durrieu, F., Martins, L.M., Nagata, S., Kaufmann, S.H., and Earnshaw, W.C., Transition from caspase-dependent to caspase-independent mechanisms at the onset of apoptotic execution, *J. Cell Biol.,* 143, 225, 1998.
8. Lazebnik, Y.A., Takahashi, A., Moir, R., Goldman, R., Poirer, G.G., Kaufmann, S.H., and Earnshaw, W.C., Studies of the lamin proteinase reveal multiple parallel biochemical pathways during apoptotic execution, *Proc. Natl. Acad. Sci. U.S.A.,* 92, 9042, 1995.
9. Ruchaud, S., Korfali, N., Villa, P., Kottke, T.J., Dingwall, C., Kaufmann, S.H., and Earnshaw, W.C., Caspase-6 gene knockout reveals a role for lamin A cleavage in apoptotic chromatin condensation, *EMBO J.,* 21, 1967, 2002.
10. Earnshaw, W.C., Nuclear changes in apoptosis, *Curr. Opin. Cell Biol.,* 7, 337, 1995.
11. Wyllie, A.H. and Duvall, E., Cell injury and death, in *Oxford Textbook of Pathology,* Vol. 1, Oxford University Press, Oxford, 1992, p. 141.
12. Koss, L.G., *Diagnostic Cytology and Its Histopathologic Bases,* Vol. 1 (4th ed.), J.B. Lippincott Company, New York, 1992.
13. Wyllie, A.H., Glucocorticoid-induced thymocyte apoptosis is associated with endogenous endonuclease activation, *Nature,* 284, 555, 1980.
14. Kornberg, R.D., Chromatin structure: a repeating unit of histones and DNA, *Science,* 184, 868, 871, 1974.

15. Hewish, D.R. and Burgoyne, L.A., Chromatin sub-structure. The digestion of chromatin DNA at regularly spaced sites by a nuclear deoxyribonuclease, *Biochem. Biophys. Res. Comm.,* 52, 504, 1973.
16. Noll, M., Subunit structure of chromatin, *Nature,* 251, 249, 1974.
17. Burgoyne, L.A., Hewish, D.R., and Mobbs, J., Mammalian chromatin substructure studies with the calcium-magnesium endonuclease and two-dimensional polyacrylamide-gel electrophoresis, *Biochem. J.,* 143, 67, 1974.
18. Peitsch, M.C., Polzar, B., Stephan, H., Crompton, T., MacDonald, H.R., Mannherz, H.G., and Tschopp, J., Characterization of the endogenous deoxyribonuclease involved in nuclear DNA degradation during apoptosis (programmed cell death), *EMBO J.,* 12, 371, 1993.
19. Polzar, B., Zanotti, S., Stephan, H., Rauch, F., Peitsch, M.C., Irmler, M., Tschopp, J., and Mannherz, H.G., Distribution of deoxyribonuclease I in rat tissues and its correlation to cellular turnover and apoptosis, *Eur. J. Cell. Biol.,* 64, 200, 1994.
20. Barry, M.A. and Eastman, A., Identification of deoxyribonuclease II as an endonuclease involved in apoptosis, *Arch. Biochem. Biophys.,* 300, 440, 1993.
21. Gottlieb, R.A., Giesing, H.A., Engler, R.L., and Babior, B.M., The acid deoxyribonuclease of neutrophils: a possible participant in apoptosis-associated genome destruction, *Blood,* 86, 2414, 1995.
22. Shiokawa, D., Ohyama, H., Yamada, T., and Tanuma, S., Purification and properties of DNase from apoptotic rat thymocytes, *Biochem. J.,* 326, 675, 1997.
23. Shiokawa, D. and Tanuma, S.-I., Molecular cloning and expression of a cDNA encoding an apoptotic endonuclease DNase, *Biochem. J.,* 332, 713, 1998.
24. Enari, M., Sakahira, H., Yokoyama, H., Okawa, K., Iwamatsu, A., and Nagata, S., A caspase-activated DNase that degrades DNA during apoptosis, and its inhibitor ICAD, *Nature,* 391, 43, 1998.
25. Sakahira, H., Enari, M., and Nagata, S., Cleavage of CAD inhibitor in CAD activation and DNA degradation during apoptosis, *Nature,* 391, 96, 1998.
26. Zhang, J., Wang, X., Bove, K.E., and Xu, M., DNA fragmentation factor 45-deficient cells are more resistant to apoptosis and exhibit different dying morphology than wild-type control cells, *J. Biol. Chem.,* 274, 37450, 1999.
27. Samejima, K., Tone, S., and Earnshaw, W.C., CAD/DFF40 nuclease is dispensable for high molecular weight DNA cleavage and stage I chromatin condensation in apoptosis, *J. Biol. Chem.,* 276, 45427, 2001.
28. Samejima, K. and Earnshaw, W.C., ICAD/DFF regulator of apoptotic nuclease is nuclear, *Exp. Cell Res.,* 243, 453, 1998.
29. Liu, X., Li, P., Widlak, P., Zou, H., Luo, X., Garrard, W.T., and Wang, X., The 40-kDa subunit of DNA fragmentation factor induces DNA fragmentation and chromatin condensation during apoptosis, *Proc. Natl. Acad. Sci. U.S.A.,* 95, 8461, 1998.
30. McIlroy, D., Sakahira, H., Talanian, R.V., and Nagata, S., Involvement of caspase 3-activated DNase in internucleosomal DNA cleavage induced by diverse apoptotic stimuli, *Oncogene,* 18, 4401, 1999.
31. Lily, L.Y., Luo, X., and Wang, X., Endonuclease G is an apoptotic DNase when released from mitochondria, *Nature,* 412, 95, 2001.
32. Igo-Kemenes, T., Greil, W., and Zachau, H.G., Preparation of soluble chromatin fractions with restriction nucleases, *Nucleic Acids Res.,* 4, 3387, 1977.
33. Kyprianou, N. and Isaacs, J.T., Activation or programmed cell death in the rat ventral prostate after castration, *Endocrinology,* 122, 552, 1988.
34. Gong, J., Traganos, F., and Darzynkiewicz, Z., A selective procedure for DNA extraction from apoptotic cells applicable for gel electrophoresis and flow cytometry, *Anal. Biochem.,* 218, 314, 1994.
35. Afanas'ev, V.N., Korol, B.A., Mantsygin, Y.A., Nelipovich, P.A., Pechantnikov, V.A., and Umansky, S.R., Flow cytometry and biochemical analysis of DNA degradation characteristic of two types of cell death, *FEBS Lett.,* 194, 347, 1986.
36. Nicoletti, I., Migliorati, G., Pagliacci, M.C., Grignani, F., and Riccardi, C., A rapid and simple method for measuring thymocyte apoptosis by propidium iodide staining and flow cytometry, *J. Immunol. Methods,* 139, 271, 1991.
37. Meng, X., Chandra, J., Loegering, D., Van Becelacre, K., Kottke, T.J., Gore, S.D., Karp, J.E., Sebolt-Leopold, J.S., and Kaufmann, S.H., Central role of FADD in apoptosis induction by the mitogen activated protein kinase inhibitor CI1040 (PD184352) in acute lymphocytic leukemia cell lines *in vitro, J. Biol. Chem.,* 278, 47236, 2003.

38. Kohn, K.W., Erickson, L.C., Ewig, R.A.G., and Friedman, C.A., Fractionation of DNA from mammalian cells by alkaline elution, *Biochemistry,* 15, 4629, 1976.

39. Kohn, K.W., DNA filter elution: a window on DNA damage in mammalian cells, *Bioessays,* 18, 505, 1996.

40. Bertrand, R., Kohn, K.W., Solary, E., and Pommier, Y., Detection of apoptosis-associated DNA fragmentation using a rapid and quantitative filter elution assay, *Drug Dev. Res.,* 34, 138, 1995.

41. Solary, E., Bertrand, R., Jenkins, J., and Pommier, Y., Radiolabeling of DNA can induce its fragmentation in HL-60 human promyelocytic leukemia cells, *Exp. Cell Res.,* 203, 495, 1992.

42. Roychoudhury, R., Jay, E., and Wu, R., Terminal labeling and addition of homopolymer tracts to duplex DNA fragments by terminal deoxynucleotidyl transferase, *Nucleic Acids Res.,* 3, 863, 1976.

43. Sambrook, J., Fritsch, E.F., and Maniatis, T., *Molecular Cloning: A Laboratory Manual,* (2nd ed.), Cold Spring Harbor Laboratory Press, Cold Spring Harbor, New York, 1989.

44. Gavrieli, Y., Sherman, Y., and Ben-Sasson, S.A., Identification of programmed cell death *in situ* via specific labeling of nuclear DNA fragmentation, *J. Cell Biol.,* 119, 493, 1992.

45. Gorczyca, W., Gong, J., and Darzynkiewicz, Z., Detection of DNA strand breaks in individual apoptotic cells by the *in situ* terminal deoxynucleotidyl transferase and nick translation assays, *Cancer Res.,* 53, 1945, 1993.

46. Jerome, K.R., Vallan, C., and Jaggi, R., The TUNEL assay in the diagnosis of graft-versus-host disease: caveats for interpretation, *Pathology,* 32, 186, 2000.

47. Gottlieb, R.A., Burleson, K.O., Kloner, R., Babior, B.M., and Engler, R.L., Reperfusion injury induces apoptosis in rabbit cardiomyocytes, *J. Clin. Invest.,* 94, 1621, 1994.

48. Grasl-Kraupp, B., Ruttkay-Nedecky, B., Koudelka, H., Bukowska, K., Bursch, W., and Schulte-Hermann, R., *In situ* detection of fragmented DNA (TUNEL assay) fails to discriminate among apoptosis, necrosis, and autolytic cell death: a cautionary note, *Hepatology,* 21, 1465, 1995.

49. Charriaut-Marlangue, C. and Ben-Ari, Y., A cautionary note on the use of the TUNEL stain to determine apoptosis, *NeuroReport,* 7, 61, 1995.

50. Yasuda, M., Umemura, S., Osamura, R.Y., Kenjo, T., and Tsutsumi, Y., Apoptotic cells in the human endometrium and placental villi: pitfalls in applying the TUNEL method, *Arch. Histol. Cytol.,* 58, 185, 1995.

51. Stadelmann, C. and Lassmann, H., Detection of apoptosis in tissue sections, *Cell Tissue Res.,* 301, 19, 2000.

52. Lecoeur, H., Nuclear apoptosis detection by flow cytometry: influence of endogenous endonucleases, *Exp. Cell Res.,* 277, 1, 2002.

53. Olive, P.L., Frazer, G., and Banath, J.P., Radiation-induced apoptosis measured in TK6 human B lymphoblast cells using the comet assay, *Radiat. Res.,* 136, 130, 1993.

54. Ostling, O. and Johanson, K.J., Microelectrophoretic study of radiation-induced DNA damages in individual mammalian cells, *Biochem. Biophys. Res. Commun.,* 123, 291, 1984.

55. Fairbairn, D.W., Olive, P.L., and O'Neill, K.L., The comet assay: a comprehensive review, *Mutat. Res.,* 339, 37, 1995.

56. Collins, A.R., The comet assay. Principles, applications, and limitations, *Meth. Mol. Biol.,* 203, 163, 2002.

57. Alnemri, E.S., Mammalian cell death proteases: a family of highly conserved aspartate specific cysteine proteases, *J. Cell. Biochem.,* 64, 33, 1997.

58. Thornberry, N.A., Caspases: key mediators of apoptosis, *Chem. Biol.,* 5, R97, 1998.

59. Earnshaw, W.C., Martins, L.M., and Kaufmann, S.H., Mammalian caspases: structure, activation, substrates and functions during apoptosis, *Ann. Rev. Biochem.,* 68, 383, 1999.

60. Kaufmann, S.H., Induction of endonucleolytic DNA cleavage in human acute myelogenous leukemia cells by etoposide, camptothecin, and other cytotoxic anticancer drugs: a cautionary note, *Cancer Res.,* 49, 5870, 1989.

61. Fischer, U., Janicke, R.U., and Schulze-Osthoff, K., Many cuts to ruin: a comprehensive update of caspase substrates, *Cell Death Diff.,* 10, 76, 2003.

62. Zheng, T.S. and Flavell, R.A., Divinations and surprises: genetic analysis of caspase function in mice, *Exp. Cell Res.,* 256, 67, 2000.

63. Patel, T., Gores, G.J., and Kaufmann, S.H., The role of proteases during apoptosis, *FASEB J.,* 10, 587, 1996.

64. Johnson, D.E., Noncaspase proteases in apoptosis, *Leukemia,* 14, 1695, 2000.
65. Lieberman, J., The ABCs of granule-mediated cytotoxicity: new weapons in the arsenal, *Nat. Rev. Immunol.,* 3, 361, 2003.
66. Jaattela, M. and Tschopp, J., Caspase-independent cell death in T lymphocytes, *Nat. Immunol.,* 4, 416, 2003.
67. Thornberry, N.A. and Molineaux, S.M., Interleukin-1β converting enzyme: a novel cysteine protease required for IL-1β production and implicated in programmed cell death, *Protein Sci.,* 4, 3, 1995.
68. Saleh, M., Vaillancourt, J.P., Graham, R., Huyck, M., Srinivasula, S.M., Alnemri, E.S., Steinberg, M.H., Nolan, V., Baldwin, C.T., Hotchkiss, R.S., Buchman, T.G., Zehnbauer, B.A., Hayden, M.R., Farrer, L.A., Roy, S., and Nicholson, D.W., Differential modulation of endotoxin responsiveness by human caspase-12 polymorphisms, *Nature,* 429, 75, 2004.
69. Kaufmann, S.H. and Hengartner, M., Programmed cell death: alive and well in the new millenium, *Trends Cell Biol.,* 11, 526, 2001.
70. Marsden, V.S., et al., Apoptosis initiated by Bcl-2-regulated caspase activation independently of the cytochrome c/Apaf-1/caspase-9 apoptosome, *Nature,* 419, 634, 2002.
71. Boatright, K.M., Renatus, M., Scott, F.L., Sperandio, S., Shin, H., Pedersen, I.M., Ricci, J.E., Edris, W.A., Sutherlin, D.P., Green, D.R., and Salvesen, G.S., A unified model for apical caspase activation, *Mol. Cell,* 11, 529, 2003.
72. Chang, D.W., Xing, Z., Capacio, V.L., Peter, M.E., and Yang, X., Interdimer processing mechanism of procaspase-8 activation, *EMBO J.,* 22, 4132, 2003.
73. Wang, X., The expanding role of mitochondria in apoptosis, *Genes Dev.,* 15, 2922, 2001.
74. Li, P., Nijhawan, D., Budihardjo, I., Srinivasula, S.M., Ahmad, M., Alnemri, E.S., and Wang, X., Cytochrome c and dATP-dependent formation of Apaf-1/caspase-9 complex initiates an apoptotic protease cascade, *Cell,* 91, 479, 1997.
75. Shi, Y., Apoptosome: the cellular engine for the activation of caspase-9, *Structure,* 10, 285, 2002.
76. Hill, M.M., Adarin, C., Duriez, P.J., Creagh, E.M., and Martin, S.J., Analysis of the composition, assembly kinetics and activity of native Apaf-1 apoptosomes, *EMBO J.,* 23, 2134, 2004.
77. Renatus, M., Stennicke, H.R., Scott, F.L., Liddington, R.C., and Salvesen, G.S., Dimer formation drives the activation of the cell death protease caspase 9, *Proc. Natl. Acad. Sci. U.S.A.,* 98, 14250, 2001.
78. Chai, J., Wu, Q., Shiozaki, E., Srinivasula, S.M., Alnemri, E.S., and Shi, Y., Crystal structure of a procaspase-7 zymogen: mechanisms of activation and substrate binding, *Cell,* 107, 399, 2001.
79. Han, Z., Hendrickson, E.A., Bremner, T.A., and Wyche, J.H., A sequential two-step mechanism for the production of the mature p17:p12 form of caspase-3 *in vitro, J. Biol. Chem.,* 272, 13432, 1997.
80. Stennicke, H.R. and Salvesen, G.S., Caspase assays, *Meth. Enzymol.,* 322, 91–100, 2000.
81. Kaufmann, S.H., Kottke, T.J., Martins, L.M., Henzing, A.J., and Earnshaw, W.C., Analysis of caspase activation during apoptosis, in *Current Protocols in Cell Biology,* Wiley Interscience, New York, 18.1.1–18.1.18, 2001.
82. Talanian, R.V., Quinlan, C. Trautz, A., Hackett, M.C., Mankovich, J.A., Banach, D., Ghayur, T. Brady, K.D., and Wong, W.W., Substrate specificities of caspase family proteases, *J. Biol. Chem.,* 272, 9677, 1997.
83. Nicholson, D.W., Ali, A., Thornberry, N.A., Vaillancourt, J.P., Ding, C.K., Gallant, M., Gareau, Y., Griffin, P.R., Labelle, M., and Lazebnik, Y.A., Identification and inhibition of the ICE/CED-3 protease necessary for mammalian apoptosis, *Nature,* 376, 37, 1995.
84. Martins, L.M., Kottke, T.J., Mesner, P.W., Basi, G.S., Sinha, S., Frignon, N., Jr., Tatar, E., Tung, J.S., Bryant, K., Takahashi, A., Svingen, P.A., Madden, B.J., McCormick, D.J., Earnshaw, W.C., and Kaufmann, S.H., Activation of multiple interleukin-1β converting enzyme homologues in cytosol and nuclei of HL-60 human leukemia cell lines during etoposide-induced apoptosis, *J. Biol. Chem.,* 272, 7421, 1997.
85. Kottke, T.J., Blajeski, A.L., Meng, X., Svingen, P.A., Ruchaud, S., Mesner, P.W., Jr., Boerner, S.A., Samejima, K., Henriquez, N.V., Chilcote, T.J., Lord, J., Salmon, M., Earnshaw, W.C., and Kaufmann, S.H., Lack of correlation between caspase activation and caspase activity assays in paclitaxel-treated MCF-7 breast cancer cells, *J. Biol. Chem.,* 277, 804, 2002.
86. Thornberry, N.A., Rano, T.A., Peterson, E.P., Rasper, D.M., Timkey, T., Garcia-Calvo, M., Houtzager, V.M., Nordstrom, P.A., Roy, S., Vaillancourt, J.P., Chapman, K.T., and Nicholson, D.W., A combinatorial approach defines specificities of members of the caspase family and granzyme B. Functional relationships established for key mediators of apoptosis, *J. Biol. Chem.,* 272, 17907, 1997.

87. Deveraux, Q.L., Leo, E., Stennicke, H.R., Welsh, K., Salvesen, G.S., and Reed, J.C., Cleavage of human inhibitor of apoptosis protein XIAP results in fragments with distinct specificities for caspases, *EMBO J.,* 18, 5242, 1999.

88. MacFarlane, M., Merrison, W. Dindale, D., and Cohen, G.M., Active caspases and cleaved cytokeratins are sequestered into cytoplasmic inclusions in TRAIL-induced apoptosis, *J. Cell Biol.,* 148, 1239, 2000.

89. Zapata, J.M., Takahashi, R. Salvesen, G.S., and Reed, J.C., Granzyme release and caspase activation in activated human T-lymphocytes, *J. Biol. Chem.,* 273, 6916, 1998.

90. Mesner, P.W., Jr., Bible, K.C., Martins, L.M., Kottke, T.J., Srinivasula, S.M., Svingen, P.A., Chilcote, T.J., Basi, G.S., Tung, J.S., Krajewski, S., Reed, J.C., Alnemri, E.S., Eranshaw, W.C., and Kaufmann, S.H., Characterization of caspase processing and activation of HL-60 cell cytosol under cell-free conditions: nucleotide requirement and inhibitor profile, *J. Biol. Chem.,* 274, 22635, 1999.

91. Rodriguez, J. and Lazebnik, Y., Caspase-9 and APAF-1 form an active holoenzyme, *Genes Dev.,* 13, 3179, 1999.

92. Stennicke, H.R., Deveraux, Q.L., Humke, E.W., Reed, J.C., Dixit, V.M., and Salvesen, G.S., Caspase-9 can be activated without proteolytic processing, *J. Biol. Chem.,* 274, 8359, 1999.

93. Huang, H.-K., Joazeiro, C.A.P., Bonfoco, E., Kamada, S., Leverson, J.D., and Hunter, T., The inhibitor of apoptosis, cIAP2, functions as a ubiquitin-protein ligase and promotes *in vitro* monoubiquitination of caspase 3 and 7, *J. Biol. Chem.,* 275, 26661, 2000.

94. Suzuki, Y., Nakabayashi, Y., and Takahashi, R., Ubiquitin-protein ligase activity of x-linked inhibitor of apoptosis protein promotes proteasomal degradation of caspase-3 and enhances its anti-apoptotic effect in Fas-induced cell death, *Proc. Natl. Acad. Sci. U.S.A.,* 98, 8662, 2001.

95. Thornberry, N.A., Peterson, E.P., Zhao, J.J., Howard, A.D., Griffin, P.R., and Chapman, K.T., Inactivation of interleukin-1 converting enzyme by peptide (acyloxy)methyl ketones, *Biochemistry,* 33, 3934, 1994.

96. Faleiro, L., Kobayashi, R., Fearnhead, H., and Lazebnik, Y., Multiple species of CPP32 and Mch2 are the major active caspases present in apoptotic cells, *EMBO J.,* 16, 2271, 1997.

97. Bedner, E., Smolewski, P., Amstad, P., and Darzynkiewicz, Z., Activation of caspases measured *in situ* by binding of fluorochrome-labeled inhibitors of caspases (FLICA): correlation with DNA fragmentation, *Exp. Cell Res.,* 259, 308, 2000.

98. Smolewski, P., Bedner, E., Du, L., Hsieh, T.C., Wu, J.M., Phelps, D.J., and Darzynkiewicz, Z., Detection of caspases activation by fluorochrome-labeled inhibitors: multiparameter analysis by laser scanning cytometry, *Cytometry,* 44, 73, 2001.

99. Slee, E.A., Adrain, C., and Martin, S.J., Ordering the cytochrome c-initiated caspase cascade: hierarchical activation of caspases-2, -3, -6, -7, -8, and -10 in a caspase-9-dependent manner, *J. Cell Biol.,* 144, 281, 1999.

100. Margolin, N., Raybuck, S.A., Wilson, K.P., Chen, W., Fox, T., Gu, Y., and Livingston, D.J., Substrate and inhibitor specificity of interleukin-1β-converting enzyme and related caspases, *J. Biol. Chem.,* 272, 7223, 1997.

101. Schotte, P., Declercq, W., Van Huffel, S., Vandenabeele, P., and Beyaert, R., Non-specific effects of methyl ketone peptide inhibitors of caspases, *FEBS Lett.,* 442, 117, 1999.

102. Pozarowski, P., Huang, X., Halicka, D.H., Lee, B., Johnson, G., and Darzynkiewicz, Z., Interactions of fluorochrome-labeled caspase inhibitors with apoptotic cells: a caution in data interpretation, *Cytometry,* 55A, 50, 2003.

103. Srinivasan, A., Roth, K.A., Sayers, R.O., Shindler, K.S., Wong, A.M., Fritz, L.C., and Tomaselli, K.J., *In situ* immunodetection of activated caspase-3 in apoptotic neurons in the developing nervous system, *Cell Death & Differentiation,* 5, 1004, 1998.

104. van Meer, G., Transport and sorting of membrane lipids, *Curr. Opin. Cell Biol.,* 5, 661, 1993.

105. Emoto, Y., Manome, Y., Meinhardt, Kisaki, H., Kharbanda, S., Robertson, M., Ghayur, T., Wong, W.W., Kamen, R., Weichselbaum, R., and Kufe, D., Proteolytic activation of protein kinase c δ by an ICE-like protease in apoptotic cells, *EMBO J.,* 14, 6148, 1995.

106. Frasch, S.C., Henson, P.M., Kailey, J.M., Richter, D.A., Janes, M.S., Fadok, V.A., and Bratton, D.L., Regulation of phospholipid scramblase activity during apoptosis and cell activation by protein kinase C, *J. Biol. Chem.,* 275, 23065, 2000.

107. Fadok, V.A., Bratton, D.L., Rose, D.M., Pearson, A., Ezekewitz, R.A., and Henson, P.M., A receptor for phosphatidylserine-specific clearance of apoptotic cells, *Nature,* 405, 85, 2000.

108. Lauber, K., Blumenthal, S.G., Waibel, M., and Wesselborg, S., Clearance of apoptotic cells: getting rid of the corpses, *Mol. Cell,* 14, 277, 2004.

109. Koopman, G., Reutelingsperger, C.P.M., Kuijten, G.A.M., Keehnen, R.M.J., Pals, S.T., and van Oers, M.H.J., Annexin V for flow cytometric detection of phosphatidylserine expression on B cells undergoing apoptosis, *Blood,* 84, 1415, 1994.

110. Martin, S.J., Reutelingsperger, C.P.M., McGahon, A.J., Rader, J.A., van Schie, R.C., LaFace, D.M., and Green, D.R., Early redistribution of plasma membrane phosphatidylserine is a general feature of apoptosis regardless of the initiating stimulus: inhibition by overexpression of Bcl-2 and Abl, *J. Exp. Med.,* 182, 1545, 1995.

111. Martin, S.J., Finucane, D.M., Amarante-Mendes, G.P., O'Brien, G.A., and Green D.R., Phosphatidylserine externalization during CD95-induced apoptosis of cells and cytoplasts requires ICE/CED-3 protease activity, *J. Biol. Chem.,* 271, 28753, 1996.

112. Vanags, D.M., Porn-Ares, M.I., Coppola, S., Burfess, D.H., and Orrenius, S., Protease involvement in fodrin cleavage and phosphatidylserine exposure in apoptosis, 271, 31075, 1996.

113. Naito, M., Nagashima, K., Mashima, T., and Tsuruo, T., Phosphatidylserine externalization is a downstsream event of interleukin-1β- converting enzyme family protease activation during apoptosis, *Blood,* 89, 2060, 1997.

114. Arends, M.J. and Wyllie, A.H., Apoptosis: mechanisms and roles in pathology, *Int. Rev. Exp. Pathol.,* 32, 223, 1991.

115. Mann, C.L., Bortner, C.D., Jewell, C.M., and Cidlowski, J.A., Glucocorticoid-induced plasma membrane depolarization during thymocyte apoptosis: association with cell shrinkage and degradation of the Na(+)/K(+)-adenosine triphosphatase, *Endocrinology,* 142, 5059, 2001.

116. Storey, N.M., Gomez-Angelats, M. Bortner, C.D., Armstrong, D.L., and Cidlowski, J.A., Stimulation of Kv1.3 potassium channels by death receptors during apoptosis in Jurkat T lymphocytes, *J. Biol. Chem.,* 278, 33319, 2003.

117. O'Brien, M.C. and Bolton, W.E., Comparison of cell viability probes compatible with fixation and permeabilization for combined surface and intracellular staining in flow cytometry, *Cytometry,* 19, 243, 1995.

118. Vayssiere, J.L., et al., Commitment to apoptosis is associated with changes in mitochondrial biogenesis and activity in cell lines conditionally immortalized with simian virus 40, *Proc. Natl. Acad. Sci. U.S.A.,* 91, 11752, 1994.

119. Zamzami, N., et al., Reduction in mitochondrial potential constitutes an early irreversible step of programmed lymphocyte death *in vivo, J. Exp. Med.,* 181, 1661, 1995.

120. Liu, X., et al., Induction of apoptotic program in cell-free extracts: requirement for dATP and cytochrome c, *Cell,* 86, 147, 1996.

121. Bossy-Wetzel, E., Newmeyer, D.D., and Green, D.R., Mitochondrial cytochrome c release in apoptosis occurs upstream of DEVD-specific caspase activation and independently of mitochondrial transmembrane depolarization, *EMBO J.,* 17, 37, 1998.

122. Ricci, J.E., et al., Disruption of mitochondrial function during apoptosis is mediated by caspase cleavage of the p75 subunit of complex I of the electron transport chain, *Cell,* 117, 773, 2004.

123. Azzone, G.F., Pietrobon, D., and Zoratti, M., Determination of proton electrochemical gradient across biological membranes, *Curr. Top. Bioenerg.,* 13, 1, 1984.

124. Zamzami, N., et al., Sequential reduction of mitochondrial transmembrane potential and generation of reactive oxygen species in early programmed cell death, *J. Exp. Med.,* 182, 367, 1995.

125. Cossarizza, A., et al., Mitochondrial modifications during rat thymocyte apoptosis: a study of the single cell level, *Exp. Cell Res.,* 214, 323, 1994.

126. Rosser, B.G. and Gores, G.J., Liver cell necrosis: cellular mechanisms and clinical implications, *Gastroenterology,* 108, 252, 1995.

127. Nad, S. and Pal, H., Electron transfer from aromatic amines to excited coumarin dyes: fluorescence quenching and picosecond transient absorption studies, *J. Phys. Chem.,* 104, 673, 2000.

4 Consequences of HIV Infection on Thymus Function and T Cell Development

Daniel B. Graham
David J. McKean

CONTENTS

INTRODUCTION

OVERVIEW

Remarkable progress has been made over the last two and one-half decades in illuminating human immunodeficiency virus (HIV) pathology and disease progression to acquired immunodeficiency syndrome (AIDS). Accordingly, the broad spectrum of immune dysfunction characteristic of AIDS has been primarily attributed to depletion of CD4+ T helper cells. It is well documented that HIV directly infects CD4 cells resulting in cell death or disruption of function; however, the virus also acts indirectly on several components of adaptive and innate immunity. This chapter will discuss the significant advancements that have been made toward understanding the mechanisms by which HIV disrupts T cell development in the thymus and the consequences pertaining to immune dysfunction in the periphery.

T CELL DEVELOPMENT IN THE THYMUS

HIV directly infects a variety of cell types in the thymus. Furthermore, the effects of HIV on thymocytes can be catastrophic to the host due to the fact that the thymus is responsible for nurturing T cell development and shaping a T cell repertoire capable of discriminating self from foreign antigens. Typically, T cell development begins with a bone marrow–derived lymphoid progenitor (Figure 4.1). These cells migrate to the thymus, where they develop into various lymphocyte lineages, including αβT cells, γT cells, NKT cells, and NK cells. The αβ T cell lineage emerges as lymphoid progenitors mature into double-negative (DN) thymocytes, named as such because

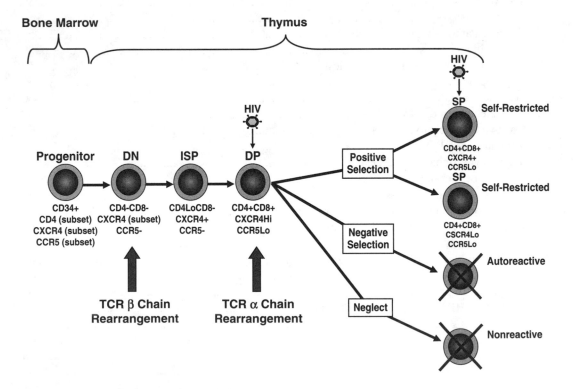

FIGURE 4.1 T cell development. During thymic selection, thymocytes undergo several phenotypic changes that render them susceptible to infection by HIV. Because DP and SP CD4 thymocytes express CD4 and co-receptors, they are preferentially infected and depleted by HIV.

they lack expression of both CD4 and CD8. At this stage, DN cells rearrange their T cell antigen receptor (TCR) β chain genes and express a clonotypic β chain on the cell surface in complex with an invariant pre-Tα chain forming the pre-TCR. DN thymocytes subsequently undergo β selection, at which point cells bearing a functional pre-TCR expand in number and mature into intermediate single-positive (ISP) thymocytes bearing the phenotype CD4^{lo}CD8. In contrast to humans, murine ISPs are CD4CD8^{lo}. The ISP stage is transient and rapidly gives rise to double-positive (DP) thymocytes expressing both CD4 and CD8. These cells rearrange their TCR α chain and express it on their surface with the β chain to form a mature TCR. At this stage of development, DP thymocytes undergo a TCR-mediated selection process to ensure self-tolerance. Autoreactive cells bearing TCRs that bind self-peptide major histocompatibility complex (MHC) with high affinity undergo negative selection culminating in apoptosis. If, however, DP thymocytes express TCRs that do not bind self-MHC, they fail to receive survival signals and die by neglect. Conversely, DP thymocytes expressing TCRs that bind self-peptide and MHC with intermediate affinity undergo positive selection and mature into either single-positive (SP) CD8 thymocytes restricted to self-MHC class I or SP CD4 thymocytes restricted to self-MHC class II. Before exiting into the periphery, SP thymocytes undergo several more rounds of division in the thymus. This tightly regulated process of T cell development is particularly vulnerable to corruption by HIV infection. In fact, HIV directly infects thymic stromal cells as well as immature T cells at various stages of development. By disrupting multiple components of the thymic microenvironment, HIV disturbs thymopoiesis and T cell homeostasis.

HIV INFECTION OF THYMOCYTE SUBSETS

THE THYMUS AS A RESERVOIR OF HIV

HIV antigens, transcripts, and genomes have been identified in the human thymus by immunohistochemistry, *in situ* hybridization, and polymerase chain reaction (PCR), respectively. Futhermore, productive infection of thymocytes was demonstrated by the observation that HIV$^+$ thymic tissue could transfer virus to permissive cell lines.[1] In addition, the thymus may serve as a reservoir of latently infected cells. In an animal model of HIV, infected severe combined immunodeficient (SCID)-hu mice bearing human fetal thymus and liver implants demonstrated that HIV infects activated DP thymocytes during selection and becomes latent as the cells enter quiescence.[2] These quiescent thymocytes exit to the periphery and produce virus upon activation. In contrast, the macaque thymus is not thought to be a reservoir for simian immunodeficiency virus (SIV).[3] These conflicting results can be explained by the fact that SIV co-receptors are not highly expressed in the macaque thymus, whereas HIV co-receptors are found at high levels in the human thymus. HIV infects cells by using the envelope protein gp120, which binds to CD4 and the co-receptors CXCR4 and CCR5 on target cells. M-tropic stains of HIV use CD4 and CCR5, and T-tropic strains use CD4 and CXCR4. Logically, different HIV strains have varying abilities to infect thymocyte subsets according to co-receptor expression. In the SCID-hu model, M-tropic strains of HIV preferentially infect thymic stromal cells, whereas T-tropic strains infect developing thymocytes.[4]

INFECTION OF THYMOCYTES

Although it is agreed that HIV infects the thymus, controversy exists over which thymocyte HIV (Table 4.1) subsets are directly infected. Some insight has been gained from studies monitoring CD4 and co-receptor expression on thymocytes.[5,6] DN thymocytes, by definition, do not express CD4. They also lack expression of CCR5, but subsets express CXCR4. Although it was suggested that HIV can infect DNs in a CD4-dependent manner,[7] these cells may have been ISPs misidentified as DNs. ISP thymocytes that express low levels of CD4 are CXCR4$^+$ and CCR5. Evidence suggests that these ISPs can be infected by HIV.[1,8] Notably, DP thymocytes are efficiently infected by HIV

TABLE 4.1
Cellular Targets of HIV in the Thymus

Cell Type	Direct Infection in vitro	Virus Present in vivo
Hematopoietic stem cell	/+	/Low
DN	+/?	+
ISP	++	++
DP	+++	+++
SP CD4	++	++
SP CD8	/?	+
TEC	+/	
Macrophage	++	
DC	++	

Notes: +, yes; , no; +/, conflicting data; ?, inconclusive.

due to the fact that they express CD4 at high levels, are CXCR4+, and express low levels of CCR5.[7] SP CD4 thymocytes also express high levels of CD4 but express less CXCR4 and CCR5. HIV can infect SP CD4 thymocytes,[1] but it is possible that infection occurs more efficiently at the DP stage and that virus persists throughout maturation. Similarly, SP CD8 thymocytes have been shown to harbor HIV,[9,10] but the majority of these cells may have become infected earlier in development. When infected SP CD4s were cocultured with uninfected SP CD8s, HIV was not transferred to the SP CD8s.[11] Because SP CD8 thymocytes lack CD4 and express CXCR4 and CCR5 at extremely low levels, direct infection of these cells by HIV may be inefficient. More comprehensive studies demonstrated that reporter genes inserted into the HIV genome can be detected in DN, DP, SP CD4, and SP CD8 thymocytes *in vitro*[9] as well as in the SCID-hu system.[10] In addition to thymocytes of the αβ T cell lineage, developing NK cells, NKT cells, and γ T cells were shown to be susceptible to HIV infection in the SCID-hu model.[12]

INFECTION OF THYMIC STROMAL CELLS

Thymic stromal cells, such as epithelia, macrophages, and dendritic cells (DCs), may also be affected during HIV infection. Thymic macrophages can be directly infected by HIV *in vitro*, and as a result, they produce excessive levels of cytokines.[13] HIV also infects human thymic DCs. After infection, these cells undergo morphological changes associated with apoptosis and produce unidentified soluble factors that are toxic to neighboring thymocytes.[14] Another study demonstrated that HIV infects thymic DCs in a CD4-dependent manner and, furthermore, showed that HIV could be transmitted to mature T cells during a mixed lymphocyte reaction.[15] The third important component of thymic stroma that is affected during infection is the epithelium, but reports disagree as to whether HIV acts directly or indirectly on these cells. Thymic epithelial monolayers cultured from human tissue can be infected by HIV independent of CXCR4 and CCR5.[16] In contrast, independent studies have suggested that thymic epithelial cells (TECs) are not infected directly by HIV but that culture supernatant from infected TECs enhances viral replication in infected T cells.[13,17] Upon analysis, the body of evidence reveals the difficulty in determining which thymic subsets are directly infected by HIV during T cell development *in vivo* and at what stage infection occurs. Furthermore, it is unclear exactly which cells may be infected by disparate strains of HIV. The strongest evidence suggests that DP, SP CD4, DCs, and macrophages are the primary targets for HIV in the thymus.

THYMIC ARCHITECTURE

THYMIC HISTOLOGY

Located in the upper anterior mediastinum, the thymus derives from epithelial outgrowths of the ventral wing of the third pharyngeal pouch. Microscopic analysis reveals that each lobe of this bilobed organ contains numerous smaller lobules. Surrounding the thymus is a collagenous capsule contiguous with capsular septa that extend inwardly into the organ, enveloping the smaller lobules. Blood vessels located along the septa nourish the thymus and provide an entry point for lymphoid progenitors. Upon histologic examination, the thymus contains two distinct regions: the cortex and the medulla. The outer cortex is highly cellular, containing a delicate network of convoluted thymic epithelium and numerous developing thymocytes. The inner medulla contains a densely packed epithelial network interspersed with fewer developing lymphocytes. Also characteristic of the medulla are degenerate epithelial cells forming spiraled structures termed *Hassal's corpuscles*. Thymic antigen presenting cells (APCs) such as macrophages and dendritic cells are also interspersed throughout the medulla and cortex. Although the functional intricacies of thymic architecture are not well understood, it is clear that differentiation of immature T cells is highly dependent on complex interactions with cortical and medullary epithelia during various stages of development.

THYMIC PATHOLOGY DURING HIV INFECTION

HIV infection has dramatic effects on thymic histology. Autopsy results from HIV-infected individuals document the presence of B cell follicles in the thymus[18] and infiltration of mature B and T cells into perivascular spaces.[19] Epithelial cell death is also observed in HIV-infected thymuses, along with disappearance of Hassall's corpuscles and disruption of the corticomedullary junction.[19] Additionally, premature thymic involution has been observed in HIV+ infants with concomitant depletion of DP and SP CD4 thymocytes.[20] In the SCID-hu model, HIV infection leads to thymic dysplasia, epithelial disorganization, and depletion of DP thymocytes.[21]

THE AGING THYMUS

In many ways, thymic pathology associated with HIV resembles the effects of aging on the thymus. In general, the thymus actively generates large numbers of T cells from birth until puberty, at which point thymic function declines throughout life. However, thymic involution begins as early as 1 year of age. During this process, the thymic epithelium folds in on itself and degenerates, as adipose tissue infiltrates the interlobular septa. Consequently, the thymus can no longer accommodate the numbers of thymocytes it once did, and T cell output declines. However, thymopoiesis continues at low levels well past the seventh decade of life.[22] Given that children exhibit such robust thymic activity, HIV may have more severe effects on the young thymus than the aged thymus. In fact, HIV+ children showing signs of premature thymic atrophy have a poor prognosis and progress rapidly to AIDS.[23]

EARLY T CELL ONTOGENY AND HEMATOPOIESIS

THYMOPOIESIS

The earliest lymphoid progenitors arise from hematopoietic stem cells in the bone marrow. During T cell development, a subset of these progenitors home to the thymus and enter the parenchyma through endothelial venules that extend into the corticomedullary junction. Although this homing process is not well understood, CXCR4, CD44, and α_6 integrin have been implicated.[24] After entry into the thymus, a fraction of these lymphoid progenitors become committed T cell precursors — they mature and subsequently undergo selection. Although the prevailing view is that hematopoietic stem cells are refractory to HIV infection,[25,26] the virus inhibits thymopoiesis at its earliest stages.

Studies performed *in vitro* using CD34+ stem cells cultured with thymic epithelial cells demonstrated that HIV inhibits T cell production.[27] Perhaps more convincing are data generated *in vivo*. In the SCID-hu model, HIV infection depleted CD34+ cells before depletion of mature thymocytes.[28] Notably, CD34+ cells were not found to be directly infected by HIV in this model. The latter studies highlight the indirect effects of HIV on early thymopoiesis; however, HIV also affects hematopoiesis on a more global scale.

HEMATOPOIESIS

In many cases, patients with AIDS exhibit pancytopenia involving anemia, thrombocytopenia, neutropenia, monocytopenia, and lymphocytopenia. Although evidence suggests that CD34+ hematopoietic stem cells are not efficiently infected *in vivo*,[25,26,29–34] an M-tropic strain of HIV has been shown to infect fetal bone marrow CD34+ cells *in vitro*.[35] These cells exhibited a reduced capacity to engraft and generate T cells in SCID-hu mice. Furthermore, the infected CD34+ cells became apoptotic and lost pluripotential characteristics *in vitro*. Interestingly, these effects were reversed by administration of a tumor necrosis factor (TNF)-α inhibitor.[35] Additional studies performed *in vitro* have confirmed that TNF-α inhibits hematopoiesis in long-term bone marrow cultures and, in addition, have shown that HIV-encoded gp120 and Nef have similar inhibitory effects.[36,37] Recent investigation *in vivo* identified a defect in the generation of myeloid and erythroid colony-forming units in human CD34+ cells recovered from HIV-infected SCID-hu mice.[38,39] Notably, this defect was transiently reversed by antiretroviral therapy.[39] Based on the observation that HIV was not detected at significant levels in CD34+ cells recovered from infected SCID-hu mice in these studies, the virus seems to indirectly inhibit hematopoiesis by an ill-defined mechanism.[28,38] One proposed model suggests that HIV indirectly disrupts the bone marrow microenvironment. Accordingly, long-term bone marrow cultures infected with HIV show reduced production of hematopoietic cytokines such as interleukin (IL)-6 and GM-CSF.[40,41,42] Additional studies with bone marrow cultures revealed that CD34+ hematopoietic progenitors from uninfected patients can develop on infected bone marrow stroma, but CD34+ cells from infected patients cannot develop on infected bone marrow stroma.[43] Overall, the indirect effects of HIV on the bone marrow microenvironment seem to be responsible for hematopoietic defects in AIDS patients.

THYMIC SELECTION AND T CELL MATURATION

POSITIVE AND NEGATIVE SELECTION

After entry into the thymus, lymphoid progenitors mature into DN thymocytes and migrate through the cortex, where they undergo selection. Developmental cues further drive DN maturation through the ISP stage and into DP thymocytes that undergo positive and negative selection at the corticomedullary junction and in the medulla. Here, medullary epithelial cells synthesize a myriad of tissue-specific self-antigens, the expressions of which are regulated by the transcription factor aire. These self-antigens are processed and presented to developing thymocytes by MHC on the surface of epithelial cells. Additionally, thymic antigen presenting cells such as macrophages and dendritic cells can present self-peptides to thymocytes. In this way, developing thymocytes are exposed to a broad spectrum of self-antigens during selection. Ultimately, autoreactive thymocytes that are activated by self-antigens are eliminated by negative selection, incompetent thymocytes die by neglect, and self-MHC-restricted thymocytes capable of recognizing foreign antigens mature as a result of positive selection (Figure 4.1). As such, central tolerance is achieved.

TOLERANCE TO HIV

Due to the fact that HIV antigens are presented to thymocytes in the thymus, it is relevant to question whether or not these antigens can be perceived as self-antigens and induce tolerance.

Indeed, functional HIV-specific T cells are present in the periphery, suggesting that tolerance to HIV is not generated during thymic selection, that tolerance is incomplete, or that HIV-specific T cells developed before infection. Additional work will be required to determine the role of HIV infection in altering central tolerance. Recent studies addressed the possibility that HIV-encoded superantigens promote clonal deletion. One study suggested that HIV superantigens were responsible for skewed TCR Vβ profiles observed in AIDS patients; however, a subsequent report suggested that this was due to cytomegalovirus (CMV) rather than HIV superantigens.[44] Also disputing the existence of HIV superantigens are data in the SCID-hu model showing that HIV infection does not preferentially deplete cells expressing specific Vβ genes.[45,46] A separate possible consequence of exposure to HIV antigen in the thymus is that thymocytes become anergized or unresponsive to the virus. Anergic T cells specific for HIV have been documented in the periphery,[47] but it is not yet clear if exposure to HIV in the thymus was necessary or sufficient for anergy induction. Alternatively, HIV-specific thymocytes that encounter antigen in the thymus may develop into regulatory T cells. Upon future encounters with HIV in the periphery, these regulatory T cells could suppress immune responses to the virus. In order to clearly elucidate the pathogenesis of HIV, it will be important to further examine the potential role of HIV in generating regulatory T cells, inducing anergy, or driving clonal deletion.

THYMOCYTE SIGNALING AND DEATH MECHANISMS

Central tolerance in the thymus is dependent on positive and negative selection, both of which are initiated by the TCR. Autoreactive thymocytes bearing TCRs with high affinity for self-peptide–MHC transmit strong TCR signals. In contrast, self-restricted thymocytes bearing TCRs with low affinity for self-peptide–MHC transmit weak TCR signals. Paradoxically, the strength of the TCR signal determines two opposite fates: death (negative selection) or life (positive selection). Specifically, strong signals trigger the negative selection pathway, weaker signals divert toward the positive selection pathway, and the absence of TCR signals leads to death by neglect. The TCR-proximal signals are graded and vary in intensity according to TCR avidity. This enables downstream signaling molecules to interpret upstream signals and initiate positive or negative selection as deemed necessary.

It is well documented that HIV proteins disrupt TCR signals in mature T cells[48]; however, little is known about how HIV influences thymocyte signaling in the context of thymic selection. Numerous mechanisms could account for HIV infection shifting TCR signals toward positive or negative selection. Specifically, CD4 engagement by HIV gp120 may activate Lck and augment TCR signals, thereby increasing negative selection. Recent studies have demonstrated that CXCR4 can co-stimulate TCR signals and augment ERK (extracellular regulated kinase) activity,[49] which regulates positive selection and maturation of DP to SP thymocytes. These studies raise additional questions as to how CXCR4 engagement by HIV gp120 may augment positive selection. Although TCR-proximal signals initiated during negative selection have been studied extensively, downstream pathways resulting in apoptosis of thymocytes largely remain a mystery. It is clear, however, that the proapoptotic Bcl-2 family members Bim, Bax, and Bak, which regulate the mitochondria-dependent apoptotic pathway, are required for negative selection[50,51] (Figure 4.2). HIV-encoded Tat and gp120 may augment negative selection by facilitating Bax insertion into the mitochondrial membrane.[52] HIV proteins may also hinder thymocyte survival by inhibiting the activity of anti-apoptotic Bcl-2 family members.[52] Overall, the possible effects of HIV on thymocyte selection are fertile ground for investigation.

TCR-initiated negative selection pathways and death by neglect are responsible for a large fraction (>95%) of apoptosis that occurs in the thymus; however, alternative apoptotic responses exist. For example, thymocytes will die after stimulation with anti-Fas antibodies or treatment with corticosteroids. Studies of TNF receptor family members (Fas, TNFR, TRAIL) in the thymus have largely focused on identifying the potential roles of these receptors in regulating negative selection. Interestingly, mice deficient in these death receptors or their ligands exhibit normal thymocyte

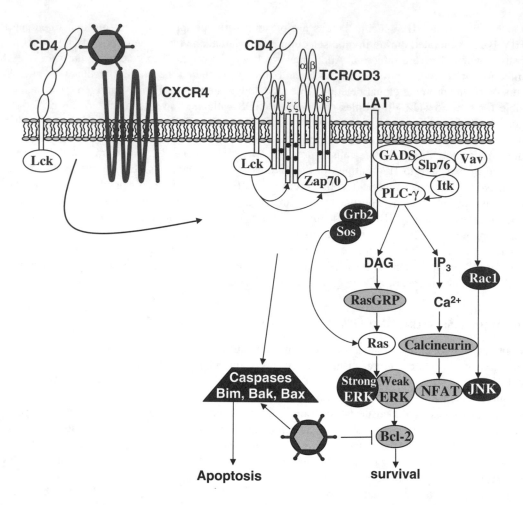

FIGURE 4.2 Thymocyte signaling. The strength of TCR signals generated in developing thymocytes dictates signaling through positive selection pathways (light gray) or negative selection pathways (dark gray). HIV may influence thymocyte survival by engaging CD4 or CXCR4 or by interfering with apoptotic regulators such as Bcl-2 family members. For a review of TCR signaling in thymocytes, see Starr, T.K., Jameson, S.C., and Hogquist, K.A., *Annu. Rev. Immunol.*, 21, 139, 2003.

development.[53,54,55] In addition, mice defective in signaling molecules downstream of death receptors (transgenic mice expressing dominant negative Fas-associated protein with death domain [FADD],[56] caspase-9 knockouts,[57] and Apaf-1 knockouts[58]) do not exhibit defects in negative selection. Together, these studies indicate that death receptors do not mediate TCR-initiated negative selection. However, HIV infection could potentially enhance thymocyte susceptibility to death receptor–initiated apoptosis. In the SIV model, infection increases Fas expression and reduces levels of Bcl-2 in thymocytes.[59] HIV proteins may also augment apoptotic signals in thymocytes by employing mechanisms similar to those used in peripheral T cells (see Chapter 5 in this text), but this is yet to be formally demonstrated.

Direct evidence for HIV-encoded proteins disrupting survival signals in thymocytes was obtained from transgenic mice. A mouse expressing HIV provirus lacking Gag, Pol, and Env exhibited thymocyte depletion despite the fact that the provirus was replication incompetent.[60] Evidence that Nef may play a pivotal role in HIV-mediated thymocyte toxicity is supported by the observation that excessive thymocyte depletion occurs in Nef transgenic mice.[61] Furthermore, the thymocytes that survived were hyperresponsive to activation by anti-CD3 antibodies.[62] Subsequent analysis of Nef transgenic mice revealed that DP thymocytes express low levels of CD4.[63] Interestingly, Nef toxicity in thymocytes correlates with reduced CD4 expression.[64] In the SCID-hu model, HIV bearing a Nef mutation that does not affect viral replication failed to decrease CD4 expression and was less toxic to thymocytes than was wild-type Nef.[65] Several additional studies confirmed that HIV Nef mutants are less pathogenic than wild-type viruses in the SCID-hu thymus.[66,67] In human thymocytes, expression of Nef impairs thymopoiesis *in vitro*[68] and leads to excessive thymocyte proliferation.[69] Ultimately, Nef may mediate its toxic effects by nonspecifically activating thymocytes and by downregulating CD4, but Nef has also been shown to kill cells by inhibiting expression of the antiapoptotic molecules Bcl-2 and Bcl-x$_L$.[52] Similarly, Tat can inhibit Bcl-2 expression, but it activates the proapoptotic factor Bax as well.[52] Although Tat expression in thymic stromal cells inhibits thymocyte maturation *in vitro*,[70] transgenic mice expressing Tat under the control of a CD2 promoter develop normally.[71] Similar to Tat, gp120 can activate Bax and inhibit Bcl-2 expression.[52]

From a broad perspective, multiple stimuli can initiate thymocyte death. In addition to negative selection, thymocytes die by neglect, death receptor engagement, exposure to corticosteroids, and cytokine deprivation. HIV disrupts the thymic microenvironment at the cellular and molecular levels, resulting in dysregulation of developmental cues and survival signals required by thymocytes.

THYMIC MICROENVIRONMENT

THYMOCYTE–STROMA CROSS TALK

Developing thymocytes and thymic stroma coexist in a complicated interdependent relationship (Figure 4.3). In addition to presenting self-peptide–MHC complexes to developing thymocytes, stromal cells express costimulatory ligands and adhesion molecules that augment TCR-initiated signals and affect both positive and negative selection. However, the functions of stromal cells during thymocyte development extend far beyond the selection phase. TECs also produce chemokines to direct thymocyte trafficking through the cortex and medulla. Furthermore, TECs secrete cytokines and growth factors, such as IL-7, SCF, and TSLP, that are critical for thymocyte survival and differentiation. The importance of TECs during T cell development is apparent in patients with DiGeorge's syndrome, who have disorganized or absent thymic epithelial tissue and, consequently, exhibit profound defects in T cell production. Conversely, thymocytes play an important role in TEC survival and organization. In transgenic mice that overexpress CD3ε early in T cell development, thymocytes die prematurely, leading to thymic atrophy and disorganization of thymic epithelial tissue.[72] HIV subverts these tenuous thymocyte–stroma interactions by directly killing thymocytes. The subsequent indirect effects of thymocyte depletion on the thymic microenvironment can further disrupt thymopoiesis.

DISRUPTION OF THE THYMIC MICROENVIRONMENT

HIV interacts with the thymic microenvironment in a bidirectional manner (Table 4.2). For example, the cytokine milieu established by TECs and thymocytes in the thymus promotes HIV replication. Studies have demonstrated that TECs are critical for HIV replication in thymocyte–TEC cocultures[73] and that TEC-derived cytokines, such as IL-1 and TNF-α, activate NFκB, resulting in HIV transcription.[74] Although TNF-α and IL-1 are primarily involved in NFκB-mediated HIV transcription, IL-7 can augment the effects of these cytokines during viral replication.[75] Additionally, TEC-derived cytokines, such as IL-6 and GM-CSF, contribute to HIV replication in cultured thymocytes.[76] Viral replication is also enhanced by IL-2, IL-4, and IL-7.[77] Furthermore, IL-2 in conjunction with IL-4 increase expression of the co-receptors CXCR4 and CCR5 on thymocytes, whereas IL-4 and IL-7 upregulate thymocyte expression of CXCR4.[78]

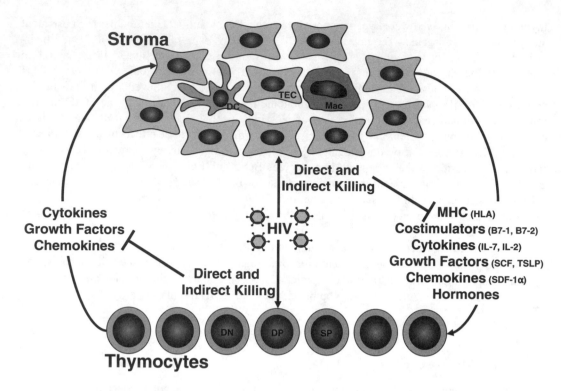

FIGURE 4.3 Thymocyte–stroma cross talk. Thymic stromal cells and thymocytes are mutually dependent on one another for survival factors and developmental cues. HIV can directly infect and kill subsets of thymocytes and stromal cells. Consequently, thymocyte–stroma cross talk is interrupted, and HIV-mediated thymic dysfunction is amplified.

TABLE 4.2
Disruption of the Thymic Microenvironment in HIV Infection

Immune Mediator	Effects of/on HIV
IL-1	Infection increases production by macrophages in thymus; increases NFκB and viral transcription *in vitro*
IL-2	Increases HIV replication in thymus *in vitro* and *in vivo*; upregulates CXCR4, CCR5 on thymocytes
IL-4	Infection increases expression in thymocytes; increases HIV replication in thymus *in vitro*; upregulates CXCR4, CCR5 on thymocytes
IL-6	Infection increases expression in thymic macrophages and thymocytes; enhances viral replication
IL-7	Infection increases levels in serum; increases HIV replication in thymus *in vitro*; cofactor for IL-1 and TNF-α in HIV transcription in thymus; upregulates CXCR4 on thymocytes
IL-10	Infection increases expression in thymocytes
GM-CSF	Enhances viral replication
TNF-α	Increases NFκB and viral transcription *in vitro*
IFN-γ	Infection increases expression in thymocytes
IFN-α	Infection increases production by DC in thymus; upregulates MHCI on thymocytes
MIP-1β	Infection reduces expression in thymocytes
Fas	Increases on thymocytes after SIV infection
Bcl-2	Decreases on thymocytes after SIV infection
MHCI	Decreases on thymocytes after SIV infection; increases in SCID-hu after infection

INDIRECT EFFECTS OF HIV ON THE THYMUS

The indirect effects of HIV on thymocyte development are exemplified by the observation that HIV infants born to HIV+ mothers exhibit reduced CD4 counts and decreased thymic output.[79] Additional evidence suggesting that HIV indirectly disrupts thymocyte development derives from the observation that the virus directly infects far fewer thymocytes than actually die during the course of infection.[80] Consistent with the latter observation, uninfected thymocytes die after contact with thymic dendritic cells induced to express FasL and TNF-α during HIV infection.[81] In addition to TNF-α, infected thymic dendritic cells produce high levels of IFN-α,[82] which can increase MHC class I expression on TECs and potentially influence selection of SP CD8 thymocytes.[83] In addition, thymic macrophages secrete IL-1 and -6 in response to HIV infection.[13] Overall, expression patterns of numerous cytokines become altered during HIV infection. For example, SCID-hu mice infected with HIV produce excessive levels of IL-10 in the thymus.[84] More detailed analysis of thymic tissue derived from infected SCID-hu mice demonstrated that thymocytes contained fewer MIP-1β transcripts and more IL-4, IFN-γ, and IL-6 transcripts than in uninfected controls.[85] Similarly, thymic stromal cells from HIV+ SCID-hu mice contained more IL-6 transcripts than controls.[85] In humans, IL-7 levels accumulate to abnormally high levels after HIV-mediated T cell depletion.[86,87] Elevated levels of TNF-α and IL-6 have also been documented in seropositive patients, whereas IL-2 and -12 levels were reduced.[88] During progression to AIDS, evidence exists for a shift in cytokine production from a Th1 profile (IL-2, -12, and IFN-γ) to a Th2 profile (IL-4, 5, 6, and -10)[89,90]; however, these observations were not confirmed in a majority of patients.[91] Although the specific consequences of cytokine dysregulation during HIV infection of the thymus are unclear, subtle changes in the thymic microenvironment can have drastic effects on thymocyte development. In addition to cytokine dysregulation, disturbances in steroid production were reported during HIV infection. Indirect effects of HIV can decrease levels of the cortisol antagonist dehydroepiandrosterone (DHEA) and may augment steroid-induced apoptosis in thymocytes.[92]

THYMIC OUTPUT AND T CELL HOMEOSTASIS

ATTRITION OF T CELLS

Thymic output as well as homeostatic expansion of the remaining T cell pool (Figure 4.4) achieves compensation for attrition of peripheral T cells. In children and young adults bearing an active thymus, the peripheral T cell pool is largely replenished by thymic output. In contrast, thymic atrophy in adults precludes thymic output as a primary mechanism for T cell replenishment, and peripheral expansion of T cells predominates in maintaining T cell numbers. The inherent drawback of peripheral expansion is that T cells with new antigen specificities cannot be generated. Rather, existing T cells merely proliferate. Thymic output, however, enriches the T cell repertoire by producing new T cells. In order to reconstitute the peripheral T cell pool after depletion by HIV, boosting thymic output could become invaluable. However, this approach has proven difficult, as the thymus may function independently of the peripheral T cell pool. When thymic tissue was implanted into mice, the size of the endogenous thymus remained unchanged.[93,94] Furthermore, cellular labeling studies performed in mice demonstrated that thymic output remains at 1 to 2% of total thymocytes regardless of the size or composition of the peripheral T cell pool.[95–97] This rate of thymic output seems to be static and does not change with age, although the total numbers of T cells generated in the thymus decline as the thymus atrophies. Depending on the age of the patient, replenishment of the T cell pool after HIV infection can occur by thymic output or peripheral expansion. However, the extent of immune reconstitution may be limited due to the fact that HIV ravages both the thymus and the peripheral T cell pool.

The natural course of HIV infection begins with an acute symptomatic phase lasting a few weeks that is typified by significant viremia and a rapid decrease in peripheral blood CD4 T cells.

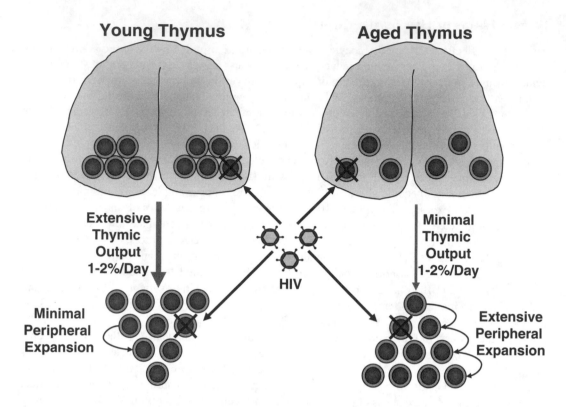

FIGURE 4.4 The impact of HIV on T cell homeostasis. HIV depletes both thymocytes and peripheral T cells. Replenishing the peripheral T cell pool can be achieved by thymic output of new T cells or peripheral expansion of existing T cells. Although young and aged thymuses produce proportionately similar numbers of T cells (1 to 2% of total thymocytes/day), output from a young thymus is greater than from an aged thymus due to the fact that a young thymus is larger and contains more total thymocytes than an aged thymus. In a young individual, T cell homeostasis is maintained by extensive thymic output and minimal peripheral expansion. Older individuals maintain T cell homeostasis by extensive peripheral expansion and minimal thymic output.

Subsequently, viremia partially subsides, and T cell counts rebound during an asymptomatic phase of chronic infection that lasts an average of 10 years. During this phase, CD4 counts gradually fall, and viremia slowly rises until the host immune system is compromised to the point at which opportunistic disease takes hold, marking the onset of AIDS. Due to technical obstacles, it has been difficult to monitor the effects of HIV on the thymus during the early stages of infection. Consequently, insight has been gained from SIV models. The initial stages of SIV infection are characterized by a reduction in peripheral blood CD4 counts correlating with suppression of thymopoiesis and increased apoptosis in the thymus.[98] Approximately 2 months after infection, a rebound in thymic function occurs that is independent of viral load and lymphopenia.[98] Subsequently, peripheral CD4 counts slowly dwindle throughout the course of disease,[98] and at the time of progression to AIDS, thymocyte numbers are significantly reduced.[99]

THYMIC OUTPUT

In human subjects, studies have unequivocally demonstrated that HIV inhibits thymic function. Frequently, phenotypic markers of naive vs. memory T cells are used to estimate thymic output. The presence of large numbers of peripheral T cells bearing the naive phenotype CD62L^{hi} CD45RA$^+$

was interpreted to indicate that the thymus is actively producing new T cells. However, phenotypic markers are not absolute determinants of recent thymic emigrants (RTEs), as naive T cells may be long-lived, and memory CD45RO[+] cells may revert back to CD45RA[+] expression. A more accurate marker of RTEs is the presence of T cell receptor excision circles (TRECs). These TRECs represent circular strands of genomic DNA generated by excision of intervening DNA sequences during rearrangement of the TCR-alpha (TCRA) locus in DP thymocytes. TRECs are remarkably stable and may be detected in RTEs by PCR. The disadvantage of using TRECs to estimate thymic output is that proliferation of RTEs results in dilution of total TRECs per cell. Consequently, TREC analysis is more appropriately used in conjunction with metabolic labeling (BrdU, deuterated glucose) or analysis of cell-cycle markers (Ki67). Consistent with this approach, TREC levels were reported to fall early in HIV infection, whereas CD4 proliferation did not rise compared with uninfected controls.[100] TREC numbers are also reduced during SIV infection[101] and in HIV[+] children.[102]

PERIPHERAL EXPANSION

In addition to suppressing thymic output, HIV hinders peripheral T cell expansion. Metabolic labeling studies have demonstrated that the half-life of circulating T cells is drastically reduced in HIV[+] patients,[103] but HIV can also counteract peripheral expansion by directly infecting proliferating T cells. A delicate equilibrium exists between HIV and the T cell pool during the asymptomatic phase of chronic infection. Because HIV inhibits thymic output as well as peripheral expansion, viremia must be completely controlled or abolished in order to replenish the T cell pool. Furthermore, achieving full reconstitution of the T cell repertoire after depletion by HIV may be greatly enhanced by robust thymic activity.

EFFECTS OF ANTI-HIV THERAPY ON THE THYMUS

HAART

Highly active antiretroviral therapy (HAART) employing a reverse transcriptase inhibitor in conjunction with two or more protease inhibitors has proven extremely effective in controlling HIV replication. Patients who respond favorably to HAART exhibit a reduction in viral load and a rebound in peripheral blood CD4 counts. Specifically, responders were found to be younger, have a larger thymus according to CT scan, and produce more naive CD45[+], CD62L[+], CD4, and CD8 T cells.[104] In addition, naive CD4 cells derived from responders contained more TRECs than nonresponders.[104] Metabolic labeling with deuterated glucose and measurement of telomere lengths in CD4 T cells demonstrated that the reduced TREC levels observed in nonresponders could not be explained by excessive T cell proliferation.[104] Several groups have reported an increase in thymic function after HAART. In these studies, RTEs identified according to phenotypic markers and the presence of TRECs were found at higher levels in HAART patients than in untreated controls.[105–107] Additional studies reported increases in thymic volume and naive T cell counts in HIV[+] patients after HAART.[105,108] Interestingly, HAART was also shown to expand the TCR repertoire in AIDS patients.[109] These results raise the question of whether or not HAART increases thymic function by reducing viremia, by directly boosting thymic function, or both. HIV protease inhibitors were shown to directly inhibit apoptosis in peripheral T cells by antagonizing the mitochondrial permeability transition pore complex.[110–113] If protease inhibitors inhibit thymocyte apoptosis by a similar mechanism, thymic output may increase, but autoreactive thymocytes could escape negative selection as well. In fact, cases of autoimmunity have been observed in AIDS patients receiving HAART.[19] Additional observations were interpreted to suggest that HAART directly promotes thymic output independent of its effects on HIV. For example, failure of HAART to reduce viremia does not always preclude increases in thymic volume, TREC levels, or naive T cell counts.[113] However, protease inhibitor–resistant strains of HIV have been shown to be less pathogenic to the thymus than protease-inhibitor-sensitive strains.[114,115] The implication is that a failure of HAART to control protease inhibitor–resistant HIV does not rule out the possibility that HAART inhibits replication of thymus-trophic HIV variants within the same individual. Thus, HAART may increase thymic output by controlling viremia and not necessarily by inhibiting thymocyte apoptosis. Furthermore, Amado et al. reported that the protease inhibitor indinavir did not increase thymic output in uninfected SCID-hu mice.[116] Studies performed in our laboratory suggest that the protease inhibitor nelfinavir does not inhibit TCR-mediated apoptosis in murine thymocytes (Graham et al., manuscript in preparation). Additional studies will be necessary to determine whether or not HAART increases thymic output by simply inhibiting HIV replication or by alternative mechanisms.

CYTOKINES

Recently, clinical trials supplementing HAART with cytokine therapy were undertaken with the intent to bolster immune reconstitution. IL-2 administration in conjunction with HAART increased CD4 counts compared with HAART alone.[117–120] Additionally, TREC levels in patients treated with IL-2 and HAART decreased or remained stable, suggesting that peripheral expansion is responsible for recovery of the CD4 T cell population.[117,119,120] In the absence of HAART, IL-2 treatment increased CD4 counts yet decreased thymic volume.[121] IL-2 treatment in SCID-hu mice infected with HIV transiently maintained relative percentages of DP thymocytes and more immature thymocytes, whereas IFN-γ treatment transiently maintained mature CD3[+]CD69[+] thymocytes.[122] The latter study also demonstrated that IL-4 or IL-7 treatment delayed thymocyte depletion but increased viral load.[122]

Several novel treatment strategies designed to augment thymic function have been attempted. Administration of growth hormone can increase thymic mass and CD4 counts in humans infected with HIV.[123] Additionally, AIDS patients have been treated with various thymic hormones in attempts to restore thymic function, but results have shown minimal to no benefit.[124–130] It has also been postulated that AIDS patients may be amenable to thymic transplants due to their immuno-suppressed condition. One study suggested that thymic grafts absent of T cells and fibroblasts were well tolerated and enhanced the clinical status of patients.[131] However, a more recent study demonstrated that allogeneic thymic tissue transplanted into HIV[+] recipients was rejected within 2 months and failed to restore T cell counts.[132] Ultimately, effective strategies to augment thymic function remain elusive.

SUMMARY

Thymic function can become severely compromised with HIV infection. The virus directly infects a variety of thymocyte subsets, including DP cells, SP CD4 cells, and thymic stromal cells, resulting in thymocyte depletion and suppression of thymic output. HIV directly kills many of the thymocytes it infects by altering intracellular signal transduction pathways; however, the virus can indirectly kill thymocytes by disrupting the thymic microenvironment. Consequently, HIV inhibits thymopoiesis by direct and indirect mechanisms. In addition, the virus indirectly inhibits hematopoiesis on a more global scale. As a result, the host's ability to replenish the peripheral T cell pool after depletion by HIV becomes compromised. With the advent of more effective antiretroviral therapies to eliminate viremia, the next immediate challenge will be to develop new strategies to promote thymic output and, thus, restore a diverse T cell repertoire to the host.

REFERENCES

1. Schmitt, N., Chene, L., Boutolleau, D., Nugeyre, M.T., Guillemard, E., Versmisse, P., Jacquemot, C., Barre-Sinoussi, F., and Israel, N., Positive regulation of CXCR4 expression and signaling by inter-leukin-7 in CD4[+] mature thymocytes correlates with their capacity to favor human immunodeficiency X4 virus replication, *J. Virol.*, 77(10), 5784, 2003.

2. Brooks, D.G., Kitchen, S.G., Kitchen, C.M., Scripture-Adams, D.D., and Zack, J.A., Generation of HIV latency during thymopoiesis, *Nat. Med.,* 7(4), 459, 2001.
3. Shen, A., Zink, M.C., Mankowski, J.L., Chadwick, K., Margolick, J.B., Carruth, L.M., Li, M., Clements, J.E., and Siliciano, R.F., Resting CD4+ T lymphocytes but not thymocytes provide a latent viral reservoir in a simian immunodeficiency virus-macaca nemestrina model of human immunodeficiency virus type 1-infected patients on highly active antiretroviral therapy, *J. Virol.,* 77(8), 4938, 2003.
4. Berkowitz, R.D., Alexander, S., and McCune, J.M., Causal relationships between HIV-1 coreceptor utilization, tropism, and pathogenesis in human thymus, *AIDS Res. Hum. Retroviruses,* 16(11), 1039, 2000.
5. Berkowitz, R.D., Beckerman, K.P., Schall, T.J., and McCune, J.M., CXCR4 and CCR5 expression delineates targets for HIV-1 disruption of T cell differentiation, *J. Immunol.,* 161(7), 3702, 1998.
6. Taylor J.R., Kimbrell, K.C., Scoggins, R., Delaney, M., Wu, L., and Camerini, D., Expression and function of chemokine receptors on human thymocytes: implication for infection by human immunodeficiency virus type 1, *J. Virol.,* 75(18), 8752, 2001.
7. Schnittman, S.M., Denning, S.M., Greenhouse, J.J., Justement, J.S., Baseler, M., Kurtzberg, J., Haynes, B.F., and Fauci, A.S., Evidence for susceptibility of intrathymic T-cell precursors and their progeny carrying T-cell antigen receptor phenotypes TCR+ and TCR+ to human immunodeficiency virus infection: a mechanism for CD4+ (T4) lymphocyte depletion, *Proc. Natl. Acad. Sci. U.S.A.,* 87(19), 7727, 1990.
8. Chene, L., Nguyen, M.T., Guillemard, E., Moulian, N., Barre-Sinoussi, F., and Israel, N.,Thymocyte-thymic epithelial cell interaction leads to high-level replication of human immunodeficiency virus exclusively in mature CD4+ CD8− CD3+ thymocytes: a critical role for tumor necrosis factor and interleukin-7, *J. Virol.,* 73(9), 7533, 1999.
9. Kitchen, S.G., and Zack, J.A., CXCR4 expression during lymphopoiesis: implications for human immunodeficiency virus type 1 infection of the thymus, *J. Virol.,* 71(9), 6928, 1997.
10. Jamieson, B.D., and Zack, J.A., *In vivo* pathogenesis of a human immunodeficiency virus type 1 reporter virus, *J. Virol.,* 72(8), 6520, 1998.
11. Lee, S., Goldstein, H., Baseler, M., Adelsberger J., and Golding, H., Human immunodeficiency virus type 1 infection of mature CD3hiCD8+ thymocytes, *J. Virol.,* 71(9), 6671, 1997.
12. Gurney, K.B., Yang, O.O., Wilson, S.B., and Uittenbogaart, C.H., TCR+ and CD161+ thymocytes express HIV-1 in the SCID-hu mouse, potentially contributing to immune dysfunction in HIV infection, *J. Immunol.,* 169(9), 5338, 2002.
13. Sandborg, C.I., Imfeld, K.L., Zaldivar, F., jr., and Berman, M.A., HIV type 1 induction of interleukin 1 and 6 production by human thymic cells, *AIDS Res. Hum. Retroviruses,* 10(10), 1221, 1994.
14. Beaulieu, S., Kessous, A., Landry, D., Montplaisir, S., Bergeron, D., and Cohen, E.A., *In vitro* characterization of purified human thymic dendritic cells infected with human immunodeficiency virus type 1, *Virology,* 222(1), 214, 1996.
15. Cameron P.U., Lowe, M.G., Sotzik, F., Coughlan, A.F., Crowe, S.M., and Shortman, K., The interaction of macrophage and non-macrophage tropic isolates of HIV-1 with thymic and tonsillar dendritic cells *in vitro, J. Exp. Med.,* 183(4), 1851, 1996.
16. Kourtis A.P., Ibegbu, C.C., Scinicarrielo, F., and Lee, F.K., Effects of different HIV strains on thymic epithelial cultures, *Pathobiology,* 69(6), 329, 2001.
17. Schnittman, S.M., Singer, K.H., Greenhouse, J.J., Stanley, S.K., Whichard, L.P., Le, P.T., Haynes, B.F., and Fauci, A.S., Thymic microenvironment induces HIV expression. Physiologic secretion of IL-6 by thymic epithelial cells up-regulates virus expression in chronically infected cells, *J. Immunol.,* 147(8), 2553, 1991.
18. Burke, A.P., Anderson, D., Benson, W., Turnicky, R., Mannan, P., Liang, Y.H., Smialek, J., and Virmani, R., Localization of human immunodeficiency virus 1 RNA in thymic tissues from asymptomatic drug addicts, *Arch. Pathol. Lab. Med.,* 119(1), 36, 1995.
19. Haynes B.F., Markert, M.L., Sempowski, G.D., Patel, D.D., and Hale, L.P., The role of the thymus in immune reconstitution in aging, bone marrow transplantation, and HIV-1 infection, *Annu. Rev. Immunol.,* 18, 529, 2000.
20. Rosenzweig, M., Clark, D.P., and Gaulton, G.N., Selective thymocyte depletion in neonatal HIV-1 thymic infection, *AIDS,* 7(12), 1601, 1993.

21. Autran, B., Guiet P., Raphael, M., Grandadam, M., Agut, H., Candotti, D., Grenot, P., Peuch, F., Debre, P., and Cesbron, J.Y., Thymocyte and thymic microenvironment alterations during a systemic HIV infection in a severe combined immunodeficient mouse model, *AIDS,* 10(7), 717, 1996.

22. Jamieson, B.D., Douek, D.C., Killian, S., Hultin, L.E., Scripture-Adams, D.D., Giorgi, J.V., Marelli D., Doup, R.A., and Zack, J.A., Generation of functional thymocytes in the human adult, *Immunity,* 10(5), 569, 1999.

23. Kourtis, A.P., Ibegbu, C., Nahmias, A.J., Lee, F.K., Clark, W.S., Sawyer, M.K., and Nesheim, S., Early progression of disease in HIV-infected infants with thymus dysfunction, *New Engl. J. Med.,* 335(19), 1431, 1996.

24. Petrie, H.T., Cell migration and the control of post-natal T-cell lymphopoiesis in the thymus, *Nat. Rev. Immunol.,* 3, 859, 2003.

25. Scadden, D.T., Shen, H., and Cheng, T., Hematopoietic stem cells in HIV disease, *J. Natl. Cancer Inst. Monogr.,* 28, 24, 2000.

26. Scadden, D.T., Stem cells and immune reconstitution in AIDS, *Blood Rev.,* 17, 227, 2003.

27. Knutsen, A.P., Roodman, S.T., Freeman, J.J., Mueller, K.R., and Bouhasin, J.D., Inhibition of thymopoiesis of CD34+ cell maturation by HIV-1 in an *in vitro* CD34+ cell and thymic epithelial organ culture model, *Stem Cells,* 17(6), 327, 1999.

28. Jenkins, M., Hanley, M.B., Moreno, M.B., Wieder, E., and McCune, J.M., Human immunodeficiency virus-1 infection interrupts thymopoiesis and multilineage hematopoiesis *in vivo, Blood,* 91(8), 2672, 1998.

29. McCune, J.M. and Kaneshima, H., The hematopathology of HIV-1 disease: experimental analysis *in vivo,* in *Human Hematopoiesis in SCID Mice,* Roncarolo, M.G., Namikawa, R., and Pault, B., Eds., R. G. Landes Co., Austin, TX, 1995.

30. Stanley, S.K., Kessler, S.W., Justement, J.S., Schnittman, S.M., Greenhouse, J.J., Brown, C.C., Musongela, L., Musey, K., Kapita, B., and Fauci, A.S., CD34+ bone marrow cells are infected with HIV in a subset of seropositive individuals, *J. Immunol.,* 149(2), 689, 1992.

31. Chelucci, C., Hassan, H.J., Locardi, C., Bulgarini, D., Pelosi, E., Mariani, G., Testa, U., Federico, M., Valtieri, M., and Peschle, C., *In vitro* human immunodeficiency virus-1 infection of purified hematopoietic progenitors in single-cell culture, *Blood,* 85(5), 1181, 1995.

32. Davis B.R., Schwartz, D.H., Marx, J.C., Johnson, C.E., Berry, J.M., Lyding J., Merigan, T.C., and Zander, A., Absent or rare human immunodeficiency virus infection of bone marrow stem/ progenitor cells *in vivo*, *J. Virol.,* 65(4), 1985, 1991.

33. Neal, T.F., Holland, H.K., Baum, C.M., Villinger, F., Ansari, A.A., Saral, R., Wingard, J.R., and Fleming, W.H., CD34+ progenitor cells from asymptomatic patients are not a major reservoir for human immunodeficiency virus-1, *Blood,* 86(5), 1749, 1995.

34. Moses, A., Nelson, J., and Bagby, G.C., The influence of human immunodeficiency virus-1 on hematopoiesis, *Blood,* 91(5), 1479, 1998.

35. Yurasov, S.V., Pettoello-Mantovani, M., Raker, C.A., and Goldstein, H., HIV type 1 infection of human fetal bone marrow cells induces apoptotic changes in hematopoietic precursor cells and suppresses their *in vitro* differentiation and capacity to engraft SCID mice, *AIDS Res. Hum. Retroviruses,* 15(18), 1639, 1999.

36. Calenda, V., Graber, P., Delamarter, J.F., and Chermann, J.C., Involvement of HIV Nef protein in abnormal hematopoiesis in AIDS: *in vitro* study on bone marrow progenitor cells, *Eur. J. Haematol.,* 52(2), 103, 1994.

37. Maciejewski, J.P., Weichold, F.F., and Young, N.S., HIV-1 suppression of hematopoiesis *in vitro* mediated by envelope glycoprotein and TNF-, *J. Immunol.,* 153(9), 4303, 1994.

38. Koka, P.S., Jamieson, B.D., Brooks, D.G., Zack, J.A., Human immunodeficiency virus type 1-induced hematopoietic inhibition is independent of productive infection of progenitor cells *in vivo, J. Virol.,* 73(11), 9089, 1999.

39. Koka, P.S., Fraser, J.K., Bryson, Y., Bristol, G.C., Aldrovandi, G.M., Daar, E.S., and Zack, J.A., Human immunodeficiency virus inhibits multilineage hematopoiesis *in vivo, J. Virol.,* 72(6), 5121, 1998.

40. Geissler, R.G., Ganser, A., Ottmann, O.G., Gute, P., Morawetz, A., Guba, P., Helm, E.B., and Hoelzer, D., *In vitro* improvement of bone marrow-derived hematopoietic colony formation in HIV-positive patients by -D-tocopherol and erythropoietin, *Eur. J. Haematol.,* 53(4), 201, 1994.

41. Geissler, R.G., Rossol, R., Mentzel, U., Ottmann, O.G., Klein, A.S., Gute, P., Helm, E.B., Hoelzer, D., and Ganser A., -T cell-receptor-positive lymphocytes inhibit human hematopoietic progenitor cell growth in HIV type 1-infected patients, *AIDS Res. Hum. Retroviruses,* 12(7), 577, 1996.

42. Moses, A.V., Williams, S., Heneveld, M.L., Strussenberg, J., Rarick, M., Loveless, M., Bagby, G., and Nelson, J.A., Human immunodeficiency virus infection of bone marrow endothelium reduces induction of stromal hematopoietic growth factors, *Blood,* 87(3), 919, 1996.

43. Gill, V., Shattock, R.J., Scopes, J., Hayes, P., Freedman, A.R., Griffin, G.E., Gordon-Smith, E.C., Gibson, and F.M., Human immunodeficiency virus infection impairs hemopoiesis in long-term bone marrow cultures: nonreversal by nucleoside analogues, *J. Infect. Dis.,* 176(6), 1510, 1997.

44. Dobrescu, D., Ursea, B., Pope, M., Asch, A.S., and Posnett, D.N., Enhanced HIV-1 replication in V 12 T cells due to human cytomegalovirus in monocytes: evidence for a putative herpesvirus superantigen, *Cell,* 82, 753, 1995.

45. Komanduri, K.V., Salha, M.D., Sekaly, R.P., and McCune, J.M., Superantigen-mediated deletion of specific T cell receptor Vβ subsets in the SCID-hu Thy/Liv mouse is induced by staphylococcal enterotoxin B, but not HIV-1, *J. Immunol.,* 158, 544, 1997.

46. Boldt-Houle, D.M., Jamieson, B.D., Aldrovandi, G.M., Rinaldo, C.R., Jr., Ehrlich, G.D., and Zack, J.A., Loss of T cell receptor V repertoires in HIV type 1-infected SCID-hu mice, *AIDS Res. Hum. Retroviruses,* 13(2), 125, 1997.

47. Bouhdoud, L., Villain, P., Merzouki, A., Arella, M., and Couture, C., T-cell receptor-mediated anergy of a human immunodeficiency virus (HIV) gp120-specific CD4$^+$ cytotoxic T-cell clone, induced by a natural HIV type 1 variant peptide, *J. Virol.,* 74(5), 2121, 2000.

48. Fackler, O.T. and Baur, A.S., Live and let die: Nef functions beyond HIV replication, *Immunity,* 16(4), 493, 2002.

49. Kremer, K.N., Humphreys, T.D., Kumar, A., Qian, N.X., and Hedin, K.E., Distinct role of ZAP-70 and Src homology 2 domain-containing leukocyte protein of 76 kDa in the prolonged activation of extracellular signal-regulated protein kinase by the stromal cell-derived factor-1/CXCL12 chemokine, *J. Immunol.,* 171(1), 360, 2003.

50. Bouillet, P., Purton, J.F., Godfrey, D.I., Zhang, L.C., Coultas, L., Puthalakath, H., Pellegrini, M., Cory, S., Adams, J.M., and Strasser, A., BH3-only Bcl-2 family member Bim is required for apoptosis of autoreactive thymocytes, *Nature,* 415, 922, 2002.

51. Rathmell, J.C., Lindsten, T., Zong, W.X., Cinalli, R.M., and Thompson, C.B., Deficiency in Bak and Bax perturbs thymic selection and lymphoid homeostasis, *Nat. Immunol.,* 3(10), 932, 2002.

52. Gougeon, M.L., Apoptosis as an HIV strategy to escape immune attack, *Nat. Rev. Immunol.,* 3, 392, 2003.

53. Page, D.M., Roberts, E.M., Peschon, J.J., and Hedrick, S.M., TNF receptor-deficient mice reveal striking differences between several models of thymocyte negative selection, *J. Immunol.,* 160, 120, 1998.

54. Adachi, M., Suematsu, S., Suda, T., Watanabe, D., Fukuyama, H., Ogasawara, J., Tanaka, T., Yoshida, N., and Nagata, S., Enhanced and accelerated lymphoproliferation in Fas-null mice, *Proc. Natl. Acad. Sci. U.S.A.,* 93, 2131, 1996.

55. Cretney, E., Uldrich, A.P., Berzins, S.P., Strasser, A., Godfrey, D.I., and Smyth, M.J., Normal thymocyte negative selection in TRAIL-deficient mice, *J. Exp. Med.,* 198(3), 491, 2003.

56. Zhang, J., Cado, D., Chen, A., Kabra, N.H., and Winoto, A., Fas-mediated apoptosis and activation-induced T-cell proliferation are defective in mice lacking FADD/Mort1, *Nature,* 392, 296, 1998.

57. Marsden V.S., O'Connor, L., O'Reilly, L.A., Stilke, J., Metcalf, D., Ekert, P.G., Huang, D.C., Cecconi, F., Kuida, K., Tomaselli, K.J., Roy, S., Nicholson, D.W., Vaux, D.L., Bouillet, P., Adams, J.M., and Strasser, A., Apoptosis initiated by Bcl-2-regulated caspase activation independently of the cytochrome c/Apaf-1/caspase-9 apoptosome, *Nature,* 419, 634, 2002.

58. Hara, H., Takeda, A., Takeuchi, M., Wakeham, A.C., Itie, A., Sasaki, M., Mak T.W., Yoshimura, A., Nomoto, K., and Yoshida, H. The apoptotic protease-activating factor 1-mediated pathway of apoptosis is dispensable for negative selection of thymocytes, *J. Immunol.,* 168, 2288, 2002.

59. Rosenzweig, M., Connole, M., Forand-Barabasz, A., Tremblay, M.P., Johnson, R.P., and Lackner, A.A., Mechanisms associated with thymocyte apoptosis induced by simian immunodeficiency virus, *J. Immunol.,* 165(6), 3461, 2000.

60. Tinkle, B.T., Ueda, H., Ngo, L., Luciw, P.A., Shaw, K., Rosen, C.A., and Jay, G., Transgenic dissection of HIV genes involved in lymphoid depletion, *J. Clin. Invest.,* 100(1), 32, 1997.

61. Lindemann, D., Wilhelm R., Renard, P., Althage, A., Zinkernagel, R., and Mous, J., Severe immunodeficiency associated with a human immunodeficiency virus 1 NEF/3-long terminal repeat transgene, *J. Exp. Med.,* 179(3), 797, 1994.

62. Hanna, Z., Kay, D.G., Rebai, N., Guimond, A., Jothy, S., and Jolicoeur, P., Nef harbors a major determinant of pathogenicity for an AIDS-like disease induced by HIV-1 in transgenic mice, *Cell,* 95(2), 163, 1998.

63. Brady, H.J., Pennington, D.J., Miles, C.G., and Dzierzak, E.A., CD4 cell surface downregulation in HIV-1 Nef transgenic mice is a consequence of intracellular sequestration, *EMBO J.,* 12(13), 4923, 1993.

64. Skowronski, J., Parks, D., and Mariani, R., Altered T cell activation and development in transgenic mice expressing the HIV-1 Nef gene, *EMBO J.,* 12(2), 703, 1993.

65. Stoddart, C.A., Geleziunas, R., Ferrell, S., Linquist-Stepps, V., Moreno, M.E., Bare, C., Xu, W., Yonemoto, W., Bresnahan, P.A., McCune, J.M., and Greene, W.C., Human immunodeficiency virus type 1 Nef-mediated downregulation of CD4 correlates with Nef enhancement of viral pathogenesis, *J. Virol.,* 77(3), 2124, 2003.

66. Duus, K.M., Miller, E.D., Smith, J.A., Kovaley, G.I., and Su, L., Separation of human immunodeficiency virus type-1 replication from Nef-mediated pathogenesis in the human thymus, *J. Virol.,* 75(8), 3916, 2001.

67. Jamieson, B.D., Aldrovandi, G.M., Planelles, V., Jowett, J.B., Gao, L., Bloch, L.M., Chen, I.S., and Zack, J.A., Requirement of human immunodeficiency virus type 1 Nef for *in vivo* replication and pathogenicity, *J. Virol.,* 68(6), 3478, 1994.

68. Stove, V., Naessens, E., Stove, C., Swigut, T., Plum, J., and Verhasselt, B., Signaling but not trafficking function of HIV-1 protein Nef is essential for Nef-induced defects in human intrathymic T-cell development, *Blood,* 102(8), 2925, 2003.

69. Verhasselt, B., Naessens, E., Verhofstede, C., De Smedt, M., Schollen, S., Kerre, T., Vanhecke, D., and Plum, J., Human immunodeficiency virus Nef gene expression affects generation and function of human T cells, but not dendritic cells, *Blood,* 94(8), 2809, 1999.

70. Maroder, M., Scarpa, S., Screpanti, I., Stigliano, A., Meco, D., Vacca, A., Stuppia, L., Frati, L., Modesti, A., and Gulino, A., Human immunodeficiency virus type 1 Tat protein modulates fibronectin expression in thymic epithelial cells and impairs *in vitro* thymocyte development, *Cell Immunol.,* 168(1), 49, 1996.

71. Brady, H.J., Abraham, D.J., Pennington, D.J., Miles, C.G., Jenkins, S., and Dzierzak, E.A., Altered cytokine expression in T lymphocytes from human immunodeficiency virus Tat transgenic mice, *J. Virol.,* 69(12), 7622, 1995.

72. Wang, B., Biron, C., She, J., Higgins, K., Sunshine, M., Lacy, E., Lonberg, N., and Terhorst, C., A block in both early T lymphocyte and natural killer cell development in transgenic mice with high-copy numbers of the human CD3E gene, *Proc. Natl. Acad. Sci. U.S.A.,* 91, 9402, 1994.

73. Braun, J., Valentin, H., Nugeyre, M-T, Ohayon, H., Gounon, P., and Barre-Sinoussi, F., Productive and persistent infection of human thymic epithelial cells *in vitro* with HIV-1, *Virology,* 225(2), 413, 1996.

74. Chene, L., Nugeyre, M.T., Guillemard, E., Moulian, N., Barre-Sinoussi, F., and Israel, N., Thymocyte–thymic epithelial cell interaction leads to high-level replication of human immunodeficiency virus exclusively in mature CD4(+) CD8(−) CD3(+) thymocytes: a critical role for tumor necrosis factor and IL-7, *J. Virol.,* 73(9), 7533, 1999.

75. Chene, L., Nugeyre, M.T., Barre-Sinoussi, F., and Israel, N., High-level replication of human immunodeficiency virus in thymocytes requires NF-B activation through interaction with thymic epithelial cells, *J. Virol.* 73(3), 2064, 1999.

76. Rothe, M., Chene, L., Nugeyre, M.T., Braun, J., Barre-Sinoussi, F., and Israel, N., Contact with thymic epithelial cells as a prerequisite for cytokine-enhanced human immunodeficiency virus type 1 replication in thymocytes, *J. Virol.,* 72(7), 5852, 1998.

77. Uittenbogaart, C.H., Anisman, D.J., Zack, J.A., Economides, A., Schmid, I., and Hays, E.F., Effects of cytokines on HIV-1 production by thymocytes, *Thymus,* 23(3–4), 155, 1994–1995.

78. Pedroza-Martins, L., Boscardin, W.J., Anisman-Posner, D.J., Schols, D., Bryson, Y.J., and Uittenbogaart, C.H., Impact of cytokines on replication in the thymus of primary human immunodeficiency virus type 1 isolates from infants, *J. Virol.,* 76(14), 6929, 2002.

79. Nielsen, S.D., Jeppesen, D.L., Kolte, L., Clark, D.R., Sorensen, T.U., Dreves, A.M., Ersboll, A.K., Ryder, L.P., Valerius, N.H., and Nielsen, J.O., Impaired progenitor cell function in HIV-negative infants of HIV-positive mothers results in decreased thymic output and low CD4 counts, *Blood,* 98(2), 398, 2001.

80. Su, L., Kaneshima, H., Bonyhadi, M., Salimi, S., Kraft, D., Rabin, L., and McCune, J.J., HIV-1-induced thymocyte depletion is associated with indirect cytopathogenicity and infection of progenitor cells *in vivo, Immunity,* 2(1), 25, 1995.

81. Beaulieu, S., Lafontaine, M., Richer, M., Courchesne, I., Cohen, E.A., and Bergeron, D., Characterization of the cytotoxic factor(s) released from thymic dendritic cells upon human immunodeficiency virus type 1 infection, *Virology,* 241(2), 285, 1998.

82. Keir, M.E., Stoddart, C.A., Linquist-Stepps, V., Moreno, M.E., and McCune, J.M., IFN- secretion by type 2 predendritic cells up-regulates MHC class I in the HIV-1- infected thymus, *J. Immunol.,* 168(1), 325, 2002.

83. Keir, M.E., Rosenberg, M.G., Sandberg, J.K., Jordan, K.A., Wiznia, A., Nixon, D.F., Stoddart, C.A., and McCune, J.M., Generation of CD3+CD8^{low} thymocytes in the HIV type 1-infected thymus, *J. Immunol.,* 169(5), 2788, 2002.

84. Kovalev, G., Duus, K., Wang, L., Lee, R., Bonyhadi, M., Ho, D., McCune, J.M., Kaneshima, H., and Su, L., Induction of MHC class I expression on immature thymocytes in HIV-1-infected SCID-hu Thy/Liv mice: evidence of indirect mechanisms, *J. Immunol.,* 162(12), 7555, 1999.

85. Koka, P.S., Brooks, D.G., Razai, A., Kitchen, C.M., and Zack, J.A., HIV type I infection alters cytokine mRNA expression in thymus, *AIDS Res. Hum. Retroviruses,* 19(1), 1, 2003.

86. Llano, A., Barretina, J., Gutierrez, A., Blanco, J., Cabrera, C., Clotet, B., and Este, J.A., Interleukin-7 in plasma correlates with CD4 T-cell depletion and may be associated with emergence of syncytium-inducing variants in human immunodeficiency virus type 1-positive individuals, *J. Virol.,* 75(21), 10319, 2001.

87. Correa, R., Resino, S., and Munoz-Fernandez, M.A., Increased interleukin-7 plasma levels are associated with recovery of CD4+ T cells in HIV-infected children, *J. Clin. Immunol.,* 23(5), 401, 2003.

88. Al-Harthi, L. and Landay, A., Immune recovery in HIV disease: role of the thymus and T cell expansion in immune reconstitution strategies, *J. Hematother. Stem Cell Res.,* 11, 777, 2002.

89. Clerici, M. and Shearer, G.M., The Th1-Th2 hypothesis of HIV infection: new insights, *Immunol. Today,* 15, 575, 1994.

90. Altfeld, M., Addo, M.M., Kreuzer, K.A., Rockstroh, J.K., Dumoulin, F.L., Schliefer, K., Leifeld, L., Sauerbruch, T., and Spengler, U., T(H)1 to T(H)2 shift of cytokines in peripheral blood of HIV-infected patients is detectable by reverse transcriptase polymerase chain reaction but not by enzyme-linked immunosorbent assay under nonstimulated conditions, *J. Acquir. Immune Defic. Syndr.,* 23, 287, 2000.

91. Alonso, K., Pontiggia, P., Medenica, R., and Rizzo, S., Cytokine patterns in adults with AIDS, *Immunol. Invest.,* 26, 341, 1997.

92. Clerici, M., Galli, M., Bosis, S., Gervasoni, C., Moroni, M., and Norbiato, G., Immunoendocrinologic abnormalities in human immunodeficiency virus infection, *Ann. NY Acad. Sci.,* 917, 956, 2000.

93. Metcalf, D., The autonomous behaviour of normal thymus grafts, *Aust. J. Exp. Biol.,* 41, 437, 1963.

94. Metcalf, D., Sparrow, N., Nakamura, K., and Ishidate, M., The behavior of thymus grafts in high and low leukemia strains of mice, *Aust. J. Exp. Biol.,* 39, 441, 1961.

95. Berzins, S.P., Boyd, R.L., and Miller, J.F., A central role for thymic emigrants in peripheral T cell homeostasis, *Proc. Natl. Acad. Sci. U.S.A.,* 96, 9787, 1999.

96. Gabor, M.J., Scollay, R., and Godfrey, D.I., Thymic T cell export is not influenced by the peripheral T cell pool, *Eur. J. Immunol.,* 27, 2986, 1997.

97. Berzins, S.P., Boyd, R.L., and Miller, J.F., The role of the thymus and recent thymic migrants in the maintenance of the adult peripheral lymphocyte pool, *J. Exp. Med.,* 187, 1839, 1998.

98. Wykrzykowska, J.J., Rosenzweig, M., Veazey, R.S., Simon, M.A., Halvorsen, K., Desrosiers, R.C., Johnson, R.P., and Lackner, A.A., Early regeneration of thymic progenitors in rhesus macaques infected with simian immunodeficiency virus, *J. Exp. Med.,* 187(11), 1767, 1998.

99. Sopper, S., Nierwetberg, D., Halbach, A., Sauer, U., Scheller, C., Stahl-Hennig, C., Matz-Rensing, K., Schafer, F., Schneider, T., ter Meulen, V., and Muller, J.G., Impact of simian immunodeficiency virus (SIV) infection on lymphocyte numbers and T-cell turnover in different organs of rhesus monkeys, *Blood,* 101(4), 1213, 2003.

100. Douek, D.C., Betts, M.R., Hill, B.J., Little, S.J., Lempicki, R., Metcalf, J.A., Casazza, J., Yoder, C., Adelsberger, J.W., Stevens, R.A., Baseler, M.W., Keiser, P., Richman, D.D., Davey, R.T., and Koup, R.A., Evidence for increased T cell turnover and decreased thymic output in HIV infection, *J. Immunol.* 167(11), 6663, 2001.

101. Sodora, D.L., Milush, J.M., Ware, F., Wozniakowski, A., Montgomery, L., McClure, H.M., Lackner, A.A., Marthas, M., Hirsch, V., Johnson, R.P., Douek, D.C., and Koup, R.A., Decreased levels of recent thymic emigrants in peripheral blood of simian immunodeficiency virus-infected macaques correlate with alteration within the thymus, *J. Virol.,* 76(19), 9981, 2002.

102. Correa, R. and Munoz-Fernandez, M.A., Production of new T cells by thymus in children: effect of HIV infection and antiretroviral therapy, *Pediat. Res.,* 52(2), 207, 2002.

103. McCune, J.M., Hanley, M.B., Cesar, D., Halvorsen, R., Hoh, R., Schmidt, D., Wieder, E., Deeks, S., Siler, S., Neese, R., and Hellerstein, M., Factors influencing T-cell turnover in HIV-1-seropositive patients, *J. Clin. Invest.,* 105(5), R1, 2000.

104. Teixeira, L., Valdez, H., McCune, J.M., Koup, R.A., Badley A.D., Hellerstein, M.K., Napolitano, L.A., Douek, D.C., Mbisa, G., Deeks, S., Harris, J.M., Barbour, J.D., Gross, B.H., Francis, I.R., Halvorsen, R., Asaad, R., and Lederman, M.M., Poor CD4 T cell restoration after suppression of HIV-1 replication may reflect lower thymic function, *AIDS,* 15(14), 1749, 2001.

105. Ruiz-Mateos, E., de la Rosa, R., Franco, J.M., Martinez-Moya, M., Rubio, A., Soriano, N., Sanchez-Quijano, A., Lissen, E., and Leal, M., Endogenous IL-7 is associated with increased thymic volume in adult HIV-infected patients under highly active antiretroviral therapy, *AIDS,* 17(7), 947, 2003.

106. De Rossi, A., Walker, A.S., Klein, N., De Forni, D., King, D., and Gibb, D.M., Increased thymic output after initiation of antiretroviral therapy in human immunodeficiency virus type 1-infected children in the Peadiatric European Network for Treatment of AIDS (PENTA) 5 Trial, *J. Infect. Dis.,* 186(3), 312, 2002.

107. Ometto, L., De Forni, D., Patiri, F., Trouplin, V., Mammano, F., Giacomet, V., Giaquinto, C., Douek, D., Koup, R., and De Rossi, A., Immune reconstitution in HIV-1-infected children on antiretroviral therapy: role of thymic output and viral fitness, *AIDS,* 16(6), 839, 2002.

108. Rubio, A., Martinez-Moya, M., Leal, M., Franco, J.M., Ruiz-Mateos, E., Merchante, E., Sanchez-Quijano, A., and Lissen, E., Changes in thymus volume in adult HIV-infected patients under HAART: correlation with the T-cell repopulation, *Clin. Exp. Immunol.,* 130(1), 121, 2002.

109. Kolte, L., Dreves, A.M., Ersboll, A.K., Strandberg, C., Jeppesen, D.L., Nielsen, J.O., Ryder, L.P., and Nielsen, S.D., Association between larger thymic size and higher thymic output in human immuno-deficiency virus-infected patients receiving highly active antiretroviral therapy, *J. Infect. Dis.,* 185(11), 1578, 2002.

110. Phenix, B.N., Lum, J.J., Nie, Z., Sanchez-Dardon, J., and Badley, A.D., Antiapoptotic mechanism of HIV protease inhibitors: preventing mitochondrial transmembrane potential loss, *Blood,* 98, 1078, 2001.

111. Phenix, B.N. and Badley, A.D., Influence of mitochondrial control of apoptosis on the pathogenesis, complications and treatment of HIV infection, *Biochimie,* 84, 251, 2002.

112. Phenix, B.N., Cooper, C., Owen, C., and Badley, A.D., Modulation of apoptosis by HIV protease inhibitors, *Apoptosis,* 7, 295, 2002.

113. Delgado, J., Leal, M., Ruiz-Mateos, E., Martinez-Moya, M., Rubio, A., Merchante E., de la Rosa, R., Sanchez-Quijano, A., and Lissen, E., Evidence of thymic function in heavily antiretroviral-treated human immunodeficiency virus type 1-infected adults with long-term virologic treatment failure, *J. Infect. Dis.,* 186(3), 410, 2002.

114. Lecossier, D., Bouchonnet, F., Schneider, P., Clavel, F., Hance, A.J., Centre de Recherche Integre sur le VIH Bichat-Claude Bernard, Discordant increases in CD4+ T cells in human immunodeficiency virus-infected patients experiencing virologic treatment failure: role of changes in thymic output and T cell death, *J. Infect. Dis.,* 183(7), 1009, 2001.

115. Stoddart, C.A., Liegler, T.J., Mammano, F., Linquist-Stepps, V.D., Hayden, M.S., Deeks, S.G., Grant, R.M., Clavel, F., and McCune, J.M., Impaired replication of protease inhibitor-resistant HIV-1 in human thymus, *Nat. Med.,* 7(6), 712, 2001.

116. Amado, R.G., Jamieson, B.D., Cortado, R., Cole, S.W., Zack, J.A., Reconstitution of human thymic implants is limited by human immunodeficiency virus breakthrough during antiretroviral therapy, *J. Virol.,* 73(8), 6361, 1999.

117. De Paoli, P., Bortolin, M.T., Zanussi, S., Monzoni, A., Pratesi, C., and Giacca, M., Changes in thymic function in HIV-positive patients treated with highly active antiretroviral therapy and interleukin-2, *Clin Exp. Immunol.*, 125(3), 440, 2001.

118. Carcelain, G., Saint-Mezard, P., Altes, H.K., Tubiana, R., Grenot, P., Rabian, C., de Boer R., Costagliola, D., Katlama, C., Debre, P., and Autran, B., IL-2 therapy and thymic production of naïve CD4 T cells in HIV-infected patients with severe CD4 lymphopenia, *AIDS*, 17(6), 841, 2003.

119. Pido-Lopez, J., Burton, C., Hardy, G., Pires, A., Sullivan, A., Gazzard, B., Aspinall, R., Gotch, F., and Imami, N., Thymic output during initial highly active antiretroviral therapy (HAART) and during HAART supplementation with interleukin 2 and/or with HIV type 1 immunogen (Remune), *AIDS Res. Hum. Retroviruses*, 19(2), 103, 2003.

120. Marchetti, G., Meroni, L., Varchetta, S., Terzieva, V., Bandera, A., Manganaro, D., Molteni, C., Trabattoni, D., Fosssati, S., Clerici, M., Galli, M., Moroni, M., Franzetti, F., and Gori, A., Low-dose prolonged intermittent interleukin-2 adjuvant therapy: results of a randomized trial among human immunodeficiency virus-positive patients with advanced immune impairment, *J. Infect. Dis.*, 186(5), 606, 2002.

121. Lu, A.C., Jones, E.C., Chow, C., Miller, K.D., Herpin, B., Rock-Kres, D., Metcalf, J.A., Lane, H.C., and Kovacs, J.A., Increases in CD4+ T lymphocytes occur without increases in thymic size in HIV-infected subjects receiving interleukin-2 therapy, *J. Acquir. Immune Defic. Syndr.*, 34(3), 299, 2003.

122. Uittenbogaart, C.H., Boscardin, W.J., Anisman-Posner, D.J., Koka, P.S., Bristol, G., and Zack, J.A., Effect of cytokines on HIV-induced depletion of thymocytes *in vivo*, *AIDS*, 14(10), 1317, 2000.

123. Napolitano, L.A., Lo, J.C., Gotway, M.B., Mulligan, K., Barbour, J.D., Schmidt, D., Grant, R.M., Halvorsen, R.A., Schambelan, M., and McCune, J.M., Increased thymic mass and circulating naïve CD4 T cells in HIV-1-infected adults treated with growth hormone, *AIDS*, 16(8), 1103, 2002.

124. Maggiolo, F., Taras, A., Pravettoni, M.G., Leone, M., Ingrosso, A., and Suter, F., Zidovudine and thymus humoral factor -2 in the treatment of HIV infection: preliminary clinical experience, *Infection*, 25(1), 35, 1997.

125. Carco, F. and Guazzotti, G., Therapeutic use of thymostimulin in HIV-seropositive subjects with lymphadenopathy syndrome, *Recenti. Prog. Med.*, 84(11), 756, 1993.

126. Silvestris, F., Gernone, A., Frassanito, M.A., and Dammacco, F., Immunologic effects of long-term thympopentin treatment in patients with HIV-induced lymphadenopathy syndrome, *J. Lab. Clin. Med.*, 113(2), 139, 1989.

127. Skotnicki, A.B., Zatz, M., Sztein, M.B., Goldstein, A.L., and Schulof, R.S., Effect of thymic hormones on interleukin 2 synthesis by lymphocytes from HIV-positive pre-AIDS subjects, *Immunol. Invest.*, 17(2), 159, 1988.

128. Valesini, G., Barnaba, V., Benvenuto, R., Balsano, F., Mazzanti, P., and Cazzola, P., A calf thymus acid lysate improves clinical symptoms and T-cell defects in the early stages of HIV infection: second report, *Eur. J. Cancer Clin. Oncol.*, 23(12), 1915, 1987.

129. Berner, Y., Handzel, Z.T., Pecht, M., Trainin, N., and Bentwich, Z., Attempted treatment of acquired immunodeficiency syndrome (AIDS) with thymic humoral factor, *Isr. J. Med. Sci.*, 20(12), 1195, 1984.

130. Beall, G., Kruger, S., Morales, F., Imagawa, D., Goldsmith, J.A., Fisher, D., Steinberg, J., Phair, J., Whaling, S., and Bitran, J., A double-blind, placebo-controlled trial of thymostimulin in symptomatic HIV-infected patients, *AIDS*, 4(7), 679, 1990.

131. Dupuy, J.M., Gilmore, N., Goldman, H., Tsoukas, C., Pekovic, D., Chausseau, J.P., Duperval, R., Joly, M., Pelletier, L., and Thibaudeau, Y., Thymic epithelial cell transplantation in patients with acquired immunodeficiency syndrome. Evidence for infection by HIV-1 of newly differentiated T cells at the site of transplantation, *Thymus*, 17(4), 205, 1991.

132. Markert, M.L., Hicks, C.B., Bartlett, J.A., Harmon, J.L., Hale, L.P., Greenberg, M.L., Ferrari, G., Ottinger, J., Boeck, A., Kloster, A.L., McLaughlin, T.M., Bleich, K.B., Ungerleider, R.M., Lyerly, H.K., Wilkinson, W.E., Rousseau, F.S., Heath-Chiozzi, M.E., Leonard, L.M., Haase, A.T., Shaw, G.M., Bucy, R.P., Douek, D.C., Koup, R.A., Haynes, B.F., Bolognesi, D.P., and Weinhold, K.J., Effect of highly active antiretroviral therapy and thymic transplantation on immunoreconstitution in HIV infection, *AIDS Res. Hum. Retroviruses*, 16(5), 403, 2000.

5 Deadly Intentions: Apoptosis in the Peripheral Immune System

Michael J. Pinkoski

CONTENTS

INTRODUCTION

A functional immune system requires tight regulation over the number and behavior of lymphocytes in the periphery. Programmed cell death, or apoptosis, is required at virtually every stage of lymphocyte development and functions to remove unwanted and potentially dangerous cells. Apoptosis is a process designed to eliminate individual cells or populations that have a specific function or defect and, therefore, it must be induced with sufficient specificity so as to remove the unwanted cells without bystander killing that could damage surrounding cells and tissue. Peripheral lymphocytes are charged with the task of immune surveillance, protecting from infected and transformed cells. Removal of activated lymphocytes is also critical for homeostasis, maintenance of lymphocyte populations, and regulation of immune responses, as the cytotoxic potential of activated lymphocytes represents a significant risk to an individual. We will address some general mechanisms of apoptosis before discussing specific scenarios involving apoptosis engaged in and by peripheral T lymphocytes.

A VIEW TO A KILL—GENERAL MECHANISMS OF APOPTOSIS

Programmed cell death is required for immune system function, tumor surveillance, and prevention of autoimmune diseases. Apoptosis was originally defined by morphological features, including cell shrinkage, membrane blebbing, nuclear condensation, and chromatin margination. Death is the consequence of an orchestrated series of biochemical events leading to the controlled destruction of the dying cell. Biochemical hallmarks, notably DNA fragmentation as seen by "laddering," loss of asymmetric distribution of phosphatidylserine (PS) in the plasma membrane, disruption of mitochondrial function, and degradation of key cellular proteins, indicate that death has occurred by an apoptotic mechanism.

initiator caspases

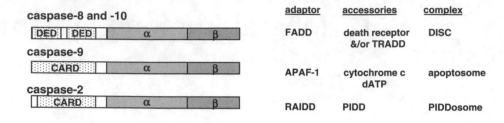

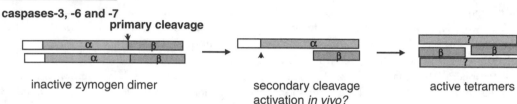

FIGURE 5.1 Overview of mammalian apoptotic caspases. The initiators are characterized by a long pro-domain that contains interaction motifs (death effector domain [DED] or caspase-recruitment domain [CARD]) responsible for intermolecular homotypic interactions with adaptor molecules, leading to enzymatic activation. Cleavage between the large (α; ~p20) and small (β; ~p10) subunits does not appear to be necessary for activation, and removal of the prodomain may serve to stabilize the active protease *in vivo*. Activation of the executioner caspases occurs following interchain cleavage between the large (α) and small (β) subunits. Subsequent autocatalytic removal of the short prodomain may be required for protease activation *in vivo*.

The pressures upon a cell to die depend on developmental stage, activation state, and environment. There is a wide variety of cell-specific and context-dependent death stimuli, and there are a number of common features shared between apoptotic mechanisms at play in the immune system. Morphological features that define apoptosis are mediated by members of the caspase family of intracellular proteases. Caspases are a family of cysteine proteinases that cleave substrates at the C-terminal side of aspartate residues (i.e., Asp at P1 of the cleavage site). Additional specificity is imparted by residues at P4, P3, and P1 of the substrate cleavage site for each respective member of the caspase family.[1,2] The best characterized caspases involved in apoptosis can be divided into two subgroups based on genetic and functional attributes: initiator and executioner caspases (see Figure 5.1). All known caspases are expressed as single polypeptide chains that are present in the cell as inactive zymogens. The initiators are the first caspases to be activated during the induction of phase of apoptosis. Initiator caspases (caspase-8, -10, -9, and -2) are characterized by long prodomains that possess CARD (caspase-recruitment domain) or DED (death-effector domain) motifs responsible for homotypic interactions with similar motifs in their corresponding adaptor partners. Initiator caspases are activated by homodimerization facilitated by an adaptor protein that acts as a scaffold upon which the initiator caspase becomes activated. Cleavage between the large and small subunits of the initiators may serve to stabilize the proteolytically active enzyme complex *in vivo*.

The executioner caspases (caspase-3, -7, and -6) are responsible for bringing about the morphological features that define apoptosis. The executioners are also expressed as inactive zymogens; however, they are activated by cleavage of the zymogen polypeptide between the large (α) and small (β) subunits. With one exception, the only known proteases known to cleave and activate executioner caspases are other caspases. The exception is the serine protease granzyme B expressed primarily by cytotoxic lymphocytes. Cleavage between the large and small subunits is followed by autoprocessing to remove the prodomain, which is substantially shorter than the prodomains found in initiator caspases and is devoid of CARD or DED motifs. The conformation of the resulting active executioner caspase is $(\alpha\beta)_2$ (see Figure 5.1).

In 1982, Wylie, an early pioneer of apoptosis research, used glucocorticoid treatment of thymocytes to demonstrate what has become recognized as a hallmark of apoptosis: DNA laddering as seen by nucleosomal-sized fragments by agarose gel electrophoresis in the final stages of cellular destruction.[3] The deoxyribonuclease (DNase) responsible for causing the signature DNA laddering was not discovered until 1998, when Nagata and colleagues identified the caspase-activated DNase (CAD).[4,5] CAD is coexpressed with its inhibitor ICAD. Cleavage of inhibitor of CAD (ICAD) by caspase-3 overcomes the inhibitory activity of ICAD, which, in turn, releases the active CAD nuclease. The coexpression of a proapoptotic protein along with its own inhibitor is a recurring theme in apoptosis. A likely explanation is that it is dangerous for a cell to express a death-inducing protein without additional regulatory constraints. Any leakiness in such a system would be rapidly selected out.

DNA fragmentation represents the end stage of cell death, as there is little chance for recovery after a cell has lost its genome. In most cases, the nucleus is not essential to initiate apoptosis (except by genotoxic stress), but because most cells contain nuclei, it is clear that degradation of DNA is necessary at some point during programmed death. It was speculated that the nucleases that degrade cellular DNA could also be co-opted to degrade DNA associated with an infecting virus in cells undergoing apoptosis induced by cytotoxic lymphocytes. Although this is an appealing notion, it is not yet supported by direct experimental evidence.

Another hallmark of apoptosis is the loss of asymmetry in the plasma membrane. Phosphatidylserine (PS) is restricted to the inner leaflet of the plasma membrane in healthy cells but becomes evenly distributed during apoptosis. Loss of membrane asymmetry, known as PS-flip, has provided a convenient indicator of apoptosis for assays in vitro.[6] More importantly, PS-flip contributes to the recognition of apoptotic cells by phagocytes in vivo, which, in turn, is involved in the immunosuppressive response observed upon macrophage engulfment of apoptotic cells. Extracellular PS exposure is one of the engulfment signals sent by apoptotic cells.[7]

Immune regulation and homeostasis rely heavily on the ability to induce death by apoptosis. Removal of the dead cell avoids the deleterious effects of a disseminated inflammatory response. Early models suggested that inflammation was avoided simply by preventing spillage of cellular contents. Recent developments, however, suggest a more complex mechanism, whereby dying cells secrete cytokines and induce the secretion of immunosuppressive cytokines from the phagocytic cells. One of the mediators of this anti-inflammatory effect has been identified as the PS-receptor (PSR). PSR is one of a growing number of recognition receptors that bind molecular flags presented on the external surface of apoptotic cells. Engagement of these receptors facilitates the uptake and removal of dying and dead cells, and the receptors also signal the production of cytokines.[8–10] These findings are consistent with earlier observations that secretion of TGF-β and IL-10 by macrophages in response to apoptotic cells contributes to the maintenance of an extracellular immunosuppressive milieu. Interestingly, dying cells can contribute to suppression of inflammation. In addition to having the decency of not dumping their contents, apoptotic cells were shown to induce expression of IL-10 and release it from existing intracellular stores.

Apoptotic and necrotic cells elicit different immunostimulatory effects on antigen-presenting cells (APCs). Engulfment of apoptotic cells resulted in better presentation of cellular antigen in the context of both major histocompatibility complex (MHC) classes I and II than did engulfment of necrotic cells, but necrotic cells elicited much stronger APC activation.[11,12] The result of this response may be that apoptotic cells are presented efficiently but elicit only a limited response due to the low level of danger, whereas a greater immune response results from the presence of necrotic cells due to the heightened possibility of danger.

OUTSIDE INFLUENCES OR AN INSIDE JOB—THE EXTRINSIC AND INTRINSIC PATHWAYS

There are two general pathways of apoptosis (and a seemingly infinite number of variations). A simple distinction by intrinsic and extrinsic induction may be made to differentiate the two. Both of these mechanisms use caspases to bring about the endpoints of death, but each relies on different initiators and engages the death pathway by different means.

The extrinsic pathway is triggered by receptors at the cell surface.[13,14] Receptor-mediated apoptosis is characterized by engagement of a death receptor from the tumor necrosis factor (TNF) family of transmembrane receptors, including Fas (CD95), TNF-R1, and DR4/5, by their corresponding ligands, Fas-ligand (CD95L; CD178), TNFα, and TRAIL, respectively. For a comprehensive review of the TNF and TNF-R family of proteins, see Locksley, Killeen, and Lenardo.[13] Perhaps the simplest example is the pathway engaged by Fas by interaction with Fas-ligand. Receptor engagement results in recruitment of the adaptor protein FADD (Fas-associated death domain), which binds Fas by intermolecular interactions between death domains of Fas and FADD. Caspase-8 (or caspase-10) is then recruited to this complex to form what is called the DISC (death-inducing signaling complex). This pathway is exemplified with Fas in Figure 5.2. TNF-R1 was recently shown to activate caspase-8 by a two-step mechanism that includes recruitment of TRADD (TNF receptor-associated protein with DD) in a multimolecular complex leading to internalization of the DISC before recruitment of FADD and caspase-8 activation.[15]

Receptor-mediated apoptosis was implicated in the regulation of peripheral lymphocytes when it was discovered that the defect responsible for the *lpr* (lymphoproliferative) and *gld* (generalized

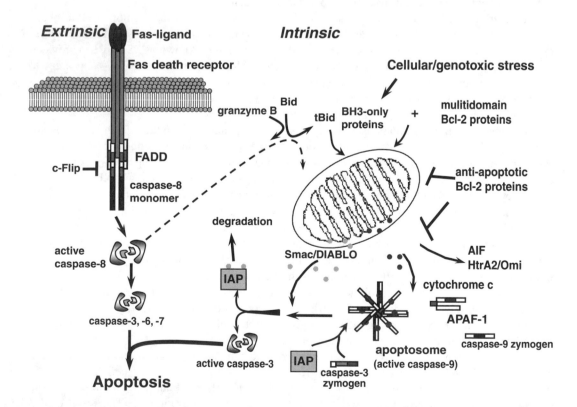

FIGURE 5.2 Overview of extrinsic and intrinsic apoptosis pathways. The extrinsic is shown with Fas as a representative member of the TNF receptor family of death receptors. Fas-mediated signaling to death is perhaps the simplest of the known death receptor–mediated pathways.

lymphoproliferative disorder) strains of mice were caused by mutations in Fas and Fas-ligand genes, respectively.[14,16,17] Mutations in the corresponding human Fas and Fas-ligand counterparts have been implicated in autoimmune lymphoproliferative syndrome (ALPS) type Ia and Ib, respectively.

The intrinsic pathway is governed by the release of proapoptotic mitochondrial proteins including cytochrome c, Smac/DIABLO, EndoG, HtrA2/Omi, and AIF.[18] After release from the mitochondria, cytochrome c associates with the cytosolic adaptor protein, apoptotic protease activating factor (Apaf-1), along with dATP to recruit procaspase-9 into a macromolecular complex known as the apoptosome. Activation of the initiator caspase-9 then results in cleavage of caspase-3, but this is not yet sufficient to cause activation of the executioner. Smac/DIABLO, another of the mitochondrial proteins released along with cytochrome c, suppresses the inhibition of caspase-3 by XIAP (X-linked inhibitor of apoptosis protein [IAP]).[19,20] De-repression of XIAP by Smac/DIABLO allows for release of active caspase-3 protease from the apoptosome into the cytosol, where it drives proteolytic degradation of the apoptotic cell.

In addition to cytochrome c, other mitochondrial proteins have been assigned apoptotic functions.[18] EndoG is a mitochondrial endonuclease that, when released, has a demonstrated activity of chromosomal DNA digestion that does not require the involvement of caspases.[21] The serine protease HtrA2/Omi was originally described as an IAP-antagonist with an effect similar to Smac/DIABLO in its ability to relieve IAP-induced repression of caspases. The IAP-inhibitory function of HtrA2/Omi is not dependent on its protease activity, which suggests an additional physiological role. Recently, a role for HtrA2/Omi as a sensor for misfolded mitochondrial proteins was proposed, providing another example of an apoptotic protein also having an important survival function.[22] Similarly, AIF (apoptosis-inducing factor) is a resident flavoprotein in the mitochondrial intermembrane space that is capable of inducing chromatin condensation and degradation of chromosomal DNA into large fragments. Knockout of the AIF gene yielded a profound phenotype resulting in death during early cavitation of embryoid bodies, suggesting a key survival role very early in embryogenesis, perhaps throughout phylogeny.[23]

The intrinsic pathway is regulated by members of the Bcl-2 protein family. Permeabilization of the mitochondrial membrane and release of apoptogenic proteins are inhibited by the pro-survival members of this family: Bcl-2, Bcl-xL, Mcl-1, A1, and Bcl-w.[24] Bcl-2, the prototypic family member, was discovered in a clinical model of B cell lymphoma in which the Bcl-2 gene translocated downstream of the immunoglobulin heavy-chain promoter becomes grossly overexpressed.[25] Dysregulated overexpression of Bcl-2 in this scenario protects immature B cells from negative selection and results in subsequent emergence of follicular lymphomas.

The release of mitochondrial proteins is mediated by two classes of proapoptotic members of the Bcl-2 family, the BH3-only proteins in concert with the multidomain family members.[26,27] The BH3-only proteins (including Bid, Bim, Bmf, Bik, BNip3, Puma, and Noxa) are viewed as cellular sensors that transmit a distress signal to effect the death program. Mitochondria form a central transmission point of the death signal, where BH3-only proteins interact with a corresponding member of the multidomain group (Bax, Bak, Bad). The biochemical mechanisms by which the proapoptotic members interact are reviewed in Scorrano and Korsmeyer.[27]

ALL REVVED UP WITH NO PLACE TO GROW—ACTIVATION-INDUCED CELL DEATH (AICD) AND PERIPHERAL DELETION

Peripheral lymphocyte populations are regulated by homeostasis, deletion, and homeostatic proliferation. Peripheral T cells undergo rapid and extensive expansion upon activation. In order to reduce lymphocyte numbers postinfection, T cells are subject to a number of selective pressures to reduce their numbers, leaving a small memory population. Some lymphocytes die following a suboptimal stimulation, either by weak MHC[Ag]:TCR interactions or lack of co-receptor involvement, and enter apoptosis during abortive proliferation. Some survive and constitute the memory pool of responders to subsequent challenge, but most activated lymphocytes undergo apoptosis in the postexpansion phase of the immune response.

Mature peripheral T cells expand upon first encounter with antigen and undergo apoptosis upon secondary stimulation by a process known as activation-induced cell death (AICD).[28] Resting lymphocytes are generally refractory to apoptotic stimuli, often because they do not express the necessary components to engage death pathways. Presumably, resting lymphocytes need not be sensitive to apoptotic stimuli, because they do not represent a significant risk, whereas activated lymphocytes, if misdirected, pose a substantial risk. The biochemical machinery necessary to drive apoptosis is expressed in peripheral T cells upon activation, but activation alone is not sufficient to drive a cell into apoptosis. AICD can be recapitulated *in vitro* by stimulating lymphocytes through the T cell receptor (TCR) in the presence of IL-2. Secondary stimulation 3 to 4 days later causes activated T cells to undergo AICD by a mechanism mediated, at least in part, by the c-myc-induced expression of Fas-ligand and engagement of the Fas-dependent extrinsic pathway. Induction of Fas-ligand is not sufficient to drive AICD; there is a corresponding sensitization of T cells to Fas-mediated apoptosis that occurs subsequent to the initial activation events. Susceptibility to receptor-mediated apoptosis includes downregulation of the Fas inhibitory protein, c-FLIP (see Figure 5.3). Induction of Fas-ligand expression and subsequent ligand–receptor interactions have also been shown to occur in lymphocyte apoptosis independent of TCR engagement: genotoxic stress such as ultraviolet (UV) radiation, etoposide, and teniposide have been shown to cause an upregulation of Fas-ligand, which leads to lymphocyte apoptosis.

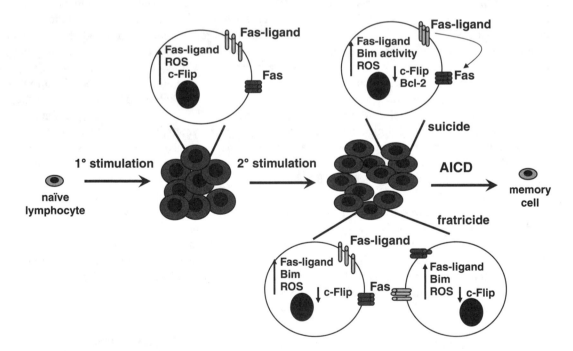

FIGURE 5.3 After a primary encounter with antigen, T lymphocytes expand; upregulate Fas, Fas-ligand, c-Flip, and Bim; and produce ROS. Upon secondary stimulation, a decrease of c-Flip levels is accompanied by an increase in sensitivity to apoptotic stimuli, notably Fas and Bim, key mediators of the extrinsic and intrinsic pathways, respectively. Cell autonomous suicide and fratricide-based mechanisms are employed to delete the majority of active lymphocytes, leaving a functional memory population. There are likely to be differences between CD4+ and CD8+ cells with respect to the contribution of each pathway, and there may also be subtleties in the employment of each mechanism, depending on the nature of the original antigenic stimulation.

Despite the demonstrated involvement of Fas-ligand and Fas in peripheral deletion and AICD, they are likely not the only mediators of apoptosis involved in this process. TNFα, a member of the same protein family of death ligands as Fas-ligand, has been implicated in both AICD and peripheral deletion.[29,30] Activated lymphocytes are also sensitive to TRAIL-induced apoptosis, and the ligand is expressed upon activation. A definitive role for TRAIL in AICD has yet to be established.

As discussed above, the intrinsic apoptotic pathway can be activated in response to cellular stress, including growth factor withdrawal. Peripheral T cells are dependent on IL-2 for continued survival, and the BH3-only protein Bim is required to signal for apoptosis from IL-2 withdrawal *in vitro*.[31] The ability of Bim to serve as a cellular sensor for IL-2 withdrawal may represent the downregulation of an immune response when growth factors become depleted due to usage or removal due to reduced production.

There are conflicting reports regarding the expression of Bim in activated T cells. It was shown that Bim levels do not change in Vβ8+ cells after challenge with superantigen SEB (*Staphylococcus* enterotoxin B), but the activity of Bim was enhanced by a decrease in Bcl-2 protein levels.[32] The net result was that Vβ8+ T cells were deleted in a Bim-dependent manner. A recent report showed differences in Bim levels in resting and activated CD4+ and CD8+ T cells[33]; CD4+ cells showed little change in Bim levels, whereas a more pronounced induction was observed in activated CD8+ cells. Therefore, it is possible that induction of Bim with or without a corresponding reduction of Bcl-2 may occur during peripheral deletion in a bulk population of SEB-responsive Vβ8+ cells.

An interesting observation arose from early studies of Fas-ligand in AICD concerning the concept of cell autonomous apoptosis. Initial studies reported that a cell could undergo AICD in conditions devoid of other cells (i.e., cell autonomous death or "suicide") (see Figure 5.3). Such a mechanism is necessary for negative selection when cells die in the absence of an appropriate signal.[28,34] In some cases of autonomous AICD *in vitro* using T cell hybridomas, apoptosis was found to be dependent on Fas–Fas–ligand interactions. AICD need not be autonomous, as there are also examples using T cell hybridomas that require cell–cell contact to undergo AICD (i.e., "fratricide").

Clearly, it would pose a problem if a lymphocyte were simply to die when activated. The coordinated expression of death proteins is tempered with the induction of death inhibitors. For example, lymphocytes express key components of the death machinery, including death ligands and receptors, notably, Fas-ligand/Fas and TRAIL/DR5, but they also induce expression of inhibitors of apoptosis, such as c-Flip, Bcl-2, and Bcl-x$_L$. Why the apoptotic proteins are expressed early is not entirely clear, but there is some suggestion that proteins, such as FADD, and caspase-8 are required for activation and proliferation of T cells as well as for apoptosis. Expression of Fas, FADD, and caspase-8 early during T cell activation may also represent a fail-safe mechanism that ensures that a cell is triggered to die if something goes awry. Presumably, it is better to remove an activated lymphocyte too soon than to have it persist as a rogue lymphocyte in the periphery, where it might cause significant harm.

Resting lymphocytes are transcriptionally and metabolically inactive. Signaling by the T cell receptor (TCR) results in a complex kinase-dependent signal transduction cascade and activation of the T cell. The transcriptional machinery becomes increasingly active, and enhanced metabolic activity supports the growing needs of the activated lymphocyte for energy and growth. Reactive oxygen species (ROS), produced as a by-product during the heightened metabolic activity, were linked in both causal and consequential roles in programmed cell death. ROS have been implicated in AICD based on the observation that MnTBAP,[35] an ROS scavenger, is capable of rescuing activated lymphocytes from death in culture *ex vivo*. Although this is a long way from proof that ROS is a causal factor in AICD, it has been since suggested that ROS produced by activated T cells are necessary for the upregulation of Fas-ligand, which was previously identified as a key factor in cell autonomous AICD.

MURDER BY PROXY—NONLYMPHOID CELLS IN PERIPHERAL LYMPHOCYTE DELETION

Activated lymphocytes are more mobile than their resting counterparts and often find their way into nonlymphoid tissues. Certain tissues possess a protective mechanism to deal with infiltrating lymphocytes. These tissues include testes, ovaries, and eye and are considered to be protected by immune privilege.[36] Foreign grafts and tumors implanted in these sites are not rejected and often flourish without eliciting an immune response. A well-characterized model of immune privilege is the eye, in which it was demonstrated that lymphocytes introduced into the anterior chamber (or recruited during an inflammatory response) are induced to die by apoptosis. Fas-ligand is constitutively expressed in the eye, and death of infiltrating lymphocytes is dependent on Fas expression by the infiltrating lymphocytes. TRAIL was also implicated in immune privilege, but its role is not yet clear.

The mechanisms involved in immune privilege have been used to explain a corresponding scenario involving lymphocyte apoptosis. Expression of Fas-ligand by transformed cells within a tumor can induce apoptosis in Fas-bearing lymphocytes by a process termed *Fas counterattack*.[37,38] Fas-ligand-expressing tumor cells avoid direct cytolytic attack and thereby minimize infiltration and potential inflammatory effects. Attempts were made to exploit features of immune privilege and tumor counterattack in experimental models of allogeneic transplantation, the idea being that Fas-ligand-expressing transplants could avoid rejection by killing infiltrating lymphocytes. In support of this possibility, Fas-ligand-expressing testes allografts were successfully protected from rejection, whereas grafts from Fas-ligand-deficient *gld* animals were not protected.[39] This model fell out of favor when it was demonstrated that, in some cases, Fas-ligand expression by the graft did not result in tolerance or protection but induced a profound granulocytosis resulting in accelerated rejection. An explanation that accommodates observations both for and against potential for practical application of the tumor counterattack model came in a report showing that Fas-ligand-expressing grafts induced apoptosis in the Fas-expressing cells of the vascular epithelium, which, in turn, led to death of the graft associated with neutrophil recruitment and infiltration.[40] Therefore, in this particular model, expression of Fas-ligand by the graft is not the direct cause of rejection but rather an indirect one. This issue is far from resolved, but clearly it should not be dismissed.

It has been shown that other nonlymphoid tissues respond to the presence of activated lymphocytes in a manner similar to inducible immune privilege.[41,42] Liver, lung, and small intestine induce expression of Fas-ligand during disseminated immune responses. Fas-ligand expressed by epithelial cells of the small intestine (IEC) confers upon IEC the capability of inducing apoptosis in Fas-bearing lymphocytes (Figure 5.4). Surprisingly, TNFα, which is secreted by activated T cells, mediates the induction of Fas-ligand expression by IEC. The phenomenon of inducible Fas-ligand expression to control lymphocyte infiltration has also been observed in the thyroid and, when dysfunctional, results in one of two disparate outcomes: Hashimoto's thyroiditis through direct destruction of thyrocytes by infiltrating lymphocytes or Graves' disease due to antibody deposition and secondary autoimmune disorder.[43,44]

Transfer of activated lymphocytes into animals lacking nonlymphoid Fas-ligand showed that there was an impairment (though not complete prevention) of peripheral lymphocyte deletion. We proposed that the response of nonlymphoid tissue to the presence of lymphocytes represents a sensing mechanism whereby involved tissues can accommodate a limited infiltration without a corresponding inflammatory response. A corollary of this model is that failure to handle the infiltrate, by defect or excessive burden, would result in either direct tissue damage caused by the infiltrating lymphocytes or a sustained inflammation. Induction of Fas-ligand and TRAIL by IEC would be preferable to constitutive expression in this case, because sustained levels may put at risk neighboring epithelial and endothelial cells that express the corresponding death receptors.

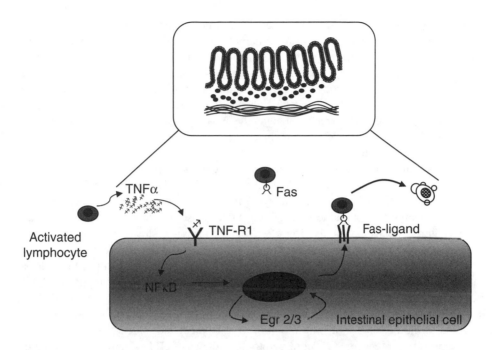

FIGURE 5.4 Involvement of nonlymphoid tissues in deletion of activated peripheral lymphocytes. In this example, activated lymphocytes that traffic through the small intestine secrete TNFα, which engages TNFR1 constitutively present on intestinal epithelial cells (IEC), resulting in upregulation of IEC Fas-ligand. IEC Fas ligand then triggers Fas-mediated apoptosis of the infiltrating lymphocytes as a means to moderate the amount and severity of allowable infiltration.

IN THE COMPANY OF KILLERS—INDUCTION OF APOPTOSIS BY CYTOTOXIC LYMPHOCYTES

The apoptosis of peripheral T lymphocytes is mirrored by the ability of cytotoxic lymphocytes (CL) to induce apoptosis. CD8+ cytotoxic T lymphocytes (CTLs) and natural killer (NK) cells are capable of inducing apoptosis by both the intrinsic and extrinsic pathways described above.[45] Targets of CL are those that are transformed (such as cells with downregulated expression of MHC class I) and cells that present appropriate "foreign" antigen in the context of MHC I for deletion by CD8+ CTLs.

CD8+ CTLs remove unwanted or dangerous cells by inducing apoptosis by one of two known mechanisms: a calcium-dependent mechanism mediated by proteins contained within cytolytic granules of the CTL, namely, the concerted action of perforin and members of the granzyme family of serine proteinases; or a triggering of the apoptotic program in target cells by CTLs' death ligand, Fas-ligand, by engagement of the Fas death receptor expressed on the surface of the target cell. Each of these triggers engages the caspase protease cascade at some point to effect the morphological changes that we observe as apoptosis, but each engages the cascade at different points of the killing process. Based on *in vitro* studies, it has long been assumed that only the Ca^{2+}-dependent mode of killing is dependent on interactions between CTL TCR and MHC class I–dependent presentation of corresponding antigen by the target cell. Fas-ligand-mediated apoptosis is viewed as a more "promiscuous" form of CTL-induced killing; this is based on the observation that many cell types that express Fas-ligand acquire the ability to kill Fas-bearing targets without the need for antigen presentation.[41,42] However, in CTLs, Fas-ligand may be compartmentalized in cytolytic granules and, therefore, it may not be deployed until favorable TCR engagement occurs.

Upon engagement of the TCR, specialized CTL granules containing granzymes, perforin, and Fas-ligand migrate toward the site of contact between the two cells.[46] Following the migration of granules to the site of contact, exocytosis of granule contents into the intercellular space between killer and target and uptake of granzymes, along with the activity of perforin, results in the generation of all biochemical features of apoptosis commonly studied *in vitro*. The kinetics of death is very rapid, within 0.25 to 1.7 h. The Ca^{2+}-dependent aspect of CTL killing enlists the concerted actions of granzymes and perforin to engage the intrinsic apoptosis pathway characterized by mitochondrial disruption, release of cytochrome *c* and Smac/DIABLO, and activation of the caspase-9 containing apoptosome.

Granzymes make up a family of serine proteases expressed primarily, but not exclusively, by CTL after activation. Granzymes B and A are the principal family members believed to be responsible for causing target cell apoptosis. Granzyme B, the only known serine protease with caspase-like substrate specificity, is capable of engaging the caspase-dependent death pathway at several points, including cleavage and activation of Bid and caspase-3. However, recent reports suggest that direct activation of caspase-3 by granzyme B requires the additional regulatory step in which inhibition of caspase-3 by XIAP must be relieved by Smac/DIABLO, which must be released from mitochondria.[47,48] Granzyme B preferentially cleaves and activates the proapoptotic Bcl-2 family member Bid,[49–51] which acts with Bax or Bak to mediate the release of mitochondrial proteins. A recent report suggests that Bax and Bak are not necessary for granzyme B-induced apoptosis,[52] which points to the possibility of multiple granzyme B–mediated effects.

Granzyme A is another constituent of cytotoxic granules of CTL. Granzyme A, when delivered with perforin, can induce target cell death, albeit with delayed kinetics and via different biochemical processes from granzyme B. CTL lacking granzyme B induced a delayed target cell apoptosis (approximately 12 h) that is attributed to granzyme A.[53] These data suggest that some compensation may occur at the level of target cell clearance, but they also indicate that the biochemical properties of the pathways engaged by each granzyme differ greatly. Whereas granzyme B engages the caspase pathway, albeit downstream of mitochondrial events, granzyme A acts in a manner independent of intracellular death proteases.[54] Known substrates of granzyme A are associated with a large multimeric complex known as the SET complex (reviewed in Lieberman and Fan[55]). It was proposed that granzyme A–mediated cleavage of the SET complex proteins may actually result in a phenotype more closely resembling necrosis than apoptosis.[56] In this model, generation of double-strand DNA breaks would result in activation of the DNA repair machinery, including poly-ADP ribose polymerase (PARP) rather than disabling it, as is often observed in apoptosis. Cleavage of PARP by caspase-3 is viewed as a hallmark of late-stage apoptosis.

Given that necrotic cells and apoptotic cells have different effects on APC activation,[11,12] it is possible that the mechanism by which a cell dies (i.e., granzyme B, granzyme A, or death ligands) will impact target cell clearance and have downstream consequences on immune cell activation and regulation. There is a strong possibility that the manner in which a target cell dies will also affect the way in which it becomes engulfed and stimulates cytokine production and activation of antigen presenting cells. Thus, the biochemical aspects of death triggered by components of CTL will contribute to an immunosuppressive or inflammatory milieu.

SUMMARY

Apoptosis is essential at virtually every stage of lymphocyte development and function. Successful resolution of a viral infection and maintenance of peripheral lymphocyte homeostasis requires the ability of cytotoxic lymphocytes to induce apoptosis in infected targets. Peripheral lymphocytes become the apoptotic targets after they have outlived their usefulness. Failure to execute the apoptotic program can result in wide-ranging and tragic consequences, from ineffective clearance of infected cells to the development of autoimmune disorders resulting from otherwise innocuous infections.

REFERENCES

1. Boatright, K.M. and Salvesen, G.S., Mechanisms of caspase activation, *Curr. Opin. Cell Biol.,* 15, 725–731, 2003.
2. Creagh, E.M., Conroy, H., and Martin, S.J., Caspase-activation pathways in apoptosis and immunity, *Immunol. Rev.,* 193, 10–21, 2003.
3. Wyllie, A.H., Glucocorticoid-induced thymocyte apoptosis is associated with endogenous endonuclease activation, *Nature,* 284, 555–556, 1980.
4. Enari, M., Sakahira, H., Yokoyama, H., Okawa, K., Iwamatsu, A., and Nagata, S., A caspase-activated DNase that degrades DNA during apoptosis, and its inhibitor ICAD, *Nature,* 391, 43–50, 1998.
5. Sakahira, H., Enari, M., and Nagata, S., Cleavage of CAD inhibitor in CAD activation and DNA degradation during apoptosis, *Nature,* 391, 96–99, 1998.
6. Martin, S.J., Reutelingsperger, C.P., McGahon, A.J., Rader, J.A., van Schie, R.C., LaFace, D.M., and Green, D.R., Early redistribution of plasma membrane phosphatidylserine is a general feature of apoptosis regardless of the initiating stimulus: inhibition by overexpression of Bcl-2 and Abl, *J. Exp. Med.,* 182, 1545–1556, 1995.
7. Henson, P.M., Bratton, D.L., and Fadok, V.A., The phosphatidylserine receptor: a crucial molecular switch? *Nat. Rev. Mol. Cell Biol.,* 2, 627–633, 2001.
8. Chen, W., Frank, M.E., Jin, W., and Wahl, S.M., TGF- released by apoptotic T cells contributes to an immunosuppressive milieu, *Immunity,* 14, 715–725, 2001.
9. Gao, Y., Herndon, J.M., Zhang, H., Griffith, T.S., and Ferguson, T.A., Anti-inflammatory effects of CD95 ligand (FasL)-induced apoptosis, *J. Exp. Med.,* 188, 887–896, 1998.
10. Huynh, M.L., Fadok, V.A., and Henson, P.M., Phosphatidylserine-dependent ingestion of apoptotic cells promotes TGF-1 secretion and the resolution of inflammation, *J. Clin. Invest.,* 109, 41–50, 2002.
11. Fonteneau, J.F., Larsson, M., and Bhardwaj, N., Interactions between dead cells and dendritic cells in the induction of antiviral CTL responses, *Curr. Opin. Immunol.,* 14, 471–477, 2002.
12. Sauter, B., Albert, M.L., Francisco, L., Larsson, M., Somersan, S., and Bhardwaj, N., Consequences of cell death: exposure to necrotic tumor cells, but not primary tissue cells or apoptotic cells, induces the maturation of immunostimulatory dendritic cells, *J. Exp. Med.,* 191, 423–434, 2000.
13. Locksley, R.M., Killeen, N., and Lenardo, M.J., The TNF and TNF receptor superfamilies: integrating mammalian biology, *Cell,* 104, 487–501, 2001.
14. Siegel, R.M., Chan, F.K., Chun, H.J., and Lenardo, M.J., The multifaceted role of Fas signaling in immune cell homeostasis and autoimmunity, *Nat. Immunol.,* 1, 469–474, 2000.
15. Micheau, O., and Tschopp, J., Induction of TNF receptor I-mediated apoptosis via two sequential signaling complexes, *Cell,* 114, 181–190, 2003.
16. Krueger, A., Fas, S.C., Baumann, S., and Krammer, P.H., The role of CD95 in the regulation of peripheral T-cell apoptosis, *Immunol. Rev.,* 193, 58–69, 2003.
17. Pinkoski, M.J. and Green, D.R., Fas ligand, death gene, *Cell Death Differ.,* 6, 1174–1181, 1999.
18. Green, D.R. and Kroemer, G., The pathophysiology of mitochondrial cell death, *Science,* 305, 626–629, 2004.
19. Du, C., Fang, M., Li, Y., Li, L., and Wang, X., Smac, a mitochondrial protein that promotes cytochrome *c*-dependent caspase activation by eliminating IAP inhibition, *Cell,* 102, 33–42, 2000.
20. Verhagen, A.M., Ekert, P.G., Pakusch, M., Silke, J., Connolly, L.M., Reid, G.E., Moritz, R.L., Simpson, R.J., and Vaux, D.L., Identification of DIABLO, a mammalian protein that promotes apoptosis by binding to and antagonizing IAP proteins, *Cell,* 102, 43–53, 2000.
21. Li, L.Y., Luo, X., and Wang, X., Endonuclease G is an apoptotic DNase when released from mitochondria, *Nature,* 412, 95–99, 2001.
22. Vaux, D.L. and Silke, J., HtrA2/Omi, a sheep in wolf's clothing, *Cell,* 115, 251–253, 2003.
23. Cande, C., Vahsen, N., Metivier, D., Tourriere, H., Chebli, K., Garrido, C., Tazi, J., and Kroemer, G., Regulation of cytoplasmic stress granules by apoptosis-inducing factor, *J. Cell Sci.,* 117, 4461–4468, 2004.
24. Kuwana, T., and Newmeyer, D.D., Bcl-2-family proteins and the role of mitochondria in apoptosis, *Curr. Opin. Cell Biol.,* 15, 691–699, 2003.
25. Lipford, E., Wright, J.J., Urba, W., Whang-Peng, J., Kirsch, I.R., Raffeld, M., Cossman, J., Longo, D.L., Bakhshi, A.M., and Korsmeyer, S.J., Refinement of lymphoma cytogenetics by the chromosome 18q21 major breakpoint region, *Blood,* 70, 1816–1823, 1987.

26. Bouillet, P. and Strasser, A. BH3-only proteins — evolutionarily conserved proapoptotic Bcl-2 family members essential for initiating programmed cell death, *J. Cell Sci.,* 115, 1567–1574, 2002.

27. Scorrano, L. and Korsmeyer, S.J., Mechanisms of cytochrome *c* release by proapoptotic BCL-2 family members, *Biochem. Biophys. Res. Commun.,* 304, 437–444, 2003.

28. Green, D.R., Droin, N., and Pinkoski, M., Activation-induced cell death in T cells, *Immunol. Rev.,* 193, 70–81, 2003.

29. Speiser, D.E., Sebzda, E., Ohteki, T., Bachmann, M.F., Pfeffer, K., Mak, T.W., and Ohashi, P.S., Tumor necrosis factor receptor p55 mediates deletion of peripheral cytotoxic T lymphocytes *in vivo, Eur. J. Immunol.,* 26, 3055–3060, 1996.

30. Sytwu, H.K., Liblau, R.S., and McDevitt, H.O., The roles of Fas/APO-1 (CD95) and TNF in antigen-induced programmed cell death in T cell receptor transgenic mice, *Immunity,* 5, 17–30, 1996.

31. Bouillet, P., Metcalf, D., Huang, D.C., Tarlinton, D.M., Kay, T.W., Kontgen, F., Adams, J.M., and Strasser, A., Proapoptotic Bcl-2 relative Bim required for certain apoptotic responses, leukocyte homeostasis, and to preclude autoimmunity, *Science,* 286, 1735–1738, 1999.

32. Hildeman, D.A., Zhu, Y., Mitchell, T.C., Bouillet, P., Strasser, A., Kappler, J., and Marrack, P., Activated T cell death *in vivo* mediated by proapoptotic Bcl-2 family member Bim, *Immunity,* 16, 759–767, 2002.

33. Sandalova, E., Wei, C.H., Masucci, M.G., and Levitsky, V., Regulation of expression of Bcl-2 protein family member Bim by T cell receptor triggering, *Proc. Natl. Acad. Sci. U.S.A.,* 101, 3011–3016, 2004.

34. Green, D.R., Overview: apoptotic signaling pathways in the immune system, *Immunol. Rev.,* 193, 5–9, 2003.

35. Hildeman, D.A., Mitchell, T., Teague, T.K., Henson, P., Day, B.J., Kappler, J., and Marrack, P.C., Reactive oxygen species regulate activation-induced T cell apoptosis, *Immunity,* 10, 735–744, 1999.

36. Green, D.R. and Ferguson, T.A., The role of Fas ligand in immune privilege, *Nat. Rev. Mol. Cell Biol.,* 2, 917–924, 2001.

37. O'Connell, J., Bennett, M.W., O'Sullivan, G.C., Collins, J.K., and Shanahan, F., The Fas counterattack: cancer as a site of immune privilege, *Immunol. Today,* 20, 46–52, 1999.

38. O'Connell, J., O'Sullivan, G.C., Collins, J.K., and Shanahan, F., The Fas counterattack: Fas-mediated T cell killing by colon cancer cells expressing Fas ligand, *J. Exp. Med.,* 184, 1075–1082, 1996.

39. Bellgrau, D., Gold, D., Selawry, H., Moore, J., Franzusoff, A., and Duke, R.C., A role for CD95 ligand in preventing graft rejection, *Nature,* 377, 630–632, 1995.

40. Janin, A., Deschaumes, C., Daneshpouy, M., Estaquier, J., Micic-Polianski, J., Rajagopalan-Levasseur, P., Akarid, K., Mounier, N., Gluckman, E., Socie, G., and Ameisen, J.C., CD95 engagement induces disseminated endothelial cell apoptosis *in vivo*: immunopathologic implications, *Blood,* 99, 2940–2947, 2002.

41. Bonfoco, E., Stuart, P.M., Brunner, T., Lin, T., Griffith, T.S., Gao, Y., Nakajima, H., Henkart, P.A., Ferguson, T.A., and Green, D.R., Inducible nonlymphoid expression of Fas ligand is responsible for superantigen-induced peripheral deletion of T cells, *Immunity,* 9, 711–720, 1998.

42. Pinkoski, M.J., Droin, N.M., Lin, T., Genestier, L., Ferguson, T.A., and Green, D.R., Nonlymphoid Fas ligand in peptide-induced peripheral lymphocyte deletion, *Proc. Natl. Acad. Sci. U.S.A.,* 99, 16174–16179, 2002.

43. Stassi, G. and De Maria, R., Autoimmune thyroid disease: new models of cell death in autoimmunity, *Nat. Rev. Immunol.,* 2, 195–204, 2002.

44. Stassi, G., Zeuner, A., Di Liberto, D., Todaro, M., Ricci-Vitiani, L., and De Maria, R., Fas-FasL in Hashimoto's thyroiditis, *J. Clin. Immunol.,* 21, 19–23, 2001.

45. Barry, M., and Bleackley, R.C., Cytotoxic T lymphocytes: all roads lead to death, *Nat. Rev. Immunol.,* 2, 401–409, 2002.

46. Trambas, C.M. and Griffiths, G.M., Delivering the kiss of death, *Nat. Immunol.,* 4, 399–403, 2003.

47. Goping, I.S., Barry, M., Liston, P., Sawchuk, T., Constantinescu, G., Michalak, K.M., Shostak, I., Roberts, D.L., Hunter, A.M., Korneluk, R., and Bleackley, R.C., Granzyme B-induced apoptosis requires both direct caspase activation and relief of caspase inhibition, *Immunity,* 18, 355–365, 2003.

48. Sutton, V.R., Wowk, M.E., Cancilla, M., and Trapani, J.A., Caspase activation by granzyme B is indirect, and caspase autoprocessing requires the release of proapoptotic mitochondrial factors, *Immunity,* 18, 319–329, 2003.

49. Barry, M., Heibein, J.A., Pinkoski, M.J., Lee, S.F., Moyer, R.W., Green, D.R., and Bleackley, R.C., Granzyme B short-circuits the need for caspase 8 activity during granule-mediated cytotoxic T-lymphocyte killing by directly cleaving Bid, *Mol. Cell Biol.,* 20, 3781–3794, 2000.
50. Pinkoski, M.J., Waterhouse, N.J., Heibein, J.A., Wolf, B.B., Kuwana, T., Goldstein, J.C., Newmeyer, D.D., Bleackley, R.C., and Green, D.R., Granzyme B-mediated apoptosis proceeds predominantly through a Bcl-2-inhibitable mitochondrial pathway, *J. Biol. Chem.,* 276, 12060–12067, 2001.
51. Sutton, V.R., Davis, J.E., Cancilla, M., Johnstone, R.W., Ruefli, A.A., Sedelies, K., Browne, K.A., and Trapani, J.A., Initiation of apoptosis by granzyme B requires direct cleavage of Bid, but not direct granzyme B-mediated caspase activation, *J. Exp. Med.,* 192, 1403–1414, 2000.
52. Thomas, D.A., Scorrano, L., Putcha, G.V., Korsmeyer, S.J., and Ley, T.J., Granzyme B can cause mitochondrial depolarization and cell death in the absence of BID, BAX, and BAK, *Proc. Natl. Acad. Sci. U.S.A.,* 98, 14985–14990, 2001.
53. Shresta, S., Graubert, T.A., Thomas, D.A., Raptis, S.Z., and Ley, T.J., Granzyme A initiates an alternative pathway for granule-mediated apoptosis, *Immunity,* 10, 595–605, 1999.
54. Beresford, P.J., Zhang, D., Oh, D.Y., Fan, Z., Greer, E.L., Russo, M.L., Jaju, M., and Lieberman, J., Granzyme A activates an endoplasmic reticulum-associated caspase-independent nuclease to induce single-stranded DNA nicks, *J. Biol. Chem.,* 276, 43285–43293, 2001.
55. Lieberman, J., and Fan, Z., Nuclear war: the granzyme A-bomb, *Curr. Opin. Immunol.,* 15, 553–559, 2003.
56. Pinkoski, M.J. and Green, D.R., Granzyme A: the road less traveled, *Nat. Immunol.,* 4, 106–108, 2003.

Section 2

Mechanisms of HIV-Associated Cell Death Viral Factors

Viral Factors

6 Cell Death in HIV Infection: gp120

Stacey R. Vlahakis

CONTENTS

INTRODUCTION

In humans, chronic infection with the human immunodeficiency virus (HIV) causes a progressive depletion of CD4+ and CD8+ T cells, resulting in the acquired immunodeficiency syndrome (AIDS). HIV-1 is a retrovirus that encodes numerous structural proteins, including the transmembrane envelope glycoprotein 160 (gp160) consisting of gp120 (the extracellular domain) and a 41 kDa anchor that holds the envelope at the cell surface. The gp120 plays an important and active role in the function and replication of HIV. The envelope gp120 recognizes and binds the CD4 and chemokine co-receptors, initiating infection of the recipient cell. The HIV gp120 is also responsible for signaling activation and death in bystander immune cells that do not become subsequently infected. Last, because the HIV gp120 is the main external identifier on the virus and activates immune cellular death in the host, the gp120 has been widely targeted for therapeutic and vaccine development. As such, substantial attention and investigation have centered on the characteristics and function of HIV-1 envelope gp120.

STRUCTURE AND EXPRESSION OF GP120

The structure of the HIV envelope gp120 is critical in dictating not only the tropism of the HIV strain but also the antigenicity of the HIV. The envelope is organized into spikes on the outside of the virion. The exterior of each spike is composed of the envelope gp120, which is noncovalently associated with each subunit of the trimeric gp41 complex.[1] The amino acid sequence of the gp120 consists of five conserved regions (C1 to C5) interspaced with five variable regions (V1 to V5). V1 to V4 create the surface-exposed loops anchored by disulfide bonds at their bases.[2] Both conserved and variable regions are highly glycosylated with carbohydrate moieties, accounting for 50% of the mass of gp120.[3] The conserved regions of the surface-exposed gp120 are involved in the interaction with the gp41 ectodomain and the host-cell receptors that recognize the virus. The envelope glycoprotein is folded into a complex three-dimensional structure such that noncontiguous regions physically associate and create binding sites for host-cell receptors.

Neutralizing antibodies isolated from HIV-infected individuals recognize both conserved and variable regions of the gp120 structure. Antienvelope antibodies that recognize specific strains of HIV are directed at epitopes in the V2 and V3 loops of the gp120. The glycosylation pattern of the envelope glycoprotein likely modulates the immunogenicity and antigenicity of the gp120, for the surface-exposed glycoprotein is the main target for neutralizing antibodies during natural infection.[4] Broadly neutralizing antienvelope antibodies recognize epitopes in discontinuous and conserved regions of the glycoproteins. The most abundant of the antibodies recovered from HIV-infected humans are directed at the CD4 receptor binding site on the gp120 envelope.[2]

HIV-1 virus has two tropisms differentiated by their ability to infect and replicate in specific cells and by the surface co-receptors used to infect host cells. Both HIV tropic strains replicate in primary CD4+ T cells, but only some strains infect and replicate in monocyte-derived macrophages or transformed T cell lines. The macrophage-tropic HIV strains (also called M-tropic or R5) can infect and replicate in macrophages but not in T cells. Similarly, T cell tropic (T-tropic or X4) HIV strains replicate in T cell lines but not in macrophages. Dual tropic isolates have also been described that can infect and replicate in both cell types. Additional work demonstrates that the gp120 on R5-tropic strains of HIV recognize and bind to the CCR5 chemokine co-receptor as well as to the CD4 receptor on a host cell. The gp120 on X4-tropic strains of HIV bind to the CXCR4 chemokine co-receptor in addition to the CD4 receptor of a host cell.[5]

The tropism of the HIV strain profoundly influences the pathogenesis of the disease. R5- or M-tropic strains are preferentially transmitted, are responsible for most cases of sexually transmitted HIV infection, and are found during the early stages of infection.[6,7] In contrast, X4 or T cell (T)-tropic strains develop in a subset of patients over time and are associated with a faster depletion in CD4+ T cells. Unlike R5 strains, X4 strains are also linked with a depletion of CD8+ T cells.[8,9] It is not clear how or why the virus transforms in an infected individual from an R5 to an X4 virus. However, the presence of dual tropic strains of virus may represent a transition or transformation of the viruses over the course of infection.[10]

Extensive studies of the virus have shown that the tropism of the virus is determined by the amino acid sequence of the envelope gp120. In particular, the amino acid sequence in the V3 loop of the envelope will determine which chemokine co-receptor the virus recognizes and, therefore, which cell type the virus can infect and subsequently replicate. Chimera studies between R5 and X4 strains have shown that mutations in the V3 loop, sometimes as little as one amino acid residue change, may alter the tropism of the virus.[11] This suggests that there is plasticity in the determinants of viral tropism. It appears that more basic residues in the V3 loop of gp120, resulting in an overall more negative charge, tend to create an X4 strain.[12]

In addition to the virion-associated form, gp120 can be detected as a soluble form in the blood of infected patients. Both forms of envelope glycoproteins are capable of binding CD4 and chemokine co-receptors, but only the whole virus can infect cells and be internalized.[13–15]

GENETIC REGULATION OF GP120

The HIV-1 envelope glycoprotein is encoded by the *env* gene, which produces a precursor protein that is 160kDa. This is proteolytically processed into the soluble (gp120) and membrane-spanning (gp41) subunits. The gene product is produced in the endoplasmic reticulum (ER) as the gp160-precursor protein. This is then folded into trimers and transported out of the ER and then to the Golgi apparatus, where it undergoes further oligosaccharide modifications.[16] During transport through the secretory pathways in human T cells, the gp160 is proteolytically cleaved to create the gp120 and gp41 subunits.[17] The envelope products are then assembled with the other viral proteins and prepared for virion budding at the cell surface.

After leaving the Golgi apparatus, the HIV envelope moves to intracellular granules that are closely regulated by the secretory pathway. Miranda et al.[18] demonstrated that large amounts of mature envelope are found stored in intracellular compartments of the cell. As part of the secretory pathway, the HIV gp120 traffics to the intracellular compartment granules that also contain CTLA-4. CTLA-4 is an important negative regulator of T cell immune responses, and its surface expression is tightly controlled as it traffics through the secretory pathways after T cell activation.[19] Moreover, the precursor gp160 is proteolytically cleaved into gp120 and gp41 by the PC6 protease, a component of the secretory pathway.

Because HIV gp120 and CTLA-4 share granules in the cell's secretory pathway on their way to the cell surface, they are expressed together. T cell activation induces cell surface expression of CTLA-4 and gp120.[18,20] It is unclear what advantage this colocalization provides HIV or the host T cell. However, it is intriguing to postulate that co-regulation of the gp120 and CTLA-4 could alert the immune system to the presence of the incipient virion and, in turn, may influence T cell activation.

GP120 RECEPTOR BINDING

CD4 BINDING

The HIV-envelope gp120 must bind two receptors in order to enter and infect a host immune cell. The gp120 binds to the CD4 receptor as well as a chemokine co-receptor, either CCR5 or CXCR4, depending on whether the viral strain is R5 or X4, respectively (Figure 6.1). Numerous

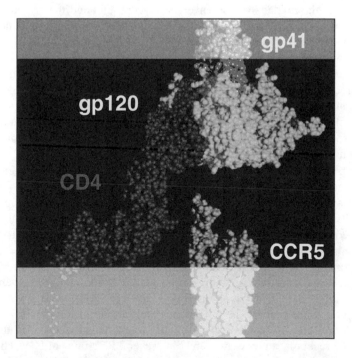

FIGURE 6.1 Structure of HIV envelope and host cell receptors. X-ray crystallographic adaptation showing HIV envelope gp120 and gp41 in proximity with the host cell receptors CD4 and chemokine co-receptor CCR5.

other G-protein-coupled chemokine receptors have been shown to function *in vitro* as co-receptors for HIV-1, HIV-2, or SIV. However, there is little evidence to suggest that any of these alternative co-receptors have a role for HIV binding or signaling *in vivo*.[21]

CD4 is the primary receptor for HIV binding and cell entry.[22] CD4 is located on the surface of many helper/inducer T cells, monocytes, and macrophages. The CD4 is a 55 kDa glycoprotein with an N-terminal extracellular domain, one transmembrane segment, and a C-terminal intracellular tail. The extracellular portion includes residues 1 to 37 that form four immunoglobulin-like domains (D1 to D4) that can be recombinantly expressed as soluble CD4 (sCD4).[23] X-ray crystallography studies of a CD4/gp120/neutralizing antibody complex have defined the exact location and nature of the HIV gp120/CD4 interaction. The gp120 binds to the D1 and D2 fragments of the CD4 receptor. CD4 is bound to gp120, with the N-terminal domain of CD4 being rigidly held in a deep-binding pocket of gp120.[23] Neither the variable loops of gp120 nor any of the carbohydrate moieties near the binding site for CD4 interfere with the orientation of binding of either partner. Direct contact is made between 22 CD4 residues and 26 gp120 residues, comprising, in total, 219 van der Waals contacts and 12 hydrogen bonds.

CHEMOKINE RECEPTOR BINDING

CXCR4 and CCR5 are members of the chemokine receptor family. They have seven transmembrane domains and are G-protein coupled receptors (GPCRs). Their functions can vary widely according to their microenvironments, possible receptor dimerization, and posttranslational modifications. The posttranslational modifications of CCR5 and CXCR4 include N-linked and O-linked glycosylation and tyrosine sulfation of the extracellular domains of the receptors, such as N-terminal, extracellular loop I (ECI), ECII, and ECIII. The intracellular loops of the receptors undergo palmtoylation, phosphorylation, and ubiquitination. Each of these modifications can play a role in receptor internalization and receptor turnover.[24]

After the gp120 binds to the CD4 receptor, it undergoes a conformational change in structure, which better exposes the residues on the envelope glycoprotein and thereby improves binding to the appropriate chemokine co-receptor. A large amount of work has been done to decipher the critical areas of the co-receptors for gp120 binding. This work has demonstrated that the N-terminal and second extracellular domain (ECD) of the co-receptor are crucial for gp120 binding and fusion. Although the interaction of HIV within these regions varies among strains, all the Cys residues on the chemokine receptors seem to be important for co-receptor function, suggesting that these residues are responsible for disulfide bonds. Deglycosylated chemokine receptors are still expressed normally on the cell surface and can still function as co-receptors, suggesting that the sugar moieties are not required for gp120 binding.[25] In fact, a deglycosylated CXCR4 has expanded tropism and can bind to dual and some R5 tropic viruses, as well as X4 viruses.[26]

Several factors determine how likely the host immune cell receptors will bind to gp120 and, therefore, allow entry and infection of the cell. The density of the CD4 receptors on the cell surface as well as CCR5 and CXCR4 conformations influences envelope binding and may explain differences in infectivity between cell types.[27] There are a number of posttranslational modifications to both chemokine co-receptors that alter their function by changing their surface density and thereby affecting the affinity of the gp120/co-receptor interactions.

The carbohydrate moieties expressed by the receptors control stability, intracellular trafficking, surface expression, and protein folding. The glycosylation sites on seven-transmembrane domain receptors are generally at the N-terminus sequence. CXCR4 contains two potential sites for N-linked glycosylation: one at the N-terminus and one in the ECII loop.[28] Removal of one or both glycosylation sites from CXCR4 results in an expanded tropism of the receptor. With both sites gone, the receptor can recognize and bind several R5 strains of HIV. The CCR5 receptor has only one potential N-linked glycosylation site, but it is not used. In multiple cell lines as well as primary macrophages, the CCR5 receptor has O-linked glycosylation at the N-terminus.

Removing these sites does not alter HIV entry but blocks the binding of its natural chemokine ligands MIP-1α and MIP-1β.[29]

Tyrosine sulfation is another posttranslational modification that is common to many proteins and is often found in tyrosine-rich regions of proteins such as the N-terminal regions of CCR5 and CXCR4. Preventing tyrosine sulfation on the chemokine co-receptors significantly decreases HIV entry but does not alter expression of the receptor on the cell surface.[30] However, the studies of tyrosine sulfation were performed in cell lines with transfected co-receptors, making it difficult to determine the physiological role of sulfation modifications and their impact on HIV infectivity in the host immune cell. It is clear that glycosylation and sulfation of the chemokine co-receptors' extracellular domains alter the overall negative charges of the receptors and thereby their specificity for binding gp120.

CCR5 and CXCR4 are internalized and recycle back to the cell surface. Multiple studies using truncated and mutated forms of the co-receptors demonstrate that receptor internalization is not required for HIV entry into the cell. The receptor turnover rate, however, affects the concentration of receptors on the cell surface that are available for viral binding. G-protein-coupled receptors internalize via agonist-induced endocytosis in clathrin-coated pits. After internalization, GPCRs follow one of two pathways. They either recycle back to the cell surface, or they move to lysosomes for degradation.[31] An immediate serine phosphorylation of the cytoplasmic tail of CXCR4 and CCR5 is needed for receptor internalization. This function seems to be modulated by GPCR kinases or protein kinase C.[32,33]

HOST IMMUNE CELL INFECTION WITH HIV

Any host immune cell that expresses CD4 and CCR5 or CXCR4 on its cell surface is susceptible to infection by HIV. The gp120 of HIV binds to the CD4 receptor on the cell in a manner previously described. This binding causes a conformational change in the HIV gp120, which allows it to bind more readily to the chemokine co-receptor.[34] After binding to either CCR5 or CXCR4, the viral envelope then fuses with the cell membrane via the transmembrane glycoprotein subunit gp41. Within the molecular structure of the gp41 are periodic hydrophobic regions in the α-helical coiled structures. Mutations in these heptad-repeat regions (HR regions) interfere with the ability of gp41 to fuse properly.[35] The gp41 rests in a high-energy configuration, and the fusion peptide is pointed inward toward the virus. After gp120 binds to CD4 and then the chemokine co-receptor, the glycoprotein undergoes another conformational change that releases the fusion peptide to spring outward into the cell membrane. The gp41 folds over into a six-helix bundle that pulls the virus into near the cell and prepares it for fusion and entry[36] (Figure 6.2).

GP120-INDUCED APOPTOSIS IN T CELLS

The hallmark of HIV infection is a progressive decline in CD4+ T cells and, subsequently, CD8+ T cells. As a result, infected individuals become susceptible to opportunistic infections as well as malignancies. There are various mechanisms that explain the decreasing CD4+ and CD8+ T cell counts. First, T cell production is impaired during the course of HIV infection, which results in fewer healthy T cells being released into the peripheral bloodstream.[37] Second, as with many viral infections, there is a generalized state of immune activation that leads to apoptosis, or programmed cell death. Third, there is now strong evidence that the HIV virus alone can induce a more specific signal to immune cells and directly cause cellular apoptosis.[38,39] The peripheral depletion of CD4+ and CD8+ T cells during HIV infection occurs occasionally in HIV-infected cells; however, more often, HIV signals cell death to uninfected bystander CD4+ and CD8+ T cells.[40]

There are several mechanisms by which HIV can directly induce cell death to host immune cells. Both CD4+ and CD8+ T cells are more susceptible to Fas-mediated apoptosis during HIV

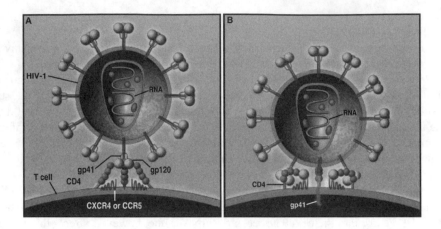

FIGURE 6.2 Mechanism of HIV cellular infection. (A) A trimer of HIV gp120 binds to a host cell CD4 receptor before changing conformation. (B) The gp120 then binds to the host cell chemokine co-receptor, allowing gp41 to spring toward the cell membrane to facilitate cellular infection.

infection.[41] T lymphocytes express higher levels of CD95 (Fas) during HIV infection, and the proportion of these Fas[hi] lymphocytes increases with disease progression.[42] FasL is also increased in peripheral blood mononuclear cells from HIV-infected individuals, and soluble FasL is increased in the plasma of HIV-infected patients.[41,43] Therefore, during HIV infection, excessive immune activation induces T cell death through Fas/FasL interactions.[44,45] The HIV proteins Nef, Env, and Tat have been shown to increase Fas and FasL levels on host immune cells and, therefore, enhance susceptibility to a Fas-mediated killing.[46–48] Several of the HIV proteins protect the virus from being removed by cytotoxic lymphocytes (CTL) that may recognize an HIV-infected cell and attempt to eliminate it. Tat and Nef proteins downregulate major histocompatibility complex class I (MHC I) expression to avoid cell recognition and the resultant death of HIV-infected cells.[49] In addition, HIV gp120, as well as Nef and Vpu, decreases CD4 expression on the cell surface.[50,51] By decreasing CD4 expression, HIV protects cells from superinfection. Taken together, the HIV proteins exert many and sometimes varying effects on the host immune cells. The end result is a state of increased immune activation that leads to increased apoptosis, increased proapoptotic proteins, decreased antiapoptotic proteins in the whole individual, and decreased CD4 receptor expression. Although more work is being done to decipher the exact effects of HIV infection on a cell, it seems that the immune cells are preparing for death while the virus is attempting to protect itself.

DEATH ASSOCIATED WITH GP120 BINDING TO CD4

The majority of the immune cells that die during HIV infection are, in fact, uninfected bystander cells. This cell death occurs through several mechanisms. When the HIV envelope glycoprotein binds to the CD4 receptor, it frequently attaches and signals the receptor without changing its conformation, binding the chemokine co-receptor and eventually infecting the cell. In fact, the envelope can bind and signal via CD4 in any of its forms, as part of the whole virion, or as a soluble glycoprotein. After binding CD4, gp120 triggers a complex cascade of intracellular signals increasing CD4+ T cell susceptibility to Fas-mediated apoptosis.[52,53] The mechanism of Fas susceptibility is complex, protein synthesis independent, involves protein kinase C activation, and results in decreased FLIP levels.[54] Thus, after priming by the HIV gp120, CD4 T cells apoptose when encountering FasL in a caspase-dependent manner that can be inhibited by blocking the caspase cascade, or by preventing the gp120 from physically interacting with the CD4 receptor, with soluble CD4 or anti-CD4 antibodies. This CD4-dependent Fas susceptibility requires CD4 signaling, because deleting the CD4 cytoplasmic tail inhibits the Fas-mediated apoptotic response.[55,56] Mutations in the V3 loop of gp120, which binds to CD4, also inhibit CD4+ T cell death.[57]

DEATH ASSOCIATED WITH GP120 BINDING FOR CCR5

Not only does gp120 cause cell death in uninfected host immune cells through Fas-dependent mechanisms as a result of binding CD4, but it also does this through binding and signaling through chemokine co-receptors alone. The gp120 can bind to CCR5 or CXCR4 without first binding the CD4 receptor, and such binding ultimately signals cell death. In the case of CCR5 binding and signaling, gp120 makes the CD4+ T cell susceptible to Fas-mediated killing. The resulting apoptosis can be abrogated by preventing gp120 from binding to CCR5 with antibodies, β-chemokines that are natural ligands for the CCR5 receptor, or nonactivating small molecule inhibitors that bind CCR5 and block gp120 binding.[9,58] Because cell death is Fas mediated and signals through the caspase cascade, the death caused by HIV signaling through CCR5 can also be blocked by inhibiting the caspase cascade. It is unclear whether the CCR5 chemokine receptor signals these messages alone or whether it acts in concert with CD4. When CCR5 is engaged by a ligand, it rafts with the CD4 receptor.[59] Therefore, it is possible that gp120 activation of CCR5 is actually signaling a caspase-dependent death to the cell via CD4 signaling.

DEATH ASSOCIATED WITH GP120 BINDING TO CXCR4

T tropic or X4 gp120 and virus may bind to and signal through the CXCR4 chemokine receptor alone without infecting the cell. Because the X4 form of the virus is present in the end stages of HIV infection and, therefore, is associated with a more rapid disease progression as well as CD8+ T cell depletion, the mechanism of the CXCR4-mediated death has been closely investigated. In contrast to CD4- or CCR5-mediated cellular killing, CXCR4 activation by gp120 causes a non-Fas-mediated caspase-independent CD4+ T cell death.[9,60] The death signal is independent of G-protein signaling.[61] The CXCR4 death is blocked by anti-CXCR4 antibodies; by SDF1α, the natural ligand of CXCR4; and by small molecule inhibitors of gp120 binding. The chemokine SDF1α signals cellular chemotaxis to the T cell but does not cause apoptosis.[62,63] The T cell death is independent of caspase activation and, therefore, is not prevented by general caspase inhibitors. Interestingly, direct contact of R5 gp120 with a CD8+ T cell does not kill the cell, but binding of X4 gp120 to the CXCR4 receptors expressed on CD8+ T cells causes a rapid caspase-independent cell death, despite equal expression of CCR5 and CXCR4 on the cell surface.[9] Therefore, CD4 and CCR5 activation by HIV gp120 causes a CD4+ T cell apoptosis by a two-step process. In the first step, the gp120 induces cell susceptibility to Fas-mediated apoptosis. In the second step, the cell dies when it engages FasL and signals death through the caspase cascade. In contrast, CXCR4 activation alone by HIV gp120 causes a direct caspase-independent cell death in both CD4+ T cells and CD8+ T cells. The mechanism of this cell death is still being investigated; however, increasing evidence suggests that the mechanism is p38 phosphorylation dependent and requires activation of caspase-9, along with a dissipation of mitochondrial transmembrane potential and, eventually, apoptotic cell death.[64,65]

Not only does HIV envelope gp120 induce cell death in immune cells, but it also does so in nonimmune cells in which the G-protein-coupled chemokine co-receptors are expressed on many cell types. Specifically, the CXCR4 receptor is expressed on the surface of human neurons. The X4 gp120 signals through the neuron CXCR4 and causes a direct caspase-independent cellular apoptosis.[66,67] This death is abrogated by blocking gp120/CXCR4 binding. It has been proposed that noninfective HIV-induced neuronal apoptosis may contribute to HIV-related dementia or peripheral neuropathy. It is also clear that HIV-infected individuals experience varying degrees of hepatotoxicity that is frequently out of proportion to, or independent of, opportunistic infections or medication-related toxicity. Interestingly, human hepatocytes also express CXCR4 on the cell surface and undergo caspase-independent apoptosis when gp120 binds the hepatocyte via CXCR4.[75] Again, this chemokine-mediated death as a result of HIV envelope gp120 signaling may explain some occurrences of liver-related disease during the course of HIV infection.

THERAPEUTIC AGENTS DIRECTED AT GP120

HIV gp120 plays a critical role in HIV viral assembly, HIV infectivity, and the signaling of host cellular death. As a result, a number of therapeutic agents have been designed to inhibit gp120 binding, signaling, and effector host-cell infection. Furthermore, agents that block gp120 binding to host T cells also block gp120-mediated T cell apoptosis.

One important area of focus is inhibiting viral entry and cellular apoptosis by blocking the interaction of gp120 with the CD4 receptor. Recombinant soluble CD4 was designed as a viral attachment decoy and has been effective at blocking HIV-1 infection *in vitro*.[69,70] However, in clinical studies, sCD4 proved ineffective except at very high doses.[71,72] Although early studies proved ineffective, gp120 and CD4 interactions are still being targeted as possible therapeutic sites. Hybrid tetramers that contain a CD4 receptor domain within an IgG2 backbone are being used as a decoy for gp120 binding.[73] Monoclonal anti-CD4 antibodies that block the CD4/gp120 interaction and can block HIV-1 replication *in vitro* are also being developed.[74] Although anti-CD4 antibodies may cause an immunosuppressive effect, clinical studies are underway to determine whether they may suppress HIV gp120–CD4 interactions. In addition, small-molecule inhibitors have been designed to compete with gp120 binding to the CD4 receptor, thereby inhibiting viral entry or signaling to the CD4+ T cell. Thus far, these compounds vary in their effect against different HIV-1 strains. Mutation in the region where the gp120 binds to the CD4 receptor confers resistance to the effects of viral inhibition and blocking.[75] Despite the limited success of the early CD4 decoys and inhibitors, it is appealing to pursue this mode of therapeutics, as inhibiting CD4 would be effective in R5 and X4 viruses and their numerous strains. Additionally, by targeting this one site, infection with the virus and HIV-induced apoptosis could be simultaneously inhibited.

Several agents have also been developed that bind to the CCR5 or CXCR4 co-receptors and block HIV-1 binding, apoptosis, and infectivity of the cell. Blocking the CCR5 receptor seems logical, because natural human mutations in the CCR5 receptor that prevent expression on the cell surface are much more resistant to infection with HIV and apoptotic signaling to the T cell, and those who do become infected have a much better outcome, with minimal immune depletion over time.[76,77] The natural ligands for CCR5 are the β-chemokines MIP 1-α, MIP 1-β, and RANTES. Incubating any of these chemokines with a CD4+ T cell will downregulate the CCR5 receptor expression on the surface of the cell and inhibit HIV replication *in vitro*.[78] As a result, several small-molecule inhibitors of CCR5 and monoclonal antibodies to CCR5 have been developed and are being used in early clinical trials alone and in conjunction with highly active antiretroviral therapy (HAART) for HIV. The preliminary results look promising, although there is significant intrapersonal variation in effect, and some small-molecule inhibitors induced viral resistance *in vitro*.[79]

Because X4 strains of HIV are present at the end stages of HIV disease and are associated with a more rapid decline in the immune cells, great effort has been made to block the gp120/CXCR4 interaction and, therefore, block infection and the rapid CXCR4-mediated apoptosis. CXCR4, however, is present on a greater number of cells than CCR5, and universally blocking this receptor may cause adverse physiological sequelae. In support of this concern, CXCR4-knockout mice die as a result of congenital defects,[80] including cardiac malformation. Several CXCR4 antagonists have been developed and brought to the phase of clinical trials, but so far, all have been difficult to administer and have had little to no effect on X4 HIV-viral load or disease progression.[81–83] Further reflecting the difficult nature of developing effective chemotherapeutics, changes in gp120 around the V3 loop render the virus resistant to the CXCR4 small-molecule inhibitor.[84]

Synthetic peptides have also been created that target the gp41 portion of the HIV envelope glycoprotein and serve to inhibit HIV viral entry. The peptides seem to mimic the HR2 segments of gp41 and bind to the coiled hydrophobic portions of the gp41.[85] Several of these agents are effective *in vitro* and are currently undergoing clinical trials. Early results are promising and demonstrate effective suppression of HIV-viral load, particularly as part of a salvage regimen.[86,87] Difficulty with these reagents has arisen with the administration of some of the agents and with the induction of viral resistance.[88]

GP120 VACCINES

The most eagerly awaited new development in the HIV community is the HIV vaccine. Because of the rapid spread of disease worldwide and, in many countries, limited resources to effectively treat the infection, a vaccine to prevent HIV infection is highly desirable. Moreover, a vaccine that would prime the host immune system to recognize and kill infected host-immune cells in an already infected individual would also be helpful, given the current issues of drug resistance and intolerance. Because the HIV envelope gp120 makes up most of the exposed portions of the HIV virus, it was one of the early focuses for HIV vaccine design.

Ideally, a vaccine would provide the following four important functions: elicit neutralizing antibodies, stimulate a T cell immune response, stimulate mucosal immunity, and stimulate the innate immune system.[89] In order to effectively achieve all four of these immune functions, a live-attenuated virus vaccine would be necessary. In the macaque model, a live-attenuated SIV vaccine caused persisting infection and even AIDS in some animals.[90,91] Therefore, the live-virus vaccine has been banned in humans. Instead, a safer killed-virus vaccine was developed using gp120 as the target. This took the form of a vaccine-stimulated neutralizing antibody but did not create a cytotoxic CD8+ T cell response. This approach was successful at producing neutralizing antibodies to laboratory-adapted strains but was ineffective at producing good antibodies against clinical isolates of viruses. Because the HIV envelope is a trimer of gp120, coated with carbohydrate, large portions of the surface are hidden from antibodies by the sugars and important functional parts of the glycoprotein are folded in and not exposed for antibody production. Although the important parts of gp120 that bind to CD4 and chemokine co-receptors are conserved, they are folded deeply into the gp120 pocket, and the outside areas are the hypervariable portions that frequently mutate and make consistent antibody production difficult.[92] Many of the important functional portions of the gp120, such as the chemokine receptor binding site, are exposed for only a fraction of a second after the gp120 binds CD4 and changes conformation before binding to the co-receptor. The gp120 envelope is, therefore, an elusive protein to target for neutralizing antibodies.[89] Several vaccines were designed to a killed-viral version of gp120 and are under study in the United States and Thailand. Great clinical success, however, is not predicted.[93] This work has suggested that a vaccine that elicits a cytotoxic CD8+ T cell response is instead preferable, using parts of the virus that are processed in the endoplasmic reticulum and are presented by MHC I.

SUMMARY

The HIV envelope gp120 is a complex glycoprotein that is crucial to the function of HIV. The glycoprotein has an intricate structure that allows for efficient binding to the host receptors CD4, CCR5, and CXCR4, yet it is protected from host-neutralizing antibodies. The gp120 seems to be designed to accommodate host cell infection with the gp41 transmembrane portion that facilitates cell entry. In addition, the gp120 can proficiently signal death to a bystander CD4+ or CD8+ T cell by simply binding only one of the receptors while not infecting the cell. Moreover, it can signal death to a bystander nonimmune cell, such as a neuron or hepatocyte, by simply binding a CXCR4 receptor in the cell's surface. Thus far, attempts to completely block the gp120 binding or stop its effects by targeted vaccines have proven ineffective in humans. HIV has clearly developed an effective shield that will require more research and time to conquer.

REFERENCES

1. Kwong, P.D., Wyatt, R. Robinson, J., Sweet, R.W., Sodroski, J., and Hendickson, W.A. (1998). Structure of an HIV gp120 envelope glycoprotein in complex with the CD4 receptor and a neutralizing human antibody. *Nature* 393(6686):648–59.

2. Wyatt, R., Kwong, P.D., Desjardins, E., Sweet, R.W., Robinson, J., Hendrickson, W.A., and Sodroski, J.G. (1998). The antigenic structure of the HIV gp120 envelope glycoprotein. *Nature* 393(6686): 705–11.

3. Hager-Braun, C., and Tomer, K.B. (2002). Characterization of the tertiary structure of soluble CD4 bound to glycosylated full-length HIVgp120 by the chemical modification of arginine residues and mass spectrometric analysis. *Biochemistry* 41(6):1759–66.

4. Profy, A.T., Salinas, P.A., Eckler, L.I., Dunlop, N.M., Nara, P.L., and Putney, S.D. (1990). Epitopes recognized by the neutralizing antibodies of an HIV-infected individual. *J Immunol* 144(12):4641–7.

5. Hoffman, T. L. and Dooms, R.W. (1999). Hiv-1 envelope determinants for cell tropism and chemokine receptor use. *Mol Membr Biol* 16(1):57–65.

6. Roos, M. T., Lange, J.M., deGoede, R.E., Coutinho, R.A., Schellenkens, P.T., Miedema, F, and Tersmette, M. (1992). Viral phenotype and immune response in primary human immunodeficiency virus type 1 infection. *J Infect Dis* 165(3):427–32.

7. Zhu, T. Mo, H., Wang, N, Nam, D.S., Cao, Y. Koup, R.A. Ho, D.D. (1993). Genotypic and phenotypic characterization of HIV-1 patients with primary infection. *Science* 261(5125):1179–81.

8. Schuitemaker, H., Koot, M., Koostra, N.A., Dercksen, M.W., deGoede, R.R., van Steenwijk, R.P., Lange, J.M., Schattenkerk, J.K., Miedema, F., and Tersmette, M. (1992). Biological phenotype of human immunodeficiency virus type 1 clones at different stages of infection: progression of disease is associated with a shift from monocytotropic to T-cell-tropic virus population. *J Virol* 66(3):1354–60.

9. Vlahakis, S.R., Algeciras-Schimnich, A., Bou, G., Heppelmann, C.J., Villasis-Keever, A., Collman, R.C., Paya, C.V. (2001). Chemokine-receptor activation by env determines the mechanism of death in HIV-infected and uninfected T lymphocytes. *J Clin Invest* 107(2):207–15.

10. Collman, R., J. W. Balliet, S. A. Gregory, H. Friedman, D. L. Kolson, N. Nathanson, and A. Srinivasan. (1992) An infectious molecular clone of an unusual macrophage-tropic and highly cytopoathic strain human immunodeficiency virus type 1. *J Virol* 66(12):7517–21.

11. Chesebro B., K. Wehrly, J. Nishio, and S. Perryman. (1996) Mapping of independent V3 envelope determinants of human immunodeficiency virus type 1 macrophage tropism and syncytium formation in lymphocytes. *J Virol* 70(12):9055–9.

12. DeJong, J.J., De Ronde, A., Keulen, W., Tersmette, M., and Goudsmit, J. (1992). Minimal requirements for the human immunodeficiency virus type 1 V3 domain to support the syncytium-inducing phenotype: analysis by single amino acid substitution. *J Virol* 66(11):6777–80.

13. Pyle, S.W., Bess, Jr., J.W., Robey, W.G., Fischinger, P.J., Gilden, R.V., and Arthur, L.O. (1987). Purification of 120,000 dalton envelope glycoprotein from culture fluids of human immunodeficiency virus (HIV)-infected H9 cells. *AIDS Res Hum Retroviruses* 3(4):387–400.

14. Willey, R. L., Bonifacino, J.S., Potts, B.J., Martin, M.A., and Klausner, R.D.. (1988). Biosynthesis, cleavage, and degradation of the human immunodeficiency virus 1 envelope glycoprotein gp160. *Proc Natl Acad USA* 85(24):9580–4.

15. McKeating, J. A. and R. L. Willey. (1989). Structure and function of the HIV envelope. *AIDS* 3 Suppl 1:S35-41.

16. Earl, P.L., Moss, B., and Doms, R.W. (1991). Folding, interaction with GRP78-BiP, assembly, and transport of the human immunodeficiency virus type 1 envelope protein. *J Virol* 65(4):2047-55.

17. Hallenberger, S., Moulard, M., Sordel, M., Klenk, H.D., and Garten, W. (1997). The role of eukaryotic subtilison-like endoproteases for the activation of human immunodeficiency virus glycoproteins in natural host cells. *J Virol* 71(2):1036–45.

18. Miranda, L.R., Schaefer, B.C., Kupfer, A., Hu, Z., and Franzusoff, A. (2002). Cell ssurface expression of the HIV-1 envelope glycoproteins is directed from intracellular CTLA-4-containing regulated secretory granules. *Proc Natl Acad USA* 99(12):8031–6.

19. Linsley, P.S., Bradshaw, J., Greene, J., Peach, R., Bennett, K.L., and Mittler, R.S. (1996). Intracellular trafficking of CTLA-4 and focal localization towards sites of TCR engagement. *Immunity* 4(6):535–43.

20. Alegre, M. L., P. J. Noel, B. J. Eisfelder, E. Chaung, M. R. Clark, S. L. Reiner, and C. B. Thompson. (1996). "Regulation of surface and intracellular expression of CTLA4 on mouse T cells." *J Immunol* 157(11):4762–70.

21. Peden, K. W. and J. M. Farber. (2000). Coreceptors for human immunodeficiency virus and simian immunodeficiency virus. *Adv Pharmacol* 48:409–78.

22. Dalgeish A. G., Beverley, P.C., Clapham, P.R., Crawford, D.H., Greaves, M.F., and Weiss, R.A.. (1984). The CD4 (T4) antigen is an essential component of the receptor for the AIDS retrovirus. *Nature* 312(5996):763–7.

23. Deen, K.C., Mc Dougal, J.S., Inacker, R., Folena-Wasserman, G., Arthos, J., Rosenberg, J., Maddon, P.J., Axel, R., and Sweet, R.W. (1988). A soluble for CD4 (T4) protein inhibits AIDS virus infection. *Nature* 331(6151):82–4.

24. Tarasova, N.I., Stauber, R.H., and Michejda, C.J. (1998). Spontaneous and ligand-induced trafficking of CXC-chemokine receptor 4. *J Biol Chem* 273(26):15883–6.

25. Zaitseva, M., Peden, K., and Golding, H. (2003). HIV coreceptors: role of structure, posttranslational modifications, and internalization in viral-cell fusion and as targets for entry inhibitors. *Biochim Biophys Acta* 1614(1):51–61.

26. Chabot, D. J., H. Chen, D. S. Dimitrov, and C. C. Broder. (2000) N-linked glycosylation of CXCR4 masks coreceptor function for CCR5-dependent human immunodeficiency virus type 1 isolates. *J Virol* 74(9):4404–13.

27. McKnight, A., Wilkinson, D., Simmons, G., Talbot, S., Picard, L., Ahuja, M., Marsh, M., Hoxie, J.A., and Clapham, P.R. (1997). Inhibition of human immunodeficiency virus fusion by a monoclonal aniotbody to a coreceptor (CXCR4) is both cell type and virus strain dependent. *J Virol* 71(2):1692–6.

28. Berson, J.F., D. Long, B.J. Doranz, J. Rucker, F.R. Jirik, and R.W. Doms. (1996). A seven transmcmbrane domain receptor involved in fusion and entry of T-cell-tropic human immunodeficiency virus type 1 strains. *J Virol* 70(9):6288–95.

29. Bannert, N., S. Craig, M. Farzan, D. Sogah, N.V. Santo, H. Chloe, and J. Sodroski. (2001). Sialylated O-glycans and sulfated tyrosincs in the NH2-terminal domain of CC chemoskine receptor 5 contribute to high affinity binding of chemokines. *J Exp Med* 194(11):1661–73.

30. Farzan, M., Mirzabekov, T., Kolchinsky, P., Wyatt, R., Cayabyab, M., Gerard, N.P., Gerard, C., Sodroski, J., and Choe, H. (1999). Tyrosine sulfation of the amino terminus of CCR5 facilitates HIV-1 entry. *Cell* 96(5):667–76.

31. Tsao, P. and M. von Zastrow. (2000). Downregulating of G protein-coupled receptors. *Curr Opin Neurobiol* 10(3):365–9.

32. Oppermann, M., Mack, M., Proudfoot, A.E., and Olbrich, H. (1999). Differential effects of CC chemokines on CC chemokine receptor 5 (CCR5) phosphorylation and identification of phosphorylation sites on the CCR5 carboxyl terminus. *J Biol Chem* 274(13):8875–85.

33. Pollok-Kopp, B., Schwarze, K., Baradari, V.K., and Oppermann, M. (2003). Analysis of ligand-stimulated CC chemokine receptor 5 (CCR5) phosphorylation in intact cells using phosphosite-specific antibodies. *J Biol Chem* 278(4):2190–8.

34. Bandres, J.C., Q. F. Wang, J. O'Leary, F. Baleaux, A. Amara, J.A. Hoxie, S. Zolla-Pazner, and M.K. Gomy. (1998). Human immunodeficiency virus (HIV) envelope binds to CXCR4 independently of CD4, and binding can be enhanced by interaction with soluble CD4 or by HIV envelope deglycosylation. *J Virol* 72(3):2500–4.

35. Dubay, J. W., Roberts, S.J., Brody, B., and Hunter, E.. (1992). Mutations in the leucine zipper of the human immunodeficiency virus type 1 transmembrane glycoprotein affect fusion and infectivity. *J Virol* 66(8):4748–56.

36. Chen C.H., T. J. Matthews, C.B. McDanal, D.P. Bolognesi, and M.L. Greenberg. (1995) A molecular clasp in the human immunodeficiency virus (HIV) type 1 TM protein determines the anti-HIV activity of gp41 derivatives: implication for viral fusion. *J Virol* 69(6):3771–7.

37. Hellerstein, M., Hanley, M.B., Cesar, D., Siler, S., Papageorgopoulos, C., Wieder, E., Schmidt, D., Hoh, R., Neese, R., Macallan, D., Deeks, S., and McCune, J.M. (1999). Directly measured kinetics of circulating T lymphocytes in normal and HIV-1-infected humans. *Nat Med* 5(1):83–9.

38. Laurent-Crawford, A.G., Krust, B., Riviere, Y., Desgranges, C., Muller, S., Kieny, M.P., Dauguet, C., and Hovanessian, A.G. (1993). Membrane expression of HIV envelope glycoproteins triggers apoptosis in CD4 cells. *AIDS Res Hum Retroviruses* 9(8):761–73.

39. Nardelli, B., Gonzalez, C.J., Scheckter, M, and Valentine, F.T. (1995). CD4+ blood lymphocytes are rapidly killed in vitro by contact with autologous human immunodeficiency virus-infected cells. *Proc Natl Acad Sci USA* 92(16):7312–6.

40. Finkel, T.H., Tudor-Williams, G., Banda, N.K., Cotton, M.F., Curiel, T., Monks, C., Baba, T.W., Ruprecht, R.M., and Kupfer, A. (1995). Apoptosis occurs predominately in bystander cells and not in productively infected cells of HIV- and SIV-infected lymph nodes. *Nat Med* 1(2):129–34.

41. Silvestris, F., Cafforio, P., Frassanito, M.A., Tucci, M., Romito, A., Nagata, S., and Dammacco, F. (1996). Overexpression of Fas antigen on T cells in advanced HIV-1 infection : differential ligation constantly induces apoptosis. *AIDS* 10(2):131–41.

42. Aries, S. P., B. Schaaf, C. Muller, R. H. Dennin, and K. Dalhoff. (1995). Fas (CD95) expression of CD4+ T cells from HIV-infected patients increase with disease progression. *J Mol Med* 73(12):59–3.

43. Hosaka, N., Oyaizu, N., Kaplan, M.H., Yagita, H., and Pahwa, S. (1998). Membrane and soluble forms of Fas (CD95) and Fas ligand in peripheral blood mononuclear cells and in plasma from human immunodeficiency virus-infected persons. *J Infect Dis* 178(4):1030–9.

44. Badley, A.D., D.H. Dockrell, A. Algeciras, S. Ziesmer, A. Landay, M.M. Ledermann, E. Connick, H. Kestler, D. Kuritzkes, D.H. Lynch, P. Roche, H. Yagita, and C.V. Paya. (1998). In vivo analysis of Fas/FasL interactions in HIV-infected patients. *J Clin Invest* 102(1):79–87.

45. Dockrell D.H., Badley, A.D., Villacian, J.S., Heppelmann, C.J., Algeciras, A., Ziesmer, S., ?Yagita, H., Lynch, D.H., Roche, P.C., Liebson, P.J., and Paya, C.V. (1998). The expression of Fas Ligand by macrophages and its upregluation by human immunodeficiency virus infection. *J Clin Invest* 101(11):2394–405.

46. Zauli, G., Gibellini, D., Secchiero, P., Durarte, H., Olive, D., Capitani, S., and Collette, Y. (1999). Human immunodeficiency virus type 1 Nef protein sensitiezes CD4(+) T lymphoid cells to apoptosis via functional upregluation of the CD95/CD95 pathway. *Blood* 93(3):1000–10.

47. Oyaizu, N., McClosky, T.W., Than, S., Hu, R., Kalyanaraman, V.S., and Pahwa, S. (1994). Cross-linking of CD4 molecules upregulates Fas antigen expression in lymphocytes by inducing interferon-gamma and tumor necrosis factor-alpha secretion. *Blood* 84(8):2622–31.

48. Westendorp, M.O., Frank, R., Ochsenbauer, C., Stricker, K., Dhein, J., Walczak, H., Debatin, K.M., and Krammer, P.H. (1995). Sensitization of T cells to CD95-mediated apoptosis by HIV-1 Tat and gp120. *Nature* 375(6531):497–500.

49. Schwartz, O., Marechal, V., LeGall, S., Lemonnier, F., and Heard, J.M. (1996). Endocytosis of major histocompatibility complex class 1 molecules is induced by the HIV-1 Nef protein. *Nat Med* 2(3):338–42.

50. Salghetti, S., Mariani, R., and Skowronski, J. (1995). Human immunodeficiency virus type 1 Nef and p56lck protein-tyrosine kinase interact with a common element in CD4 cytoplasmic tail. *Proc Natl Acad Sci USA* 92(2):349–53.

51. Willey, R.L., Maldarelli, F., Martin, M.A., and Strebel, K. (1992). Human immunodefifiency virus type 1 Vpu protein induces rapid degradation of CD4. *J Virol* 66(12):7193–200.

52. Banda, N.K., J. Bernier, D.K. Kurahara, R. Kurrle, N. Haigwood, R.P. Sekaly, and T.H. Finkel. (1992). Crosslinking CD4 by human immunodeficiency virus gp120 primes T cells for activation-induced apoptosis. *J Exp Med* 176(4):1099–106.

53. Algeciras, A., D.H. Dockrell, D.H. Lynch, and C.V. Paya. (1998). CD4 regulates susceptibility to Fas ligand- and tumor necrosis factor-mediated apoptosis. *J Exp Med* 187(5):711–20.

54. Algeciras-Schimnich, A., T.S. Griffith, D.H. Lynch, and C.V. Paya. (1999). Cell cycle-dependent regulation of FLIP levels and susceptibility to Fas-mediated apoptosis. *J Immunol* 162(9):5205–11.

55. Moutouh, L., Estaquier, J., Richman, D.D., and Corbeil, J. (1998). Molecular and cellualr analysis of human immunodeficiency virus-induced apoptosis in lymphoblastoid T-cell-line-expressing wild-type and mutated CD4 receptors. *J Virol* 72(10):8061–72.

56. Guillerm, C., Coudronniere, N., Robert-Hebmann, V., and Devaux, C. (1998). Delayed human immunodeficiency virus type 1-induced apoptosis in cells expressing truncated forms of CD4. *J Virol* 72(3):1754–61.

57. Laurent-Crawford, A.G., Coccia, E., Krust, B., and Hovanessian, A.G. (1995). Membrane-expressed HIV envelope glycoprotein heterodimer is a powerful inducer of cell death in uninfected CD4+ target cells. *Res Virol* 146(1):5–17.

58. Algeciras-Schimnich, A., S.R. Vlahakis, A. Villasis-Keever, T. Gomez, C.J. Heppelmann, C. Bou, and C.V. Paya. (2002). CCR5 mediates Fas- and caspase 8 dependent apoptosis of both uninfected and HIV infected primary human CD4 T cells. *AIDS* 16(11):1467–78.

59. Singer, I.I., S. Scott, D.W. Kawka, J. Chin, B.L. Daugherty, J.A. DeMartino, J. DiSalvo, S.L. Gould, J.E. Lineberger, L. Malkowitz, M.D. Miller, L. Mitnaul, S.J. Siciliano, M.J. Staruch, H.R. Williams,

H.J. Zeerink, and M.S. Springer. (2001). CCR5, CXCR4, and CD4 are clustered and closely opposed on microvilli of human macrophages and T cells. *J Virol* 75(8):3779–90.

60. Berndt, C., B. Mopps, S. Angermuller, P. Gierschik, and P.H. Krammer. (1998). CXCR4 and CD4 mediate a rapid CD95-independent cell death in CD4(+) T cells. *Proc Natl Acad Sci USA* 95(21): 12556–61.

61. Blanco, J., E. Jacotot, C. Cabrera, A. Cardona, B. Clotet, E. De Clercq, and J.A. Este. (1999). The implications of the chemokine recptor CXCR 4 in HIV-1 envelope protein-induced apoptosis is independent of the G proetin-mediated signalling. *AIDS* 13(8):909–17.

62. Vlahakis, S.R., Villasis-Keever, A., Gomez, T., Vanegas, M., Vlahakis, N. and Paya C.V. (2002). G Protein-coupled chemokine receptors induce both survival and apoptotic signaling pathways. *J Immunol* 169(10):5546–54.

63. Blanco, J., J. Barretina, G. Henson, G. Bridger, E. De Clercq, B. Clotet, and J.A. Este. (2000). The CXCR4 antagonist AMD3100 efficiently inhibits cell-surface-expressed human immunodeficiency virus type 1 envelope-induced apoptosis. *Antimicrob Agents Chemother* 44(1):51–6.

64. Kaul, M. and S.A. Lipton. (1999). Chemokines and activated macrophages in HIV gp120-induced neuronal apoptosis. *Proc Natl Acad Sci USA* 96(14):8212–6.

65. Roggero, R., V. Robert-Hebmann, S. Harrington, J. Roland, L. Vergne, S. Jaleco, C. Devaux, and M. Biard-Piecharaczyk. (2001). Binding of human immunodeficiency virus type 1 gp120 CXCR4 induces mitochondrial transmembrane depolarization and cytochrome c-mediated apoptosis independently of Fas signaling. *J Virol* 75(16):7637–50.

66. Hesselgesser, J., Taub, D., Baskar, P., Greenberg, M., Hoxie, J., Kolson, D.L., and Horuk, R. (1998). Neuronal apoptosis induced by HIV-1 gp120 and chemokine SDF-1 alpha is mediated by the chemokine receptor CXCR4. *Curr Biol* 8(10):595–8.

67. Herzberg, U. and J. Sagen. (2001). Peripheral nerve exposure to HIV viral envelope protein gp120 induces neuropathic pain and spinal gliosis. *J Neuroimmunol* 116(1):29–39.

68. Vlahakis, S.R., Villasis-Keever, A., Gomez, T., Bren, G.D., and Paya, C.V. (2003). Human immunodeficiency virus-induced apoptosis of human hepatocytes via CXCR4. *J Infect Dis* 188(10): 1455–60.

69. Fisher, R.A., Bertonis, J.M., Meier, W., Johnson, V.A., Costopoulos, D.S., Liu, T., Tizard, R., Walker, B.D., Hirsch, M.S., and Schooley, R.T. (1988). HIV infection blocked in vitro by recombinant soluble CD4. *Nature* 331(6151):76–8.

70. Hussey, R.E., and Richardson, N.E. (1998). A soluble CD4 protein selectively inhibits HIV replication and syncytium formation. *Nature* 331(6151):78–81.

71. Schooley, R.T., Merigan, T., Gaut, P. Hirsch, M.S., Holodniy, M., Flynn, T., Liu, S., Byington, R.E., Henochowicz, and Gubish, S.. (1990). Recombinant soluble CD4 therapy in patients with the acquired immunodeficiency syndrome (AIDS) and AIDS-related comoplex. A phase I-II escalating dosage trial. *Ann Intern Med* 112(4)247–53.

72. Moore, J.P., McKeating, J.A., Huang, Y.X., Ashkenazi, A., and Ho, D.D. (1992). Virions of primary human immunodeficiency virus type 1 isolates resistant to soluble CD4 (sCD4) neutralization differ in sCD4 binding and glycoprotein gp120 retention from sCD4-sensitive isolates. *J Virol* 66(1):235–43.

73. Trkola, A., Pomales, A.B., Yuan, H., Korber, B., Maddon, P.J., Allaway, G.P., Katinger, H., Barbas, 3rd, C.F., Burton, D.R., and Ho, D.D. (1995). Cross-clade neutralization of primary isolates of human immunodeficiency virus type 1 by human monoclonal antibodies and tetrameric CD4-IgG. *J Virol* 69(11):6609–17.

74. Shearer, M.H., Timanus, D.K., Benton, P.A., Lee, D.R., and Kennedy, R.C. (1998). Cross-clade inhibition of human immunodeficiency virus type 1 primary isolates by monoclonal anti-CD4. *J Infect Dis* 177(6):1727–9.

75. Lin, C.M. and F.H. Wang. (2002). Selective modification of antigen-specific CD4(+) T cells by retroviral-mediated gene transfer and in vitro sensitization with dendritic cells. Clin Immunol 104(1):58–66.

76. Dean, M., Carrington, M., Winkler, C., Huttley, G.A., Smith, M.W., Allikmets, R., Goedert, J.J., Buchbinder, S.P., Vittinghoff, E., Gomperts, E., Donfiedls, S., Vlahov, D., Kaslow, R., Saah, A, Rinaldo, Detels, R., and O'Biren, S.J. (1996). Genetic restriction of HIV-1 infection and progression to AIDS by a deletion allele of the CKR5 structural gene. Hemophilia Growth and Development Study, Multicenter Hemophilia Cohort Study, San Francisco City Cohort, ALIVE Study. *Science* 273(5283):1856–62.

77. Liu, R., Paxton, W.A., Choe, S., Ceradini, D., Martin, S.R., Horuk, R., MacDonald, M.E., Stuhlmann, H, Koup, R.A., and Landau, N.R.. (1996). Homozygous defect in HIV-1 coreceptor accounts of resistance of some multiply-exposed individuals to HIV-1 infection. *Cell* 86(3):367–77.

78. Mosier, D.E., Picchio, G.R., Gulizia, R.J., Sabbe, R., Poignard, P., Picard, L., Offord, R.E., Thompson, D.A., Wilken, J. (1999). Highly potent RANTES analouges either prevent CCR5-using human immunodeficiency virus type 1 infection in vivo or rapidly select for CXCR4-using variants. *J Virol* 73(5):3544–50.

79. Trkola, A., Kuhmann, S.E., Strizki, J.M., Maxwell, E., Ketas, T., Morgan, T., Pugach, P., Xu, S., Wojcik, L., Tagat, J., Palani, A., Shapiro, S., Clader, J.W., McCombie, S., Reyes, G.R., Baroudy, B.M., and Moore, J.P. (2002). HIV-1 escape from a small molecule, CCr5-specific entry inhibitor does not involve CXCR4 use. *Proc Natl Acad Sci USA* 99(1):395–400.

80. Nagasawa, T., Hirota, S., Tachibana, K., Takakura, N., Nishikawa, S., Kitamura, Y., Yoshida, N., Kikutani, H., and Kishimoto, T. (1996). Defects of B-cell lymphopoiesis and bone-marrow myelopoiesis in mice lacking the CXC chemokine PBSF/SDF-1. *Nature* 382(6592):635–8.

81. Donzella, G.A., Schols, D., Lin, S.W., Este, J.A., Nagashima, K.A., Maddon, P.J., Allaway, G.P., Sakmar, T.P., Henson, G. DeClercq, E., and Moore, J.P. (1998). AMD3100, a small molecule inhibitor of HIV-1 entry via the CXCR4 co-receptor. *Nat Med* 4(1):72–7.

82. Hendrix, C.W., Flexner, C., MacFarland, R.T., Giandomenico, C., Fuchs, E.J., Redpath, E., Bridger, G., and Henson, G.W. (2000). Pharmacokinetics and safety of AMD-3100, a novel antagonist of the CXCR4 chemokine receptor, in human volunteers. *Antimicrob Agents Chemother* 44(6):1667–73.

83. Doranz, B.J., Filion, L.G., Diaz-Mitoma, F., Sitar, D.S., Sahai, J., Baribaud, F., Orsini, M.J., Benovic, J.L., Caeron, W., and Doms, R.W. (2001). Safe use of the CXCR4 inhibitor ALX40-4C in humans. *AIDS Res Hum Retroviruses* 17(6):475–86.

84. de Vreese, K., Kofler-Mongold, V., Leutgeb, C., Weber, V., Vermeire, K., Schacht, S., Anne, J., de Clercq, E., Datema, R., and Werner, G. (1996). The molecular target of bicyclams, potent inhibitors of human immunodeficiency virus replication. *J Virol* 70(2):689–96.

85. Wild, C., Greenwell, T., and Matthews, T. (1993). A synthetic peptide from HIV-1 gp41 is a potent inhibitor of virus-mediated cell-cell fusion. *AIDS Res Hum Retroviruses* 9(11):1051–3.

86. Lalezari, J. P., Henry, K., O'Hearn, M., Montaner, J.S., Piliero, P.J., Trottier, B., Walmsley, S., Cohen, C., Kuritzkes, D.R., Eron, Jr., J.J., Chung, J., DeMasi, R., Donatacci, L., Drobnes, C., Delehanty, J., and Salgo, M. (TORO1 Study Group). (2003). Enfuvirtide, and HIV-1 infusion inhibitor, for drug-resistant HIV infection in North and South America. *N Engl J Med* 348(22):2175–85.

87. Lazzaarin, A., Clotet, B., Cooper, D., Reynes, J., Arasteh, K., Nelson, M., Katlama, C., Stellbrink, H.J., Delfraissy, J.F., Lange, J., Huson, L., DeMasi, R., Wat, C., Delehanty, J., Drobnes, C., and Salgo, M. (TORO2 Study Group). (2003). Efficacy of enfuvirtide in patients infected with drug-resistant HIV-1 in Europe and Australia. *N Engl J Med* 348(22):2186–95.

88. Wei, X., Decker, J.M., Liu, H., Zhang, Z., Arani, R.B., Kilby, J.M., Saag, M.S., Wu, X., Shaw, G.M., and Kappes, J.C. (2002). Emergence of resistant human immunodeficiency virus type 1 in patients receiving fusion inhibitor (T-20) monotherapy. *Antimicrob Agents Chemother* 46(6):1896–905.

89. McMichael, A., Mwau, M., and Hanke, T. (2002). Design and tests of an HIV vaccine. *Br Med Bull* 62:87–98.

90. Baba, T.W., Y.S. Jeong, D. Pennick, R. Bronson, M.F. Greene, and R.M. Ruprecht. (1995). Pathogenicity of live, attenuated SIV after mucosal infection of neonatal macaques. *Science* 267(5205): 1820-5.

91. Daniel, M.D., Kirchhoff, F., Czajak, S.C., Sehgal, P.K., and Desrosiers, R.C. (1992). Protective effects of a live attentuated SIV vaccine with a delection in the nef gene." *Science* 258(5090):1938–41.

92. Moore, J.P. and D.R. Burton. (1999). HIV-1 neutralizing antibodies: how full is the bottle? *Nat Med* 5(2):142–4.

93. Connor, R.I., Korber, B.T., Graham, B.S., Hahn, B.H., Ho, D.D., Walker, B.D., Neumann, A.U., Vermund, S.H., Mestecky, J., Jackson, S., Fenamore, E., Cao, Y., Gao, F., Kalams, S., Kunstman, K.J., McDonald, D., McWilliams, N., Trkola, Moore, J.P., and Wolinsky, S.M. (1998). Immunological and virological analyses of persons infected by human immunodeficiency virus type 1 while participating in trials of recombinat gp120 subunit vaccines. *J Virol* 72(2):1552–76.

7 Vpr

Julian J. Lum
Andrew D. Badley

CONTENTS

INTRODUCTION

The life or death of the host cell determines the fate of the virus. Cell lysis and target cell destruction too early would terminate progeny virion production, whereas appropriate cell death enhances viral dissemination. A growing body of literature indicates that Vpr impinges on the cell cycle and the general machinery that regulates programmed cell death, but exactly how this occurs is still largely a matter of intense investigation and debate. This chapter will highlight the key roles of Vpr in human immunodeficiency virus (HIV)-1 biology, its host interactions, and the mechanisms of Vpr-induced cell death. Finally, some discussion will be given on strategies that may be used to neutralize the actions of Vpr.

VIRAL PROTEIN R (VPR): STRUCTURE AND FUNCTION IN THE VIRAL LIFE CYCLE

GENOMIC STRUCTURE OF VPR IN HIV-1

Vpr constitutes one of the four HIV accessory proteins that include Nef, Vpu, and Vif. Its expression late in the viral life cycle requires the specific interaction with the C-terminal domain of p55 Gag19, and approximately 100 copies are incorporated into each virion.[1] Vpr is highly conserved among HIV-1, HIV-2, and simian immunodeficiency virus (SIV).[2] The full structure of Vpr has not been determined given its tendency to aggregate and multimerize in solution. As an alternative approach,

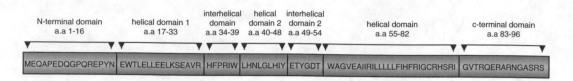

FIGURE 7.1 Amino acid structure and domains of HIV-1 Vpr.

synthetic peptides were used to analyze several Vpr domains by circular dichorism and homonuclear and heteronuclear nuclear magnetic resonance (NMR) methods.[3–6] Its structure is characterized by several tandem repeats of γ-turns and α-helices. Mutagenesis studies indicate that the C-terminal domain is responsible for cell-cycle arrest and protein stability,[7,8] and it contains several acidic motifs that play a role in its cytotoxicity.[9–11] The N-terminal domain is required for nuclear localization, virion packaging, and binding of a number of viral (nucleocapsid protein NCp7, p17) and cellular proteins (RIP1, Sp1).[12,13] Based on this information, Vpr domains mediate distinct functions and may explain the pleiotropic effects of its action during the viral life cycle.

Amino Acid Sequence

Vpr is a 14 kDa protein composed of 96 amino acids divided into several domains. The key domains are shown in Figure 7.1. The N-terminal (amino acids 1 to 16) forms the first γ-turn. Sequence variations in this region among HIV-1 clades were noted.[14] This is followed by a helical domain (amino acids 17 to 33), which by specific site-directed mutagenesis was shown to be important for subcellular localization, virion incorporation, stability, and oligomerization.[15–19] In particular, an amino acid substitution at residue 33 was isolated in naturally occurring viruses that dramatically affects viral subcellular localization and virion incorporation.[20] A second γ-turn (amino acids 34 to 39) is found between the first and second helical domains. Although some natural mutations were isolated, further characterization is needed to determine their functional consequences on Vpr activity.[21] The second helical domain forms a highly hydrophobic region that significantly abrogates virion incorporation when mutated.[16] The second interhelical domain spans the third γ-turn comprising amino acids 49 to 54. Although naturally occurring polymorphisms are found in this region, its contribution to Vpr function is not known. The last helical domain (amino acids 55 to 82) was implicated in both virion incorporation and subcellular localization. This region contains an arginine- rich motif that, when expressed in yeast or mammalian cells, exhibits potent apoptotic-inducing activity.[7,20,22–24] The C-terminal region of Vpr participates in cell-cycle arrest with four serine residues that are targets for phosphorylation.[25]

Lessons from Yeast

A large number of cellular functions are highly conserved in both yeast and mammalian cells. This has led to a number of focused studies on Vpr function using fission yeast *Schizosaccharomyces pombe* and budding yeast *Saccharomyces cerevisiae* (reviewed in Zhao and Elder[2]). In yeast, progression through G2 requires phosphorylation of Cdc2 on tyrosine 15 by the Wee1 and Mik1 kinases.[26,27] Genetic studies suggest that Vpr induces G2 arrest by inhibiting phosphorylation of Cdc2 kinase.[26,28–31] This effect is distinct from physiological or pharmacological forms of DNA damage or replication checkpoint pathways. DNA damage in yeast activates the Cdc2 pathway, whereas the checkpoint pathway involves Rad1, Rad3, Rad9, and Rad17, which lead to phosphorylation of Chk1 kinase. Chk1 directly phosphorylates Cdc25 phosphatase,[31] allowing the transport of Cdc25 out of the nucleus, rendering it inactive.[26] Vpr induces G2 arrest in *chk1/cds1* double-yeast mutants, supporting the hypothesis that the mechanism of Vpr action is distinct from DNA damage or checkpoint regulation. Interestingly, another protein that regulates cell-cycle progression, PP2A, plays a major role in G2 arrest.[28] PP2A interacts with both Wee1 and Cdc25 in *S. pombe*, and its overexpression results in G2 arrest. The PP2A inhibitor okadaic acid suppresses Vpr-induced G2 arrest, and deletion of one of the catalytic PP2A subunits in *S. pombe* has a similar negative effect.[32]

The cytotoxic domain of Vpr (HFRIGCRHSRIG) was originally described in *S. cerevisiae* and *S. pombe* systems.[23] Using synthetic peptides, the authors showed that the cytotoxic domain induced cell-cycle arrest and loss of mitochondria membrane potential. Only recently was a similar comprehensive study performed in mammalian cells. In fact, the data on cell-cycle targets recapitulated what was found in the yeast system. Because yeast lacks the core apoptotic components of the mitochondria-induced cell death pathway, the more recent work provided important data regarding the aspects of Vpr-induced T cell death.

BIOLOGY OF VPR

NUCLEAR ENTRY BY PREINTEGRATION COMPLEXES

An unsolved puzzle for many years is how HIV causes productive infection in macrophages when nuclear breakdown does not occur as part of ongoing cell division. In contrast to retrovirus, such as murine moloney leukemia virus, a unique feature of lentiviruses such as SIV and HIV is the ability to permit productive infection in cells that do not exhibit breakdown of the nuclear membrane. One of the first activities of Vpr was the recognition of its involvement in transport of the HIV-1 preintegration complex (PIC) to the nucleus.[33–35] To a large extent, this discovery helped clarify the essential role of Vpr for productive infection of macrophages,[36,37] because at the time, it was reported that Vpr was dispensable for viral replication in T cells. A number of studies identified features of Vpr that are related to HIV-1 nuclear translocation. One possibility may be that Vpr can enter the nucleus and localize to the nuclear envelope.[21] Alternatively, disruption of nuclear envelope integrity by dynamic herniations may allow entry of large PIC complexes.[38]

However, for most proteins targeted for nuclear import, the universal feature is the use of the classical nuclear localization signal (NLS). The NLS is recognized by a family of karyopherin proteins known as importins. Importin-α binds to the NLS, which, together, further interact with importin-β. The importin-α–β complex docks to a nuclear pore complex and is then shuttled across the nuclear pore via a number of energy-dependent steps. Because Vpr does not possess a conventional NLS, it was suggested that it binds directly to importin-α[21,35] and, therefore, imitates the action of importin-β.[39,40] This form of nuclear import mimickery was described in the transport of proteins that bypass the requirement for importin-α, using importin-β directly. The precise targeting sequence for this form of transport was not mapped, although sequences rich in arginine residues seem to be a common feature.[41] In addition, Vpr's access to the nucleus is not inhibited by traditional NLS blockers, consistent with the possibility that Vpr may use this alternative form of transport that is independent of the known importin pathway. Controversy surrounding this model has been proposed, because Vpr was shown to be small enough to passively diffuse across the nuclear pore complex.[42] Recent work also indicates that Vpr possesses a nuclear export signal (NES)[43] to allow exit of Vpr back into the cytoplasm for subsequent PIC incorporation. This property is thought to facilitate infection of macrophages and other nondividing cells. However, one group reported that Vpr was dispensable for *ex vivo* productive infection of resting T cells.[44] These confounding data indicated that the effect of PIC import may differ depending on the target cell type.

INDUCTION OF CELL-CYCLE ARREST

Control of cell-cycle progression is a complex process involved in growth, proliferation, development, DNA damage, and repair. Numerous regulatory proteins ensure that specific events are coordinated to guarantee genomic stability for the proper production of two daughter cells. The major checkpoint regulators in the G2–mitosis interface are the cyclin-dependent kinases, cdc2, and cyclin B1. Activation of cdc2–cyclin B1 permits progression from G2 to mitosis, whereas inactivation of the complex results in G2 cell-cycle arrest. In the inactive cdc2–cyclin B1 complex, cdc2 is phosphorylated on three residues, Thr14, Tyr15, and Thr161. The kinases responsible for Thr14 and Th15 phosphorylation are Wee1 and Myt1, respectively, and the cdk activation complex is responsible for phosphorylation at Thr161. When activation of cdc2–cyclin B1 is required, Thr14 and Tyr15 are dephosphorylated by the cdc25c phosphatase. Similar to yeast, the activities of Wee1 and cdc25c are also governed by their phosphorylation status, such that Wee1 phosphorylation renders it inactive and, conversely, cdc25c active. Cdc2–cyclin B1 kinase targets both Wee1 and cdc2, which provides a positive feedback amplification that reinforces cdc2 dephosphorylation and irreversible entry into mitosis (reviewed in Amini, Khalili, and Sawaya[45]).

Precisely how Vpr mediates cell-cycle arrest is under debate. Are the effects of Vpr epistatic to cell-cycle regulators, or are their effects secondary to a DNA damaging response? In overexpression experiments, a correlation was observed between the level of Vpr and the hyperphosphorylation state of cdc2 and hypophosphorylation state of cdc25c.[30,46] Another possible explanation of how Vpr induces G2 arrest comes from one study in which overexpression of Vpr caused marked alternations and breaks in the nuclear lamin structure. This loss of nuclear envelope integrity would conceivably disrupt the segregation of the cdc2 and cyclin B1 in the nucleus and cytoplasm. Physical stress associated with nuclear disruption may also trigger a DNA damage response that results from incomplete DNA synthesis.[47] Mutational analysis demonstrated that an intact C-terminal Vpr domain is essential to mediate G2 arrest, as truncations or point mutations in this domain fail to induce cell-cycle arrest.[7] A summary of Vpr's action on the nuclear function during the cell cycle is depicted in Figure 7.2.

MOLECULAR TARGETS OF VPR

Vpr has multiple activities and interacts with host machinery involved in these processes, such as cell-cycle regulators and proteins that control the apoptotic response. Here, we describe a set of potential interactors of Vpr and discuss the relevance of such interactions.

Early reports indicate that activation of HIV-1 long terminal repeat (LTR) or other heterologous viral promoters requires the activity of Vpr (reviewed in Cohen[48]), in association with the host transcription factor SP1. Although a direct interaction between Vpr and SP1 was not found, Vpr associates with SP1 responsive elements in coimmunoprecipitation and gel shift assays.[49] This association is mediated by the second -helix of Vpr. Other transcription factors that can associate with Vpr include TFIIB, as shown by GST pull-down assays[50] and by acting as an adaptor molecular to bridge the co-activators p300/cAMP-responsive element-binding (CREB) protein.[51] Both *in vitro* and *in vivo* binding assays confirm that the interaction of Vpr and p300 depends on the third -helix of Vpr.[51]

In a large-scale yeast two-hybrid screen with wild-type or mutant Vpr molecules as bait, several additional host proteins were found to interact with Vpr. Among these proteins was a kinase, hVIP/mov34, a homologue of yeast mov34 that belongs to a family of proteosomal and transcriptional regulators.[52] The binding of hVIP/mov34 requires the C-terminal portion of Vpr. However, the relevance of this interaction on cell-cycle arrest is not known. In another yeast two-hybrid screen, two groups, the human homologue of yeast rad23 and HHR23A, bound to Vpr through its C-terminal,[53–55] which when overexpressed suppressed the cell-cycle effect of Vpr.[55] However, using mutational analysis, other investigators did not observe this effect of HH23RA on binding and cell-cycle arrest.[56] Several other Vpr-interacting molecules were described. For example, IκB and STAT5β binds Vpr and may have consequences on immune activation and its ensuing humoral and innate responses.[56–58]

Using a mutant form of Vpr, L64Λ, as bait, several human 14-3-3 proteins were identified as potential Vpr-binding partners. The 14-3-3 proteins comprise a family of nine isotypes that bind to phosphorylated serine/threonine residues and regulate the activities of their target through subcellular

A

1. disruption of nuclear envelope
2. nuclear import (PIC)
3. Wee1 phosphorylation

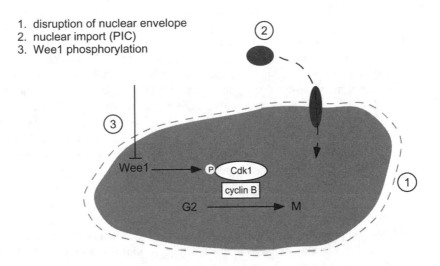

B

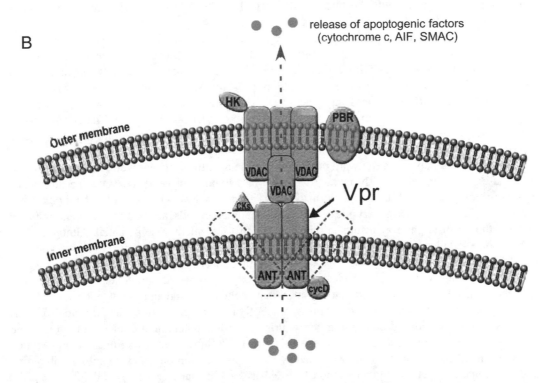

FIGURE 7.2 Major nuclear and mitochondrial targets of Vpr. (A) Vpr disrupts nuclear envelope structure by invoking dynamic herniations, allowing direct import of preintegration complexes and disturbing cellular localization of cell-cycle regulators. Vpr also targets nuclear pore complexes by mimicking importin proteins and bypassing the requirement for a nuclear localization signal. Wee1 kinase is regulated by its level of phosphorylation and its subsequent activity on cdc2. Expression of Vpr correlates with hyperphosphorylated Wee1 and G2 arrest. (B) Vpr induces cell death through its interaction with the adenine nucleotide translocator on the inner mitochondrial membrane. This leads to release of apoptogenic factors including cytochrome *c*, AIF, and SMAC.

localization and stability. The role of 14-3-3 proteins in modulating cell-cycle progression was well documented. They regulate the activity of Cdc25C activity[59–62] and bind to Wee1 kinases to increase stability. In addition, Chk1/2 kinases are activated by 14-3-3 by sequestering in the nucleus.[63] Conversely, 14-3-3 proteins inactivate phosphorylated cdc2 and cyclin B1 by exporting them into the cytoplasm.[64] The 14-3-3- attenuated Vpr induced cell-cycle arrest, and lack of the σ-isotype reduced the effect of Vpr. This effect was dependent on binding to the C-terminal portion of Vpr and correlates with the role of Vpr in modulating the activities of cell-cycle regulators. Therefore, the reported effect of Vpr on cell-cycle arrest may, in part, be the result of its association with 14-3-3.

In addition, Vpr was found to interact with a number of multifunctional cellular proteins. The molecular basis and biological consequences of these interactions were not clarified, but they offer new insight into the action of Vpr during HIV infection. The conservation of a proline-rich motif at residues 14 and 35 in the N-terminal of Vpr suggests the possibility of protein–protein interactions. This domain exhibits a high degree of *cis*-conformations of the imidic bond. One study suggested that a host peptidyl-prolyl isomerase (PPIase) regulates the interconversion of the imidic bond in the N-terminal of Vpr.[65] Cyclophilin A belongs to the PPIase family and is a highly abundant protein found to be associated with capsid proteins of HIV-1 virions. Previous work supported that cyclophilin A functions to support the formation of infectious virions.[66] This study demonstrated, in pull-down experiments, that Vpr was found to interact with cyclophilin A and, importantly, that cyclophilin A was required for the *de novo* expression of Vpr. This interplay was dependent on the N-terminal proline residues. In the absence of cyclophilin A, Vpr incorporation into virions is decreased, and cell-cycle arrest is impaired.

Vpr was reported to interact with the DNA repair enzyme, uracil DNA glycosylase (UNG). This interaction was identified in a yeast two-hybrid screen. UNG can be incorporated into viral particles, and this correlates with a reduced rate of reverse transcriptase mutation in the viral genome.[67,68] This may select for increased fitness of progeny. Another way in which Vpr is thought to disrupt the DNA/RNA processing machinery is through the lys-tRNA synthetase product, tRNA[lys]. tRNA[lys] functions as the primer for the reverse transcriptase enzyme. The N-terminal of Vpr is thought to interact with lys-tRNA synthetase and thereby disrupt initiation of reverse transcription of the viral genome.[69]

An important step forward toward understanding the apoptosis-inducing effects of Vpr was the observation that Vpr can directly interact with the mitochondria in yeast. However, because yeast lack core components of the apoptotic machinery, it was unclear what the role of mitochondria played during Vpr-induced eukaryotic cell death. In mammalian cells, an apoptotic response initiated through the intrinsic pathway results in mitochondria that lose membrane potential. Although there is still considerable debate as to whether the mitochondria permeability transition pore complex (PTPC) is in an open or closed conformation, the resulting release of cytochrome *c* induces a cascade of events leading to the eventual cleavage of effector caspases and subsequent cell death. The mitochondria PTPC is a multiprotein complex that includes the adenine nucleotide translocator (ANT) found on the inner mitochondria membrane. Synthetic Vpr peptides directly bind to ANT, and this correlates with a decrease in mitochondria membrane potential and release of cytochrome *c*, whereas Vpr mutant peptides do not have this effect.[11,70] Subsequent experiments demonstrated that the Vpr-induced apoptosis can be blocked by HIV-1 protease inhibitors, cyclophilin A inhibitors, and bonkregic acid, laying further support for ANT as a molecular target for Vpr.[11,70,71]

VPR MEDIATES CELL DEATH

APOPTOTIC VS. NECROTIC DEATH

Maintenance of proper cell number requires a balance between proliferation, production, and programmed cell death (PCD). Inhibition of apoptosis leads to autoimmune disease and development of malignancies. Conversely, enhanced apoptosis can cause diseases associated with neurodegeneration, ischemia–reperfusion damage, and immunodeficiency. The progressive decline in CD4[+] T cells is the central feature of HIV infection and was proposed to be the result of increased levels of PCD (reviewed in Badley, Pilon, Landay, and Lynch[72]). The mechanism by which HIV infection causes lymphocyte destruction continues to be a matter of great debate. The earliest studies examining the mode of T cell death proposed that apoptosis was the predominant pathway involved. However, a recent study indicates that T cell destruction and the pathogenesis of HIV may involve an alternative form of necrotic cell death that does not rely on caspase activity.[73,74] A large part of the disparate data stems from the highly divergent tissues, cells, models, and assays used to analyze PCD. For example, HIV replication was shown to lead to rapid death of peripherally infected T cells, thereby resulting in a constant turnover of lymphocytes.[75,76] In contrast, analysis of lymph nodes showed that cell death occurs mainly in uninfected cells.[77]

Almost all the HIV-1 proteins were implicated in modulating the apoptotic response of T cells. However, genetic analyses that directly correlate with the apoptosis-inducing activities of these proteins are lacking and raise significant concern as to the physiological involvement of these proteins as mediators of T cell death. Vpr is not a unique protein in this regard, although an emerging pool of literature has documented naturally occurring mutations in Vpr that directly abrogate cell death both *in vitro* and *in vivo*. There are also conflicting reports that suggest that Vpr may serve as both an antiapoptotic and a proapoptotic regulator of T cell death. In this section, we will discuss the current models of apoptotic and nonapoptotic modes of Vpr-induced T cell death and new modes for inhibiting the pathophysiological effects of Vpr.

Early studies on Vpr function initially found that expression of Vpr-induced myocytes to differentiate[78] and led to reports demonstrating that the Vpr's action was mediated by the glucocorticoid receptor complex.[79,80] Given that glucocorticoids inhibit the action of NF-κB, Vpr was proposed to modify cellular functions by inducing a glucocorticoid effect, perhaps through modulation of T cell death or survival. Answers to how Vpr induced cell death came from studies transfecting Vpr into cells. In these cells, Vpr potently suppressed immune activation through upregulation of IκBα, which inhibited NF-κB transcriptional activity and resulted in the induction of cell death.[81] These initial results led several investigators to examine details of Vpr-induced cell death. As noted previously, Vpr induces cell-cycle arrest at the G2 stage of the cell cycle and subsequently results in apoptosis of a number of human cells, including fibroblasts, T cell lines, and primary cells.[9–11,82–85] These observations were universally validated in a number of different systems. For example, in one study, adenovirus was used to introduce Vpr into HeLa cells in the absence of other HIV gene products.[10] Adenovirus infection with Vpr led to G2/M cell-cycle arrest within 24 h after infection, and staining with annexin V revealed significant apoptosis with increased activation of caspase-9. Moreover, mice expressing a transgene for Vpr on a CD4 promoter exhibited enhanced apoptotic cell death with a marked decrease in T cells of the lymphoid organs and peripheral blood.[86]

The major components of apoptosis initiated via the intrinsic mitochondrial pathway were largely clarified in recent years (reviewed in Green and Reed[87] and Green and Kroemer[88]). A more detailed review of this pathway can be found in Chapter 2. Execution of cell death through this route occurs by causing the translocation and insertion of Bax or Bak into the mitochondria membrane, loss in membrane potential, and release of cytochrome *c* into the cytosol. After being released, cytochrome *c* binds to form a complex consisting of procaspase-9 and apoptotic activating factor 1 (Apaf-1). The formation of this multimeric complex results in cleavage of procaspase-9 to an active caspase-9. This pathway requires the continued presence of adenosine triphosphate(ATP) in order to carry out these energetic dependent reactions (reviewed in Green and Reed[87] and Green and Kroemer[88]).

Using synthetic Vpr peptides, it was demonstrated that a major target of Vpr is the ANT, a component of the inner mitochondrial permeability transition pore complex.[70] Vpr binding to ANT causes loss of membrane potential, release of cytochrome *c*, and activation of caspase-3 and caspase-9.[9,10,70,85,89] The ability of Vpr to induce cell death seems to be dependent on the C-terminal H(S/F)RIG motif, as specific mutations in this region abrogate Vpr's ability to bind ANT and also reduce the capacity to induce cell death.[11,70,89]

Given the important role of the mitochondria in the production of ATP, the nucleotide pool within the cell provides an indirect measure of cellular bioenergetics. NMR analysis of α, β, and γ^{31}P in HeLa cells infected with adenovirus containing Vpr demonstrated a lower density of α^{31}P than that of β^{31}P. When this analysis was used to compare β^{31}P in necrotic cells vs. adenovirus Vpr-infected cells, necrotic cells had no detectable β^{31}P, whereas adenoviral-infected cells contained significant levels (reviewed in Muthumani et al.[90]). Because the apoptotic cascade is an energy-dependent process and necrosis is associated with energy collapse, this result is consistent with Vpr-infected cells maintaining active metabolism and cytoplasmic integrity and providing further evidence that apoptosis is the main pathway in which Vpr executes cell death.

The mechanism of HIV-1 induced death was called into question based on studies demonstrating that infection of Jurkat and H9 T cell lines failed to exhibit characteristics of classical apoptosis.[73] This included assays that measured caspase activity and activation, DNA fragmentation, annexin V exposure, and loss of mitochondrial membrane potential. Cells overexpressing the antiapoptotic proteins Bcl-xL or v-FLIP failed to prevent death. Furthermore, no difference in cytopathogenicity to HIV-1 infection was observed in Jurkat cells deficient for RIP, caspase-8, and FADD, or when caspase inhibitors were used. Therefore, the authors concluded that a caspase and a mitochondrial-independent mechanism were responsible for the observed cell death. In a concurrent article, the same authors used electron microscopy to distinguish apoptotic vs. necrotic cell death based on cell morphology. In this case, VSV-G pseudotype viruses were engineered to contain various HIV-1 gene products and were then used to infected T cell lines. In line with the notion that nonapoptotic pathways are induced by HIV-1, the authors found that Env was not required for HIV-1-induced T cell death.[74] The C-terminal of Vpr was found to induce cell death even in the absence of caspase-9, AIF, or Apaf-1.[83] Also the overexpression of Bcl-2 does not protect against Vpr-induced death, lending support to the notion that T cell death may occur through multiple pathways.

There were also several reports that Vpr may function as an antiapoptotic protein. HepG2 cells undergo apoptosis after treatment with sorbitol, but cells expressing Vpr were largely protected against these effects.[91] In another study, low levels of Vpr stably expressed in Jurkat cells were shown to have an impaired capacity to undergo apoptosis by cycloheximide, TNF-α, anti-Fas, or serum starvation. Furthermore, cells transfected with antisense Vpr oligonucleotides were more sensitive to these apoptotic stimuli.[92] In a later study, the same group found that the protective effect of Vpr could be recapitulated even when cells were infected *in vitro* with HIV-1, although this was observed only during early time points after infection.[93] At later time points, these infected cells ultimately underwent apoptosis. Further evidence suggesting that Vpr prevents cell death comes from a study examining the effects of Jurkat cells expressing Vpr on cell adhesion properties.[94] Both cell–cell and cell–matrix interactions are important for regulating locomotion, differentiation, homing, growth, and cell death. However, lack of matrix attachment can induce apoptosis through a process termed "anoikis." Along these lines, constitutive expression of Vpr in Jurkat cells exhibits features distinct from mock infected cells. In particular, there was a dramatic increase in polymerized F-actin and improved rearrangement of filamentous stress fibers. In addition, surface expression of α5 and α6 integrins as well as cadherins was significantly increased compared with controls. Stimulation of cells with Fas or cycloheximide/TNFα triggered a lower degree of apoptosis in Vpr-expression cells. Therefore, enhanced cell–matrix, adhesion, and spreading of Jurkat cells by Vpr favor increased integrin signaling associated with inhibition of apoptosis and cell survival.[94]

Based on this evidence, Vpr may have a dual role in modulating the apoptotic response that is dependent on several factors, including the time of infection and the level of Vpr. It may be possible that low levels of Vpr early in infection results in protection from cell death that facilitates maximal viral replication, persistence, and viral survival. However, ultimately, at later time points, high levels of Vpr mediate the death of the host cell to support viral spreading and dissemination.

Vpr Polymorphisms

Given the multitude of functions attributed to Vpr, numerous studies have attempted to identify naturally occurring polymorphisms within Vpr that are associated with enhanced or attenuated Vpr function. Studies in macaques, chimpanzees, and human subjects underscore the important contribution of Vpr to disease progression. For example, macaques infected with Vpr-defective SIV strains do not exhibit progression to AIDS or have severely attenuated disease.[95-97] In a human HIV-1 infected patient and chimpanzees infected with nonfunctional Vpr, slower disease progression was observed.[98] Therefore, it is reasonable to suggest that disease progression could be attributed to Vpr function. A unique group of patients collectively referred to as long-term nonprogressors (LTNPs) were identified as being HIV-1 infected, with high viral titers but low levels of CD4+ T cell death. Importantly, these patients did not progress to AIDS, although some reports suggested that several of these patients ultimately showed disease progression. A number of these patient cohorts exist and were studied extensively to isolate both host and viral genetic aberrations that may account for their favorable clinical outcome. Many Vpr mutations were documented that interfere with cell-cycle arrest, decreased phosphorylation, and nuclear localization (Figure 7.3). A major portion of Vpr's cytotoxicity is found clustered within the C-terminal, consisting of a dodecapeptide domain (residues 71 to 82). Several studies have demonstrated that polymorphisms within this arginine-rich C-terminal are frequently observed in LTNPs.[13,99,100] In these cases, such mutations abolish Vpr function on mitochondria-induced apoptosis, cell-cycle arrest, and nuclear localization, and they may explain decreased pathogenicity of viruses harboring these alleles. One study found a unique polymorphism, Q3R, in a LTNP. This mutation resulted in a premature stop codon that correlated with reduced ability to induce T cell apoptosis.[101] A more direct role of Vpr in disease progression was demonstrated in a molecular analysis of an HIV-1-infected mother–child LTNP pair. Clustering of amino acid substitutions in the cytotoxic domain of Vpr correlated with

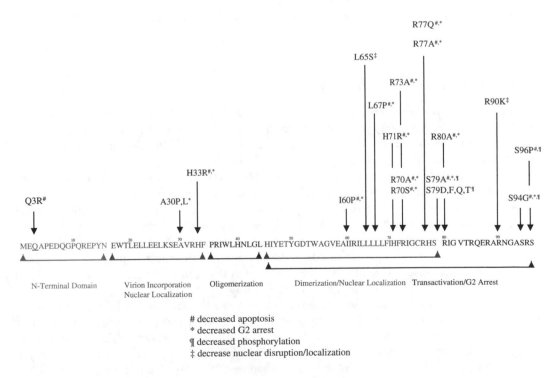

FIGURE 7.3 Mutations in Vpr domain affect its cellular activity.

high T cell count.[13] However, reversion to wild-type Vpr resulted in a progressive decline in T cell numbers, strongly indicating that Vpr is an important determinant in viral pathogenesis and disease progression. In another study, arginine substitution at position 77 to glutamine was found at high frequency among LTNPs.[89] This mutation was also observed in a larger sample of LTNP patients identified through the Los Alamos database and was associated with the reduced ability of Vpr to induce T cell death *in vitro* and *in vivo*. The incidence of this mutation was independently confirmed in another report.[102] However, conflicting data were recently published as part of a larger clinical study examining the clinical outcomes of patients treated with a single- vs. a double-nucleoside reverse-transcriptase inhibitor.[103] In this report, the authors found no correlation between the R77Q mutation and disease severity, progression, or absolute CD4+ T cell counts. Therefore, the role of Vpr mutations, specifically R77Q, was questioned as a predictor of clinical outcome. Several explanations may account for the contrasting observations. First, this study was designed to assess whether HAART masks any benefit toward disease progression that may have otherwise resulted from C-terminal mutations within the Vpr, whereas the original report studied only drug-naive HIV-1-infected patients with defined nondisease progression. Second, other viral or host factors that were suggested to play unique roles in disease progression were not defined in the latter study, whereas nef was found to be intact in all enrolled individuals, and no patient that was homozygous or heterozygous for the Δ32bp mutation in the chemokine co-receptor CCR5 was included in the former study. Third, viral load information was not available in the second study and may account for the lack of protective effects of the R77Q mutation. In addition, without viral load data, it would not be possible to speculate severity of pre-existing disease before treatment. Despite these discrepancies, 41% of the drug naive patients had viruses containing the R77Q mutation in Vpr, a value similar to that found for nearly 300 LTNPs in the Los Alamos database but lower than the 80% cited in the initial report. It is difficult to extract from the data in the second study whether patients who experienced an increase in CD4+ T cell counts were also drug naive. Perhaps the most compelling evidence will require the isolation of clinical viruses from both cohorts of patients so that careful *in vitro* analysis can be performed to assess the virulence of these mutant Vpr-containing isolates.

NOVEL STRATEGIES TO INHIBIT VPR FUNCTION

The multifunctional role of Vpr in the viral life cycle and its interaction with a number of host targets allow for the possibility of designing specific inhibitors to block Vpr function. Several line studies have demonstrated robust inhibition of Vpr activity in *in vitro* systems, and it will be important to validate these potential agents in animal models.

Cyclophilin blockers such as cyclosporine A and NIM811 were shown to have potent antiretroviral activity. For example, NIM811 interferes with nuclear localization of preintegration complexes. Cyclosporine was shown to block expression of interleukin-2, thereby exhibiting indirect antiviral activity through reducing T cell activation and release of infectious virions.[104] In rhesus monkeys acutely infected with SIV, cyclosporine A showed some benefit in preventing T cell loss,[105] but human trials in advanced stages of AIDS showed only a mild effect in patients treated with cyclosporine A. Because cyclophilin A inhibitors are less toxic, these drugs need to be evaluated clinically for and in combination with current HIV therapy.

Vpr-mediated cell-cycle arrest was observed in fission yeast systems and acts through a pathway involving PP2A, Wee1, and cdc25c.[29,106] In addition, Vpr experimental evidence demonstrates that Vpr induces cell killing of budding yeast through its effect on the mitochondrial transition pore complex.[70] Therefore, an elegant genetic system using the budding yeast *S. cerevisiae* was employed to screen for GST-fusion hexamic peptides, which suppressed the growth arrest phenotype of HIV-1 Vpr.[107] In this system, yeast expression plasmids were generated, encoding Vpr under the control of galactose-inducible GAL1 promoter. When grown in Vpr-inducible medium, cells transformed with wild-type Vpr underwent growth arrest compared with control. When a hexamic peptide library consisting of 1×10^8 GST peptides was transformed into yeast, 15 clones were found to exhibit strong growth defects. Sequence analysis revealed that all GST-inhibiting peptides contained a conserved diW motif (WXXF) and a stretch of hydrophobic residues at their C-terminals. This WXXF motif was previously shown to interact with Vpr,[108] but the basis of this interaction is not known. Furthermore, these peptides interfered with the ability of VSV-G pseudotype viruses to induce G2 arrest and apoptosis. It was concluded that these diW-containing peptides may be capable of preventing protein–protein interactions between Vpr and host targets or may act as dissociative inhibitors. Along similar lines, other investigators examined the effects of dominant negative Vpr mutants on HIV-1 replication. Mutation of arginine residue 73 to serine in Vpr reduced the transactivation and G2 arrest by wild-type Vpr.[109,110] However, the ability of this dominant negative R73S to block apoptosis was not investigated. Other approaches were to use antagonists of the glucocorticoid receptor (RU486) and pentoxifylline.[79]

Using another similar yeast genetic approach, one group screened for suppressors of G2 arrest (reviewed in Bukrinsky and Zhao[111]). Out of more than 2.0×10^4 transformants, two small heat-shock proteins, Hsp27 and Hsp70, were identified to inhibit G2 arrest by Vpr. When transfected into 293T cells, Hsp27 strongly suppressed G2 arrest induced by VSV-G pseudotype HIV-1 infection, whereas HSP70 had similar suppressive properties. The ability of HIV-1 to replicate in nonproliferating macrophages is dependent on the function of Vpr. It is believed that Vpr assists in the translocation of viral PICs via the nuclear envelope through interactions with a number of host proteins, including nucleoporins, importins, and heat-shock proteins. Hsp70 was found to stimulate nuclear import in a cell-free system.[112] More recently, Hsp70 expression was induced following HIV-1 infection of macrophages.[113] Interestingly, in the absence of Vpr, Hsp70 was able to stimulate import and replication in macrophages. However, Hsp70 reduced replication and reversed G2 arrest in Vpr sufficient viruses. The authors concluded that Hsp70 and Vpr act in parallel pathways when expressed separately but inhibit each other's activities when dually expressed. These results may provide an opportunity to modulate Hsp70 function during HIV-1 infection.

Recent work showed that the use of RNAi technology may have potential in inhibiting HIV-1 replication and other viral functions. Small interfering RNAs (siRNAs) are RNA duplexes of 21 to 23 nucleotides that induce specific degradation of a homologous mRNA that results in gene silencing.[114–116] One of the earliest findings that HIV-1 infection could be inhibited was the discovery that homozymous mutations in the chemokine co-receptors CCR5 conferred protection against HIV-1 infection.[117] At least one other study showed that knockdown of CCR5 provided strong protection against infection *in vitro*,[118] another potential target of RNAi during the viral uncoating and entry process of the target cell.

Another approach was to develop a system to deliver exogenous siRNAs through endogenous expression via DNA plasmids. This method would bypass the need to produce siRNA in cells, which have low expression of Dicer.[119,120] Still, the requirement for stem–loop structure processing by Dicer or another endonuclease was problematic. Recent work showed that these stem–loop structures can be processed in a manner similar to micro-RNA precursors, and anti-HIV siRNAs processing from micro-RNA precursors can occur efficiently.[121] Both HIV-1 Rev and Tat can be inhibited by these approaches in both transient transfection or from a lentiviral transduction of hematopoietic progenitor cells.[122,123] In these studies, cotransfection of siRNA directed against Rev and Tat reduced the level of HIV-1 NL4-3 proviral DNA and inhibition of p24 expression by 4 logs[123] in human T cell lines and primary lymphocytes.[124] In similar work, silencing of CD4 blocked viral entry and syncytia formation.[125] Given the proof of concept in these studies, it will be of interest to determine whether HIV-1 Vpr may be a potential target for siRNA silencing. Conceivably, the knockdown of Vpr expression would prevent infection of nondividing cells as well as abrogate its cell-cycle-inducing functions. Also, decreased Vpr expression may mimic naturally occurring mutations in Vpr, whereby these alleles have reduced capacity to induce cell death of target cells.

The advances in siRNA technology and continued improvement in its efficiency provide a unique opportunity to exploit delivery of anti-HIV treatments, including those that target Vpr. Loss-of-function Vpr studies indicate that inhibition of Vpr activity on cell cycle, apoptosis, or nuclear integration may be an alternative strategy to curb the deleterious effects of HIV-1 infection in humans.

REFERENCES

1. Cohen, E.A., Dehni, G., Sodroski, J.G., and Haseltine, W.A., Human immunodeficiency virus Vpr product is a virion-associated regulatory protein, *J. Virol.,* 64(6), 3097–3099, 1990.
2. Zhao, Y. and Elder, R.T., Yeast perspectives on HIV-1 VPR, *Front Biosci.,* 5(1), D905–D916, 2000.
3. Morellet, N., Bouaziz, S., Petitjean, P., and Roques, B.P., NMR structure of the HIV-1 regulatory protein VPR, *J. Mol. Biol.,* 327(1), 215–227, 2003.
4. Engler, A., Stangler, T., and Willbold, D., Structure of human immunodeficiency virus type 1 Vpr(34–51) peptide in micelle containing aqueous solution, *Eur. J. Biochem.,* 269(13), 3264–3269, 2002.
5. Luo, Z., Butcher, D.J., Murali, R., Srinivasan, A., and Huang, Z., Structural studies of synthetic peptide fragments derived from the HIV-1 Vpr protein, *Biochem. Biophys. Res. Commun.,* 244(3), 732–736, 1998.
6. Wecker, K., Morellet, N., Bouaziz, S., and Roques, B.P., NMR structure of the HIV-1 regulatory protein Vpr in H2O/trifluoroethanol. Comparison with the Vpr N-terminal (1–51) and C-terminal (52–96) domains, *Eur. J. Biochem.,* 269(15), 3779–3788, 2002.
7. Zhou, Y., Lu, Y., and Ratner, L., Arginine residues in the C-terminus of HIV-1 Vpr are important for nuclear localization and cell cycle arrest, *Virology,* 242(2), 414–424, 1998.
8. Zhou, Y. and Ratner, L., Phosphorylation of human immunodeficiency virus type 1 Vpr regulates cell cycle arrest, *J. Virol.,* 74(14), 6520–6527, 2000.
9. Muthumani, K., Hwang, D.S., Desai, B.M., Zhang, D., Dayes, N., Green, D.R., and Weiner, D.B., HIV-1 Vpr induces apoptosis through caspase 9 in T cells and peripheral blood mononuclear cells, *J. Biol. Chem.,* 277(40), 37820–37831, 2002.
10. Muthumani, K., Zhang, D., Hwang, D.S., Kudchodkar, S., Dayes, N.S., Desai, B.M., Malik, A.S., Yang, J.S., Chattergoon, M.A., Maguire, H.C., Jr., and Weiner, D.B., Adenovirus encoding HIV-1 Vpr activates caspase 9 and induces apoptotic cell death in both p53 positive and negative human tumor cell lines, *Oncogene,* 21(30), 4613–4625, 2002.
11. Jacotot, E., Ravagnan, L., Loeffler, M., Ferri, K.F., Vieira, H.L., Zamzami, N., Costantini, P., Druillennec, S., Hoebeke, J., Briand, J.P., Irinopoulou, T., Daugas, E., Susin, S.A., Cointe, D., Xie, Z.H., Reed, J.C., Roques, B.P., and Kroemer, G., The HIV-1 viral protein R induces apoptosis via a direct effect on the mitochondrial permeability transition pore, *J. Exp. Med.,* 191(1), 33–46, 2000.
12. Wecker, K. and Roques, B.P., NMR structure of the (1–51) N-terminal domain of the HIV-1 regulatory protein Vpr, *Eur. J. Biochem.,* 266(2), 359–369, 1999.
13. Wang, B., Ge, Y.C., Palasanthiran, P., Xiang, S.H., Ziegler, J., Dwyer, D.E., Randle, C., Dowton, D., Cunningham, A., and Saksena, N.K., Gene defects clustered at the C-terminus of the Vpr gene of HIV-1 in long-term nonprogressing mother and child pair: *in vivo* evolution of Vpr quasispecies in blood and plasma, *Virology,* 223(1), 224–232, 1996.
14. Tungaturthi, P.K., Sawaya, B.E., Ayyavoo, V., Murali, R., and Srinivasan, A., HIV-1 Vpr: genetic diversity and functional features from the perspective of structure, *DNA Cell Biol.,* 23(4), 207–222, 2004.
15. Bukrinsky, M. and Adzhubei, A., Viral protein R of HIV-1, *Rev. Med. Virol.,* 9(1), 39–49, 1999.
16. Singh, S.P., Tomkowicz, B., Lai, D., Cartas, M., Mahalingam, S., Kalyanaraman, V.S., Murali, R., and Srinivasan, A., Functional role of residues corresponding to helical domain II (amino acids 35 to 46) of human immunodeficiency virus type 1 Vpr, *J. Virol.,* 74(22), 10650–10657, 2000.
17. Mahalingam, S., Khan, S.A., Murali, R., Jabbar, M.A., Monken, C.E., Collman, R.G., and Srinivasan, A., Mutagenesis of the putative -helical domain of the Vpr protein of human immunodeficiency virus type 1: effect on stability and virion incorporation, *Proc. Natl. Acad. Sci. U.S.A.,* 92(9), 3794–3798, 1995.

18. Mahalingam, S., Khan, S.A., Jabbar, M.A., Monken, C.E., Collman, R.G., and Srinivasan, A., Identification of residues in the N-terminal acidic domain of HIV-1 Vpr essential for virion incorporation, *Virology,* 207(1), 297–302, 1995.

19. Mahalingam, S., Ayyavoo, V., Patel, M., Kieber-Emmons, T., and Weiner, D.B., Nuclear import, virion incorporation, and cell cycle arrest/differentiation are mediated by distinct functional domains of human immunodeficiency virus type 1 Vpr, *J. Virol.,* 71(9), 6339–6347, 1997.

20. Chen, M., Elder, R.T., Yu, M., O'Gorman, M.G., Selig, L., Benarous, R., Yamamoto, A., and Zhao, Y., Mutational analysis of Vpr-induced G2 arrest, nuclear localization, and cell death in fission yeast, *J. Virol.,* 73(4), 3236–3245, 1999.

21. Vodicka, M.A., Koepp, D.M., Silver, P.A., and Emerman, M., HIV-1 Vpr interacts with the nuclear transport pathway to promote macrophage infection, *Genes Dev.,* 12(2), 175–185, 1998.

22. Nishizawa, M., Kamata, M., Katsumata, R., and Aida, Y., A carboxy-terminally truncated form of the human immunodeficiency virus type 1 Vpr protein induces apoptosis via G(1) cell cycle arrest, *J. Virol.,* 74(13), 6058–6067, 2000.

23. Macreadie, I.G., Arunagiri, C.K., Hewish, D.R., White, J.F., and Azad, A.A., Extracellular addition of a domain of HIV-1 Vpr containing the amino acid sequence motif H(S/F)RIG causes cell membrane permeabilization and death, *Mol. Microbiol.,* 19(6), 1185–1192, 1996.

24. Macreadie, I.G., Castelli, L.A., Hewish, D.R., Kirkpatrick, A., Ward, A.C., and Azad, A.A., A domain of human immunodeficiency virus type 1 Vpr containing repeated H(S/F)RIG amino acid motifs causes cell growth arrest and structural defects, *Proc. Natl. Acad. Sci. U.S.A.,* 92(7), 2770–2774, 1995.

25. Agostini, I., Popov, S., Hao, T., Li, J.H., Dubrovsky, L., Chaika, O., Chaika, N., Lewis, R., and Bukrinsky, M., Phosphorylation of Vpr regulates HIV type 1 nuclear import and macrophage infection, *AIDS Res. Hum. Retroviruses,* 18(4), 283–288, 2002.

26. Lopez-Girona, A., Furnari, B., Mondesert, O., and Russell, P., Nuclear localization of Cdc25 is regulated by DNA damage and a 14-3-3 protein, *Nature,* 397(6715), 172–175, 1999.

27. Krek, W. and Nigg, E.A., Differential phosphorylation of vertebrate p34cdc2 kinase at the G1/S and G2/M transitions of the cell cycle: identification of major phosphorylation sites, *EMBO J.,* 10(2), 305–316, 1991.

28. Zhao, Y., Cao, J., O'Gorman, M.R., Yu, M., and Yogev, R., Effect of human immunodeficiency virus type 1 protein R (Vpr) gene expression on basic cellular function of fission yeast *Schizosaccharomyces pombe, J. Virol.,* 70(9), 5821–5826, 1996.

29. Elder, R.T., Yu, M., Chen, M., Edelson, S., and Zhao, Y., Cell cycle G2 arrest induced by HIV-1 Vpr in fission yeast (*Schizosaccharomyces pombe*) is independent of cell death and early genes in the DNA damage checkpoint, *Virus Res.,* 68(2), 161–173, 2000.

30. He, J., Choe, S., Walker, R., Di Marzio, P., Morgan, D.O., and Landau, N.R., Human immunodeficiency virus type 1 viral protein R (Vpr) arrests cells in the G2 phase of the cell cycle by inhibiting p34cdc2 activity, *J. Virol.,* 69(11), 6705–6711, 1995.

31. Furnari, B., Rhind, N., and Russell, P., Cdc25 mitotic inducer targeted by Chk1 DNA damage checkpoint kinase, *Science,* 277(5331), 1495–1497, 1997.

32. Masuda, M., Nagai, Y., Oshima, N., Tanaka, K., Murakami, H., Igarashi, H., and Okayama, H., Genetic studies with the fission yeast *Schizosaccharomyces pombe* suggest involvement of Wee1, Ppa2, and Rad24 in induction of cell cycle arrest by human immunodeficiency virus type 1 Vpr, *J. Virol.,* 74(6), 2636–2646, 2000.

33. Heinzinger, N.K., Bukinsky, M.I., Haggerty, S.A., Ragland, A.M., Kewalramani, V., Lee, M.A., Gendelman, H.E., Ratner, L., Stevenson, M., and Emerman, M., The Vpr protein of human immunodeficiency virus type 1 influences nuclear localization of viral nucleic acids in nondividing host cells, *Proc. Natl. Acad. Sci. U.S.A.,* 91(15), 7311–7315, 1994.

34. Popov, S., Rexach, M., Ratner, L., Blobel, G., and Bukrinsky, M., Viral protein R regulates docking of the HIV-1 preintegration complex to the nuclear pore complex, *J. Biol. Chem.,* 273(21), 13347–13352, 1998.

35. Popov, S., Rexach, M., Zybarth, G., Reiling, N., Lee, M.A., Ratner, L., Lane, C.M., Moore, M.S., Blobel, G., and Bukrinsky, M., Viral protein R regulates nuclear import of the HIV-1 pre-integration complex, *EMBO J.,* 17(4), 909–917, 1998.

36. Subbramanian, R.A., Kessous-Elbaz, A., Lodge, R., Forget, J., Yao, X.J., Bergeron, D., and Cohen, E.A., Human immunodeficiency virus type 1 Vpr is a positive regulator of viral transcription and infectivity in primary human macrophages, *J. Exp. Med.,* 187(7), 1103–1111, 1998.

37. Emerman, M., HIV-1, Vpr and the cell cycle, *Curr. Biol.,* 6(9), 1096–1103, 1996.
38. de Noronha, C.M., Sherman, M.P., Lin, H.W., Cavrois, M.V., Moir, R.D., Goldman, R.D., and Greene, W.C., Dynamic disruptions in nuclear envelope architecture and integrity induced by HIV-1 Vpr, *Science,* 294(5544), 1105–1108, 2001.
39. Gallay, P., Stitt, V., Mundy, C., Oettinger, M., and Trono, D., Role of the karyopherin pathway in human immunodeficiency virus type 1 nuclear import, *J. Virol.,* 70(2), 1027–1032, 1996.
40. Gallay, P., Hope, T., Chin, D., and Trono, D., HIV-1 infection of nondividing cells through the recognition of integrase by the importin/karyopherin pathway, *Proc. Natl. Acad. Sci. U.S.A.,* 94(18), 9825–9830, 1997.
41. Moore, J.D., Yang, J., Truant, R., and Kornbluth, S., Nuclear import of Cdk/cyclin complexes: identification of distinct mechanisms for import of Cdk2/cyclin E and Cdc2/cyclin B1, *J. Cell Biol.,* 144(2), 213–224, 1999.
42. Lu, Y.L., Spearman, P., and Ratner, L., Human immunodeficiency virus type 1 viral protein R localization in infected cells and virions, *J. Virol.,* 67(11), 6542–6550, 1993.
43. Sherman, M.P., de Noronha, C.M., Eckstein, L.A., Hataye, J., Mundt, P., Williams, S.A., Neidleman, J.A., Goldsmith, M.A., and Greene, W.C., Nuclear export of Vpr is required for efficient replication of human immunodeficiency virus type 1 in tissue macrophages, *J. Virol.,* 77(13), 7582–7589, 2003.
44. Eckstein, D.A., Sherman, M.P., Penn, M.L., Chin, P.S., De Noronha, C.M., Greene, W.C., and Goldsmith, M.A., HIV-1 Vpr enhances viral burden by facilitating infection of tissue macrophages but not nondividing CD4+ T cells, *J. Exp. Med.,* 194(10), 1407–1419, 2001.
45. Amini, S., Khalili, K., and Sawaya, B.E., Effect of HIV-1 Vpr on cell cycle regulators, *DNA Cell Biol.,* 23(4), 249–260, 2004.
46. Jowett, J.B., Planelles, V., Poon, B., Shah, N.P., Chen, M.L., and Chen, I.S., The human immunodeficiency virus type 1 Vpr gene arrests infected T cells in the G2 + M phase of the cell cycle, *J. Virol.,* 69(10), 6304–6313, 1995.
47. Moir, R.D., Spann, T.P., Herrmann, H., and Goldman, R.D., Disruption of nuclear lamin organization blocks the elongation phase of DNA replication, *J. Cell Biol.,* 149(6), 1179–1192, 2000.
48. Cohen, J., New role for HIV: a vehicle for moving genes into cells, *Science,* 272(5259), 195. 1996.
49. Wang, L., Mukherjee, S., Jia, F., Narayan, O., and Zhao, L.J., Interaction of virion protein Vpr of human immunodeficiency virus type 1 with cellular transcription factor Sp1 and trans-activation of viral long terminal repeat, *J. Biol. Chem.,* 270(43), 25564–25569, 1995.
50. Agostini, I., Navarro, J.M., Rey, F., Bouhamdan, M., Spire, B., Vigne, R., and Sire, J., The human immunodeficiency virus type 1 Vpr transactivator: cooperation with promoter-bound activator domains and binding to TFIIB, *J. Mol. Biol.,* 261(5), 599–606, 1996.
51. Kino, T., Gragerov, A., Slobodskaya, O., Tsopanomichalou, M., Chrousos, G.P., and Pavlakis, G.N., Human immunodeficiency virus type 1 (HIV-1) accessory protein Vpr induces transcription of the HIV-1 and glucocorticoid-responsive promoters by binding directly to p300/CBP coactivators, *J. Virol.,* 761(19), 9724–9734, 2002.
52. Mahalingam, S., Ayyavoo, V., Patel, M., Kieber-Emmons, T., Kao, G.D., Muschel, R.J., and Weiner, D.B., HIV-1 Vpr interacts with a human 34-kDa mov34 homologue, a cellular factor linked to the G2/M phase transition of the mammalian cell cycle, *Proc. Natl. Acad. Sci. U.S.A.,* 95(7), 3419–3424, 1998.
53. Withers-Ward, E.S., Jowett, J.B., Stewart, S.A., Xie, Y.M., Garfinkel, A., Shibagaki, Y., Chow, S.A., Shah, N., Hanaoka, F., Sawitz, D.G., Armstrong, R.W., Souza, L.M., and Chen, I.S., Human immunodeficiency virus type 1 Vpr interacts with HHR23A, a cellular protein implicated in nucleotide excision DNA repair, *J. Virol.,* 71(12), 9732–9742, 1997.
54. Dieckmann, T., Withers-Ward, E.S., Jarosinski, M.A., Liu, C.F., Chen, I.S., and Feigon, J., Structure of a human DNA repair protein UBA domain that interacts with HIV-1 Vpr, *Nat. Struct. Biol.,* 5(12), 1042–1047, 1998.
55. Gragerov, A., Kino, T., Ilyina-Gragerova, G., Chrousos, G.P., and Pavlakis, G.N., HHR23A, the human homologue of the yeast repair protein RAD23, interacts specifically with Vpr protein and prevents cell cycle arrest but not the transcriptional effects of Vpr, *Virology,* 245(2), 323–330, 1998.
56. Mansky, L.M., Preveral, S., Le Rouzic, E., Bernard, L.C., Selig, L., Depienne, C., Benarous, R., and Benichou, S., Interaction of human immunodeficiency virus type 1 Vpr with the HHR23A DNA repair protein does not correlate with multiple biological functions of Vpr, *Virology,* 282(1), 176–185, 2001.

57. Roux, P., Alfieri, C., Hrimech, M., Cohen, E.A., and Tanner, J.E., Activation of transcription factors NF-B and NF-IL-6 by human immunodeficiency virus type 1 protein R (Vpr) induces interleukin-8 expression, *J. Virol.,* 74(10), 4658–4665, 2000.

58. Muthumani, K., Hwang, D.S., Dayes, N.S., Kim, J.J., and Weiner, D.B., The HIV-1 accessory gene Vpr can inhibit antigen-specific immune function, *DNA Cell Biol.,* 21(9), 689–695, 2002.

59. Peng, C.Y., Graves, P.R., Thoma, R.S., Wu, Z., Shaw, A.S., and Piwnica-Worms, H., Mitotic and G2 checkpoint control: regulation of 14-3-3 protein binding by phosphorylation of Cdc25C on serine-216, *Science,* 277(5331), 1501–1505, 1997.

60. Sanchez, Y., Wong, C., Thoma, R.S., Richman, R., Wu, Z., Piwnica-Worms, H., and Elledge, S.J., Conservation of the Chk1 checkpoint pathway in mammals: linkage of DNA damage to Cdk regulation through Cdc25, *Science,* 277(5331), 1497–1501, 1997.

61. Zeng, Y., Forbes, K.C., Wu, Z., Moreno, S., Piwnica-Worms, H., and Enoch, T., Replication checkpoint requires phosphorylation of the phosphatase Cdc25 by Cds1 or Chk1, *Nature,* 395(6701), 507–510, 1998.

62. Graves, P.R., Lovly, C.M., Uy, G.L., and Piwnica-Worms, H., Localization of human Cdc25C is regulated both by nuclear export and 14-3-3 protein binding, *Oncogene,* 20(15), 1839–1851, 2001.

63. Chen, L., Liu, T.H., and Walworth, N.C., Association of Chk1 with 14-3-3 proteins is stimulated by DNA damage, *Genes Dev.,* 13(6), 675–685, 1999.

64. Chan, T.A., Hermeking, H., Lengauer, C., Kinzler, K.W., and Vogelstein, B., 14-3-3 is required to prevent mitotic catastrophe after DNA damage, *Nature,* 401(6753), 616–620, 1999.

65. Zander, K., Sherman, M.P., Tessmer, U., Bruns, K., Wray, V., Prechtel, A.T., Schubert, E., Henklein, P., Luban, J., Neidleman, J., Greene, W.C., and Schubert, U., Cyclophilin A interacts with HIV-1 Vpr and is required for its functional expression, *J. Biol. Chem.,* 278(44), 43202–43213, 2003.

66. Luban, J., Absconding with the chaperone: essential cyclophilin-Gag interaction in HIV-1 virions, *Cell,* 87(7), 1157–1159, 1996.

67. Bouhamdan, M., Benichou, S., Rey, F., Navarro, J.M., Agostini, I., Spire, B., Camonis, J., Slupphaug, G., Vigne, R., Benarous, R., and Sire, J., Human immunodeficiency virus type 1 Vpr protein binds to the uracil DNA glycosylase DNA repair enzyme, *J. Virol.,* 70(2), 697–704, 1996.

68. Selig, L., Benichou, S., Rogel, M.E., Wu, L.I., Vodicka, M.A., Sire, J., Benarous, R., and Emerman, M., Uracil DNA glycosylase specifically interacts with Vpr of both human immunodeficiency virus type 1 and simian immunodeficiency virus of sooty mangabeys, but binding does not correlate with cell cycle arrest, *J. Virol.,* 71(6), 4842–4846, 1997.

69. Stark, L.A. and Hay, R.T., Human immunodeficiency virus type 1 (HIV-1) viral protein R (Vpr) interacts with Lys-tRNA synthetase: implications for priming of HIV-1 reverse transcription, *J. Virol.,* 72(4), 3037–3044, 1998.

70. Jacotot, E., Ferri, K.F., El Hamel, C., Brenner, C., Druillennec, S., Hoebeke, J., Rustin, P., Metivier, D., Lenoir, C., Geuskens, M., Vieira, H.L., Loeffler, M., Belzacq, A.S., Briand, J.P., Zamzami, N., Edelman, L., Xie, Z.H., Reed, J.C., Roques, B.P., and Kroemer, G., Control of mitochondrial membrane permeabilization by adenine nucleotide translocator interacting with HIV-1 viral protein rR and Bcl-2, *J. Exp. Med.,* 193(4), 509–519, 2001.

71. Phenix, B.N., Lum, J.J., Nie, Z., Sanchez-Dardon, J., and Badley, A.D., Antiapoptotic mechanism of HIV protease inhibitors: preventing mitochondrial transmembrane potential loss, *Blood,* 98(4), 1078–1085, 2001.

72. Badley, A.D., Pilon, A.A., Landay, A., and Lynch, D.H., Mechanisms of HIV-associated lymphocyte apoptosis, *Blood,* 96(9), 2951–2964, 2000.

73. Lenardo, M.J., Angleman, S.B., Bounkeua, V., Dimas, J., Duvall, M.G., Graubard, M.B., Hornung, F., Selkirk, M.C., Speirs, C.K., Trageser, C., Orenstein, J.O., and Bolton, D.L., Cytopathic killing of peripheral blood CD4(+) T lymphocytes by human immunodeficiency virus type 1 appears necrotic rather than apoptotic and does not require Env, *J. Virol.,* 76(10), 5082–5093, 2002.

74. Bolton, D.L., Hahn, B.I., Park, E.A., Lehnhoff, L.L., Hornung, F., and Lenardo, M.J., Death of CD4(+) T-cell lines caused by human immunodeficiency virus type 1 does not depend on caspases or apoptosis, *J. Virol.,* 76(10), 5094–5107, 2002.

75. Wei, X., Ghosh, S.K., Taylor, M.E., Johnson, V.A., Emini, E.A., Deutsch, P., Lifson, J.D., Bonhoeffer, S., Nowak, M.A., Hahn, B.H., Saag, M.S., and Shaw, G.M., Viral dynamics in human immunodeficiency virus type 1 infection, *Nature,* 373(6510), 117–122, 1995.

76. Ho, D.D., HIV-1 dynamics *in vivo, J. Biol. Regul. Homeost. Agents,* 9(3), 76–77, 1995.
77. Finkel, T.H., Tudor-Williams, G., Banda, N.K., Cotton, M.F., Curiel, T., Monks, C., Baba, T.W., Ruprecht, R.M., and Kupfer, A., Apoptosis occurs predominantly in bystander cells and not in productively infected cells of HIV- and SIV-infected lymph nodes, *Nat. Med.,* 1(2), 129–134, 1995.
78. Levy, D.N., Fernandes, L.S., Williams, W.V., and Weiner, D.B., Induction of cell differentiation by human immunodeficiency virus 1 Vpr, *Cell,* 72(4), 541–550, 1993.
79. Kino, T., Gragerov, A., Kopp, J.B., Stauber, R.H., Pavlakis, G.N., and Chrousos, G.P., The HIV-1 virion-associated protein Vpr is a coactivator of the human glucocorticoid receptor, *J. Exp. Med.,* 189(1), 51–62, 1999.
80. Refaeli, Y., Levy, D.N., and Weiner, D.B., The glucocorticoid receptor type II complex is a target of the HIV-1 Vpr gene product, *Proc. Natl. Acad. Sci. U.S.A.,* 92(8), 3621–3625, 1995.
81. Ayyavoo, V., Mahboubi, A., Mahalingam, S., Ramalingam, R., Kudchodkar, S., Williams, W.V., Green, D.R., and Weiner, D.B., HIV-1 Vpr suppresses immune activation and apoptosis through regulation of nuclear factor B [see comments], *Nat. Med.,* 3(10), 1117–1123, 1997.
82. Levy, D.N., Refaeli, Y., MacGregor, R.R., and Weiner, D.B., Serum Vpr regulates productive infection and latency of human immunodeficiency virus type 1, *Proc. Natl. Acad. Sci. U.S.A.,* 91(23), 10873–10877, 1994.
83. Roumier, T., Vieira, H.L., Castedo, M., Ferri, K.F., Boya, P., Andreau, K., Druillennec, S., Joza, N., Penninger, J.M., Roques, B., and Kroemer, G., The C-terminal moiety of HIV-1 Vpr induces cell death via a caspase-independent mitochondrial pathway, *Cell Death Differ.,* 9(11), 1212–1219, 2002.
84. Stewart, S.A., Poon, B., Jowett, J.B., and Chen, I.S., Human immunodeficiency virus type 1 Vpr induces apoptosis following cell cycle arrest, *J. Virol.,* 71(7), 5579–5592, 1997.
85. Stewart, S.A., Poon, B., Song, J.Y., and Chen, I.S., Human immunodeficiency virus type 1 Vpr induces apoptosis through caspase activation, *J. Virol.,* 74(7), 3105–3111, 2000.
86. Yasuda, J., Miyao, T., Kamata, M., Aida, Y., and Iwakura, Y., T cell apoptosis causes peripheral T cell depletion in mice transgenic for the HIV-1 Vpr gene, *Virology,* 285(2), 181–192, 2001.
87. Green, D.R. and Reed, J.C., Mitochondria and apoptosis, *Science,* 281(5381), 1309–1312, 1998.
88. Green, D. and Kroemer, G., The central executioners of apoptosis: caspases or mitochondria?, *Trends Cell Biol.,* 8(7), 267–271, 1998.
89. Lum, J.J., Cohen, O.J., Nie, Z., Weaver, J.G., Gomez, T.S., Yao, X.J., Lynch, D., Pilon, A.A., Hawley, N., Kim, J.E., Chen, Z., Montpetit, M., Sanchez-Dardon, J., Cohen, E.A., and Badley, A.D., Vpr R77Q is associated with long-term nonprogressive HIV infection and impaired induction of apoptosis, *J. Clin. Invest.,* 111(10), 1547–1554, 2003.
90. Muthumani, K., Choo, A.Y., Hwang, D.S., Chattergoon, M.A., Dayes, N.N., Zhang, D., Lee, M.D., Duvvuri, U., and Weiner, D.B., Mechanism of HIV-1 viral protein R-induced apoptosis, *Biochem. Biophys. Res. Commun.,* 304(3), 583–592, 2003.
91. Fukumori, T., Akari, H., Iida, S., Hata, S., Kagawa, S., Aida, Y., Koyama, A.H., and Adachi, A., The HIV-1 Vpr displays strong anti-apoptotic activity, *FEBS Lett.,* 432(1–2), 17–20, 1998.
92. Conti, L., Rainaldi, G., Matarrese, P., Varano, B., Rivabene, R., Columba, S., Sato, A., Belardelli, F., Malorni, W., and Gessani, S., The HIV-1 Vpr protein acts as a negative regulator of apoptosis in a human lymphoblastoid T cell line: possible implications for the pathogenesis of AIDS, *J. Exp. Med.,* 187(3), 403–413, 1998.
93. Conti, L., Matarrese, P., Varano, B., Gauzzi, M.C., Sato, A., Malorni, W., Belardelli, F., and Gessani, S., Dual role of the HIV-1 Vpr protein in the modulation of the apoptotic response of T cells, *J. Immunol.,* 165(6), 3293–3300, 2000.
94. Matarrese, P., Conti, L., Varano, B., Gauzzi, M.C., Belardelli, F., Gessani, S., and Malorni, W., The HIV-1 Vpr protein induces anoikis-resistance by modulating cell adhesion process and microfilament system assembly, *Cell Death Differ.,* 7(1), 25–36, 2000.
95. Gibbs, J.S., Lackner, A.A., Lang, S.M., Simon, M.A., Sehgal, P.K., Daniel, M.D., and Desrosiers, R.C., Progression to AIDS in the absence of a gene for Vpr or Vpx, *J. Virol.,* 69(4), 2378–2383, 1995.
96. Hoch, J., Lang, S.M., Weeger, M., Stahl-Hennig, C., Coulibaly, C., Dittmer, U., Hunsmann, G., Fuchs, D., Muller, J., Sopper, S., Fleckenstein, B., and Ubuda, K.T., Vpr deletion mutant of simian immunodeficiency virus induces AIDS in rhesus monkeys, *J. Virol.* 69 (8), 4807–13, 1995.
97. Lang, S.M., Weeger, M., Stahl-Hennig, C., Coulibaly, C., Hunsmann, G., Muller, J., Muller-Hermelink, H., Fuchs, D., Wachter, H., Daniel, M.M., Duresiens, R.C., and Fleckenstein, B., Importance

of Vpr for infection of rhesus monkeys with simian immunodeficiency virus, *J. Virol.,* 67(2), 902–912, 1993.

98. Goh, W.C., Rogel, M.E., Kinsey, C.M., Michael, S.F., Fultz, P.N., Nowak, M.A., Hahn, B.H., and Emerman, M., HIV-1 Vpr increases viral expression by manipulation of the cell cycle: a mechanism for selection of Vpr *in vivo, Nat. Med.,* 4(1), 65–71, 1998.

99. Saksena, N.K., Ge, Y.C., Wang, B., Xiang, S.H., Dwyer, D.E., Randle, C., Palasanthiran, P., Ziegler, J., and Cunningham, A.L., An HIV-1 infected long-term non-progressor (LTNP): molecular analysis of HIV-1 strains in the Vpr and Nef genes, *Ann. Acad. Med. Singapore,* 25(6), 848–854, 1996.

100. Yamada, T. and Iwamoto, A., Comparison of proviral accessory genes between long-term nonprogressors and progressors of human immunodeficiency virus type 1 infection, *Arch. Virol.,* 145(5), 1021–1027, 2000.

101. Somasundaran, M., Sharkey, M., Brichacek, B., Luzuriaga, K., Emerman, M., Sullivan, J.L., and Stevenson, M., Evidence for a cytopathogenicity determinant in HIV-1 Vpr, *Proc. Natl. Acad. Sci. U.S.A.,* 99(14), 9503–9508, 2002.

102. Rodes, B., Toro, C., Paxinos, E., Poveda, E., Martinez-Padial, M., Benito, J.M., Jimenez, V., Wrin, T., Bassani, S., and Soriano, V., Differences in disease progression in a cohort of long-term non-progressors after more than 16 years of HIV-1 infection, *AIDS,* 18(8), 1109–1116, 2004.

103. Cavert, W., Webb, C.H., and Balfour, H.H., Jr., Alterations in the C-terminal region of the HIV-1 accessory gene Vpr do not confer clinical advantage to subjects receiving nucleoside antiretroviral therapy, *J. Infect. Dis.,* 189(12), 2181–2184, 2004.

104. Emmel, E.A., Verweij, C.L., Durand, D.B., Higgins, K.M., Lacy, E., and Crabtree, G.R., Cyclosporin A specifically inhibits function of nuclear proteins involved in T cell activation, *Science,* 246(4937), 1617–1620, 1989.

105. Martin, L.N., Murphey-Corb, M., Mack, P., Baskin, G.B., Pantaleo, G., Vaccarezza, M., Fox, C.H., and Fauci, A.S., Cyclosporin A modulation of early virologic and immunologic events during primary simian immunodeficiency virus infection in rhesus monkeys, *J. Infect. Dis.,* 176(2), 374–383, 1997.

106. Matsuda, Z., Yu, X., Yu, Q.C., Lee, T.H., and Essex, M., A virion-specific inhibitory molecule with therapeutic potential for human immunodeficiency virus type 1, *Proc. Natl. Acad. Sci. U.S.A.,* 90(8), 3544–3548, 1993.

107. Yao, X.J., Lemay, J., Rougeau, N., Clement, M., Kurtz, S., Belhumeur, P., and Cohen, E.A., Genetic selection of peptide inhibitors of human immunodeficiency virus type 1 Vpr, *J. Biol. Chem.,* 277(50), 48816–48826, 2002.

108. Gu, J., Emerman, M., and Sandmeyer, S., Small heat shock protein suppression of Vpr-induced cytoskeletal defects in budding yeast, *Mol. Cell Biol.,* 17(7), 4033–4042, 1997.

109. Sawaya, B.E., Khalili, K., Rappaport, J., Serio, D., Chen, W., Srinivasan, A., and Amini, S., Suppression of HIV-1 transcription and replication by a Vpr mutant, *Gene Ther.,* 6(5), 947–950, 1999.

110. Sawaya, B.E., Khalili, K., Gordon, J., Srinivasan, A., Richardson, M., Rappaport, J., and Amini, S., Transdominant activity of human immunodeficiency virus type 1 Vpr with a mutation at residue R73, *J. Virol.,* 74(10), 4877–4881, 2000.

111. Bukrinsky, M. and Zhao, Y., Heat-shock proteins reverse the G2 arrest caused by HIV-1 viral protein R, *DNA Cell Biol.,* 23(4), 223–225, 2004.

112. Agostini, I., Popov, S., Li, J., Dubrovsky, L., Hao, T., and Bukrinsky, M., Heat-shock protein 70 can replace viral protein R of HIV-1 during nuclear import of the viral preintegration complex, *Exp. Cell Res.,* 259(2), 398–403, 2000.

113. Iordanskiy, S., Zhao, Y., DiMarzio, P., Agostini, I., Dubrovsky, L., and Bukrinsky, M., Heat-shock protein 70 exerts opposing effects on Vpr-dependent and Vpr-independent HIV-1 replication in macrophages, *Blood,* 104(6), 1367–1372, 2004.

114. Hutvagner, G. and Zamore, P.D., RNAi: nature abhors a double-strand, *Curr. Opin. Genet. Dev.,* 12(2), 225–232, 2002.

115. Elbashir, S.M., Lendeckel, W., and Tuschl, T., RNA interference is mediated by 21- and 22-nucleotide RNAs, *Genes Dev.,* 15(2), 188–200, 2001.

116. Elbashir, S.M., Harborth, J., Lendeckel, W., Yalcin, A., Weber, K., and Tuschl, T., Duplexes of 21-nucleotide RNAs mediate RNA interference in cultured mammalian cells, *Nature,* 411(6836), 494–498, 2001.

117. Doranz, F., Rucker, B.I., Liesnard, J., Farber, C., Saragosti, C.M., Lapoumeroulie, S., Samson, C., Liebert, M., Cognaux, J., Forceille, C., Muyldermans, G., Verhofstede, C., Burtonboy, G., Georges, M., Imai, T., Rana, S., Yi, Y., Smyth, R.J., Collman, R.G., Doms, R.W., Vassart, G., and Parmentier, M., Resistance to HIV-1 infection in caucasion individuals bearing mutant alleles of the CCR5 receptor gene, *Nature,* 382(6593), 722–725, 1996.

118. Martinez, M.A., Gutierrez, A., Armand-Ugon, M., Blanco, J., Parera, M., Gomez, J., Clotet, B., and Este, J.A., Suppression of chemokine receptor expression by RNA interference allows for inhibition of HIV-1 replication, *Aids,* 16(18), 2385–2390, 2002.

119. Kitabwalla, M. and Ruprecht, R.M., RNA interference—a new weapon against HIV and beyond, *N. Engl. J. Med.,* 347(17), 1364–1367, 2002.

120. Jacque, J.M., Triques, K., and Stevenson, M., Modulation of HIV-1 replication by RNA interference, *Nature,* 418(6896), 435–438, 2002.

121. Zeng, Y. and Cullen, B.R., RNA interference in human cells is restricted to the cytoplasm, *RNA,* 8(7), 855–860, 2002.

122. Banerjea, A., Li, M.J., Bauer, G., Remling, L., Lee, N.S., Rossi, J., and Akkina, R., Inhibition of HIV-1 by lentiviral vector-transduced siRNAs in T lymphocytes differentiated in SCID-hu mice and CD34+ progenitor cell-derived macrophages, *Mol. Ther.,* 8(1), 62–71, 2003.

123. Lee, N.S., Dohjima, T., Bauer, G., Li, H., Li, M.J., Ehsani, A., Salvaterra, P., and Rossi, J., Expression of small interfering RNAs targeted against HIV-1 Rev transcripts in human cells, *Nat. Biotechnol.,* 20(5), 500–505, 2002.

124. Coburn, G.A. and Cullen, B.R., Potent and specific inhibition of human immunodeficiency virus type 1 replication by RNA interference, *J. Virol.,* 76(18), 9225–9231, 2002.

125. Novina, C.D., Murray, M.F., Dykxhoorn, D.M., Beresford, P.J., Riess, J., Lee, S.K., Collman, R.G., Lieberman, J., Shankar, P., and Sharp, P.A., siRNA-directed inhibition of HIV-1 infection, *Nat. Med.,* 8(7), 681–686, 2002.

8 Interference of the Nef Protein of HIV-1 with Pro- and Antiapoptotic Pathways of T Cells

Andreas Baur

CONTENTS

INTRODUCTION

The Nef protein of the human immunodeficiency virus (HIV) is as important for HIV disease progression as it is perplexing in its number of target molecules and functions. In the early years of acquired immunodeficiency syndrome (AIDS) and Nef research, Nef was thought to be a negative factor (Nef for negative), moderately repressing HIV transcription.[1] This view changed completely in 1991, when Kestler and colleagues reported that Nef is essential for the maintenance of highly viral loads and, thus, progression to AIDS in the rhesus macaque animal model.[2] Their finding was subsequently confirmed in numerous reports using animal models and was corroborated by follow-up studies of individuals who were infected with Nef-defective viruses but did not progress in the disease.[3,4] Today, it is generally accepted that Nef quite drastically increases the viral load, but, surprisingly, after almost two decades of research, the underlying molecular mechanisms that would explain that phenomenon are still not understood. Nevertheless, there seems to be a consensus that several effects and functions of Nef contribute to the overall increase of viral load in the infected host. From a molecular point of view, Nef has two functions. First, it stimulates T cell activation pathways, and second, it interferes

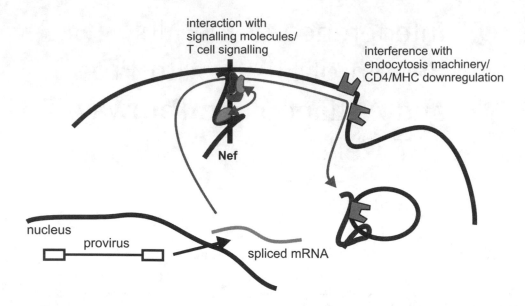

FIGURE 8.1 The two molecular functions of Nef. Nef is expressed early in the viral life cycle from spliced messenger RNA. After protein translation, the protein is translocated to the plasma membrane. There, upon anchoring into the membrane via the myristoylated N-terminus, possibly a conformational change occurs, enabling the protein to associate with signaling molecules from the T cell receptor environment and leading to the assembly of the Nef signaling complex. Subsequently, Nef interacts or binds cell surface receptors like CD4 and MHC class I, causing their internalization and degradation.

with cellular trafficking and the endocytosis machinery (Figure 8.1). Surprisingly, these molecular functions of Nef do not seem to stimulate viral replication directly. Rather, particle production and release from infected cells are increased by indirect means. Despite circumstantial evidence that Nef expression might stimulate replication directly, it has never been formally proven.

Additional indirect effects of Nef are mechanisms that both induce and block cellular death programs. At first glance, it seems to be contradictory that one protein induces opposite effects to support one function, which is the increase of viral particle production. However, as will be discussed below, these effects may not occur at the same time in the viral life cycle and are, therefore, not mutually exclusive. Interestingly, both molecular functions of Nef, T cell signaling and interference with endocytosis, converge to manipulate cellular death programs. The co-evolution of two obviously different molecular mechanisms in one small viral protein is remarkable and again demonstrates the complexity and sophistication of HIV and simian immunodeficiency virus (SIV).

It is generally believed that the increase of viral load is the only contribution of Nef to disease progression. In the mouse model, however, Nef expression alone is sufficient to induce an AIDS-like disease with T cell depletion and immunodeficiency.[5] Although there are many differences between transgenic expression of Nef in mice and natural infection of humans and monkeys, it cannot be ruled out that Nef has direct detrimental effects on the immune system in the absence of any viral replication. Such effects, however, have not yet been sufficiently described.

INTERACTION OF NEF WITH THE HOST T CELL

NEF STRUCTURE AND INTERACTION WITH CELLULAR PROTEINS

Nef is a small viral regulatory protein of approximately 27 kDa that is expressed abundantly in the early stages of viral replication. In fact, it is expressed even before the virus is integrated into the genome.[6] Nef is posttranslationally modified by the irreversible attachment of myristic acid to its

N-terminus, which targets Nef to the cellular membrane. So far, it seems that membrane targeting is absolutely essential for most, if not all, functions of Nef.

Nef achieves its biochemical effect by requisite interactions with cellular components. More than 30 putative Nef targets have been reported, and the number is still growing.[7] Structurally, HIV-1 Nef has three distinct features. The first 70 amino acids constitute a genetically diverse and structurally flexible N-terminal arm that is followed by a well-conserved and folded core domain of approximately 120 residues. Out of the core domain projects a 30-amino-acid loop, which is again highly flexible. Only HIV-2 and SIV have an additional C-terminal tail of 10 to 30 amino acids. The core domain is the only part of Nef that adopts a stable tertiary fold. Its high-sequence conservation in all Nef alleles implies that this region has the same structure in HIV-2 and SIV Nef. Each of the described domains has interaction sites for the binding and interaction of and with cellular proteins. For example, mutational analysis of the core domain identified several residues relevant for the binding of SH3 domains of src kinases and the Pak kinase. Also, the C-terminal flexible loop contains three defined protein motifs that interact with components of the endocytosis machinery, an example of which would be the regulatory subunit of the v-ATPase.[8] In addition to these better-defined interactions, most interacting proteins associate with Nef in the context of one or more protein complexes. In these instances, several domains of Nef seem to be involved in the association with those complexes. For example, the N-terminus of Nef interacts with a protein complex that contains Lck and PKC and the Polycomb protein Eed.[9] Surprisingly, domains in the far C-terminus contribute to the binding of this complex. Some interactions cannot occur simultaneously, as their binding sites on Nef are overlapping. For example, the association of Nef with Pak, the human thioesterase, involve the same hydrophobic patch on the core domain, forming a collar around F121.

Given the dynamic interaction of Nef with its cellular environment, it is likely that Nef is sequentially directed to distinct cellular locations and different sets of interacting effectors. Therefore, Nef may be viewed as an adapter protein, assembling and connecting several protein and signaling complexes in a timely, defined fashion. The fairly extensive unstructured regions of Nef provide a vast accessible surface for serving this function.[7]

INTERACTION OF NEF WITH THE ENDOCYTOSIS MACHINERY

The accelerated endocytosis and lysosomal degradation of CD4 was one of the first functions attributed to Nef.[10] Meanwhile, additional factors that are required were identified. These include cellular proteins generally involved in trafficking, such as adapter protein complexes (AP-complexes), ß-COP, the vacuolar ATPase, thioesterase, and the PACS-1 protein.[11] To recruit CD4 into the endocytotic pathway, Nef binds CD4 and acts as an adapter between the receptor and components of clathrin-coated pits. This involves the interaction of Nef with AP-complexes and the regulatory unit of the vacuolar proton pump of the v-ATPase.[12] The precise meaning of the latter interaction is not obvious and awaits further clarification. Subsequently, Nef binds to the -subunit of COPI coatomers (β-COP) to direct CD4 to a degradation pathway.[13]

In subsequent studies, the downregulation of other surface receptors was reported, including CD28[14] and CD3 (only SIV).[15] However, there are more receptors that are affected (O. Keppler, personal communication). Functionally, it was shown that downregulation of CD4 increases viral infectivity by preventing the formation of complexes between gp120 and CD4 on viral particles, as well as intracellularly.[16,17] It was further argued that CD4 downregulation may also prevent reinfection (also called superinfection), in order not to overload the biosynthetic capability of the cell.[18] On the other hand, it is known that HIV genomes recombine within an infected cell; therefore, the latter argument is on shaky grounds. Taken together, CD4 downregulation by Nef may have an additional role that has not yet been described. Unlike CD4, other receptors that are downregulated do not bind directly to Nef, perhaps with the exception of CD3 in SIV Nef. Therefore, the rather promiscuous downregulation of a seemingly unrelated set of surface receptors may point to a general interference with trafficking pathways that is not fully explained or understood until now.

Surprisingly, Nef uses a different mechanism and also different protein domains to downregulate another cell surface protein: the major histocompatibility complex (MHC) class I, A and B.[19] The mechanism was recently described in more detail. Nef binds sequentially a PACS-1/AP 1 complex and PI3 kinase before reaching the plasma membrane. There, the ARF6 endocytic pathway is activated, leading to increased downregulation of MHC class I. In addition, Nef seems to block the recycling of MHC molecules back to the cell surface, which ultimately leads to an accumulation of MHC complexes in the trans-Golgi network (TGN).[20] Downregulation of MHC class I by Nef is not as efficient as CD4 downregulation and seems to require higher expression levels of Nef. The effect, however, has been shown to be sufficient to diminish recognition of HIV-infected cells by cytotoxic T lymphocyte (CTL).[21] Therefore, it is a classical viral immune escape mechanism[22] (see below). In summary, at least the downmodulation of MHC class I by Nef clearly is a function that increases viral particle production.

Interaction of Nef with Signaling Proteins of the T Cell Receptor (TCR) Environment

The ability of Nef to activate T cells was originally described in 1994 using Jurkat cells stably transfected with a CD8-Nef chimeric protein.[23] Cells that had higher concentrations of CD8-Nef at the membrane induced tyrosine phosphorylation and upregulation of activation markers such as CD69. In addition, these cells underwent activation-induced apoptosis, indirectly confirming that Nef activated these cells. In 2001, these results were confirmed by gene expression profiling of inducible T cell lines, showing that Nef and anti-CD3-mediated T cell activation overlap by 97%.[24] As much as these results imply that Nef activates T cells, there are recognizable differences between T cell receptor (TCR)–mediated activation by antigen or antibodies and Nef expression with respect to the phenotype of activated T cells (see below). The use of tumor cell lines, rather than the primary cells, and chimerical proteins may mask subtle differences of classical T cell activation and Nef-induced signaling. Hence, we still do not know which T cell signaling pathways are, and which are not, activated by Nef.

On the other hand, the ability of Nef to induce T cell activation pathways was confirmed by a number of publications, showing the direct interaction of Nef with both the T cell receptor and its immediate downstream effectors.[25] Functional as well as binding studies analyzed the interaction of Nef with the T cell receptor -chain[26] and proteins of the T cell receptor environment, including adaptor proteins Vav[27] and LAT,[5] the tyrosine kinase Lck,[9] the serine kinase Pak,[28] PKC,[29] the DOCK2-ELMO1 complex,[30] the map kinases ERK1 and ERK2,[31] and membrane micro domains (rafts).[32] Although the picture is still not clear, these reports suggest that Nef is part of and probably acts through a T cell receptor–associated multiprotein complex.[25] In line with these findings, development of an AIDS-like disease in an HIV transgenic mouse model correlated with Nef-mediated activation of mouse T cells.[5] How Nef activates T cells is not known. Because it has no intrinsic enzymatic activity, it likely functions as an adapter protein. By binding to signaling molecules of different compartments/lipid-rich micro domains (rafts) and possibly by forming oligomers, Nef may function as an intracellular cross-linker.

From the sheer number of interacting molecules in the Nef signaling complex, it may be assumed that several distinct signaling pathways are activated, similar as seen, for example, when the T cell receptor complex is activated through a cross-linking antibody. Although detailed individual pathways have not yet been sufficiently documented, at least three different functional consequences of Nef signaling have been found:

1. Nef signaling and viral replication: Nef signaling or Nef-mediated T cell activation seems to perfectly fill the needs of HIV, as one of the very early findings was that T cells had to be activated for HIV replication to start.[33] T cell activation leads to the nuclear translocation of transcription factors such as NFAT and NFkB, probably priming

the viral promoter or establishing a basal viral transcription that would lead to the expression of more Tat protein.[34] Upon infection of resting primary T cells, HIV does, in fact, express Nef, even before the virus is integrated,[6] but viral replication remains very low or not detectable. Conversely, stimulation of the T cell receptor by antibodies potently activates replication. Therefore, Nef- and antibody-mediated T cell receptor activation clearly differs, at least with respect to viral replication. In other words, Nef-mediated T cell activation does not start viral replication. On the other hand, there is little doubt that Nef supports HIV replication. This was shown in a number of studies using primary cells. Particularly revealing were experiments using co-cultures of resting T cells and immature dendritic cells.[35–37] Immature dendritic cells replicate HIV at a very low level. However, upon co-culture with resting T cells, a significant increase of viral replication is observed in the T cell compartment. This, however, requires the presence of a functional Nef gene. These results point to a role for dendritic cell surface receptors as co-stimulators of viral replication. Therefore, besides Nef, a second, thus far not characterized, signal (receptor, cytokine, or chemokine) is required to start HIV replication.

2. Nef signaling and cytoskeleton rearrangement/chemotaxis: More recently, at least two publications have demonstrated in detail that Nef signaling leads to cytoskeleton rearrangement in the infected cell. Both reports analyzed the Nef-induced pathways leading to this effect, but came up with different results. Fackler and colleagues[27] found that the PxxP domain of Nef binds and activates the Vav protein, leading further downstream to the activation of Rac1 and the Pak kinase. The authors speculated that this Nef effect could play a major role in viral morphogenesis and budding, similar to that seen with other viruses and bacteria. This mechanism would basically lead to an increased release of viral particles. Janardhan and colleagues[30] confirmed that Nef activates Rac1 but suggested a different mechanism. They found that Nef associates with the DOCK2-ELMO1 proteins and showed that this complex is required to activate downstream Rac1 and the Pak kinase. They further found that the chronic activation of DOCK2-ELMO1 leads to impaired chemotaxis of Nef-expressing cells. They speculated and suggested that this would affect the ordered migration patterns of infected T cells and, therefore, the generation and maturation of the immune response to viral antigens. Although both studies come to different conclusions, it seems clear that Nef activates a pathway generally associated with cytoskeleton rearrangement and chemotaxis. How these molecular mechanisms support the viral life cycle remains to be elucidated.

3. Nef signaling and immune evasion/antiapoptosis: The third consequence of Nef signaling in T cells is the stimulation of antiapoptotic and, possibly later in the viral life cycle, proapoptotic signaling pathways. To date, only the antiapoptotic mechanisms are well described. A logical consequence of T cell signaling from the T cell receptor is in the understandable interest of the virus. Much less defined are the proapoptotic effects of Nef, as will be discussed below. In addition, Nef signaling from the T cell receptor leads to the upregulation of Fas ligand, which may be the only real but indirect proapoptotic signaling function of Nef. It is most likely a mechanism of viral immune evasion (see below).

TCR SIGNALS THAT STIMULATE ANTI- AND PROAPOPTOTIC PATHWAYS

The T cell receptor is one of the best studied of the cellular surface receptors. When pro- and antiapoptotic signaling pathways were detected and described, it was soon learned that this receptor is not only a key regulator of proliferation but also of cellular death programs in T cells.[38] This is because both T cell activation/cellular proliferation and T cell death are intimately correlated.

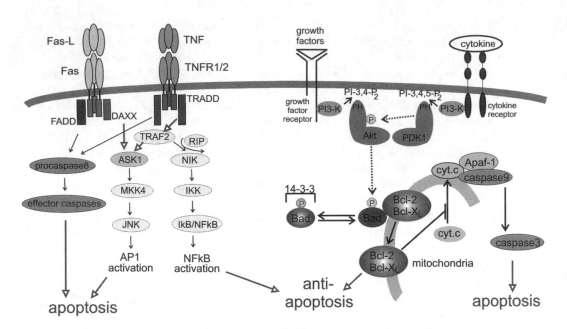

FIGURE 8.2 The "intrinsic" and "extrinsic" apoptotic pathways. Apoptotic cell death is principally induced by two different molecular mechanisms. The extrinsic pathway transmits extracellular signals through death receptors such as Fas and TNFR, as depicted on the left side of the figure. The intrinsic pathway is activated through internal sensors (for example, p53, not shown), leading to the release of cytochrome c from mitochondria and subsequent activation of caspases (right lower part of the figure). The release of cytochrome c is regulated through members of the Bcl-2 family, which again are under the influence of signaling pathways emanating from surface receptor and PI3 kinase (upper right part of the figure).

In recent years, the PI3 kinase–Akt signaling pathways (Figure 8.2) have emerged as an important and general mechanism of antiapoptosis.[39] In T cells, this pathway is activated through stimulation, for example, through the T cell receptor or growth factor receptors. Beyond that, the constitutive upregulation of PI3 kinase–Akt survival signaling has also been implicated in oncogenic transformation and tumor development, as it prevents apoptotic cell death during uncontrolled proliferation. The phosphatidylinositol-3 kinases (PI3 kinases) comprise a family of heterodimeric proteins that produce lipid second messengers by phosphorylation of plasma membrane phospho-inositides at the 3 OH of the inositol ring. The products of this enzymatic reaction (3,4-biphosphate and 3,4,5-triphosphate) are anchored at the plasma membrane and subsequently recruit cellular signaling molecules that contain a lipid binding — a so-called pleckstrin homology (PH) domain.[40] PI3 kinase is activated through the T cell receptor or CD28. A precise mechanism is not clear but likely involves other proteins, such as the tyrosine kinase Lck and the adapter proteins Gab 1 and 2.[38] A wide variety of signal transduction proteins interacts with PI3 kinase products, one of which is the Akt kinase. Akt was identified as an enzyme similar in many respects to protein kinase A (PKA) and protein kinase C (PKC), and it is, therefore, also referred to as protein kinase B (PKB). After it is recruited to the plasma membrane, Akt is activated by phosphorylation at threonine 308 by another PH-domain-encoding kinase, PDK-1. After its activation, Akt phosphorylates a number of cellular signaling proteins, including the proapoptotic Bad protein (Figure 8.2), a member of the Bcl-2 family that regulates cell death through mitochondria-associated proteins.[41] Akt-phosphorylated Bad is inactivated through association with 14-3-3 proteins, which prevents Bad from dimerizing with and inactivating pro-survival members of the same family, for example, Bcl-2. However, in T cells, this mechanism or pathway may not play a major role in survival, as Akt transgenic T cells do not display increased phosphorylation of Bad.[42] Rather, it has been suggested that Akt activation leads to the activation of NF-kB, which, in turn, would lead to transcriptional upregulation of Bcl-XL that protects the cell from proapoptotic effects of Bad (also see discussion below). Potential mechanisms of Akt-mediated NFkB activation are controversial and are not understood at this time.

In addition to antiapoptotic signals, the T cell receptor regulates proapoptotic signals through activation-induced cell death (AICD), a mechanism that is thought to limit clonal expansion of T cells and cause the deletion of T cell clones.[43] The latter mechanism is particularly important in protecting the host from autoimmune aggression, inflammation, tissue damage, and T cell transformation. *In vitro*, AICD can be mimicked by stimulating T cells through ligation of the T cell receptor and addition of interleukin-2 and then by reactivation of the T cells. Various mechanisms have been described that induce AICD. In principle, chronic T cell activation leads to the upregulation of death receptors and their ligands, leading to cell death in an autocrine fashion (suicide, autonomous AICD) or by cell–cell contact (fratricide). Although several death receptors have been implicated in this mechanism, Fas and Fas ligand seem to be the most important receptor/ligand couple involved. For example, immune-privileged tissues as well as certain tumors express Fas ligand in order to avoid cell death induced by attacking CD8 cells. There are two possible ways that activated T cells can upregulate Fas ligand. First, T cell activation increases Fas ligand transcription through the induction of a number of transcription factors, including NFAT, NF-kB, and Egr1 and 2.[43] In addition, there are preformed Fas-ligand-containing vesicles in the cell that are released into the extracellular space and membrane upon T cell stimulation.[44]

NEF SIGNALS THAT STIMULATE ANTI- AND PROAPOPTOTIC PATHWAYS

SURVIVAL STRATEGIES OF VIRUSES: ANTIAPOPTOSIS AND IMMUNE EVASION

In recent years, we have started to understand how important survival strategies are for invading pathogenic viruses, in particular, when they intend to establish a chronic infection.[22,45,46] In fact, in a chronic phase of a viral infection, when viral replication is often barely detectable, these mechanisms may be even more important than viral replication. Recent work has shown that most viruses have a bivalent defense strategy. The first arm of this strategy relies on numerous mechanisms of immune evasion, which include change of protein conformation and sequence, interference with MHC antigen presentation, modulation or mimicry of cytokine activity, and induction of apoptosis by Fas ligand.[22] HIV is equipped with several of these mechanisms, two of which are exerted by the Nef protein (see below). Viral-induced immune evasion alone will not ensure survival of an infected cell until the next virus generation is ready to leave. In expecting a viral infection, the cell has learned to self-destruct by apoptosis as soon as something unexpected occurs (i.e., unscheduled DNA synthesis).

Cellular death is induced, in principle, by two different routes. Internal sensors (for example, p53) activate the so-called intrinsic pathway, which initiates a process leading to the ultimate loss of mitochondrial integrity and apoptosis. Proteins of the Bcl-2 family (Figure 8.2) regulate this process. Proapoptotic members of this family (Bad, Bax, Bak, Bid, and others) form heterodimers with and thereby inactivate pro-survival members of the same family (Bcl-2, Bcl-Xl, Bcl-w, and others). In antiapoptotic signaling, the proapoptotic effectors such as Bad are phosphorylated on specific serine residues that release Bcl-2 for pro-survival activity by blocking Apaf-1-mediated activation of initiator caspase-9. The signaling pathway that leads to Bad phosphorylation usually starts with the activation of PI3 kinase, through ligation of cytokine or growth factor receptors. Downstream, PI3 kinase activates the Akt serine/threonine kinase that directly phosphorylated the Bad protein. As discussed above, this mechanism may not apply to T cells (see also below). The middle T antigen of polyomavirus was found to activate this pathway for antiapoptotic signaling.[47] Other viruses, particularly lymphotropic herpesviruses such as Epstein–Barr virus (EBV), overexpress a Bcl-2/Bcl-XL homologue to bind and neutralize death-inducing members of the same family. Bcl-2 and Bcl-2 XL prevent the release of mitochondrial cytochrome c and subsequent activation of caspases, presumably by interacting with Apaf-1.[48]

The second, an extrinsic pathway, is initiated through death receptors of the tumor necrosis family (TNF) of receptors transmitting external signals usually provided by immune effector cells (Figure 8.2). So far, five death receptors have been identified: Fas, TNF receptor 1, TRAMP, TRAIL-R1, and TRAIL-R2. Activation of these receptors leads to the recruitment of a death-inducing signaling complex. This complex contains adapter proteins, such as the initiator caspase-8/FLICE, with protein-binding motifs called death domains (DD) and death effector domains (DED). Ultimately, the death-signaling complex leads to the activation of effector caspases and, subsequently, cell death. Viruses, and again the lymphotropic herpesviruses such as Kaposi's sarcoma-associated herpesvirus (KSHV), overexpress proteins, the so-called viral FLIPs (FLICE inhibitory proteins), which block the recruitment of FLICE and thus the formation of a functional death-inducing signaling complex.[48]

INTERFERENCE OF NEF WITH THE DEATH RECEPTOR/EXTRINSIC PATHWAY

Signals from the surface through death receptors TNFR1/2 and Fas could potentially stimulate death programs. In HIV infection, this is a realistic scenario.[25] First, HIV-mediated upregulation of Fas-ligand (see below) could kill the infected cell in an autocrine fashion through Fas ligation (Figure 8.3).

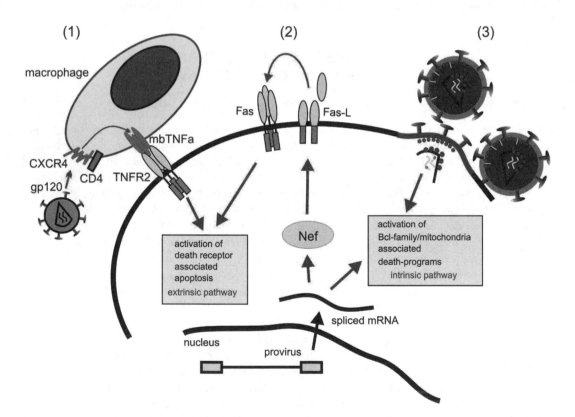

FIGURE 8.3 Apoptosis-inducing signals that threaten the infected cell. Several signals can potentially induce apoptotic cell death in the infected cell. (1) Cross-linking of CXCR4 by HIV gp120 envelope induces the upregulation of membrane-bound TNF on macrophages. This could potentially stimulate the TNF receptor on the infected cells. (2) Nef expression leads to the upregulation of Fas ligand, which could potentially stimulate the Fas receptor in an autocrine fashion. Such a mechanism would lead to the self-destruction of the infected cell. (3) The invading virus during cellular infection likely activates internal sensors that would lead to the activation of the intrinsic death program regulated through the Bcl-2 family.

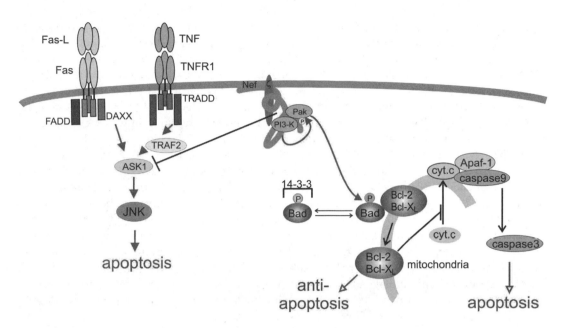

FIGURE 8.4 Nef blocks the extrinsic and intrinsic apoptotic pathways. Nef blocks death receptor–mediated signals through inhibition of ASK1. The serine/threonine kinase ASK1 is an upstream effector of the JNK/p38 pathway and a positive mediator of apoptotic signals. Nef associates with and activates the PI3 and Pak kinases. Subsequently, Pak phosphorylates and inactivates the Bad protein. This liberates Bcl-2 and Bcl-XL to block the release of cytochrome *c*.

Second, HIV gp120 ligation of CXCR4 on macrophages induces the upregulation of membrane-bound TNF. This has been shown to induce cell death via the TNFR in CD8 cells (Figure 8.3), which may lead to T cell depletion.[49] Because both apoptotic signals, "outside in" by death receptors and "inside in" by unbalanced cellular homeostasis, are regulated by different signaling pathways, the invading virus has to block both routes early and efficiently.

HIV seems to manage this dilemma through functions of the Nef protein. Geleziunas and coworkers[50] found that Nef associates with and blocks the activity of apoptosis signal-regulating kinase 1 (ASK1) (Figure 8.4). ASK1, a MAPKKK, links both the Fas- and the TNFR-mediated signals (by Fas-ligand and TNF) to the downstream JNK/p38 pathways (Figure 8.2). In both instances, ASK1 interacts with the death-signaling complex through association with TRAF2 (TNFR-1/2) and Daxx (Fas).[51,52] Although overexpression of ASK induces apoptosis, a transdominant negative ASK1 mutant will block receptor-induced death signals. ASK1 kinase activity is inhibited by thioredoxin (Trx), a redox regulator protein. Only a reduced form of Trx associates with ASK1 and keeps the kinase inactive. A precise mechanism of how Nef activates ASK1 is lacking. However, it seems that Nef blocks the stimulus-dependent release of Trx from ASK1. At this point, it is not clear how this signaling function of Nef connects to the Nef signaling or T cell receptor complex. On the other hand, it has been shown that Nef recruits the Vav protein (see above), which causes the activation and nuclear translocation of the JNK kinase. Therefore, it is conceivable that Nef and death receptor–mediated activation of JNK converge at a level that remains to be determined.

INTERFERENCE OF NEF WITH THE MITOCHONDRIA-ASSOCIATED/INTRINSIC PATHWAY

Another report has demonstrated that Nef signaling also interferes with regulation of the intrinsic, mitochondria-associated cellular death program regulated by the Bcl family of proteins. In order to block proapoptotic members of this family, Wolf et al.[53] demonstrated that HIV/Nef uses the PI3 kinase pathway, however, in a surprising new variation. Like mTAg, Nef was found to associate with and activate PI3 kinase but not to stimulate Akt, as one would expect, but, rather, to activate the Nef-associated serine kinase Pak. The association and activation of Pak is one of the hallmark signaling functions of Nef that was described early.[28] Up to that time, a functional consequence for this interaction was lacking or at least not clearly evident. The Nef-PI3 kinase–PAK complex was able to phosphorylate Bad on serine residues (Figure 8.4), resulting in a block of apoptosis induced by serum starvation and, more importantly, by HIV replication. As would have been predicted, Nef antiapoptotic signaling increased viral particle release between five- and tenfold from infected Jurkat cells *in vitro*, which indicated that this mechanism was of particular importance for viral replication. Beyond that, the phosphorylation of Bad through PI3 kinase and Pak suggested an intriguing new signaling pathway emanating from the TCR independent of the Akt kinase.

By interfering with the Bcl family, Nef follows the common path of other viruses (e.g., herpesviruses). This shows that this is a particularly efficient way to secure viral interests. Herpesviruses, however, which have large DNA genomes, "bring" and overexpress their own Bcl homologues. Conversely, HIV and polyomavirus, due to their small-sized genomes, had to evolve a different strategy. They manage to induce phosphorylation of Bad, which seems to be comparably efficient. Because polyomavirus and HIV infect different target cells, the signaling pathways leading to Bad inactivation are slightly different. For the same reason, Nef may interfere with ASK1 kinase activity rather than overexpress its "own" protein to block early death reporter signaling (e.g., through vFLIPs).

NEF AND IMMUNE EVASION OF HIV

The control and eradication of a viral infection normally require virus-specific cytotoxic T lymphocytes (CTLs). Professional antigen-presenting cells induce CTLs, for example, dendritic cells, which process viral proteins and present MHC-associated specific peptides. CTLs recognize virally infected cells through viral peptides associated with MHC class I molecules and kill them through perforin- or Fas-dependent pathways before new virus particles are made and released. After recognition, the elimination of infected cells takes less than 4 h. To counteract this mechanism, viruses developed tricks to avoid recognition by CTLs. Herpesviruses, such as herpes simplex virus (HSV) or cytomegalovirus (CMV) avoid recognition by inhibiting the transport of viral peptides from the cytosol to the endoplasmic reticulum (ER).[22] HIV developed a similar idea but changed the execution.

Nef-dependent downmodulation of MHC class I, A and B was shown to diminish recognition of HIV-infected cells by cytotoxic T cells[21] (Figure 8.5). Surprisingly, the effect is not as efficient as CD4 downregulation and seems to require higher expression levels of Nef. More importantly, HLA C and HLA B molecules are not downregulated by Nef, probably because they are able to bind to inhibitory receptors of natural killer (NK) cells. Taken together, Nef-mediated MHC downmodulation may delay or avoid recognition by specific CTLs as well as nonspecific NK cells. The latter could give the infected cell a deadly advantage. Possibly, in a timely connected process, Nef interacts with the -chain of the T cell receptor, leading to the upregulation of Fas-ligand[26] (Figure 8.5). HIV-specific cytotoxic T cells that screen T cells for viral antigen may not recognize the infected cell quickly enough to react. Instead, they might be stimulated for their own death program through Fas-ligand on the infected cell. Although not formally proven, there is evidence, *in vivo* as well as *in vitro*, that it is the infected cell that is protected, while bystander cells (as seen by immunostainings[54]) and attacking CTLs (as seen in monkey experiments[55]) undergo Fas/Fas ligand-mediated apoptosis. The immune evasion function of Nef is a good example of how different molecular mechanisms of this protein (T cell receptor–associated signaling and interference with cellular trafficking) converge in order to make up for one function. Upregulation of Fas-ligand clearly is a consequence of Nef signaling and, more specifically, the activation of the T cell receptor

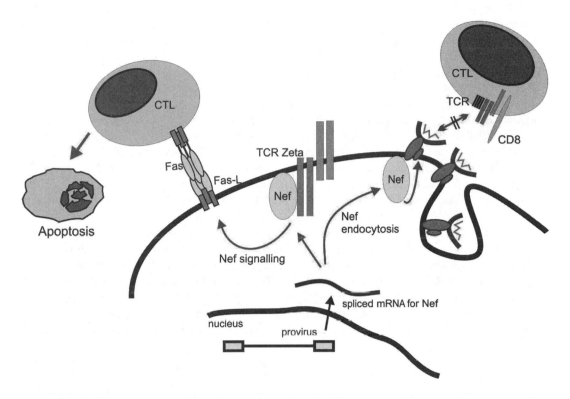

FIGURE 8.5 Immune evasion of HIV-infected cells through Nef. At the plasma membrane, Nef interacts with TCR-, which stimulates FasL expression. This may protect infected cells from attacking CTLs through Fas/Fas ligand ligation. Nef also downregulates MHC class I expression and, therefore, avoids specific recognition of the infected cells by HIV-specific CTLs.

ζ-chain through Nef. The latter occurs in the context of the Nef-mediated assembly of signaling molecules from the T cell receptor environment. In this manner, for example, Nef could bring together Nef-associated Lck and TCR-ζ, which, as in physiological TCR stimulation, could lead to tyrosine phosphorylation and, thus, activation of ζ. In line with this finding, the ζ chain was found to be required for the upregulation of Fas ligand in a physiological setting.[56]

PROAPOPTOTIC EFFECTS OF NEF

HIV, similar to other viruses, dedicates a great deal of its genetic information to survival strategies, leaving the impression that it is a master of survival. Surprisingly, the virus completely changes the strategy in the late phase of the replication cycle. At least four viral proteins (Vpu, Vpr, Env, and Tat) have been reported to induce apoptosis in the infected cell (for review, see Ross[57]). In addition, Nef was shown to induce apoptosis in stably transfected T cell lines when targeted to the plasma membrane. This effect was interpreted as activation-induced cell death due to the prolonged and increased membrane targeting of Nef using a CD8 tag.[23] Because Nef is rapidly internalized, it is at least doubtful that prolonged membrane targeting and, therefore, activation-induced cell death occur in primary lymphocytes. The latter may have been an artifact of that particular system *in vitro*, which showed that Nef is able to activate T cells. However, it cannot be ruled out that a continuous expression of Nef will lead to a fast depletion of signaling intermediates, such as phosphoinositols, that cannot be replenished by the cell. In addition, Nef may sequester important signaling molecules from the plasma membrane. Both effects would render the cell refractory to certain stimuli, which has been described by a number of reports. In addition, these cells may then be more susceptible to apoptotic cell death. For example, if PI3 kinase cannot be activated after prolonged Nef expression, this important survival pathway could no longer inhibit proapoptotic molecules of the Bcl-2 family.

But why would the virus induce apoptosis of the infected cell toward the end of the viral replication cycle? A simple explanation may be that this would increase the release of newly generated viral particles. Apoptotic, in contrast to necrotic, cells are phagocytosed by macrophages and dendritic cells, leading to cross-presentation of viral antigens. This could possibly induce T cell tolerance rather than immunity.[58] The latter would constitute yet another and perhaps important mechanism of immune evasion. It should be noted, however, that cross-presentation could also lead to increased immunity. The latter may depend on the cell that presents the antigen.

In recent years, several reports have demonstrated that a soluble Nef protein induces apoptotic cell death in noninfected human lymphocytes through a Fas-independent mechanism.[59,60] Furthermore, it has been shown that extracellular Nef targets the CXCR4 surface receptor to induce apoptosis.[61] However, this mechanism awaits a more detailed analysis and confirmation through different groups. Also, whether it constitutes an important mechanism of CD4/CD8 cell depletion *in vivo* remains to be determined. In this context, Mueller and colleagues[62] reported that in HIV infection, the HIV-specific CD8 immune response shows a higher susceptibility to apoptotic cell death but not for the CMV-specific immune response. This implies a more specific cellular immune depletion than anticipated from a broad and general killing effect of extracellular Nef protein.

CONCLUSION

The more we look into Nef functions, the more details we find, seemingly confusing our understanding of the actions of this small viral protein. It is likely that the picture is even more complex than anticipated. For example, we still know little regarding Nef functions in macrophages and dendritic cells and, possibly, in HIV latency. The antiapoptotic, more than the proapoptotic, effects of Nef are likely a very important function of this protein. They are easily explained through the molecular interactions of Nef, namely, its ability to associate with signaling molecules and the endocytosis machinery of the cell. However, Nef is likely not the only weapon of HIV in the struggle for survival in the infected host. We have learned that Nef-negative viruses are able to maintain a low-replication capacity and, more importantly, cannot be eradicated by the immune system. Although Nef is not absolutely required for the virus *in vitro* or *in vivo*, it may be responsible for the usually fast progression of the disease. In this context, Nef may also have direct detrimental effects on the immune system, as suggested, in part, by transgenic mouse models. For the foreseeable future, Nef remains one of the most enigmatic proteins of HIV.

REFERENCES

1. Ahmad, N. and Venkatesan, S., Nef protein of HIV-1 is a transcriptional repressor of HIV-1 LTR, *Science,* 241, 1481–1485, 1988.
2. Kestler, H.W., III, Ringler, D.J., Mori, K., Panicali, D.L., Sehgal, P.K., Daniel, M.D., and Desrosiers, R.C., Importance of the Nef gene for maintenance of high virus loads and for development of AIDS, *Cell,* 65, 651–662, 1991.
3. Deacon, N.J., Tsykin, A., Solomon, A., Smith, K., Ludford-Menting, M., Hooker, D.J., McPhee, D.A., Greenway, A. L., Ellett, A., Chatfield, C., Lawson, V.A., Crowe, S., Maerz, A., Souza, S., Learmont, J., Sullivan, J.S., Cunningham, A., Dwyer, D., Druton, D., and Mills, J., Genomic structure of an attenuated quasi species of HIV-1 from a blood transfusion donor and recipients, *Science,* 270, 988–991, 1995.
4. Kirchhoff, F., Greenough, T.C., Brettler, D.B., Sullivan, J.L., and Desrosiers, R.C., Brief report: absence of intact Nef sequences in a long-term survivor with nonprogressive HIV-1 infection, *N. Engl. J. Med.,* 332, 228–232, 1995.

5. Hanna, Z., Kay, D.G., Rebai, N., Guimond, A., Jothy, S., and Jolicoeur, P., Nef harbors a major determinant of pathogenicity for an AIDS-like disease induced by HIV-1 in transgenic mice, *Cell,* 95, 163–175, 1998.

6. Wu, Y. and Marsh, J.W., Selective transcription and modulation of resting T cell activity by preintegrated HIV DNA, *Science,* 293, 1503–1506, 2001.

7. Arold, S.T. and Baur, A.S., Dynamic Nef and Nef dynamics: how structure could explain the complex activities of this small HIV protein, *Trends Biochem. Sci.,* 26, 356–363, 2001.

8. Lu, X., Yu, H., Liu, S.H., Brodsky, F.M., and Peterlin, B.M., Interactions between HIV1 Nef and vacuolar ATPase facilitate the internalization of CD4, *Immunity,* 8, 647–656, 1998.

9. Baur, A.S., Sass, G., Laffert, B., Willbold, D., Cheng-Mayer, C., and Peterlin, B.M., The N-terminus of Nef from HIV-1/SIV associates with a protein complex containing Lck and a serine kinase, *Immunity,* 6, 283–291, 1997.

10. Garcia, J.V. and Miller, A.D., Serine phosphorylation-independent downregulation of cell-surface CD4 by Nef, *Nature,* 350, 508–511, 1991.

11. Doms, R.W., and Trono, D., The plasma membrane as a combat zone in the HIV battlefield, *Genes Dev.,* 14, 2677–2688, 2000.

12. Geyer, M., Yu, H., Mandic, R., Linnemann, T., Zheng, Y.H., Fackler, O.T., and Peterlin, B.M., Subunit H of the V-ATPase binds to the medium chain of adaptor protein complex 2 and connects Nef to the endocytic machinery, *J. Biol. Chem.,* 277, 28521–28529, 2002.

13. Piguet, V., Gu, F., Foti, M., Demaurex, N., Gruenberg, J., Carpentier, J.L., and Trono, D., Nef-induced CD4 degradation: a diacidic-based motif in Nef functions as a lysosomal targeting signal through the binding of -COP in endosomes, *Cell,* 97, 63–73, 1999.

14. Swigut, T., Shohdy, N., and Skowronski, J., Mechanism for down-regulation of CD28 by Nef, *EMBO J.,* 20, 1593–1604, 2001.

15. Swigut, T., Greenberg, M., and Skowronski, J., Cooperative interactions of simian immunodeficiency virus Nef, AP-2, and CD3- mediate the selective induction of T-cell receptor-CD3 endocytosis, *J. Virol.,* 77, 8116–8126, 2003.

16. Lama, J., Mangasarian, A., and Trono, D., Cell-surface expression of CD4 reduces HIV-1 infectivity by blocking Env incorporation in a Nef- and Vpu-inhibitable manner, *Curr. Biol.,* 9, 622–631, 1999.

17. Ross, T.M., Oran, A.E., and Cullen, B.R., Inhibition of HIV-1 progeny virion release by cell-surface CD4 is relieved by expression of the viral Nef protein, *Curr. Biol.,* 9, 613–621, 1999.

18. Lama, J., The physiological relevance of CD4 receptor down-modulation during HIV infection, *Curr. HIV Res.,* 1, 167–184, 2003.

19. Schwartz, O., Marechal, V., Le Gall, S., Lemonnier, F., and Heard, J.M., Endocytosis of major histocompatibility complex class I molecules is induced by the HIV-1 Nef protein, *Nat. Med.,* 2, 338–342, 1996.

20. Blagoveshchenskaya, A.D., Thomas, L., Feliciangeli, S.F., Hung, C.H., and Thomas, G., HIV-1 Nef downregulates MHC-I by a PACS-1- and PI3K-regulated ARF6 endocytic pathway, *Cell,* 111, 853–866, 2002.

21. Collins, K.L., Chen, B.K., Kalams, S.A., Walker, B.D., and Baltimore, D., HIV-1 Nef protein protects infected primary cells against killing by cytotoxic T lymphocytes, *Nature,* 391, 397–401, 1998.

22. Xu, X.N., Screaton, G.R., and McMichael, A.J., Virus infections: escape, resistance, and counterattack, *Immunity,* 15, 867–870, 2001.

23. Baur, A.S., Sawai, E.T., Dazin, P., Fantl, W.J., Cheng-Mayer, C., and Peterlin, B.M., HIV-1 Nef leads to inhibition or activation of T cells depending on its intracellular localization, *Immunity,* 1, 373–384, 1994.

24. Simmons, A., Aluvihare, V., and McMichael, A., Nef triggers a transcriptional program in T cells imitating single-signal T cell activation and inducing HIV virulence mediators, *Immunity,* 14, 763–777, 2001.

25. Fackler, O.T., and Baur, A.S., Live and let die: Nef functions beyond HIV replication, *Immunity,* 16, 493–497, 2002.

26. Xu, X.N., Laffert, B., Screaton, G.R., Kraft, M., Wolf, D., Kolanus, W., Mongkolsapay, J., McMichael, A.J., and Baur, A.S., Induction of Fas ligand expression by HIV involves the interaction of Nef with the T cell receptor chain, *J. Exp. Med.,* 189, 1489–1496, 1999.

27. Fackler, O.T., Luo, W., Geyer, M., Alberts, A.S., and Peterlin, B.M., Activation of Vav by Nef induces cytoskeletal rearrangements and downstream effector functions, *Mol. Cell,* 3, 729–739, 1999.

28. Sawai, E.T., Baur, A., Struble, H., Peterlin, B.M., Levy, J.A., and Cheng-Mayer, C., Human immunodeficiency virus type 1 Nef associates with a cellular serine kinase in T lymphocytes, *Proc. Natl. Acad. Sci. U.S.A.,* 91, 1539–1543, 1994.

29. Smith, B.L., Krushelnycky, B.W., Mochly-Rosen, D., and Berg, P., The HIV Nef protein associates with protein kinase C, *J. Biol. Chem.,* 271, 16753–16757, 1996.

30. Janardhan, A., Swigut, T., Hill, B., Myers, M.P., and Skowronski, J., HIV-1 Nef binds the DOCK2-ELMO1 complex to activate Rac and inhibit lymphocyte chemotaxis, *PLoS Biol.,* 2, 2(1):E6, 2004.

31. Schrager, J.A., Der Minassian, V., and Marsh, J.W., HIV Nef increases T cell ERK MAP kinase activity, *J. Biol. Chem.,* 277, 6137–6142, 2002.

32. Wang, J.K., Kiyokawa, E., Verdin, E., and Trono, D., The Nef protein of HIV-1 associates with rafts and primes T cells for activation, *Proc. Natl. Acad. Sci. U.S.A.,* 97, 394–399, 2000.

33. Zack, J.A., Arrigo, S.J., Weitsman, S.R., Go, A.S., Haislip, A., and Chen, I.S., HIV-1 entry into quiescent primary lymphocytes: molecular analysis reveals a labile, latent viral structure, *Cell,* 61, 213–222, 1990.

34. Kinoshita, S., Chen, B.K., Kaneshima, H., and Nolan, G.P., Host control of HIV-1 parasitism in T cells by the nuclear factor of activated T cells, *Cell,* 95, 595–604, 1998.

35. Fackler, O.T., Wolf, D., Weber, H.O., Laffert, B., DíAloja, P., Schuler-Thurner, B., Geffin, R., Saksela, K., Geyer, M., Peterlin, B.M., Schuler, G. and Baver, A.S., A natural variability in the proline-rich motif of Nef modulates HIV-1 replication in primary T cells, *Curr. Biol.,* 11, 1294–1299, 2001.

36. Messmer, D., Jacque, J.M., Santisteban, C., Bristow, C., Han, S.Y., Villamide-Herrera, L., Mehlhop, E., Marx, P.A., Steinman, R.M., Gettie, A., and Pope, M., Endogenously expressed Nef uncouples cytokine and chemokine production from membrane phenotypic maturation in dendritic cells, *J. Immunol.,* 169, 4172–4182. 2002.

37. Petit, C., Buseyne, F., Boccaccio, C., Abastado, J.P., Heard, J.M., and Schwartz, O., Nef is required for efficient HIV-1 replication in cocultures of dendritic cells and lymphocytes, *Virology,* 286, 225–236, 2001.

38. Kane, L.P. and Weiss, A., The PI-3 kinase/Akt pathway and T cell activation: pleiotropic pathways downstream of PIP3, *Immunol. Rev.,* 192, 7–20, 2003.

39. Cooray, S., The pivotal role of phosphatidylinositol 3-kinase-Akt signal transduction in virus survival, *J. Gen. Virol.,* 85, 1065–1076, 2004.

40. Fruman, D.A., Meyers, R.E., and Cantley, L.C., Phosphoinositide kinases, *Annu. Rev. Biochem.,* 67, 481–507, 1998.

41. Marsden, V.S. and Strasser, A., Control of apoptosis in the immune system: Bcl-2, BH3-only proteins and more, *Annu. Rev. Immunol.,* 21, 71–105, 2003.

42. Jones, R.G., Parsons, M., Bonnard, M., Chan, V.S., Yeh, W.C., Woodgett, J.R., and Ohashi, P.S., Protein kinase B regulates T lymphocyte survival, nuclear factor B activation, and Bcl-X(L) levels *in vivo, J. Exp. Med.,* 191, 1721–1734, 2000.

43. Green, D.R., Droin, N., and Pinkoski, M., Activation-induced cell death in T cells, *Immunol. Rev.,* 193, 70–81, 2003.

44. Andreola, G., Rivoltini, L., Castelli, C., Huber, V., Perego, P., Deho, P., Squarcina, P., Accornero, P., Lozupone, F., Lugini, L., Stringaro, A., Molinari, A., Arancia, G., Gentile, M., Parmiani, G., and Fais, S., Induction of lymphocyte apoptosis by tumor cell secretion of FasL-bearing microvesicles, *J. Exp. Med.,* 195, 1303–1316, 2002.

45. Benedict, C.A., Norris, P.S., and Ware, C.F., To kill or be killed: viral evasion of apoptosis, *Nat. Immunol.,* 3, 1013–1018, 2002.

46. Tschopp, J., Thome, M., Hofmann, K., and Meinl, E., The fight of viruses against apoptosis, *Curr. Opin. Genet. Dev.,* 8, 82–87, 1998.

47. Dahl, J., Jurczak, A., Cheng, L.A., Baker, D.C., and Benjamin, T.L., Evidence of a role for phosphatidylinositol 3-kinase activation in the blocking of apoptosis by polyomavirus middle T antigen, *J. Virol.,* 72, 3221–3226, 1998.

48. Meinl, E., Fickenscher, H., Thome, M., Tschopp, J., and Fleckenstein, B., Anti-apoptotic strategies of lymphotropic viruses, *Immunol. Today,* 19, 474–479, 1998.

49. Herbein, G., Mahlknecht, U., Batliwalla, F., Gregersen, P., Pappas, T., Butler, J., O'Brien, W.A., and Verdin, E., Apoptosis of CD+ T cells is mediated by macrophages through interaction of HIV gp120 with chemokine receptor CXCR4, *Nature,* 395, 189–194, 1998.

50. Geleziunas, R., Xu, W., Takeda, K., Ichijo, H., and Greene, W.C., HIV-1 Nef inhibits ASK1-dependent death signalling providing a potential mechanism for protecting the infected host cell, *Nature,* 410, 834–838, 2001.
51. Chang, H.Y., Nishitoh, H., Yang, X., Ichijo, H., and Baltimore, D., Activation of apoptosis signal-regulating kinase 1 (ASK1) by the adapter protein Daxx, *Science,* 281, 1860–1863, 1998.
52. Tobiume, K., Matsuzawa, A., Takahashi, T., Nishitoh, H., Morita, K., Takeda, K., Minowa, O., Miyazono, K., Noda, T., and Ichijo, H., ASK1 is required for sustained activations of JNK/p38 MAP kinases and apoptosis, *EMBO Rep.,* 2, 222–228, 2001.
53. Wolf, D., Witte, V., Laffert, B., Blume, K., Stromer, E., Trapp, S., d'Aloja, P., Schurmann, A., and Baur, A.S., HIV-1 Nef associated PAK and PI3-kinases stimulate Akt-independent Bad-phosphorylation to induce anti-apoptotic signals, *Nat. Med.,* 7, 1217–1224, 2001.
54. Finkel, T.H., Tudor-Williams, G., Banda, N.K., Cotton, M.F., Curiel, T., Monks, C., Baba, T.W., Ruprecht, R.M., and Kupfer, A., Apoptosis occurs predominantly in bystander cells and not in productively infected cells of HIV- and SIV-infected lymph nodes, *Nat. Med.,* 1, 129–134, 1995.
55. Xu, X.N., Screaton, G.R., Gotch, F.M., Dong, T., Tan, R., Almond, N., Walker, B., Stebbings, R., Kent, K., Nagata, S., Stott, J.E., and McMichael, A.J., Evasion of cytotoxic T lymphocyte (CTL) responses by Nef-dependent induction of Fas ligand (CD95L) expression on simian immunodeficiency virus-infected cells, *J. Exp. Med.,* 186, 7–16, 1997.
56. Combadiere, B., Freedman, M., Chen, L., Shores, E.W., Love, P., and Lenardo, M.J., Qualitative and quantitative contributions of the T cell receptor chain to mature T cell apoptosis, *J. Exp. Med.,* 183, 2109–2117, 1996.
57. Ross, T.M., Using death to one's advantage: HIV modulation of apoptosis, *Leukemia,* 15, 332–341, 2001.
58. Steinman R.M., DC-SIGN: a guide to some mysteries of dendritic cells. *Cell,* 2000 Mar 3;100(5): 491–4.
59. Fujii, Y., Otake, K., Tashiro, M., and Adachi, A., Soluble Nef antigen of HIV-1 is cytotoxic for human CD4+ T cells, *FEBS Lett.,* 393, 93–96, 1996.
60. Okada, H., Takei, R., and Tashiro, M., Nef protein of HIV-1 induces apoptotic cytolysis of murine lymphoid cells independently of CD95 (Fas) and its suppression by serine/threonine protein kinase inhibitors, *FEBS Lett.,* 417, 61–64, 1997.
61. James, C.O., Huang, M.B., Khan, M., Garcia-Barrio, M., Powell, M.D., and Bond, V.C., Extracellular Nef protein targets CD4+ T cells for apoptosis by interacting with CXCR4 surface receptors, *J. Virol.,* 78, 3099–3109, 2004.
62. Mueller, Y.M., De Rosa, S.C., Hutton, J.A., Witek, J., Roederer, M., Altman, J.D., and Katsikis, P.D., Increased CD95/Fas-induced apoptosis of HIV-specific CD8(+) T cells, *Immunity,* 15, 871–882, 2001.

9 HIV-1 Tat and Apoptotic Death

Sylviane Muller
Claude Desgranges

CONTENTS

INTRODUCTION

Human immunodeficiency virus type 1 (HIV-1) infection is characterized by a progressive loss of CD4+ T cells, and apoptosis has been proposed to be the primary mechanism involved in this process.[1,2] A growing body of literature suggests that the HIV-1 transactivator protein, Tat, plays a key role in HIV-1-mediated cell death and induces apoptosis in T lymphocyte cell lines as well as in peripheral blood mononuclear cells (PBMCs). Tat has also been shown to possess pro-survival properties. Other HIV-1 proteins have been implicated in regulating apoptosis namely Env, Vpr, and Nef.[3–5] In this chapter, we will focus our attention on Tat, a small regulatory protein with pleiotropic properties that regulates apoptosis of infected and noninfected cells by a variety of possible mechanisms.

HIV-1 Tat is a well-conserved protein. Although the laboratory HIV strains produce an active 86-amino acid Tat protein, most Tat proteins from primary isolates contain an additional 15- to 16-residue-long sequence at their C-terminus. Tat is encoded by two exons. The first encodes residues 1 to 72 and is classically described as a modular protein (Figure 9.1). Tat contains six regions, namely, the proline-rich N-terminus (aa 1 to aa 21), the cysteine-rich region (aa 22 to aa 37), the hydrophobic core region (aa 38 to aa 48), the basic region (aa 49 to aa 59), the glutamine-rich region (aa 60 to aa 72), and the C-terminus encoded by the second exon and with size that corresponds to the isolate. Tat displays sequence variability from isolate to isolate but contains relatively conserved residues or sequences, including a conserved tryptophane residue at position 11; seven cysteine residues at positions 22, 25, 27, 30, 31, 34, and 37 in the cysteine-rich region; the sequence [43]LGISYG[48] in the core region; and the sequence [49]RKKRRQRRR[57] in the basic region.

The functions of several regions of the protein have been well characterized.[6–13] The N-terminal region 1-9 is involved in the binding of Tat to CD26 (DP IV[8,11]), and the region 49-57 is involved in Tat uptake. Albini et al.[9] demonstrated that the synthetic peptide 24-51, which encompassed the "chemokine-like" region of Tat, induces the rapid and transient Ca++ influx in monocytes and macrophages, analogous to β-chemokines. This peptide is angiogenic *in vivo*.[14] Boykins et al.[10] showed that the fragment 21-41 is sufficient to transactivate, induce HIV replication, and trigger angiogenesis. Mutations at the positions 22 or 22/37 lead to a protein that lacks transactivation capacities and that displays a transdominant negative phenotype.[15] Acetylation of Tat at positions

```
              N-terminal        Cys-rich      Core   K,R-rich(LNS)    Gln-rich    RGD

Tat Lai    (B)   MEPVDPRLEPWKHPGSQPKTACTTCYCKKCCFHCQVCFTTKALGISYGRKKRRQRRRPPQGSQTHQVSLSKQPTSQPRGDPTGPKE                86

Tat HXB2   (B)   .................N..........I.............AH.N...A...........-KKKVERETETDPFD                100

Tat CM240  (A/E) ..L...N....N......T...SK.......W...L..LK.G.........KH..GT..S.KD..NPIP...LPII.RN..D...S..E.ASKA...QC.    101

Tat 92Br   (C)   ......N...N..........NN....R.SY..L...Q..G.............SA.PS.ED..NPIP...LP.T...Q..SE.S.....SK.....F.    101

Tat Eli    (D)   .D....N...N......R.P.NK.H....Y..P...LN.G.............G...G.A..PIP...S............Q.....S.A....         99

Tat Eg11RP (A)   .D....NI...N.......TP.NK....V..Y..L...QS.G.........K...GPT.SNKQ..NPIP...IPRTQ.IS...E.S.....DK....RR.   101
```

FIGURE 9.1 Sequences and modular structure of Tat proteins. HXB2 HIV-1 strain was isolated in France, *CM240* in Thailand, *92Br* in Brazil, *Eli* in the Democratic Republic of Congo, and Ug11RP HIV-1 strain in Uganda. (Adapted from Opi, S., Peloponese, J.M. Jr, Esquieu, D., Campbell, G., de Mareuil, J., Walburger, A., Solomiac, M., Gregoire, C., Bouveret, E., Yirrell, D.L., and Loret E.P., *J. Biol. Chem.*, 277, 35915–35919, 2002. With permission.)

K^{28} and K^{50} is particularly important for its transactivation activity.[16–18] Mutation and competition experiments, as well as ribonuclease protection assays, demonstrated that the arginine-rich motif of Tat (aa 49 to aa 57), and especially the arginine residue 52 (Figure 9.1), is essential for its high-affinity binding to transactivation response (TAR) element RNA, a 59-nucleotide stem–loop structure located at the 5′ end of all transcripts.[19] It is through its interaction with TAR that Tat acts as a powerful transcriptional activator. It seems that in addition to residues present in the basic domain, the N-terminus region of Tat is also important for Tat–TAR interaction *in vitro*.[20] *In vivo*, the N-terminal domain of Tat (aa 1 to aa 48) interacts directly with cyclin T1, a regulatory partner of cyclin-dependent kinase 9 (Cdk9) in the positive transcription elongation factor (P-TEFb) complex, and Tat/cyclinT1 in the P-TEFb complex binds TAR RNA cooperatively. The assembly of the Tat/TAR/P-TEFb complex to the HIV promoter activates the Cdk9 activities, which further auto-phosphorylate the P-TEFb complex and hyperphosphorylate the C-terminal domain of RNA polymerase II, which leads to the formation of processive elongation complexes that synthesize full-length HIV viral mRNA.[21,22] Recent studies have shown that the nascent P-TEFb complex contains other RNA and proteins of known or unknown identity.

Regarding the C-terminal domain of Tat, Poggi et al.[13] reported that HIV-1 Tat and, particularly, the peptide 65-80 inhibit natural killer (NK) cell–mediated lysis of dendritic cells (DCs) and interferon (INF)-γ release by NK cells after their contact with DC. Finally, the RGD motif present in residues 78 to 80 binds the integrins α5β1 and αvβ3, the receptors of fibronectin and vibronectin, respectively.[23,24]

A panoply of peptide analogues or molecules targeting specific regions of Tat as well as TAR and cyclin T1 was used as possible blocking agents in therapeutic strategies (see Hamy et al.[25]; Fulcher and Jans[26]; Holmes, Arzumanov, and Gait[27]; Bai et al.[28]; and Tikhonov et al.[29] for examples).

HIV-1 TAT AND APOPTOSIS

Tat is produced early after infection and is essential for viral gene expression and virus production. Work conducted in the last few years by several laboratories has shed light on the numerous interconnected functions of this striking small protein, both intra- and extracellularly. Its extracellular functions play an important role in viral infection. Significant levels of extracellular Tat are found in the serum from HIV-1-infected patients and in Kaposi's sarcoma lesions, in which it is associated with endothelial spindle cells. It is also released in the culture supernatants of cells infected with HIV-1.[30,31] It has been shown with Tat-transfected COS-1 cells that Tat is released in the absence of cell death or changes in cell permeability via a leaderless exocytosic pathway.[32,33] During acute infection of T cells by HIV, extracellular Tat released in the stromal microenvironment of infected cells can bind or can be efficiently taken up by most cell types.[30,32–35] In infected cells, Tat will then transactivate virus gene expression and replication, and in uninfected cells it will favor the transmission of both macrophage-tropic and T lymphocyte-tropic HIV-1 strains by inducing the expression of the chemokine receptors (and HIV-1 co-receptors) CCR5 and CXCR4.[36–38] Specific motifs are engaged in Tat internalization, namely, the basic domain, which interacts with negatively charged sulfated groups of free heparin and heparan sulfates and to cell-surface heparan sulfate proteoglycans, and the RGD motif. Thus, Tat protein, which is readily released from infected cells and taken up by neighboring (infected or uninfected) cells, is implicated in HIV-1 pathogenesis not only by its indispensability for virus replication but also by its numerous other control functions that ultimately affect cell survival, growth, and angiogenesis.

By the early 1990s, Tat was shown to inhibit proliferation and induce apoptosis in activated uninfected cells.[31,39–41] Depending on the cell type and conditions (in particular, Tat concentrations and source), apoptosis of uninfected "bystander" cells was also described. Tat-dependent apoptosis was described in several cell types, including T and B cells, erythroid cells, endothelial cells, and neuronal cells. For example, Tat was found to display apoptotic properties on CD4[+] Jurkat T lymphoblastoid cells but was ineffective when unstimulated PBMCs or PBMCs activated by anti-CD3 antibodies (OKT3) were incubated with up to 5 μM synthetic Tat Lai or Tat fragments (Figure 9.2). In the case of Jurkat cells, apoptosis was demonstrated by different techniques, including flow

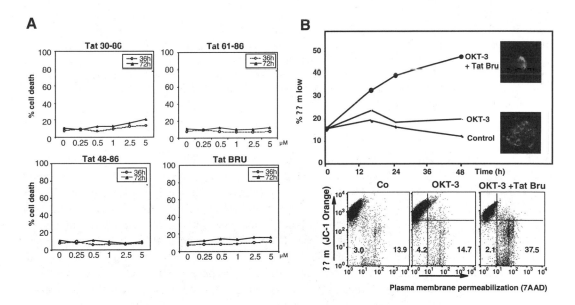

FIGURE 9.2 Tat promotes apoptosis of CD3-stimulated Jurkat cells but does not kill PBMCs. (A) PBMCs were exposed up to 72 h to indicated doses of synthetic Tat protein or Tat fragments, labeled with 7 amino-actinomycin D (7-AAD), and submitted to FACS analysis. Curves are the means of three independent experiments (SD < 5%). Note that, in these experimental conditions, there is no significant cellular uptake of 7-AAD, thus indicating the absence of Tat-induced PBMC death. (B) Jurkat cells were stimulated for 24 h in the presence of OKT3 (anti-CD3), were then submitted or not to synthetic Tat Lai, were colabeled with the $\Delta\Psi$m-sensitive probe JC-1 and with the DNA-interacting dye Hoechst 33342 and with 7-AAD, and were submitted to both FACS analysis and fluorescence microscopy observations. A representative $\Delta\Psi$m time response (after OKT3 plus Tat treatment) is shown in the upper panel. Inserted fluorescent micrographs show Jurkat cells incubated (upper insert) or not (lower insert) with synthetic Tat (5 μM) for 24 h and stained with 1 μM $\Delta\Psi_m$-sensitive dye JC-1 (light dots in bottom image fluorescence shows mitochondria with high $\Delta\Psi_m$, light area on the upper right side of the top image fluorescence shows mitochondria with low $\Delta\Psi_m$) and with Hoechst 33342 (base color of top and bottom images fluorescence). Dot plot was obtained by FACS analysis after costaining with JC-1 and 7-AAD.

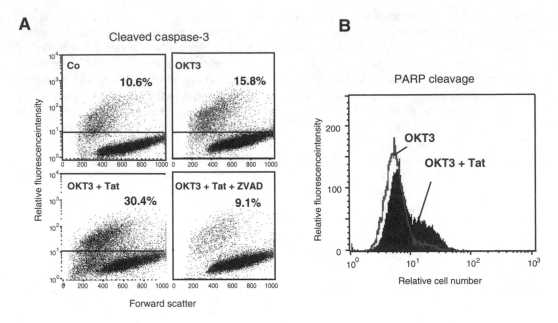

FIGURE 9.3 Tat enhances caspase-3 activation and PARP cleavage in CD3-stimulated Jurkat cells. Experimental conditions were similar to those described in the legend of Figure 9.2. At 24 h post-Tat treatment, cells were subjected to permeabilization and labeled with either a polyclonal FITC-tagged antibody directed to activated caspase-3 (A) or a polyclonal FITC-tagged antibody directed to cleaved PARP (B). A representative experiment is shown. Note that pretreatment with the pan-caspase inhibitor zVAD-fmk (50 μM) fully inhibited Tat-induced caspase-3 activation.

cytometry after propidium iodide labeling of treated cells, the TUNEL (terminal deoxynucleotidyltransferase-mediated dUTP-biotin nick-end labeling) method, the cleavage of caspase-3 and poly(ADP ribose polymer)ase (PARP) revealed by FITC-labeled specific antibodies and FACS analysis, and the evaluation of mitochondrial transmembrane potential ($\Delta\Psi m$) state (Figure 9.2 and Figure 9.3)[42] (Lecoeur and Jacotot, personal communication). Tat was also shown to induce cell death by apoptosis in primary cortical neurons (Lecoeur, Chauvier et al.[43]) and cultured human fetal neurons[44] and rat neurons. For example, it was shown recently with striatal neurons that both HIV-1 gp120 and Tat fragment 1-72 significantly increased caspase-3 activation in a concentration-dependent manner, but that even though both components induced significant neurotoxicity, the nature of the apoptotic pathways preceding death differed.[45]

Mechanisms by which Tat was reported to induce apoptosis include upregulation of the apoptotic effector molecule Fas ligand in a pathway that might involve NF-κB,[31,46] inhibition of the expression of manganese-dependent superoxide dismutase (Mn-SOD),[47] and the activation of cyclin-dependent kinases.[41] These studies were performed in the context of uninfected cells. Thus, in both T cells and HeLa cells, it was shown, for example, that Tat suppressed the expression of Mn-SOD, a mitochondrial enzyme that is part of the cellular defense system against oxidative stress.[47] The truncated Tat fragment 1-72, known to transactivate the HIV-1 long terminal repeat (LTR), was unable to affect Mn-SOD expression, the cellular redox state, or TNF-mediated cytotoxicity. (TNF is known to stimulate HIV-1 replication through activation of NF-κB.) This result highlights the importance of the C-terminal end of Tat protein in this process.

Tat binds tubulin and polymerized microtubules through a four-amino-acid residue subdomain located in the core region, leading to the alteration of microtubule dynamics and activation of a mitochondria-dependent apoptotic pathway.[48]

Tat-dependent apoptosis may also be explained by the effects of Tat on Bcl-2 and related proteins. Tat-mediated downregulation of Bcl-2 was effectively observed at both the transcriptional and translational levels. Also, Tat-transfected cells expressed increased amounts of Bax, a Bcl-2 family protein known to induce apoptosis.[49] In contrast, however, stable and transient transfections of Jurkat cells with the cDNA of Tat and a plasmid containing Bcl-2 promoter in front of CAT (Bcl-2 Pr/CAT) were found to stimulate CAT activity and lead to an increase of Bcl-2 mRNA and protein expression. This effect was specifically related to Tat, because Jurkat cells transfected with the cDNA of Tat in antisense orientation, Tat carrying a mutation at the position 22 (mutant C22G), or the control vector alone (pRPneo-SL3) did not show any significant difference in Bcl-2 promoter activity with respect to parental Jurkat cells.[50]

In the context of infected cells, Tat was found to induce apoptosis by upregulating caspase-8, a protease that cleaves downstream caspases such as caspase-3, thus initiating the caspase cascade and increasing the sensitivity of cells to apoptotic signals.[51] The induction of apoptosis could be prevented by treating cells with the caspase-8 inhibitor 2-IETD-fmk. The capacity of Tat to upregulate caspase-8 was shown in both CD4+ T cells and T cell lines, and it was found that Tat does not require Fas–Fas ligand interaction to induce apoptosis. It was demonstrated further that Tat transactivation mutants, namely, Tat C22G and Tat K41A mutants, induced apoptosis, whereas a truncated Tat fragment 1-72 did not. This result supports the fact that the C-terminal of the protein (expressed by the second exon) is required to induce apoptosis in the context of a viral infection and that induction of apoptosis by Tat is a function independent from transactivation.[51] One report proposed that Tat may accumulate at the mitochondria in stressed Tat-expressing cells, and this mitochondrial localization was found to be correlated with the $\Delta\Psi$m disruption detected in these cells.[52]

Controversies remain, however, because Tat can either induce apoptosis[31,39–49,51–53] or inhibit apoptosis, depending on context.[50,54,55] Furthermore, certain cells, such as chimpanzee T cells, have been shown to be resistant to Tat-enhanced apoptosis. A number of studies were devoted to the molecular mechanisms by which Tat might directly or indirectly regulate apoptosis. Several cofactors mediating Tat-induced apoptosis were characterized, and models of Tat-mediated apoptosis were proposed. At this stage, however, it seems apparent that the observed ability of Tat to activate or inhibit apoptosis is highly dependent on the cell type, its stage of activation and infection, the use of tumoral or primary cells, their environments (presence of co-stimulatory signals and other cells), and the method used to reveal apoptosis.[54,56,57] Significant differences were also observed in the presence of recombinant and synthetic Tat according to their integrity, their degree of purity, and their proper folding. The length (86 versus 106 residues) of the protein might also be important. This suggests that the results of some studies would need to be revisited.

MODULATING APOPTOSIS WITH TAT ANTIBODIES

Tat antibodies are detected in the serum of HIV-infected patients, and a number of short linear Tat epitopes recognized by IgM or IgG patients' antibodies have been characterized in several regions of the protein.[58–62] Independent groups showed that, in contrast to antibodies to Env, Nef, and Gag, the level of Tat antibodies is elevated in long-term non- or slow-progressors and is low in fast progressors; their decline in asymptomatic individuals would be related to the imminence of acquired immunodeficiency syndrome (AIDS).[62–66] It was also reported that the presence of Tat natural IgM antibodies in normal individuals (and chimpanzees, but not in other primates and mice) and reacting with functional Tat regions might have some influence on the course of AIDS progression.[64,67] Taken together, this body of evidence suggests that immunization with Tat or Tat fragments might block certain extracellular deleterious functions of Tat and, consequently, may considerably delay the emergence and spread of infection. Until now, however, different groups have reported controversial results describing partial[68–71] or no protection[72–74] in monkeys vaccinated with Tat protein, Tat toxoid, or short Tat peptides.

B cell epitopes important for Tat neutralization were located in the N-terminus and the basic region of the Tat protein.[60–62,74–78] These regions of Tat, in particular, the basic region, are relatively well conserved among all viral subtypes, except for the O subtype.[42,79] Recently, we showed that only sera-containing antibodies reacting with both regions were able to inhibit extracellular Tat-dependent transactivation *in vitro*.[80] This feature was found in mice, rabbits, macaques, and humans vaccinated with recombinant or synthetic Tat, Tat toxoid, and Tat peptides. Furthermore, human monoclonal antibodies as well as human single-chain fragment variable (scFv) antibodies specific for Tat were found to modulate HIV-1 replication in infected cell cultures. Interestingly, the two IgGs reacted with the basic region 44-61 of Tat, and the three scFvs were all directed to the N-terminus region 1-20 of the protein.[81] These results suggest that inhibition of secreted Tat contributes significantly to hampering viral replication *in vitro*.

Specific antibodies generated in rabbits and macaques against Tat peptides were also shown to inhibit Tat-induced apoptosis of T cells. This was demonstrated, for example, with OKT3-activated Jurkat T cells.[42] Such antibodies, either produced actively by vaccinated individuals or administrated passively at the very beginning of infection, might play a critical role in controlling Tat-induced death of uninfected cells. They might hamper the effect of Tat on critical immune cells and particularly on dendritic cells that are targeted and activated by Tat.[82,83] Several independent clinical trials have demonstrated the safety of Tat toxoid,[84–86] and Ensoli and colleagues recently started preventive and therapeutic phase I clinical trials in Italy to evaluate the effectiveness of a Tat-based vaccine strategy.[87] Intense research is also devoted to the production of Tat-specific human intrabodies that may be useful for passive immunization.[81,88,89]

CONCLUDING REMARKS

Tat displays pleiotropic properties, and blocking exogenous Tat might interfere with the development of immunodeficiency. Several studies, however, have shown controversial effects, and it is sometimes difficult to reconcile the experimental observations. As described above, according to the types of cells and their environments, and according also to the type of Tat used for testing apoptosis (purified, recombinant, or synthetic protein), Tat may induce or inhibit apoptosis. Moreover, Tat influences different pathways that can have ultimately contrasting effects on cell survival. Finally, an important question concerns the effect of extracellular Tat *in vivo*. The amount of Tat released locally from infected cells is difficult to evaluate. Its conformation when it is secreted from these cells is unknown. Is this conformation compatible with its rapid penetration into neighboring cells *in vivo*? Is the secreted Tat in a biologically active state? Are the numerous effects of Tat demonstrated *in vitro* relevant *in vivo*? Thus, several questions remain unanswered in the absence of an *in vivo* study model. As is often the case in science, it is possible that researchers will provide successful protection studies with Tat before having understood all the secrets of this striking small molecule.

ACKNOWLEDGMENTS

We are grateful to Hervé Lecoeur and Etienne Jacotot (Theraptosis, Paris) for helpful comments on the manuscript and for generating the results shown in Figure 9.2 and Figure 9.3.

REFERENCES

1. Ameisen, J.C., Estaquier, J., Idziorek, T., and De Bels, F., Programmed cell death and AIDS pathogenesis: significance and potential mechanisms, *Curr. Top. Microbiol. Immunol.*, 200, 195–211, 1995.
2. Roshal, M., Zhu, Y., and Planelles, V., Apoptosis in AIDS, *Apoptosis*, 6, 103–116, 2001.
3. Dianzani, U., Bensi, T., Savarino, A., Sametti, S., Indelicato, M., Mesturini, R., and Chiocchetti, A., Role of FAS in HIV infection, *Curr. HIV Res.* 1, 405–417, 2003.

4. Castedo, M., Perfettini, J.L., Andreau, K., Roumier, T., Piacentini, M., and Kroemer, G., Mitochondrial apoptosis induced by the HIV-1 envelope, *Ann. N.Y. Acad. Sci.*, 1010, 19–28, 2003.
5. Seelamgari, A., Maddukuri, A., Berro, R., de la Fuente, C., Kehn, K., Deng, L., Dadgar, S., Bottazzi, M.E., Ghedin, E., Pumfery, A., and Kashanchi, F., Role of viral regulatory and accessory proteins in HIV-1 replication, *Front. Biosci.* 9, 2388–2413, 2004.
6. Opi, S., Peloponese, J.M. Jr, Esquieu, D., Campbell, G., de Mareuil, J., Walburger, A., Solomiac, M., Gregoire, C., Bouveret, E., Yirrell, D.L., and Loret E.P., Tat HIV-1 primary and tertiary structures critical to immune response against non-homologous variants, *J. Biol. Chem.*, 277, 35915–35919, 2002.
7. Frankel, A.D., Biancalana, S., and Hudson, D., Activity of synthetic peptides from the Tat protein of human immunodeficiency virus type 1, *Proc. Natl. Acad. Sci. U.S.A.*, 86, 7397–7401, 1989.
8. Wrenger, S., Hoffmann, T., Faust, J., Mrestani-Klaus, C., Brandt, W., Neubert, K., Kraft, M., Olek, S., Frank, R., Ansorge, S., and Reinhold, D., The N-terminal structure of HIV-1 Tat is required for suppression of CD26-dependent T cell growth, *J. Biol. Chem.*, 272, 30283–30288, 1997.
9. Albini, A., Ferrini, S., Benelli, R., Sforzini, S., Giunciuglio, D., Aluigi, M.G., Proudfoot, A.E., Alouani, S., Wells, T.N., Mariani, G., Rabin, R.L., Farber, J.M., and Noonan, D.M., HIV-1 Tat protein mimicry of chemokines, *Proc. Natl. Acad. Sci. U.S.A.*, 95, 13153–13158, 1998.
10. Boykins, R.A., Mahieux, R., Shankavaram, U.T., Gho, Y.S., Lee, S.F., Hewlett, I.K., Wahl, L.M., Kleinman, H.K., Brady, J.N., Yamada, K.M., and Dhawan, S., A short polypeptide domain of HIV-1-Tat protein mediates pathogenesis, *J. Immunol.*, 163, 15–20, 1999.
11. Wrenger, S., Reinhold, D., Faust, J., Mrestani-Klaus, C., Brandt, W., Fengler, A., Neubert, K., and Ansorge, S., Effects of nonapeptides derived from the N-terminal structure of human immunodeficiency virus-1 (HIV-1) Tat on suppression of CD26-dependent T cell growth, *Adv. Exp. Med. Biol.*, 477, 161–165, 2000.
12. Betti, M., Voltan, R., Marchisio, M., Mantovani, I., Boarini, C., Nappi, F., Ensoli, B., and Caputo, A., Characterization of HIV-1 Tat proteins mutated in the transactivation domain for prophylactic and therapeutic application, *Vaccine*, 19, 3408–3419, 2001.
13. Poggi, A., Carosio, R., Spaggiari, G.M., Fortis, C., Tambussi, G., Dell'Antonio, G., Dal Cin, E., Rubartelli, A., and Zocchi, M.R., NK cell activation by dendritic cells is dependent on LFA-1-mediated induction of calcium-calmodulin kinase II: inhibition by HIV-1 Tat C-terminal domain, *J. Immunol.*, 168, 95–101, 2002.
14. Benelli, R., Barbero, A., Ferrini, S., Scapini, P., Cassatella, M., Bussolino, F., Tacchetti, C., Noonan, D.M., and Albini, A., Human immunodeficiency virus transactivator protein (Tat) stimulates chemotaxis, calcium mobilization, and activation of human polymorphonuclear leukocytes: implications for Tat-mediated pathogenesis, *J. Infect. Dis.*, 182, 1643–1651, 2000.
15. Rossi, C., Balboni, P.G., Betti, M., Marconi, P.C., Bozzini, R., Grossi, M.P., Barbanti-Brodano, G., and Caputo, A., Inhibition of HIV-1 replication by a Tat transdominant negative mutant in human peripheral blood lymphocytes from healthy donors and HIV-1-infected patients, *Gene Ther.*, 4, 1261–1269, 1997.
16. Kiernan, R.E., Vanhulle, C., Schiltz, L., Adam, E., Xiao, H., Maudoux, F., Calomme, C., Burny, A., Nakatani, Y., Jeang, K.T., Benkirane, M., and Van Lint, C., HIV-1 tat transcriptional activity is regulated by acetylation, *EMBO J.*, 18, 6106–6118, 1999.
17. Ott, M., Schnolzer, M., Garnica, J., Fischle, W., Emiliani, S., Rackwitz, H.R., and Verdin, E., Acetylation of the HIV-1 Tat protein by p300 is important for its transcriptional activity, *Curr. Biol.*, 9, 1489–1492, 1999.
18. Kaechlcke, K., Dorr, A., Hetzer-Egger, C., Kiermer, V., Henklein, P., Schnoelzer, M., Loret, E., Cole, P.A., Verdin, E., and Ott, M., Acetylation of Tat defines a cyclin T1-independent step in HIV transactivation, *Mol. Cell*, 12, 167–176, 2003.
19. Green, M., Ishino, M., and Loewenstein, P.M., Mutational analysis of HIV-1 Tat minimal domain peptides: identification of trans-dominant mutants that suppress HIV-LTR-driven gene expression, *Cell*, 58, 215–223, 1989.
20. Chaloin, O., Peter, J.-C., Briand, J.-P., Masquida, B., Desgranges, C., Muller, S., and Hoebeke, J., The N-terminus of HIV-1 Tat protein is essential for Tat-TAR RNA interaction, *Cell. Mol. Life Sci.*, 62, 355–361, 2005.
21. Brigati, C., Giacca, M., Noonan, D.M., and Albini, A., HIV Tat, its TARgets and the control of viral gene expression, *FEMS Microbiol. Lett.*, 220, 57–65, 2003.

22. Huigen, M.C., Kamp, W., and Nottet, H.S., Multiple effects of HIV-1 trans-activator protein on the pathogenesis of HIV-1 infection, *Eur. J. Clin. Invest.,* 34, 57–66, 2004.

23. Barillari, G., Gendelman, R., Gallo, R.C., and Ensoli, B., The Tat protein of human immunodeficiency virus type 1, a growth factor for AIDS Kaposi sarcoma and cytokine-activated vascular cells, induces adhesion of the same cell types by using integrin receptors recognizing the RGD amino acid sequence, *Proc. Natl. Acad. Sci. U.S.A.,* 90, 7941–7945, 1993.

24. Zauli, G., Gibellini, D., Celeghini, C., Mischiati, C., Bassini, A., La Placa, M., and Capitani, S., Pleiotropic effects of immobilized vs. soluble recombinant HIV-1 Tat protein on CD3-mediated activation, induction of apoptosis, and HIV-1 long terminal repeat transactivation in purified CD4+ T lymphocytes, *J. Immunol.,* 157, 2216–2224, 1996.

25. Hamy, F., Gelus, N., Zeller, M., Lazdins, J.L., Bailly, C., and Klimkait, T., Blocking HIV replication by targeting Tat protein, *Chem. Biol.,* 7, 669–676, 2000.

26. Fulcher, A.J. and Jans, D.A., The HIV-1 Tat transactivator protein: a therapeutic target? *IUBMB Life,* 55, 669–680, 2003.

27. Holmes, S.C., Arzumanov, A.A., and Gait, M.J., Steric inhibition of human immunodeficiency virus type-1 Tat-dependent trans-activation *in vitro* and in cells by oligonucleotides containing 2-O-methyl G-clamp ribonucleoside analogues, *Nucleic Acids Res.,* 31, 2759–2768, 2003.

28. Bai, J., Sui, J., Zhu, R.Y., Tallarico, A.S., Gennari, F., Zhang, D., and Marasco, W.A., Inhibition of Tat-mediated transactivation and HIV-1 replication by human anti-hCyclinT1 intrabodies, *J. Biol. Chem.,* 278, 1433–1442, 2003.

29. Tikhonov, I., Ruckwardt, T.J., Berg, S., Hatfield, G.S., and Pauza, C.D., Furin cleavage of the HIV-1 Tat protein, *FEBS Lett.,* 565, 89–92, 2004.

30. Ensoli, B., Buonaguro, L., Barillari, G., Fiorelli, V., Gendelman, R., Morgan, R.A., Wingfield, P., and Gallo, R.C., Release, uptake, and effects of extracellular human immunodeficiency virus type 1 Tat protein on cell growth and viral transactivation, *J. Virol.,* 67, 277–287, 1993.

31. Westendorp, M.O., Frank, R., Ochsenbauer, C., Stricker, K., Dhein, J., Walczak, H., Debatin, K.M., and Krammer, P.H., Sensitization of T cells to CD95-mediated apoptosis by HIV-1 Tat and gp120, *Nature,* 375, 497–500, 1995.

32. Chang, H.C., Samaniego, F., Nair, B.C., Buonaguro, L., and Ensoli, B., HIV-1 Tat protein exits from cells via a leaderless secretory pathway and binds to extracellular matrix-associated heparan sulfate proteoglycans through its basic region, *AIDS,* 11, 1421–1431, 1997.

33. Frankel, A.D. and Pabo, C.O., Cellular uptake of the Tat protein from human immunodeficiency virus, *Cell,* 55, 1189–1193, 1988.

34. Westendorp, M.O., Li-Weber, M., Frank, R.W., and Krammer, P.H., Human immunodeficiency virus type 1 Tat upregulates interleukin-2 secretion in activated T cells, *J. Virol.,* 68, 4177–4185, 1994.

35. Rusnati, M. and Presta, M., HIV-1 Tat protein and endothelium: from protein/cell interaction to AIDS-associated pathologies, *Angiogenesis,* 5, 141–151, 2002.

36. Li, C.J., Ueda, Y., Shi, B., Borodyansky, L., Huang, L., Li, Y.Z., and Pardee, A.B., Tat protein induces self-perpetuating permissivity for productive HIV-1 infection, *Proc. Natl. Acad. Sci. U.S.A.,* 94, 8116–8120, 1997.

37. Huang, L., Bosch, I., Hofmann, W., Sodroski, J., and Pardee, A.B., Tat protein induces human immunodeficiency virus type 1 (HIV-1) coreceptors and promotes infection with both macrophage-tropic and T-lymphotropic HIV-1 strains, *J. Virol.,* 72, 8952–8960, 1998.

38. Secchiero, P., Zella, D., Capitani, S., Gallo, R.C., and Zauli, G., Extracellular HIV-1 Tat protein up-regulates the expression of surface CXC-chemokine receptor 4 in resting CD4+ T cells, *J. Immunol.,* 162, 2427–2431, 1999.

39. Viscidi, R.P., Mayur, K., Lederman, H.M., and Frankel, A.D., Inhibition of antigen-induced lymphocyte proliferation by Tat protein from HIV-1, *Science,* 246, 1606–1608, 1989.

40. Meyaard, L., Otto, S.A., Schuitemaker, H., and Miedema, F., Effects of HIV-1 Tat protein on human T cell proliferation, *Eur. J. Immunol.,* 22, 2729–2732, 1992.

41. Li, C.J., Friedman, D.J., Wang, C., Metelev, V., and Pardee, A.B., Induction of apoptosis in uninfected lymphocytes by HIV-1 Tat protein, *Science,* 268, 429–431, 1995.

42. Belliard, G., Romieu, A., Zagury, J.-F., Dali, H., Chaloin, O., Le Grand, R., Loret, E., Briand, J.-P., Roques, B., Desgranges, C., and Muller, S., Specificity and effect on apoptosis of Tat antibodies from vaccinated and SHIV-infected rhesus macaques and HIV-infected individuals, *Vaccine,* 21, 3186–3199, 2003.

43. Lecoeur, H., Chalouin, O., Briere, J.-J., Rebouillat, D., Langonne, A., Longin, C., Deniaud, A., Brenner, C., Briand, J.P., Muller, S., Rustin, P., and Jacoto, E., HIV-1 Tat induces inner and out mitochondrial membrane permeabilization in isolated mitochondria and inactivates cytochrome C oxydase. Submitted.

44. Nath, A. and Geiger, A., Neurobiological aspects of human immunodeficiency virus infection: neurotoxic mechanisms, *Prog. Neurobiol.*, 54, 19–33, 1998.

45. Singh, I.N., Goody, R.J., Dean, C., Ahmad, N.M., Lutz, S.E., Knapp, P.E., Nath, A., and Hauser, K.F., Apoptotic death of striatal neurons induced by human immunodeficiency virus-1 Tat and gp120: differential involvement of caspase-3 and endonuclease G, *J. Neurovirol.*, 10, 141–151, 2004.

46. Li-Weber, M., Laur, O., Dern, K., and Krammer, P.H., T cell activation-induced and HIV tat-enhanced CD95(APO-1/Fas) ligand transcription involves NF-kappaB, *Eur. J. Immunol.*, 30, 661–670, 2000.

47. Westendorp, M.O., Shatrov, V.A., Schulze-Osthoff, K., Frank, R., Kraft, M., Los, M., Krammer, P.H., Droge, W., and Lehmann, V., HIV-1 Tat potentiates TNF-induced NF-B activation and cytotoxicity by altering the cellular redox state, *EMBO J.*, 14, 546–554, 1995.

48. Chen, D., Wang, M., Zhou, S., and Zhou, Q., HIV-1 Tat targets microtubules to induce apoptosis, a process promoted by the pro-apoptotic Bcl-2 relative Bim, *EMBO J.*, 21, 6801–6810, 2002.

49. Sastry, K.J., Marin, M.C., Nehete, P.N., McConnell, K., el-Naggar, A.K., and McDonnell, T.J., Expression of human immunodeficiency virus type I Tat results in down-regulation of Bcl-2 and induction of apoptosis in hematopoietic cells, *Oncogene,* 13, 487–493, 1996.

50. Zauli, G., Gibellini, D., Caputo, A., Bassini, A., Negrini, M., Monne, M., Mazzoni, M., and Capitani, S., The human immunodeficiency virus type-1 Tat protein upregulates Bcl-2 gene expression in Jurkat T-cell lines and primary peripheral blood mononuclear cells, *Blood,* 86, 3823–3834, 1995.

51. Bartz, S.R. and Emerman, M., Human immunodeficiency virus type 1 Tat induces apoptosis and increases sensitivity to apoptotic signals by up-regulating FLICE/caspase-8, *J. Virol.*, 73, 1956–1963, 1999.

52. Macho, A., Calzado, M.A., Jimenez-Reina, L., Ceballos, E., Leon, J., and Munoz, E., Susceptibility of HIV-1-TAT transfected cells to undergo apoptosis. Biochemical mechanisms, *Oncogene,* 18, 7543–7551, 1999.

53. Purvis, S.F., Jacobberger, J.W., Sramkoski, R.M., Patki, A.H., and Lederman, M.M., HIV type 1 Tat protein induces apoptosis and death in Jurkat cells, *AIDS Res. Hum. Retroviruses,* 11, 443–450, 1995.

54. McCloskey, T.W., Ott, M., Tribble, E., Khan, S.A., Teichberg, S., Paul, M.O., Pahwa, S., Verdin, E., and Chirmule, N., Dual role of HIV Tat in regulation of apoptosis in T cells, *J. Immunol.*, 158, 1014–1019, 1997.

55. Gibellini, D., Re, M.C., Ponti, C., Maldini, C., Celeghini, C., Cappellini, A., La Placa, M., and Zauli, G., HIV-1 Tat protects CD4+ Jurkat T lymphoblastoid cells from apoptosis mediated by TNF-related apoptosis-inducing ligand, *Cell Immunol.*, 207, 89–99, 2001.

56. Katsikis, P.D., Garcia-Ojeda, M.E., Torres-Roca, J.F., Greenwald, D.R., Herzenberg, L.A., and Herzenberg, L.A., HIV type 1 Tat protein enhances activation—but not Fas (CD95)-induced peripheral blood T cell apoptosis in healthy individuals, *Int. Immunol.*, 9, 835–841, 1997.

57. Park, I.W., Ullrich, C.K., Schoenberger, E., Ganju, R.K., and Groopman, J.E., HIV-1 Tat induces microvascular endothelial apoptosis through caspase activation, *J. Immunol.*, 167, 2766–2771, 2001.

58. Krone, W.J., Debouck, C., Epstein, L.G., Heutink, P., Meloen, R., and Goudsmit, J., Natural antibodies to HIV-Tat epitopes and expression of HIV-1 genes *in vivo, J. Med. Virol.*, 26, 261–270, 1988.

59. McPhee, D.A., Kemp, B.E., Cumming, S., Stapleton, D., Gust, I.D., and Doherty, R.R., Recognition of envelope and Tat protein synthetic peptide analogs by HIV positive sera or plasma, *FEBS Lett.*, 233, 393–396, 1988.

60. Tähtinen, M., Ranki, A., Valle, S.L., Ovod, V., and Krohn, K., B-cell epitopes in HIV-1 Tat and Rev proteins colocalize with T-cell epitopes and with functional domains, *Biomed. Pharmacother.*, 51, 480–487, 1997.

61. Demirhan, I., Chandra, A., Hasselmayer, O., Biberfeld, P., and Chandra, P., Detection of distinct patterns of anti-Tat antibodies in HIV-infected individuals with or without Kaposi's sarcoma, *J. Acquir. Immune Defic. Syndr.*, 22, 364–368, 1999.

62. Re, M.C., Vignoli, M., Furlini, G., Gibellini, D., Colangeli, V., Vitone, F., and La Placa, M., Antibodies against full-length Tat protein and some low-molecular-weight Tat-peptides correlate with low or undetectable viral load in HIV-1 seropositive patients, *J. Clin. Virol.*, 21, 81–89, 2001.

63. Rodman, T.C., Pruslin, F.H., To, S.E., and Winston, R., Human immunodeficiency virus (HIV) Tat-reactive antibodies present in normal HIV-negative sera and depleted in HIV-positive sera. Identification of the epitope, *J. Exp. Med.,* 175, 1247–1253, 1992.

64. Rodman, T.C., To, S.E., Hashish, H., and Manchester, K., Epitopes for natural antibodies of human immunodeficiency virus (HIV)-negative (normal) and HIV-positive sera are coincident with two key functional sequences of HIV Tat protein, *Proc. Natl. Acad. Sci. U.S.A.,* 90, 7719–7723, 1993.

65. Re, M.C., Furlini, G., Vignoli, M., Ramazzotti, E., Zauli, G., and La Placa, M., Antibody against human immunodeficiency virus type 1 (HIV-1) Tat protein may have influenced the progression of AIDS in HIV-1-infected hemophiliac patients, *Clin. Diagn. Lab. Immunol.,* 3, 230–232, 1996.

66. Zagury, J.F., Sill, A., Blattner, W., Lachgar, A., Le Buanec, H., Richardson, M., Rappaport, J., Hendel, H., Bizzini, B., Gringeri, A., Carcagno, M., Criscuolo, M., Burny, A., Gallo, R.C., and Zagury, D., Antibodies to the HIV-1 Tat protein correlated with nonprogression to AIDS: a rationale for the use of Tat toxoid as an HIV-1 vaccine, *J. Hum. Virol.,* 1, 282–292, 1998.

67. Rodman, T.C., Sullivan, J.J., Bai, X., and Winston, R., The human uniqueness of HIV: innate immunity and the viral Tat protein, *Hum. Immunol.,* 60, 631–639, 1999.

68. Cafaro, A., Caputo, A., Fracasso, C., Maggiorella, M.T., Goletti, D., Baroncelli, S., Pace, M., Sernicola, L., Koanga-Mogtomo, M.L., Betti, M., Borsetti, A., Belli, R., Akerblom, L., Corrias, F., Butto, S., Heeney, J., Verani, P., Titti, F., and Ensoli, B., Control of SHIV-89.6P-infection of cynomolgus monkeys by HIV-1 Tat protein vaccine, *Nat. Med.,* 5, 643–650, 1999.

69. Cafaro, A., Caputo, A., Maggiorella, M.T., Baroncelli, S., Fracasso, C., Pace, M., Borsetti, A., Sernicola, L., Negri, D.R., Ten Haaft, P., Betti, M., Michelini, Z., Macchia, I., Fanales-Belasio, E., Belli, R., Corrias, F., Butto, S., Verani, P., Titti, F., and Ensoli, B., SHIV89.6P pathogenicity in cynomolgus monkeys and control of viral replication and disease onset by human immunodeficiency virus type 1 Tat vaccine, *J. Med. Primatol.,* 29, 193–208, 2000.

70. Goldstein, G., Manson, K., Tribbick, G., and Smith, R., Minimization of chronic plasma viremia in rhesus macaques immunized with synthetic HIV-1 Tat peptides and infected with a chimeric simian/human immunodeficiency virus (SHIV33), *Vaccine,* 18, 2789–2795, 2000.

71. Pauza, C.D., Trivedi, P., Wallace, M., Ruckwardt, T.J., Le Buanec, H., Lu, W., Bizzini, B., Burny, A., Zagury, D., and Gallo, R.C., Vaccination with Tat toxoid attenuates disease in simian/ HIV-challenged macaques, *Proc. Natl. Acad. Sci. U.S.A.,* 97, 3515–3519, 2000.

72. Richardson, M.W., Mirchandani, J., Silvera, P., Regulier, E.G., Capini, C., Bojczuk, P.M., Hu, J., Gracely, E.J., Boyer, J.D., Khalili, K., Zagury, J.F., Lewis, M.G., and Rappaport, J., Immunogenicity of HIV-1 IIIB and SHIV 89.6P Tat and Tat toxoids in rhesus macaques: induction of humoral and cellular immune responses, *DNA Cell. Biol.,* 21, 637–651, 2002.

73. Silvera, P., Richardson, M.W., Greenhouse, J., Yalley-Ogunro, J., Shaw, N., Mirchandani, J., Khalili, K., Zagury, J.F., Lewis, M.G., and Rappaport, J., Outcome of simian-human immunodeficiency virus strain 89.6p challenge following vaccination of rhesus macaques with human immunodeficiency virus Tat protein, *J. Virol.,* 76, 3800–3809, 2002.

74. Belliard, G., Hurtrel, B., Moreau, E., Lafont, B.A.P., Monceaux, V., Roques, B., Desgranges, C. Aubertin, A.-M., Le Grand, R., and Muller, S., Tat-neutralizing vs. Tat-protecting antibodies in Rhesus macaques vaccinated with Tat peptides, *Vaccine,* 23, 1399-407, 2005.

75. Tosi, G., Meazza, R., De Lerma Barbaro, A., D'Agostino, A., Mazza, S., Corradin, G., Albini, A., Nooman, D.M., Ferrini, S., and Accolla, R.S., Highly stable oligomerization forms of HIV-1 Tat detected by monoclonal antibodies and requirement of monomeric forms for the transactivating function on the HIV-1 LTR, *Eur. J. Immunol.,* 30, 1120–1126, 2000.

76. Marinaro, M., Riccomi, A., Rappuoli, R., Pizza, M., Fiorelli, V., Tripiciano, A., Cafaro, A., Ensoli, B., and De Magistris, M.T., Mucosal delivery of the human immunodeficiency virus-1 Tat protein in mice elicits systemic neutralizing antibodies cytotoxic T lymphocytes and mucosal IgA, *Vaccine,* 21, 3972–3981, 2003.

77. Tikhonov, I., Ruckwardt, T.J., Hatfield, G.S., and Pauza, C.D., Tat-neutralizing antibodies in vaccinated macaques, *J. Virol.,* 77, 3157–3166, 2003.

78. Ruckwardt, T.J., Tikhonov, I., Berg, S., Hatfield, G.S., Chandra, A., Chandra, P., Gilliam, B., Redfield, R.R., Gallo, R.C., and Pauza, C.D., Sequence variation within the dominant amino terminus epitope affects antibody binding and neutralization of human immunodeficiency virus type 1 Tat protein, *J. Virol.,* 78, 13190–13196, 2004.

79. Goldstein, G., Tribbick, G., and Manson, K., Two B cell epitopes of HIV-1 Tat protein have limited antigenic polymorphism in geographically diverse HIV-1 strains, *Vaccine,* 19, 1738–1746, 2001.
80. Moreau, E., Belliard, G., Partidos, C.D., Pradezinsky, F., Le Buanec, H., Muller, S., and Desgranges, C., Important B cell epitopes for Tat-neutralisation in the serum samples of humans and different animal species immunised with Tat protein or Tat peptides, *J. Gen. Virol.,* 85, 2893–2901, 2004.
81. Moreau, E., Hoebeke, J., Zagury, D., Muller, S., and Desgranges, C., Generation and characterization of neutralizing human monoclonal antibodies against human immunodeficient virus type 1 Tat antigen, *J. Virol.,* 78, 3792–3796, 2004.
82. Fanales-Belasio, E., Moretti, S., Nappi, F., Barillari, G., Micheletti, F., Cafaro, A., and Ensoli, B., Native HIV-1 Tat protein targets monocyte-derived dendritic cells and enhances their maturation, function, and antigen-specific T cell responses, *J. Immunol.,* 168, 197–206, 2002.
83. Gavioli, R., Gallerani, E., Fortini, C., Fabris, M., Bottoni, A., Canella, A., Bonaccorsi, A., Marastoni, M., Micheletti, F., Cafaro, A., Rimessi, P., Caputo, A., and Ensoli, B., HIV-1 Tat protein modulates the generation of cytotoxic T cell epitopes by modifying proteasome composition and enzymatic activity, *J. Immunol.,* 173, 3838–3843, 2004.
84. Gringeri, A., Santagostino, E., Muca-Perja, M., Mannucci, P.M., Zagury, J.-F., Bizzini, B., Lachgar, A., Carcagno, M., Rappaport, J., Criscuolo, M., Blattner, W., Burny, A., Gallo, R.C., and Zagury, D., Safety and immunogenicity of HIV-1 Tat toxoid in immunocompromised HIV-1-infected patients, *J. Hum. Virol.,* 1, 293–298, 1998.
85. Gringeri, A., Santagostino, E., Muca-Perja, M., Le Buanec, H., Bizzini, B., Lachgar, A., Zagury, J.-F., Rappaport, J., Burny, A., Gallo, R.C., and Zagury, D., Tat toxoid as a component of a preventive vaccine in seronegative subjects, *J. Acquir. Immune Defic. Syndr. Hum. Retrovirol.,* 20, 371–375, 1999.
86. Fanales-Belasio, E., Cafaro, A., Cara, A., Negri, D.R., Fiorelli, V., Butto, S., Moretti, S., Maggiorella, M.T., Baroncelli, S., Michelini, Z., Tripiciano, A., Sernicola, L., Scoglio, A., Borsetti, A., Ridolfi, B., Bona, R., Ten Haaft, P., Macchia, I., Leone, P., Pavone-Cossut, M.R., Nappi, F., Vardas, E., Magnani, M., Laguardia, E., Caputo, A., Titti, F., and Ensoli, B., HIV-1 Tat-based vaccines: from basic science to clinical trials, *DNA Cell. Biol.,* 21, 599–610, 2002.
87. Caputo, A., Gavioli, R., and Ensoli, B., Recent advances in the development of HIV-1 Tat-based vaccines, *Curr. HIV Res.,* 2, 357–376, 2004.
88. Marasco, W.A., LaVecchio, J., and Winkler, A., Human anti-HIV-1 Tat sFv intrabodies for gene therapy of advanced HIV-1-infection and AIDS, *J. Immunol. Methods,* 231, 223–238, 1999.
89. Poznansky, M.C., La Vecchio, J., Silva-Arietta, S., Porter-Brooks, J., Brody, K., Olszak, I.T., Adams, G.B., Ramstedt, U., Marasco, W.A., and Scadden, D.T., Inhibition of human immunodeficiency virus replication and growth advantage of CD4+ T cells and monocytes derived from CD34+ cells transduced with an intracellular antibody directed against human immunodeficiency virus type 1 Tat, *Hum. Gene Ther.,* 10, 2505–2514, 1999.

10 HIV Protease (PR) and Cell Death

Zilin Nie
David R. McNamara
Andrew D. Badley

CONTENTS

PROTEASE (PR) IN VIRUS LIFE CYCLE AND AIDS PATHOGENESIS

The human immunodeficiency virus (HIV) genome consists of a single-stranded positive sense ribonucleic acid (RNA) molecule that is organized into three major coding elements: Gag, Pol, and Env. Enzymes necessary for viral replication are included within the Pol gene and include protease, reverse transcriptase, and integrase. During the viral assembly and maturation, the P55 Gag precursor and P160 Gag–Pol precursors are protealitically cleaved by preformed protease, which is contained within the HIV virion and introduced into the cell at the time of initial infection. The preformed protease consequently generates new protease by virtue of new cleavages between the protease and reverse transcriptase regions of the Pol gene. HIV protease is critical to the HIV life cycle, as demonstrated by observations that mutation or inhibition of HIV protease results in formation of structurally disorganized noninfectious viral particles.

HIV protease (PR) is also a cytotoxic protein that leads to the death of human and bacterial cells after transfection. These findings may suggest that HIV PR contributes to the death of HIV-infected T cells by directing induced cell death. This concept is further supported by the recent

155

finding that inducible expression of HIV protease causes growth arrest, loss of plasma membrane, and lysis, both in *Saccharomyces cerevisiae* and in human cell lines. Although morphologically the mode of death in these cells is consistent with necrosis (as determined by electron microscopy), other features of apoptosis, such as caspase activation, mitochrondrial transmembrane potential change, and DNA fragmentation, were not directly assessed. Thus, although it is clear that HIV PR is cytotoxic, the mechanisms by which HIV PR induces cell death are unknown.

There are two subcellular locations of PR within infected cells that correspond to the two locations where viral polyproteins are processed: first, as membrane-associated polyproteins that are cleaved for virus assembly and maturation, and second, as free polyproteins processed within cytosols. Active PR within the cytosolic fraction of infected cells cleaves not only the viral Gag–Pol precursor but also nonviral proteins, including both structural and cell-death regulatory proteins. This chapter will review the science concerning the cytotoxic effects of HIV PR putative mechanisms responsible for cell death induced by PR and will review evidence that PR-induced killing might contribute to the T cell depletion seen in HIV infection.

EFFECT OF PROTEASE EXPRESSION IN TRANSGENIC MODELS

Although it is well known that small animals such as mice are not optimal models for the study of AIDS pathogenesis, transgenic mice that harbor defective viral genomes that contain a PR sequence or single PR gene have been developed.[1–4] The main pathological effect of these transgenic mice is cytotoxicity in tissues in which PR is expressed; for example, targeted ocular expression of PR results in lens fiber cell damage, which causes cataracts. In particular, the expression of HIV-1 protease driven by an A-crystallin promoter causes proteolysis of the aquaporin protein MIP26 and the lens crystallin proteins, resulting in cataracts. However, the proteolysis patterns of lens proteins in HIV-1 PR transgenic mice suggest that these proteins are not directly processed by HIV-1 PR[5]; rather, HIV-1 PR activates an endogenous protease, probably calpain, that cleaves the proteins of the lens in HIV-1 PR transgenic mice.[4,6] In particular, these studies demonstrate conclusively that HIV PR is active within mammalian cells, that the principal effect of PR expression is cell loss, that the degree of cell loss varies discretely with levels of PR expression,[5] and that PR selectively targets host cell proteins.[6]

HIV PR AND CELL DEATH

EVIDENCE OF CELL DEATH CAUSED BY HIV PR

Necrosis

There is some evidence that expression of HIV-1 PR can cause cell death through necrosis. In yeast, the expression of HIV PR causes yeast growth arrest and cell lysis.[7] The growth arrest of yeast-expressing HIV-1 PR occurs 14 to 18 h after induction, which is coincident with the peak of HIV-1 PR expression as detected by Western blotting. The lytic activity of HIV-1 PR in yeast is evidenced by the release of intracellular protein into the culture medium from damaged cells. Ultrastructural examination by electron microscopy shows morphological alterations of the cells expressing HIV-1 PR including altered cell size, detachment of the plasma membrane from the cell wall, and a normal appearance of the nucleus. COS cells that express HIV PR through infection using a hybrid vaccinia virus-T7 RNA polymerase expression system exhibit a necrotic phenotype associated with increased plasma membrane permeability and cell lysis.[7] The molecular events responsible for initiating death, however, were not investigated in these studies; consequently, whether this form of cell death shares some features of apoptosis (e.g., mitochondrial depolarization, caspase activation) remains to be determined.

Mechanism of the Cell Death Caused by HIV PR

Bcl-2

In 1996, Strack and colleagues first identified Bcl-2 as an apoptosis regulatory protein that is cleaved by HIV PR and suggested that this event results in the infected cell's death.[8,9] Bcl-2 is an important cellular survival protein that inhibits programmed cell death via homo- or heterodimerization with other Bcl-2 family members. This antiapoptotic molecule contains four functional BCR-homology GTPase activation domains (BH-domains) and one membrane target domain (transmembrane [TM] domain) (see Figure 10.1).[10,11] All four BH domains are required for the antiapoptotic effects of Bcl-2 to form homodimers and heterodimers. Specifically, the BH1 to BH4 domains mediate interactions with other Bcl-2 members in which the BH1 to BH3 domains form a hydrophobic groove, and the N-terminal BH4 domain stabilizes this structure.[11] The HIV PR cleavage site within Bcl-2 is located between the BH3 and BH1 domains (between phenylalanine 112 and alanine 113).[8] This cleavage separates Bcl-2 into two parts: an N-terminal part containing the BH4 and BH3 domains and the C-terminal part containing the BH1, BH2, and TM domains (Figure 10.1). Although the C-terminal part still contains Bcl-2 functional motifs -helix 5 and 6 (pore formation), this fragment is unlikely to confer cell death protection, as it lacks the BH3 domain (ligand domain) and BH4 domain (cell-death protecting domain). The BH3 and BH4 domains are important for the ability of the full length of Bcl-2 to protect cells from death. Specifically, the BH3 domain is necessary for Bcl-2 to bind BH3-containing proapoptotic proteins,[12] whereas the BH4 domain is required for interaction with Raf kinases.[13] During caspase-3-dependent apoptosis, active caspase-3 removes the BH4 domain of Bcl-2, converting it to a Bax-like proapoptotic effector.[14] Therefore, removal of the BH3 and BH4 domains of Bcl-2 by HIV PR cleavage is likely to abrogate the antiapoptotic function of Bcl-2 and may even promote proapoptotic signaling.

Procaspase-8

Caspase-8 is an initiator caspase that is required for death receptor–initiated death signaling.[15,16] Procaspase-8 is activated after its interaction with the adaptor proteins FADD (Fas-associated protein with death domain) or TRADD (tumor necrosis factor receptor–associated protein with death domain), resulting in the removal of the procaspase N-terminal prodomain by caspase-8 autoprocessing. The activation of this initiator caspase is a key step in physiological apoptosis, and

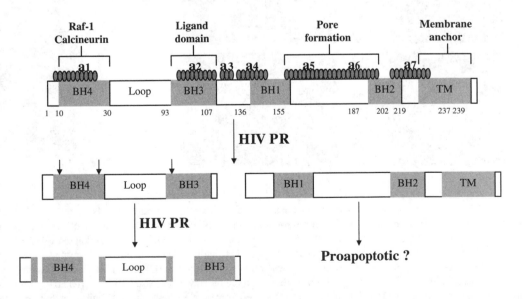

FIGURE 10.1 Human immunodeficiency virus (HIV) protease inhibitor (PR) cleaves Bcl-2.

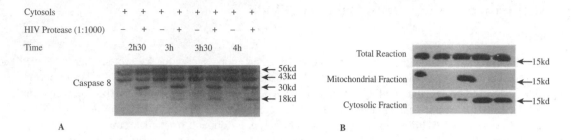

Cytosols	+	+	+	+	+	+	+	+
HIV Protease (1:1000)	–	+	–	+	–	+	–	+
Time		2h30		3h		3h30		4h

Caspase 8 ← 56kd / ← 43kd / ← 30kd / ← 18kd

Total Reaction ←15kd
Mitochondrial Fraction ←15kd
Cytosolic Fraction ←15kd

A   B

FIGURE 10.2 HIV-1 PR processes and activates procaspase 8 (A) and causes cytochrome c release from mitochondria (B) in cell-free system.

the activation of procaspase-8 may also be modified during virus infections. Some lytic viruses (e.g., Sendai virus) induce apoptosis through activation of caspase-8,[17] but several latent viruses, such as -herpesviruses (including Kaposi's sarcoma–associated human herpesvirus-8) directly inhibit the maturation of procaspase-8 through the death-effector domain (DED) containing viral protein vFlip.[18] Recent data suggest that HIV PR may also process procaspase-8. HIV PR processes recombinant procaspase-8 *in vitro* and cellular procaspase-8 in a cell free system.[19] The processing of procaspase-8 by HIV-1 PR leads to cytochrome *c* release from mitochondria into the cytoplasmic compartment in a comparable manner to the release of cytochrome *c* seen with recombinant active caspase-8 or Granzyme B (Figure 10.2A, B), resulting in the activation of the downstream caspase cascade. Co-incubation of HeLa nuclei with the activated cytoplasm extracts from Jurkat cells causes HeLa nuclei fragmentation and internucleosomal DNA cleavage (as determined by DNA ladder analysis), which are the classical markers of cell death through apoptosis (Figure 10.3A–E).

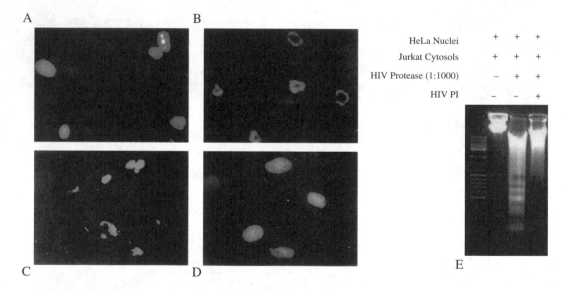

A   B

HeLa Nuclei	+	+	+
Jurkat Cytosols	+	+	+
HIV Protease (1:1000)	–	+	+
HIV PI	–	–	+

C   D   E

FIGURE 10.3 HIV-1 protease induces the nuclear changes of apoptosis. Jurkat cytosols (1 mg) were treated with recombinant HIV-1 protease at 30°C for 3 hours and then co-incubated with HeLa nuclei. Treated nuclei were imaged under microscopy by Hoechst 33342 staining. (A) Nuclei incubated with Jurkat cytosols without HIV-1 protease treatment. (B) Nuclei incubated with Jurkat cytosols treated with HIV-1 protease, resulting in fragmentation of the nuclear membrane and chromatin condensation or (C) margination of chromatin. (D) The induction of apoptotic changes were completely inhibited by HIV-1 protease inhibitor. (E) HIV-1 protease induces internucleosomal DNA fragmentation. DNA gel from nuclei incubated with Jurkat cytosols in the presence or absence of HIV-1 protease with or without HIV-1 protease inhibitor (PI).

Taken together, HIV-1 PR could activate an apoptotic-signaling pathway via direct targeting of procaspase-8 to induce apoptosis in HIV-infected cells.

INITIATION FACTOR OF TRANSLATION EIF4GI AND OTHER CELLULAR PROTEINS

HIV PR also cleaves the initiation factor of translation eIF4GI and other cellular structure proteins such as actin, laminin B, desmin, and vimentin.[20–27] Although these cellular proteins are not intrinsically apoptotic regulatory molecules, it is likely that they are involved in the decision of cells to die or remain viable. For example, the morphological changes associated with apoptosis are controlled by cytoskeletal reorganization,[28–30] as most cytoskeleton proteins are cleaved by activated caspases in the processes of apoptosis.[16] Therefore, PR-mediated cleavage of these proteins may contribute to the phenotypic effects of apoptosis. Moreover, the cleavage of eIF4GI by HIV PR drastically impairs the translation of capped and uncapped mRNAs,[25,26,31] which, in turn, leads to inhibition of protein synthesis, resulting in metabolic arrest, impaired cellular respiration, and consequent death. There are at least two HIV-1 PR cleavage sites on eIF4GI: one located at the N-terminus between positions 718 and 719 and another located at the C-terminus between positions 1126 and 1127.[32] Teleogically, eIF4GI cleavage by PR, resulting in inhibition of protein synthesis, is consistent with the concept that HIV PR is cytotoxic.

MODEL FOR APOPTOSIS VS. NECROSIS

The expression of HIV PR in transfected cells, therefore, can cause death through apoptosis or necrosis. This seeming paradox may be reconciled by observations that stimuli other than PR may, in different contexts, lead to apoptosis or necrosis. For example, caspase inhibitors can switch cell death from apoptosis phenotype to necrosis.[33] Another example shows two functional domains of FADD point cell death signals to different directions — to DED induction of apoptosis through interaction with procaspase-8 and death domain (DD) activation resulting in necrosis via binding RIP.[34] Therefore, it is possible that HIV PR induces cell death either through apoptosis or necrosis, potentially depending on PR expression levels. Such a model is reminiscent of lysosomal proteases, for example, cathepsin B. When small amounts of cathepsin B are in the cytosol, caspases are activated, inducing apoptosis, but when larger amounts are released from ruptured lysosomes, the cell death phenotype resembles necrosis.[35]

CLINICAL SIGNIFICANCE OF THE CELL DEATH INDUCED BY PR

DISCORDANCE

The initiation of highly active antiretroviral therapy (HAART) in HIV-infected patients most commonly results in a decrease in plasma HIV viral load (a virologic response) with a corresponding increase in CD4 T cell count and CD4 percentage (an immunologic response). Within 2 years of the introduction of protease inhibitor (PI)–based antiretroviral therapy, discordant responses between the immunologic and virologic effects of antiretroviral therapy were reported.[36] Discordant responses can involve a virologic reponse with immunologic failure (inhibition of viral replication and decrease in HIV plasma viral load, with no increase in CD4 count or percentage). Conversely, virologic failure with immunologic response (ongoing viral replication accompanied by lack of CD4 T cell loss or even CD4 T cell increase) may also occur.

Soon after the initial recognition of discordant virologic responses, improved clinical outcomes were observed in patients with virologic failure but immunologic response to antiretroviral therapy. One retrospective analysis of a cohort of patients with an earlier nucleoside reverse transcriptase inhibitor (NRTI) treatment and CD4 counts <50 cells/μL revealed similar clinical outcomes between two groups with similar immunologic response but substantially different virologic responses.[37] More than 2500 Swiss patients with HIV were followed prospectively for 24 months after initiation of antiretroviral therapy. Among those with an initial virologic response (viral load <400 copies/ml within 12 months of initiation of therapy) followed by either maintenance of HIV viral load <500 copies/mL or development of viral rebound, there was no significant difference in the incidence of death or development of a new AIDS-defining condition. Conversely, patients who never achieved a virologic response had a significantly higher risk of death or new diagnosis of an AIDS-defining condition.[38] However, a prospective cohort study of more than 2000 PI-naive patients in France displayed favorable clinical outcomes in patients with immunologic responses after 6 months of PI-based antiretroviral therapy (ART), regardless of virologic response.[39] These and other studies reporting sustained increases in CD4 T cells and significant clinical benefit despite this ongoing detectable viral replication[37–42] generated important data challenging the view that the beneficial immunological effect of PI-based antiretroviral therapy was solely due to the suppression of ongoing viral replication.

Understanding the mechanisms behind these discordant responses to PI-based ART may provide important insight into T cell biology and the causes of CD4 cell loss during HIV infection. A variety of reasons behind discordant responses to ART has been proposed, including impaired replicative (and possibly pathogenic) fitness of a resistant virus; direct cytotoxic effects of HIV protease; viral escape, leading to autoimmunization and enhanced cytotoxic T cell function; and possibly other unidentified mechanisms.

An emerging body of literature indicates that a drug-resistant virus (especially a PI-resistant virus) has impaired replicative potential compared with a wild-type virus.[43–49] Such studies use laboratory strains of HIV with point mutations in PR or clinical isolates containing PR mutations and compare replicative capacity between the mutant and wild-type virusus, either in simple replication assays or in competition experiments. More recently, these experiments have been performed in severe combined immunodeficient (SCID)–hu Thy/LIV mice, activated peripheral blood mononuclear cells, and cultures of thymic organ specimens from humans. In the overwhelming majority of cases, an impaired replicative fitness of mutant virus was shown.[50] It is, therefore, reasonable to suspect that such mutations, which result in diminished replicative capacity, might also be associated with impaired pathogenic capacity, which manifests clinically as decreased propensity for CD4 T cell depletion.

PROTEASE INHIBITOR RESISTANCE

HIV protease is a homodimer of two identical 99 amino acid chains with an active site at positions 25 to 27 in each chain. This enzyme cleaves the Gag (p55) and Gag–Pol (p160) polyproteins to yield the HIV core proteins p6, p7, p17, and p24 and the enzymes reverse transcriptase, integrase, and protease.[51] The critical role of HIV protease in the replication cycle of HIV via the production of mature virions was recognized as a target for antiretroviral drug therapy. In 1995, HIV protease inhibitors, the third class of antiretroviral agents for treatment of HIV infection, were introduced.[52] PIs exert their therapeutic effect by selectively binding to and inhibiting the catalytic activity of HIV protease, thereby preventing cleavage of the large Gag and Gag–Pol polyprotein precursors. Shortly after introduction of PIs into routine use for the care of HIV-infected individuals, antiretroviral combination therapy that included PIs was recognized as having a profound ability to alter the progressive course of HIV infection. Clinical research analyzing the survival of patients treated with these drugs confirmed these observations.[53] In the 9 years after their introduction, PIs have remained potent agents against HIV and, when used in combination with other drug classes, form the cornerstone of current ART against HIV.

Similar to the previously recognized phenomenon of viral resistance to NRTIs and non-NRTIs (NNRTIs) development of resistance to PIs was soon recognized to occur commonly and to adversely affect the susceptibility of HIV to PI therapy.[54] High levels of ongoing viral replication and turnover in the steady state of HIV infection, as well as the highly error-prone mechanism of HIV reverse transcriptase that lacks a proofreading mechanism, combine to produce frequent mutations in the population of newly produced virions. In addition, recombination of genetic material among HIV strains increases the genetic diversity of the HIV population within a host.[55]

Currently available combination ART does not fully suppress viral replication in all patients. Ongoing viral replication during periods of drug exposure may be associated with drug concentration sufficient to exert selective pressure on the population of HIV quasispecies present. In this environment, viral quasispecies with resistance to antiretrovirals emerge.[56] Extensive cross-resistance among PIs also commonly occurs and has been associated with a genotypic basis for class resistance to available PIs.[57] Ultimately, an HIV mutational variant is selected that exhibits a survival advantage over wild-type virus, at least when that virus remains in the continued presence of the antiretroviral selective process. In addition to altering substrate specificities, these resistance-associated mutations typically confer altered enzyme kinetics due to their location in or around the active site of the HIV PR homodimer.

In a variety of different T cell lines, viral replication of virions containing mutant protease (assessed using indirect immunofluorescence assay) was impaired tenfold in the mutant virus compared to that in the wild-type virus; processing of the pr55 Gag polyprotein precursor and p24 production were similarly impaired. Of particular interest was that the ability of mutant protease to kill MT4 cells *in vitro* was also impaired.[58] Other groups subsequently demonstrated similar impairment in HIV PR activity associated with resistance mutations using techniques such as dual infection competition assays,[47,49] pr55 Gag polyprotein processing, single-cell infectivity assays, viral RNA and p24 production rates, and Gag cleavage kinetics. These altered outcomes (compared with wild-type virus) were observed using patient viral isolates, *in vitro*–selected protease mutations, site-directed protease mutations, and recombinant HIV-1 vectors using laboratory strains of HIV deleted for HIV PR or reverse transcriptase, into which patient-derived protease or reverse transcriptase is inserted.[43,46,47,49,59]

It now seems that at least three classes of mutations may result from subinhibitory HIV PI selective pressure: primary HIV resistance mutations that impair viral replicative potential *in vitro*, secondary protease resistance mutations that may compensate for the impairment associated with primary mutations, and Gag–Pol cleavage site mutations. Examples include the primary resistance mutations D30N and L90M, both of which significantly impact viral replicative fitness compared with wild-type virus, whereas the secondary mutation, L63P, does not independently impact viral fitness.[60] However, the addition of L63P to either of the primary mutations compensates for their impaired fitness — significantly in the case of the L90M isolate and slightly for the D30N isolate.[46] Alternately, defects within Gag–Pol processing may also impact replicative capacity, presumably by altering the efficacy of viral polyprotein processing, and virion packaging and progeny release. For example, paired pretherapy and postresistance viral isolates assayed for infectivity and Gag–Pol precursor cleavage demonstrated that impaired Gag–Pol cleavage (assessed by the accumulation of cleavage intermediates) is associated with impaired viral infectivity.[45] Impaired Gag–Pol processing can be caused by HIV PR or Gag–Pol cleavage site mutations.[44,61]

Although the majority of PR mutations impair protease activity, viral infectivity, packaging, and polyprotein processing, some mutations may actually enhance fitness. For example, viral evolution in one patient receiving protease inhibitor monotherapy demonstrated mutant viruses that initially had reduced protease activity and, consequently, reduced viral replicative capacity. *In vivo* replication of these viral isolates in the presence of drug selected for viral populations with more protease mutations eventually acquired enhanced replicative potential compared with the wild-type viral strain.[62] Insertions of up to six amino acids were also associated with enhanced viral replicative fitness compared with wild-type controls.[63] Mutational improvements in viral replicative capacity may also enhance viral susceptibility to some therapeutic PIs. For example, the Gag polyprotein–processing mutations at positions 449 and 353 are associated with improved viral fitness, despite impaired sensitivity to antiretroviral agent amprenavir. Similarly, the N88 mutation in HIV PR is associated with impaired viral fitness and hypersusceptibility to amprenavir.[64]

Overall, the interrelationships between viral replication, viral drug sensitivity, and Gag–Pol processing are complex and remain poorly defined. It is likely that each of these measures confers different information regarding the clinical impact of such mutations. At a molecular level, because protease inhibitor–resistant isolates have impaired catalytic activity, other viral functions will also likely be impaired. For example, the impaired ability of HIV PR to cleave the Gag–Pol polyprotein precursor may result in impaired production of reverse transcriptase (RT) and, therefore, reduced level of RT activity in infected cell cultures.[65,66] Moreover, such mutants are likely to have an impaired ability to cleave host cell proteins.

One example concerns the PR active site mutation (T26S), which has approximately five- to tenfold lower activity on HIV-1 polyprotein processing.[67,68] Of interest is that viruses containing this mutation also have impaired CD4 T cell killing potential. They did not induce toxicity upon bacterial expression, and their cleavage of cytoskeletal proteins *in vitro* was significantly impaired in infected T cell lines.[67] Another PR mutant (V32I/A71V) grows only slightly more slowly than wild-type virus on peripheral blood mononuclear cells (PBMCs)[43]; however, it cannot recognize and cleave several cellular proteins, including vimentin, focal adhesion plaque kinase (FAK), fimbrin, and integrins.[27] Altogether, these observations demonstrate that substrates of host cell origin can have different affinities for HIV PR than does the HIV polyprotein, supporting the hypothesis that mutations within HIV PR can, and do, impact the ability of HIV to kill target T cells.

ASSOCIATIONS BETWEEN PROTEASE MUTATIONS AND CLINICAL OR CD4 OUTCOME

The impact of protease mutations was studied extensively from the viral standpoint, but the effect of protease mutations upon the clinical course of disease received less attention. An emerging literature suggests that mutations in PR may be associated with a slower course of disease progression. Much of this literature is embedded in the literature concerning discordant response to therapy, in which the majority (if not all) of patients with immunologic discordance have viral isolates with PR resistance mutations, impaired viral fitness, and, therefore, impaired protease enzyme catalytic activity.

An association between infection with PI-resistant HIV and decreased CD4 cell loss could be envisioned to be the result of impaired catalytic activity of mutant protease (Figure 10.4). These protease mutations, which are selected for in the presence of PI, confer survival benefit in the presence of PI and possess decreased catalytic ability. The evidence linking HIV protease with mechanisms of cellular apoptosis suggests that mutant protease, with its decreased catalytic activity, also may have impaired ability to initiate apoptosis, thereby explaining an association between protease mutations and improved immunologic and clinical outcomes.

Because there is strong evidence that HIV PR induces cell death through apopotosis or necrosis, we hypothesize that PR inhibitor–resistant mutations might attenuate or enhance HIV protease ability to induce cell death. Indeed, some data suggest that PR mutation may slow disease progression.[49] Recently, the hypothesis that impaired PR catalytic activity will lead to enhanced CD4 T cell counts was tested. In 191 HIV-infected patients, viral sequences corresponding to the terminal 18 amino acids of Gag, all of PR, and a proximal segment of RT were cloned from patient isolates. This sequence was inserted into an expression vector, and cleavage kinetics was compared with WT HIV (strain NL4-3). Impaired fitness of PR was significantly associated with improved CD4 T cell counts by both cross-sectional and longitudinal analyses.[69]

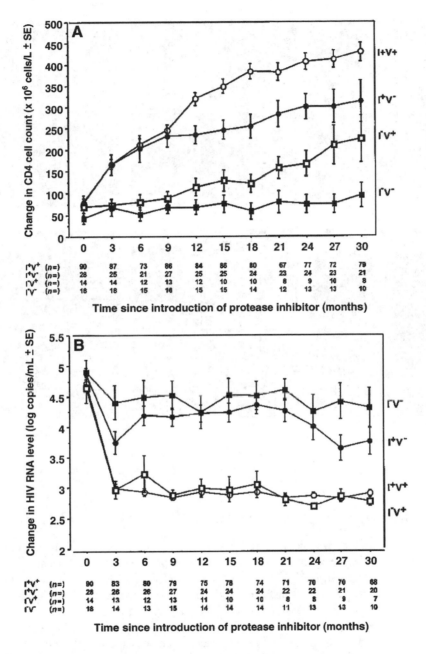

FIGURE 10.4 Changes in CD4 cell counts and plasma human immunodeficiency virus (HIV) RNA levels, according to immunologic and virologic status within the first 12 months of highly active antiretroviral treatment. (A) CD4 cell count changes ±SE. (B) Mean plasma HIV RNA level changes ±SE. Number of patients evaluated at each time point is shown. I+V+, patients with an immunologic and virologic response to therapy; I+V−, patients with a sustained increase in CD4 cells in the absence of a significant decrease in plasma HIV RNA or without a transient decrease in plasma HIV RNA; I−V+, patients with a decreased plasma HIV load without an increase in CD4 cell count; I−V−, patients without an immunologic or virologic response to therapy.

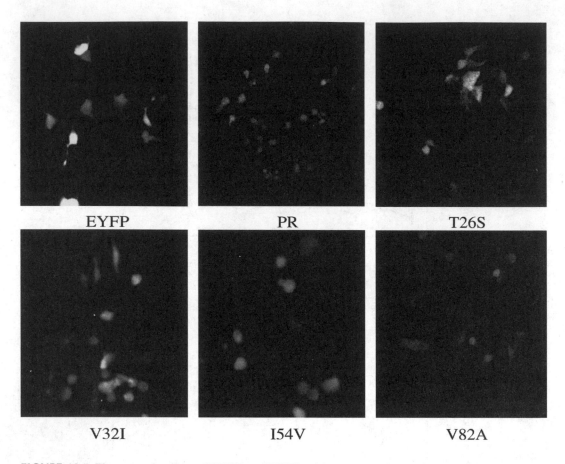

FIGURE 10.5 The cytotoxic effects of YFPPR and YFPPR primary mutants in transiently transfused 293 cells.

IN VITRO EFFECTS OF PR MUTATIONS ON CELL KILLING

It is important to discern whether different protease mutations may result in differences in the cytotoxic effects of drug-resistant PR mutants. We developed a yellow fluorescence protein (YFP) reporter system in which an HIV-1 PR precursor is fused with YFP. After autocatalytic cleavage of the fusion protein by PR in the transfected cells, a free and toxic PR is formed to induce cytotoxic effects. The toxicity of PR is measured by the YFP density and cell morphology changes of the transfected cells. A panel of PR mutants, which often appear as primary and secondary mutations in PR inhibitor therapy, is generated by site-directed mutagenesis. Our data show that even some single amino acid substitutions are enough to allow PR to lose or reduce its cytotoxic effects in 293 cells (Figure 10.5). These data demonstrate proof of the concept that HIV PR mutations can, independently, alter protease-mediated cytotoxicity.

CONCLUSIONS

Although it is understood that HIV PR is cytotoxic in systems in which it is expressed independently, it is less certain whether HIV PR can and does induce cell death in the context of viral infection. Moreover, the molecular mechanisms governing the initiation of PR-induced cell death remain controversial; however, PR-mediated cleavage of procaspase-8, Bcl-2, eIF4G, and cytoskeletal proteins is likely involved. We propose that low-level expression of PR favors the selective cleavage of these regulatory proteins favoring an apoptotic phenotype, whereas higher levels of expression may favor more extensive proteolysis, favoring an autophagocytic or necrotic phenotype. Whether PR-mediated cell death plays a clinically significant role in patients also remains controversial; however, indirect evidence suggests that in patients with immunologic discordance, mutations within PR may be contributory. Additional research directed at each of these controversies will undoubtedly resolve such controversies.

REFERENCES

1. Iwakura, Y., Shioda, T., Tosu, M., Yoshida, E., Hayashi, M., Nagata, T., and Shibuta, H. The induction of cataracts by HIV-1 in transgenic mice, *AIDS,* 6, 1069–1075, 1992.
2. Bettelheim, F.A., Zeng, F.F., Bia, Y., Tumminia, S.J., and Russell, P., Lens hydration in transgenic mice containing HIV-1 protease linked to the lens -A-crystallin promoter, *Arch. Biochem. Biophys.,* 20(324), 2, 1995.
3. Bettelheim, F.A., Churchill, A.C., Siew, E.L., Tumminia, S.J., and Russell, P., Light scattering and morphology of cataract formation in transgenic mice containing the HIV-1 protease linked to the lens -A-crystallin promoter, *Exp. Eye Res.,* 64(5), 667–674, 1997.
4. Tumminia, S.J., Clark, J.I., Richiert, D.M., Mitton, K.P., Duglas-Tabor, Y., Kowalak, J.A., Garland, D.L., and Russell, P., Three distinct stages of lens opacification in transgenic mice expressing the HIV-1 protease, *Exp. Eye Res.,* 72(2), 115–121, 2001.
5. Tumminia, S.J., Jonak, G.J., Focht, R.J., Cheng, Y.S., and Russell, P., Cataractogenesis in transgenic mice containing the HIV-1 protease linked to the lens -A-crystallin promoter, *J. Biol. Chem.,* 271(1), 425–431, 1996.
6. Mitton, K.P., Kamiya, T., Tumminia, S.J., and Russell, P., Cysteine protease activated by expression of HIV-1 protease in transgenic mice. MIP26 (aquaporin-0) cleavage and cataract formation *in vivo* and *ex vivo, J. Biol. Chem.,* 271(50), 31803–31806, 1996.
7. Blanco, R., Carrasco, L., and Ventoso, I., Cell killing by HIV-1 protease, *J. Biol. Chem.,* 278(2), 1086–1093, 2003.
8. Strack, P.R., Frey, M.W., Rizzo, C.J., Cordova, B., George, H.J., Meade, R., Ho, S.P., Corman, J., Tritch, R., and Korant, B.D., Apoptosis mediated by HIV protease is preceded by cleavage of Bcl-2, *Proc. Natl. Acad. Sci. U.S.A.,* 93(18), 9571–9576, 1996.
9. Korant, B.D., Strack, P.R., Frey, M.W., and Rizzo, C.J., A cellular anti-apoptosis protein is cleaved by the HIV-1 protease, *Adv. Exp. Med. Biol.,* 436, 27–29, 1998.
10. Reed, J.D., Double identity for proteins of the Bcl-2 family, *Nature,* 387(6635), 773–776, 1997.
11. Kirkin, V., Joos, S., and Zornig, M., The role of Bcl-2 family members in tumorigenesis, *Biochim. Biophys. Acta,* 1644(2–3), 229–249, 2004.
12. Cory, S., Huang, D.C., and Adams, J.M., The Bcl-2 family: roles in cell survival and oncongenesis, *Oncogene,* 22(53), 8590 8607, 2003.
13. Rapp, U.R., Rennefahrt, U., and Troppmair, J., Bcl-2 proteins: master switches at the intersection of death signaling and the survival control by Raf kinases, *Biochim. Biophys. Acta,* 1644(2–3), 149–158, 2004.
14. Cheng, E.H., Kirsch, D.G., Clem, R.J., Ravi, R., Kastan, M.B., Bedi, A., Ueno, K., and Wardwick, J.M., Conversion of Bcl-2 to a Bax-like death effector by caspases, *Science,* 278(5345), 1966–1968, 1997.
15. Cohen, G.M., Caspases: the executioners of apoptosis, *Biochem. J.,* 326(Pt. 1), 1–16, 1997.
16. Chang, H.Y. and Yang, X., Proteases for cell suicide: functions and regulation of caspases, *Microbiol. Mol. Biol. Rev.,* 64(4), 821–846, 2000.
17. Bitzer, M., Prinz, F., Bauer, M., Spiegel, M., Neubert, W.J., Gregor, M., Schulze-Osthoff, K., and Lauer, U., Sendai virus infection induces apoptosis through activation of caspase-8 (FLICE) and caspase-3 (CPP32), *J. Virol.,* 73(1), 702–708, 1999.
18. Thome, M., Schneider, P., Hofmann, K., Fickenscher, H., Meinl, E., Neipel, F., Mattmann, C., Burns, K., Bodmer, J.L., Schroter, M., Scaffidi, C., Krammer, P.H., Peter, M.E., and Tschopp, J., Viral FLICE-inhibitory proteins (FLIPs) prevent apoptosis induced by death receptors, *Nature,* 386(6624), 517–521, 1997.

19. Nie, Z., Phenix, B.N., Lum, J.J., Alam, A., Lynch, D.H., Beckett, B., Krammer, P.H., Sekaly, R.P., and Badley, A.D., HIV-1 protease processes procaspase-8 to cause mitochondrial release of cytochrome c, caspase cleavage and nuclear fragmentation, *Cell Death Differ.,* 9(11), 1172–1184, 2002.

20. Shoeman, R.L., Honer, B., Stoller, T.J., Kesselmeier, C., Miedel, M.C., Traub, P., and Graves, M.C., Human immunodeficiency virus type 1 protease cleaves the intermediate filament proteins vimentin, desmin, and glial fibrillary acidic protein, *Proc. Natl. Acad. Sci. U.S.A.,* 87(16), 6336–6340, 1990.

21. Tomasselli, A.G., Hui, J.O., Adams, L., Chosay, J. Lowery, D., Greenberg, B., Yem, A., Deibel, M.R., Zurcher-Neely, H., and Heinrikson, R.L., Actin, troponin C, Alzheimer amyloid precursor protein and pro-interleukin 1 as substrates of the protease from human immunodeficiency virus, *J. Biol. Chem.,* 266(22), 14548–14553, 1991.

22. Adams, L.D., Tommasselli, A.G., Robbins, P., Moss, B., and Heinrikson, R.L., HIV-1 protease cleaves actin during acute infection of human T-lymphocytes, *AIDS Res. Hum. Retroviruses,* 8(2), 291–295, 1992.

23. Shoeman, R.L., Sachse, C., Honer, B., Mothes, E., Kaufmann, M., and Traub, P., Cleavage of human and mouse cytoskeletal and sarcomeric proteins by human immunodeficiency virus type 1 protease. Actin, desmin, myosin, and tropomyosin, *Am. J. Pathol.,* 142(1), 221–230, 1993.

24. Shoeman, R.L., Huttermann, C., Hartig, R., and Traub, P., Amino-terminal polypeptides of vimentin are responsible for the changes in nuclear architecture associated with human immunodeficiency virus type 1 protease activity in tissue culture cells, *Mol. Biol. Cell,* 12(1), 143–154, 2001.

25. Ventoso, I., Blanco, R., Perales, C., and Carrasco, L., HIV-1 protease cleaves eukaryotic initiation factor 4G and inhibits cap-dependent translation, *Proc. Natl. Acad. Sci. U.S.A.,* 98(23), 12966–12971, 2001.

26. Ohlmann, T., Prevot, D., Decimo, D., Roux, F., Garin, J., Morley, S.J., and Darlix, J.L., *In vitro* cleavage of eIF4GI but not eIF4GII by HIV-1 protease and its effects on translation in the rabbit reticulocyte lysate system, *J. Mol. Biol.,* 318(1), 9–20, 2002.

27. Shoeman, R.L., Hartig, R., Hauses, C., and Traub, P., Organization of focal adhesion plaques is disrupted by action of the HIV-1 protease, *Cell Biol. Int.,* 26(6), 529–539, 2002.

28. Coleman, M.L., Sahai, E.A., Yeo, M., Bosch, M., Dewar, A., and Olson, M.F., Membrane blebbing during apoptosis results from caspase-mediated activation of ROCK I, *Nat. Cell Biol.,* 3(4), 339–345, 2001.

29. Coleman, M.L. and Olson, M.F., Rho GTPase signalling pathways in the morphological changes associates with apoptosis, *Cell Death Differ.,* 9(5), 493–504, 2002.

30. Grzanka, A., Grzanka, D., and Orlikowska, M., Cytoskeletal reorganization during process of apoptosis induced by cytostatic drugs in K-562 and HL-60 leukemia cell lines, *Biochem. Pharmacol.,* 66(8), 1611–1617, 2003.

31. Perales, C., Carrasco, L., and Ventoso, I., Cleavage of eIF4G by HIV-1 protease: effects on translation, *FEBS Lett.,* 533(1–3), 89–94, 2003.

32. Alvarez, E., Menendez-Arias, L., and Carrasco, L., The eukaryotic translation initiation factor 4GI is cleaved by different retroviral proteases, *J. Virol.,* 77(23), 12392–400, 2003.

33. Cauwels, A., Janssen, B., Waeytens, A., Cuvelier, C., and Brouckaert, P., Caspase inhibition causes hyperacute tumor necrosis factor-induced shock via oxidative stress and phospholipase A2, *Nat. Immunol.,* 4(4), 387–393, 2003.

34. Vanden Berghe, T., van Loo, G., Saelens, X., Van Gurp, M., Brouckaert, G., Kalai, M., Declercq, W., and Vandenabeele, P., Differential signaling to apoptotic and necrotic cell death by Fas-associated death domain protein FADD, *J. Biol. Chem.,* 279(9), 7925–7933, 2004.

35. Li, W., Yuan, X., Nordgren, G., Dalen, H., Dubowchik, G.M., Firestone, R.A., and Brunk, U.T., Induction of cell death by the lysosomotropic detergent MSDH, *FEBS Lett.,* 470(1), 35–39, 2000.

36. Piketty, C., Castiel, P., Belec, L., Batisse, D., Si Mohamed, A., Gilquin, J., Gonzalez-Canali, G., Jayle, D., Karmochkine, M., Weiss, L., Aboulker, J.P., and Kazatchkine, M.D., Discrepant responses to triple combination antiretroviral therapy in advanced HIV disease, *AIDS,* 12(7), 745–750, 1998.

37. Mezzaroma, I., Carlesimo, M., Pinter, E., Muratori, D.S., Di Sora, F., Chiarotti, F., Cunsolo, M.G., Sacco, G., and Aiuti, F., Clinical and immunologic response without decrease in virus load in patients with AIDS after 24 months of highly active antiretroviral therapy, *Clin. Infect. Dis.,* 29(6), 1423–1430, 1999.

38. Ledergerber, B., Egger, M., Erard, V., Weber, R., Hirschel, B., Furrer, H., Battegay, M., Vernazza, P., Bernasconi, E., Opravil, M., Kaufmann, D., Sudre, P., Francioli, P., and Telenti, A., Clinical progression

and virological failure on highly active antiretroviral therapy in HIV-1 patients: a prospective cohort study. Swiss HIV Cohort Study, *Lancet*, 353(9156), 863–868, 1999.

39. Grabar, S., Le Moing, V., Goujard, C., Leport, C., Kazatchkine, M.D., Costagliola, D., and Weiss, L., Clinical outcome of patients with HIV-1 infection according to immunologic and virologic response after 6 months of highly active antiretroviral therapy, *Ann. Intern. Med.,* 133(6), 401–410, 2000.

40. Autran, B., Carcelain, G., Li, T.S., Blanc, C., Mathez, D., Tubiana, R., Katlama, C., Debre, P., and Leibowitch, J., Positive effects of combined antiretroviral therapy on CD4+ T cell homeostasis and function in advanced HIV disease, *Science,* 277(5322), 112–116, 1997.

41. Kaufmann, D., Pantaleo, G., Sudre, P., and Telenti, A., CD4-cell count in HIV-1-infected individuals remaining viraemic with highly active antiretroviral therapy (HAART). Swiss HIV Cohort Study, *Lancet,* 351(9104), 723–724, 1998.

42. Deeks, S.G., Hecht, F.M., Swanson, M., Elbeik, T., Loftus, R., Cohen, P.T., and Grant, R.M., HIV RNA and CD4 cell count response to protease inhibitor therapy in an urban AIDS clinic: response to both initial and salvage therapy, *AIDS,* 13(6), F35–F43, 1999.

43. Croteau, G., Doyon, L., Thibeault, D., McKercher, G., Pilote, L., and Lamarre, D., Impaired fitness of human immunodeficiency virus type 1 variants with high-level resistance to protease inhibitors, *J. Virol.,* 71(2), 1089–1096, 1997.

44. Mammano, F., Petit, C., and Clavel, F., Resistance-associated loss of viral fitness in human immunodeficiency virus type 1: phenotypic analysis of protease and Gag coevolution in protease inhibitor-treated patients: loss of viral fitness associated with multiple Gag and Gag-Pol processing defects in human immunodeficiency virus type 1 variants selected for resistance to protease inhibitors *in vivo,* *J. Virol.,* 72(9), 7632–7637, 1998.

45. Zennou, V., Mammano, F., Paulous, S., Mathez, D., and Clavel, F., Loss of viral fitness associated with multiple Gag and Gag-Pol processing defects in human immunodeficiency virus type 1 variants selected for resistance to protease inhibitors *in vivo,* *J. Virol.,* 72(4), 3300–3306, 1998.

46. Martinez-Picado, J., Savara, A.V., Sutton, L., and D'Aquila, R.T., Replicative fitness of protease inhibitor-resistant mutants of human immunodeficiency virus type 1, *J. Virol.,* 73(5), 3744–3752, 1999.

47. Mammano, F., Trouplin, V., Zennous, V., and Clavel, F., Retracing the evolutionary pathways of human immunodeficiency virus type 1 resistance to protease inhibitors: virus fitness in the absence and in the presence of drug, *J. Virol.,* 74(18), 8524–8531, 2000.

48. Maree, A.F., Keulen, W., Boucher, C.A., and De Boer, R.J., Estimating relative fitness in viral competition experiments, *J. Virol.,* 74(23), 11067–11072, 2000.

49. Quinones-Mateu, M.E., Ball, S.C., Marozsan, A.J., Torre, V.S., Albright, J.L., Vanham, G., van Der Groen, G., Colebunders, R.L., and Arts, E.J., A dual infection/competition assay shows a correlation between *ex vivo* human immunodeficiency virus type 1 fitness and disease progression, *J. Virol.,* 74(19), 9222–9233, 2000.

50. Stoddart, C.A., Liegler, T.J., Mammano, F., Linquist-Stepps, V.D., Hayden, M.S., Deeks, S.G., Grant, R.M., Clavel, F., and McCune, J.M., Impaired replication of protease inhibitor-resistant HIV-1 in human thymus, *Nat. Med.,* 7(6), 712–718, 2001.

51. Peng, C., Ho, B.K., Chang, T.W., and Chang, N.T., Role of human immunodeficiency virus type 1-specific protease in core protein maturation and viral infectivity, *J. Virol.,* 63(6), 2550–2556, 1989.

52. Pomerantz, R.J. and Horn, D.L., Twenty years of therapy for HIV-1 infection, *Nat. Med.,* 9(7), 867–873, 2003.

53. Palella, F.J., Jr., Delaney, K.M., Moorman, A.C., Loveless, M.O., Fuhrer, J., Satten, G.A., Aschman, D.J., and Holmberg, S.D., Declining morbidity and mortality among patients with advanced human immunodeficiency virus infection. HIV Outpatient Study Investigators, *N. Engl. J. Med.,* 338(13), 853–860, 1998.

54. Boden, D. and Markowitz, M., Resistance to human immunodeficiency virus type 1 protease inhibitors, *Antimicrob. Agents Chemother.,* 42(11), 2775–2783, 1998.

55. Cornelissen, M., Kampinga, G., Zorgdrager, F., and Goudsmit, J., Human immunodeficiency virus type 1 subtypes defined by Env show high frequency of recombinant gag genes. The UNAIDS Network for HIV Isolation and Characterization, *J. Virol.,* 70(11), 8209–8212, 1996.

56. Clavel, F. and Hance, A.J., HIV drug resistance, *N. Engl. J. Med.,* 350(10), 1023–1035, 2004.

57. Hertogs, K., Bloor, S., Kemp, S.D., Van den Eynde, C., Alcorn, T.M., Pauwels, R., Van Houtte, M., Staszewski, S., Miller, V., and Larder, B.A., Phenotypic and genotypic analysis of clinical HIV-1

isolates reveals extensive protease inhibitor cross-resistance: a survey of over 6000 samples, *AIDS,* 14(9), 1203–1210, 2000.

58. Kuroda, M.J., el-Farrash, M.A., Choudhury, S., and Harada, S., Impaired infectivity of HIV-1 after a single point mutation in the POL gene to escape the effect of a protease inhibitor *in vitro, Virology,* 210(1), 212–216, 1995.

59. Martinez-Picado, J., Savara, A.V., Shi, L., Sutton, L., and D'Aquila, R.T., Fitness of human immunodeficiency virus type 1 protease inhibitor-selected single mutants, *Virology,* 275(2), 318–322, 2000.

60. Devereux, H.L., Emery, V.C., Johnson, M.A., and Loveday, C., Replicative fitness *in vivo* of HIV-1 variants with multiple drug resistance-associated mutations, *J. Med. Virol.,* 65(2), 218–224, 2001.

61. Bally, F., Martinez, R., Peters, S., Sudre, P., and Telenti, A., Polymorphism of HIV type 1 Gag p7/p1 and p1/p6 cleavage sites: clinical significance and implications for resistance to protease inhibitors, *AIDS Res. Hum. Retroviruses,* 16(13), 1209–1213, 2000.

62. Nijhuis, M., Schuurman, R., de Jong, D., Erickson, J., Gustchina, E., Albert, J., Schipper, P., Gulnik, S., and Boucher, C.A., Increased fitness of drug resistant HIV-1 protease as a result of acquisition of compensatory mutations during suboptimal therapy, *AIDS,* 13(17), 2349–2359, 1999.

63. Kim, E.Y., Winters, M.A., Kagan, R.M., and Merigan, T.C., Functional correlates of insertion mutations in the protease gene of human immunodeficiency virus type 1 isolates from patients, *J. Virol.,* 75(22), 11227–11233, 2001.

64. Maguire, M.F., Guinea, R., Griffin, P., Macmanus, S., Elston, R.C., Wolfram, J., Richards, N., Hanlon, M.H., Porter, D.J., Wrin, T., Parkin, N., Tisdale, M., Furfine, E., Petropoulos, C., Snowden, B.W., and Kleim, J.P., Changes in human immunodeficiency virus type 1 Gag at positions L449 and P453 are linked to I50V protease mutants *in vivo* and cause reduction of sensitivity to amprenavir and improved viral fitness *in vitro, J. Virol.,* 76(15), 7398–7406, 2002.

65. de la Carriere, L.C., Paulous, S., Clavel, F., and Mammano, F., Effects of human immunodeficiency virus type 1 resistance to protease inhibitors on reverse transcriptase processing, activity, and drug sensitivity, *J. Virol.,* 73(4), 3455–3459, 1999.

66. Bleiber, G., Munoz, M., Ciuffi, A., Meylan, P., and Telenti, A., Individual contributions of mutant protease and reverse transcriptase to viral infectivity, replication, and protein maturation of antiretroviral drug-resistant human immunodeficiency virus type 1, *J. Virol.,* 75(7), 3291–3300, 2001.

67. Konvalinka, J., Litterst, M.A., Welker, R., Kottler, H., Rippmann, F., Heuser, A.M., and Krausslich, H.G., An active-site mutation in the human immunodeficiency virus type 1 proteinase (PR) causes reduced PR activity and loss of PR-mediated cytotoxicity without apparent effect on virus maturation and infectivity, *J. Virol.,* 69(11), 7180–7186, 1995.

68. Rose, J.R., Babe, L.M., and Craik, C.S., Defining the level of human immunodeficiency virus type 1 (HIV-1) protease activity required for HIV-1 particle maturation and infectivity, *J. Virol.,* 69(5), 2751–2758, 1995.

69. Barbour, J.D., Hecht, F.M., Wrin, T., Segal, M.R., Ramstead, C.A., Liegler, T.J., Busch, M.P., Petropoulos, C.J., Hellmann, N.S., Kahn, J.O., and Grant, R.M. Higher CD4+ T cell counts associated with low viral pol replication capacity among treatment-naïve adults in early HIV-1 infection. *J. Infect. Dis.,* 190, 251–256, 2004.

Immune Mechanisms

11 Immune Activation

Barbara N. Phenix
Savita Pahwa

CONTENTS

INTRODUCTION

Immune activation is a characteristic feature of primary viral infections and is associated with marked expansion of CD8 T cells.[1,2] In the case of human immunodeficiency virus (HIV), immune activation occurs during the chronic phase of the infection as well. Not all of the immune activation is HIV specific, and "bystander activation" of multiple cell types has been documented.[3,4] Although pathogen-specific immune activation is a necessary host requirement for self-protection, heightened generalized immune activation such as that which occurs with HIV has dire consequences. It is known to accelerate CD4 T cell depletion, increase viral replication, and decrease host survival.[5–8] The cause for heightened immune activation in HIV infection is not fully understood and is attributed to many factors, including effects of virus gene products,[9–16] coinfection with other pathogens,[17,18] and reactivation of latent viruses.[3,4] Increases in immune activation generally precede the inflection point of CD4 T cell depletion, and HIV-1 viral replication increases, suggesting a role for immune activation in the pathogenesis of HIV infection.

Aberrant immune activation dampens HIV-specific immunity, and emerging data support this contention, although it has never been conclusively proven. One of the most compelling arguments for the adverse prognostic significance of immune activation may be made on the basis of observations in the sooty mangabeys, the natural hosts of simian immunodeficiency virus (SIV), and in rhesus macaques inoculated with SIV.[19] Sooty mangabeys do not develop progressive disease despite ongoing viral replication, whereas rhesus macaques develop progressive acquired immunodeficiency syndrome (AIDS)–like disease. The major distinguishing characteristic between the mangabeys and macaques is the low level of markers of immune activation in the mangabeys and high-immune activation in the macaques. This chapter will discuss markers of immune activation, possible causes of immune activation, and its consequences, particularly in reference to events leading to activation-induced cell death (AICD) of CD4 and CD8 T cells.

MARKERS OF IMMUNE ACTIVATION

Several studies have demonstrated an increase in CD38 and HLA-DR activation antigens on CD4 and CD8 cells in HIV-infected individuals.[5,6] CD38 antigen is broadly distributed on both immature and activated lymphocytes, and its increased expression has been linked to poor prognosis of subjects with HIV disease. The median density of CD38 molecules on CD4 and CD8 T cells is, in some studies, a more reliable measure of activated T cells than percentage of CD38 cells.[20] The T cell activation in HIV is generalized, involving cells of antigen specificities other than HIV, as documented by CD38 upregulation in CD8 T cells specific for Epstein–Barr virus (EBV), cytomegalovirus, and influenza virus during primary HIV infection.[3,4]

Although CD38 and DR are two markers that are used reliably to indicate immune activation, other markers that are indicative of ongoing immune activation include expression of HLA-DR and CD25 in CD4 T cells, CD57, and CD71 on CD8 T cells, and CD71 on B cells.[8] Polyclonal hypergammaglobulinemia is another characteristic feature of HIV infection, which results from immune activation of B cells.[21,22] CD4 T cells from HIV-1-infected patients express tumor necrosis factor (TNF)α, the TNF receptor (TNFR) ligand, and can polyclonally activate B cells.[23] Levels of sCD8, sCD14, TNFα, neopterin, β_2-microglobulin, sTNFR2, and sCD25 are also reported to be increased in serum of patients with progressive HIV infection.[8]

Cytokine dysregulation[24,25] is a characteristic feature of HIV infection; excessive immune activation is associated with an increase of proinflammatory cytokines, interleukin (IL)-4, IL-5, IL-6, TNFα, and interferon (IFN) γ, and with a decline in IL-2 and IL-12. Increased lymphocyte apoptosis involving CD4 and CD8 T cells[10,26,27] is regularly seen in HIV-infected individuals and reflects ongoing immune activation.

CAUSES AND MECHANISMS OF IMMUNE ACTIVATION IN HIV INFECTION

Although considerable strides have been made in this field, the mechanism of aberrant immune activation in HIV infection remains incompletely understood. Immune activation, per se, is an essential component of an immune response when it involves activation of antigen-specific T cells in response to an antigen, usually in the context of an infectious organism. However, this type of immune activation can also be deleterious to the host if the inflammatory component exceeds the cytolytic activity[28] or if generalized activation occurs, which encompasses cells of multiple specificities. In these situations, the immune response is potentially deleterious to the host. Cytokine dysregulation that characterizes HIV infection is a direct reflection of the overt immune activation that mediates much of the immunologic dysfunction.

Many factors can contribute to the heightened immune activation in HIV infection. Many of the virally encoded proteins, particularly, Nef, Env, Tat, and Vpr, can induce chronic immune activation,[9–16] thereby creating an environment that favors the ability of the virus to persist in the host. For example, Nef has the capacity to modify the plasma membrane signaling by regulation of receptor/ligand endocytosis and to promote lymphocyte activation and apoptosis.[9,13,15] HIV Env was shown in numerous studies to induce CD4 T cell activation via CD4 cross-linking and binding to chemokine receptors, which, in turn, can prime the cells to undergo AICD.[10,14] HIV-envelope interaction with CD4 T cells results in activation of NF-κB transcription factors, which amplifies the host inflammatory response and activates the HIV-1 long terminal repeat (LTR) sequences. HIV Tat, when presented to T cells exogenously, can promote their activation and apoptosis.[11] HIV infection in humans and SIV infection in rhesus macaques interfere with cell-cycle control in T lymphocytes in association with chronic immune activation and excess apoptosis.[29] HIV Vpr can cause cell-cycle arrest in G_2 and lymphocyte apoptosis, which can occur independently of the cell-cycle dysfunction.[12]

Independently of the effect of HIV viral gene products, another important cause of immune activation stems from co-infections with non-HIV pathogens. In many regions of the world, the co-infection of HIV-infected patients with other pathogens is highly prevalent, and these co-infections often persist as chronic and recurrent acute infections. Common co-infections[17,18] include protozoan parasites (e.g., *Leishmania donovani, Toxoplasma gondii, Plasmodium falciparum*) and bacteria (e.g., *Mycobacterium* sp. and *Neisseria gonorrhea*). Co-infections with viruses such as human herpesvirus (HHV)-6, human simplex virus (HSV)-1, cytomegalovirus, and hepatitis B and C viruses are also prevalent worldwide. Certain co-infections, in particular, parasitic infections, are believed to contribute significantly to HIV-associated morbidity and mortality by a mechanism of microbial-induced immune activation and its consequences. The mechanism of immune activation in these situations has been attributed to signaling via the Toll-like receptor (TLR) family, which is the major innate recognition system for microbial invaders in vertebrates.[30] There is now evidence to link the microbial-stimulated TLR signaling with HIV LTR induction and viral expression.[18] Events triggered upon ligation of TLR lead to the transcription of many of the genes involved in immune activation and activation of NFκ-B and mitogen-activated protein (MAP) kinases, which are the main transcriptional regulators for the genes encoding proinflammatory cytokines, TNFα and IL-1.

In the healthy state, upon completion of an immune effector function, the immune response comes to an end through an incompletely understood process. Recently, regulatory T cells (T regs) were identified, both natural and induced, that play important roles in maintaining immunologic balance.[31–34] The forkhead/winged helix transcription factor FoxP3 is a marker for T regs, and the best-studied regulatory T cells are CD4+ CD25+ T cells that express FoxP3. The original role of T reg was in mediating tolerance, because deficiencies of T regs are associated with autoimmunity; however, several subsets of T reg with distinct phenotypes and functions were identified. Increasingly, in addition to the T regs that originate in the thymus, antigen-specific regulatory cells that are induced in the periphery are being recognized as important participants in adaptive immune responses, thereby suggesting that T regs maintain tolerance and also control immune responses. Thus, a lack of T regs may theoretically lead to excessive immune activation in situations of chronic exposure antigenic burden. The work of Kinter et al., demonstrating a better outcome of HIV-infected patients who had higher T regs, supports this view.[35] The role of regulatory T cells was not investigated in depth in the pathogenesis of immune activation in HIV infection.

The generation of T regs in the adaptive immune response is dependent on dendritic cells (DC).[32–34] The differential roles of myeloid vs. plasmacytoid dendritic cell (DC) in the induction of T regs are under investigation. Moreover, the roles of immature vs. mature DC, and their ability to generate as well as overcome the suppressive properties of T regs, are being identified. In one model, DC blocked normal CD4+ CD25+ T-reg-mediated suppression, partially via IL-6 secretion. DC can thus control the suppressive ability, expansion, and differentiation of T reg cells *in vivo*. Thus, in addition to the definitive role of DC in the pathogenesis of immune activation secondary to infection with a co-pathogen, these cells likely play important and even essential independent roles in immune activation. The absence of immune activation in the nonpathogenic natural infection of sooty mangabeys with SIV seems to be linked to the function of plasmacytoid DCs.[36] The mechanism of action of regulatory T cells is mediated by TGF-β or IL-10 in some models, whereas others have failed to show a need for these cytokines.

It is possible that other "regulatory" cells are also important in the immune activation of HIV infection. It was demonstrated that the function of natural killer (NK) cells is compromised in HIV infection[37] and that γδ T cells are rapidly and permanently depleted in HIV infection.[38] Whether or not these cells also play a role in the regulation of immune activation merits exploration.

CONSEQUENCES OF IMMUNE ACTIVATION

A heightened state of chronic immune activation is expected to be deleterious to the host on many fronts. In a prospective cohort of acutely infected adults, Deeks et al. demonstrated that an immunologic activation set point is established early in the natural history of HIV disease and that this set point determines the rate at which CD4 T cells are lost over time.[39] In chronic HIV infection, increased T cell activation is associated with shorter duration of viral suppression, frequent low-level viremia, and lower nadir CD4+ T cell counts.[40] Immune activation accelerates viral replication[7] and leads to depletion of naive cells and exhaustion of the memory T cell pool,[41–45] rendering the host unable to generate new primary immune responses. The enhanced viral replication is facilitated by an increased pool of susceptible target cells and the altered cytokine milieu favoring virus replication. The activation of HIV antigen–specific CD4 T cells that occurs during a natural host antiviral response makes them susceptible to HIV infection.[46] CD4 T cell activation also affects surface expression of HIV-1 co-receptors by uninfected CD4 T cells and pushes the balance of naive and memory cells in favor of memory cells.[41] These are the preferential targets of HIV, as HIV-1 replicates preferentially in CD45RO-memory T lymphocytes compared with CD45RA T cells, in which reverse transcriptase is impaired. The proinflammatory aspects of HIV infection also result in the activation and proliferation of other CD4 T cells, including those of specificities other than HIV, thus favoring both proinflammatory cytokines as well as more HIV-susceptible T cells.[3,4] Viral gene products interact with infected CD4 T cells to induce LTR activation, as described earlier, and augment the proinflammatory aspects of HIV infection, influencing uninfected CD4 T cells.[47]

Another consequence of generalized immune activation is to dampen antigen-specific immune responses to HIV and other antigens. Cellular immunity mediated by memory and effector CD8 T cells is considered an essential component of antiviral immune responses, although characteristics of the response in different viral infections seem to be different.[42–44] The CD8 T cell memory and effector subsets can be distinguished phenotypically on the basis of expression of maturation antigens on their cell surfaces.[48] Heightened activation leads to increased lymphocyte turnover[49] and promotes premature T cell replicative senescence with a consequent decline in immune competence.[4,50] Observations of skewed maturation of virus-specific cells with impaired effector function and low perforin levels in HIV-specific CD8 T cells support the deleterious effects of activation in HIV-infected patients.[41,51] These data were generated in limited data sets, and a direct relationship between immune activation and immune function and T cell maturation was not performed in a systematic manner.

The hallmark of HIV disease is the progressive loss of CD4 T cells. One of the important mechanisms that contributes to this CD4 loss is lymphocyte apoptosis, and immune activation is a major underlying cause of the increased lymphocyte apoptosis.[52] Increased lymphocyte apoptosis is believed to involve not only CD4 T cells but also CD8 T cells and possibly other cells, such as γδ T cells, B cells, and NK cells.[38–40] In persons infected with HIV, both infected and uninfected cells undergo accelerated apoptosis. Apoptosis of cells directly infected with HIV results in only a small percentage of the apoptosis observed, as the vast majority of cells undergoing apoptosis are uninfected.[53,54] One of the mechanisms proposed to mediate apoptosis of uninfected cells in HIV is AICD. Normally, AICD is a physiological process of cell death that occurs in a controlled manner for the central and peripheral deletion of cells in the context of tolerance and homeostasis.[55] In HIV infection, lymphocyte AICD is exaggerated because of the heightened state of immune activation.

Supportive evidence that apoptosis contributes to loss of CD4 T cells in HIV-1 infection of humans comes from the normal levels of apoptosis in apathogenic lentivirus infections of nonhuman primates, including HIV-1 infection of chimpanzees.[56–58] Chimpanzees infected with the LAI/LAV-1b strain of HIV develop progressive loss of CD4 T cells with high viral burdens and increased levels of CD4 T cell apoptosis, both *in vitro* and *in vivo*.[59] In contrast, SIV-infected chimpanzees do not lose CD4 T cells and have no increases in apoptosis compared with control groups.

Spontaneous apoptosis is greater in patients with progressive HIV disease than in long-term nonprogressors (LTNPs). The frequency of T cells exhibiting loss of mitochondrial transmembrane potential and increased reactive oxygen species (ROS) production is also lower in LTNPs.[60] These findings indicate that the degree of apoptosis correlates inversely with CD4 T cell depletion and suggest that apoptosis contributes to HIV-induced immunodeficiency.

DEATH RECEPTOR/LIGANDS AND INTRINSIC APOPTOTIC MACHINERY IN HIV INFECTION

The death program can be set in motion by the binding of ligands to members of the TNF superfamily of death receptors. AICD occurs as a result of sequential activation of T cells. The death receptors/ligands best studied in HIV infection are depicted in Figure 11.1. This includes the Fas/FasL, TNFR/TNF, and tumor necrosis factor–related apoptosis-inducing ligand–receptor (TRAILR)/tumor necrosis factor–related apoptosis-inducing ligand (TRAIL) pathways of apoptosis. Upon binding of their respective ligands, the death receptors recruit adaptors and initiator caspases sequentially through specific interaction domains to form the death-inducing signaling complex (DISC) at the plasma membrane.[55]

The Fas/FasL death signaling pathway constitutes one of the major apoptotic pathways in lymphocytes.[61] Although Fas is expressed in resting lymphocytes, the sensitivity to apoptosis by Fas ligation is induced only upon cell activation, which also upregulates Fas expression.[62] Whether the primary stimulus involves the CD3/T cell receptor (TCR),[63] sole stimulation through the CD4 receptor,[64] activation without costimulation,[65] or infection with HIV,[66] stimulation results in an increase in the expression of Fas and increased susceptibility to Fas-mediated apoptosis.[67–71] Phenotypic analysis of cells undergoing apoptosis has revealed that a population of activated lymphocytes with the memory phenotype lacking the co-stimulatory molecule CD28 is especially prone

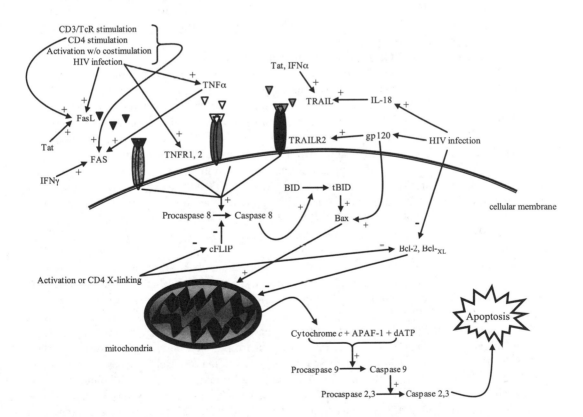

FIGURE 11.1 The direct and indirect influences of HIV on pathways of AICD, showing both the extrinsic pathway, mediated at the level of the cellular membrane by members of the TNF family of death-inducing receptors, and the intrinsic pathway of AICD, involving targeting of mitochondria.

to undergoing apoptosis. Evidence also exists for both Fas-dependent and Fas-independent pathways of apoptosis in HIV disease in children.[72]

Regulation of Fas and FasL expression can be affected by a variety of different factors within the context of HIV infection. CD4 cross-linking by viral glycoprotein (gp)120 triggers increases in Fas expression, upregulation of FasL, and CD4 and CD8 T cell apoptosis.[73,74] Both IFNγ and TNFα were found to contribute to Fas upregulation, suggesting that aberrant cytokine secretion induced by CD4 cross-linking contributes to HIV disease pathogenesis.[73] However, it was also suggested that it is the lateral association of CD4 with Fas after gp120 binding that triggers apoptosis.[75] Thus, an early FasL-independent effect of gp120 could lead to Fas triggering, followed by a secondary upregulation of FasL. Of note is the fact that gp120s from diverse HIV-1 strains display different capacities to induce cell death.[75] Tat and Nef, two virally-encoded proteins, were also shown to upregulate FasL expression,[60,77] the former via effects on nuclear factor (NF) κB,[76] and the latter via activation of the T cell receptor (reviewed in Fackler, 2002).

In addition to upregulating FasL expression, the expression of other death receptors, including TNFR1 and TNFR2, is also dysregulated in HIV infection. CD8+ bronchioalveolar (BAL) T cells accumulating in the alveolar spaces of patients infected with HIV express more phenotypic markers of activation (CD45RA, CD69) and demonstrate increased expression of TNFR2 than do BAL T cells from uninfected donors.[78] In one study, both TNFR2 and TNFR1 were found to be expressed at higher levels in lymphocytes from HIV-infected persons, although expression of the former was higher than the latter. These and other observations suggest that apoptosis mediated by the TNFRs may play a substantial role in the AICD of T cells observed in HIV.[79] Increased production of TNFα occurs early in HIV infection and may further increase with the development of opportunistic infections. A correlation exists between serum TNFα levels and progression to AIDS, suggesting a role for this cytokine in HIV pathogenesis.[80]

In addition to Fas and TNFR, apoptosis mediated by TRAIL also plays a role in the AICD associated with HIV infection. TRAIL expression is increased in peripheral blood mononuclear cells (PBMCs) from HIV-infected persons experiencing virologic failure and has been attributed to persistently high-serum IL-18 levels.[81] HIV-infected monocytes are activated indirectly via IFNα or directly via extracellular Tat protein[82–85] and express several soluble mediators of apoptosis, including TRAIL.[86] Research into potential dysregulation of TRAIL receptor expression identified increases in TRAILR2 in both CD4 and CD8 T cells from HIV-infected individuals. A similar increase was observed *in vitro* after gp120 stimulation of uninfected Jurkat and primary cells, indicating a potential mechanism for upregulation. Although TRAIL does not induce apoptosis in normal peripheral blood T and B cells, a neutralizing monoclonal antibody to TRAIL effectively inhibits AICD in some, but not all, HIV-infected patients. TRAIL agonists also induce apoptosis of *in vitro* HIV-infected monocyte-derived macrophages (MDMs) as well as peripheral blood lymphocytes (PBLs) from HIV-infected patients.[87]

Members of the Bcl-2 family of molecules are key regulators of apoptosis and include both pro- and antiapoptotic molecules. It is the ratio between these two subsets that helps determine, in part, the susceptibility of a cell to undergo apoptosis.[88] Regulation of Bcl-2, an antiapoptotic protein, seems to be altered in CD8 T cells in HIV infection.[89] Downmodulation of Bcl-2 expression in CD4 T cells occurs after CD4 cross-linking.[90] The important role of immune activation in altering the ratio of pro- and antiapoptotic molecules is evident from the expression of activation markers CD45RO, HLA-DR, and CD38 and increased levels of Fas in the low Bcl-2 expressing fraction of CD8 T cells.[89] Induction of Bcl-xL, another antiapoptotic member of the Bcl-2 family, is also impaired in HIV-infected individuals.[91] In HIV-specific CD8 T cells, expression of both Bcl-2 and Bcl-xL is markedly reduced.[92] In addition to downmodulating Bcl-2 expression, engagement of CD4 by HIV gp120 leads to an increase in the expression of Bax, a proapoptotic member of the Bcl-2 family,[93] further biasing the intracellular milieu toward promotion of apoptosis. Altered expression of Bcl-2 may account for the increased sensitivity of T cells to TNFR1- and TNFR2-mediated apoptosis.[79]

MODULATION OF HIV-MEDIATED IMMUNE ACTIVATION, AICD, AND EFFECT OF ANTIRETROVIRAL THERAPY

Interventions that directly target T cell activation or the determinants of activation are being investigated as adjuvants to antiretroviral therapy. In an interesting *in vivo* study,[94] Rizzardi et al. gave the immune-modulating drug cyclosporin A (CsA) in combination with highly active antiretroviral therapy (HAART) to nine adults during primary HIV-1 infection. The CsA was discontinued at week 8, and patients were maintained on HAART alone. Although this was not a controlled trial, the results were impressive. CsA restored normal CD4 T cell levels, and at week 48, the proportion of IFN-γ secreting CD4 T cells was significantly higher in the CsA + HAART cohort than in the HAART-alone cohort. Thus, the early shutdown of T cell activation in primary HIV infection seemed to have a beneficial effect. It had a more favorable immunologic set point than that in the absence of CsA intervention. This concept is currently undergoing investigation in a larger controlled multicenter study.

Another strategy was tested in the acute SIV infection of rhesus macaques.[95] *In vivo* blockade of CD28 and CD40 T cell co-stimulation pathways during acute infection by the transient administration of CTLA4-Ig and anti-CD40L mAb resulted in dramatic inhibition of SIV-specific cellular and humoral immune responses, implying that proliferating CD4+ T cells function both as sources of virus production and as antiviral effectors. Thus, further studies are needed to determine the best approach to reducing immune activation without impairing the antigen-specific prospective immune response. It is of the authors' opinion that approaches that can manipulate DC to induce T regs to keep the immune activation under control need to be researched.

Modulation of AICD presents itself as a potential strategy to slow the progressive loss of T cells in HIV-infected subjects. Because apoptosis is regulated at a variety of different levels, there are many targets that offer themselves as attractive candidates, and insights into possible approaches are evolving. Cytokines, in particular IL-2 and IL-15, can influence apoptosis.[96] Although IL-2 can both protect and augment apoptosis depending on the state of activation of target cells, IL-15 is characteristically antiapoptotic. Exogenous addition of IL-2 can partially prevent Bcl-2 downregulation,[97,98] and IL-15 is able to induce expression of both Bcl-2 and Bcl-xL in HIV-specific CD8 T cells.[93,99] The chemoattractant cytokine IL-16, which is a ligand for CD4, was reported to suppress lymphocyte activation[100] and was found to significantly reduce the percentage of lymphocytes undergoing AICD, and it also caused a decrease in Fas expression in CD4 T cells from HIV-infected patients.[101]

Antiretroviral therapy (ART) of HIV-infected patients also results in a decrease in immune activation and in apoptosis, but these remain elevated in comparison to uninfected controls.[102–105] The decrease in apoptosis in response to ART has been attributed to a decrease in viremia, to a decrease in immune activation, and also to the effects of the treatment drugs. Institution of ART significantly reduces the proportion of low Bcl-2-expressing lymphocytes in patients' peripheral blood.[97] Treatment of uninfected lymphocytes with the HIV protease inhibitor Ritonavir leads to decreases in FasL expression, suggesting that these drugs may function to directly and indirectly influence the Fas/FasL pathway in patients.[104] A cysteine protease inhibitor was also shown to inhibit FasL expression in PBLs from HIV-infected persons.[106] Incubation of PBMCs from HIV-infected individuals with all-trans retinoic acid *in vitro* results in decreased AICD,[107] presumably via downmodulation of FasL.[108] Chemokines also exert an antiapoptoic effect on lymphocytes *in vitro*.[109]

CONCLUDING REMARKS

Progression of HIV is intimately related to the loss of lymphocytes. Because these cells play vital roles in maintaining and augmenting immune responses, any decline in their numbers results in deficits in both humoral and cell-mediated immunity. AICD of lymphocytes has specific consequences relating to the pathogenesis of HIV disease, including loss of memory responses, impaired

CTL function, and cytokine dysregulation. The immune activation that characterizes the chronic state of HIV infection occurs as the result of various virus-specific and nonspecific causes. Attrition of the memory T cell pools by repeated immune activation events and depletion of naive T cells may account for the increased susceptibility to opportunistic infections and malignancies observed with HIV infection. Preferential targeting of CD4 T cells for AICD may explain, in part, some of the CTL dysfunction observed in HIV. The cytokine dysregulation that occurs in HIV disease has the potential to play a pivotal role in pathogenesis, as type-1 and type-2 cytokines have the potential to either inhibit or enhance AICD, respectively. Additional evidence that immune activation is linked to apoptosis in HIV infection stems from recent studies that suggest that increases in immune activation observed in other pathogenic states also correlate with increases in cell death, as, for example, is seen in certain septic states.[110] An understanding of the means to control immune activation has the potential of converting a lethal progressive disease into one in which the host and the pathogen can live in harmony, as in the sooty mangabeys and some cases of LTNPs.

REFERENCES

1. Butz, E.A. and Bevan, M.J., Massive expansion of antigen-specific CD8+ T cells during an acute virus infection, *Immunity*, 8(2), 167–175, 1998.
2. Pantaleo, G., Demarest, J.F., Soudeyns, H., Graziosi, C., Denis, F., Adelsberger, J.W., Borrow, P., Saag, M.S., Shaw, G.M., Sekaly, R.P., Major expansion of CD8+ T cells with a predominant V usage during the primary immune response to HIV, *Nature*, 370(6489), 463–467, 1994.
3. Doisne, J.M., Urrutia, A., Lacabaratz-Porret, C., Goujard, C., Meyer, L., Chaix, M.L., Sinet, M., Venet, A., CD8+ T cells specific for EBV, cytomegalovirus, and influenza virus are activated during primary HIV infection, *J. Immunol.*, 173(4), 2410–2418, 2004.
4. Papagno, L., Spina, C.A., Marchant, A., Salio, M., Rufer, N., Little, S., Dong, T., Chesney, G., Waters, A., Easterbrook, P., Dunbar, P.R., Shepherd, D., Cerundolo, V., Emery, V., Griffiths, P., Conlon, C., McMichael, A.J., Richman, D.D., Rowland-Jones, S.L., Appay, V., Immune activation and CD8(+) T-cell differentiation towards senescence in HIV-1 infection. *PLoS Biol.*, 2(2), E20, 2004.
5. Giorgi, J.V., Fahey, J.L., Smith, D.C., Hultin, L.E., Cheng, H.L., Mitsuyasu, R.T., Detels, R., Early effects of HIV on CD4 lymphocytes *in vivo*, *J. Immunol.*, 138(11), 3725–3730, 1987.
6. Chun, T.W., Justement, J.S., Sanford, C., Hallahan, C.W., Planta, M.A., Loutfy, M., Kottilil, S., Moir, S., Kovacs, C., Fauci, A.S., Relationship between the frequency of HIV-specific CD8+ T cells and the level of CD38+CD8+ T cells in untreated HIV-infected individuals, *Proc. Natl. Acad. Sci. U.S.A.*, 101(8), 2464–2469, 2004.
7. Sousa, A.E., Carneiro, J., Meier-Schellersheim, M., Grossman, Z., Victorino, R.M., CD4 T cell depletion is linked directly to immune activation in the pathogenesis of HIV-1 and HIV-2 but only indirectly to the viral load, *J. Immunol.*, 169(6), 3400–3406, 2002.
8. Mildvan, D., Bosch, R.J., Kim, R.S., Spritzler, J., Haas, D.W., Kuritzkes, D., Kagan, J., Nokta, M., DeGruttola, V., Moreno, M., Landay, A., Immunophenotypic markers and antiretroviral therapy (IMART): T cell activation and maturation help predict treatment response, *J. Infect. Dis.*, 189(10), 1811–1820, 2004.
9. Tolstrup, M., Ostergaard, L., Laursen, A.L., Pedersen, S.F., Duch, M., HIV/SIV escape from immune surveillance: focus on Nef, *Curr. HIV Res.*, 2(2), 141–151, 2004.
10. Oyaizu, N., McCloskey, T.W., Coronesi, M., Chirmule, N., Kalyanaraman, V.S., Pahwa, S., Accelerated apoptosis in peripheral blood mononuclear cells (PBMCs) from human immunodeficiency virus type-1 infected patients and in CD4 cross-linked PBMCs from normal individuals, *Blood*, 82(11), 3392–3400, 1993.
11. McCloskey, T.W., Ott, M., Tribble, E., Khan, S.A., Teichberg, S., Paul, M.O., Pahwa, S., Verdin, E., Chirmule, N., Dual role of HIV Tat in regulation of apoptosis in T cells, *J. Immunol.*, 158(2), 1014–1019, 1997.
12. Gaynor, E. and Chen, I.S., Analysis of apoptosis induced by HIV-1 Vpr and examination of the possible role of the hHR23A protein, *Exp. Cell Res.*, 267(2), 243–257, 2001.

13. Simmons, A., Aluvihare, V., and McMichael, A., Nef triggers a transcriptional program in T cells imitating single-signal T cell activation and inducing HIV virulence mediators, *Immunity*, 14(6), 763–777, 2001.

14. Kinter, A.L., Craig, A., Umscheid, J.A., Claudia C., Yin L., Robert J., Eileen D., Linda E., Joseph A., Ronald L.R., and Anthony S.F., HIV envelope induces virus expression from resting CD4$^+$ T cells isolated from HIV-infected individuals in the absence of markers of cellular activation or apoptosis, *J. Immunol.*, 170(5), 2449–2455, 2003.

15. Stoddart, C.A., Geleziunas, R., Ferrell, S., Linquist-Stepps, V., Moreno, M.E., Bare, C., Xu, W., Yonemoto, W., Bresnahan, P.A., McCune, J.M., Greene, W.C., Human immunodeficiency virus type 1 Nef-mediated downregulation of CD4 correlates with Nef enhancement of viral pathogenesis, *J. Virol.*, 77(3), 2124–2133, 2003.

16. Lahti, A.L., Manninen, A., and Saksela, K., Regulation of T cell activation by HIV-1 accessory proteins: Vpr acts via distinct mechanisms to cooperate with Nef in NFAT-directed gene expression and to promote transactivation by CREB, *Virology*, 310(1), 190–196, 2003.

17. Bentwich, Z., Kalinkovich, A., and Weisman, Z., Immune activation is a dominant factor in the pathogenesis of African AIDS, *Immunol. Today*, 16(4), 187–191, 1995.

18. Bafica, A., Scanga, C.A., Schito, M., Chaussabel, D., Sher, A., Influence of coinfecting pathogens on HIV expression: evidence for a role of Toll-like receptors, *J. Immunol.*, 172(12), 7229–7234, 2004.

19. Silvestri, G., Sodora, D.L., Koup, R.A., Paiardini, M., O'Neil, S.P., McClure, H.M., Staprans, S.I., Feinberg, M.B., Nonpathogenic SIV infection of sooty mangabeys is characterized by limited bystander immunopathology despite chronic high-level viremia, *Immunity*, 18(3), 441–452, 2003.

20. Giorgi, J.V., Lyles, R.H., Matud, J.L., Yamashita, T.E., Mellors, J.W., Hultin, L.E., Jamieson, B.D., Margolick, J.B., Rinaldo, C.R. Jr, Phair, J.P., Detels, R., Multicenter AIDS Cohort Study., Predictive value of immunologic and virologic markers after long or short duration of HIV-1 infection, *J. Acquir. Immune Defic. Syndr.*, 29(4), 346–355, 2002.

21. Pahwa, S., Pahwa, R., Saxinger, C., Gallo, R.C., Good, R.A., Influence of the human T-lymphotropic virus/lymphadenopathy-associated virus on functions of human lymphocytes: evidence for immuno-suppressive effects and polyclonal B-cell activation by banded viral preparations, *Proc. Natl. Acad. Sci. U.S.A.*, 82(23), 8198–8202, 1985.

22. Lane, H.C., Masur, H., Edgar, L.C., Whalen, G., Rook, A.H., Fauci, A.S., Abnormalities of B-cell activation and immunoregulation in patients with the acquired immunodeficiency syndrome, *N. Engl. J. Med.*, 309(8), 453–458, 1983.

23. Johansson, C.C., Bryn, T., Yndestad, A., Eiken, H.G., Bjerkeli, V., Froland, S.S., Aukrust, P., Tasken, K., Cytokine networks are pre-activated in T cells from HIV-infected patients on HAART and are under the control of cAMP, *AIDS*, 18(2), 171–179, 2004.

24. Poli, G., T lymphocytes of HIV-positive individuals: preloaded guns in spite of highly active antiretroviral therapy? *AIDS*, 18(2), 327–328, 2004.

25. Moir, S., Malaspina, A., Ogwaro, K.M., Donoghue, E.T., Hallahan, C.W., Ehler, L.A., Liu, S., Adelsberger, J., Lapointe, R., Hwu, P., Baseler, M., Orenstein, J.M., Chun, T.W., Mican, J.A., Fauci, A.S., HIV-1 induces phenotypic and functional perturbations of B cells in chronically infected individuals, *Proc. Natl. Acad. Sci. U.S.A.*, 98(18), 10362–10367, 2001.

26. Gougeon, M.-L., Lecoeur, H., Dulioust, A., Enouf, M.G., Crouvoiser, M., Goujard, C., Debord, T., Montagnier, L., Programmed cell death in peripheral lymphocytes from HIV-infected persons: increased susceptibility to apoptosis of CD4 and CD8 T cells correlates with lymphocyte activation and with disease progression, *J. Immunol.*, 156,(9): 3509–3521, 1996.

27. Miedema, F., Petit, A.J., Terpstra, F.G., Schattenkerk, J.K., de Wolf, F., Al, B.J., Roos, M., Lange, J.M., Danner, S.A., Goudsmit, J., Immunological abnormalities in human immunodeficiency virus (HIV)-infected asymptomatic homosexual men. HIV affects the immune system before CD4$^+$ T helper cell depletion occurs, *J. Clin. Invest.*, 82(6), 1908–1914, 1988.

28. Deeks, S.G. and Walker, B.D., The immune response to AIDS virus infection: good, bad, or both? *J. Clin. Invest.*, 113(6), 808–810, 2004.

29. Paiardini, M., Cervasi, B., Dunham, R., Sumpter, B., Radziewicz, H., Silvestri, G., Cell-cycle dysregulation in the immunopathogenesis of AIDS, *Immunol. Res.*, 29(1–3), 253–268, 2004.

30. Takeda, K., Kaisho, T., and Akira, S., Toll-like receptors, *Annu. Rev. Immunol.*, 21, 335–376, 2003.

31. Wraith, D.C., Nicolson, K.S., and Whitley, N.T., Regulatory CD4$^+$ T cells and the control of autoimmune disease, *Curr. Opin. Immunol.*, 16(6), 695–701, 2004.

32. Fehervari, Z. and Sakaguchi, S., Control of Foxp3+ CD25+CD4+ regulatory cell activation and function by dendritic cells, *Int. Immunol.*, 16(12), 1769–1780, 2004.

33. Chen, T.C., Cobbold, S.P., Fairchild, P.J., Waldmann, H., Generation of anergic and regulatory T cells following prolonged exposure to a harmless antigen, *J. Immunol.*, 172(10), 5900–5907, 2004.

34. Groux, H., Fournier, N., and Cottrez, F., Role of dendritic cells in the generation of regulatory T cells, *Semin. Immunol.*, 16(2), 99–106, 2004.

35. Kinter, A.L., Hennessey, M., Bell, A., Kern, S., Lin, Y., Daucher, M., Planta, M., McGlaughlin, M., Jackson, R., Ziegler, S.F., Fauci, A.S., CD25+CD4+ regulatory T cells from the peripheral blood of asymptomatic HIV-infected individuals regulate CD4+ and CD8+ HIV-specific T cell immune responses *in vitro* and are associated with favorable clinical markers of disease status, *J. Exp. Med.*, 200(3), 331–343, 2004.

36. Barry, A.P., Chavan, R., Wernett, M., McCausland, M., Staprans, S., Silvestri, G., and Feinberg, M.B., Differences in Dendritic Cell Populations and Signaling May Account for Divergent SIV Infection Outcomes in Rhesus Macaques and Sooty Mangabeys, paper presented at 11th Conference on Retroviruses and Opportunistic Infections, San Francisco, CA, 2004.

37. Kottilil, S., Chun, T.W., Moir, S., Liu, S., McLaughlin, M., Hallahan, C.W., Maldarelli, F., Corey, L., Fauci, A.S., Innate immunity in human immunodeficiency virus infection: effect of viremia on natural killer cell function, *J. Infect. Dis.*, 187(7), 1038–1045, 2003.

38. Poles, M.A., Barsoum, S., Yu, W., Yu, J., Sun, P., Daly, J., He, T., Mehandru, S., Talal, A., Markowitz, M., Hurley, A., Ho, D., Zhang, L., Human immunodeficiency virus type 1 induces persistent changes in mucosal and blood T cells despite suppressive therapy, *J. Virol.*, 77(19), 10456–10467, 2003.

39. Deeks, S.G., Kitchen, C.M., Liu, L., Guo, H., Gascon, R., Narvaez, A.B., Hunt, P., Martin, J.N., Kahn, J.O., Levy, J., McGrath, M.S., Hecht, F.M., Immune activation set point during early HIV infection predicts subsequent CD4+ T-cell changes independent of viral load, *Blood*, 104(4), 942–947, 2004.

40. Hunt, P.W., Martin, J.N., Sinclair, E., Bredt, B., Hagos, E., Lampiris, H., Deeks, S.G., T cell activation is associated with lower CD4+ T cell gains in human immunodeficiency virus-infected patients with sustained viral suppression during antiretroviral therapy, *J. Infect. Dis.*, 187(10), 1534–1543, 2003.

41. McMichael, A.J. and Rowland-Jones, S.L., Cellular immune responses to HIV, *Nature*, 410(6831), 980–987, 2001.

42. Wherry, E.J., Teichgraber, V., Becker, T.C., Masopust, D., Kaech, S.M., Antia, R., von Andrian, U.H., Ahmed, R., Lineage relationship and protective immunity of memory CD8 T cell subsets, *Nat. Immunol.*, 4(3), 225–234, 2003.

43. Appay, V., Dunbar, P.R., Callan, M., Klenerman, P., Gillespie, G.M., Papagno, L., Ogg, G.S., King, A., Lechner, F., Spina, C.A., Little, S., Havlir, D.V., Richman, D.D., Gruener, N., Pape, G., Waters, A., Easterbrook, P., Salio, M., Cerundolo, V., McMichael, A.J., Rowland-Jones, S.L., Memory CD8+ T cells vary in differentiation phenotype in different persistent virus infections, *Nat. Med.*, 8(4), 379–385, 2002.

44. van Lier, R.A., ten Berge, I.J., and Gamadia, L.E., Human CD8(+) T-cell differentiation in response to viruses, *Nat. Rev. Immunol.*, 3(12), 931–939, 2003.

45. Champagne, P., Ogg, G.S., King, A.S., Knabenhans, C., Ellefsen, K., Nobile, M., Appay, V., Rizzardi, G.P., Fleury, S., Lipp, M., Forster, R., Rowland-Jones, S., Sekaly, R.P., McMichael, A.J., Pantaleo, G., Skewed maturation of memory HIV-specific CD8 T lymphocytes, *Nature*, 410(6824), 106–111, 2001.

46. Douek, D.C., Brenchley, J.M., Betts, M.R., Ambrozak, D.R., Hill, B.J., Okamoto, Y., Casazza, J.P., Kuruppu, J., Kunstman, K., Wolinsky, S., Grossman, Z., Dybul, M., Oxenius, A., Price, D.A., Connors, M,. Koup, R.A., HIV preferentially infects HIV-specific CD4+ T cells, *Nature*, 417(6884), 95–98, 2002.

47. Speth, C., Joebstl, B., Barcova, M., Dierich, M.P., HIV-1 envelope protein gp41 modulates expression of interleukin-10 and chemokine receptors on monocytes, astrocytes and neurones, *AIDS*, 14(6), 629–636, 2000.

48. Sallusto, F., Lenig, D., Forster, R., Lipp, M., Lanzavecchia, A., Two subsets of memory T lymphocytes with distinct homing potentials and effector functions, *Nature*, 401(6754), 708–712, 1999.

49. Silvestri, G. and Feinberg, M.B., Turnover of lymphocytes and conceptual paradigms in HIV infection, *J. Clin. Invest.*, 112(6), 821–824, 2003.

50. Effros, R.B., Telomeres and HIV disease, *Microbes Infect.*, 2(1), 69–76, 2000.

51. Migueles, S.A., Laborico, A.C., Shupert, W.L., Sabbaghian, M.S., Rabin, R., Hallahan, C.W., Van Baarle, D., Kostense, S., Miedema, F., McLaughlin, M., Ehler, L., Metcalf, J., Liu, S., Connors, M.,

HIV-specific CD8[+] T cell proliferation is coupled to perforin expression and is maintained in nonprogressors, *Nat. Immunol.,* 3(11), 1061–1068, 2002.

52. Hunt, P.W., Martin, J.N., Sinclair, E., Bredt, B., Hagos, E., Lampiris, H., Deeks, S.G., T cell activation is associated with lower CD4[+] T cell gains in human immunodeficiency virus-infected patients with sustained viral suppression during antiretroviral therapy, *J. Infect. Dis.,* 187(10), 1534–1543, 2003.

53. Finkel, T.H., Tudor-Williams, G., Banda, N.K., Cotton, M.F., Curiel, T., Monks, C., Baba, T.W., Ruprecht, R.M., Kupfer, A., Apoptosis occurs predominantly in bystander cells and not in productively infected cells of HIV- and SIV-infected lymph nodes, *Nat. Med.,* 1(2), 129–134, 1995.

54. Muro-Cacho, C.A., Pantaleo, G., and Fauci, A.S., Analysis of apoptosis in lymph nodes of HIV-infected persons. Intensity of apoptosis correlates with the general state of activation of the lymphoid tissue and not with stage of disease or viral burden, *J. Immunol.,* 154(10), 5555–5566, 1995.

55. Benedict, C.A., Norris, P.S., and Ware, C.F., To kill or be killed: viral evasion of apoptosis, *Nat. Immunol.,* 3(11), 1013–1018, 2002.

56. Estaquier, J., Idziorek, T., de Bels, F., Barre-Sinoussi, F., Hurtrel, B., Aubertin, A.M,. Venet, A., Mehtali, M., Muchmore, E., Michel, P., Programmed cell death and AIDS: significance of T-cell apoptosis in pathogenic and nonpathogenic primate lentiviral infections, *Proc. Natl. Acad. Sci. U.S.A.,* 91(20), 9431–9435, 1994.

57. Gougeon, M.-L., Garcia, S., Heeney, J., Tschopp, R., Lecoeur, H., Guetard, D., Rame, V., Dauguet, C., Montagnier, L., Programmed cell death in AIDS-related HIV and SIV infections, *AIDS Res. Hum. Retroviruses,* 9(6), 553–563, 1993.

58. Gougeon, M.-L. , Lecoeur, H., Boudet, F., Ledru, E., Marzabal, S., Boullier, S., Roue, R., Nagata, S., Heeney, J., Lack of chronic immune activation in HIV-infected chimpanzees correlates with the resistance of T cells to Fas/Apo-1 (CD95)-induced apoptosis and preservation of a T helper 1 phenotype, *J. Immunol.,* 158(6), 2964–2976, 1997.

59. Davis, I.C., Girard, M., and Fultz, P.N., Loss of CD4[+] T cells in human immunodeficiency virus type 1-infected chimpanzees is associated with increased lymphocyte apoptosis, *J. Virol.,* 72(6), 4623–4632, 1998.

60. Moretti, S., Marcellini, S., Boschini, A., Famularo, G., Santini, G., Alesse, E., Steinberg, S.M., Cifone, M.G., Kroemer, G., De Simone, C., Apoptosis and apoptosis-associated perturbations of peripheral blood lymphocytes during HIV infection: comparison between AIDS patients and asymptomatic long-term non-progressors, *Clin. Exp. Immunol.,* 122(3), 364–373, 2000.

61. Lynch, D.H., Ramsdell, F., and Alderson, M.R., Fas and FasL in the homeostatic regulation of immune responses, *Immunol. Today,* 16(12), 569–574, 1995.

62. Suda, T., Okazaki, T., Naito, Y., Yokota, T., Arai, N., Ozaki, S., Nakao, K., Nagata, S., Expression of the Fas ligand in cells of T cell lineage, *J. Immunol.,* 154(8), 3806–3813, 1995.

63. Kabelitz, D., Pohl, T., and Pechhold, K., T cell apoptosis triggered via the CD3/T cell receptor γ complex and alternative activation pathways, *Curr. Top. Microbiol. Immunol.,* 200, 1–14, 1995.

64. Banda, N.K., Bernier, J., Kurahara, D.K., Kurrle, R., Haigwood, N., Sekaly, R.P., Finkel, T.H., Crosslinking CD4 by human immunodeficiency virus gp120 primes T cells for activation-induced apoptosis, *J. Exp. Med.,* 176(4), 1099–1106, 1992.

65. Borthwick, N.J., Lowdell, M., Salmon, M., Akbar, A.N., Loss of CD28 expression on CD8(+) T cells is induced by IL-2 receptor γ-chain signalling cytokines and type I IFN, and increases susceptibility to activation-induced apoptosis, *Int. Immunol.,* 12(7), 1005–1013, 2000.

66. Algeciras, A., Dockrell, D.H., Lynch, D.H., Paya, C.V., CD4 regulates susceptibility to Fas ligand- and tumor necrosis factor-mediated apoptosis, *J. Exp. Med.,* 187(5), 711–720, 1998.

67. Debatin, K.M., Fahrig-Faissner, A., Enenkel-Stoodt, S,, Kreuz, W., Benner, A., Krammer, P.H., High expression of APO-1 (CD95) on T lymphocytes from human immunodeficiency virus-1-infected children, *Blood,* 83(10), 3101–3103, 1994.

68. Katsikis, P.D., Wunderlich, E.S., Smith, C.A., Herzenberg, L.A., Herzenberg, L.A., Fas antigen stimulation induces marked apoptosis of T lymphocytes in human immunodeficiency virus-infected individuals, *J. Exp. Med.,* 181(6), 2029–2036, 1995.

69. Baumler, C.B., Bohler, T., Herr, I., Benner, A., Krammer, P.H., Debatin, K.M., Activation of the CD95 (APO-1/Fas) system in T cells from human immunodeficiency virus type-1-infected children, *Blood,* 88(5), 1741–1746, 1996.

70. McCloskey, T.W., Oyaizu, N., Bakshi, S., Kowalski, R., Kohn, N., Pahwa, S., CD95 expression and apoptosis during pediatric HIV infection: early upregulation of CD95 expression, *Clin. Immunol. Immunopathol.*, 87(1), 33–41, 1998.

71. McCloskey, T.W., Oyaizu, N., Kaplan, M., Pahwa, S., Expression of the Fas antigen in patients infected with human immunodeficiency virus, *Cytometry*, 22(2), 111–114, 1995.

72. McCloskey, T.W., Bakshi, S., Than, S., Arman, P., Pahwa, S., Immunophenotypic analysis of peripheral blood mononuclear cells undergoing *in vitro* apoptosis after isolation from human immunodeficiency virus-infected children, *Blood*, 92(11), 4230–4237, 1998.

73. Oyaizu, N., McCloskey, T.W., Than, S., Hu, R., Kalyanaraman, V.S., Pahwa, S., Cross-linking of CD4 molecules upregulates Fas antigen expression in lymphocytes by inducing interferon- and tumor necrosis factor- secretion, *Blood*, 84(8), 2622–2631, 1994.

74. Desbarats, J., Freed, J.H., Campbell, P.A., Newell, M.K., Fas (CD95) expression and death-mediating function are induced by CD4 cross-linking on CD4+ T cells, *Proc. Natl. Acad. Sci. U.S.A.*, 93(20), 11014–11018, 1996.

75. Agostini, C., Siviero, M., Facco, M., Carollo, D., Binotto, G., Tosoni, A., Cattelan, A.M., Zambello, R., Trentin, L., Semenzato, G., Antiapoptotic effects of IL-15 on pulmonary Tc1 cells of patients with human immunodeficiency virus infection, *Am. J. Respir. Crit. Care. Med.*, 163(2), 484–489, 2001.

76. Kalinkovich, A., Geleziunas, R., Kemper, O., Belenki, D., Wallach, D., Wainberg, M.A., Bentwich, Z., Increased soluble tumor necrosis factor receptor expression and release by human immunodeficiency virus type 1 infection, *J. Interferon Cytokine Res.*, 15(9), 749–757, 1995.

77. Fackler, O.T. and Baur, A.S., Live and let die: Nef functions beyond HIV replication, *Immunity*, 16(4), 493–497, 2002.

78. Hestdal, K., Aukrust, P., Muller, F., Lien, E., Bjerkeli, V., Espevik, T., Froland, S.S., Dysregulation of membrane-bound tumor necrosis factor- and tumor necrosis factor receptors on mononuclear cells in human immunodeficiency virus type 1 infection: low percentage of p75-tumor necrosis factor receptor positive cells in patients with advanced disease and high viral load, *Blood*, 90(7), 2670–2679, 1997.

79. de Oliveira Pinto, L.M., Garcia, S., Lecoeur, H., Rapp, C., Gougeon, M.L., Increased sensitivity of T lymphocytes to tumor necrosis factor receptor 1 (TNFR1)- and TNFR2-mediated apoptosis in HIV infection: relation to expression of Bcl-2 and active caspase-8 and caspase-3, *Blood*, 99(5), 1666–1675, 2002.

80. Lahdevirta, J., Maury, C.P., Teppo, A.M., Repo, H., Elevated levels of circulating cachectin/tumor necrosis factor in patients with acquired immunodeficiency syndrome, *Am. J. Med.*, 85(3), 289–291, 1988.

81. Stylianou, E., Bjerkeli, V., Yndestad, A., Heggelund, L., Waehre, T., Damas, J.K., Aukrust, P., Froland, S.S., Raised serum levels of interleukin-18 is associated with disease progression and may contribute to virological treatment failure in HIV-1-infected patients, *Clin. Exp. Immunol.*, 132(3), 462–466, 2003.

82. Zagury, D., Lachgar, A., Chams, V., Fall, L.S., Bernard, J., Zagury, J.F., Bizzini, B., Gringeri, A., Santagostino, E., Rappaport, J., Feldman, M., Burny, A., Gallo, R.C., Interferon- and Tat involvement in the immunosuppression of uninfected T cells and C-C chemokine decline in AIDS, *Proc. Natl. Acad. Sci. U.S.A.*, 95(7), 3851–3856, 1998.

83. Fanales-Belasio, E., Cafaro, A., Cara, A., Negri, D.R., Fiorelli, V., Butto, S., Moretti, S., Maggiorella, M.T., Baroncelli, S., Michelini, Z., Tripiciano, A., Sernicola, L., Scoglio, A., Borsetti, A., Ridolfi, B., Bona, R., Ten Haaft, P., Macchia, I., Leone, P., Pavone-Cossut, M.R., Nappi, F., Vardas, E., Magnani, M., Laguardia, E., Caputo, A., Titti, F., Ensoli, B., Native HIV-1 Tat protein targets monocyte-derived dendritic cells and enhances their maturation, function, and antigen-specific T cell responses, *J. Immunol.*, 168(1), 197–206, 2002.

84. Huang, L., Bosch, I., Hofmann, W., Sodroski, J., Pardee, A.B., Tat protein induces human immunodeficiency virus type 1 (HIV-1) coreceptors and promotes infection with both macrophage-tropic and T-lymphotropic HIV-1 strains, *J. Virol.*, 72(11), 8952–8960, 1998.

85. Zhang, M., Li, X., Pang, X., Ding, L., Wood, O., Clouse, K., Hewlett, I., Dayton, A.I., Identification of a potential HIV-induced source of bystander-mediated apoptosis in T cells: upregulation of TRAIL in primary human macrophages by HIV-1 Tat, *J. Biomed. Sci.*, 8(3), 290–296, 2001.

86. Yang, Y., Tikhonov, I., Ruckwardt, T.J., Djavani, M., Zapata, J.C., Pauza, C.D., Salvato, M.S., Monocytes treated with human immunodeficiency virus Tat kill uninfected CD4(+) cells by a tumor necrosis factor-related apoptosis-induced ligand-mediated mechanism, *J. Virol.,* 77(12), 6700–6708, 2003.

87. Lum, J.J., Pilon, A.A., Sanchez-Dardon, J., Phenix, B.N., Kim, J.E., Mihowich, J., Jamison, K., Hawley-Foss, N., Lynch, D.H., Badley, A.D., Induction of cell death in human immunodeficiency virus-infected macrophages and resting memory CD4 T cells by TRAIL/Apo2l, *J. Virol.,* 75(22), 11128–11136, 2001.

88. Oltvai, Z.N., Milliman, C.L., and Korsmeyer, S.J., Bcl-2 heterodimerizes *in vivo* with a conserved homolog, Bax, that accelerates programmed cell death, *Cell,* 74(4), 609–619, 1993.

89. Boudet, F., Lecoeur, H., and Gougeon, M.L., Apoptosis associated with *ex vivo* down-regulation of Bcl-2 and up-regulation of Fas in potential cytotoxic CD8+ T lymphocytes during HIV infection, *J. Immunol.,* 156(6), 2282–2293, 1996.

90. Hashimoto, F., Oyaizu, N., Kalyanaraman, V.S., Pahwa, S., Modulation of Bcl-2 protein by CD4 cross-linking: a possible mechanism for lymphocyte apoptosis in human immunodeficiency virus infection and for rescue of apoptosis by interleukin-2, *Blood,* 90(2), 745–753, 1997.

91. Blair, P.J., Boise, L.H., Perfetto, S.P., Levine, B.L., McCrary, G., Wagner, K.F., St Louis, D.C., Thompson, C.B., Siegel, J.N., June, C.H., Impaired induction of the apoptosis-protective protein Bcl-xL in activated PBMC from asymptomatic HIV-infected individuals, *J. Clin. Immunol.,* 17(3), 234–246, 1997.

92. Petrovas, C., Mueller, Y.M., Dimitriou, I.D., Bojczuk, P.M., Mounzer, K.C., Witek, J., Altman, J.D., Katsikis, P.D., HIV-specific CD8(+) T cells exhibit markedly reduced levels of Bcl 2 and Bcl-x(L), *J. Immunol.,* 172(7), 4444–4453, 2004.

93. Somma, F., Tuosto, L., Gilardini Montani, M.S., Di Somma, M.M., Cundari, E., Piccolella, E., Engagement of CD4 before TCR triggering regulates both Bax- and Fas (CD95)-mediated apoptosis, *J. Immunol.,* 164(10), 5078–5087, 2000.

94. Rizzardi, G.P., Harari, A., Capiluppi, B., Tambussi, G., Ellefsen, K., Ciuffreda, D., Champagne, P., Bart, P.A., Chave, J.P., Lazzarin, A., Pantaleo, G., Treatment of primary HIV-1 infection with cyclosporin A coupled with highly active antiretroviral therapy, *J. Clin. Invest.,* 109(5), 681–688, 2002.

95. Garber, D.A., Silvestri, G., Barry, A.P., Fedanov, A., Kozyr, N., McClure, H., Montefiori, D.C., Larsen, C.P., Altman, J.D., Staprans, S.I., Feinberg, M.B., Blockade of T cell costimulation reveals interrelated actions of CD4+ and CD8+ T cells in control of SIV replication, *J. Clin. Invest.,* 113(6), 836–845, 2004.

96. Waldmann, T.A., Dubois, S., and Tagaya, Y., Contrasting roles of IL-2 and IL-15 in the life and death of lymphocytes: implications for immunotherapy, *Immunity,* 14(2), 105–110, 2001.

97. Regamey, N., Harr, T., Battegay, M., Erb, P., Downregulation of Bcl-2, but not of Bax or Bcl-x, is associated with T lymphocyte apoptosis in HIV infection and restored by antiretroviral therapy or by interleukin 2, *AIDS Res. Hum. Retroviruses,* 15(9), 803–810, 1999.

98. Adachi, Y., Oyaizu, N., Than, S., McCloskey, T.W., Pahwa, S., IL-2 rescues *in vitro* lymphocyte apoptosis in patients with HIV infection: correlation with its ability to block culture-induced down-modulation of Bcl-2, *J. Immunol.,* 157(9), 4184–4193, 1996.

99. Mueller, Y.M., Bojczuk, P.M., Halstead, E.S., Kim, A.H., Witek, J., Altman, J.D., Katsikis, P.D., IL-15 enhances survival and function of HIV-specific CD8+ T cells, *Blood,* 101(3), 1024–1029, 2003.

100. Cruikshank, W.W., Lim, K., Theodore, A.C., Cook, J., Fine, G., Weller, P.F., Center, D.M., IL-16 inhibition of CD3-dependent lymphocyte activation and proliferation, *J. Immunol.,* 157(12), 5240–5248, 1996.

101. Idziorek, T., Khalife, J., Billaut-Mulot, O., Hermann, E., Aumercier, M., Mouton, Y., Capron, A., Bahr, G.M., Recombinant human IL-16 inhibits HIV-1 replication and protects against activation-induced cell death (AICD), *Clin. Exp. Immunol.,* 112(1), 84–91, 1998.

102. Badley, A.D., Parato, K., Cameron, D.W., Kravcik, S., Phenix, B.N., Ashby, D., Kumar, A., Lynch, D.H., Tschopp, J., Angel, J.B., Dynamic correlation of apoptosis and immune activation during treatment of HIV infection, *Cell Death Differ.,* 6(5), 420–432, 1999.

103. Chavan, S., Kodoth, S., Pahwa, R., Pahwa, S., The HIV protease inhibitor Indinavir inhibits cell-cycle progression *in vitro* in lymphocytes of HIV-infected and uninfected individuals, *Blood,* 98(2), 383–389, 2001.

104. Sloand, E.M., Kumar, P.N., Kim, S., Chaudhuri, A., Weichold, F.F., Young, N.S., Human immunodeficiency virus type 1 protease inhibitor modulates activation of peripheral blood CD4(+) T cells and decreases their susceptibility to apoptosis *in vitro* and *in vivo, Blood,* 94(3), 1021–1027, 1999.

105. Groux, H., Torpier, G., Monte, D., Mouton, Y., Capron, A., Ameisen JC., Activation-induced death by apoptosis in CD4⁺ T cells from human immunodeficiency virus-infected asymptomatic individuals, *J. Exp. Med.,* 175(2), 331–340, 1992.

106. Yang, Y., Liu, Z.H., Ware, C.F., Ashwell, J.D., A cysteine protease inhibitor prevents activation-induced T-cell apoptosis and death of peripheral blood cells from human immunodeficiency virus-infected individuals by inhibiting upregulation of Fas ligand, *Blood,* 89(2), 550–557, 1997.

107. Yang, Y., Bailey, J., Vacchio, M.S., Yarchoan, R., Ashwell, J.D., Retinoic acid inhibition of *ex vivo* human immunodeficiency virus-associated apoptosis of peripheral blood cells, *Proc. Natl. Acad. Sci. U.S.A.,* 92(7), 3051–3055, 1995.

108. Szondy, Z., Lecoeur, H., Fesus, L., Gougeon, M.L., All-trans retinoic acid inhibition of anti-CD3-induced T cell apoptosis in human immunodeficiency virus infection mostly concerns CD4 T lymphocytes and is mediated via regulation of CD95 ligand expression, *J. Infect. Dis.,* 178(5), 1288–1298, 1998.

109. Pinto, L.A., Williams, M.S., Dolan, M.J., Henkart, P.A., Shearer, G.M., Beta-chemokines inhibit activation-induced death of lymphocytes from HIV-infected individuals, *Eur. J. Immunol.,* 30(7), 2048–2055, 2000.

110. Roth, G., Moser, B., Krenn, C., Brunner, M., Haisjackl, M., Almer, G., Gerlitz, S., Wolner, E., Boltz-Nitulescu, G., Ankersmit, H.J., Susceptibility to programmed cell death in T-lymphocytes from septic patients: a mechanism for lymphopenia and Th2 predominance, *Biochem. Biophys. Res. Commun.,* 308(4), 840–846, 2003.

12 Impairment of HIV-Specific Immune Effector Cell Function

Demetre C. Daskalakis
Eric S. Rosenberg

CONTENTS

INTRODUCTION

The human immune system is exceedingly complex in its response to viral pathogens. In the process of resolving an acute illness, some viruses, such as influenza, are eradicated by the immune system. Other viruses, such as members of the Herpesviridae family, are controlled but not eradicated, resulting in chronic but typically clinically latent infection.

Latency is defined as a state in which the viral genome is present within a cell without production of detectable virus.[1] During this phase of the host–viral interaction, there is a delicate balance between viral persistence, pressure to replicate, and host immunity. The intricate framework of latency is established during early infection through mechanisms that modulate the immune response, which allows for containment of replication but not eradication of virus. These mechanisms include inhibition by natural killer cells, secretion of viral products that mimic chemokines and cytokines, inhibition of apoptosis, and interference with antigen presentation.[1] After establishment of a latently infected reservoir in various sanctuary sites, these viruses alter their expression of genes, seemingly to a limited repertoire that allows the viral genome to exist undisturbed.[1] Very low-level viral replication may occur, perhaps below the threshold of immune activation, preventing a full-blown immune response. Through a mechanism that is not well understood, these viruses may reactivate, often in the setting of waning immunity.[1]

Similar to the Herpesviridae family, infection with the human immunodeficiency virus (HIV) causes chronic infection; however, HIV replication is typically never fully controlled and is, at best, partially contained. The persistence of viremia inevitably results in a period of clinical latency that lasts between 8 and 10 years before progression to acquired immunodeficiency syndrome (AIDS).[2] During this time, the infected host may have no symptoms referable to HIV infection, but virus continues to replicate, as evidenced by the presence of HIV-1 ribonucleic acid (RNA) detectable in plasma.[3] Several studies demonstrate that individuals with high levels of viremia have more rapid progression of disease than those with lower viral loads.[3-6] The level of viremia in this clinically quiescent phase of infection correlates with a decline in $CD4^+$ count and progression to symptomatic disease. Within 6 to 12 months after acute infection, viremia reaches an equilibrium commonly referred to as the viral load set point. For some individuals, the viral load set point may be very low and is associated with slow or nonprogressive disease. In some, the infection is controlled so efficiently that HIV remains undetectable in plasma without therapy. Many of these HIV controllers still harbor virus in their lymph nodes.[7] The existence of these individuals raises the possibility that HIV viremia could be controlled as a latent infection without complete viral eradication.

Several viral, host, and immunologic factors have been described that may influence the viral load set point and rate of progression to disease. The precise correlates of immune protection, however, are still unknown. In this chapter, we will review these factors and focus on potential mechanisms used by HIV to evade, delete, or attenuate HIV-specific host immunity.

VIRAL FACTORS THAT INFLUENCE LEVEL OF VIREMIA

One potential factor that influences level of viremia is viral fitness. There are several examples of specific strains of HIV that are associated with less virulent infection. Early data for this were found in simian models. SIVmac239, a variant of simian immunodeficiency virus (SIV) with a premature stop signal in its Nef coding sequence, demonstrated the ability to replicate in cell culture, but disease was severely attenuated in infected rhesus monkeys.[8] Similarly, a deletion in Nef causing attenuated disease was demonstrated in humans through accidental transmission of HIV via blood transfusion.

The Sydney Blood Bank Cohort is a group of eight blood transfusion recipients who acquired HIV from a single donor, the ninth member of the cohort. These patients were described in 1992 after infection with HIV-contaminated blood products administered between 1980 and 1984. Despite documented HIV infection, disease progression in the majority of these individuals was noted to be very slow, leading to the theory that the virus with which they were infected was somehow attenuated.[9] Sequence analysis revealed that their virus harbored a deletion of 150 base pairs in the Nef-LTR (long terminal repeats) overlap region as well as duplications and rearrangements of the Sp1 transcription factor binding site and the nuclear factor κ-B binding site within the LTR.[10] Twenty or more years after their HIV infection, the living members of this cohort continue to have averted any AIDS-defining illnesses. One of the nine patients died due to *Pneumocystis jerovecii* pneumonia. It is unclear, however, if the immunocompromise that led to this opportunistic infection was secondary to HIV infection or other therapies for systemic lupus erythematosus. Two of the nine died of causes unrelated to HIV infection. Three recipients and the donor have low but detectable viremia, with slowly declining $CD4^+$ counts. The other two recipients have undetectable viral loads but still demonstrate slowly declining $CD4^+$ counts.[11] Other examples of HIV mutations that attenuate virulence have been described but are beyond the scope of this chapter.[12]

HOST GENETIC FACTORS THAT INFLUENCE DISEASE PROGRESSION

There are several host genetic factors that are postulated to influence viral load set point and rate of disease progression. Host-specific alterations in soluble mediators or their receptors, host human leukocyte antigen (HLA)-type, and gender are genetic factors presumably involved in influencing the rate of disease progression.

CHEMOKINE RECEPTOR MUTATIONS

After identification of HIV as the causative agent of AIDS, it became evident that the presence of CD4 molecules on target cell surfaces was necessary but not sufficient by itself for viral entry, raising the possibility that a second receptor may be needed for HIV to enter a cell. The identification of HIV-inhibitory chemokines, such as regulated upon activation normal T cell expressed and secreted (RANTES), macrophage inflammatory protein (MIP)-1α, MIP-1β, and stromal derived factor (SDF)-1, as well as their natural receptors CCR5 and CXCR4, led to the observation that these chemokine receptors served an important co-receptor function for HIV that allowed infection of CD4+ cells.[13–15]

Investigators identified two men who were exposed to HIV on repeated occasions but remained uninfected. Examination of these two subjects led to the identification of a 32-base pair deletion in the CCR5 chemokine receptor referred to as the CCR5-Δ32 mutation that explained their resistance to infection by certain strains of HIV. The two subjects were found to be homozygotes for this mutation, resulting in alteration of CCR5 such that viruses using this co-receptor could not infect CD4+ cells.[16] A subsequent study of 1252 men who have sex with men (MSM) enrolled in the Multicenter AIDS Cohort Study (MACS) revealed that the mutant allele had a relatively high frequency in the general Caucasian population (0.08%) but was comparatively overrepresented in the at-risk, HIV-uninfected MSM population—3.6% of them were homozygotes. In combination with the observation that none of the HIV-infected MSM were homozygotes, CCR5-Δ32 homozygosity was postulated to be protective against strains of HIV infection requiring CCR5 for a co-receptor (R5 strain).[17] Cases of HIV infection in individuals homozygous at this allele are caused by viruses that use receptors other than CCR5, such as CXCR4, CCR2, and CCR3.[18]

Heterozygotes with the CCR5-Δ32 mutation also seem to have a slower rate of disease progression, despite only being partially protected against infection with the R5-strain HIV.[19] As expected, the protection afforded by heterozygosity for CCR5-Δ32 is lost if the infecting virus uses CXCR4 (X4 strain) as its co-receptor.[20] It is important to note that there are individuals who are highly exposed to HIV and remain uninfected who do not harbor any known co-receptor polymorphisms. The operative mechanism providing protection from infection in these cases is unknown.

Mutations in SDF-1 (the natural ligand of CXCR4), specific polymorphisms of CCR5, and homozygosity for a specific haplotype of a newly described receptor CX3CR1 are additional putative host genetic characteristics that may influence the immunologic and virologic course of HIV infection.[21]

Conversely, differential expression of certain proinflammatory cytokines such as tumor necrosis factor (TNF)α or interleukin (IL)-1β, as well as nonspecific immune activation by other infections such as helminths, seems to correlate with enhanced replication and higher viral loads in infected individuals.[22,23] Interestingly, treatment of helminthic infections in a cohort of patients in Ethiopia resulted in a significant decrease in viral load, although there was no difference in initial viral load between helminth-infected and helminth-free subjects.[24]

HUMAN LEUKOCYTE ANTIGEN (HLA)

The HLA genes are located on the short arm of chromosome 6 and make up the human major histocompatibility complex (MHC). The two classes of MHC are coded by HLA class I and II genes. HLA-A, HLA-B, and HLA-C encode the MHC class I molecules expressed on all nucleated cells that are responsible for presenting antigen to CD8+ cytotoxic T cells. HLA-DR, HLA-DQ, and HLA-DP encode the MHC class II molecules expressed on the surface of antigen-presenting cells that are responsible for displaying antigen to CD4+ helper T cells. These six HLA loci are considered the most polymorphic in the human genome, allowing humans to respond to multiple and diverse infectious challenges.[25] Because the cellular immune system relies on presentation of viral antigen within the context of HLA molecules, it is conceivable that HLA molecules and their interactions with viral peptides would influence the efficiency and effectiveness of the cellular immune response.

Several associations have been made between HLA genotype and HIV disease progression and susceptibility. Certain class I alleles, such as B57, B27, and B35, are associated with variable rates of disease progression. Additionally, increasing heterozygosity at the three HLA class I loci, irrespective of specific allele, seems to confer some level of protection from disease progression.[25] This observation is thought to confirm the theory of overdominant selection of MHC molecules. This hypothesis posits that HLA heterozygotes have more combinations of MHC molecules and, therefore, have an enhanced ability to present a greater diversity of epitopes than homozygotes. Therefore, diversity at HLA alleles may be expressed phenotypically by a broadened ability to respond to varying pathogens, conferring a selective advantage.[26] In one study, 498 individuals from five AIDS cohorts were recruited and HLA typed. Those homozygous at one HLA class I locus progressed to AIDS more rapidly than heterozygotes. More loci for which individuals were homozygous resulted in an outcome that was worse. Heterozygotes fared better than single-locus homozygotes, who, in turn, fared better than individuals homozygous at two or three of their HLA loci.[26] Similar data were obtained in Dutch and Rwandan cohorts.[27]

Though diversity seems important in immunologic control of HIV infection, specific alleles have also been characterized as being protective or deleterious in HIV progression. The three best-studied HLA class I alleles are HLA-B57, HLA-B27, and HLA-B35. Less is known about specific HLA II alleles.

Both HLA-B57 and HLA-B27 are associated with slower disease progression. One hypothesis is that HLA-B57 is able to present multiple HIV epitopes, providing a broad response as well as limiting the possibility of viral escape.[25] HLA-B27 seems to recognize and present a HIV-1 Gag epitope that is under evolutionary pressure to remain stable, limiting variation of that epitope and eventual immunologic escape.[25] It seems, however, that whole alleles may not be the only prognosticators of disease progression and immune control.

HLA-B alleles may be divided into two groups based on differential amino acid sequences at the carboxyl ends of their α1 helices. The Bw4 group includes HLA-B57, HLA-B27, and several other loci. The Bw6 group includes, among others, HLA-B35. Individuals heterozygous for an HLA-B allele may still be considered homozygous for their Bw group. For instance, a heterozygote expressing HLA-B57 and HLA-B27 has two Bw4 alleles, making them Bw4 homozygotes. One study compared HLA types of 39 HIV-positive individuals, 20 of whom were characterized as "controllers of viremia." Analysis revealed that Bw4 epitope group homozygosity was associated with control of viremia and slowed disease progression, whereas the presence of a Bw6 allele, such as HLA-B35, was not.[28] Interestingly, epitopes presented in the context of the Bw4 motif are involved in blocking natural killer cell inhibition, whereas peptides presented in the context of the Bw6 motif are not. This observation led to the conclusion that all HLA homozygosity may not be deleterious in the immune handling of HIV.[28]

The Bw6 allele HLA-B35 is associated with increased susceptibility to progression of HIV infection to AIDS. Differential rates of progression have been noted in Caucasian and African-American patients who carried the HLA-B35 allele. In one study, it seemed that HLA-B35 was correlated with worse outcome in both groups, but more so in Caucasians.[26] Further study revealed that the HLA-B35 allele could be divided into subtypes HLAB35-Px and HLAB35-Py based on peptide specificity at epitope positions 2 and 9. Analysis of a cohort of Caucasian and African-American patients revealed that the HLA-B35-Px subtype was associated with faster disease progression, whereas the HLA-B35-Py was not. The lower frequency of HLAB35Px subtype alleles in association with the higher occurrence of HLAB35Py alleles in the African-Americans studied was postulated to explain the previous racial differences in disease progression noted among individuals who carried the HLAB35 allele.[29] Interestingly, the same study demonstrated that the HLA-Cw4, previously considered an allele that portended rapid progression to HIV, did not do so independently of HLA-B35-Px. Its postulated role as a marker of rapid disease progression is likely attributed to linkage.[29]

Significantly fewer durable associations have been made between disease progression and HLA-II alleles. HLADRB1*13 is frequently cited with mixed disease progression outcomes, some due to race, raising the possibility of further complexity of this allele and its role in HIV disease progression.[25] One study noted better virologic outcome with highly active antiretroviral therapy (HAART) in subjects with the HLA-DR1*13 haplotype. This improved outcome was attributed to optimized MHCII HIV epitope presentation, leading to a more effective anti-HIV Th2 effector response as demonstrated by lymphocyte proliferation and interferon-γ secretion.[30] Further scrutiny of HLA II alleles will be necessary to better understand their role in progression of asymptomatic HIV infection to AIDS.

GENDER

Gender seems to be another genetic factor that may impact viral load set point and, ultimately, disease progression. Several studies have been published comparing differences in viremia between male and female subjects infected with HIV. A meta-analysis of these studies revealed that women have a 41% lower viral load set point yet progress at the same rate for a given CD4 count and disease stage than men. This observation raises the possibility that women may progress more rapidly than men to AIDS at a given viral load set point. Women, therefore, may need to be treated more aggressively given the gender-specific magnification of the viral load's impact on progression to AIDS.[31] The mechanism of this phenomenon is not known, and no specific guidelines exist for differential interpretation of viral load in men and women.

IMMUNOLOGIC FACTORS IMPACTING DISEASE PROGRESSION

Unlike the Herpesviridae family, HIV produces chronic active, rather than latent, infection. Several observations regarding HIV and the immune response in infected subjects provide insight into the lack of persistent immune control of HIV. A group of HIV-infected subjects was described with members who have vigorous HIV-specific immune responses and spontaneously control viremia in the absence of antiviral therapy. Many of these patients harbor robust cytotoxic and T helper cell responses, a finding that implies that for some, cellular immunity may limit HIV to a latent infection. These patients are the exception, however, rather than the rule.

The vast majority of HIV-infected individuals only partially contain viral replication and often progress to AIDS. Although the precise mechanisms for immunologic success or failure remain to be elucidated, there are several observations that shed insight into HIV pathogenesis. Impairment of HIV-specific T helper cell function is one of the most notable defects in the immunologic repertoire of infected hosts. Early in HIV infection, the initial viremia triggers a robust immune response that leads to the activation of as many as 10% of CD8+, cytotoxic T lymphocytes (CTLs).[32] These cytotoxic T cells are postulated to be the mechanism behind the initial decline in viremia and establishment of a viral load set point. After the initial viremia begins to decline, HIV-specific neutralizing antibody becomes detectable, lending supportive evidence for the primary role of CTLs rather than antibodies in viral containment. After 6 to 12 months of infection, a viral set point is reached that influences the rate of disease progression. Higher viral loads are correlated with more rapid decline in CD4 cell count and more rapid disease progression.

Individuals with long-term nonprogressing infection and chronically infected persons with undetectable plasma viral loads on therapy harbor latent HIV provirus in their lymph nodes and spleen.[33] Mathematical models estimate that a 60- to 73-year course of therapy would be necessary to eradicate the latent reservoir of HIV, suggesting that the eradication of HIV with potent antiviral therapy is essentially unachievable.[33] Some evidence exists that early initiation of antiretroviral therapy in the setting of recent infection may lower levels of latent viral reservoirs.[34]

Three key immune responses are postulated to be important in control of HIV infection. These are neutralizing antibodies, CD8+ cytotoxic T lymphocytes, and CD4+ T helper cells. Impairment of HIV-specific T helper cell responses in chronic infection is likely to cause a significant immunologic defect in both humoral and cellular immune responses. First, we will discuss the role of HIV-specific neutralizing antibodies and cytotoxic T cells. Second, we will review the role of CD4+ T helper cells in HIV-specific immunity and discuss potential mechanisms involved in the success or failure of immunologic control of HIV infection.

ANTIBODY RESPONSE

Within 1 to 3 months after infection, HIV antibodies are detected. Nonneutralizing antibody production begins to appear 2 to 3 weeks after infection, although these antibodies do not seem to play a role in the initial decline in viremia.[35] These antibodies are directed against HIV-1 structural proteins such as p24 and p17 from Gag. Subsequently, antibodies against HIV-1 Env and Pol as well as other accessory and regulatory proteins develop over time.[36] It is not well understood, however, what role these early antibodies play in the overall immunologic control of HIV infection, because they may not target neutralizing epitopes. Neutralizing antibodies targeting the viral envelope are formed in early infection and may be able to bind HIV before entry into target cells. One potential limitation of detected antibodies is that they may be directed against viral debris rather than true functional epitopes.[37–39] The role of neutralizing antibody in control of viremia remains unclear. For instance, some studies indicate that the presence of neutralizing maternal antibodies may decrease perinatal HIV infection.[40] Other studies, however, provide conflicting evidence that the presence of neutralizing maternal antibody does not protect from transmission.[41] Another study of health care workers exposed to HIV-positive blood through needlestick injury correlated the presence of early neutralizing antibody with decreased viral load compared with workers who did not develop neutralizing antibody.[42] In another study, some subjects treated during primary HIV who subsequently underwent treatment interruption seemed to develop neutralizing antibody responses temporally associated with a decline in viremia after viral rebound.[43] The role of HIV-specific antibody responses in immunologic control of HIV infection is further complicated by three observations:

1. Some HIV-infected individuals with long-term nonprogressive infection maintain strong neutralizing antibody responses, whereas others do not.[44]
2. Viremia during acute HIV infection declines before neutralizing antibodies are detectable, providing evidence that the neutralizing antibody response may not be the primary mechanism in controlling viral replication.[45]
3. Neutralizing antibodies through passive immunization protects monkeys from challenge infection with laboratory strains of virus, but high titers of these antibodies are required to prevent infection by primary isolates from humans or monkeys not passed through cell culture.[46]

Though not all antibodies are neutralizing, they may serve other functions in controlling HIV, such as complement-directed lysis of infected cells as well as antibody-dependent cell-mediated cytotoxicity (ADCC).[47]

HIV may evade an antibody response by several mechanisms. Given the poor proofreading mechanisms of HIV, rapid viral diversification may result in escape from neutralizing antibody. This antibody escape was observed in a laboratory worker infected with a strain of virus known to be susceptible to neutralizing antibody. The infecting virus ultimately developed variants that were resistant to antibody that previously had been neutralizing.[48] Another mechanism postulated to be involved in escape from neutralization is glycosylation. Subjects studied after acute seroconversion developed successive strains of neutralization-resistant virus.[49] Sequence analysis revealed very few mutations of the Env gene. Additionally, these mutations were not in areas that coded sequences known to be important targets of HIV-neutralizing antibody. It was observed that increased glycosylation developed on these neutralization-resistant viruses, raising the possibility that they develop a "glycan shield" that sterically inhibits neutralizing antibody from binding to virus.[49] Steric hindrance of antibody attachment by overlapping hypervariable loops may also be responsible for failure of neutralization.[50] Neutralizing antibody continues to be the subject of intense research scrutiny given its possible application in HIV vaccine development.

CYTOTOXIC T LYMPHOCYTES

The first effector cell to play a role in viral containment is the CD8+ cytotoxic T lymphocyte. Initial viremia triggers a robust immune response that leads to the activation of as many as 10% of the infected host's CD8+, cytotoxic T lymphoctes.[32] This response is temporally associated with the decline of viremia observed during primary HIV. Multiple lines of evidence in both humans and monkeys indicate the importance of this response in the immunologic control of HIV. Experimental depletion of CD8+ cells in SIV-infected monkeys results in a rebound in viremia. As CD8+ cells are replenished, viremia declines. If CD8 depletion is sustained, so, in turn, is the increase in the SIV load.[51]

Tetramer studies confirm that cytotoxic T lymphocyte activity coincides with peak viremia.[32] Peak in HIV-specific CTLs occurs before the development of neutralizing antibodies, supporting the role of virus-specific CTLs in control of HIV. Initially, the CTL response is limited to a few epitopes in most acutely infected individuals. This narrowly directed response may reflect the fact that the virus is relatively homogenous in early infection. Some theorize that the diversity in CTL response is important prognostically. HIV-infected individuals may generate a very narrow CTL response to HIV in acute infection that is associated with poor immune control of the virus. This narrow response may lead to overexpansion of the CTL clones and immunologic exhaustion.[52]

HIV-specific CTLs in chronic stages of infection persist over time. Tetramer staining reveals that even in later stages of infection, 0.1 to 2% of circulating CD8 CTLs are HIV specific.[53] The persistence of the HIV-specific CTL response seems to require ongoing stimulation with HIV antigens. Tetramer staining of lymphocytes in patients on antiretroviral therapy reveals a steady decline in the number of circulating HIV-specific CTLs.[54,55] In untreated disease, the high number of HIV-specific CTLs persists as disease progression to AIDS occurs. The precise mechanisms of failure of cytotoxic T cell immunologic control of HIV are not well understood.

T HELPER CELLS

CD4+ T helper (Th) cells are critical in coordinating and maintaining host immune responses to viral infection. One role of CD4+ cells is to prime cellular immunity. A major function of Th cells is to stimulate antigen-presenting cells (APCs), leading to the activation of CD8+ effector cells. In a murine model of cytotoxic T cell activation, CD4+ lymphocytes were necessary to activate the antigen-presenting dendritic cells of immunized mice, ultimately allowing MHC I restricted CD8+ cells to kill cells presenting foreign antigen on their surfaces. In the absence of priming of APCs by epitope-specific CD4+ cells, surrogate anti-CD40 antibody acting in lieu of the CD4+ CD40+ cells, or helper-independent viral infection of the dendritic cell, the CD8 effector response was ineffective.[56] One can postulate, therefore, that the loss of Th activity through direct depletion or lack of appropriate activation would result in a suboptimal CD8+ response and ineffective clearance of virus.

In addition to priming, Th cell function has been postulated, through *in vitro* and *in vivo* evidence, to be necessary to maintain robust virus-specific cytotoxic T cell responses necessary to contain viral replication. One example of the dependence of CTL on CD4[+] T cell help was observed in human cytomegalovirus (CMV) infection. Clones of CMV-specific CTLs were adoptively transferred into bone marrow transplant patients at high risk for CMV infection during the course of their procedure. The cytotoxic T cell activity was maintained in subjects with sufficient CMV-specific Th cell responses but rapidly waned in persons without adequate virus-specific Th cell activity. The level of cytotoxic activity waned at a much faster rate in recipients who did not demonstrate a CMV-specific CD4[+] response than in those patients who did.[57] Adoptive transfer of HIV-specific CTL similarly resulted in waning of activity, presumably due to the lack of appropriate CD4[+] T cell help needed to maintain their response.[58]

Adoptive transfer of HIV-specific CTL accompanied by large doses of IL-2 resulted in maintenance of the HIV-specific CD8[+] response. This cytokine infusion was postulated to provide "help" to the transferred CTL via an IL-2 dependent pathway, potentially mimicking the activity of T helper cells. This population of CTL effectively cleared viremia and, by so doing, was thought to have exerted an evolutionary pressure on HIV. Ultimately, sequence variants were generated that escaped the clonal specificity of the infused cytotoxic lymphocytes, resulting in recurrent viremia with a mutant virus.[59]

Additional support for the role of CD4[+] T helper cells in generating and maintaining cellular immune responses is provided by a murine model of viral infection with lymphocytic choriomeningitis virus (LCMV).[60] In this model, CD4[+] knockout mice were compared to their CD4[+] wild-type counterparts. When infected with the indolent Armstrong strain of LCMV, both the knockout and wild-type mice produced a vigorous CD8[+] CTL response at the onset of infection, achieving clearance of the LCMV viremia. When the same populations of mice were exposed to more virulent LCMV strains, several differences in the responses were observed. Both strains of mice were able to produce CD8[+] cells, but the CD4[+] knockout mice had waning virus-specific CTL responses and eventually lost control of viral replication. Antibody-mediated depletion of CD4[+] cells in the wild-type mice yielded results similar to those noted in the CD4[+] knockouts. Replenishing the depleted CD4[+] cells in these mice did not result in a "wild-type" LCMV-specific immune response.[60] Similar data in a murine model of Epstein–Barr virus (EBV) infection also demonstrated that CD4[+] cell activity is likely involved in maintaining CD8[+] effector cell control of viremia.[61]

Extrapolating from these models, CD4[+] Th responses also seem to be important in the control of HIV replication and disease pathogenesis. An inverse correlation between HIV-specific CD4[+] Th response and viremia in chronic infection has been demonstrated.[62] Similarly, some persons with long-term nonprogressive infection who are able to spontaneously control viremia in the absence of antiretroviral therapy typically have robust CD4[+] and CD8[+] responses. Additionally, in some subjects treated during acute infection who generated and maintained an HIV-specific T cell response, viremia was transiently controlled even when therapy was interrupted. Similar to the LCMV murine model described above, *in vitro* depletion of CD4[+] cells in the PBMC of a patient with known CTL responses against HIV epitopes results in loss of these responses.[62,63] Murine models of retrovirus infection with Friend virus (FV) also seem to indicate that the CD4[+] compartment is necessary in maintaining control of viral infection. Depletion of CD4[+] cells in these models resulted in FV replication, viral activation from reservoir B cells, as well as signs and symptoms of active disease.[64]

In HIV-infected individuals, virus-specific cellular immune responses seem to be maintained by treatment with HAART early in primary infection.[65] Comparisons of subjects initiated during early or later phases of infection demonstrate differences in *in vitro* HIV-specific immune responses. In chronic HIV infection, virus-specific proliferation is typically weak or absent. Despite effective therapy with HAART, most chronically infected individuals do not restore HIV-specific proliferative responses. In persons treated during acute HIV infection, subjects maintained on HAART had persistent HIV-specific Th responses as well as persistent CTL activity. Similarly, subjects treated early, but in an interrupted fashion, displayed durability of their HIV-specific CD4+ and CD8+ proliferative responses.[66] Transient immunologic and virologic control in HIV-infected individuals was also reported in the context of supervised treatment interruptions (STI) of HAART in small cohorts treated early in the course of acute/early HIV infection. Both CD4+ and CD8+ responses are preserved in this group, with some evidence that the immunologic repertoire may become more diverse.[65]

Chronically infected individuals not treated during early infection lack HIV-specific CD4+ proliferative responses. More recently, evidence has emerged that the lack of CD4 proliferation does not correlate with an absolute absence of HIV-specific CD4 cells or Th1 cytokine production in chronically infected patients.[67] Furthermore, there is a dichotomous response observed in which there is cytokine secretion but no proliferation.[68] A recent study demonstrated that these cells are present in chronically infected individuals and are able to produce intracellular cytokines such as IFN- but do not proliferate in the presence of HIV-specific CD4 antigens.[68] The lack of proliferation implies that the response exists in these individuals but that it may somehow be impaired or down- regulated.

CD4+ T CELL IMPAIRMENT

A defect of the HIV-specific CD4+ response results in significant immunologic impairment. One hypothesis is that during primary HIV infection, virus-specific T helper cells are activated and made more susceptible to infection and HIV-mediated cell death.[69]

Several mechanisms have been postulated to be involved in the impairment of CD4+ T helper responses. All of these pathways to CD4 attenuation have been characterized through *in vitro* evidence, so the significance of these pathways is unclear in the infected host. These theories include the following: apoptosis, HIV-mediated cell killing, syncytia formation, antibody-dependent cellular cytotoxicity, and deletion by HIV-specific cytolytic T lymphocytes (Figure 12.1).[70]

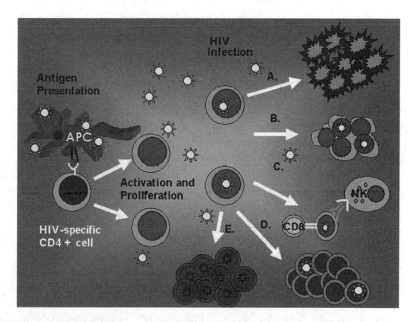

FIGURE 12.1 Antigen is presented to HIV-specific CD4+ cells causing them to activate and proliferate. Infection of CD4 cells with HIV is postulated to trigger several pathways involved in attenuation of the HIV-specific CD4 response. These include direct HIV-induced cytotoxicity and apoptosis (A.), syncytia formation (B.), cell-mediated cytotoxicity by means of CD8+ CTL and natural killer cells (C.), and HIV-induced CD4 dysfunction.[70] Additionally, some infected cells that do not die revert to resting CD4+ cells and reservoirs for latent HIV (E.).

Apoptosis

Apoptosis is one mechanism postulated in the HIV-mediated impairment of HIV-specific T helper cells. Several mechanisms are thought to be involved in HIV-induced apoptosis of effector cells. Apoptosis involves several classes of proteins that regulate the process of programmed cell death. The proteins that are involved in initiating apoptosis are cysteine-dependent aspartate-specific proteases or caspases. When signaled to undergo programmed death, the cell's caspases are triggered, changing the activation state of multiple intracellular proteins to result in mitochondrial, nuclear, and membrane changes characteristic of apoptosis. Terminal commitment to apoptosis is characterized by activation of caspase-3.[71]

HIV is postulated to enhance programmed cell death in both infected and uninfected cells through several pathways. Among these mechanisms, CD4+ T lymphocytes in HIV-infected subjects express higher levels of Fas and Fas ligand than similar cells in uninfected controls, and the number of the Fas-enhanced cells seems to increase in later stages of infection.[72,73] One potential mechanism for increased Fas/Fas ligand expression is thought to be induction by the HIV proteins Nef, Env, and Tat. HIV proteins may also alter the balance of regulatory proteins favoring programmed cell death in the cells it infects.[71] It is not, however, optimal for HIV's survival to immediately induce apoptosis in infected cells. Infection with the virus shortens the life of the CD4 cell and promotes apoptosis, but it seems as if uninfected bystander cells may undergo apoptosis more readily than their infected counterparts.[74,75] Uninfected cells are thought to undergo apoptosis due to HIV proteins released from infected cells or by activation-induced cell death.[71]

Several soluble HIV proteins have been found to cause uninfected cells to undergo programmed cell death. The gp120 causes cross-linking of CD4 molecules without a co-stimulatory signal, resulting in apoptosis.[76] Tat upregulates caspase-8 and Fas-ligand in uninfected cells.[76,77] Activation-induced cell death (AICD) is also thought to participate in HIV-induced CD4 apoptosis through upregulation of Fas and Fas-ligand.[78] Alternatively, AICD may be secondary to the other mechanisms that HIV uses to upregulate Fas and Fas-ligand.[79] All of these mechanisms may be further amplified by HIV-mediated cytokine dysregulation.[71]

HIV-Mediated Cell Killing (Cytopathic Death)

Several theories exist as to how HIV exerts its cytopathic effect on CD4+ cells, leading to direct induction of cell death. Most experts agree that when active HIV replication occurs, the host cell dies. It is still not completely clear how HIV causes this cytopathic effect. Accumulation of HIV DNA in the infected cell is one mechanism postulated to be involved in the direct killing of CD4. During the replication cycle of HIV and other retroviruses, the RNA genome is transcribed into an unintegrated DNA form by a reverse transcriptase. This sequence is usually integrated into host DNA. In animal models of other retroviral infections, unintegrated viral DNA rarely accumulates in infected cells. When DNA accumulates, its presence correlates with the cytopathic effect of the infecting retroviruses. Early evaluations of cells infected by HIV demonstrated that, unlike other retroviruses, accumulation of unintegrated viral DNA seemed to be a more common event. This observation prompted the theory that like other retroviral systems, this DNA accumulation could be responsible for CD4+ cell killing by HIV.[80,81] Subsequent studies, however, indicated that accumulation of unintegrated DNA is not directly coupled to cytopathic effect of HIV on CD4+ cells. This was demonstrated by introducing zidovudine or neutralizing antibody against HIV into *in vitro* systems, where it was observed that CD4+ cell death still occurred despite cessation of accumulation of unintegrated viral DNA genome, effectively demonstrating that unintegrated DNA accumulation was not necessary to induce direct cell death.[82,83]

Disruption of cellular integrity by HIV has also been postulated to play a role in the direct killing of CD4+ cells. Cellular permeability may be increased and the integrity of the CD4+ cells' membrane may be compromised during budding of HIV virions from infected CD4+ cells.[81] It is also believed that viral products participate in the disruption of the CD4[+] membrane. The VPU protein product of HIV causes increased permeability of cells and may also participate in membrane disruption and cell death.[84]

Interactions between the CD4[+] receptor and viral envelope may also participate in the direct cytopathic effect of HIV on T lymphocytes. Two theories exist as to how the envelope protein of HIV interacts with CD4[+] receptors to cause cell death. Both CD4[+] and the gp160 protein of HIV that codes for the envelope proteins of the virus are processed in the endoplasmic reticulum. This localization of receptor and antigen in the endoplasmic reticulum allows for binding of CD4[+] and the envelope protein intracellularly. This receptor–ligand complex may be directly cytopathic.[85] Alternatively, the binding of CD4[+] in the endoplasmic reticulum decreases the surface expression of CD4[+], the receptor for the MHC class II molecule that presents antigen to CD4[+] cells to allow activation.[86]

The HIV envelope protein may directly cause cell death.[87] Alterations in the viral envelope were demonstrated to decrease the direct cytopathic effect of the virus without altering the ability of the virus to infect CD4[+] cells.[88,89] A subsequent study challenged the role of the envelope protein in T cell destruction by engineering a virus with and without the envelope protein. CD4[+] cells infected with either of these viruses demonstrated significant cell death independent of the presence or absence of the envelope protein. Though both viruses caused a cytopathic effect, cell death was accelerated more in envelope-positive virus-infected cells than in envelope-negative infected cells.[90]

Syncytia Formation

One consequence of HIV infection is the formation of CD4[+] cell syncytia. Specifically, these syncytia are defined as the fusion of infected and uninfected CD4[+] cells to form multinucleated giant cells that ultimately result in cell death and depletion of CD4[+] T cells. Early work showed that CD4 molecules were important in syncytia formation by demonstrating decreased formation of these multinucleated cells in the presence of antibody against CD4. Additionally, CD4[+] cells do not have to be infected by HIV to be involved in syncytia formation.[91]

The role of CD4 expression in the formation of syncytia is likely related to the envelope protein (gp120) of HIV. The expression of gp120 on the surface of infected cells may allow for interactions with the CD4 receptor on uninfected cells, leading to formation of syncytia and accelerated CD4[+] cell depletion.[92] In addition, interaction between cell surface CD4 and HIV gp120 requires other cell adhesion molecules, which may also play a role in the formation of these multinucleated complexes. One such adhesion molecule is leukocyte adhesion receptor-1 (LFA-1). *In vitro* introduction of an antibody targeting this adhesion molecule inhibits production of HIV-induced multinucleated cells.[93] Additional confirmation that LFA-1 is necessary in syncytia formation was supported by identification of a subject with LFA-1-deficient T lymphocytes. In this system, the lack of expression of LFA-1 resulted in the lack of formation of syncytia. Infection of these LFA-1-deficient CD4[+] lymphocytes by HIV, however, was unaffected by the lack of expression of this protein on the surface of these CD4[+] T cells.[94]

The strain of HIV involved may also influence the frequency and occurrence of syncytia. A shift toward the CXCR4 co-receptor-requiring T lymphotropic strains of HIV, also known as syncytium-inducing (SI) virus, in later stages of disease may favor formation of multinucleated cells and enhance depletion of T helper cells. In simian models, shifts from the macrophage-tropic strain of SIV to more T lymphotropic strains of the virus were demonstrated in later stages of disease. This observation raises the possibility that phenotypic shifts in human virus late in infection may accelerate immunodeficiency, perhaps by the participation of the lymphotropic strain in formation of syncytia, depletion of CD4[+] cells, and disease progression.[95] Clinically, though not an independent predictor of death, the presence of the syncytium-inducing (SI) phenotype of HIV may be a strong predictor of decline in CD4[+] cell count and, ultimately, disease progression.[96] Despite these observations, syncytia are rarely seen in lymphoid tissue of infected patients, calling to question the *in vivo* significance of this mechanism of CD4[+] cell depletion.[96]

Antibody-Dependent Cell-Mediated Cytotoxicity

Antibody-dependent cell-mediated cytotoxicity (ADCC) is a process that links the humoral and cellular immune responses. Antibodies directed against the surface of a cell serve as the target of natural killer (NK) cells that carry receptors for the Fc portion of the antibody and are able to then kill the antibody-coated target cell. ADCC may be one of the mechanisms of CD4[+] T helper cell depletion in HIV-infected individuals.

In one *in vitro* study, both infected and uninfected CD4[+] cells were coated with gp120 and were exposed to gp120-reactive human sera. The human sera were able to direct an ADCC response against both infected and uninfected CD4[+] cells coated with gp120.[97] This observation raised the possibility that ADCC may be a pathway that allows for clearance of HIV-infected cells but, by so doing, may also participate in the depletion of HIV-specific CD4[+] cells. Excess soluble gp120 produced from infected cells may be able to coat uninfected CD4[+] cells, making them targets of *in vivo* ADCC. An association was also observed between an antibody response against the HIV protein Nef and disease progression. Similar to gp120 discussed above, efficient ADCC was demonstrated against Nef, potentially linking the presence of a Nef antibody response with increased ADCC-mediated CD4[+] cell death and HIV disease progession.[98]

Cytotoxic T Lymphocyte Directed CD4[+] Lymphocyte Depletion

HIV-specific cytotoxic T lymphocytes are postulated to play a very important role in controlling HIV during acute infection and during the more latent phases of HIV progression.[99] One important function of CD8[+] CTLs is the killing of cells infected with HIV. It is, therefore, possible that the CD8[+] CTL response thought to be important in controlling infection may lyse HIV-specific CD4[+] cells. They may also play a role in dismantling the delicate architecture of the lymphoid organs that allow for an orchestrated immune response to occur against a given antigen.[100] There is additional evidence that HIV-infected individuals harbor a population of CD8[+] cytotoxic T lymphocytes that participate in depletion of non-HIV-infected "bystander" CD4[+] T cells.[101]

Other Mechanisms of CD4 Depletion and Viral Evasion

Although CD4 depletion likely plays a role in the loss of immune activity against HIV, MHC class II– restricted epitope variation may also play an important role in HIV pathogenesis. In one study, four variant gp120 HIV Th cell epitopes were presented to cells that previously demonstrated a proliferative response against wild-type sequence. Three of the tested variants did not induce proliferation. The fourth variant stimulated attenuated proliferation compared with the wild-type epitope. Pre-exposure of the previously responsive CD4[+] T cell clone with variant peptide followed by stimulation with wild-type peptide resulted in inhibition of proliferation as well as cytotoxicity. Inhibition was reversed or prevented by exposing the CD4 clone cells to IL-2.[102] The characterization of these epitope variants and their roles in T helper cell immune response is limited. Other models of infection, both human and murine, have demonstrated that sequence variation may alter cellular immune response by decreasing recognition or presentation of MHC class II–restricted epitopes or shifting T helper cell phenotypes from the more protective Th1 response to the less protective Th2 response (Table 12.1).[103–105]

CONCLUSION

The immune response to HIV in both the acute and chronic phases of infection is extraordinarily complex. HIV-specific immunity seems to play a significant role in controlling virus and perhaps slowing disease progression. However, HIV is able to impair the immune system through deletion or evasion of the specific immune effectors activated to control infection. Several lines of evidence seemingly indicate the importance of the cellular immune response to control of viremia in HIV

TABLE 12.1
Potential Mechanisms of HIV-mediated CD4 impairment

Direct HIV cytotoxicity
HIV mediated syncytia formation
Cell-mediated cytotoxicity
CD8-mediated
NK cell ADDC
HIV-mediated CD4 dysfunction
Other:
Superantigen-mediated
Other autoimmune mechanisms
Anergy

---

as well as other viral infections. HIV is unique in that its target, the T helper cell, may play a critical role in the coordination of important virus-specific immune responses. The loss of this coordination probably results in impairment of both cellular and humoral immune responses, giving the virus an opportunity to replicate. This unchecked replication creates a cycle of viral replication, immune destruction, and further loss of immune control. A better understanding of the mechanisms employed by both host and virus will provide important insights into the pathogenesis of HIV infection and possibly stimulate future development of novel immune-based therapies and vaccines.

REFERENCES

1. Redpath, S., Angulo, A., Gascoigne, N.R., and Ghazal, P., Immune checkpoints in viral latency, *Annu. Rev. Microbiol.,* 55, 531–560, 2001.
2. Vergis, E.N., and Mellors, J.W., Natural history of HIV-1 infection, *Infect. Dis. Clin. North Am.,* 14, 809–825, v–vi, 2000.
3. Piatak, M., Jr., Saag, M.S., Yang, L.C., Clark, S.J., Kappes, J.C., Luk, K.C., Hahn, B.H., Shaw, G.M., and Lifson, J.D., High levels of HIV-1 in plasma during all stages of infection determined by competitive PCR, *Science,* 259, 1749–1754, 1993.
4. Mellors, J.W., Kingsley, L.A., Rinaldo, C.R., Jr., Todd, J.A., Hoo, B.S., Kokka, R.P., and Gupta, P., Quantitation of HIV-1 RNA in plasma predicts outcome after seroconversion, *Ann. Intern. Med.,* 122, 573–579, 1995.
5. Mellors, J.W., Munoz, A., Giorgi, J.V., Margolick, J.B., Tassoni, C.J., Gupta, P., Kingsley, L.A., Todd, J.A., Saah, A.J., Detels, R., Phair, J.P., and Rinaldo, C.R., Jr., Plasma viral load and CD4+ lymphocytes as prognostic markers of HIV-1 infection, *Ann. Intern. Med.,* 126, 946–954, 1997.
6. Mellors, J.W., Rinaldo, C.R., Jr., Gupta, P., White, R.M., Todd, J.A., and Kingsley, L.A., Prognosis in HIV-1 infection predicted by the quantity of virus in plasma, *Science,* 272, 1167–1170, 1996.
7. Pantaleo, G., Menzo, S., Vaccarezza, M., Graziosi, C., Cohen, O.J., Demarest, J.F., Montefiori, D., Orenstein, J.M., Fox, C., and Schrager, L.K., Studies in subjects with long-term nonprogressive human immunodeficiency virus infection, *N. Engl. J. Med.,* 332, 209–216, 1995.
8. Kestler, H.W., III, Ringler, D.J., Mori, K., Panicali, D.L., Sehgal, P.K., Daniel, M.D., and Desrosiers, R.C., Importance of the Nef gene for maintenance of high virus loads and for development of AIDS, *Cell,* 65, 651–662, 1991.
9. Learmont, J., Tindall, B., Evans, L., Cunningham, A., Cunningham, P., Wells, J., Penny, R., Kaldor, J., and Cooper, D.A., Long-term symptomless HIV-1 infection in recipients of blood products from a single donor, *Lancet,* 340, 863–867, 1992.
10. Deacon, N.J., Tsykin, A., Solomon, A., Smith, K., Ludford-Menting, M., Hooker, D.J., McPhee, D.A., Greenway, A.L., Ellett, A., and Chatfield, C., Genomic structure of an attenuated quasi species of HIV-1 from a blood transfusion donor and recipients, *Science,* 270, 988–991, 1995.

11. Learmont, J.C., Geczy, A.F., Mills, J., Ashton, L.J., Raynes-Greenow, C.H., Garsia, R.J., Dyer, W.B., McIntyre, L., Oelrichs, R.B., Rhodes, D.I., Deacon, N.J., and Sullivan, J.S., Immunologic and virologic status after 14 to 18 years of infection with an attenuated strain of HIV-1. A report from the Sydney Blood Bank Cohort, *N. Engl. J. Med.,* 340, 1715–1722, 1999.

12. Kirchhoff, F., Greenough, T.C., Brettler, D.B., Sullivan, J.L., and Desrosiers, R.C., Brief report: absence of intact Nef sequences in a long-term survivor with nonprogressive HIV-1 infection, *N. Engl. J. Med.,* 332, 228–232, 1995.

13. Cocchi, F., DeVico, A.L., Garzino-Demo, A., Arya, S.K., Gallo, R.C., and Lusso, P., Identification of RANTES, MIP-1, and MIP-1 as the major HIV-suppressive factors produced by CD8+ T cells, *Science,* 270, 1811–1815, 1995.

14. Deng, H., Liu, R., Ellmeier, W., Choe, S., Unutmaz, D., Burkhart, M., Di Marzio, P., Marmon, S., Sutton, R.E., Hill, C.M., Davis, C.B., Peiper, S.C., Schall, T.J., Littman, D.R., and Landau, N.R., Identification of a major co-receptor for primary isolates of HIV-1, *Nature,* 381, 661–666, 1996.

15. Feng, Y., Broder, C.C., Kennedy, P.E., and Berger, E.A., HIV-1 entry cofactor: functional cDNA cloning of a seven-transmembrane, G protein-coupled receptor, *Science,* 272, 872–877, 1996.

16. Liu, R., Paxton, W.A., Choe, S., Ceradini, D., Martin, S.R., Horuk, R., MacDonald, M.E., Stuhlmann, H., Koup, R.A., and Landau, N.R., Homozygous defect in HIV-1 coreceptor accounts for resistance of some multiply-exposed individuals to HIV-1 infection, *Cell,* 86, 367–377, 1996.

17. Huang, Y., Paxton, W.A., Wolinsky, S.M., Neumann, A.U., Zhang, L., He, T., Kang, S., Ceradini, D., Jin, Z., Yazdanbakhsh, K., Kunstman, K., Erickson, D., Dragon, E., Landau, N.R., Phair, J., Ho, D.D., and Koup, R.A., The role of a mutant CCR5 allele in HIV-1 transmission and disease progression, *Nat. Med.,* 2, 1240–1243, 1996.

18. Michael, N.L., Nelson, J.A., KewalRamani, V.N., Chang, G., O'Brien, S.J., Mascola, J.R., Volsky, B., Louder, M., White, G.C., 2nd, Littman, D.R., Swanstrom, R., and O'Brien, T.R., Exclusive and persistent use of the entry coreceptor CXCR4 by human immunodeficiency virus type 1 from a subject homozygous for CCR5 32, *J. Virol.,* 72, 6040–6047, 1998.

19. Huang, W., De Gruttola, V., Fischl, M., Hammer, S., Richman, D., Havlir, D., Gulick, R., Squires, K., and Mellors, J., Patterns of plasma human immunodeficiency virus type 1 RNA response to antiretroviral therapy, *J. Infect. Dis.,* 183, 1455–1465, 2001.

20. Michael, N.L., Chang, G., Louie, L.G., Mascola, J.R., Dondero, D., Birx, D.L., and Sheppard, H.W., The role of viral phenotype and CCR-5 gene defects in HIV-1 transmission and disease progression, *Nat. Med.,* 3, 338–340, 1997.

21. Hogan, C.M. and Hammer, S.M., Host determinants in HIV infection and disease. Part 2: genetic factors and implications for antiretroviral therapeutics, *Ann. Intern. Med.,* 134, 978–996, 2001.

22. Kinter, A.L., Ostrowski, M., Goletti, D., Oliva, A., Weissman, D., Gantt, K., Hardy, E., Jackson, R., Ehler, L., and Fauci, A.S., HIV replication in CD4+ T cells of HIV-infected individuals is regulated by a balance between the viral suppressive effects of endogenous beta-chemokines and the viral inductive effects of other endogenous cytokines, *Proc. Natl. Acad. Sci. U.S.A.,* 93, 14076–14081, 1996.

23. Bentwich, Z., Kalinkovich, A., and Weisman, Z., Immune activation is a dominant factor in the pathogenesis of African AIDS, *Immunol. Today,* 16, 187–191, 1995.

24. Wolday, D., Mayaan, S., Mariam, Z.G., Berhe, N., Seboxa, T., Britton, S., Galai, N., Landay, A., and Bentwich, Z., Treatment of intestinal worms is associated with decreased HIV plasma viral load, *J. Acquir. Immune Defic. Syndr.,* 31, 56–62, 2002.

25. Carrington, M. and O'Brien, S.J., The influence of HLA genotype on AIDS, *Annu. Rev. Med.,* 54, 535–551, 2003.

26. Carrington, M., Nelson, G.W., Martin, M.P., Kissner, T., Vlahov, D., Goedert, J.J., Kaslow, R., Buchbinder, S., Hoots, K., and O'Brien, S.J., HLA and HIV-1: heterozygote advantage and B*35-Cw*04 disadvantage, *Science,* 283, 1748–1752, 1999.

27. Tang, J., Costello, C., Keet, I.P., Rivers, C., Leblanc, S., Karita, E., Allen, S., and Kaslow, R.A., HLA class I homozygosity accelerates disease progression in human immunodeficiency virus type 1 infection, *AIDS Res. Hum. Retroviruses,* 15, 317–324, 1999.

28. Flores-Villanueva, P.O., Yunis, E.J., Delgado, J.C., Vittinghoff, E., Buchbinder, S., Leung, J.Y., Uglialoro, A.M., Clavijo, O.P., Rosenberg, E.S., Kalams, S.A., Braun, J.D., Boswell, S.L., Walker, B.D., and Goldfeld, A.E., Control of HIV-1 viremia and protection from AIDS are associated with HLA-Bw4 homozygosity, *Proc. Natl. Acad. Sci. U.S.A.,* 98, 5140–5145, 2001.

29. Gao, X., Nelson, G.W., Karacki, P., Martin, M.P., Phair, J., Kaslow, R., Goedert, J.J., Buchbinder, S., Hoots, K., Vlahov, D., O'Brien, S.J., and Carrington, M., Effect of a single amino acid change in MHC class I molecules on the rate of progression to AIDS, *N. Engl. J. Med.,* 344, 1668–1675, 2001.

30. Malhotra, U., Holte, S., Dutta, S., Berrey, M.M., Delpit, E., Koelle, D.M., Sette, A., Corey, L., and McElrath, M.J., Role for HLA class II molecules in HIV-1 suppression and cellular immunity following antiretroviral treatment, *J. Clin. Invest.,* 107, 505–517, 2001.

31. Napravnik, S., Poole, C., Thomas, J.C., and Eron, J.J., Jr., Gender difference in HIV RNA levels: a meta-analysis of published studies, *J. Acquir. Immune. Defic. Syndr.,* 31, 11–19, 2002.

32. Wilson, J.D., Ogg, G.S., Allen, R.L., Davis, C., Shaunak, S., Downie, J., Dyer, W., Workman, C., Sullivan, S., McMichael, A.J., and Rowland-Jones, S.L., Direct visualization of HIV-1-specific cytotoxic T lymphocytes during primary infection, *AIDS,* 14, 225–233, 2000.

33. Finzi, D., Blankson, J., Siliciano, J.D., Margolick, J.B., Chadwick, K., Pierson, T., Smith, K., Lisziewicz, J., Lori, F., Flexner, C., Quinn, T.C., Chaisson, R.E., Rosenberg, E., Walker, B., Gange, S., Gallant, J., and Siliciano, R.F., Latent infection of CD4+ T cells provides a mechanism for lifelong persistence of HIV-1, even in patients on effective combination therapy, *Nat. Med.,* 5, 512–517, 1999.

34. Pires, A., Hardy, G., Gazzard, B., Gotch, F., and Imami, N., Initiation of antiretroviral therapy during recent HIV-1 infection results in lower residual viral reservoirs, *J. Acquir. Immune. Defic. Syndr.,* 36, 783–790, 2004.

35. Hogan, C.M. and Hammer, S.M., Host determinants in HIV infection and disease. Part 1: cellular and humoral immune responses, *Ann. Intern. Med.,* 134, 761–776, 2001.

36. Gandhi, R.T. and Walker, B.D., Immunologic control of HIV-1, *Annu. Rev. Med.,* 53, 149–172, 2002.

37. Parren, P.W., Burton, D.R., and Sattentau, Q.J., HIV-1 antibody—debris or virion? *Nat. Med.,* 3, 366–367, 1997.

38. Pilgrim, A.K., Pantaleo, G., Cohen, O.J., Fink, L.M., Zhou, J.Y., Zhou, J.T., Bolognesi, D.P., Fauci, A.S., and Montefiori, D.C., Neutralizing antibody responses to human immunodeficiency virus type 1 in primary infection and long-term-nonprogressive infection, *J. Infect. Dis.,* 176, 924–932, 1997.

39. Moog, C., Fleury, H.J., Pellegrin, I., Kirn, A., and Aubertin, A.M., Autologous and heterologous neutralizing antibody responses following initial seroconversion in human immunodeficiency virus type 1-infected individuals, *J. Virol.,* 71, 3734–3741, 1997.

40. Scarlatti, G., Albert, J., Rossi, P., Hodara, V., Biraghi, P., Muggiasca, L., and Fenyo, E.M., Mother-to-child transmission of human immunodeficiency virus type 1: correlation with neutralizing antibodies against primary isolates, *J. Infect. Dis.,* 168, 207–210, 1993.

41. Mabondzo, A., Narwa, R., Roques, P., Gras, G.S., Herve, F., Parnet-Mathieu, F., Lasfargues, G., Courpotin, C., and Dormont, D., Lack of correlation between vertical transmission of HIV-1 and maternal antibody titers against autologous virus in human monocyte-derived macrophages, *J. Acquir. Immune Defic. Syndr. Hum. Retrovirol.,* 17, 92–94, 1998.

42. Lathey, J.L., Pratt, R.D., and Spector, S.A., Appearance of autologous neutralizing antibody correlates with reduction in virus load and phenotype switch during primary infection with human immunodeficiency virus type 1, *J. Infect. Dis.,* 175, 231–232, 1997.

43. Montefiori, D.C., Hill, T.S., Vo, H.T., Walker, B.D., and Rosenberg, E.S., Neutralizing antibodies associated with viremia control in a subset of individuals after treatment of acute human immunodeficiency virus type 1 infection, *J. Virol.,* 75, 10200–10207, 2001.

44. Harrer, T., Harrer, E., Kalams, S.A., Elbeik, T., Staprans, S.I., Feinberg, M.B., Cao, Y., Ho, D.D., Yilma, T., Caliendo, A.M., Johnson, R.P., Buchbinder, S.P., and Walker, B.D., Strong cytotoxic T cell and weak neutralizing antibody responses in a subset of persons with stable nonprogressing HIV type 1 infection, *AIDS Res. Hum. Retroviruses,* 12, 585–592, 1996.

45. Koup, R.A., Safrit, J.T., Cao, Y., Andrews, C.A., McLeod, G., Borkowsky, W., Farthing, C., and Ho, D.D., Temporal association of cellular immune responses with the initial control of viremia in primary human immunodeficiency virus type 1 syndrome, *J. Virol.,* 68, 4650–4655, 1994.

46. Baba, T.W., Liska, V., Hofmann-Lehmann, R., Vlasak, J., Xu, W., Ayehunie, S., Cavacini, L.A., Posner, M.R., Katinger, H., Stiegler, G., Bernacky, B.J., Rizvi, T.A., Schmidt, R., Hill, L.R., Keeling, M.E., Lu, Y., Wright, J.E., Chou, T.C., and Ruprecht, R.M., Human neutralizing monoclonal antibodies of the IgG1 subtype protect against mucosal simian-human immunodeficiency virus infection, *Nat. Med.,* 6, 200–206, 2000.

47. Bender, B.S., Davidson, B.L., Kline, R., Brown, C., and Quinn, T.C., Role of the mononuclear phagocyte system in the immunopathogenesis of human immunodeficiency virus infection and the acquired immunodeficiency syndrome, *Rev. Infect. Dis.,* 10, 1142–1154, 1988.

48. Beaumont, T., van Nuenen, A., Broersen, S., Blattner, W.A., Lukashov, V.V., and Schuitemaker, H., Reversal of human immunodeficiency virus type 1 IIIB to a neutralization-resistant phenotype in an accidentally infected laboratory worker with a progressive clinical course, *J. Virol.,* 75, 2246–2252, 2001.

49. Wei, X., Decker, J.M., Wang, S., Hui, H., Kappes, J.C., Wu, X., Salazar-Gonzalez, J.F., Salazar, M.G., Kilby, J.M., Saag, M.S., Komarova, N.L., Nowak, M.A., Hahn, B.H., Kwong, P.D., and Shaw, G.M., Antibody neutralization and escape by HIV-1, *Nature,* 422, 307–312, 2003.

50. Parren, P.W., Moore, J.P., Burton, D.R., and Sattentau, Q.J., The neutralizing antibody response to HIV-1: viral evasion and escape from humoral immunity, *AIDS,* 13 Suppl. A, S137–S162, 1999.

51. Jin, X., Bauer, D.E., Tuttleton, S.E., Lewin, S., Gettie, A., Blanchard, J., Irwin, C.E., Safrit, J.T., Mittler, J., Weinberger, L., Kostrikis, L.G., Zhang, L., Perelson, A.S., and Ho, D.D., Dramatic rise in plasma viremia after CD8(+) T cell depletion in simian immunodeficiency virus-infected macaques, *J. Exp. Med.,* 189, 991–998, 1999.

52. Pantaleo, G., Demarest, J.F., Schacker, T., Vaccarezza, M., Cohen, O.J., Daucher, M., Graziosi, C., Schnittman, S.S., Quinn, T.C., Shaw, G.M., Perrin, L., Tambussi, G., Lazzarin, A., Sekaly, R.P., Soudeyns, H., Corey, L., and Fauci, A.S., The qualitative nature of the primary immune response to HIV infection is a prognosticator of disease progression independent of the initial level of plasma viremia, *Proc. Natl. Acad. Sci. U.S.A.,* 94, 254–258, 1997.

53. Altman, J.D., Moss, P.A., Goulder, P.J., Barouch, D.H., McHeyzer-Williams, M.G., Bell, J.I., McMichael, A.J., and Davis, M.M., Phenotypic analysis of antigen-specific T lymphocytes, *Science,* 274, 94–96, 1996.

54. Ogg, G.S., Jin, X., Bonhoeffer, S., Moss, P., Nowak, M.A., Monard, S., Segal, J.P., Cao, Y., Rowland-Jones, S.L., Hurley, A., Markowitz, M., Ho, D.D., McMichael, A.J., and Nixon, D.F., Decay kinetics of human immunodeficiency virus-specific effector cytotoxic T lymphocytes after combination anti-retroviral therapy, *J. Virol.,* 73, 797–800, 1999.

55. Kalams, S.A., Goulder, P.J., Shea, A.K., Jones, N.G., Trocha, A.K., Ogg, G.S., and Walker, B.D., Levels of human immunodeficiency virus type 1-specific cytotoxic T-lymphocyte effector and memory responses decline after suppression of viremia with highly active antiretroviral therapy, *J. Virol.,* 73, 6721–6728, 1999.

56. Ridge, J.P., Di Rosa, F., and Matzinger, P., A conditioned dendritic cell can be a temporal bridge between a CD4+ T-helper and a T-killer cell, *Nature,* 393, 474–478, 1998.

57. Walter, E.A., Greenberg, P.D., Gilbert, M.J., Finch, R.J., Watanabe, K.S., Thomas, E.D., and Riddell, S.R., Reconstitution of cellular immunity against cytomegalovirus in recipients of allogeneic bone marrow by transfer of T-cell clones from the donor, *N. Engl. J. Med.,* 333, 1038–1044, 1995.

58. Tan, R., Xu, X., Ogg, G.S., Hansasuta, P., Dong, T., Rostron, T., Luzzi, G., Conlon, C.P., Screaton, G.R., McMichael, A.J., and Rowland-Jones, S., Rapid death of adoptively transferred T cells in acquired immunodeficiency syndrome, *Blood,* 93, 1506–1510, 1999.

59. Koenig, S., Conley, A.J., Brewah, Y.A., Jones, G.M., Leath, S., Boots, L.J., Davey, V., Pantaleo, G., Demarest, J.F., and Carter, C., Transfer of HIV-1-specific cytotoxic T lymphocytes to an AIDS patient leads to selection for mutant HIV variants and subsequent disease progression, *Nat. Med.,* 1, 330–336, 1995.

60. Zajac, A.J., Blattman, J.N., Murali-Krishna, K., Sourdive, D.J., Suresh, M., Altman, J.D., and Ahmed, R., Viral immune evasion due to persistence of activated T cells without effector function, *J. Exp. Med.,* 188, 2205–2213, 1998.

61. Cardin, R.D., Brooks, J.W., Sarawar, S.R., and Doherty, P.C., Progressive loss of CD8+ T cell-mediated control of a gamma-herpesvirus in the absence of CD4+ T cells, *J. Exp. Med.,* 184, 863–871, 1996.

62. Kalams, S.A., Buchbinder, S.P., Rosenberg, E.S., Billingsley, J.M., Colbert, D.S., Jones, N.G., Shea, A.K., Trocha, A.K., and Walker, B.D., Association between virus-specific cytotoxic T-lymphocyte and helper responses in human immunodeficiency virus type 1 infection, *J. Virol.,* 73, 6715–6720, 1999.

63. Rosenberg, E.S., Billingsley, J.M., Caliendo, A.M., Boswell, S.L., Sax, P.E., Kalams, S.A., and Walker, B.D., Vigorous HIV-1-specific CD4+ T cell responses associated with control of viremia, *Science,* 278, 1447–1450, 1997.

64. Hasenkrug, K.J., Brooks, D.M., and Dittmer, U., Critical role for CD4(+) T cells in controlling retrovirus replication and spread in persistently infected mice, *J. Virol.,* 72, 6559–6564, 1998.

65. Rosenberg, E.S., Altfeld, M., Poon, S.H., Phillips, M.N., Wilkes, B.M., Eldridge, R.L., Robbins, G.K., D'Aquila, R.T., Goulder, P.J., and Walker, B.D., Immune control of HIV-1 after early treatment of acute infection, *Nature,* 407, 523–526, 2000.

66. Oxenius, A., Price, D.A., Easterbrook, P.J., O'Callaghan, C.A., Kelleher, A.D., Whelan, J.A., Sontag, G., Sewell, A.K., and Phillips, R.E., Early highly active antiretroviral therapy for acute HIV-1 infection preserves immune function of CD8+ and CD4+ T lymphocytes, *Proc. Natl. Acad. Sci. U.S.A.,* 97, 3382–3387, 2000.

67. Pitcher, C.J., Quittner, C., Peterson, D.M., Connors, M., Koup, R.A., Maino, V.C., and Picker, L.J., HIV-1-specific CD4+ T cells are detectable in most individuals with active HIV-1 infection, but decline with prolonged viral suppression, *Nat. Med.,* 5, 518–525, 1999.

68. Wilson, J.D., Imami, N., Watkins, A., Gill, J., Hay, P., Gazzard, B., Westby, M., and Gotch, F.M., Loss of CD4+ T cell proliferative ability but not loss of human immunodeficiency virus type 1 specificity equates with progression to disease, *J. Infect. Dis.,* 182, 792–798, 2000.

69. Pope, M., Gezelter, S., Gallo, N., Hoffman, L., and Steinman, R.M., Low levels of HIV-1 infection in cutaneous dendritic cells promote extensive viral replication upon binding to memory CD4+ T cells, *J. Exp. Med.,* 182, 2045–2056, 1995.

70. Pantaleo, G., Graziosi, C., and Fauci, A.S., New concepts in the immunopathogenesis of human immunodeficiency virus infection, *N. Engl. J. Med.,* 328, 327–335, 1993.

71. Alimonti, J.B., Ball, T.B., and Fowke, K.R., Mechanisms of CD4+ T lymphocyte cell death in human immunodeficiency virus infection and AIDS, *J. Gen. Virol.,* 84, 1649–1661, 2003.

72. Silvestris, F., Cafforio, P., Frassanito, M.A., Tucci, M., Romito, A., Nagata, S., and Dammacco, F., Overexpression of Fas antigen on T cells in advanced HIV-1 infection: differential ligation constantly induces apoptosis, *AIDS,* 10, 131–141, 1996.

73. Aries, S.P., Schaaf, B., Muller, C., Dennin, R.H., and Dalhoff, K., Fas (CD95) expression on CD4+ T cells from HIV-infected patients increases with disease progression, *J. Mol. Med.,* 73, 591–593, 1995.

74. Finkel, T.H., Tudor-Williams, G., Banda, N.K., Cotton, M.F., Curiel, T., Monks, C., Baba, T.W., Ruprecht, R.M., and Kupfer, A., Apoptosis occurs predominantly in bystander cells and not in productively infected cells of HIV- and SIV-infected lymph nodes, *Nat. Med.,* 1, 129–134, 1995.

75. Ho, D.D., Neumann, A.U., Perelson, A.S., Chen, W., Leonard, J.M., and Markowitz, M., Rapid turnover of plasma virions and CD4 lymphocytes in HIV-1 infection, *Nature,* 373, 123–126, 1995.

76. Banda, N.K., Bernier, J., Kurahara, D.K., Kurrle, R., Haigwood, N., Sekaly, R.P., and Finkel, T.H., Crosslinking CD4 by human immunodeficiency virus gp120 primes T cells for activation-induced apoptosis, *J. Exp. Med.,* 176, 1099–1106, 1992.

77. Li-Weber, M., Laur, O., Dern, K., and Krammer, P.H., T cell activation-induced and HIV Tat-enhanced CD95(APO-1/Fas) ligand transcription involves NF-B, *Eur. J. Immunol.,* 30, 661–670, 2000.

78. Badley, A.D., Parato, K., Cameron, D.W., Kravcik, S., Phenix, B.N., Ashby, D., Kumar, A., Lynch, D.H., Tschopp, J., and Angel, J.B., Dynamic correlation of apoptosis and immune activation during treatment of HIV infection, *Cell Death Differ.,* 6, 420–432, 1999.

79. Yang, Y., Dong, B., Mittelstadt, P.R., Xiao, H., and Ashwell, J.D., HIV Tat binds Egr proteins and enhances Egr-dependent transactivation of the Fas ligand promoter, *J. Biol. Chem.,* 277, 19482–19487, 2002.

80. Shaw, G.M., Hahn, B.H., Arya, S.K., Groopman, J.E., Gallo, R.C., and Wong-Staal, F., Molecular characterization of human T-cell leukemia (lymphotropic) virus type III in the acquired immune deficiency syndrome, *Science,* 226, 1165–1171, 1984.

81. Fauci, A.S., The human immunodeficiency virus: infectivity and mechanisms of pathogenesis, *Science,* 239, 617–622, 1988.

82. Bergeron, L. and Sodroski, J., Dissociation of unintegrated viral DNA accumulation from single-cell lysis induced by human immunodeficiency virus type 1, *J. Virol.,* 66, 5777–5787, 1992.

83. Laurent-Crawford, A.G., Krust, B., Muller, S., Riviere, Y., Rey-Cuille, M.A., Bechet, J.M., Montagnier, L., and Hovanessian, A.G., The cytopathic effect of HIV is associated with apoptosis, *Virology,* 185, 829–839, 1991.

84. Gonzalez, M.E. and Carrasco, L., Human immunodeficiency virus type 1 VPU protein affects Sindbis virus glycoprotein processing and enhances membrane permeabilization, *Virology,* 279, 201–209, 2001.

85. Koga, Y., Sasaki, M., Yoshida, H., Wigzell, H., Kimura, G., and Nomoto, K., Cytopathic effect determined by the amount of CD4 molecules in human cell lines expressing envelope glycoprotein of HIV, *J. Immunol.*, 144, 94–102, 1990.

86. Kawamura, I., Koga, Y., Oh-Hori, N., Onodera, K., Kimura, G., and Nomoto, K., Depletion of the surface CD4 molecule by the envelope protein of human immunodeficiency virus expressed in a human CD4+ monocytoid cell line, *J. Virol.*, 63, 3748–3754, 1989.

87. Cao, J., Park, I.W., Cooper, A., and Sodroski, J., Molecular determinants of acute single-cell lysis by human immunodeficiency virus type 1, *J. Virol.*, 70, 1340–1354, 1996.

88. Kowalski, M., Bergeron, L., Dorfman, T., Haseltine, W., and Sodroski, J., Attenuation of human immunodeficiency virus type 1 cytopathic effect by a mutation affecting the transmembrane envelope glycoprotein, *J. Virol.*, 65, 281–291, 1991.

89. Stevenson, M., Haggerty, S., Lamonica, C., Mann, A. M., Meier, C., and Wasiak, A., Cloning and characterization of human immunodeficiency virus type 1 variants diminished in the ability to induce syncytium-independent cytolysis, *J. Virol.*, 64, 3792–3803, 1990.

90. Lenardo, M.J., Angleman, S.B., Bounkeua, V., Dimas, J., Duvall, M.G., Graubard, M.B., Hornung, F., Selkirk, M.C., Speirs, C.K., Trageser, C., Orenstein, J.O., and Bolton, D.L., Cytopathic killing of peripheral blood CD4(+) T lymphocytes by human immunodeficiency virus type 1 appears necrotic rather than apoptotic and does not require Env, *J. Virol.*, 76, 5082–5093, 2002.

91. Lifson, J.D., Reyes, G.R., McGrath, M.S., Stein, B.S., and Engleman, E.G., AIDS retrovirus induced cytopathology: giant cell formation and involvement of CD4 antigen, *Science*, 232, 1123–1127, 1986.

92. Sodroski, J., Haseltine, W., and Kowalski, M., Role of the human immunodeficiency virus type I envelope glycoprotein in cytopathic effect, *Adv. Exp. Med. Biol.*, 300, 193–199; discussion 200–201, 1991.

93. Hildreth, J.E. and Orentas, R.J., Involvement of a leukocyte adhesion receptor (LFA-1) in HIV-induced syncytium formation, *Science*, 244, 1075–1078, 1989.

94. Pantaleo, G., Poli, G., Butini, L., Fox, C., Dayton, A.I., and Fauci, A.S., Dissociation between syncytia formation and HIV spreading. Suppression of syncytia formation does not necessarily reflect inhibition of HIV infection, *Eur. J. Immunol.*, 21, 1771–1774, 1991.

95. Rudensey, L.M., Kimata, J.T., Benveniste, R.E., and Overbaugh, J., Progression to AIDS in macaques is associated with changes in the replication, tropism, and cytopathic properties of the simian immunodeficiency virus variant population, *Virology*, 207, 528–542, 1995.

96. Richman, D.D. and Bozzette, S.A., The impact of the syncytium-inducing phenotype of human immunodeficiency virus on disease progression, *J. Infect. Dis.*, 169, 968–974, 1994.

97. Lyerly, H.K., Matthews, T.J., Langlois, A.J., Bolognesi, D.P., and Weinhold, K.J., Human T-cell lymphotropic virus IIIB glycoprotein (gp120) bound to CD4 determinants on normal lymphocytes and expressed by infected cells serves as target for immune attack, *Proc. Natl. Acad. Sci. U.S.A.*, 84, 4601–4605, 1987.

98. Yamada, T., Watanabe, N., Nakamura, T., and Iwamoto, A., Antibody-dependent cellular cytotoxicity via humoral immune epitope of Nef protein expressed on cell surface, *J. Immunol.*, 172, 2401–2406, 2004.

99. Borrow, P., Lewicki, H., Hahn, B.H., Shaw, G.M., and Oldstone, M.B., Virus-specific CD8+ cytotoxic T-lymphocyte activity associated with control of viremia in primary human immunodeficiency virus type 1 infection, *J. Virol.*, 68, 6103–6110, 1994.

100. Zinkernagel, R.M. and Hengartner, H., T-cell-mediated immunopathology versus direct cytolysis by virus: implications for HIV and AIDS, *Immunol. Today*, 15, 262–268, 1994.

101. Miskovsky, E.P., Liu, A. Y., Pavlat, W., Viveen, R., Stanhope, P.E., Finzi, D., Fox, W.M., 3rd, Hruban, R.H., Podack, E.R., and Siliciano, R.F., Studies of the mechanism of cytolysis by HIV-1-specific CD4+ human CTL clones induced by candidate AIDS vaccines, *J. Immunol.*, 153, 2787–2799, 1994.

102. Bouhdoud, L., Villain, P., Merzouki, A., Arella, M., and Couture, C., T-cell receptor-mediated anergy of a human immunodeficiency virus (HIV) gp120-specific CD4(+) cytotoxic T-cell clone, induced by a natural HIV type 1 variant peptide, *J. Virol.*, 74, 2121–2130, 2000.

103. Plebanski, M., Flanagan, K.L., Lee, E.A., Reece, W.H., Hart, K., Gelder, C., Gillespie, G., Pinder, M., and Hill, A.V., Interleukin 10-mediated immunosuppression by a variant CD4 T cell epitope of *Plasmodium falciparum*, *Immunity*, 10, 651–660, 1999.

104. Wang, J.H., Layden, T.J., and Eckels, D.D., Modulation of the peripheral T-cell response by CD4 mutants of hepatitis C virus: transition from a Th1 to a Th2 response, *Hum. Immunol.*, 64, 662–673, 2003.
105. Ciurea, A., Hunziker, L., Martinic, M.M., Oxenius, A., Hengartner, H., and Zinkernagel, R.M., CD4+ T-cell-epitope escape mutant virus selected *in vivo*, *Nat. Med.*, 7, 795–800, 2001.

Infected vs. Uninfected Death

13 Mechanisms of HIV-Infected vs. Uninfected T Cell Killing

Jason F. Kreisberg
Gilad Doitsh
Warner C. Greene

CONTENTS

INTRODUCTION

The progressive loss of CD4+ T cells is a defining clinical feature of AIDS resulting from infection with the human immunodeficiency virus (HIV). CD4+ T cell depletion in HIV-infected individuals almost certainly is a multifactorial process. Remarkably, the mechanisms underlying this phenomenon remain poorly understood. In this chapter, we will review the mechanisms implicated in both the direct death of HIV-infected cells and the indirect HIV-mediated death of uninfected CD4+ T cells, often termed "bystander killing."

On the basis of morphological changes, two types of cell death have been described: apoptosis and necrosis. In necrotic cells, swelling of the cytoplasm leads to destruction of organelles and eventual rupture of the plasma membrane. The mechanism of cellular necrosis is not well understood and is often evoked as a mechanism only after a role for apoptosis has been excluded. Morphologically, apoptotic cells are characterized by cell shrinkage, intact organelles, and the condensation and fragmentation of DNA. Many of the molecular mechanisms underlying apoptosis have been identified and shown to be regulated at multiple levels.[1–3] Please note that although the terms apoptosis and necrosis are useful descriptions, several recent papers have emphasized that these pathways may overlap, making their distinction more difficult.[4,5]

The apoptotic signaling cascade can be initiated by extracellular or intracellular signals (Figure 13.1). Extracellular signals frequently involve the binding of members of the tumor necrosis

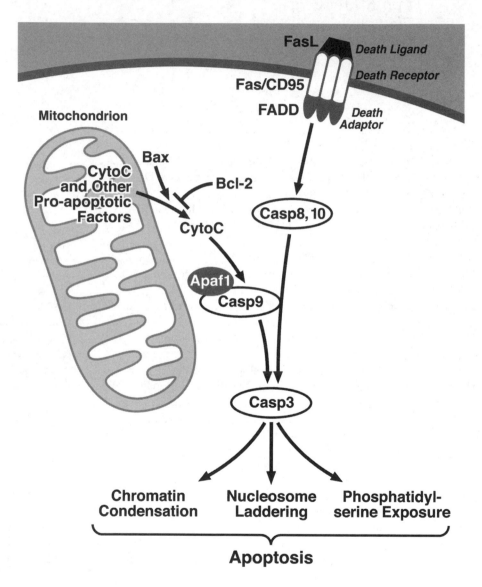

FIGURE 13.1 Extrinsic and intrinsic signaling pathways underlying apoptosis. Extrensic signals from death ligands are transmitted by trimerized death receptors such as Fas promoting recruitment of death adaptors culminating in the activation of the initiator caspases, caspase-8, and -10. This leads to the cleavage of the executor caspase, caspase-3, which either directly or indirectly underlies many of the phenotypes observed during apoptosis. Caspase-3 also can be activated by a complex of caspase-9 and Apaf1, which is activated when cytochrome-C is released from the mitochondria. Release of cytochrome-C and other apoptosis-inducing molecules from the mitochondria is both negatively and positively regulated by various Bcl-2 family members including Bcl-2 and Bax, respectively.

factor (TNF) superfamily to a set of death domain–containing receptors.[6] The death domain motifs in the cytoplasmic tail of these receptors inducibly bind to a set of adaptors that, in turn, mediate the activation of a family of cysteine proteases, known as caspases.[7] In particular, caspase-8 and caspase-10 interact with and are activated by death receptor adaptors that initiate a cascade of proteases, culminating in the activation caspase-3. By direct or indirect mechanisms, active caspase-3 causes many of the features of apoptosis. Thus, activation of caspase-3 is viewed as "a point of no return" in the apoptotic pathway.

Intracellular proapoptotic signals, such as DNA damage, are primarily coordinated by factors residing in the mitochondria and lead to the activation of caspase-3. Key among these mitochondrial factors is cytochrome c. Release of cytochrome c from the mitochondria promotes activation of caspase-9 involving the action of Apaf-1. Caspase-9 activation leads, in turn, to the activation of caspase-3. Release of cytochrome c and other mitochondrial proteins is controlled by Bcl-2 family members[8,9] that can function as either pro- or antiapoptotic factors. Proapoptotic Bcl-2 family members, such as Bax and Bak, are inhibited by heterodimerization with antiapoptotic Bcl-2 family members, such as Bcl-2 and Bcl-xL. The relative expression levels of these proteins within a cell establish a threshold for apoptosis. NF-κB is one of the key transcription factors that controls the levels of antiapoptotic molecules.[10]

HIV KILLING OF INFECTED T CELLS

How do HIV-infected cells die? A number of general pathways promoting cell death have been dissected in detail. We will discuss how HIV infection, as well as the actions of isolated viral factors, influences these cell death pathways.

DIRECT KILLING BY APOPTOSIS AND OTHER MECHANISMS

Precise characterization of HIV-mediated cell death has proved rather difficult. Although many studies have presented evidence for direct HIV-mediated death in T cell lines or primary T cells, it is often hard to ascertain whether the dead cells correspond to infected cells, bystander cells, or a combination of both. Not surprisingly, the literature is replete with conflicting reports. The few studies that have characterized HIV-infected cells at the single-cell level with reporter genes and flow cytometry have implicated apoptosis as the major mechanism of cell death, including hallmarks of this death pathway such as loss of mitochondrial membrane potential,[11] phosphatidylserine accumulation on the exterior plasma membrane,[12] and fragmentation of DNA.[13,14] These observations are supported by some studies in HIV-infected cultures[15–17] but not by others.[11,18–21] The reasons for these experimental disparities remain unclear. These studies do agree that death receptor–mediated signaling through death adaptors involving caspase-8 activation does not seem to play a major role and that broadly acting caspase inhibitors often fail to protect infected cells from death.[11,14,19]

In summary, HIV infection in many cellular systems leads to phenotypic changes consistent with apoptosis, yet classic caspase cascades may not play a prominent role. HIV-induced cytotoxicity probably depends on the expression levels of various viral and host proteins that establish a cellular threshold for apoptosis. In cells with a high threshold, HIV infection may be so damaging that death results, regardless of whether the classical apoptotic signaling pathways are detectably induced.

CYTOPATHICITY OF THE HIV ENVELOPE PROTEIN INVOLVES BOTH CD4 AND THE CHEMOKINE CO-RECEPTORS

The HIV envelope (Env) protein is a type I transmembrane protein, initially translated in the endoplasmic reticulum (ER) as a 160 kDa molecule (gp160). In the ER, gp160 is cleaved by the cellular protease furin. The resulting transmembrane gp41 protein contains the fusion peptide, whereas the noncovalently associated surface gp120 protein contains domains that sequentially interact with CD4 and one of two principal HIV co-receptors, CCR5 or CXCR4. These gp120 interactions trigger a conformational change in gp41 structure that exposes the fusion peptide, promoting its insertion into the target cell membrane. Importantly, gp120 and gp160 associate with CD4 at the plasma membrane and within the endoplasmic reticulum, respectively. In fact, interactions in the ER are important for the viral life cycle: they retard surface expression of CD4 maturation and act in concert with a second HIV protein, Vpu, which promotes polyubiquitylation of CD4 and its degradation by the 26S proteasome. Because the interaction between gp120 and gp41 is noncovalent, gp120 can be shed from the surface of either HIV-infected cells or virions.[22,23]

Env was the first viral protein shown to promote cell death.[24,25] Three models emerged to explain its cytopathic effects. The first proposes that gp120 transiently interacts with CD4 and the HIV co-receptor on target cells and triggers a set of intracellular signaling events that promotes cell death. The source of gp120 can be the virion surface, infected cells, or shed gp120. Because CD4 is effectively downregulated from the surface of infected cells by the combined action of the HIV proteins Env,[26,27] Nef,[28,29] and Vpu,[30,31] this mechanism would presumably target bystander cells expressing CD4 (Figure 13.3 and Figure 13.4).

A second model, also principally related to bystander killing, holds that expression of gp120/41 on the surface of infected cells can mediate fusion between infected and uninfected cells, leading to syncytia formation (Figure 13.3).[24,25] Of note is that syncytia appear to occur principally in T cell lines *in vitro*,[32] although some evidence for syncytia has been documented *in vivo*.[33–37]

In the third model, Env mediates "single-cell killing,"[38–40] a phrase used to describe cell death occurring in HIV-infected cultures in the absence of syncytia formation. In fact, single-cell killing, syncytia formation, and the promotion of viral replication are genetically separable events within Env.[32,38] In HIV-infected primary cell cultures and some T cell lines, single-cell killing is the most prominent form of cell death.[32]

Three models have also been proposed to explain how Env mediates single-cell killing (Figure 13.2A). They all agree that Env must interact with CD4 but disagree over the consequences of this interaction or where in the cell this interplay occurs. Nomoto and colleagues hypothesized that unprocessed gp160 interacts with CD4 in the ER, leading to apoptosis. Although the regions of gp160 and CD4 necessary for this action, as well as the mechanism, are unclear, complexes of unprocessed gp160 and CD4 were correlated with a disruption of nuclear pores and disturbances in calcium signaling.[41–45] Sodroski and colleagues favor a second model that suggests that intracellular fusion events may be facilitated by interactions between gp120/41, CD4, and the appropriate co-receptor in the secretory pathway.[39,46–49] These intracellular fusion events damage cellular membranes and lead to cell death reminiscent of necrosis. This phenotype was observed in HIV-infected cultures and in cells expressing the HIV envelope together with CD4 and the appropriate co-receptor. Two key findings support this model: single-cell killing is resistant to soluble CD4, and the viral envelope must not only be able to interact with CD4 but must also possess a functional fusion peptide.[39,46–49] A third model supported by Corbeil and colleagues states that surface-expressed complexes of gp120/41 and CD4 are toxic and that the cytoplasmic tail of CD4, but not its ability to signal through the tyrosine kinase Lck, is important for this response.[50–52] In contrast to the findings of the Sodroski group, these authors report that Env-mediated killing can be inhibited by soluble CD4.[46,50–52] Thus, in summary, although Env is one of the most potent mediators of HIV-induced cell death, the molecular mechanism underlying its action remains highly controversial.

VPR MEDIATES APOPTOSIS BY DIRECTLY TARGETING THE MITOCHONDRIA

Expression of viral protein R (Vpr) in a variety of systems induces apoptosis. Perhaps most telling is that genetic studies[53–55] and studies of Vpr alleles from long-term nonprogressors[56,57] have demonstrated that Vpr is a key determinant of cytopathicity. Expressed alone, Vpr is cytotoxic in systems as disparate as yeast,[58] murine T cells,[59] and human T cells,[60–63] although some studies

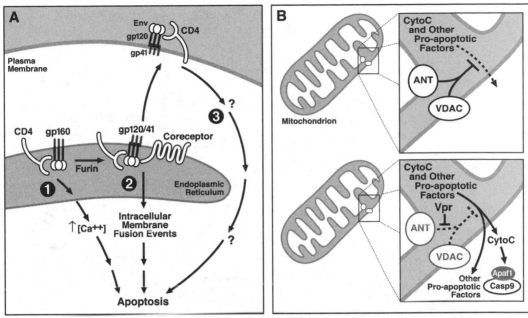

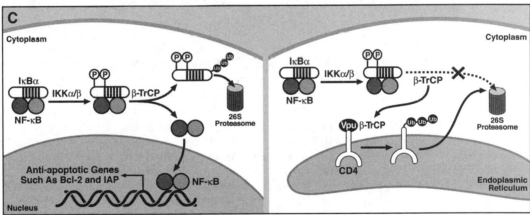

FIGURE 13.2 Models underlying direct killing by HIV Env, Vpr, and Vpu. (A) The three models of Env-mediated cytotoxicity, including (1) gp160 binding to CD4 in the ER leading to cytotoxic signaling, (2) intracellular fusion events involving Env/CD4–co-receptor interactions, and (3) gp120 signaling from the cell surface through HIV receptors, are outlined and discussed in the text. (B) Vpr directly interacts with and inhibits the activity of ANT, a key component of the machinery needed to maintain the integrity of the mitochondria. (C) In the absence of Vpu, β-TrCP plays a key role in degrading IκBα, which allows for NF-κB mediated activation of antiapoptotic genes. In the presence of Vpu, β-TrCP binds to and is sequestered by Vpu, thereby inhibiting the normal degradation of IκBα and activation of NF-κB. NF-κB regulates the expression of several key antiapoptotic genes.

have suggested that this phenotype depends on either the levels of Vpr[64,65] or the activation state of the cell.[66] This is an important point, because Vpr is detectable in patient serum,[67] in part, because it can cross biological membranes in an energy-independent manner.[68] Thus, at sufficient concentrations, Vpr could mediate the killing of infected and bystander cells (Figure 13.4). Vpr-induced cell death seems to be apoptotic: it typically correlates with activation of caspases, loss of mitochondrial membrane potential, release of cytochrome *c* into the cytoplasm, phosphatidylserine exteriorization, and hypodiploidy.[57,60,62,63,69–72] Activation of caspases seems to play a significant

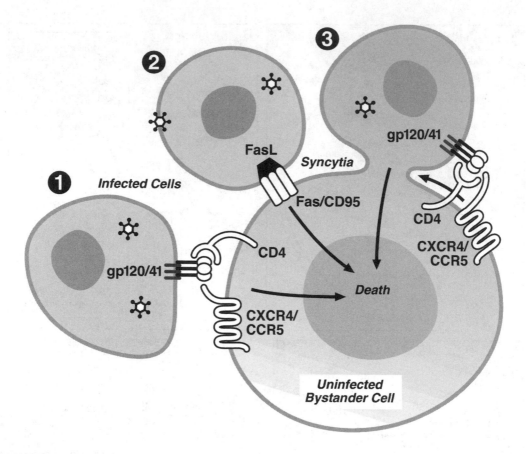

FIGURE 13.3 Possible mechanisms of bystander CD4+ T cell killing by direct cell–cell interaction. (1) Interaction of Env expressed on the surface of HIV-infected cells with CD4 and a co-receptor expressed on the surface of the uninfected T cells. (2) Engagement of Fas receptors on uninfected cells by FasL expressed on the surface of the infected cells. (3) Syncytia formation induced by the interaction of HIV Env expressed on the surface of infected cells and the CD4/co-receptors present on bystander cells promoting cell-to-cell fusion.

role in Vpr-mediated toxicity, because Vpr-mediated killing can be inhibited by broadly acting caspase inhibitors.[57,61,62,72]

Although the connection between most other cytotoxic HIV genes and the cellular death machinery may be indirect, Vpr directly targets one of two key proteins that regulate mitochondrial membrane potential (MMP). In a cell-free system, Vpr effectively disrupts MMP, facilitating the release of cytochrome c.[72] MMP is regulated by the permeability transition pore complex that comprises the voltage-dependent anion channel (VDAC) in the outer mitochondrial membrane leaflet and the adenine nucleotide translocator (ANT) located on the inner leaflet. These proteins interact at sites where the two membranes are juxtaposed. Vpr physically interacts with ANT with nanomolar affinity, and the interaction leads to changes in membrane permeability.[72,73] Consistent with the finding that Bcl-2 antagonizes the proapoptotic affects of Vpr, Bcl-2 decreases the affinity of the interaction of Vpr with ANT. These studies suggest that Vpr directly targets and destabilizes a key component of the mitochondria responsible for regulating membrane potential (Figure 13.2B).[73] Surprisingly, a recent study by the same group indicates that Vpr-induced toxicity occurs independently of the caspases and other catabolic hydrolases that are normally activated after cytochrome c release from the mitochondria.[74] As noted earlier for HIV-mediated direct killing, this is probably another example in which, even though the dominant death pathway is inhibited, sufficient toxic effects are generated, perhaps as a result of bioenergetic catastrophe, and cell death ensues.

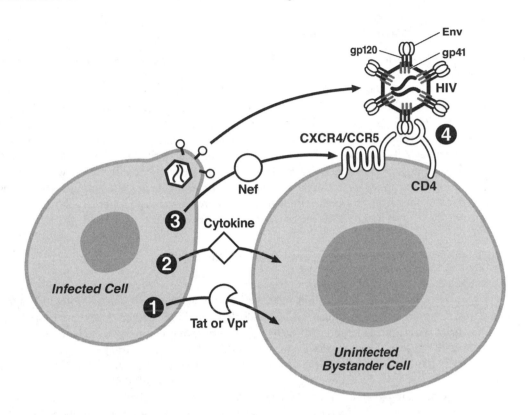

FIGURE 13.4 Possible mechanisms of bystander CD4⁺ T cell killing by soluble factors. (1) Cytopathic effect of soluble viral factors (Tat or Vpr) entering bystander cells due to their protein transduction properties. (2) Production of cytotoxic cytokines by infected cells, altering viability of bystander cells. (3) Binding of extracellular Nef to CXCR4, triggering cytotoxic effects in the bystander cells. (4) Interaction of gp120 presented on HIV virions budding from infected cells with CD4 and co-receptors present on the surface of the bystander CD4⁺ T cells. The HIV virion is not depicted to scale.

VPU CONTRIBUTES TO APOPTOSIS BY INHIBITING THE ANTIAPOPTOTIC EFFECTS OF NF-κB

Although far fewer studies have reported proapoptotic effects of Vpu than those of Env and Vpr, a recent genetic screen has yielded intriguing results. Strebel and colleagues identified Vpu as a proapoptotic gene after screening HIV mutants lacking individual genes for their ability to induce apoptosis in Jurkat T cells. They further confirmed the role of Vpu as a proapoptotic gene during infection of primary CD4⁺ T cells.[53] Although it is clear in these experiments that Vpu functions as a proapoptotic gene, infections with HIV viruses lacking Vpu still induce cytopathic effects, supporting a role for other HIV genes as well.[54,75]

The ability of Vpu to promote apoptosis likely involves inhibition of NF-κB signaling (Figure 13.2C). Activation of NF-κB requires phosphorylation, ubiquitylation, and degradation of the NF-κB inhibitor IκBα.[76] IκBα is ubiquitylated by the same E3 ubiquitin ligase, β-TrCP,[77–82] employed by Vpu to ubiquitylate CD4.[83] By competing for β-TrCP, Vpu inhibits IκBα degradation, which, in turn, prevents NF-κB from promoting the expression of various antiapoptotic genes. The result is that Vpu is proapoptotic.[84,85] What is unclear from this study is why the proapoptotic potential of Vpu emerged only recently and not in similar previous studies.[54,75] Perhaps the endogenous levels of β-TrCP, IκBα, NF-κB, or NF-κB-induced antiapoptotic genes in some cell types confer greater or lesser sensitivity to this pathway.

HIV PROTEASE AS A MEDIATOR OF DIRECT CELL KILLING

HIV protease (PR) is toxic when expressed in *Escherichia coli*, *Saccharomyces cerevisiae*, or mammalian fibroblast cell lines.[86-88] Although PR cleaves a variety of host substrates *in vitro*,[89-96] few have been demonstrated to be responsible for the toxicity of PR. Two possible exceptions are Bcl-2[97] and caspase-8.[98] Both are cleaved in infected T cell lines, although neither study has convincingly demonstrated similar effects in primary CD4+ T cells. Perhaps this result is not altogether surprising, as enzymatically active PR in an HIV-infected cell would most likely be restricted to the plasma membrane, whereas Bcl-2 is principally localized at the mitochondria. Of note is that a recently described PR mutant supports normal viral replication but lacks cytotoxic effects.[99] It will be interesting to investigate the ability of this mutant to inactivate Bcl-2 and to activate caspase-8 as well as other proapoptotic factors in the cell. Although PR displays cytotoxic properties, the biological significance of these findings in HIV-infected cells remains unclear.

TAT INCREASES THE SENSITIVITY OF INFECTED CELLS TO APOPTOSIS

Tat has attracted considerable interest as a proapoptotic factor for two reasons. First, like Vpr, Tat contains a basic protein transduction domain that can promote its transit across biological membranes. Thus, Tat can promote the death of both infected and uninfected cells (Figure 13.4).[100-103] Second, Tat has been reported to interact directly with a number of proteins involved in transcription, including the RNA Pol II elongation factor, pTEF-b[104-106], the acetyltransferases p300[107-109], pCAF[108], CREB-binding protein[107], and the transcription factor Egr.[110] Consequently, Tat may exert a wide range of effects on host gene expression that will make it difficult to identify a single proapoptotic mechanism. Whereas other HIV proteins directly interact with the apoptotic machinery of the host, Tat may modulate apoptosis by altering the balance of pro- and antiapoptotic gene expression.

A number of different approaches, including treatment of cells with soluble Tat or stable expression of Tat within cells, have shown that Tat either is a proapoptotic factor or renders cells more sensitive to other proapoptotic factors. However, no clear consensus on the underlying mechanism has emerged. One model suggests that Tat-mediated activation of cyclin-dependent kinases underlies its proapoptotic activity. Inhibition of these kinases reduced Tat-mediated apoptosis.[111] Another model suggests that the proapoptotic activity of Tat is associated with its ability to alter the sensitivity of cells to external death signals. This notion is controversial. Specifically, although various groups have documented increased levels of CD95L expression[112,113] or increased sensitivity to Fas-mediated death,[75,114] others have failed to detect these changes.[115,116] In one report, cell death in HIV-infected T cell cultures was prevented by neutralizing CD95 or Tat.[113] The ability of Tat to increase CD95L expression levels was associated with the activation of the transcription factors NF-κB[117] or Egr.[110] Another potentially key change elicited by Tat is increased expression of caspase-8.[75] The relevance of these findings to HIV-infected T cells remains unclear, because increases in CD95L expression or Fas sensitivity were not detected in such cultures.[14,118]

In light of the controversy surrounding these findings, the proposed actions of Tat further downstream merit consideration. Tat can indirectly alter the redox state of the cell by downregulating Mn superoxide dismutase.[119] This leads to increased levels of reactive oxygen intermediates[115] that may exert proapoptotic effects. In addition, Tat has been proposed to alter the regulation of Bcl-2 family members, although both increases[120-122] and decreases[116,123] in the expression of the antiapoptotic Bcl-2 protein have been described. Levels of the proapoptotic proteins Bax[116] and Bim[124] are increased. These findings are consistent with increased levels of activated caspase-9[124] and caspase-3[123] in Tat-expressing cells. Finally, the use of pan-caspase inhibitors or a caspase-3 inhibitor prevents Tat-induced cell death, implicating activation of a classical cysteine protease cascade in Tat-induced cell death.[124]

Of note, Tat has also been reported to exert antiapoptotic effects. This claim is based on functional studies[120,122,125–127] and increases in the expression of Bcl-2.[120,122,128] The root of the discrepancy in these contrasting results remains unclear but likely reflects differences in the cell types analyzed and perhaps in the levels of Tat expression achieved.

KILLING OF UNINFECTED (BYSTANDER) T CELLS BY HIV

The fact that only a small fraction of the CD4+ T cells in HIV-infected persons harbors the virus at any given time points to an important role for the death of uninfected bystander cells in the clinical evolution of AIDS.[129] However, as with direct killing, the underlying mechanism for bystander cell death remains poorly understood. Possible mechanisms include expression of death-inducing ligands on infected cells that engage cognate death receptors on uninfected cells; the interaction of uninfected cells with viral products such as gp120, Tat, Vpr, or Nef; the formation of syncytia between infected and uninfected cells; and paracrine secretion of cytokines by productively infected cells that induce apoptosis in neighboring uninfected cells (Figure 13.3 and Figure 13.4). Additionally, changes in the overall production of CD4+ T lymphocytes and the death of CD4+ T cells as a result of a generalized state of immune activation driven by the specific immune response against HIV could play central roles in the inexorable decline in CD4+ T cells occurring in untreated HIV-infected patients. As with direct cell killing, it seems possible that multiple mechanisms may underlie bystander cell death.

CD4+ T CELL APOPTOSIS MEDIATED BY CELL–CELL INTERACTIONS: DEATH RECEPTORS

One mechanism implicated in the HIV-induced apoptosis of bystander cells involves the upregulation of FasL expression on infected cells. Subsequent interaction with Fas on neighboring cells triggers bystander cell apoptosis (Figure 13.3). In contrast to the Fas receptor, which is widely distributed throughout the body, FasL expression is principally restricted to lymphocytes and natural killer (NK) cells. FasL expression is activated by the transcription factors NF-κB, AP-1, nuclear factor of activated T cells (NF-AT), and Sp-1.[130] Additionally, FasL is stored in the cytoplasm in secretory vesicles and expressed as a 40 kDa membrane-bound protein that can be cleaved by matrix metalloproteinase to release a 28 kDa soluble ligand (sFasL).[131]

In HIV-infected persons, FasL is upregulated on peripheral blood mononuclear cells (PBMCs), and increased levels of sFasL can be detected in the plasma.[132,133] This increase in FasL expression on PBMCs seems to correlate with disease progression, because it occurs at higher levels in patients with CD4+ T cell counts of fewer than 200 cells/μL. The interplay of FasL with Fas may be a principal mediator of the death of PBMCs isolated from HIV-infected patients. The Fas/FasL interaction can be blocked with recombinant soluble Fas or antagonistic anti-Fas antibodies. Apoptosis in these cultures is greatly reduced in the presence of these agents.[134–136] Furthermore, T cells from HIV-infected subjects express higher levels of the Fas receptor than do cells from uninfected donors and are more sensitive to Fas engagement.[137] Increased FasL sensitivity can be induced by cross-linking CD4 or by treating lymphocytes with purified gp120. Activated T cells also upregulate FasL mRNA when treated with HIV Tat, an inductive process that seems to depend on NF-κB activation. Interestingly, several studies have indicated that nonlymphoid cells, including macrophages, express FasL and participate in the killing of uninfected T lymphocytes.[138–140] In accord with these studies, monocytes upregulate FasL after CD4 cross-linking and induce Fas-mediated bystander killing of CD4+ T cells by apoptosis *in vivo*.[141]

Other death receptors and ligands, such as the TNF-α receptor and TRAIL and its receptors, may also play a role in HIV-induced bystander killing (Figure 13.4). TNF-α is a multifunctional cytokine produced primarily by activated monocytes and macrophages. Although TNF-α can contribute to T cell apoptosis in the presence of HIV-infected macrophages,[142,143] the importance of increased TNF-α expression in the death of directly infected vs. bystander cells remains to be established. T cells from HIV-infected subjects seem to be more sensitive to TRAIL compared with activated T cells from normal donors.[144] Anti-TRAIL antibodies also partially block activation-induced apoptosis of T cells from asymptomatic HIV-positive patients.[145] Again, as with TNF-α, firm evidence implicating TRAIL as a major mediator of bystander T cell apoptosis in HIV-infected patients is lacking.

THE HIV ACCESSORY GENES TAT, VPR, AND NEF AS MEDIATORS OF BYSTANDER CELL DEATH

Several HIV-encoded gene products have been implicated as initiators of cell-death signaling involving uninfected bystander cells. We will focus on Tat, Vpr, and Nef in this discussion. As discussed above, Tat kills cells either by direct or indirect mechanisms, whereas Vpr acts directly through targeting of the MMP. Due to the protein-transducing properties of these proteins, it is possible that these same processes extend to bystander cells (Figure 13.4). Several lines of evidence also suggest a role for Nef as a mediator of bystander T cell apoptosis. Nef is present in significant quantities in the sera of HIV-infected patients.[146] Soluble Nef has been shown to cause the death of uninfected CD4+ T cells *in vitro* by a process termed apoptotic cytolysis.[147–149] A recent study highlighted a direct role for Nef in bystander killing, showing that soluble Nef protein targets CD4+ T cells for apoptosis through a direct interaction with the CXCR4 co-receptor (Figure 13.4).[150]

Nef might also play a role in bystander killing independent of its presence in the extracellular media. Within infected cells, Nef interacts with and modulates the activity of a number of kinases that play a role in apoptosis: the Ser/Thr p21-activated kinase (PAK),[151] the apoptosis signal-regulating kinases (ASK1),[152] and a number of the Src-family kinases (Hck, Lck, and Lyn). Gele-ziunas et al. showed that the assembly with Nef inhibits the catalytic activity of ASK1, which, in turn, inhibits both Fas and TNF-α mediated apoptosis.[152] The authors suggest a model by which Nef promotes the killing of bystander cells through the induction of FasL[153–155] and TNF-α production while simultaneously protecting the HIV-infected host from these same proapoptotic signals through its interference with ASK1 function. These antiapoptotic effects of Nef may allow the infected cell to survive for a sufficient period of time to produce new virions, whereas bystander cells succumb.

ACCELERATED T CELL TURNOVER: THE EFFECTS OF HIV ON THE THYMUS AND BONE MARROW

T cell dynamics in HIV-infected patients have been studied extensively using *in vivo* methods of DNA labeling with both radioactive and nonradioactive labels. After administration, these labels are metabolized into nucleotides that are incorporated in a stable manner into dividing cells during DNA synthesis. During administration of the label, uptake can be employed as a marker of cell division, and, after administration, the rate at which the label disappears within specific cellular subsets provides information on the half-life of labeled cells. The fractional turnover rate of each cell population can then be delineated.

During HIV infection, 4×10^7 to 2×10^9 CD4+ T cells are replaced in the blood each day.[156] Initially, it was thought that this upper estimate exceeded the proliferative capacity of the CD4+ T cells. However, recent studies revealed that HIV-infected patients treated with potent drug regimens produce new CD4+ and CD8+ T cells at abnormally high rates.[157–160] In addition to its ability to promote high turnover of peripheral CD4+ T cells, HIV can infect thymic precursor cells, leading to greatly impaired production of new CD4+ T cells. This process, termed "regenerative failure," further contributes to profound loss of CD4+ T cells occurring in untreated HIV infection.[161–164]

It is worth considering this phenomenon in greater detail. The primary supply route for naive T cells before they settle in the lymph node involves the bone marrow–thymus axis, where there are several precursor cells that are susceptible to HIV infection. Although the thymus involutes over time and intrathymic production of T cells diminishes progressively with increasing age, the residual functioning thymic tissue present in adults might be quickly eliminated by infection with HIV.[165,166] In particular, thymocytes are susceptible to infection with X4-tropic strains of HIV.[167–171] There is mounting evidence that viral infection is associated with thymic pathology in HIV-infected humans,[172–176] severe combined immunodeficient (SCID)-hu mice,[177–179] and simian immunodeficiency virus-infected nonhuman primates.[180,181] A number of different thymocyte populations are sensitive to HIV infection, including CD3-CD4+CD8- intrathymic progenitor cells, CD3+CD4+CD8+ cortical thymocytes, and CD3+CD4+CD8- or CD3-CD4-CD8+ medullary thymocytes.[165,177–179] Because the thymus is more active in children and young adults, these individuals may be less sensitive to the peripheral destruction of CD4+ T lymphocytes. However, this advantage may be forfeited if the intrathymic T cell progenitor pool is infected by HIV,[182] leading to greatly diminished T cell production. Such a scenario may underlie the higher set point for viral loads observed in children and the observation that HIV-infected children often progress to AIDS more rapidly than adults.[183–186] This possibility is countered by other investigations indicating that the inhibitory effect of HIV does not involve the direct infection of the T-progenitor cells.[187] Rather, stromal cells in the bone marrow may form the target of HIV infection, leading to their dysfunction. Infection of these cells leads to a progressive decline in T cell production involving multiple cellular subsets. As noted, it is likely that CD4+ T cell loss in HIV infection is multifactorial and certainly could involve accelerated peripheral destruction coupled with diminished central production.

Another hypothesis that has received great interest of late is the idea that activation-induced cell death is an underlying mechanism of CD4+ T cell depletion. In this model, the loss of T cell homeostasis is due principally to an increased state of T cell activation leading to activation-induced apoptosis. The driving force behind the immune activation is likely the immune response to the virus and consequent release of cytokines. Thus, this loss of T cell homeostasis is a consequence of the immune response to the virus rather than infection of cells by the virus.[161] This hypothesis clearly merits further investigation.

The Role of Chemokine Receptors in HIV-Mediated Bystander Killing

As discussed previously, Env plays a role in both direct cell killing and bystander cell killing (Figure 13.2A, Figure 13.3, and Figure 13.4). Again, this activity depends on the ability of Env to interact with both CD4 and the chemokine co-receptor. The mechanism by which engagement of the CD4 receptor by HIV Env leads to accelerated T cell death is poorly understood. Accumulating evidence indicates that signaling through the CD4 receptor alone is insufficient. For example, cross-linking of CD4 leads to activation of the tyrosine kinase Lck[188], but cells expressing a truncated form of CD4 and thus failing to activate Lck continue to undergo gp120-induced apoptosis.[52] Similarly, it has been suggested that CD4 cross-linking leads to Fas/FasL upregulation, which may mediate the cytotoxic response.[189–193] However, treatment with both anti-CD4 and anti-CXCR4 antibodies induces death in CD4+ T cells independent of caspase activation, arguing against a key role for the Fas/FasL circuit, which operates in a caspase-dependent manner.[193,194]

It seems evident then that Env-mediated bystander killing involves not only CD4 binding but also co-receptor engagement. Using an *ex vivo* human lymphoid culture system that preserves the natural lymphoid environment, Jekle and colleagues showed that bystander apoptosis occurs when cultures are infected with CXCR4-dependent strains of HIV.[195] Only minimal cell death is observed with CCR5-tropic viruses in this system, perhaps reflecting the relative paucity of CCR5-expressing cells in these unstimulated cultures. Of note, bystander apoptosis was not detected in CXCR4-expressing CD8+ or B cells, implicating a key role for CD4 in the response. However, interaction with CXCR4 is also required, as indicated by a blockade of the cytopathic response with the CXCR4 antagonist AMD3100 or with anti-CXCR4 monoclonal antibodies. A recent study showed that both CCR5 and CXCR4 activation by gp120 on the surface of cells can induce the death of uninfected CD4[+] and CD8[+] T cells in a different culture system (Figure 13.4).[196] Interestingly, engagement of CCR5 by HIV gp120 induces caspase-8-dependent cell death, whereas engagement of CXCR4 induces a caspase-independent form of cell death.

Finally, as discussed earlier, Env can mediate bystander killing due to syncytia formation (Figure 13.3). Using an experimental system based principally on transfected fibroblast cell lines, Kroemer and colleagues showed that in syncytia-induced cell death, cdk1 activity is necessary for p53 phosphorylation involving mTOR and that activated phospho-p53 subsequently upregulates expression of the Bax proapoptotic factor.[197–199] Intriguingly, activated forms of mTOR and p53 are found in lymph node biopsies from HIV-infected patients.[198]

SUMMARY AND PERSPECTIVES

Even after more than 20 years of intensive investigation, the fundamental question of how HIV kills CD4[+] T lymphocytes has defied a clear answer. Although a number of HIV proteins are cytotoxic in a variety of cellular systems, an open question remains as to which, if any, play a significant role in killing infected T cells *in vivo*. It is also unclear what contribution the various mechanisms posited to explain bystander killing play *in vivo*. Understanding not only the possible mechanisms of killing that exist but also, more importantly, identifying those that play a significant role *in vivo* are crucial for the development of new therapeutic strategies to prevent the death of CD4[+] T cells in HIV-infected patients.

REFERENCES

1. Tibbetts, M.D., Zheng, L., and Lenardo, M.J., The death effector domain protein family: regulators of cellular homeostasis, *Nat. Immunol.,* 4, 404–409, 2003.
2. Budihardjo, I., Oliver, H., Lutter, M., Luo, X., and Wang, X., Biochemical pathways of caspase activation during apoptosis, *Annu. Rev. Cell Dev. Biol.,* 15, 269–290, 1999.
3. Schultz, D.R. and Harrington, W.J., Jr., Apoptosis: programmed cell death at a molecular level, *Semin. Arthritis Rheum.,* 32, 345–369, 2003.
4. Lockshin, R.A. and Zakeri, Z., Caspase-independent cell death? *Oncogene,* 23, 2766–2773, 2004.
5. Jaattela, M. and Tschopp, J., Caspase-independent cell death in T lymphocytes, *Nat. Immunol.,* 4, 416–423, 2003.
6. Aggarwal, B.B., Signaling pathways of the TNF superfamily: a double-edged sword, *Nat. Rev. Immunol.,* 3, 745–756, 2003.
7. Barnhart, B.C., Lee, J.C., Alappat, E.C., and Peter, M.E., The death effector domain protein family, *Oncogene,* 22, 8634–8644, 2003.
8. Droin, N.M. and Green, D.R., Role of Bcl-2 family members in immunity and disease, *Biochim. Biophys. Acta,* 1644, 179–188, 2004.
9. Petros, A.M., Olejniczak, E.T., and Fesik, S.W., Structural biology of the Bcl-2 family of proteins, *Biochim. Biophys. Acta,* 1644, 83–94, 2004.
10. Kucharczak, J., Simmons, M.J., Fan, Y., and Gelinas, C., To be, or not to be: NF-B is the answer—role of Rel/NF-B in the regulation of apoptosis, *Oncogene,* 22, 8961–8982, 2003.
11. Petit, F., Arnoult, D., Lelievre, J.D., Moutouh-de Parseval, L., Hance, A.J., Schneider, P., Corbeil, J., Ameisen, J.C., and Estaquier, J., Productive HIV-1 infection of primary CD4+ T cells induces mitochondrial membrane permeabilization leading to a caspase-independent cell death, *J. Biol. Chem.,* 277, 1477–1487, 2002.
12. Noraz, N., Gozlan, J., Corbeil, J., Brunner, T., and Spector, S.A., HIV-induced apoptosis of activated primary CD4+ T lymphocytes is not mediated by Fas-Fas ligand, *AIDS,* 11, 1671–1680, 1997.
13. Herbein, G., Van Lint, C., Lovett, J.L., and Verdin, E., Distinct mechanisms trigger apoptosis in human immunodeficiency virus type 1-infected and in uninfected bystander T lymphocytes, *J. Virol.,* 72, 660–670, 1998.

14. Gandhi, R.T., Chen, B.K., Straus, S.E., Dale, J.K., Lenardo, M.J., and Baltimore, D., HIV-1 directly kills CD4+ T cells by a Fas-independent mechanism, *J. Exp. Med.,* 187, 1113–1122, 1998.

15. Terai, C., Kornbluth, R.S., Pauza, C.D., Richman, D.D., and Carson, D.A., Apoptosis as a mechanism of cell death in cultured T lymphoblasts acutely infected with HIV-1, *J. Clin. Invest.,* 87, 1710–1715, 1991.

16. Laurent-Crawford, A.G., Krust, B., Muller, S., Riviere, Y., Rey-Cuille, M.A., Bechet, J.M., Montagnier, L., and Hovanessian, A.G., The cytopathic effect of HIV is associated with apoptosis, *Virology,* 185, 829–839, 1991.

17. Genini, D., Sheeter, D., Rought, S., Zaunders, J.J., Susin, S.A., Kroemer, G., Richman, D.D., Carson, D.A., Corbeil, J., and Leoni, L.M., HIV induces lymphocyte apoptosis by a p53-initiated, mitochondrial-mediated mechanism, *FASEB J.,* 15, 5–6, 2001.

18. Lenardo, M.J., Angleman, S.B., Bounkeua, V., Dimas, J., Duvall, M.G., Graubard, M.B., Hornung, F., Selkirk, M.C., Speirs, C.K., Trageser, C., Orenstein, J.O., and Bolton, D.L., Cytopathic killing of peripheral blood CD4(+) T lymphocytes by human immunodeficiency virus type 1 appears necrotic rather than apoptotic and does not require Env, *J. Virol.,* 76, 5082–5093, 2002.

19. Bolton, D.L., Hahn, B.I., Park, E.A., Lehnhoff, L.L., Hornung, F., and Lenardo, M.J., Death of CD4(+) T-cell lines caused by human immunodeficiency virus type 1 does not depend on caspases or apoptosis, *J. Virol.,* 76, 5094–5107, 2002.

20. Kolesnitchenko, V., King, L., Riva, A., Tani, Y., Korsmeyer, S.J., and Cohen, D.I., A major human immunodeficiency virus type 1-initiated killing pathway distinct from apoptosis, *J. Virol.,* 71, 9753–9763, 1997.

21. Kolesnitchenko, V., Wahl, L.M., Tian, H., Sunila, I., Tani, Y., Hartmann, D.P., Cossman, J., Raffeld, M., Orenstein, J., Samuelson, L.E., Human immunodeficiency virus 1 envelope-initiated G2-phase programmed cell death, *Proc. Natl. Acad. Sci. U.S.A.,* 92, 11889–11893, 1995.

22. McKeating, J.A., McKnight, A., and Moore, J.P., Differential loss of envelope glycoprotein gp120 from virions of human immunodeficiency virus type 1 isolates: effects on infectivity and neutralization, *J. Virol.,* 65, 852–860, 1991.

23. Schneider, J., Kaaden, O., Copeland, T.D., Oroszlan, S., and Hunsmann, G., Shedding and interspecies type sero-reactivity of the envelope glycopolypeptide gp120 of the human immunodeficiency virus, *J. Gen. Virol.,* 67 (Pt. 11), 2533–2538, 1986.

24. Sodroski, J., Goh, W.C., Rosen, C., Campbell, K., and Haseltine, W.A., Role of the HTLV-III/LAV envelope in syncytium formation and cytopathicity, *Nature,* 322, 470–474, 1986.

25. Lifson, J.D., Reyes, G.R., McGrath, M.S., Stein, B.S., and Engleman, E.G, AIDS retrovirus induced cytopathology: giant cell formation and involvement of CD4 antigen, *Science,* 232, 1123–1127, 1986.

26. Buonocore, L. and Rose, J.K., Prevention of HIV-1 glycoprotein transport by soluble CD4 retained in the endoplasmic reticulum, *Nature,* 345, 625–628, 1990.

27. Crise, B., Buonocore, L., and Rose, J.K., CD4 is retained in the endoplasmic reticulum by the human immunodeficiency virus type 1 glycoprotein precursor, *J. Virol.,* 64, 5585–5593, 1990.

28. Garcia, J.V. and Miller, A.D., Serine phosphorylation-independent downregulation of cell-surface CD4 by Nef, *Nature,* 350, 508–511, 1991.

29. Aiken, C., Konner, J., Landau, N.R., Lenburg, M.E., and Trono, D., Nef induces CD4 endocytosis: requirement for a critical dileucine motif in the membrane-proximal CD4 cytoplasmic domain, *Cell,* 76, 853–864, 1994.

30. Willey, R.L., Maldarelli, F., Martin, M.A., and Strebel, K., Human immunodeficiency virus type 1 Vpu protein regulates the formation of intracellular gp160-CD4 complexes, *J. Virol.,* 66, 226–234, 1992.

31. Willey, R.L., Maldarelli, F., Martin, M.A., and Strebel, K., Human immunodeficiency virus type 1 Vpu protein induces rapid degradation of CD4, *J. Virol.,* 66, 7193–7200, 1992.

32. Somasundaran, M. and Robinson, H.L., A major mechanism of human immunodeficiency virus-induced cell killing does not involve cell fusion, *J. Virol.,* 61, 3114–3119, 1987.

33. Dargent, J.L., Lespagnard, L., Kornreich, A., Hermans, P., Clumeck, N., and Verhest, A., HIV-associated multinucleated giant cells in lymphoid tissue of the Waldeyer's ring: a detailed study, *Mod. Pathol.,* 13, 1293–1299, 2000.

34. Lewin-Smith, M., Wahl, S.M., and Orenstein, J.M., Human immunodeficiency virus-rich multinucleated giant cells in the colon: a case report with transmission electron microscopy, immunohistochemistry, and *in situ* hybridization, *Mod. Pathol.,* 12, 75–81, 1999.

35. Li, S.L., Kaaya, E.E., Feichtinger, H., Putkonen, P., Parravicini, C., Bottiger, D., Biberfeld, G., and Biberfeld, P., Monocyte/macrophage giant cell disease in SIV-infected cynomolgus monkeys, *Res. Virol.*, 142, 173–182, 1991.

36. Maier, H., Budka, H., Lassmann, H., and Pohl, P., Vacuolar myelopathy with multinucleated giant cells in the acquired immune deficiency syndrome (AIDS). Light and electron microscopic distribution of human immunodeficiency virus (HIV) antigens, *Acta Neuropathol. (Berl.)*, 78, 497–503, 1989.

37. Ryzhova, E.V., Crino, P., Shawver, L., Westmoreland, S.V., Lackner, A.A., and Gonzalez-Scarano, F., Simian immunodeficiency virus encephalitis: analysis of envelope sequences from individual brain multinucleated giant cells and tissue samples, *Virology*, 297, 57–67, 2002.

38. Stevenson, M., Haggerty, S., Lamonica, C., Mann, A.M., Meier, C., and Wasiak, A., Cloning and characterization of human immunodeficiency virus type 1 variants diminished in the ability to induce syncytium-independent cytolysis, *J. Virol.*, 64, 3792–3803, 1990.

39. Cao, J., Vasir, B., and Sodroski, J.G., Changes in the cytopathic effects of human immunodeficiency virus type 1 associated with a single amino acid alteration in the ectodomain of the gp41 transmembrane glycoprotein, *J. Virol.*, 68, 4662–4668, 1994.

40. Kowalski, M., Bergeron, L., Dorfman, T., Haseltine, W., and Sodroski, J., Attenuation of human immunodeficiency virus type 1 cytopathic effect by a mutation affecting the transmembrane envelope glycoprotein, *J. Virol.*, 65, 281–291, 1991.

41. Sasaki, M., Uchiyama, J., Ishikawa, H., Matsushita, S., Kimura, G., Nomoto, K., and Koga, Y., Induction of apoptosis by calmodulin-dependent intracellular Ca2+ elevation in CD4+ cells expressing gp 160 of HIV, *Virology*, 224, 18–24, 1996.

42. Lu, Y.Y., Koga, Y., Tanaka, K., Sasaki, M., Kimura, G., and Nomoto, K., Apoptosis induced in CD4+ cells expressing gp160 of human immunodeficiency virus type 1, *J. Virol.*, 68, 390–399, 1994.

43. Koga, Y., Nakamura, K., Sasaki, M., Kimura, G., and Nomoto, K., The difference in gp160 and gp120 of HIV type 1 in the induction of CD4 downregulation preceding single-cell killing, *Virology*, 201, 137–141, 1994.

44. Koga, Y., Sasaki, M., Nakamura, K., Kimura, G., and Nomoto, K., Intracellular distribution of the envelope glycoprotein of human immunodeficiency virus and its role in the production of cytopathic effect in CD4+ and CD4 human cell lines, *J. Virol.*, 64, 4661–4671, 1990.

45. Koga, Y., Sasaki, M., Yoshida, H., Oh-Tsu, M., Kimura, G., and Nomoto, K., Disturbance of nuclear transport of proteins in CD4+ cells expressing gp160 of human immunodeficiency virus, *J. Virol.*, 65, 5609–5612, 1991.

46. Bergeron, L. and Sodroski, J., Dissociation of unintegrated viral DNA accumulation from single-cell lysis induced by human immunodeficiency virus type 1, *J. Virol.*, 66, 5777–5787, 1992.

47. LaBonte, J.A., Madani, N., and Sodroski, J., Cytolysis by CCR5-using human immunodeficiency virus type 1 envelope glycoproteins is dependent on membrane fusion and can be inhibited by high levels of CD4 expression, *J. Virol.*, 77, 6645–6659, 2003.

48. LaBonte, J.A., Patel, T., Hofmann, W., and Sodroski, J., Importance of membrane fusion mediated by human immunodeficiency virus envelope glycoproteins for lysis of primary CD4-positive T cells, *J. Virol.*, 74, 10690–10698, 2000.

49. Park, I.W., Kondo, E., Bergeron, L., Park, J., and Sodroski, J., Effects of human immunodeficiency virus type 1 infection on programmed cell death in the presence or absence of Bcl-2, *J. Acquir. Immune Defic. Syndr. Hum. Retrovirol.*, 12, 321–328, 1996.

50. Corbeil, J., Tremblay, M., and Richman, D.D., HIV-induced apoptosis requires the CD4 receptor cytoplasmic tail and is accelerated by interaction of CD4 with p56lck, *J. Exp. Med.*, 183, 39–48, 1996.

51. Corbeil, J. and Richman, D.D., Productive infection and subsequent interaction of CD4-gp120 at the cellular membrane is required for HIV-induced apoptosis of CD4+ T cells, *J. Gen. Virol.*, 76 (Pt. 3), 681–690, 1995.

52. Moutouh, L., Estaquier, J., Richman, D.D., and Corbeil, J., Molecular and cellular analysis of human immunodeficiency virus-induced apoptosis in lymphoblastoid T-cell-line-expressing wild-type and mutated CD4 receptors, *J. Virol.*, 72, 8061–8072, 1998.

53. Akari, H., Bour, S., Kao, S., Adachi, A., and Strebel, K., The human immunodeficiency virus type 1 accessory protein Vpu induces apoptosis by suppressing the nuclear factor B-dependent expression of antiapoptotic factors, *J. Exp. Med.*, 194, 1299–1311, 2001.

54. Mustafa, F. and Robinson, H.L., Context-dependent role of human immunodeficiency virus type 1 auxiliary genes in the establishment of chronic virus producers, *J. Virol.,* 67, 6909–6915, 1993.

55. Rogel, M.E., Wu, L.I., and Emerman, M., The human immunodeficiency virus type 1 Vpr gene prevents cell proliferation during chronic infection, *J. Virol.,* 69, 882–888, 1995.

56. Somasundaran, M., Sharkey, M., Brichacek, B., Luzuriaga, K., Emerman, M., Sullivan, J.L., and Stevenson, M., Evidence for a cytopathogenicity determinant in HIV-1 Vpr, *Proc. Natl. Acad. Sci. U.S.A.,* 99, 9503–9508, 2002.

57. Lum, J.J., Cohen, O.J., Nie, Z., Weaver, J.G., Gomez, T.S., Yao, X. J., Lynch, D., Pilon, A.A., Hawley, N., Kim, J.E., Chen, Z., Montpetit, M., Sanchez-Dardon, J., Cohen, E.A., and Badley, A.D., Vpr R77Q is associated with long-term nonprogressive HIV infection and impaired induction of apoptosis, *J. Clin. Invest.,* 111, 1547–1554, 2003.

58. Macreadie, I.G., Castelli, L.A., Hewish, D.R., Kirkpatrick, A., Ward, A.C., and Azad, A.A., A domain of human immunodeficiency virus type 1 Vpr containing repeated H(S/F)RIG amino acid motifs causes cell growth arrest and structural defects, *Proc. Natl. Acad. Sci. U.S.A.,* 92, 2770–2774, 1995.

59. Yasuda, J., Miyao, T., Kamata, M., Aida, Y., and Iwakura, Y., T cell apoptosis causes peripheral T cell depletion in mice transgenic for the HIV-1 Vpr gene, *Virology,* 285, 181–192, 2001.

60. Stewart, S.A., Poon, B., Jowett, J.B., and Chen, I.S., Human immunodeficiency virus type 1 Vpr induces apoptosis following cell cycle arrest, *J. Virol.,* 71, 5579–5592, 1997.

61. Stewart, S.A., Poon, B., Song, J.Y., and Chen, I.S., Human immunodeficiency virus type 1 Vpr induces apoptosis through caspase activation, *J. Virol.,* 74, 3105–3111, 2000.

62. Muthumani, K., Hwang, D.S., Desai, B.M., Zhang, D., Dayes, N., Green, D.R., and Weiner, D.B., HIV-1 Vpr induces apoptosis through caspase 9 in T cells and peripheral blood mononuclear cells, *J. Biol. Chem.,* 277, 37820–37831, 2002.

63. Muthumani, K., Zhang, D., Hwang, D.S., Kudchodkar, S., Dayes, N.S., Desai, B.M., Malik, A.S., Yang, J.S., Chattergoon, M.A., Maguire, H.C., Jr., and Weiner, D.B., Adenovirus encoding HIV-1 Vpr activates caspase 9 and induces apoptotic cell death in both p53 positive and negative human tumor cell lines, *Oncogene,* 21, 4613–4625, 2002.

64. Conti, L., Matarrese, P., Varano, B., Gauzzi, M.C., Sato, A., Malorni, W., Belardelli, F., and Gessani, S., Dual role of the HIV-1 Vpr protein in the modulation of the apoptotic response of T cells, *J. Immunol.,* 165, 3293–3300, 2000.

65. Conti, L., Rainaldi, G., Matarrese, P., Varano, B., Rivabene, R., Columba, S., Sato, A., Belardelli, F., Malorni, W., and Gessani, S., The HIV-1 Vpr protein acts as a negative regulator of apoptosis in a human lymphoblastoid T cell line: possible implications for the pathogenesis of AIDS, *J. Exp. Med.,* 187, 403–413, 1998.

66. Ayyavoo, V., Mahboubi, A., Mahalingam, S., Ramalingam, R., Kudchodkar, S., Williams, W.V., Green, D.R., and Weiner, D.B., HIV-1 Vpr suppresses immune activation and apoptosis through regulation of nuclear factor B, *Nat. Med.,* 3, 1117–1123, 1997.

67. Levy, D.N., Refaeli, Y., MacGregor, R.R., and Weiner, D.B., Serum Vpr regulates productive infection and latency of human immunodeficiency virus type 1, *Proc. Natl. Acad. Sci. U.S.A.,* 91, 10873–10877, 1994.

68. Sherman, M.P., Schubert, U., Williams, S.A., de Noronha, C.M., Kreisberg, J.F., Henklein, P., and Greene, W.C., HIV-1 Vpr displays natural protein-transducing properties: implications for viral pathogenesis, *Virology,* 302, 95–105, 2002.

69. Stewart, S.A., Poon, B., Jowett, J.B., Xie, Y., and Chen, I.S., Lentiviral delivery of HIV-1 Vpr protein induces apoptosis in transformed cells, *Proc. Natl. Acad. Sci. U.S.A.,* 96, 12039–12043, 1999.

70. Yao, X.J., Mouland, A.J., Subbramanian, R.A., Forget, J., Rougeau, N., Bergeron, D., and Cohen, E.A., Vpr stimulates viral expression and induces cell killing in human immunodeficiency virus type 1-infected dividing Jurkat T cells, *J. Virol.,* 72, 4686–4693, 1998.

71. Nishizawa, M., Kamata, M., Mojin, T., Nakai, Y., and Aida, Y., Induction of apoptosis by the Vpr protein of human immunodeficiency virus type 1 occurs independently of G(2) arrest of the cell cycle, *Virology,* 276, 16–26, 2000.

72. Jacotot, E., Ravagnan, L., Loeffler, M., Ferri, K.F., Vieira, H.L., Zamzami, N., Costantini, P., Druillennec, S., Hoebeke, J., Briand, J.P., Irinopoulou, T., Daugas, E., Susin, S.A., Cointe, D., Xie, Z.H., Reed, J.C., Roques, B.P., and Kroemer, G., The HIV-1 viral protein R induces apoptosis via a direct effect on the mitochondrial permeability transition pore, *J. Exp. Med.,* 191, 33–46, 2000.

73. Jacotot, E., Ferri, K.F., El Hamel, C., Brenner, C., Druillennec, S., Hoebeke, J., Rustin, P., Metivier, D., Lenoir, C., Geuskens, M., Vieira, H.L., Loeffler, M., Belzacq, A.S., Briand, J.P., Zamzami, N., Edelman, L., Xie, Z.H., Reed, J.C., Roques, B.P., and Kroemer, G., Control of mitochondrial membrane permeabilization by adenine nucleotide translocator interacting with HIV-1 viral protein rR and Bcl-2, *J. Exp. Med.*, 193, 509–519, 2001.

74. Roumier, T., Vieira, H.L., Castedo, M., Ferri, K.F., Boya, P., Andreau, K., Druillennec, S., Joza, N., Penninger, J.M., Roques, B., and Kroemer, G., The C-terminal moiety of HIV-1 Vpr induces cell death via a caspase-independent mitochondrial pathway, *Cell Death Differ.*, 9, 1212–1219, 2002.

75. Bartz, S.R. and Emerman, M., Human immunodeficiency virus type 1 Tat induces apoptosis and increases sensitivity to apoptotic signals by up-regulating FLICE/caspase-8, *J. Virol.*, 73, 1956–1963, 1999.

76. Ghosh, S. and Karin, M., Missing pieces in the NF-B puzzle, *Cell,* 109 (Suppl.), S81–S96, 2002.

77. Suzuki, H., Chiba, T., Kobayashi, M., Takeuchi, M., Suzuki, T., Ichiyama, A., Ikenoue, T., Omata, M., Furuichi, K., and Tanaka, K., IB ubiquitination is catalyzed by an SCF-like complex containing Skp1, cullin-1, and two F-box/WD40-repeat proteins, TrCP1 and TrCP2, *Biochem. Biophys. Res. Commun.*, 256, 127–132, 1999.

78. Spencer, E., Jiang, J., and Chen, Z.J., Signal-induced ubiquitination of IB by the F-box protein Slimb/-TrCP, *Genes Dev.*, 13, 284–294, 1999.

79. Wu, C. and Ghosh, S., -TrCP mediates the signal-induced ubiquitination of IB, *J. Biol. Chem.*, 274, 29591–29594, 1999.

80. Hatakeyama, S., Kitagawa, M., Nakayama, K., Shirane, M., Matsumoto, M., Hattori, K., Higashi, H., Nakano, H., Okumura, K., Onoe, K., and Good, R.A., Ubiquitin-dependent degradation of IB is mediated by a ubiquitin ligase Skp1/Cul 1/F-box protein FWD1, *Proc. Natl. Acad. Sci. U.S.A.*, 96, 3859–3863, 1999.

81. Kroll, M., Margottin, F., Kohl, A., Renard, P., Durand, H., Concordet, J.P., Bachelerie, F., Arenzana-Seisdedos, F., and Benarous, R., Inducible degradation of IB by the proteasome requires interaction with the F-box protein h-TrCP, *J. Biol. Chem.*, 274, 7941–7945, 1999.

82. Yaron, A., Hatzubai, A., Davis, M., Lavon, I., Amit, S., Manning, A.M., Andersen, J.S., Mann, M., Mercurio, F., and Ben-Neriah, Y., Identification of the receptor component of the IB-ubiquitin ligase, *Nature,* 396, 590–594, 1998.

83. Margottin, F., Bour, S.P., Durand, H., Selig, L., Benichou, S., Richard, V., Thomas, D., Strebel, K., and Benarous, R., A novel human WD protein, h-TrCp, that interacts with HIV-1 Vpu connects CD4 to the ER degradation pathway through an F-box motif, *Mol. Cell,* 1, 565–574, 1998.

84. Bour, S., Perrin, C., Akari, H., and Strebel, K., The human immunodeficiency virus type 1 Vpu protein inhibits NF-B activation by interfering with TrCP-mediated degradation of IB, *J. Biol. Chem.*, 276, 15920–15928, 2001.

85. Besnard-Guerin, C., Belaidouni, N., Lassot, I., Segeral, E., Jobart, A., Marchal, C., and Benarous, R., HIV-1 Vpu sequesters -transducin repeat-containing protein (TrCP) in the cytoplasm and provokes the accumulation of -catenin and other SCFTrCP substrates, *J. Biol. Chem.*, 279, 788–795, 2004.

86. Krausslich, H.G., Human immunodeficiency virus proteinase dimer as component of the viral polyprotein prevents particle assembly and viral infectivity, *Proc. Natl. Acad. Sci. U.S.A.*, 88, 3213–3217, 1991.

87. Blanco, R., Carrasco, L., and Ventoso, I., Cell killing by HIV-1 protease, *J. Biol. Chem.*, 278, 1086–1093, 2003.

88. Buttner, J., Dornmair, K., and Schramm, H.J., Screening of inhibitors of HIV-1 protease using an *Escherichia coli* cell assay, *Biochem. Biophys. Res. Commun.*, 233, 36–38, 1997.

89. Schlick, P. and Skern, T., Eukaryotic initiation factor 4GI is a poor substrate for HIV-1 proteinase, *FEBS Lett.*, 529, 337–340, 2002.

90. Alvarez, E., Menendez-Arias, L., and Carrasco, L., The eukaryotic translation initiation factor 4GI is cleaved by different retroviral proteases, *J. Virol.*, 77, 12392–12400, 2003.

91. Ventoso, I., Blanco, R., Perales, C., and Carrasco, L., HIV-1 protease cleaves eukaryotic initiation factor 4G and inhibits cap-dependent translation, *Proc. Natl. Acad. Sci. U.S.A.*, 98, 12966–12971, 2001.

92. Shoeman, R.L., Huttermann, C., Hartig, R., and Traub, P., Amino-terminal polypeptides of vimentin are responsible for the changes in nuclear architecture associated with human immunodeficiency virus type 1 protease activity in tissue culture cells, *Mol. Biol. Cell,* 12, 143–154, 2001.

93. Honer, B., Shoeman, R.L., and Traub, P., Human immunodeficiency virus type 1 protease micro-injected into cultured human skin fibroblasts cleaves vimentin and affects cytoskeletal and nuclear architecture, *J. Cell Sci.,* 100, 799–807, 1991.

94. Shoeman, R.L., Honer, B., Stoller, T.J., Kesselmeier, C., Miedel, M.C., Traub, P., and Graves, M.C., Human immunodeficiency virus type 1 protease cleaves the intermediate filament proteins vimentin, desmin, and glial fibrillary acidic protein, *Proc. Natl. Acad. Sci. U.S.A.,* 87, 6336–6340, 1990.

95. Riviere, Y., Blank, V., Kourilsky, P., and Israel, A., Processing of the precursor of NF-B by the HIV-1 protease during acute infection, *Nature,* 350, 625–626, 1991.

96. Tomasselli, A.G., Hui, J.O., Adams, L., Chosay, J., Lowery, D., Greenberg, B., Yem, A., Deibel, M.R., Zurcher-Neely, H., and Heinrikson, R.L., Actin, troponin C, Alzheimer amyloid precursor protein and pro-interleukin 1 as substrates of the protease from human immunodeficiency virus, *J. Biol. Chem.,* 266, 14548–14553, 1991.

97. Strack, P.R., Frey, M.W., Rizzo, C.J., Cordova, B., George, H.J., Meade, R., Ho, S.P., Corman, J., Tritch, R., and Korant, B.D., Apoptosis mediated by HIV protease is preceded by cleavage of Bcl-2, *Proc. Natl. Acad. Sci. U.S.A.,* 93, 9571–9576, 1996.

98. Nie, Z., Phenix, B.N., Lum, J.J., Alam, A., Lynch, D.H., Beckett, B., Krammer, P.H., Sekaly, R.P., and Badley, A.D., HIV-1 protease processes procaspase 8 to cause mitochondrial release of cytochrome *c,* caspase cleavage and nuclear fragmentation, *Cell Death Differ.,* 9, 1172–1184, 2002.

99. Konvalinka, J., Litterst, M.A., Welker, R., Kottler, H., Rippmann, F., Heuser, A.M., and Krausslich, H.G., An active-site mutation in the human immunodeficiency virus type 1 proteinase (PR) causes reduced PR activity and loss of PR-mediated cytotoxicity without apparent effect on virus maturation and infectivity, *J. Virol.,* 69, 7180–7186, 1995.

100. Ensoli, B., Buonaguro, L., Barillari, G., Fiorelli, V., Gendelman, R., Morgan, R.A., Wingfield, P., and Gallo, R.C., Release, uptake, and effects of extracellular human immunodeficiency virus type 1 Tat protein on cell growth and viral transactivation, *J. Virol.,* 67, 277–287, 1993.

101. Frankel, A.D. and Pabo, C.O., Cellular uptake of the Tat protein from human immunodeficiency virus, *Cell,* 55, 1189–1193, 1988.

102. Mann, D.A. and Frankel, A.D., Endocytosis and targeting of exogenous HIV-1 Tat protein, *EMBO J.,* 10, 1733–1739, 1991.

103. Helland, D.E., Welles, J.L., Caputo, A., and Haseltine, W.A., Transcellular transactivation by the human immunodeficiency virus type 1 Tat protein, *J. Virol.,* 65, 4547–4549, 1991.

104. Wei, P., Garber, M.E., Fang, S.M., Fischer, W.H., and Jones, K.A., A novel CDK9-associated C-type cyclin interacts directly with HIV-1 Tat and mediates its high-affinity, loop-specific binding to TAR RNA, *Cell,* 92, 451–462, 1998.

105. Peng, J., Marshall, N.F., and Price, D.H., Identification of a cyclin subunit required for the function of *Drosophila* P-TEFb, *J. Biol. Chem.,* 273, 13855–13860, 1998.

106. Mancebo, H.S., Lee, G., Flygare, J., Tomassini, J., Luu, P., Zhu, Y., Peng, J., Blau, C., Hazuda, D., Price, D., and Flores, O., P-TEFb kinase is required for HIV Tat transcriptional activation *in vivo* and *in vitro,* *Genes Dev.,* 11, 2633–2644, 1997.

107. Marzio, G., Tyagi, M., Gutierrez, M.I., and Giacca, M., HIV-1 Tat transactivator recruits p300 and CREB-binding protein histone acetyltransferases to the viral promoter, *Proc. Natl. Acad. Sci. U.S.A.,* 95, 13519–13524, 1998.

108. Benkirane, M., Chun, R.F., Xiao, H., Ogryzko, V.V., Howard, B.H., Nakatani, Y., and Jeang, K.T., Activation of integrated provirus requires histone acetyltransferase. p300 and P/CAF are coactivators for HIV-1 Tat, *J. Biol. Chem.,* 273, 24898–24905, 1998.

109. Hottiger, M.O. and Nabel, G.J., Interaction of human immunodeficiency virus type 1 Tat with the transcriptional coactivators p300 and CREB binding protein, *J. Virol.,* 72, 8252–8256, 1998.

110. Yang, Y., Dong, B., Mittelstadt, P.R., Xiao, H., and Ashwell, J.D., HIV Tat binds Egr proteins and enhances Egr-dependent transactivation of the Fas ligand promoter, *J. Biol. Chem.,* 277, 19482–19487, 2002.

111. Li, C.J., Friedman, D.J., Wang, C., Metelev, V., and Pardee, A.B., Induction of apoptosis in uninfected lymphocytes by HIV-1 Tat protein, *Science,* 268, 429–431, 1995.

112. Ehret, A., Westendorp, M.O., Herr, I., Debatin, K.M., Heeney, J.L., Frank, R., and Krammer, P.H., Resistance of chimpanzee T cells to human immunodeficiency virus type 1 Tat-enhanced oxidative stress and apoptosis, *J. Virol.,* 70, 6502–6507, 1996.

113. Westendorp, M.O., Frank, R., Ochsenbauer, C., Stricker, K., Dhein, J., Walczak, H., Debatin, K.M., and Krammer, P.H., Sensitization of T cells to CD95-mediated apoptosis by HIV-1 Tat and gp120, *Nature,* 375, 497–500, 1995.

114. McCloskey, T.W., Ott, M., Tribble, E., Khan, S. A., Teichberg, S., Paul, M.O., Pahwa, S., Verdin, E., and Chirmule, N., Dual role of HIV Tat in regulation of apoptosis in T cells, *J. Immunol.,* 158, 1014–1019, 1997.

115. Katsikis, P.D., Garcia-Ojeda, M.E., Torres-Roca, J.F., Greenwald, D.R., and Herzenberg, L.A., HIV type 1 Tat protein enhances activation—but not Fas (CD95)-induced peripheral blood T cell apoptosis in healthy individuals, *Int. Immunol.,* 9, 835–841, 1997.

116. Sastry, K.J., Marin, M.C., Nehete, P.N., McConnell, K., el-Naggar, A.K., and McDonnell, T.J., Expression of human immunodeficiency virus type I Tat results in down-regulation of Bcl-2 and induction of apoptosis in hematopoietic cells, *Oncogene,* 13, 487–493, 1996.

117. Li-Weber, M., Laur, O., Dern, K., and Krammer, P.H., T cell activation-induced and HIV Tat-enhanced CD95(APO-1/Fas) ligand transcription involves NF-B, *Eur. J. Immunol.,* 30, 661–670, 2000.

118. Yagi, T., Sugimoto, A., Tanaka, M., Nagata, S., Yasuda, S., Yagita, H., Kuriyama, T., Takemori, T., and Tsunetsugu-Yokota, Y., Fas/FasL interaction is not involved in apoptosis of activated CD4+ T cells upon HIV-1 infection *in vitro, J. Acquir. Immune Defic. Syndr. Hum. Retrovirol.,* 18, 307–315, 1998.

119. Westendorp, M.O., Shatrov, V.A., Schulze-Osthoff, K., Frank, R., Kraft, M., Los, M., Krammer, P.H., Droge, W., and Lehmann, V., HIV-1 Tat potentiates TNF-induced NF-B activation and cytotoxicity by altering the cellular redox state, *EMBO J.,* 14, 546–554, 1995.

120. Zauli, G., Gibellini, D., Caputo, A., Bassini, A., Negrini, M., Monne, M., Mazzoni, M., and Capitani, S., The human immunodeficiency virus type-1 Tat protein upregulates Bcl-2 gene expression in Jurkat T-cell lines and primary peripheral blood mononuclear cells, *Blood,* 86, 3823–3834, 1995.

121. Wang, Z., Morris, G.F., Reed, J.C., Kelly, G.D., and Morris, C.B., Activation of Bcl-2 promoter-directed gene expression by the human immunodeficiency virus type-1 Tat protein, *Virology,* 257, 502–510, 1999.

122. Corallini, A., Sampaolesi, R., Possati, L., Merlin, M., Bagnarelli, P., Piola, C., Fabris, M., Menegatti, M.A., Talevi, S., Gibellini, D., Rocchetti, R., Caputo, A., and Barbanti-Brodano, G., Inhibition of HIV-1 Tat activity correlates with down-regulation of Bcl-2 and results in reduction of angiogenesis and oncogenicity, *Virology,* 299, 1–7, 2002.

123. Macho, A., Calzado, M.A., Jimenez-Reina, L., Ceballos, E., Leon, J., and Munoz, E., Susceptibility of HIV-1-TAT transfected cells to undergo apoptosis. Biochemical mechanisms, *Oncogene,* 18, 7543–7551, 1999.

124. Chen, D., Wang, M., Zhou, S., and Zhou, Q., HIV-1 Tat targets microtubules to induce apoptosis, a process promoted by the pro-apoptotic Bcl-2 relative Bim, *EMBO J.,* 21, 6801–6810, 2002.

125. Gibellini, D., Re, M.C., Ponti, C., Celeghini, C., Melloni, E., La Placa, M., and Zauli, G., Extracellular Tat activates c-fos promoter in low serum-starved CD4+ T cells, *Br. J. Haematol.,* 112, 663–670, 2001.

126. Gibellini, D., Re, M.C., Ponti, C., Maldini, C., Celeghini, C., Cappellini, A., La Placa, M., and Zauli, G., HIV-1 Tat protects CD4+ Jurkat T lymphoblastoid cells from apoptosis mediated by TNF-related apoptosis-inducing ligand, *Cell. Immunol.,* 207, 89–99, 2001.

127. Gibellini, D., Caputo, A., Celeghini, C., Bassini, A., La Placa, M., Capitani, S., and Zauli, G., Tat-expressing Jurkat cells show an increased resistance to different apoptotic stimuli, including acute human immunodeficiency virus-type 1 (HIV-1) infection, *Br. J. Haematol.,* 89, 24–33, 1995.

128. Wang, L., Klimpel, G.R., Planas, J.M., Li, H., and Cloyd, M.W., Apoptotic killing of CD4+ T lymphocytes in HIV-1-infected PHA-stimulated PBL cultures is mediated by CD8+ LAK cells, *Virology,* 241, 169–180, 1998.

129. Finkel, T.H., Tudor-Williams, G., Banda, N.K., Cotton, M.F., Curiel, T., Monks, C., Baba, T.W., Ruprecht, R.M., and Kupfer, A., Apoptosis occurs predominantly in bystander cells and not in productively infected cells of HIV- and SIV-infected lymph nodes, *Nat. Med.,* 1, 129–134, 1995.

130. Dockrell, D.H., The multiple roles of Fas ligand in the pathogenesis of infectious diseases, *Clin. Microbiol. Infect.,* 9, 766–779, 2003.

131. Kayagaki, N., Kawasaki, A., Ebata, T., Ohmoto, H., Ikeda, S., Inoue, S., Yoshino, K., Okumura, K., and Yagita, H., Metalloproteinase-mediated release of human Fas ligand, *J. Exp. Med.,* 182, 1777–1783, 1995.

132. Mitra, D., Steiner, M., Lynch, D.H., Staiano-Coico, L., and Laurence, J., HIV-1 upregulates Fas ligand expression in CD4+ T cells *in vitro* and *in vivo:* association with Fas-mediated apoptosis and modulation by aurintricarboxylic acid, *Immunology,* 87, 581–585, 1996.

133. Hosaka, N., Oyaizu, N., Kaplan, M.H., Yagita, H., and Pahwa, S., Membrane and soluble forms of Fas (CD95) and Fas ligand in peripheral blood mononuclear cells and in plasma from human immunodeficiency virus-infected persons, *J. Infect. Dis.,* 178, 1030–1039, 1998.

134. Katsikis, P.D., Wunderlich, E.S., Smith, C.A., and Herzenberg, L.A., Fas antigen stimulation induces marked apoptosis of T lymphocytes in human immunodeficiency virus-infected individuals, *J. Exp. Med.,* 181, 2029–2036, 1995.

135. Yang, Y., Liu, Z.H., Ware, C.F., and Ashwell, J.D., A cysteine protease inhibitor prevents activation-induced T-cell apoptosis and death of peripheral blood cells from human immunodeficiency virus-infected individuals by inhibiting upregulation of Fas ligand, *Blood,* 89, 550–557, 1997.

136. Baumler, C.B., Bohler, T., Herr, I., Benner, A., Krammer, P.H., and Debatin, K.M., Activation of the CD95 (APO-1/Fas) system in T cells from human immunodeficiency virus type-1-infected children, *Blood,* 88, 1741–1746, 1996.

137. Debatin, K.M., Fahrig-Faissner, A., Enenkel-Stoodt, S., Kreuz, W., Benner, A., and Krammer, P.H., High expression of APO-1 (CD95) on T lymphocytes from human immunodeficiency virus-1-infected children, *Blood,* 83, 3101–3103, 1994.

138. Badley, A.D., McElhinny, J.A., Leibson, P.J., Lynch, D.H., Alderson, M.R., and Paya, C.V., Upregulation of Fas ligand expression by human immunodeficiency virus in human macrophages mediates apoptosis of uninfected T lymphocytes, *J. Virol.,* 70, 199–206, 1996.

139. Dockrell, D.H., Badley, A.D., Villacian, J.S., Heppelmann, C.J., Algeciras, A., Ziesmer, S., Yagita, H., Lynch, D.H., Roche, P.C., Leibson, P.J., and Paya, C.V., The expression of Fas ligand by macrophages and its upregulation by human immunodeficiency virus infection, *J. Clin. Invest.,* 101, 2394–2405, 1998.

140. Cohen, S.S., Li, C., Ding, L., Cao, Y., Pardee, A.B., Shevach, E.M., and Cohen, D.I., Pronounced acute immunosuppression *in vivo* mediated by HIV Tat challenge, *Proc. Natl. Acad. Sci. U.S.A.,* 96, 10842–10847, 1999.

141. Oyaizu, N., Adachi, Y., Hashimoto, F., McCloskey, T.W., Hosaka, N., Kayagaki, N., Yagita, H., and Pahwa, S., Monocytes express Fas ligand upon CD4 cross-linking and induce CD4+ T cells apoptosis: a possible mechanism of bystander cell death in HIV infection, *J. Immunol.,* 158, 2456–2463, 1997.

142. Han, X., Becker, K., Degen, H.J., Jablonowski, H., and Strohmeyer, G., Synergistic stimulatory effects of tumour necrosis factor and interferon on replication of human immunodeficiency virus type 1 and on apoptosis of HIV-1-infected host cells, *Eur. J. Clin. Invest.,* 26, 286–292, 1996.

143. Badley, A.D., Dockrell, D., Simpson, M., Schut, R., Lynch, D.H., Leibson, P., and Paya, C.V., Macrophage-dependent apoptosis of CD4+ T lymphocytes from HIV-infected individuals is mediated by FasL and tumor necrosis factor, *J. Exp. Med.,* 185, 55–64, 1997.

144. Jeremias, I., Herr, I., Boehler, T., and Debatin, K.M., TRAIL/Apo-2-ligand-induced apoptosis in human T cells, *Eur. J. Immunol.,* 28, 143–152, 1998.

145. Katsikis, P.D., Garcia-Ojeda, M.E., Torres-Roca, J.F., Tijoe, I.M., Smith, C.A., and Herzenberg, L.A., Interleukin-1 converting enzyme-like protease involvement in Fas-induced and activation-induced peripheral blood T cell apoptosis in HIV infection. TNF-related apoptosis-inducing ligand can mediate activation-induced T cell death in HIV infection, *J. Exp. Med.,* 186, 1365–1372, 1997.

146. Fujii, Y., Otake, K., Tashiro, M., and Adachi, A., Soluble Nef antigen of HIV-1 is cytotoxic for human CD4+ T cells, *FEBS Lett.,* 393, 93–96, 1996.

147. Okada, H., Takei, R., and Tashiro, M., HIV-1 Nef protein-induced apoptotic cytolysis of a broad spectrum of uninfected human blood cells independently of CD95(Fas), *FEBS Lett.,* 414, 603–606, 1997.

148. Okada, H., Takei, R., and Tashiro, M., Nef protein of HIV-1 induces apoptotic cytolysis of murine lymphoid cells independently of CD95 (Fas) and its suppression by serine/threonine protein kinase inhibitors, *FEBS Lett.,* 417, 61–64, 1997.

149. Okada, H., Takei, R., and Tashiro, M., Inhibition of HIV-1 Nef-induced apoptosis of uninfected human blood cells by serine/threonine protein kinase inhibitors, fasudil hydrochloride and M3, *FEBS Lett.,* 422, 363–367, 1998.

150. James, C.O., Huang, M.B., Khan, M., Garcia-Barrio, M., Powell, M.D., and Bond, V.C., Extracellular Nef protein targets CD4+ T cells for apoptosis by interacting with CXCR4 surface receptors, *J. Virol.,* 78, 3099–3109, 2004.

151. Manninen, A., Hiipakka, M., Vihinen, M., Lu, W., Mayer, B.J., and Saksela, K., SH3-Domain binding function of HIV-1 Nef is required for association with a PAK-related kinase, *Virology,* 250, 273–282, 1998.

152. Geleziunas, R., Xu, W., Takeda, K., Ichijo, H., and Greene, W.C., HIV-1 Nef inhibits ASK1-dependent death signaling providing a potential mechanism for protecting the infected host cell, *Nature,* 410, 834–838, 2001.

153. Hodge, S., Novembre, F.J., Whetter, L., Gelbard, H.A., and Dewhurst, S., Induction of Fas ligand expression by an acutely lethal simian immunodeficiency virus, SIVsmmPBj14, *Virology,* 252, 354–363, 1998.

154. Xu, X.N., Laffert, B., Screaton, G.R., Kraft, M., Wolf, D., Kolanus, W., Mongkolsapay, J., McMichael, A.J., and Baur, A.S., Induction of Fas ligand expression by HIV involves the interaction of Nef with the T cell receptor chain, *J. Exp. Med.,* 189, 1489–1496, 1999.

155. Xu, X.N., Screaton, G.R., Gotch, F.M., Dong, T., Tan, R., Almond, N., Walker, B., Stebbings, R., Kent, K., Nagata, S., Stott, J.E., and McMichael, A.J., Evasion of cytotoxic T lymphocyte (CTL) responses by Nef-dependent induction of Fas ligand (CD95L) expression on simian immunodeficiency virus-infected cells, *J. Exp. Med.,* 186, 7–16, 1997.

156. Flint, S.J., *Principles of Virology: Molecular Biology, Pathogenesis, and Control,* ASM Press, Washington, DC, 2000.

157. Gougeon, M.L., Apoptosis as an HIV strategy to escape immune attack, *Nat. Rev. Immunol.,* 3, 392–404, 2003.

158. Hellerstein, M., Hanley, M.B., Cesar, D., Siler, S., Papageorgopoulos, C., Wieder, E., Schmidt, D., Hoh, R., Neese, R., Macallan, D., Deeks, S., and McCune, J.M., Directly measured kinetics of circulating T lymphocytes in normal and HIV-1-infected humans, *Nat. Med.,* 5, 83–89, 1999.

159. Lempicki, R.A., Kovacs, J.A., Baseler, M.W., Adelsberger, J.W., Dewar, R.L., Natarajan, V., Bosche, M.C., Metcalf, J.A., Stevens, R.A., Lambert, L.A., Alvord, W.G., Polis, M.A., Davey, R.T., Dimitrov, D.S., and Lane, H.C., Impact of HIV-1 infection and highly active antiretroviral therapy on the kinetics of CD4+ and CD8+ T cell turnover in HIV-infected patients, *Proc. Natl. Acad. Sci. U.S.A.,* 97, 13778–13783, 2000.

160. Mohri, H., Perelson, A.S., Tung, K., Ribeiro, R.M., Ramratnam, B., Markowitz, M., Kost, R., Hurley, A., Weinberger, L., Cesar, D., Hellerstein, M.K., and Ho, D.D., Increased turnover of T lymphocytes in HIV-1 infection and its reduction by antiretroviral therapy, *J. Exp. Med.,* 194, 1277–1287, 2001.

161. Grossman, Z., Meier-Schellersheim, M., Sousa, A.E., Victorino, R.M., and Paul, W.E., CD4+ T-cell depletion in HIV infection: are we closer to understanding the cause? *Nat. Med.,* 8, 319–323, 2002.

162. Ho, D.D., Neumann, A.U., Perelson, A.S., Chen, W., Leonard, J.M., and Markowitz, M., Rapid turnover of plasma virions and CD4 lymphocytes in HIV-1 infection, *Nature,* 373, 123–126, 1995.

163. Mohri, H., Bonhoeffer, S., Monard, S., Perelson, A.S., and Ho, D.D., Rapid turnover of T lymphocytes in SIV-infected rhesus macaques, *Science,* 279, 1223–1227, 1998.

164. Perelson, A.S., Neumann, A.U., Markowitz, M., Leonard, J.M., and Ho, D.D., HIV-1 dynamics *in vivo:* virion clearance rate, infected cell life-span, and viral generation time, *Science,* 271, 1582–1586, 1996.

165. McCune, J.M., Loftus, R., Schmidt, D.K., Carroll, P., Webster, D., Swor-Yim, L.B., Francis, I.R., Gross, B.H., and Grant, R.M., High prevalence of thymic tissue in adults with human immunodeficiency virus-1 infection, *J. Clin. Invest.,* 101, 2301–2308, 1998.

166. McCune, J.M., Thymic function in HIV-1 disease, *Semin. Immunol.,* 9, 397–404, 1997.

167. Chene, L., Nugeyre, M.T., Guillemard, E., Moulian, N., Barre-Sinoussi, F., and Israel, N., Thymocyte–thymic epithelial cell interaction leads to high-level replication of human immunodeficiency virus exclusively in mature CD4(+) CD8(−) CD3(+) thymocytes: a critical role for tumor necrosis factor and interleukin-7, *J. Virol.,* 73, 7533–7542, 1999.

168. Pedroza-Martins, L., Boscardin, W.J., Anisman-Posner, D.J., Schols, D., Bryson, Y.J., and Uittenbogaart, C.H., Impact of cytokines on replication in the thymus of primary human immunodeficiency virus type 1 isolates from infants, *J. Virol.,* 76, 6929–6943, 2002.

169. Wolf, S.S. and Cohen, A., Expression of cytokines and their receptors by human thymocytes and thymic stromal cells, *Immunology,* 77, 362–368, 1992.

170. Napolitano, L.A., Grant, R.M., Deeks, S.G., Schmidt, D., De Rosa, S.C., Herzenberg, L.A., Herndier, B.G., Andersson, J., and McCune, J.M., Increased production of IL-7 accompanies HIV-1-mediated T-cell depletion: implications for T-cell homeostasis, *Nat. Med.,* 7, 73–79, 2001.

171. Napolitano, L.A., Lo, J.C., Gotway, M.B., Mulligan, K., Barbour, J.D., Schmidt, D., Grant, R.M., Halvorsen, R.A., Schambelan, M., and McCune, J.M., Increased thymic mass and circulating naive CD4 T cells in HIV-1-infected adults treated with growth hormone, *AIDS,* 16, 1103–1111, 2002.

172. Burke, A.P., Anderson, D., Benson, W., Turnicky, R., Mannan, P., Liang, Y.H., Smialek, J., and Virmani, R., Localization of human immunodeficiency virus 1 RNA in thymic tissues from asymptomatic drug addicts, *Arch. Pathol. Lab. Med.,* 119, 36–41, 1995.

173. Joshi, V.V. and Oleske, J.M., Pathologic appraisal of the thymus gland in acquired immunodeficiency syndrome in children. A study of four cases and a review of the literature, *Arch. Pathol. Lab. Med.,* 109, 142–146, 1985.

174. Joshi, V.V., Oleske, J.M., Saad, S., Gadol, C., Connor, E., Bobila, R., and Minnefor, A.B., Thymus biopsy in children with acquired immunodeficiency syndrome, *Arch. Pathol. Lab. Med.,* 110, 837–842, 1986.

175. Prevot, S., Audouin, J., Andre-Bougaran, J., Griffais, R., Le Tourneau, A., Fournier, J.G., and Diebold, J., Thymic pseudotumorous enlargement due to follicular hyperplasia in a human immunodeficiency virus sero-positive patient. Immunohistochemical and molecular biological study of viral infected cells, *Am. J. Clin. Pathol.,* 97, 420–425, 1992.

176. Linder, J., The thymus gland in secondary immunodeficiency, *Arch. Pathol. Lab. Med.,* 111, 1118–1122, 1987.

177. Stanley, S.K., McCune, J.M., Kaneshima, H., Justement, J.S., Sullivan, M., Boone, E., Baseler, M., Adelsberger, J., Bonyhadi, M., Orenstein, J., Human immunodeficiency virus infection of the human thymus and disruption of the thymic microenvironment in the SCID-hu mouse, *J. Exp. Med.,* 178, 1151–1163, 1993.

178. Aldrovandi, G.M., Feuer, G., Gao, L., Jamieson, B., Kristeva, M., Chen, I.S., and Zack, J.A., The SCID-hu mouse as a model for HIV-1 infection, *Nature,* 363, 732–736, 1993.

179. Bonyhadi, M.L., Rabin, L., Salimi, S., Brown, D.A., Kosek, J., McCune, J.M., and Kaneshima, H., HIV induces thymus depletion *in vivo, Nature,* 363, 728–732, 1993.

180. Baskin, G.B., Murphey-Corb, M., Martin, L.N., Davison-Fairburn, B., Hu, F.S., and Kuebler, D., Thymus in simian immunodeficiency virus-infected rhesus monkeys, *Lab. Invest.,* 65, 400–407, 1991.

181. Li, S.L., Kaaya, E.E., Ordonez, C., Ekman, M., Feichtinger, H., Putkonen, P., Bottiger, D., Biberfeld, G., and Biberfeld, P., Thymic immunopathology and progression of SIVsm infection in cynomolgus monkeys, *J. Acquir. Immune Defic. Syndr. Hum. Retrovirol.,* 9, 1–10, 1995.

182. Su, L., Kaneshima, H., Bonyhadi, M., Salimi, S., Kraft, D., Rabin, L., and McCune, J.M., HIV-1-induced thymocyte depletion is associated with indirect cytopathogenicity and infection of progenitor cells *in vivo, Immunity,* 2, 25–36, 1995.

183. Levy, J.A., Pathogenesis of human immunodeficiency virus infection, *Microbiol. Rev.,* 57, 183–289, 1993.

184. Kourtis, A.P., Ibegbu, C., Nahmias, A.J., Lee, F.K., Clark, W.S., Sawyer, M.K., and Nesheim, S., Early progression of disease in HIV-infected infants with thymus dysfunction, *N. Engl. J. Med.,* 335, 1431–1436, 1996.

185. Frederick, T., Mascola, L., Eller, A., O'Neil, L., and Byers, B., Progression of human immunodeficiency virus disease among infants and children infected perinatally with human immunodeficiency virus or through neonatal blood transfusion. Los Angeles County Pediatric AIDS Consortium and the Los Angeles County–University of Southern California Medical Center and the University of Southern California School of Medicine, *Pediatr. Infect. Dis. J.,* 13, 1091–1097, 1994.

186. Luzuriaga, K., Wu, H., McManus, M., Britto, P., Borkowsky, W., Burchett, S., Smith, B., Mofenson, L., and Sullivan, J.L., Dynamics of human immunodeficiency virus type 1 replication in vertically infected infants, *J. Virol.,* 73, 362–367, 1999.

187. Douek, D.C., Picker, L.J., and Koup, R.A., T cell dynamics in HIV-1 infection, *Annu. Rev. Immunol.,* 21, 265–304, 2003.

188. Luo, K.X. and Sefton, B.M., Cross-linking of T-cell surface molecules CD4 and CD8 stimulates phosphorylation of the Lck tyrosine protein kinase at the autophosphorylation site, *Mol. Cell Biol.,* 10, 5305–5313, 1990.

189. Oyaizu, N., McCloskey, T.W., Than, S., Hu, R., Kalyanaraman, V.S., and Pahwa, S., Cross-linking of CD4 molecules upregulates Fas antigen expression in lymphocytes by inducing interferon- and tumor necrosis factor- secretion, *Blood,* 84, 2622–2631, 1994.

190. Hashimoto, F., Oyaizu, N., Kalyanaraman, V.S., and Pahwa, S., Modulation of Bcl-2 protein by CD4 cross-linking: a possible mechanism for lymphocyte apoptosis in human immunodeficiency virus infection and for rescue of apoptosis by interleukin-2, *Blood,* 90, 745–753, 1997.

191. Algeciras, A., Dockrell, D.H., Lynch, D.H., and Paya, C.V., CD4 regulates susceptibility to Fas ligand- and tumor necrosis factor-mediated apoptosis, *J. Exp. Med.,* 187, 711–720, 1998.

192. Nagata, S. and Golstein, P., The Fas death factor, *Science,* 267, 1449–1456, 1995.

193. Yang, Y. and Ashwell, J.D., Exploiting the apoptotic process for management of HIV: are we there yet? *Apoptosis,* 6, 139–146, 2001.

194. Berndt, C., Mopps, B., Angermuller, S., Gierschik, P., and Krammer, P.H., CXCR4 and CD4 mediate a rapid CD95-independent cell death in CD4(+) T cells, *Proc. Natl. Acad. Sci. U.S.A.,* 95, 12556–12561, 1998.

195. Jekle, A., Keppler, O.T., De Clercq, E., Schols, D., Weinstein, M., and Goldsmith, M.A., *In vivo* evolution of human immunodeficiency virus type 1 toward increased pathogenicity through CXCR4-mediated killing of uninfected CD4 T cells, *J. Virol.,* 77, 5846–5854, 2003.

196. Vlahakis, S.R., Algeciras-Schimnich, A., Bou, G., Heppelmann, C.J., Villasis-Keever, A., Collman, R.C., and Paya, C.V., Chemokine-receptor activation by Env determines the mechanism of death in HIV-infected and uninfected T lymphocytes, *J. Clin. Invest.,* 107, 207–215, 2001.

197. Roumier, T., Castedo, M., Perfettini, J.L., Andreau, K., Metivier, D., Zamzami, N., and Kroemer, G., Mitochondrion-dependent caspase activation by the HIV-1 envelope, *Biochem. Pharmacol.,* 66, 1321–1329, 2003.

198. Castedo, M., Roumier, T., Blanco, J., Ferri, K.F., Barretina, J., Tintignac, L.A., Andreau, K., Perfettini, J.L., Amendola, A., Nardacci, R., Leduc, P., Ingber, D.E., Druillennec, S., Roques, B., Leibovitch, S.A., Vilella-Bach, M., Chen, J., Este, J.A., Modjtahedi, N., Piacentini, M., and Kroemer, G., Sequential involvement of Cdk1, mTOR and p53 in apoptosis induced by the HIV-1 envelope, *EMBO J.,* 21, 4070–4080, 2002.

199. Castedo, M., Ferri, K.F., Blanco, J., Roumier, T., Larochette, N., Barretina, J., Amendola, A., Nardacci, R., Metivier, D., Este, J.A., Piacentini, M., and Kroemer, G., Human immunodeficiency virus 1 envelope glycoprotein complex-induced apoptosis involves mammalian target of rapamycin/FKBP12-rapamycin-associated protein-mediated p53 phosphorylation, *J. Exp. Med.,* 194, 1097–1110, 2001.

14 Elevated Apoptosis of CD8+ T Lymphocytes during HIV-1 Infection

John Zaunders
Jérôme Estaquier
Jacques Corbeil

CONTENTS

CD8⁺ T LYMPHOCYTES IN HIV-1 INFECTION

Many studies have indicated the importance of CD8⁺ T lymphocytes in limiting viral replication. During primary infection, most individuals develop human immunodeficiency virus (HIV)-specific cytotoxic T lymphocytes (CTLs), which are believed to contribute to a significant decrease in viremia.[1] However, disease progression has been associated with either loss of these cells *in vivo*[2–5] or impaired function.[6,7] Similarly, subjects with asymptomatic HIV-1 infection maintain CD8⁺ T cells that suppress HIV-1 replication without lysis of the infected cells, but this activity is also apparently lost in the course of disease progression.[8] Therefore, an understanding of the maintenance of antiviral CD8⁺ T cells is critical to an understanding of HIV-related pathogenesis as well as guiding the design of vaccines and adjuvants to maximize cell-mediated immunity. Inappropriate apoptosis of HIV-specific CD8⁺ T cells may underpin loss of their activity.

Apoptosis exerts tight control over the number of circulating CD8⁺ T lymphocytes throughout life, particularly after the expansion of effector cells during acute viral infections, such as in the dramatic response to Epstein–Barr virus (EBV) infection in adolescents or adults.[9–11] However, the rate of apoptosis of CD8⁺ T lymphocytes remains elevated throughout the course of HIV-1 infection and, therefore, the question arises whether this is abnormal and contributes to the pathogenesis of HIV-associated immunodeficiency.

EXPANSION AND CONTRACTION OF CD8⁺ T LYMPHOCYTES DURING RESPONSES TO VIRAL INFECTION

To maintain homeostasis, the increase in cell numbers due to antigen-specific proliferation must be balanced by death of an equal number of cells. For example, in one study of a mouse model of acute viral infection, there was a 10,000-fold increase in CD8⁺ T cells specific for immunodominant viral antigens in just 8 days.[12] As the viral infection clears, 95% of these cells disappear from the spleen[13] in association with elevated rates of apoptosis. However, it must be noted that many CD8⁺ effector T cells migrate from the spleen to nonlymphoid tissues.[14] Up to 50% of the expanded antigen-specific CD8⁺ T cells may be present in nonlymphoid tissues for at least 4 weeks after infection,[15] for example, the persistence of antigen-specific CD8⁺ T cells that protect lung tissue from viral reinfection.[16] However, a number of changes in the activated effector cells are believed to predispose these cells to apoptosis after resolution of viral infection (as detailed below), and further apoptosis may occur after these cells traffic to the liver.[17]

The cytokine milieu during the initial activation is believed to influence the fate of the antigen-specific CD8⁺ T cells. Live bacteria are superior to heat-killed antigen in generating memory CD8⁺ T cells.[18] Similarly, absence of CD4⁺ T cell help via CD40L[19,20] and antigen-presenting cell (APC) interactions via 4-1BB/4-1BBL[21] leads to impaired generation of memory CD8⁺ T cells, as does an absence of interleukin (IL)-15 signaling.[22] Also, the conversion rate of effector memory cells to resting central memory cells is determined during the activation stage, and these latter cells are capable of extensive proliferation and are better able to control viral replication *in vivo*.[23]

The total number of memory CD8⁺ T cells is relatively stable throughout much of adult life.[24] Therefore, the viral antigen-specific cells that are retained after the peak of expansion must replace preexisting cells in the total CD8⁺ T lymphocyte population.[25] The preexisting cells may be primed for apoptosis by IFN type I,[26] and it is also possible that newly generated memory CD8⁺ T cells may outcompete older memory cells for essential survival factors.[27]

These results suggest that long-term memory CD8⁺ T cell survival is programmed during the initial differentiation, involving helper CD4⁺ T cells and antigen-presenting cells, and the regulation of subsequent apoptosis is central to the maintenance of antiviral CD8⁺ T lympho-cytes.

ROLE OF TNF/TNFR SUPERFAMILY MEMBERS IN HOMEOSTASIS OF CD8+ T LYMPHOCYTES

Activated effector cells exhibit increased expression of CD95 and tumour necrosis factor receptor 1 (TNFR1) and TNFR2, which are known to mediate induction of apoptosis *in vitro*.[28] However, knockout studies in mice showed that deletion of CD95 or either tumor necrosis factor (TNF)-α or TNFRp55 did not affect apoptosis of effector cells.[29,30] Nevertheless, in some reports, a contribution of TNF-α was observed in the deletion of activated peripheral CD8 T cells.[31,32] Loss of Fas/FasL function does not lead to increases in CD8+ T cell numbers,[33] and loss of TNF/TNFR (TNF receptor) function reduces inflammation rather than disturbing CD8+ T cell homeostasis.[28] Also, caspase-8 links CD95 and TNFR ligation to apoptosis,[28] but human caspase-8 deficiency leads to immunodeficiency rather than unrestrained lymphoproliferation.[34] Therefore, the physiological role of apoptosis mediated by these ligands, at least in viral infections, remains unclear.

ROLE OF BCL-2 FAMILY MEMBERS IN CD8+ T CELL APOPTOSIS

Many different studies in mice and humans have shown that expanded CD8+ T cells during acute viral infection, and antigen specific cells in particular, reduce intracellular levels of Bcl-2.[9,10,35] The decrease in Bcl-2 in effector cells is associated with a decrease in IL-7 receptor (IL-7R), whereas long-term memory cells regain expression of IL-7R and Bcl-2.[36,37] This situation is similar to early T cell progenitors, for which expression of IL-7R is required for upregulation of Bcl-2 and survival during ontogeny.[38,39]

However, it was found in two studies that overexpression of Bcl-2 did not prevent apoptosis of excess effector cells after viral infections in murine models.[40,41] A recent study showed that a knockout dose of proapoptotic Bim, a BH3-only member of the Bcl-2 family that antagonizes the effect of Bcl-2,[42] prevented apoptosis of effector cells.[43] These results suggest that an alternative antiapoptotic member of the Bcl-2 family may be important in effector T cell survival.

In this regard, a recent study showed that MCL-1, another antiapoptotic member of the Bcl-2 family, is essential for peripheral T cell survival, is induced by IL-7, and interacts strongly with Bim.[44] MCL-1 also antagonizes the proapoptotic effect of Bak.[45] Current data suggest that Bak combines with Bax to mediate lymphocyte apoptosis via the intrinsic mitochondrial pathway.[46,47] This is normally actively prevented by Bcl-2 and MCL-1, which are, in turn, counteracted by apoptotic signals leading to transcriptional or posttranslational changes in the BH3-only proteins Bim, Bid, and Bad.[48,49]

Studies of lymphoid tumorigenesis suggest there is a reciprocal interaction between upregulation of c-Myc, which promotes both proliferation and apoptosis, and Bcl-2, which promotes quiescence and survival.[48,50] Proliferation of effector T cells, which is regulated by c-Myc,[51] has long been known to involve cytokines that use the common γ-chain as part of their receptors, especially IL-2.[52] However, IL-2 also seems to regulate Bcl-2 expression and cell survival, in addition to proliferation, as detailed below.

ROLE OF COMMON φ-CHAIN CYTOKINES IN CD8+ T CELL APOPTOSIS

In vitro, IL-2 is a potent growth factor for CD8+ T cells, but *in vivo*, its role as a growth and survival factor[53,54] is much more enigmatic, even though IL-2 is able to induce upregulation of Bcl-2.[55] IL-2 may actually inhibit the survival of memory CD8+ T cells *in vivo*,[56] and a knockout dose of IL-2 or IL-2R results in T cell lymphocytosis[57,58] rather than leading to a deficit of peripheral T cells. The inhibitory effects of exposure to IL-2 *in vivo* may be due to its induction of FasL expression and possible fratricide[59] as well as its major role in generating regulatory T cells.[60]

However, IL-7 and IL-15 are unequivocally survival factors for mature CD8+ T cells *in vivo*.[61,62] In particular, expression of the IL-7Rα chain is closely correlated with Bcl-2 expression and long-term survival.[36,37] Also, IL-15 is required for CD8[+] memory cell generation and survival,[22,63–65] and IL-15Rα is required for maintenance of Bcl-2 levels in the periphery.[66,67] IL-15 seems to counteract a proapoptotic effect of IL-2 and increase the survival of memory CD8[+] T cells.[56] In addition to direct survival effects, IL-15 induces low-level turnover of memory CD8[+] T cells[68] as well as mediates homeostatic proliferation in response to lymphopenia.[69]

Therefore, the common γ-chain cytokines, IL-7, IL-2, and IL-15, seem to be central to CD8[+] effector and memory homeostasis, controlling proliferation as well as survival via regulation of Bcl-2 and possibly MCL-1.

ROLE OF IFN TYPE I IN CD8[+] T CELL APOPTOSIS

Another important factor in CD8 responses is IFN type I, which was originally described to induce nonspecific proliferation *in vivo*.[70] Subsequently, this effect of IFN type I was shown to be mediated via induction of IL-15 synthesis.[68] Production of IFN type I is a hallmark of viral infections and is mainly synthesized *in vivo* by plasmacytoid dendritic cells.[71] IFN-β was shown to rescue activated T cells from apoptosis,[72,73] possibly via activation of protein kinase activity.[74]

IFN type I, at the same time, may lead to attrition of older memory cells.[26] IFN-α has also been shown to be involved in inducing apoptosis of B cells via translocation of Daxx to PML bodies[75] as well as apoptosis of multiple myeloma cells via tumor necrosis factor–related apoptosis-inducing ligand (TRAIL).[76] Therefore, again, a regulator of apoptosis may have complex, contradictory effects, but in the case of viral infections, it should act to inhibit CD8[+] T effector cell apoptosis through the action of IL-15.

ROLE OF MOLECULES PRODUCED BY EFFECTOR T CELLS IN CD8[+] T CELL APOPTOSIS

Recently, it was shown that disruption of the perforin gene leads to impaired contraction of primary CD8[+] T cell expansion[77] and unrestrained secondary CD8[+] T cell responses.[78] Similarly, a knockout dose of the IFN-γ gene delays contraction of CD8[77] and CD4 responses.[79]

An emerging concept is that many products of an activated T cell, which are believed to mediate important antiviral effector functions as well as immune-activating functions, also have a feedback role in homeostasis. In addition to perforin and IFN-γ, these products include TNF-α, FasL, IL-2, and, possibly, CD70 (via CD27).[80]

ACTIVATION, PROLIFERATION, AND APOPTOSIS OF CD8[+] T LYMPHOCYTES DURING HIV INFECTION

HIV-1 infection typically results in a long-term doubling of the number of circulating CD8[+] T lymphocytes[81] as well as an increased number of effector phenotype CD8[+] T cells in lymphoid tissue.[5] Paradoxically, as detailed below, there is an increased rate of apoptosis of these cells *in vivo*, and in cultured cells *ex vivo,* compared with CD8[+] T cells from healthy uninfected individuals. Clearly, then, there is not a simple relationship between the rate of cell death and the total number of cells in the circulation, but we must take into account the rate of production of new cells, their states of differentiation, and their partitioning between blood and tissues.

DYNAMICS OF CD8[+] T LYMPHOCYTES IN CHRONIC HIV-1 INFECTION

HIV-1 infection is characterized by an elevated rate of proliferation of CD8[+] T cells, approximately five- to eightfold, as determined by [2]H-glucose[82] or BrdU[83] incorporation into DNA or expression of the cell cycle antigen Ki-67.[84] CD8[+] T cells also exhibit elevated expression of activation antigens, especially CD38 and HLA-DR.[85] Both activation and proliferation are decreased after initiation of potent antiretroviral therapy,[83,86–89] demonstrating their dependence on viral replication. In particular, the level of expression of CD38 is highly correlated with plasma HIV RNA viral load[90,91] in both treated and untreated individuals and is closely aligned with the rate of decline of CD4+ T cells.[92]

The observed discrepancy between the several-fold elevation in activation and proliferation, but only a net doubling in cell number, suggests that the steady state is balanced by increased apoptosis of CD8+ T lymphocytes. Furthermore, mathematical analysis of BrdU incorporation suggests that HIV-1 or SIV infection enlarges a pool of high-turnover CD8+ T cells that are both produced and disappear at an increased rate, compared with CD8+ T cells in uninfected controls.[89,93] It is usually assumed that decay of labeled cells equates to cell death, but this was not found to be significantly elevated in one study,[89] and an alternative explanation may be trafficking of labeled cells[14] to tissues that are not sampled.

The cause of the increased activation and proliferation of CD8+ T cells is not clear but may be critical to the later survival of the cells. Studies in mouse models of acute viral infection would suggest that a large proportion of the activated, proliferating CD8+ T cells are antigen specific.[12,13] Also, in the case of acute EBV infection, up to 50% of the CD8+ T cells have been shown to be antigen specific.[94] HIV-1 antigen-specific CD8+ T cells have an activated CD38+ phenotype[95,96] but numerically seem to be a small subset of the population of activated CD8+ T cells, even when comprehensive overlapping peptide sets for all HIV-1 proteins are used to stimulate peripheral blood mononuclear cells (PBMCs).[97,98]

It is possible that antigen-specific cells may be underestimated due to a number of factors: rapid divergence of autologous virus epitopes from consensus sequences[99,100]; responses to cryptic, frame-shift epitopes[101]; or antigen-specific cells' functions other than IFN-γ secretion.[102,103] It is plausible that the increasing heterogeneity of viral quasispecies during the asymptomatic stage of HIV-1 infection[104,105] continually drives constant recruitment of new antigen-specific CD8+ T cells.

Alternatively, nonspecific effects of cytokines may drive CD8 activation and proliferation. Previously, it was shown that IFN type I nonspecifically induced CD8+ T cell turnover in mice, and this was dependent on IL-15 production.[68] IFN type I production is typically elevated in viral infections, and there is a reported increase of mRNA for IFN type I target genes in lymph nodes during SIV infection,[106] consistent with recruitment of activated plasmacytoid dendritic cells[71] to these sites.[107] There are two reports that the plasma level of IL-15 is increased in HIV-1 infection,[108,109] but other reports have suggested that production of IL-15 may actually be decreased.[110,111] Consistent with the latter possibility, it has been found that the number of circulating, IFN type I–producing plasmacytoid dendritic cells is decreased in progressive HIV-1 infection.[112]

Overall, the increased activation and proliferation of CD8+ T cells during chronic HIV-1 infection may be a composite of antigen- and cytokine-driven processes.

DYNAMICS OF CD8+ T LYMPHOCYTES IN NONPATHOGENIC CHRONIC HIV-1 AND SIV INFECTION

Long-term nonprogressors[113,114] and subjects infected with less pathogenic HIV-2[115] exhibit reduced activation of CD8+ T cells. Simlarly, sooty mangabeys and chimpanzees, which are relatively unaffected by simian immunodeficiency virus (SIV) and HIV infection, respectively, have less activation of CD8+ T cells than rhesus macaques, which rapidly lose their CD4+ T cells when infected with SIV.[116–118] Importantly, in the case of sooty mangabeys, the lower levels of CD8 activation and stable CD4 cell counts are found despite relatively high plasma viral loads.[117,119] Altogether, these observations suggest that CD8 T cell activation may be an important part of the pathogenesis of HIV-1 infection.

DYNAMICS OF CD8+ T LYMPHOCYTES IN PRIMARY HIV-1 INFECTION

After exposure to HIV-1 and an incubation period of 2 to 3 weeks, the acute onset of symptoms is associated with a peak of viral load and commonly a brief lymphopenia at presentation,[120] involving both CD4[+] and CD8[+] T cells. The lymphopenia may result from shutdown of efferent lymph,[121] because there is a rapid increase in the lymphocyte count in the days afterward.[120,122] The CD4[+] T cell count exhibits a nadir coinciding with the peak of viremia and remains depressed in approximately 50% of subjects, but the CD8[+] T cell count progressively increases to a peak approximately 30 days after the onset of symptoms.[122] Thereafter, the CD8[+] T cell count remains elevated, resulting in the typical long-term inversion of the CD4:CD8 ratio, although the CD8[+] T cell count decreases just before the onset of AIDS.[123,81] In nonpathogenic primate models, despite intense viral replication during the acute phase, there are no major long-term changes in CD4[+] or CD8[+] T cell counts in blood or lymph nodes,[117,124] compared with pathogenic primate models.[125]

Symptomatic acute EBV infection in adolescents and adults also results in a very large increase in the CD8[+] T cell count. Unlike HIV-1 infection, the activation of CD8[+] T cells returns to normal in the vast majority of patients for the next 16 weeks of follow-up.[126]

In both of these acute infections, the increase in CD8[+] T cells is due to the appearance of highly activated, proliferating memory effector phenotype cells.[11,127,128] However, in acute HIV-1 infection, only a small proportion of the increased CD8[+] T cells seems to be antigen specific in most subjects.[129,130] Whether the activation and proliferation are antigen specific or cytokine driven is unclear, similar to the situation in chronic infection, as described above.

DYSFUNCTION OF CD8[+] T CELLS IN HIV-1 INFECTION

The impairment of HIV-specific CD4[+] T cell helper cells late in acute infection[131] probably affects the state of differentiation of antigen-specific CD8[+] T cells. The evidence suggests that CD8[+] effector T cells control high levels of viremia during resolution of acute HIV[132,133] and SIV[134] infections, but they are typically unable to completely prevent ongoing replication.

It has been suggested that HIV-specific CD8[+] CTLs are functionally impaired in chronic HIV infection, with low expression of perforin and possible alterations in T cell antigen receptor (TCR) signaling[6,7] as well as a state of differentiation of HIV-specific CD8[+] T cells that is quite different to comparable cytomegalovirus (CMV)- and EBV-specific cells.[135,136] The production of IL-2 by CD4[+] T cells is characteristically reduced in untreated HIV-1 infection,[137] particularly in antigen-specific CD4[+] T cells,[138,139] and lack of IL-2 may be the cause of impaired CTL differentiation[135,136] and perforin expression.[140]

Consistent with this possibility, long-term nonprogressors maintain antigen-specific CD4[+] T cell responses,[141] and their CD8[+] T cells proliferate and express perforin, in response to HIV antigen, unlike cells from progressing subjects.[142] However, treatment of HIV-infected subjects with IL-2 does not lead to an increase in circulating CD8[+] T lymphocytes or an increase in HIV antigen-specific responses.[143,144]

In the absence of CD4[+] T cell help, short-term CD8[+] effector cells may be constantly generated in HIV-1 infection, but these cells may be preprogrammed to be senescent.[145,146] In fact, early HIV-1 infection is characterized by an increased proportion of CD8[+] T cells that are CD28-negative CD57[+],[147] but such cells do not proliferate in vitro and are particularly prone to apoptosis.[148] Therefore, apoptosis as a result of incomplete differentiation during acute infection may be responsible for inadequate antiviral CD8[+] T cell activity during HIV and SIV infections.

INCREASED RATE OF CD8[+] T CELL APOPTOSIS IN HIV-1 INFECTION

ELEVATED CD8[+] T CELL APOPTOSIS IN VIVO

An increase in the number of apoptotic CD8[+] T cells has been observed in lymph nodes from chronically infected HIV[+] subjects, and has been related to the state of activation of the lymph nodes[149] and the reduction in Bcl-2 expression.[150] Importantly, in one study, apoptosis was found to mostly involve uninfected T cells.[151] These results suggest that activation of CD8+ T cells during HIV-1 infection predisposes the CD8+ T cells to an indirect apoptotic effect *in vivo*.

In acute SIV infection, increased apoptosis of T cells in lymphoid tissue has been observed.[125,152,153] In one of these studies, much greater lymphocyte apoptosis was observed very early in infection with pathogenic virus compared with infection with nonpathogenic virus strain.[152] In another study, a greater ratio of apoptotic to cycling (Ki-67+) lymphocytes in the T cell areas of lymph nodes was observed in macaques that went on to progress rapidly, compared with a lower ratio of apoptotic cells to cycling cells in the lymph nodes of macaques that subsequently progressed more slowly.[125] Unfortunately, no further phenotyping of the apoptotic cells was possible.

These results suggest that early apoptosis may influence later pathogenesis, particularly if antiviral CD8+ T lymphocytes inappropriately die. It is important to consider that HIV-1 infection is most active in lymphoid tissues, and CD8+ T cells undergoing programmed cell death in these tissues will be rapidly phagocytosed[154] and will not return to the circulation. Therefore, investigation of CD8+ T cell apoptosis in lymphoid tissues may be important, even though most studies have examined the propensity of peripheral blood CD8+ T cells to undergo apoptosis in *ex vivo* culture.

Increased Rate of Spontaneous Apoptosis *Ex Vivo*

Several early studies described abnormal levels of apoptosis in cultures of CD4+ and CD8+ T cells from HIV-1-infected individuals *in vitro*.[116,155–161] In particular, increased spontaneous cell death was associated with decreased expression of Bcl-2,[162,163] as previously reported for spontaneous apoptosis of CD8+ T cells during acute EBV infection.[10] Furthermore, incubation of cells with IL-2[162] or IL-15[164,165] simultaneously increased intracellular Bcl-2 levels and led to reduced spontaneous apoptosis.

The phenotype of the cells undergoing apoptosis showed that they were highly activated *ex vivo*.[161,163,166] Similarly, during acute HIV-1 infection, incubation with IL-2 or IL-15 increased Bcl-2 expression and reduced spontaneous apoptosis of highly activated CD8+ T cells.[167] Furthermore, in this study, decreased Bcl-2 levels were associated with decreased expression of the IL-7 receptor, and both changes were observed in Ki-67+ CD8+ T cells.[167] A recent study directly addressed the question of whether antigen-specific CD8+ T lymphocytes undergo spontaneous apoptosis *ex vivo*.[168] It was found that Bcl-2 was decreased in antigen-specific CD8+ T cells, which was associated with relatively increased spontaneous apoptosis—this could be inhibited by incubation with IL-15, which also induced an increase in intracellular Bcl-2.

Interestingly, a recent study showed an increase in the intracellular concentration of the proapoptotic molecules Bim and Bak in T cells from SIV+ macaques.[169] In previous studies, it was found that proapoptotic Bax was not significantly increased in circulating T cells from HIV-infected individuals.[170,171] The decrease in Bcl-2, combined with an increase in Bim and its association with spontaneous apoptosis,[169] is consistent with studies showing the nonredundant role of Bim in the decrease of effector CD8+ T cells after viral infection.[43] Disruption to mitochondrial membrane potential, where Bim and Bak have their effects,[49] was widely reported in PBMC during chronic HIV-1[172,173] and SIV infection[169] and during acute HIV-1 infection.[174]

Taken together, these results suggest that spontaneous apoptosis is primarily related to cytokine regulation of Bcl-2 family members and involves the intrinsic mitochondrial pathway, which is consistent with the cytokine control of antigen-specific effector and memory CD8+ T cell numbers observed in murine models.[62] However, the observation that forced expression of Bcl-2 in transgenic mice did not prevent apoptosis of effector CD8+ T cells[40,41] does not fit this hypothesis. It is possible that Mcl-1 expression[44] may be more important than Bcl-2, although this has not yet been studied.

The continued spontaneous apoptosis after resolution of acute infection seems to reflect continued activation of CD8+ T cells, which is characteristic of untreated HIV-1 infection but is not observed after acute EBV infection.[126] It is not clear whether the spontaneous apoptosis observed *in vitro* is an exact correlate of the apoptosis observed in lymph nodes, but it is clearly related to the increased activation of CD8[+] T cells, which is, in turn, clearly associated with CD4[+] T cell decline (see above).

APOPTOSIS IN NONPATHOGENIC HIV-1 AND SIV INFECTIONS AND AFTER HAART

The level of spontaneous apoptosis and mitochondrial dysfunction of CD4[+] and CD8[+] T cells is lower in HIV[+] long-term nonprogressors than in subjects with rapidly progressive disease.[173,175] Similarly, spontaneous apoptosis of CD8[+] T cells in nonpathogenic SIV infection in sooty mangabeys or HIV-1 infection in chimpanzees is not significantly different compared with uninfected controls.[116,117,152,176]

Similarly, initiation of HAART leads to a reduction in the level of spontaneous apoptosis as well as CD95-induced apoptosis (see below).[177–179] These results are consistent with the observed decrease in activation of CD8[+] T cells observed after antiretroviral therapy[86,87] and is associated with increases in Bcl-2 expression.[171]

ROLE OF CD95/CD95L INTERACTIONS IN ELEVATED APOPTOSIS OF CD8[+] T CELLS IN HIV-1 INFECTION

Many studies have demonstrated an increased expression of CD95 on the surface of circulating CD8[+] T cells during HIV-1 infection and have demonstrated that these cells are highly sensitive to treatment with anti-CD95 antibodies and readily undergo apoptosis, compared to cells from uninfected controls.[180–183] However, memory CD8[+] T cells in uninfected controls also express CD95 and, therefore, some additional change must account for the increased sensitivity during HIV-1 infection. A possible candidate is FLICE inhibitory protein (FLIP), which is believed to inhibit CD95-induced apoptosis,[28] but it has been reported that there is no difference in its expression between HIV[+] subjects and uninfected controls.[169,184]

Several studies have suggested that, as with spontaneous apoptosis, increased sensitivity to CD95 ligation is related to decreased intracellular Bcl-2 and can be counteracted by incubation with IL-2[185] or IL-15.[186] IL-12 prevents Fas-mediated apoptosis of CD4[+] T cells from HIV-infected persons,[181] but IL-10 prevents Fas-mediated apoptosis of CD8[+] T cells from HIV-infected persons and has no preventive effect on CD4 T cell death.[185] These observations are reminiscent of previous studies that show that IL-10 can rescue CD8[+] T cells from apoptotic cell death.[187,188] Altogether, these results are consistent with the suggestion that the intrinsic and extrinsic pathways cooperate in terminating immune responses.[49]

The signaling cascades activated by IL-2 and IL-15 differ from that used by IL-10. IL-2 and IL-15 receptors share a common γ-chain and use a common JAK3/STAT5 signaling pathway.[189] Moreover, the activation of STAT5 by IL-2 has been reported to be involved in T cell proliferation and activation-induced cell death (AICD), whereas the IL-2Rß chain has been reported to be involved in the activation of the protein kinase Akt and Bcl-2 production.[55] The binding of IL-10 to its receptor complex activates the JAK1 and Tyk2 kinases and involves STAT3 and STAT1, similar to interferon receptors.[190–192]

Serum levels of soluble CD95L are elevated in HIV-1 infection compared to uninfected controls,[193] but it is not known if this translates into increased apoptosis of CD95[+] CD8[+] T cells in vivo.[194] Soluble CD95L is less able to induce apoptosis than the membrane-bound form,[195] and this may simply reflect increased expression of CD95L in HIV-1 and SIV infection.[169,196] HIV- and SIV-infected cells reportedly have increased expression of CD95L,[169,197–199] and this may be important in protecting infected cells from lysis by CD95[+] CD8[+] CTL. However, back-killing of CTLs has not yet been reported, but new methods of determining CTL function by using flow cytometric detection of intracellular caspase activation[200] may be informative.

ROLE OF TNF/TNFR INTERACTIONS AND TRAIL IN APOPTOSIS OF CD8+ T LYMPHOCYTES IN HIV-1 INFECTION

In HIV-infected individuals, it has been found that the addition of TNF-α does not enhance spontaneous cell death.[185] A recent report described increased apoptosis of CD8+ T cells from HIV+ subjects mediated by antibodies to TNFR1 and TNFR2.[201] Decreased levels of intracellular Bcl-2 and increased levels of activated caspases were also associated with this activity.

Some reports also suggest a role for TRAIL,[202–204] although this has not been observed in all studies.[205]

The level of TNF-α in serum is characteristically increased in HIV-1 infection[206] and is implicated in viral replication.[207] Furthermore, decreasing CD4+ T cell counts are most closely correlated with plasma TNFR2 levels, which are correlated with CD8+ T cell activation.[208] Therefore, TNF-α seems to play a central part in HIV-1 pathogenesis, although treatment with inhibitors of TNF-α activity did not show any effect on plasma HIV RNA,[209] and an effect on apoptosis was not reported in this study.

ROLE OF AICD IN APOPTOSIS OF CD8+ T LYMPHOCYTES IN HIV-1 INFECTION

The original studies of apoptosis in HIV-1 infection demonstrated increased AICD,[116,156,160] possibly resulting from upregulation of CD95L expression.[59] Whereas CD95 ligation–mediated CD8 T cell death can be prevented by pretreatment with synthetic peptide inhibitors caspases, AICD mediated by TCR stimulation was not found to be prevented by an antagonistic Fas mAb or by a pan caspase inhibitor, zVAD-fmk,[185,210] suggesting that TCR-mediated CD8 T cell death occurs through Fas- and caspase-independent pathways. Moreover, the addition of antibodies to IL-10 or the addition of IL-12 has a preventive effect on abnormal programmed cell death induction in response to *in vitro* stimulation in PBMC from HIV-infected subjects.[159,181]

It was recently found that stimulation through the TCR leads to increased expression of proapoptotic Bim,[211] which may be responsible for the ensuing apoptosis.[49] Therefore, it seems that AICD may be regulated by exogenous cytokines and the intrinsic mitochondrial pathway, consistent with a role for Bcl-2 in regulation of AICD in murine models.[41]

ROLE OF HIV-1 PROTEINS IN ELEVATED APOPTOSIS OF CD8+ T CELLS IN HIV-1 INFECTION

The envelope glycoprotein complex (Env) seems to be one of the dominant apoptosis-inducing molecules encoded by the HIV-1 genome. Incubation of resting PBMCs from healthy donors with HIV, even in the presence of an inhibitor of the viral replication, has been sufficient to prime both CD4+ and CD8+ T cells to die by apoptosis in response to Fas ligation.[212–214] Therefore, as most of the HIV particles produced are noninfectious, the simple fixation or penetration of viruses, without integration, may be sufficient to prime T cells for Fas-mediated apoptosis in quiescent T cells. Furthermore, virions also contain host membrane proteins with immunomodulatory activity.[215] However, it has been reported that increased Fas sensitivity of CD8+ T cells from HIV-infected individuals and SIV-infected macaques did not correlate with plasma viral load.[169,216]

HIV Env binding to CXCR4 has been reported to induce apoptosis of uninfected CD8+ T cells *in vitro*,[217] and infected macrophages have been reported to be capable of inducing CD8+ T cell apoptosis through a pathway involving TNF/TNFR.[218] However, the binding to a co-receptor by Env glycoproteins from clinical isolates is rarely CD4 independent.[219] Also, most HIV quasispecies, especially during the early phases of infection, use CCR5 as co-receptor.[104,220] Therefore, a significant direct effect of Env on CD8+ T cells *in vivo*, particularly during acute infection, may not be plausible. It must be remembered, though, that a feature of activated effector CD8+ T cells during primary HIV-1 infection is upregulation of CCR5, and these are the cells that undergo spontaneous apoptosis *in vitro*.[167]

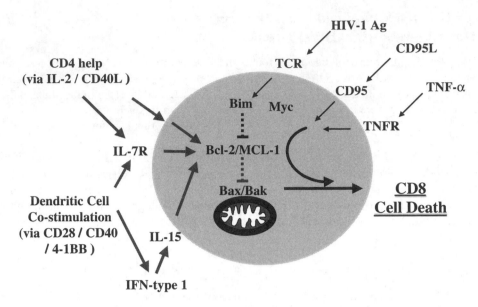

FIGURE 14.1 Proposed model for elevated apoptosis of CD8+ T lymphocytes in HIV-1 infection. Apoptosis is regulated by the balance between the anti-apoptotic effect of γ-chain cytokines and co-stimulation of up-regulation of antiapoptotic Bcl-2 family members versus the pro-apoptotic effect of antigen-driven up-regulation of Bim and Myc and down-regulation of Bcl-2. Chronic antigenic stimulation also down-regulates the IL-7R and up-regulates CD95, CD95L, TNFR1, and TNF-α. Processes that increased during HIV-1 infection (HIV-1 Ag, CD95L, TNF-α, Bim, Myc, CD95, and TNFR) are shown in the upper right quadrant of the figure, whereas processes are shown in left half of the figure.

OTHER HIV-RELATED MECHANISMS

Most reported effects of HIV-1 proteins on apoptosis apply to infected cells, and CD8[+] T cells do not seem to be significantly infected *in vivo*,[221] presumably due to lack of CD4 expression. However, there may be upregulation of CD4 after activation of CD8[+] T cells *in vitro*[222]; *in vivo*, it was shown that CD8[+] T cells, which also express CD4, are infected early in the course of infection.[223]

Exogenous Tat reportedly increases apoptosis of bystander cells[224,225] but at relatively high concentrations.[224] However, at such concentrations, Tat also was reported to have antiapoptotic effects.[226] Therefore, the overall effect of Tat remains to be defined (Figure 14.1).

SUMMARY

Survival of CD8[+] T cells under the normal conditions of homeostasis is primarily regulated by the effect of common γ-chain cytokines on expression of Bcl-2 family members, especially Bcl-2 itself. Antigenic activation in HIV-1 infection seems to upregulate Bim and decrease both IL-7R and Bcl-2, leading to a state in which CD8[+] T cells are particularly sensitive to proapoptotic signals. A feature of chronic HIV-1 infection is an increase in plasma levels of soluble forms of CD95, CD95L, TNF-α, TNFR2, and IFN-γ, all of which may also be involved in feedback control of immune responses.

The resulting increased rate of apoptosis is directly related to increased turnover of activated, proliferating cells, which is also seen in other viral infections at acute stages but usually resolves as the virus is cleared. HIV antigen–driven chronic activation and proliferation occurs in an environment of greatly reduced CD4[+] T cell help, as well as decreased dendritic cell numbers and function. Because longevity of memory CD8[+] T cells is determined during the activation phase, HIV-specific CD8[+] T cells may be programmed to contain lower amounts of antiapoptotic Bcl-2 family members and proceed rapidly toward senescence and apoptosis. Loss of effector CD8$^+$ T cells with antiviral activity probably contributes significantly to disease progression.

ACKNOWLEDGMENTS

JC holds a Canada Research Chair in Medical Genomics. This research is supported by the San Diego Center for AIDS Research and the Veterans Medical Research Foundation.

REFERENCES

1. McMichael, A.J. and Rowland-Jones, S.L., Cellular immune responses to HIV, *Nature,* 410 (6831), 980–987, 2001.
2. Klein, M.R., van Baalen, C.A., Holwerda, A.M., Kerkhof Garde, S.R., Bende, R.J., Keet, I.P.M., Eeftinck-Schattenkerk, J.-K.M., Osterhaus, A.D.M.E., Schuitemaker, H., and Miedema, F., Kinetics of Gag-specific cytotoxic T lymphocyte responses during the clinical course of HIV-1 infection: a longitudinal analysis of rapid progressors and long-term asymptomatics, *J. Exp. Med.,* 181 (4), 1365–1372, 1995.
3. Pantaleo, G. and Fauci, A.S., Immunopathogenesis of HIV infection, *Annu. Rev. Microbiol.,* 50 (18), 825–854, 1996.
4. Pantaleo, G., Soudeyns, H., Demarest, J.F., Vaccarezza, M., Graziosi, C., Paolucci, S., Daucher, M., Cohen, O.J., Denis, F., Biddison, W.E., Sekaly, R.P., and Fauci, A.S., Evidence for rapid disappearance of initially expanded HIV-specific CD8+ T cell clones during primary HIV infection, *Proc. Natl. Acad. Sci. U.S.A.,* 94 (18), 9848–9853, 1997.
5. Tenner-Racz, K., Racz, P., Thome, C., Meyer, C.G., Anderson, P.J., Schlossman, S.F., and Letvin, N.L., Cytotoxic effector cell granules recognized by the monoclonal antibody TIA-1 are present in CD8+ lymphocytes in lymph nodes of human immunodeficiency virus-1-infected patients, *Am. J. Pathol.,* 142 (6), 1750–1758, 1993.
6. Appay, V., Nixon, D.F., Donahoe, S.M., Gillespie, G.M., Dong, T., King, A., Ogg, G.S., Spiegel, H.M., Conlon, C., Spina, C.A., Havlir, D.V., Richman, D.D., Waters, A., Easterbrook, P., McMichael, A.J., and Rowland-Jones, S.L., HIV-specific CD8(+) T cells produce antiviral cytokines but are impaired in cytolytic function, *J. Exp. Med.,* 192 (1), 63–75, 2000.
7. Lieberman, J., Shankar, P., Manjunath, N., and Andersson, J., Dressed to kill? A review of why antiviral CD8 T lymphocytes fail to prevent progressive immunodeficiency in HIV-1 infection, *Blood,* 98 (6), 1667–1677, 2001.
8. Landay, A.L., Mackewicz, C.E., and Levy, J.A., An activated CD8+ T cell phenotype correlates with anti-HIV activity and asymptomatic clinical status, *Clin. Immunol. and Immunopathol.,* 69 (1), 106–116, 1993.
9. Uehara, T., Miyawaki, T., Ohta, K., Tamaru, Y., Yokoi, T., Nakamura, S., and Taniguchi, N., Apoptotic cell death of primed CD45RO+ T lymphocytes in Epstein–Barr virus-induced infectious mononucleosis, *Blood,* 80 (2), 452–458, 1992.
10. Akbar, A.N., Borthwick, N., Salmon, M., Gombert, W., Bofill, M., Shamsadeen, N., Pilling, D., Pett, S., Grundy, J.E., and Janossy, G., The significance of low Bcl-2 expression by CD45RO T cells in normal individuals and patients with acute viral infections. The role of apoptosis in T cell memory, *J. Exp. Med.,* 178 (2), 427–438, 1993.
11. Callan, M.F., Fazou, C., Yang, H., Rostron, T., Poon, K., Hatton, C., and McMichael, A.J., CD8(+) T-cell selection, function, and death in the primary immune response *in vivo, J. Clin. Invest.,* 106 (10), 1251–1261, 2000.
12. Butz, E.A. and Bevan, M.J., Massive expansion of antigen-specific CD8+ T cells during an acute virus infection, *Immunity,* 8 (2), 167–175, 1998.
13. Murali-Krishna, K., Altman, J.D., Suresh, M., Sourdive, D.J., Zajac, A.J., Miller, J.D., Slansky, J., and Ahmed, R., Counting antigen-specific CD8 T cells: a reevaluation of bystander activation during viral infection, *Immunity,* 8 (2), 177–187, 1998.
14. Masopust, D., Vezys, V., Marzo, A.L., and Lefrancois, L., Preferential localization of effector memory cells in nonlymphoid tissue, *Science,* 291 (5512), 2413–2417, 2001.

15. Marshall, D.R., Turner, S.J., Belz, G.T., Wingo, S., Andreansky, S., Sangster, M.Y., Riberdy, J.M., Liu, T., Tan, M., and Doherty, P.C., Measuring the diaspora for virus-specific CD8+ T cells, *Proc. Natl. Acad. Sci. U.S.A.,* 98 (11), 6313–6318, 2001.

16. Roberts, A.D. and Woodland, D.L., Cutting edge: effector memory CD8+ T cells play a prominent role in recall responses to secondary viral infection in the lung, *J. Immunol.,* 172 (11), 6533–6537, 2004.

17. Huang, L., Soldevila, G., Leeker, M., Flavell, R., and Crispe, I.N., The liver eliminates T cells undergoing antigen-triggered apoptosis *in vivo, Immunity,* 1 (9), 741–749, 1994.

18. Lauvau, G., Vijh, S., Kong, P., Horng, T., Kerksiek, K., Serbina, N., Tuma, R.A., and Pamer, E.G., Priming of memory but not effector CD8 T cells by a killed bacterial vaccine, *Science,* 294 (5547), 1735–1739, 2001.

19. Sun, J.C. and Bevan, M.J., Defective CD8 T cell memory following acute infection without CD4 T cell help, *Science,* 300 (5617), 339–342, 2003.

20. Shedlock, D.J. and Shen, H., Requirement for CD4 T cell help in generating functional CD8 T cell memory, *Science,* 300 (5617), 337–339, 2003.

21. Cooper, D., Bansal-Pakala, P., and Croft, M., 4-1BB (CD137) controls the clonal expansion and survival of CD8 T cells *in vivo* but does not contribute to the development of cytotoxicity, *Eur. J. Immunol.,* 32 (2), 521–529, 2002.

22. Schluns, K.S., Williams, K., Ma, A., Zheng, X.X., and Lefrancois, L., Cutting edge: requirement for IL-15 in the generation of primary and memory antigen-specific CD8 T cells, *J. Immunol.,* 168 (10), 4827–4831, 2002.

23. Wherry, E.J., Teichgraber, V., Becker, T.C., Masopust, D., Kaech, S.M., Antia, R., von Andrian, U.H., and Ahmed, R., Lineage relationship and protective immunity of memory CD8 T cell subsets, *Nat. Immunol.,* 4 (3), 225–234, 2003.

24. Sprent, J. and Surh, C.D., T cell memory, *Annu. Rev. Immunol.,* 20, 551–579, 2002.

25. Selin, L.K., Lin, M.Y., Kraemer, K.A., Pardoll, D.M., Schneck, J.P., Varga, S.M., Santolucito, P.A., Pinto, A.K., and Welsh, R.M., Attrition of T cell memory: selective loss of LCMV epitope-specific memory CD8 T cells following infections with heterologous viruses, *Immunity,* 11 (6), 733–742, 1999.

26. McNally, J.M., Zarozinski, C.C., Lin, M.Y., Brehm, M.A., Chen, H.D., and Welsh, R.M., Attrition of bystander CD8 T cells during virus-induced T-cell and interferon responses, *J. Virol.,* 75 (13), 5965–5976, 2001.

27. Freitas, A.A. and Rocha, B., Population biology of lymphocytes: the flight for survival, *Annu. Rev. Immunol.,* 18, 83–111, 2000.

28. Wallach, D., Varfolomeev, E.E., Malinin, N.L., Goltsev, Y.V., Kovalenko, A.V., and Boldin, M.P., Tumor necrosis factor receptor and Fas signaling mechanisms, *Annu. Rev. Immunol.,* 17, 331–367, 1999.

29. Reich, A., Korner, H., Sedgwick, J.D., and Pircher, H., Immune down-regulation and peripheral deletion of CD8 T cells does not require TNF receptor–ligand interactions nor CD95 (Fas, APO-1), *Eur. J. Immunol.,* 30 (2), 678–682, 2000.

30. Nguyen, L.T., McKall-Faienza, K., Zakarian, A., Speiser, D.E., Mak, T.W., and Ohashi, P.S., TNF receptor 1 (TNFR1) and CD95 are not required for T cell deletion after virus infection but contribute to peptide-induced deletion under limited conditions, *Eur. J. Immunol.,* 30 (2), 683–688, 2000.

31. Sytwu, H.K., Liblau, R.S., and McDevitt, H.O., The roles of Fas/APO-1 (CD95) and TNF in antigen-induced programmed cell death in T cell receptor transgenic mice, *Immunity,* 5 (1), 17–30, 1996.

32. Turner, S.J., La Gruta, N.L., Stambas, J., Diaz, G., and Doherty, P.C., Differential tumor necrosis factor receptor 2-mediated editing of virus-specific CD8+ effector T cells, *Proc. Natl. Acad. Sci. U.S.A.,* 101 (10), 3545–3550, 2004.

33. Fisher, G.H., Rosenberg, F.J., Straus, S.E., Dale, J.K., Middleton, L.A., Lin, A.Y., Strober, W., Lenardo, M.J., and Puck, J.M., Dominant interfering Fas gene mutations impair apoptosis in a human autoimmune lymphoproliferative syndrome, *Cell,* 81 (6), 935–946, 1995.

34. Chun, H.J., Zheng, L., Ahmad, M., Wang, J., Speirs, C.K., Siegel, R.M., Dale, J.K., Puck, J., Davis, J., Hall, C.G., Skoda-Smith, S., Atkinson, T.P., Straus, S.E., and Lenardo, M.J., Pleiotropic defects in lymphocyte activation caused by caspase-8 mutations lead to human immunodeficiency, *Nature,* 419 (6905), 395–399, 2002.

35. Grayson, J.M., Zajac, A.J., Altman, J.D., and Ahmed, R., Cutting edge: increased expression of Bcl-2 in antigen-specific memory CD8+ T cells, *J. Immunol.,* 164 (8), 3950–3954, 2000.

36. Grayson, J.M., Murali-Krishna, K., Altman, J.D., and Ahmed, R., Gene expression in antigen-specific CD8+ T cells during viral infection, *J. Immunol.*, 166 (2), 795–799, 2001.

37. Kaech, S.M., Tan, J.T., Wherry, E.J., Konieczny, B.T., Surh, C.D., and Ahmed, R., Selective expression of the interleukin 7 receptor identifies effector CD8 T cells that give rise to long-lived memory cells, *Nat. Immunol.*, 4 (12), 1191–1198, 2003.

38. von Freeden-Jeffry, U., Solvason, N., Howard, M., and Murray, R., The earliest T lineage-committed cells depend on IL-7 for Bcl-2 expression and normal cell cycle progression, *Immunity*, 7 (1), 147–154, 1997.

39. Strasser, A., O'Connor, L., and Dixit, V.M., Apoptosis signaling, *Annu. Rev. Biochem.*, 69, 217–245, 2000.

40. Razvi, E.S., Jiang, Z., Woda, B.A., and Welsh, R.M., Lymphocyte apoptosis during the silencing of the immune response to acute viral infections in normal, Lpr, and Bcl-2-transgenic mice, *Am. J. Pathol.*, 147 (1), 79–91, 1995.

41. Petschner, F., Zimmerman, C., Strasser, A., Grillot, D., Nunez, G., and Pircher, H., Constitutive expression of Bcl-xL or Bcl-2 prevents peptide antigen-induced T cell deletion but does not influence T cell homeostasis after a viral infection, *Eur. J. Immunol.*, 28 (2), 560–569, 1998.

42. O'Connor, L., Strasser, A., O'Reilly, L.A., Hausmann, G., Adams, J.M., Cory, S., and Huang, D.C., Bim: a novel member of the Bcl-2 family that promotes apoptosis, *EMBO J.*, 17 (2), 384–395, 1998.

43. Pellegrini, M., Belz, G., Bouillet, P., and Strasser, A., Shutdown of an acute T cell immune response to viral infection is mediated by the proapoptotic Bcl-2 homology 3-only protein Bim, *Proc. Natl. Acad. Sci. U.S.A.*, 100 (24), 14175–14180, 2003.

44. Opferman, J.T., Letai, A., Beard, C., Sorcinelli, M.D., Ong, C.C., and Korsmeyer, S.J., Development and maintenance of B and T lymphocytes requires antiapoptotic MCL-1, *Nature*, 426 (6967), 671–676, 2003.

45. Leu, J.I., Dumont, P., Hafey, M., Murphy, M.E., and George, D.L., Mitochondrial p53 activates Bak and causes disruption of a Bak–Mcl1 complex, *Nat. Cell. Biol.*, 6 (5), 443–450, 2004.

46. Lindsten, T., Ross, A.J., King, A., Zong, W.X., Rathmell, J.C., Shiels, H.A., Ulrich, E., Waymire, K.G., Mahar, P., Frauwirth, K., Chen, Y., Wei, M., Eng, V.M., Adelman, D.M., Simon, M.C., Ma, A., Golden, J.A., Evan, G., Korsmeyer, S.J., MacGregor, G.R., and Thompson, C.B., The combined functions of proapoptotic Bcl-2 family members Bak and Bax are essential for normal development of multiple tissues, *Mol. Cell*, 6 (6), 1389–1399, 2000.

47. Wei, M.C., Zong, W.X., Cheng, E.H., Lindsten, T., Panoutsakopoulou, V., Ross, A.J., Roth, K.A., MacGregor, G.R., Thompson, C.B., and Korsmeyer, S.J., Proapoptotic BAX and BAK: a requisite gateway to mitochondrial dysfunction and death, *Science*, 292 (5517), 727–730, 2001.

48. Cory, S., Huang, D.C., and Adams, J.M., The Bcl-2 family: roles in cell survival and oncogenesis, *Oncogene*, 22 (53), 8590–8607, 2003.

49. Marsden, V.S. and Strasser, A., Control of apoptosis in the immune system: Bcl-2, BH3-only proteins and more, *Annu. Rev. Immunol.*, 21, 71–105, 2003.

50. Evan, G. and Littlewood, T., A matter of life and cell death, *Science*, 281 (5381), 1317–1322, 1998.

51. Bouchard, C., Staller, P., and Eilers, M., Control of cell proliferation by Myc, *Trends Cell Biol.*, 8 (5), 202–206, 1998.

52. Smith, K.A., Interleukin 2, *Annu. Rev. Immunol.*, 2, 319–333, 1984.

53. D'Souza, W.N., Schluns, K.S., Masopust, D., and Lefrancois, L., Essential role for IL-2 in the regulation of antiviral extralymphoid CD8 T cell responses, *J. Immunol.*, 168 (11), 5566–5572, 2002.

54. D'Souza, W.N. and Lefrancois, L., IL-2 is not required for the initiation of CD8 T cell cycling but sustains expansion, *J. Immunol.*, 171 (11), 5727–5735, 2003.

55. Van Parijs, L., Refaeli, Y., Lord, J.D., Nelson, B.H., Abbas, A.K., and Baltimore, D., Uncoupling IL-2 signals that regulate T cell proliferation, survival, and Fas-mediated activation-induced cell death, *Immunity*, 11 (3), 281–288, 1999.

56. Ku, C.C., Murakami, M., Sakamoto, A., Kappler, J., and Marrack, P., Control of homeostasis of CD8+ memory T cells by opposing cytokines, *Science*, 288 (5466), 675–678, 2000.

57. Sadlack, B., Merz, H., Schorle, H., Schimpl, A., Feller, A.C., and Horak, I., Ulcerative colitis-like disease in mice with a disrupted interleukin-2 gene, *Cell*, 75 (2), 253–261, 1993.

58. Willerford, D.M., Chen, J., Ferry, J.A., Davidson, L., Ma, A., and Alt, F.W., Interleukin-2 receptor chain regulates the size and content of the peripheral lymphoid compartment, *Immunity*, 3 (4), 521–530, 1995.

59. Lenardo, M., Chan, K.M., Hornung, F., McFarland, H., Siegel, R., Wang, J., and Zheng, L., Mature T lymphocyte apoptosis — immune regulation in a dynamic and unpredictable antigenic environment, *Annu. Rev. Immunol.,* 17, 221–253, 1999.

60. Shevach, E.M., CD4+ CD25+ suppressor T cells: more questions than answers, *Nat. Rev. Immunol.,* 2 (6), 389–400, 2002.

61. Schluns, K.S. and Lefrancois, L., Cytokine control of memory T-cell development and survival, *Nat. Rev. Immunol.,* 3 (4), 269–279, 2003.

62. Marrack, P. and Kappler, J., Control of T cell viability, *Annu. Rev. Immunol.,* 22, 765–787, 2004.

63. Vella, A.T., Dow, S., Potter, T.A., Kappler, J., and Marrack, P., Cytokine-induced survival of activated T cells *in vitro* and *in vivo, Proc. Natl. Acad. Sci. U.S.A.,* 95 (7), 3810–3815, 1998.

64. Lodolce, J.P., Boone, D.L., Chai, S., Swain, R.E., Dassopoulos, T., Trettin, S., and Ma, A., IL-15 receptor maintains lymphoid homeostasis by supporting lymphocyte homing and proliferation, *Immunity,* 9 (5), 669–676, 1998.

65. Kennedy, M.K., Glaccum, M., Brown, S.N., Butz, E.A., Viney, J.L., Embers, M., Matsuki, N., Charrier, K., Sedger, L., Willis, C.R., Brasel, K., Morrissey, P.J., Stocking, K., Schuh, J.C., Joyce, S., and Peschon, J.J., Reversible defects in natural killer and memory CD8 T cell lineages in interleukin 15-deficient mice, *J. Exp. Med.,* 191 (5), 771–780, 2000.

66. Li, X.C., Demirci, G., Ferrari-Lacraz, S., Groves, C., Coyle, A., Malek, T.R., and Strom, T.B., IL-15 and IL-2: a matter of life and death for T cells *in vivo, Nat. Med.,* 7 (1), 114–118, 2001.

67. Wu, T.S., Lee, J.M., Lai, Y.G., Hsu, J.C., Tsai, C.Y., Lee, Y.H., and Liao, N.S., Reduced expression of Bcl-2 in CD8+ T cells deficient in the IL-15 receptor -chain, *J. Immunol.,* 168 (2), 705–712, 2002.

68. Zhang, X., Sun, S., Hwang, I., Tough, D.F., and Sprent, J., Potent and selective stimulation of memory-phenotype CD8+ T cells *in vivo* by IL-15, *Immunity,* 8 (5), 591–599, 1998.

69. Goldrath, A.W., Sivakumar, P.V., Glaccum, M., Kennedy, M.K., Bevan, M.J., Benoist, C., Mathis, D., and Butz, E.A., Cytokine requirements for acute and basal homeostatic proliferation of naive and memory CD8+ T cells, *J. Exp. Med.,* 195 (12), 1515–1522, 2002.

70. Tough, D.F., Borrow, P., and Sprent, J., Induction of bystander T cell proliferation by viruses and type 1 interferon *in vivo, Science,* 272, 1947–1950, 1996.

71. Siegal, F.P., Kadowaki, N., Shodell, M., Fitzgerald-Bocarsly, P.A., Shah, K., Ho, S., Antonenko, S., and Liu, Y.J., The nature of the principal type 1 interferon-producing cells in human blood, *Science,* 284 (5421), 1835–1837, 1999.

72. Marrack, P., Kappler, J., and Mitchell, T., Type I interferons keep activated T cells alive, *J. Exp. Med.,* 189 (3), 521–530, 1999.

73. Pilling, D., Akbar, A.N., Girdlestone, J., Orteu, C.H., Borthwick, N.J., Amft, N., Scheel-Toellner, D., Buckley, C.D., and Salmon, M., Interferon- mediates stromal cell rescue of T cells from apoptosis, *Eur. J. Immunol.,* 29 (3), 1041–1050, 1999.

74. Scheel-Toellner, D., Pilling, D., Akbar, A.N., Hardie, D., Lombardi, G., Salmon, M., and Lord, J.M., Inhibition of T cell apoptosis by IFN- rapidly reverses nuclear translocation of protein kinase C-, *Eur. J. Immunol.,* 29 (8), 2603–2612, 1999.

75. Shimoda, K., Kamesaki, K., Numata, A., Aoki, K., Matsuda, T., Oritani, K., Tamiya, S., Kato, K., Takase, K., Imamura, R., Yamamoto, T., Miyamoto, T., Nagafuji, K., Gondo, H., Nagafuchi, S., Nakayama, K., and Harada, M., Cutting edge: Tyk2 is required for the induction and nuclear translocation of Daxx which regulates IFN–induced suppression of B lymphocyte formation, *J. Immunol.,* 169 (9), 4707–4711, 2002.

76. Chen, Q., Gong, B., Mahmoud-Ahmed, A.S., Zhou, A., Hsi, E.D., Hussein, M., and Almasan, A., Apo2L/TRAIL and Bcl-2-related proteins regulate type I interferon-induced apoptosis in multiple myeloma, *Blood,* 98 (7), 2183–2192, 2001.

77. Badovinac, V.P., Tvinnereim, A.R., and Harty, J.T., Regulation of antigen-specific CD8+ T cell homeostasis by perforin and interferon-, *Science,* 290 (5495), 1354–1358, 2000.

78. Badovinac, V.P., Hamilton, S.E., and Harty, J.T., Viral infection results in massive CD8+ T cell expansion and mortality in vaccinated perforin-deficient mice, *Immunity,* 18 (4), 463–474, 2003.

79. Dalton, D.K., Haynes, L., Chu, C.Q., Swain, S.L., and Wittmer, S., Interferon eliminates responding CD4 T cells during mycobacterial infection by inducing apoptosis of activated CD4 T cells, *J. Exp. Med.,* 192 (1), 117–122, 2000.

80. Tesselaar, K., Arens, R., van Schijndel, G.M., Baars, P.A., van der Valk, M.A., Borst, J., van Oers, M.H., and van Lier, R.A., Lethal T cell immunodeficiency induced by chronic costimulation via CD27–CD70 interactions, *Nat. Immunol.*, 4 (1), 49–54, 2003.

81. Margolick, J.B., Munoz, A., Donnenberg, A.D., Park, L.P., Galai, N., Giorgi, J.V., Gorman, M.R.G., and Ferbas, J., for the Multicenter AIDS cohort study, Failure of T-cell homeostasis preceding AIDS in HIV-1 infection, *Nat. Med.*, 1 (7), 674–680, 1995.

82. Hellerstein, M., Hanley, M.B., Cesar, D., Siler, S., Papageorgopoulos, C., Wieder, E., Schmidt, D., Hoh, R., Neese, R., Macallan, D., Deeks, S., and McCune, J.M., Directly measured kinetics of circulating T lymphocytes in normal and HIV-1-infected humans, *Nat. Med.*, 5 (1), 83–89, 1999.

83. Lempicki, R.A., Kovacs, J.A., Baseler, M.W., Adelsberger, J.W., Dewar, R.L., Natarajan, V., Bosche, M.C., Metcalf, J.A., Stevens, R.A., Lambert, L.A., Alvord, W.G., Polis, M.A., Davey, R.T., Dimitrov, D.S., and Lane, H.C., Impact of HIV-1 infection and highly active antiretroviral therapy on the kinetics of CD4+ and CD8+ T cell turnover in HIV-infected patients, *Proc. Natl. Acad. Sci. U.S.A.*, 97 (25), 13778–13783, 2000.

84. Sachsenberg, N., Perelson, A.S., Yerly, S., Schockmel, G.A., Leduc, D., Hirschel, B., and Perrin, L., Turnover of CD4+ and CD8+ T lymphocytes in HIV-1 infection as measured by Ki-67 antigen, *J. Exp. Med.*, 187 (8), 1295–1303, 1998.

85. Giorgi, J.V. and Detels, R., T-cell subset alterations in HIV-infected homosexual men: NIAID Multicenter AIDS Cohort Study, *Clin. Immunol. Immunopathol.*, 52 (1), 10–18, 1989.

86. Kelleher, A.D., Carr, A., Zaunders, J., and Cooper, D.A., Alterations in the immune response of human immunodeficiency virus (HIV)-infected subjects treated with an HIV-specific protease inhibitor, ritonavir, *J. Infect. Dis.*, 173 (2), 321–329, 1996.

87. Gray, C.M., Schapiro, J.M., Winters, M.A., and Merigan, T.C., Changes in CD4+ and CD8+ T cell subsets in response to highly active antiretroviral therapy in HIV type 1-infected patients with prior protease inhibitor experience, *AIDS Res. Hum. Retroviruses*, 14 (7), 561–569, 1998.

88. Hazenberg, M.D., Stuart, J.W., Otto, S.A., Borletts, J.C., Boucher, C.A., de Boer, R.J., Miedema, F., and Hamann, D., T-cell division in human immunodeficiency virus (HIV)-1 infection is mainly due to immune activation: a longitudinal analysis in patients before and during highly active antiretroviral therapy (HAART), *Blood*, 95 (1), 249–255, 2000.

89. Mohri, H., Perelson, A.S., Tung, K., Ribeiro, R.M., Ramratnam, B., Markowitz, M., Kost, R., Hurley, A., Weinberger, L., Cesar, D., Hellerstein, M.K., and Ho, D.D., Increased turnover of T lymphocytes in HIV-1 infection and its reduction by antiretroviral therapy, *J. Exp. Med.*, 194 (9), 1277–1287, 2001.

90. Bouscarat, F., Levacher-Clergeot, M., Dazza, M.-C., Strauss, K.W., Girard, P.-M., Ruggeri, C., and Sinet, M., Correlation of CD8 lymphocyte activation with cellular viremia and plasma HIV RNA levels in asymptomatic patients infected by human immunodeficiency virus type 1, *AIDS Res. Hum. Retroviruses*, 12 (1), 17–24, 1996.

91. Zaunders, J.J., Cunningham, P.H., Kelleher, A.D., Kaufmann, G.R., Jaramillo, A.B., Wright, R., Smith, D., Grey, P., Vizzard, J., Carr, A., and Cooper, D.A., Potent antiretroviral therapy of primary human immunodeficiency virus type 1 (HIV-1) infection: partial normalization of T lymphocyte subsets and limited reduction of HIV-1 DNA despite clearance of plasma viremia, *J. Infect. Dis.*, 180 (2), 320–329, 1999.

92. Liu, Z., Cumberland, W.G., Hultin, L.E., Kaplan, A.H., Detels, R., and Giorgi, J.V., CD8+ T-lymphocyte activation in HIV-1 disease reflects an aspect of pathogenesis distinct from viral burden and immunodeficiency, *J. Acquir. Immune Defic. Syndr. Hum. Retrovirol.*, 18 (4), 332–340, 1998.

93. Kovacs, J.A., Lempicki, R.A., Sidorov, I.A., Adelsberger, J.W., Herpin, B., Metcalf, J.A., Sereti, I., Polis, M.A., Davey, R.T., Tavel, J., Falloon, J., Stevens, R., Lambert, L., Dewar, R., Schwartzentruber, D.J., Anver, M.R., Baseler, M.W., Masur, H., Dimitrov, D.S., and Lane, H.C., Identification of dynamically distinct subpopulations of T lymphocytes that are differentially affected by HIV, *J. Exp. Med.*, 194 (12), 1731–1741, 2001.

94. Callan, M.F., Tan, L., Annels, N., Ogg, G.S., Wilson, J.D., O'Callaghan, C.A., Steven, N., McMichael, A.J., and Rickinson, A.B., Direct visualization of antigen-specific CD8+ T cells during the primary immune response to Epstein–Barr virus *in vivo*, *J. Exp. Med.*, 187 (9), 1395–1402, 1998.

95. Ho, H.-N., Hutlin, L.E., Mitsuyasu, R.T., Matud, J.L., Hausner, M.A., Bockstoce, D., Chou, C.-C., O'Rourke, S., Taylor, J.M.G., and Giorgi, J.V., Circulating HIV-specific CD8+ cytotoxic T cells express CD38 and HLA-DR antigens, *J. Immunol.*, 150 (7), 3070–3079, 1993.

96. Ogg, G.S., Jin, X., Bonhoeffer, S., Moss, P., Nowak, M.A., Monard, S., Segal, J.P., Cao, Y., Rowland-Jones, S.L., Hurley, A., Markowitz, M., Ho, D.D., McMichael, A.J., and Nixon, D.F., Decay kinetics of human immunodeficiency virus-specific effector cytotoxic T lymphocytes after combination anti-retroviral therapy, *J. Virol.,* 73, 797–800, 1999.

97. Betts, M.R., Ambrozak, D.R., Douek, D.C., Bonhoeffer, S., Brenchley, J.M., Casazza, J.P., Koup, R.A., and Picker, L.J., Analysis of total human immunodeficiency virus (HIV)-specific CD4(+) and CD8(+) T-cell responses: relationship to viral load in untreated HIV infection, *J. Virol.,* 75 (24), 11983–11991, 2001.

98. Addo, M.M., Yu, X.G., Rathod, A., Cohen, D., Eldridge, R.L., Strick, D., Johnston, M.N., Corcoran, C., Wurcel, A.G., Fitzpatrick, C.A., Feeney, M.E., Rodriguez, W.R., Basgoz, N., Draenert, R., Stone, D.R., Brander, C., Goulder, P.J., Rosenberg, E.S., Altfeld, M., and Walker, B.D., Comprehensive epitope analysis of human immunodeficiency virus type 1 (HIV-1)-specific T-cell responses directed against the entire expressed HIV-1 genome demonstrate broadly directed responses, but no correlation to viral load, *J. Virol.,* 77 (3), 2081–2092, 2003.

99. Allen, T.M., O'Connor, D.H., Jing, P., Dzuris, J.L., Mothe, B.R., Vogel, T.U., Dunphy, E., Liebl, M.E., Emerson, C., Wilson, N., Kunstman, K.J., Wang, X., Allison, D.B., Hughes, A.L., Desrosiers, R.C., Altman, J.D., Wolinsky, S.M., Sette, A., and Watkins, D.I., Tat-specific cytotoxic T lymphocytes select for SIV escape variants during resolution of primary viraemia, *Nature,* 407 (6802), 386–390, 2000.

100. Altfeld, M., Addo, M.M., Shankarappa, R., Lee, P.K., Allen, T.M., Yu, X.G., Rathod, A., Harlow, J., O'Sullivan, K., Johnston, M.N., Goulder, P.J., Mullins, J.I., Rosenberg, E.S., Brander, C., Korber, B., and Walker, B.D., Enhanced detection of human immunodeficiency virus type 1-specific T-cell responses to highly variable regions by using peptides based on autologous virus sequences, *J. Virol.,* 77 (13), 7330–7340, 2003.

101. Cardinaud, S., Moris, A., Fevrier, M., Rohrlich, P.S., Weiss, L., Langlade-Demoyen, P., Lemonnier, F.A., Schwartz, O., and Habel, A., Identification of cryptic MHC I-restricted epitopes encoded by HIV-1 alternative reading frames, *J. Exp. Med.,* 199 (8), 1053–1063, 2004.

102. Roederer, M., Brenchley, J.M., Betts, M.R., and De Rosa, S.C., Flow cytometric analysis of vaccine responses: how many colors are enough? *Clin. Immunol.,* 110 (3), 199–205, 2004.

103. Lichterfeld, M., Yu, X.G., Waring, M.T., Mui, S.K., Johnston, M., Cohen, D., Addo, M.M., Zaunders, J., Alter, G., Pae, E., Strick, D., Allen, T.M., Rosenberg, E.S., Walker, B.D., and Altfeld, M., HIV-1-specific cytotoxicity is preferentially mediated by a subset of CD8+ T cells producing both interferon- and tumor-necrosis factor-, *Blood,* April 1 (vol 104 (2), 487–494), 2004.

104. Shankarappa, R., Margolick, J.B., Gange, S.J., Rodrigo, A.G., Upchurch, D., Farzadegan, H., Gupta, P., Rinaldo, C.R., Learn, G.H., He, X., Huang, X.L., and Mullins, J.I., Consistent viral evolutionary changes associated with the progression of human immunodeficiency virus type 1 infection, *J. Virol.,* 73 (12), 10489–10502, 1999.

105. Wei, X., Decker, J.M., Wang, S., Hui, H., Kappes, J.C., Wu, X., Salazar-Gonzalez, J.F., Salazar, M.G., Kilby, J.M., Saag, M.S., Komarova, N.L., Nowak, M.A., Hahn, B.H., Kwong, P.D., and Shaw, G.M., Antibody neutralization and escape by HIV-1, *Nature,* 422 (6929), 307–312, 2003.

106. Abel, K., Alegria-Hartman, M.J., Rothaeusler, K., Marthas, M., and Miller, C.J., The relationship between simian immunodeficiency virus RNA levels and the mRNA levels of / interferons (IFN-/) and IFN-/-inducible Mx in lymphoid tissues of rhesus macaques during acute and chronic infection, *J. Virol.,* 76 (16), 8433–8445, 2002.

107. Lore, K., Sonnerborg, A., Brostrom, C., Goh, L.E., Perrin, L., McDade, H., Stellbrink, H.J., Gazzard, B., Weber, R., Napolitano, L.A., van Kooyk, Y., and Andersson, J., Accumulation of DC-SIGN+CD40+ dendritic cells with reduced CD80 and CD86 expression in lymphoid tissue during acute HIV-1 infection, *AIDS,* 16 (5), 683–692, 2002.

108. Kacani, L., Stoiber, H., and Dierich, M.P., Role of IL-15 in HIV-1-associated hypergammaglobuli-naemia, *Clin. Exp. Immunol.,* 108 (1), 14–18, 1997.

109. Amicosante, M., Poccia, F., Gioia, C., Montesano, C., Topino, S., Martini, F., Narciso, P., Pucillo, L.P., and D'Offizi, G., Levels of interleukin-15 in plasma may predict a favorable outcome of structured treatment interruption in patients with chronic human immunodeficiency virus infection, *J. Infect. Dis.,* 188 (5), 661–665, 2003.

110. d'Ettorre, G., Forcina, G., Lichtner, M., Mengoni, F., D'Agostino, C., Massetti, A.P., Mastroianni, C.M., and Vullo, V., Interleukin-15 in HIV infection: immunological and virological interactions in antiretroviral-naive and -treated patients, *AIDS,* 16 (2), 181–188, 2002.

111. Ahmad, R., Sindhu, S.T., Toma, E., Morisset, R., and Ahmad, A., Studies on the production of IL-15 in HIV-infected/AIDS patients, *J. Clin. Immunol.*, 23 (2), 81–90, 2003.

112. Donaghy, H., Pozniak, A., Gazzard, B., Qazi, N., Gilmour, J., Gotch, F., and Patterson, S., Loss of blood CD11c(+) myeloid and CD11c(−) plasmacytoid dendritic cells in patients with HIV-1 infection correlates with HIV-1 RNA virus load, *Blood,* 98 (8), 2574–2576, 2001.

113. Giorgi, J.V., Ho, N.-H., Hirji, K., Chou, C.-C., Hultin, L.E., O'Rourke, S., Park, L., Margolick, J.B., Ferbas, J., Phair, J.P., and The Multicenter AIDS Cohort Study Group, CD8+ lymphocyte activation at human immunodeficiency virus type 1 seroconversion: development of HLA-DR+CD38-CD8+ cells is associated with subsequent stable CD4+ cell levels, *J. Infect. Dis.,* 170 (4), 775–781, 1994.

114. Zaunders, J.J., Geczy, A.F., Dyer, W.B., McIntyre, L.B., Cooley, M.A., Ashton, L.J., Raynes-Greenow, C.H., Learmont, J., Cooper, D.A., and Sullivan, J.S., Effect of long-term infection with Nef-defective attenuated HIV type 1 on CD4+ and CD8+ T lymphocytes: increased CD45RO+CD4+ T lymphocytes and limited activation of CD8+ T lymphocytes, *AIDS Res. Hum. Retroviruses,* 15 (17), 1519–1527, 1999.

115. Sousa, A.E., Carneiro, J., Meier-Schellersheim, M., Grossman, Z., and Victorino, R.M., CD4 T cell depletion is linked directly to immune activation in the pathogenesis of HIV-1 and HIV-2 but only indirectly to the viral load, *J. Immunol.,* 169 (6), 3400–3406, 2002.

116. Gougeon, M.L., Garcia, S., Heeney, J., Tschopp, R., Lecoeur, H., Guetard, D., Rame, V., Dauguet, C., and Montagnier, L., Programmed cell death in AIDS-related HIV and SIV infections, *AIDS Res. Hum. Retroviruses,* 9 (6), 553–563, 1993.

117. Silvestri, G., Sodora, D.L., Koup, R.A., Paiardini, M., O'Neil, S.P., McClure, H.M., Staprans, S.I., and Feinberg, M.B., Nonpathogenic SIV infection of sooty mangabeys is characterized by limited bystander immunopathology despite chronic high-level viremia, *Immunity,* 18 (3), 441–452, 2003.

118. Monceaux, V., Ho Tsong Fang, R., Cumont, M.C., Hurtrel, B., and Estaquier, J., Distinct cycling CD4(+)- and CD8(+)-T-cell profiles during the asymptomatic phase of simian immunodeficiency virus SIVmac251 infection in rhesus macaques, *J. Virol.,* 77 (18), 10047–10059, 2003.

119. Chakrabarti, L.A., Lewin, S.R., Zhang, L., Gettie, A., Luckay, A., Martin, L.N., Skulsky, E., Ho, D.D., Cheng-Mayer, C., and Marx, P.A., Normal T-cell turnover in sooty mangabeys harboring active simian immunodeficiency virus infection, *J. Virol.,* 74 (3), 1209–1223, 2000.

120. Tindall, B. and Cooper, D.A., Primary HIV infection: host responses and intervention strategies, *AIDS,* 5 (1), 1–14, 1991.

121. Mackay, C.R., Marston, W., and Dudler, L., Altered patterns of T cell migration through lymph nodes and skin following antigen challenge, *Eur. J. Immunol.,* 22 (9), 2205–2210, 1992.

122. Kaufmann, G.R., Cunningham, P., Zaunders, J., Law, M., Carr, A., Cooper, D.A., and Group, T.S.P.H.I.S., Impact of early HIV-1 RNA and T-lymphocyte dynamics during primary HIV-1 infection on the subsequent course of HIV-1 RNA and CD4+ T-lymphocyte counts in the first year of HIV-1 infection, *J. Acquir. Immun. Defic. Syndr.,* 22 (5), 437–444, 1999.

123. Roederer, M., Dubs, J.G., Anderson, M.T., Raju, P.A., Herzenberg, L.A., and Herzenberg, L.A., CD8 naive T cell counts decrease progressively in HIV-infected adults, *J. Clin. Invest.,* 95 (5), 2061–2066, 1995.

124. Onanga, R., Kornfeld, C., Pandrea, I., Estaquier, J., Souquiere, S., Rouquet, P., Mavoungou, V.P., Bourry, O., M'Boup, S., Barre-Sinoussi, F., Simon, F., Apetrei, C., Roques, P., and Muller-Trutwin, M.C., High levels of viral replication contrast with only transient changes in CD4(+) and CD8(+) cell numbers during the early phase of experimental infection with simian immunodeficiency virus SIVmnd-1 in *Mandrillus sphinx, J. Virol.,* 76 (20), 10256–10263, 2002.

125. Monceaux, V., Estaquier, J., Fevrier, M., Cumont, M.C., Riviere, Y., Aubertin, A.M., Ameisen, J.C., and Hurtrel, B., Extensive apoptosis in lymphoid organs during primary SIV infection predicts rapid progression towards AIDS, *AIDS,* 17 (11), 1585–1596, 2003.

126. Lynne, J.E., Schmid, I., Matud, J.L., Hirji, K., Buessow, S., Shlian, D.M., and Giorgi, J.V., Major expansions of select CD8+ subsets in acute Epstein–Barr virus infection: comparison with chronic human immunodeficiency virus disease, *J. Infect. Dis.,* 177 (4), 1083–1087, 1998.

127. Roos, M.T., van Lier, R.A., Hamann, D., Knol, G.J., Verhoofstad, I., van Baarle, D., Miedema, F., and Schellekens, P.T., Changes in the composition of circulating CD8+ T cell subsets during acute Epstein–Barr and human immunodeficiency virus infections in humans, *J. Infect. Dis.,* 182 (2), 451–458, 2000.

128. Zaunders, J.J., Kaufmann, G.R., Cunningham, P.H., Smith, D., Grey, P., Suzuki, K., Carr, A., Goh, L.E., and Cooper, D.A., Increased turnover of CCR5+ and redistribution of CCR5 CD4 T lymphocytes during primary human immunodeficiency virus type 1 infection, *J. Infect. Dis.,* 183 (5), 736–743, 2001.

129. Altfeld, M., Rosenberg, E.S., Shankarappa, R., Mukherjee, J.S., Hecht, F.M., Eldridge, R.L., Addo, M.M., Poon, S.H., Phillips, M.N., Robbins, G.K., Sax, P.E., Boswell, S., Kahn, J.O., Brander, C., Goulder, P.J., Levy, J.A., Mullins, J.I., and Walker, B.D., Cellular immune responses and viral diversity in individuals treated during acute and early HIV-1 infection, *J. Exp. Med.,* 193 (2), 169–180, 2001.

130. Alter, G., Merchant, A., Tsoukas, C.M., Rouleau, D., LeBlanc, R.P., Cote, P., Baril, J.G., Thomas, R., Nguyen, V.K., Sekaly, R.P., Routy, J.P., and Bernard, N.F., Human immunodeficiency virus (HIV)-specific effector CD8 T cell activity in patients with primary HIV infection, *J. Infect. Dis.,* 185 (6), 755–765, 2002.

131. Rosenberg, E.S., Altfeld, M., Poon, S.H., Phillips, M.N., Wilkes, B.M., Eldridge, R.L., Robbins, G.K., D'Aquila, R.T., Goulder, P.J., and Walker, B.D., Immune control of HIV-1 after early treatment of acute infection, *Nature,* 407 (6803), 523–526, 2000.

132. Koup, R.A., Safrit, J.T., Cao, Y., Andrews, C.A., McLeod, G., Borkowsky, W., Farthing, C., and Ho, D.D., Temporal association of cellular immune responses with the initial control of viremia in primary human immunodeficiency virus type 1 syndrome, *J. Virol.,* 68 (7), 4650–4655, 1994.

133. Borrow, P., Lewicki, H., Hahn, B.H., Shaw, G.M., and Oldstone, M.B.A., Virus-specific CD8+ cytotoxic T-lymphocyte activity associated with control of viremia in primary human immunodeficiency virus type 1 infection, *J. Virol.,* 68 (9), 6103–6110, 1994.

134. Schmitz, J.E., Kuroda, M.J., Santra, S., Sasseville, V.G., Simon, M.A., Lifton, M.A., Racz, P., Tenner-Racz, K., Dalesandro, M., Scallon, B.J., Ghrayeb, J., Forman, M.A., Montefiori, D.C., Rieber, E.P., Letvin, N.L., and Reimann, K.A., Control of viremia in simian immunodeficiency virus infection by CD8+ lymphocytes, *Science,* 283 (5304), 857–860, 1999.

135. Appay, V., Dunbar, P.R., Callan, M., Klenerman, P., Gillespie, G.M., Papagno, L., Ogg, G.S., King, A., Lechner, F., Spina, C.A., Little, S., Havlir, D.V., Richman, D.D., Gruener, N., Pape, G., Waters, A., Easterbrook, P., Salio, M., Cerundolo, V., McMichael, A.J., and Rowland-Jones, S.L., Memory CD8+ T cells vary in differentiation phenotype in different persistent virus infections, *Nat. Med.,* 8 (4), 379–385, 2002.

136. Champagne, P., Ogg, G.S., King, A.S., Knabenhans, C., Ellefsen, K., Nobile, M., Appay, V., Rizzardi, G.P., Fleury, S., Lipp, M., Forster, R., Rowland-Jones, S., Sekaly, R.P., McMichael, A.J., and Pantaleo, G., Skewed maturation of memory HIV-specific CD8 T lymphocytes, *Nature,* 410 (6824), 106–111, 2001.

137. Clerici, M. and Shearer, G.M., TH-1 to TH2 switch is a critical step in the etiology of HIV infection, *Immunol. Today,* 14 (3), 107–111, 1993.

138. Younes, S.A., Yassine-Diab, B., Dumont, A.R., Boulassel, M.R., Grossman, Z., Routy, J.P., and Sekaly, R.P., HIV-1 viremia prevents the establishment of interleukin 2-producing HIV-specific memory CD4+ T cells endowed with proliferative capacity, *J. Exp. Med.,* 198 (12), 1909–1922, 2003.

139. Iyasere, C., Tilton, J.C., Johnson, A.J., Younes, S., Yassine-Diab, B., Sekaly, R.P., Kwok, W.W., Migueles, S.A., Laborico, A.C., Shupert, W.L., Hallahan, C.W., Davey, R.T., Jr., Dybul, M., Vogel, S., Metcalf, J., and Connors, M., Diminished proliferation of human immunodeficiency virus-specific CD4+ T cells is associated with diminished interleukin-2 (IL-2) production and is recovered by exogenous IL-2, *J. Virol.,* 77 (20), 10900–10909, 2003.

140. Zhou, J., Zhang, J., Lichtenheld, M.G., and Meadows, G.G., A role for NF-B activation in perforin expression of NK cells upon IL-2 receptor signaling, *J. Immunol.,* 169 (3), 1319–1325, 2002.

141. Picker, L.J. and Maino, V.C., The CD4(+) T cell response to HIV-1, *Curr. Opin. Immunol.,* 12 (4), 381–386, 2000.

142. Migueles, S.A., Laborico, A.C., Shupert, W.L., Sabbaghian, M.S., Rabin, R., Hallahan, C.W., Van Baarle, D., Kostense, S., Miedema, F., McLaughlin, M., Ehler, L., Metcalf, J., Liu, S., and Connors, M., HIV-specific CD8+ T cell proliferation is coupled to perforin expression and is maintained in nonprogressors, *Nat. Immunol.,* 3 (11), 1061–1068, 2002.

143. Kelleher, A.D., Roggensack, M., Emery, S., Carr, A., French, M.A., and Cooper, D.A., Effects of IL-2 therapy in asymptomatic HIV-infected individuals on proliferative responses to mitogens, recall antigens and HIV-related antigens, *Clin. Exp. Immunol.,* 113 (1), 85–91, 1998.

144. Sereti, I., Anthony, K.B., Martinez-Wilson, H., Lempicki, R., Adelsberger, J., Metcalf, J.A., Hallahan, C.W., Follmann, D., Davey, R.T., Kovacs, J.A., and Lane, H.C., IL-2 induced CD4+ T cell expansion in HIV-infected patients is associated with long-term decreases in T cell proliferation, *Blood,* 104 (3), 775–780, 2004.

145. Effros, R.B., Allsopp, R., Chiu, C.P., Hausner, M.A., Hirji, K., Wang, L., Harley, C.B., Villeponteau, B., West, M.D., and Giorgi, J.V., Shortened telomeres in the expanded CD28-CD8+ cell subset in HIV disease implicate replicative senescence in HIV pathogenesis, *AIDS,* 10 (8), F17–F22, 1996.

146. Papagno, L., Spina, C.A., Marchant, A., Salio, M., Rufer, N., Little, S., Dong, T., Chesney, G., Waters, A., Easterbrook, P., Dunbar, P.R., Shepherd, D., Cerundolo, V., Emery, V., Griffiths, P., Conlon, C., McMichael, A.J., Richman, D.D., Rowland-Jones, S.L., and Appay, V., Immune activation and CD8(+) T-cell differentiation towards senescence in HIV-1 infection, *PLoS Biol.,* 2 (2), E20, 2004.

147. Lieberman, J., Trimble, L.A., Friedman, R.S., Lisziewicz, J., Lori, F., Shankar, P., and Jessen, H., Expansion of CD57 and CD62L-CD45RA+ CD8 T lymphocytes correlates with reduced viral plasma RNA after primary HIV infection, *AIDS,* 13 (8), 891–899, 1999.

148. Brenchley, J.M., Karandikar, N.J., Betts, M.R., Ambrozak, D.R., Hill, B.J., Crotty, L.E., Casazza, J.P., Kuruppu, J., Migueles, S.A., Connors, M., Roederer, M., Douek, D.C., and Koup, R.A., Expression of CD57 defines replicative senescence and antigen-induced apoptotic death of CD8+ T cells, *Blood,* 101 (7), 2711–2720, 2003.

149. Muro-Cacho, C.A., Pantaleo, G., and Fauci, A.S., Analysis of apoptosis in lymph nodes of HIV-infected persons. Intensity of apoptosis correlates with the general state of activation of the lymphoid tissue and not with stage of disease or viral burden, *J. Immunol.,* 154 (10), 5555–5566, 1995.

150. Bofill, M., Gombert, W., Borthwick, N.J., Akbar, A.N., McLaughlin, J.E., Lee, C.A., Johnson, M.A., Pinching, A.J., and Janossy, G., Presence of CD3+CD8+Bcl-2(low) lymphocytes undergoing apoptosis and activated macrophages in lymph nodes of HIV-1+ patients, *Am. J. Pathol.,* 146 (6), 1542–1555, 1995.

151. Finkel, T.H., Tudor-Williams, G., Banda, N.K., Cotton, M.F., Curiel, T., Monks, C., Baba, T.W., Ruprecht, R.M., and Kupfer, A., Apoptosis occurs predominantly in bystander cells and not in productively infected cells of HIV- and SIV-infected lymph nodes, *Nat. Med.,* 1 (2), 129–134, 1995.

152. Iida, T., Ichimura, H., Ui, M., Shimada, T., Akahata, W., Igarashi, T., Kuwata, T., Ido, E., Yonehara, S., Imanishi, J., and Hayami, M., Sequential analysis of apoptosis induction in peripheral blood mononuclear cells and lymph nodes in the early phase of pathogenic and nonpathogenic SIVmac infection, *AIDS Res. Hum. Retroviruses,* 15 (8), 721–729, 1999.

153. Wallace, M., Waterman, P.M., Mitchen, J.L., Djavani, M., Brown, C., Trivedi, P., Horejsh, D., Dykhuizen, M., Kitabwalla, M., and Pauza, C.D., Lymphocyte activation during acute simian/human immunodeficiency virus SHIV(89.6PD) infection in macaques, *J. Virol.,* 73 (12), 10236–10244, 1999.

154. Savill, J., Dransfield, I., Gregory, C., and Haslett, C., A blast from the past: clearance of apoptotic cells regulates immune responses, *Nat. Rev. Immunol.,* 2 (12), 965–975, 2002.

155. Ameisen, J.C., Estaquier, J., and Idziorek, T., From AIDS to parasite infection: pathogen-mediated subversion of programmed cell death as a mechanism for immune dysregulation, *Immunol. Rev.,* 142, 9–51, 1994.

156. Meyaard, L., Otto, S.A., Jonker, R.R., Mijnster, M.J., Keet, R.P., and Miedema, F., Programmed death of T cells in HIV-1 infection, *Science,* 257 (5067), 217–219, 1992.

157. Sarin, A., Clerici, M., Blatt, S.P., Hendrix, C.W., Shearer, G.M., and Henkart, P.A., Inhibition of activation-induced programmed cell death and restoration of defective immune responses of HIV+ donors by cysteine protease inhibitors, *J. Immunol.,* 153 (2), 862–872, 1994.

158. Oyaizu, N., McCloskey, T.W., Coronesi, M., Chirmule, N., Kalyanaraman, V.S., and Pahwa, S., Accelerated apoptosis in peripheral blood mononuclear cells (PBMCs) from human immunodeficiency virus type-1 infected patients and in CD4 cross-linked PBMCs from normal individuals, *Blood,* 82 (11), 3392–3400, 1993.

159. Clerici, M., Sarin, A., Coffman, R.L., Wynn, T.A., Blatt, S.P., Hendrix, C.W., Wolf, S.F., Shearer, G.M., and Henkart, P.A., Type 1/type 2 cytokine modulation of T-cell programmed cell death as a model for human immunodeficiency virus pathogenesis, *Proc. Natl. Acad. Sci. U.S.A.,* 91 (25), 11811–11815, 1994.

160. Estaquier, J., Idziorek, T., de Bels, F., Barre-Sinoussi, F., Hurtrel, B., Aubertin, A.M., Venet, A., Mehtali, M., Muchmore, E., Michel, P., Mouton, Y., Girard, M., and Ameisen, J.C., Programmed cell death and AIDS: significance of T-cell apoptosis in pathogenic and nonpathogenic primate lentiviral infections, *Proc. Natl. Acad. Sci. U.S.A.,* 91 (20), 9431–9435, 1994.

161. Lewis, D.E., Tang, D.S., Adu-Oppong, A., Schober, W., and Rodgers, J.R., Anergy and apoptosis in CD8+ T cells from HIV-infected persons, *J. Immunol.,* 153 (1), 412–420, 1994.

162. Adachi, Y., Oyaizu, N., Than, S., McCloskey, T.W., and Pahwa, S., IL-2 rescues *in vitro* lymphocyte apoptosis in patients with HIV infection: correlation with its ability to block culture-induced down-modulation of Bcl-2, *J. Immunol.,* 157, 4184–4193, 1996.

163. Boudet, F., Lecoeur, H., and Gougeon, M.L., Apoptosis associated with *ex vivo* down-regulation of Bcl-2 and up-regulation of Fas in potential cytotoxic CD8+ T lymphocytes during HIV infection, *J. Immunol.,* 156 (6), 2282–2293, 1996.

164. Chehimi, J., Marshall, J.D., Salvucci, O., Frank, I., Chehimi, S., Kawecki, S., Bacheller, D., Rifat, S., and Chouaib, S., IL-15 enhances immune functions during HIV infection, *J. Immunol.,* 158 (12), 5978–5987, 1997.

165. Naora, H. and Gougeon, M.L., Interleukin-15 is a potent survival factor in the prevention of sponta-neous but not CD95-induced apoptosis in CD4 and CD8 T lymphocytes of HIV-infected individuals. Correlation with its ability to increase BCL-2 expression, *Cell Death Differ.,* 6 (10), 1002–1011, 1999.

166. McCloskey, T.W., Bakshi, S., Than, S., Arman, P., and Pahwa, S., Immunophenotypic analysis of peripheral blood mononuclear cells undergoing *in vitro* apoptosis after isolation from human immu-nodeficiency virus-infected children, *Blood,* 92 (11), 4230–4237, 1998.

167. Zaunders, J.J., Moutouh-de Parseval, L., Kitada, S., Reed, J.C., Rought, S., Genini, D., Leoni, L., Kelleher, A., Cooper, D.A., Smith, D.E., Grey, P., Estaquier, J., Little, S., Richman, D.D., and Corbeil, J., Polyclonal proliferation and apoptosis of CCR5+ T lymphocytes during primary human immuno-deficiency virus type 1 infection: regulation by interleukin (IL)-2, IL-15, and Bcl-2, *J. Infect. Dis.,* 187 (11), 1735–1747, 2003.

168. Petrovas, C., Mueller, Y.M., Dimitriou, I.D., Bojczuk, P.M., Mounzer, K.C., Witek, J., Altman, J.D., and Katsikis, P.D., HIV-specific CD8(+) T cells exhibit markedly reduced levels of Bcl-2 and Bcl-x(L), *J. Immunol.,* 172 (7), 4444–4453, 2004.

169. Arnoult, D., Petit, F., Lelievie, J.D., Lecossier, D., Hance, A., Monceaux, V., Ho Tsong Fang, R., Huntrel, B., Ameisen, J.C., and Estaquier, J., Caspase-dependent and -independent T-cell death pathways in pathogenic simian immunodeficiency virus infection: relationship to disease progression, *Cell Death Differ.,* 10 (11), 1240–1252, 2003.

170. Borthwick, N.J., Bofill, M., Hassan, I., Panayiotidis, P., Janossy, G., Salmon, M., and Akbar, A.N., Factors that influence activated CD8+ T-cell apoptosis in patients with acute herpesvirus infections: loss of costimulatory molecules CD28, CD5 and CD6 but relative maintenance of Bax and Bcl-X expression, *Immunology,* 88 (4), 508–515, 1996.

171. Regamey, N., Harr, T., Battegay, M., and Erb, P., Downregulation of Bcl-2, but not of Bax or Bcl-x, is associated with T lymphocyte apoptosis in HIV infection and restored by antiretroviral therapy or by interleukin 2, *AIDS Res. Hum. Retroviruses,* 15 (9), 803–810, 1999.

172. Macho, A., Castedo, M., Marchetti, P., Aguilar, J.J., Decaudin, D., Zamzami, N., Girard, P.M., Uriel, J., and Kroemer, G., Mitochondrial dysfunctions in circulating T lymphocytes from human immun-odeficiency virus-1 carriers, *Blood,* 86 (7), 2481–2487, 1995.

173. Moretti, S., Marcellini, S., Boschini, A., Famularo, G., Santini, G., Alesse, E., Steinberg, S.M., Cifone, M.G., Kroemer, G., and De Simone, C., Apoptosis and apoptosis-associated perturbations of peripheral blood lymphocytes during HIV infection: comparison between AIDS patients and asymptomatic long-term non-progressors, *Clin. Exp. Immunol.,* 122 (3), 364–373, 2000.

174. Cossarizza, A., Mussini, C., Mongiardo, N., Borghi, V., Sabbatini, A., De Rienzo, B., and Franceschi, C., Mitochondria alterations and dramatic tendency to undergo apoptosis in peripheral blood lympho-cytes during acute HIV syndrome, *AIDS,* 11 (1), 19–26, 1997.

175. Liegler, T.J., Yonemoto, W., Elbeik, T., Vittinghoff, E., Buchbinder, S.P., and Greene, W.C., Diminished spontaneous apoptosis in lymphocytes from human immunodeficiency virus-infected long-term non-progressors, *J. Infect. Dis.,* 178 (3), 669–679, 1998.

176. Davis, I.C., Girard, M., and Fultz, P.N., Loss of CD4+ T cells in human immunodeficiency virus type 1-infected chimpanzees is associated with increased lymphocyte apoptosis, *J. Virol.,* 72 (6), 4623–4632, 1998.

177. Badley, A.D., Dockrell, D.H., Algeciras, A., Ziesmer, S., Landay, A., Lederman, M.M., Connick, E., Kessler, H., Kuritzkes, D., Lynch, D.H., Roche, P., Yagita, H., and Paya, C.V., *In vivo* analysis of Fas/FasL interactions in HIV-infected patients, *J. Clin. Invest.,* 102 (1), 79–87, 1998.

178. Gougeon, M.L., Lecoeur, H., and Sasaki, Y., Apoptosis and the CD95 system in HIV disease: impact of highly active anti-retroviral therapy (HAART), *Immunol. Lett.,* 66 (1–3), 97–103, 1999.

179. Badley, A.D., Pilon, A.A., Landay, A., and Lynch, D.H., Mechanisms of HIV-associated lymphocyte apoptosis, *Blood,* 96 (9), 2951–2964, 2000.

180. Katsikis, P.D., Wunderlich, E.S., Smith, C.A., and Herzenberg, L.A., Fas antigen stimulation induces marked apoptosis of T lymphocytes in human immunodeficiency virus-infected individuals, *J. Exp. Med.,* 181 (6), 2029–2036, 1995.

181. Estaquier, J., Idziorek, T., Zou, W., Emilie, D., Farber, C.M., Bourez, J.M., and Ameisen, J.C., T helper type 1/T helper type 2 cytokines and T cell death: preventive effect of interleukin 12 on activation-induced and CD95 (FAS/APO-1)-mediated apoptosis of CD4+ T cells from human immunodeficiency virus-infected persons, *J. Exp. Med.,* 182 (6), 1759–1767, 1995.

182. Sloand, E.M., Young, N.S., Kumar, P., Weichold, F.F., Sato, T., and Maciejewski, J.P., Role of Fas ligand and receptor in the mechanism of T-cell depletion in acquired immunodeficiency syndrome: effect on CD4+ lymphocyte depletion and human immunodeficiency virus replication, *Blood,* 89 (4), 1357–1363, 1997.

183. Gehri, R., Hahn, S., Rothen, M., Steuerwald, M., Nuesch, R., and Erb, P., The Fas receptor in HIV infection: expression on peripheral blood lymphocytes and role in the depletion of T cells, *AIDS,* 10 (1), 9–16, 1996.

184. Badley, A.D., Parato, K., Cameron, D.W., Kravcik, S., Phenix, B.N., Ashby, D., Kumar, A., Lynch, D.H., Tschopp, J., and Angel, J.B., Dynamic correlation of apoptosis and immune activation during treatment of HIV infection, *Cell Death Differ.,* 6 (5), 420–432, 1999.

185. Estaquier, J., Tanaka, M., Suda, T., Nagata, S., Golstein, P., and Ameisen, J.C., Fas-mediated apoptosis of CD4+ and CD8+ T cells from human immunodeficiency virus-infected persons: differential *in vitro* preventive effect of cytokines and protease antagonists, *Blood,* 87 (12), 4959–4966, 1996.

186. Mueller, Y.M., Bojczuk, P.M., Halstead, E.S., Kim, A.H., Witek, J., Altman, J.D., and Katsikis, P.D., IL-15 enhances survival and function of HIV-specific CD8+ T cells, *Blood,* 101 (3), 1024–1029, 2003.

187. Cohen, S.B., Crawley, J.B., Kahan, M.C., Feldmann, M., and Foxwell, B.M., Interleukin-10 rescues T cells from apoptotic cell death: association with an upregulation of Bcl-2, *Immunology,* 92 (1), 1–5, 1997.

188. Taga, K., Cherney, B., and Tosato, G., IL-10 inhibits apoptotic cell death in human T cells starved of IL-2, *Int. Immunol.,* 5 (12), 1599–1608, 1993.

189. Waldmann, T., Tagaya, Y., and Bamford, R., Interleukin-2, interleukin-15, and their receptors, *Int. Rev. Immunol.,* 16 (3–4), 205–226, 1998.

190. Tan, J.C., Indelicato, S.R., Narula, S.K., Zavodny, P.J., and Chou, C.C., Characterization of interleukin-10 receptors on human and mouse cells, *J. Biol. Chem.,* 268 (28), 21053–21059, 1993.

191. Kotenko, S.V., Krause, C.D., Izotova, L.S., Pollack, B.P., Wu, W., and Pestka, S., Identification and functional characterization of a second chain of the interleukin-10 receptor complex, *EMBO J.,* 16 (19), 5894–5903, 1997.

192. Ho, A.S., Liu, Y., Khan, T.A., Hsu, D.H., Bazan, J.F., and Moore, K.W., A receptor for interleukin 10 is related to interferon receptors, *Proc. Natl. Acad. Sci. U.S.A.,* 90 (23), 11267–11271, 1993.

193. Hosaka, N., Oyaizu, N., Kaplan, M.H., Yagita, H., and Pahwa, S., Membrane and soluble forms of Fas (CD95) and Fas ligand in peripheral blood mononuclear cells and in plasma from human immunodeficiency virus-infected persons, *J. Infect. Dis.,* 178 (4), 1030–1039, 1998.

194. Kaplan, D. and Sieg, S., Role of the Fas/Fas ligand apoptotic pathway in human immunodeficiency virus type 1 disease, *J. Virol.,* 72 (8), 6279–6282, 1998.

195. Jang, S., Krammer, P.H., and Salgame, P., Lack of proapoptotic activity of soluble CD95 ligand is due to its failure to induce CD95 oligomers, *J. Interferon Cytokine Res.,* 23 (8), 441–447, 2003.

196. Dockrell, D.H., Badley, A.D., Algeciras-Schimnich, A., Simpson, M., Schut, R., Lynch, D.H., and Paya, C.V., Activation-induced CD4+ T cell death in HIV-positive individuals correlates with Fas susceptibility, CD4+ T cell count, and HIV plasma viral copy number, *AIDS Res. Hum. Retroviruses,* 15 (17), 1509–1518, 1999.

197. Badley, A.D., McElhinny, J.A., Leibson, P.J., Lynch, D.H., Alderson, M.R., and Paya, C.V., Upregulation of Fas ligand expression by human immunodeficiency virus in human macrophages mediates apoptosis of uninfected T lymphocytes, *J. Virol.,* 70 (1), 199–206, 1996.

198. Mitra, D., Steiner, M., Lynch, D.H., Staiano-Coico, L., and Laurence, J., HIV-1 upregulates Fas ligand expression in CD4+ T cells *in vitro* and *in vivo*: association with Fas-mediated apoptosis and modulation by aurintricarboxylic acid, *Immunology,* 87 (4), 581–585, 1996.

199. Xu, X.-N., Screaton, G.R., Gotch, F.M., Dong, T., Tan, R., Almond, N., Walker, B., Stebbings, R., Kent, K., Nagata, S., Stott, J.E., and McMichael, A.J., Evasion of cytotoxic T lymphocyte (CTL) responses by *Nef*-dependent induction of Fas ligand (CD95L) expression on simian immunodeficiency virus-infected cells, *J. Exp. Med.,* 186 (1), 7–16, 1997.

200. Chahroudi, A., Silvestri, G., and Feinberg, M.B., Measuring T cell-mediated cytotoxicity using fluorogenic caspase substrates, *Methods,* 31 (2), 120–126, 2003.

201. de Oliveira Pinto, L.M., Garcia, S., Lecoeur, H., Rapp, C., and Gougeon, M.L., Increased sensitivity of T lymphocytes to tumor necrosis factor receptor 1 (TNFR1)- and TNFR2-mediated apoptosis in HIV infection: relation to expression of Bcl-2 and active caspase-8 and caspase-3, *Blood,* 99 (5), 1666–1675, 2002.

202. Katsikis, P.D., Garcia-Ojeda, M.E., Torres-Roca, J.F., Tijoe, I.M., Smith, C.A., and Herzenberg, L.A., Interleukin-1 converting enzyme-like protease involvement in Fas-induced and activation-induced peripheral blood T cell apoptosis in HIV infection. TNF-related apoptosis-inducing ligand can mediate activation-induced T cell death in HIV infection, *J. Exp. Med.,* 186 (8), 1365–1372, 1997.

203. Jeremias, I., Herr, I., Boehler, T., and Debatin, K.M., TRAIL/Apo-2-ligand-induced apoptosis in human T cells, *Eur. J. Immunol.,* 28 (1), 143–152, 1998.

204. Miura, Y., Misawa, N., Maeda, N., Inagaki, Y., Tanaka, Y., Ito, M., Kayagaki, N., Yamamoto, N., Yagita, H., Mizusawa, H., and Koyanagi, Y., Critical contribution of tumor necrosis factor-related apoptosis-inducing ligand (TRAIL) to apoptosis of human CD4+ T cells in HIV-1-infected hu-PBL-NOD-SCID mice, *J. Exp. Med.,* 193 (5), 651–660, 2001.

205. Lecossier, D., Bouchonnet, F., Schneider, P., Clavel, F., and Hance, A.J., Discordant increases in CD4+ T cells in human immunodeficiency virus-infected patients experiencing virologic treatment failure: role of changes in thymic output and T cell death, *J. Infect. Dis.,* 183 (7), 1009–1016, 2001.

206. von Sydow, M., Sonnerborg, A., Gaines, H., and Strannegard, O., Interferon- and tumor necrosis factor- in serum of patients in various stages of HIV-1 infection, *AIDS Res. Hum. Retroviruses,* 7 (4), 375–380, 1991.

207. Poli, G., Kinter, A., Justement, J.S., Kehrl, J.H., Bressler, P., Stanley, S., and Fauci, A.S., Tumor necrosis factor functions in an autocrine manner in the induction of human immunodeficiency virus expression, *Proc. Natl. Acad. Sci. U.S.A.,* 87 (2), 782–785, 1990.

208. Lederman, M.M., Kalish, L.A., Asmuth, D., Fiebig, E., Mileno, M., and Busch, M.P., "Modeling" relationships among HIV-1 replication, immune activation and CD4+ T-cell losses using adjusted correlative analyses, *AIDS,* 14 (8), 951–958, 2000.

209. Wallis, R.S., Kyambadde, P., Johnson, J.L., Horter, L., Kittle, R., Pohle, M., Ducar, C., Millard, M., Mayanja-Kizza, H., Whalen, C., and Okwera, A., A study of the safety, immunology, virology, and microbiology of adjunctive etanercept in HIV-1-associated tuberculosis, *AIDS,* 18 (2), 257–264, 2004.

210. Katsikis, P.D., Garcia-Ojeda, M.E., Wunderlich, E.S., Smith, C.A., Yagita, H., Okumura, K., Kayagaki, N., Alderson, M., and Herzenberg, L.A., Activation-induced peripheral blood T cell apoptosis is Fas independent in HIV-infected individuals, *Int. Immunol.,* 8 (8), 1311–1317, 1996.

211. Sandalova, E., Wei, C.H., Masucci, M.G., and Levitsky, V., Regulation of expression of Bcl-2 protein family member Bim by T cell receptor triggering, *Proc. Natl. Acad. Sci. U.S.A.,* 101 (9), 3011–3016, 2004.

212. Petit, F., Corbeil, J., Lelievre, J.D., Parseval, L.M., Pinon, G., Green, D.R., Ameisen, J.C., and Estaquier, J., Role of CD95-activated caspase-1 processing of IL-1 in TCR-mediated proliferation of HIV-infected CD4(+) T cells, *Eur. J. Immunol.,* 31 (12), 3513–3524, 2001.

213. Estaquier, J., Lelievre, J.D., Petit, F., Brunner, T., Moutouh-De Parseval, L., Richman, D.D., Ameisen, J.C., and Corbeil, J., Effects of antiretroviral drugs on human immunodeficiency virus type 1-induced CD4(+) T-cell death, *J. Virol.,* 76 (12), 5966–5973, 2002.

214. Esser, M.T., Bess, J.W., Jr., Suryanarayana, K., Chertova, E., Marti, D., Carrington, M., Arthur, L.O., and Lifson, J.D., Partial activation and induction of apoptosis in CD4(+) and CD8(+) T lymphocytes by conformationally authentic noninfectious human immunodeficiency virus type 1, *J. Virol.,* 75 (3), 1152–1164, 2001.

215. Tremblay, M.J., Fortin, J.F., and Cantin, R., The acquisition of host-encoded proteins by nascent HIV-1, *Immunol. Today,* 19 (8), 346–351, 1998.

216. Mueller, Y.M., De Rosa, S.C., Hutton, J.A., Witek, J., Roederer, M., Altman, J.D., and Katsikis, P.D., Increased CD95/Fas-induced apoptosis of HIV-specific CD8(+) T cells, *Immunity,* 15 (6), 871–882, 2001.

217. Vlahakis, S.R., Algeciras-Schimnich, A., Bou, G., Heppelmann, C.J., Villasis-Keever, A., Collman, R.C., and Paya, C.V., Chemokine-receptor activation by Env determines the mechanism of death in HIV-infected and uninfected T lymphocytes, *J. Clin. Invest.,* 107 (2), 207–215, 2001.

218. Herbein, G., Van Lint, C., Lovett, J.L., and Verdin, E., Distinct mechanisms trigger apoptosis in human immunodeficiency virus type 1-infected and in uninfected bystander T lymphocytes, *J. Virol.,* 72 (1), 660–670, 1998.

219. Berger, E.A., Murphy, P.M., and Farber, J.M., Chemokine receptors as HIV-1 coreceptors: roles in viral entry, tropism, and disease, *Annu. Rev. Immunol.,* 17, 657–700, 1999.

220. Connor, R.I., Sheridan, K.E., Ceradini, D., Choe, S., and Landau, N.R., Change in coreceptor use correlates with disease progression in HIV-1-infected individuals, *J. Exp. Med.,* 185 (4), 621–628, 1997.

221. Brenchley, J.M., Hill, B.J., Ambrozak, D.R., Price, D.A., Guenaga, F.J., Casazza, J.P., Kuruppu, J., Yazdani, J., Migueles, S.A., Connors, M., Roederer, M., Douek, D.C., and Koup, R.A., T-cell subsets that harbor human immunodeficiency virus (HIV) *in vivo*: implications for HIV pathogenesis, *J. Virol.,* 78 (3), 1160–1168, 2004.

222. Kitchen, S.G., Korin, Y.D., Roth, M.D., Landay, A., and Zack, J.A., Costimulation of naive CD8(+) lymphocytes induces CD4 expression and allows human immunodeficiency virus type 1 infection, *J. Virol.,* 72 (11), 9054–9060, 1998.

223. Khatissian, E., Monceaux, V., Cumont, M.C., Ho Tsong Fang, R., Estaquier, J., and Hurtrel, B., Simian immunodeficiency virus infection of CD4+CD8+ T cells in a macaque with an unusually high peripheral CD4+CD8+ T lymphocyte count, *AIDS Res. Hum. Retroviruses,* 19 (4), 267–274, 2003.

224. Westendorp, M.O., Frank, R., Oschenbauer, C., Stricker, K., Dheln, J., Walczak, H., Debatin, K.-M., and Krammer, P.H., Sensitisation of T cells to CD95-mediated apoptosis by HIV-1 Tat and gp120, *Nature,* 375, 497–500, 1995.

225. Li, C.J., Friedman, D.J., Wang, C., Metelev, V., and Pardee, A.B., Induction of apoptosis in uninfected lymphocytes by HIV-1 Tat protein, *Science,* 268 (5209), 429–431, 1995.

226. Li, C.J., Wang, C., Friedman, D.J., and Pardee, A.B., Reciprocal modulations between p53 and Tat of human immunodeficiency virus type 1. *Proc. Natl. Acad. Sci. USA,* 92 (12), 5461–5464, 1995.

15 Autologous Cell-Mediated Killing

Georges Herbein

CONTENTS

INTRODUCTION

Human immunodeficiency virus type 1 (HIV-1) infection causes functional impairment and progressive loss of T lymphocytes, leading to immunodeficiency and AIDS. Direct lysis of infected T lymphocytes and apoptosis or programmed cell death of bystander cells occur during HIV-1 replication *in vitro*. The relative importance of direct cytopathic effects compared with apoptosis of uninfected bystander cells remains a subject of controversy. Most of the apoptotic T lymphocytes in the peripheral blood and lymphoid tissues of HIV-infected subjects are uninfected. Increased apoptosis of other types of uninfected cells, B cells, thymocytes, and neurons is also detected in HIV-infected subjects. Among mechanisms that are involved in HIV-associated apoptosis of bystander cells are the death of bystander cells by proapoptotic virus proteins or cytotoxic factors that are released by infected cells, and altered expression of cellular apoptosis regulatory molecules by lymphocytes and antigen-presenting cells (APC) as a consequence of HIV-mediated immune activation. Macrophages (MØ), dendritic cells (DCs), CD4+ T cells, and CD8+ T cells derived from HIV-infected patients may induce the death of uninfected T lymphocytes, indicating that apoptosis of uninfected bystander cells plays an important role in AIDS pathogenesis.

AUTOLOGOUS CELL-MEDIATED KILLING VIA
INTERCELLULAR CONTACTS

MONOCYTE/MACROPHAGE-MEDIATED T CELL APOPTOSIS

CD4+ T Cell Apoptosis

In the lymph nodes of HIV-seropositive individuals, HIV-mediated immune activation is observed,[1,2] and apoptosis occurs predominantly in the bystander uninfected cell populations, especially in the presence of APCs, such as MØ.[3-13] The key role of HIV-infected MØ in causing CD4+ T cell depletion was first suggested by results obtained from the severe combined immunodeficient (SCID)-HIV mouse model.[14] It was observed that monocytotropic viruses resulted in enhanced and accelerated apoptosis of the repopulated CD4+ T cell population.[14] Also, monkeys infected with HIV-1, which does not infect simian monocytes/MØ, show no signs of disease, have stable CD4+ T cell counts, and do not show accelerated T cell apoptosis.[15,16] These observations suggested that MØ might play a crucial role in CD4+ T cell depletion after HIV infection.

CD4 cross-linking by the HIV envelope glycoprotein (gp)120 was shown to prime T cells for activation-induced apoptosis, via the upregulation of expression of Fas and Fas ligand (FasL) on uninfected T cells and MØ, respectively (Figure 15.1).[3,17] Macrophages express basal levels of FasL that are significantly upregulated after infection with HIV,[3] and FasL expression is enhanced compared with monocytes from HIV-negative controls.[18] HIV-infected MØ (and, to a lesser extent, uninfected MØ) were shown to kill Fas-sensitive T cell targets[3] in a major histocompatibility complex–unrestricted and Fas/tumor necrosis factor (TNF)-dependent manner.[4] FasL is elevated in peripheral blood mononuclear cells (PBMCs, which contain monocytes)[19-21] from HIV-infected patients. The plasma level of soluble FasL is increased in HIV-positive patients, correlates with HIV RNA burden, and can be used as a predictive marker for the progression to acquired immunodeficiency syndrome (AIDS).[22,23] A marked increase in MØ-associated FasL was also detected in lymphoid tissue from HIV-positive subjects, which correlates with the level of tissue apoptosis.[24] Dexamethasone inhibits CD4+ T cell deletion in a dose-dependent manner. The deletion of normal CD4+ T cells by MØ from HIV-infected patients was also inhibited by dexamethasone. Furthermore, upregulation of Fas expression on T cells exposed to anti-CD4 and gp120/IgG, which predisposes T cells to Fas-mediated apoptosis, is inhibited by dexamethasone in a dose-dependent fashion. Dexamethasone inhibits the MØ-mediated deletion of CD4+ T lymphocytes in HIV-infected persons.[25] MØ-mediated CD4+ T cell apoptosis has implications *in vivo*, because levels of tissue apoptosis directly correlate with levels of MØ-associated FasL.[24] Thus, FasL may be one of the main mediators of uninfected CD4+ T cell death by monocytes/MØ.

As well as the involvement of Fas–FasL interactions, TNF-α was reported to trigger apoptosis in uninfected CD4+ T cells upon HIV infection.[4] HIV-infected MØ-mediated killing of uninfected CD4+ T cell blasts can be partially reduced by the administration of soluble TNF-receptor (TNFR) decoys,[26] and TNF may contribute to apoptosis induced by gp120-mediated cross-linking of CD4.[27]

Although FasL and, to a lesser extent, TNF seem to be the mediators of MØ-mediated apoptosis of CD4+ T cells, antigen-induced apoptosis of CD4+ T cells from HIV-infected individuals can be mediated by other members of the TNFR family known to induce apoptosis, such as the TNF-related apoptosis-inducing ligand (TRAIL/APO2L).[28] T cells from HIV-positive individuals are susceptible to TRAIL-mediated killing, in contrast to cells from control donors, and activation-induced cell death (AICD) is partially inhibited by antagonistic TRAIL-specific antibodies.[29] The fact that AICD in patients with HIV infection may be partially inhibited using antagonistic TRAIL/APO2L-specific antibodies[28] suggests that TRAIL/APO2L and TRAIL/APO2L receptor dysfunction may contribute to HIV pathogenesis. After infection with HIV, TRAIL is released by MØ, and exogenous HIV-encoded Tat protein upregulates the production of TRAIL by primary MØ *in vitro*,[30] indicating that a TRAIL-dependent mechanism of destruction of bystander CD4+ T cells may occur *in vivo*, which might be triggered by Tat produced by HIV-infected cells. Tat induces TRAIL in PBMCs, and the

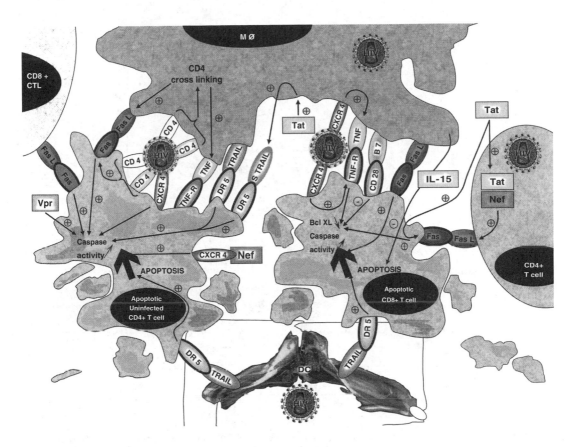

FIGURE 15.1 Intercellular contacts and soluble factors involved in autologous cell-mediated killing during HIV infection. Uninfected CD4+ T cells in contact with MØ undergo apoptosis following CD4 cross-linking and the resultant upregulation of expression of Fas and FasL on the surface of CD4+ T cells and MØ, respectively. Fas FasL interaction triggers the apoptosis of uninfected CD4+ T cells via activation of the caspase pathway. CD8+ T cells in contact with MØ undergo apoptosis after CXCR4 stimulation by X4 viruses. TNF–TNFR2 interaction triggers apoptosis of CD8+ T cells via downregulation of antiapoptotic proteins of the Bcl-2 family, such as Bcl-xL, and via caspase activation. As CD28 stimulation increases the levels of antiapoptotic members of the Bcl-2 family, such as Bcl-xL, B7–CD28 interaction might rescue CD8+ T cells from MØ-mediated CD8+ T cell apoptosis. DC could kill both CD4+ T cells and CD8+ T cells via TRAIL-DR5-mediated apoptosis.

majority of TRAIL is produced by the monocyte subset of PBMCs. A slight increase in cell surface expression of TRAIL was observed on monocytes, and sufficient TRAIL was secreted to be toxic for T cells. Remarkably, uninfected T cells are more susceptible to TRAIL than are HIV-infected T cells. The production of TRAIL by Tat-stimulated monocytes provides a mechanism by which HIV infection can destroy uninfected bystander cells.[31]

HIV-1-infected antigen-presenting monocytes are able to prime *in vitro* non-HIV-infected antigen-specific CD4+ T cells or peripheral blood CD4+ T cells to undergo apoptosis after antigen-specific restimulation. Priming for apoptosis required two concomitant signals present on the same APC, an antigenic stimulus and a second signal provided by the HIV gp120 protein.[6]

Because HIV gp120 binds to both CD4 and the chemokine receptors CXCR4 and CCR5 on the cell surface, it was suggested that the binding of gp120 not only to CD4 but also to the chemokine receptors can trigger apoptosis of uninfected T cells (Figure 15.2). Monoclonal blocking antibodies directed against the gp120 binding site of CD4 and against CXCR4 prevent the apoptotic signal

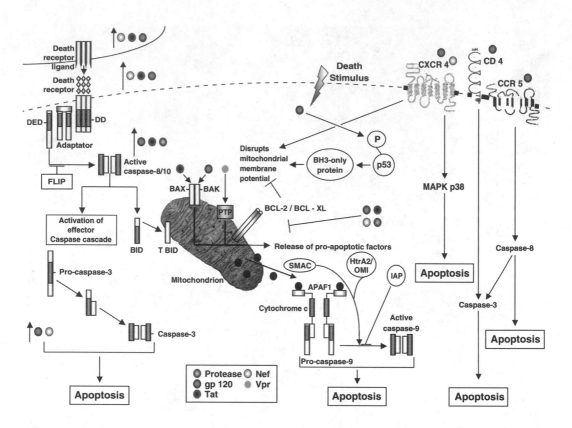

FIGURE 15.2 Molecular mechanisms involved in autologous cell-mediated killing during HIV infection. The extrinsic pathway is initiated by the binding of TNF-family death-receptor ligands to their cognate receptors. Through their death domains (DDs), multimerized receptors interact with the DDs adaptor proteins, which also contain death-effector domains (DEDs) that facilitate binding to procaspase-8 and procaspase-10 to form the death-inducing signal complex (DISC). As part of the DISC, the procaspases are cleaved into their active forms and initiate the intrinsic pathway of apoptosis. BID (BH3-interacting domain death agonist) is then cleaved to produce truncated (T)BID, and the effector caspase cascade is activated. Death-receptor-induced apoptosis can be blocked by FLIP (FLICE-like inhibitory protein), which inhibits the proteolytic processing of caspase-8. HIV-encoded proteins gp120, Nef, and Tat upregulate the expression of Fas and FasL. In addition, Tat upregulates the expression of TRAIL; gp120, Tat, and protease upregulate caspase-8; and gp120 and Nef increase the activity of caspase-3. The intrinsic apoptotic pathway is initiated by internal sensors, such as p53, that activate BH3-containing proteins and mediate the assembly of the proapoptotic members of the Bcl-2 family, including BAX and BAK, into hetero-oligomeric pores in the mitochondrial membrane. This results in the release of proapoptotic factors—cytochrome c, SMAC (second mitochondria-derived activator of caspases), and HtrA2/OMI—into the cytoplasm. This is associated with the loss of mitochondrial potential, which can be blocked by antiapoptotic proteins Bcl-2/Bcl-xL. Release of cytochrome c promotes the formation of the apoptosome, which includes Apaf-1 (apoptotic protease activating factor 1) and procaspase-9. Autolytic activation of caspase-9 initiates the effector caspase cascade. Caspase activation is negatively regulated by IAPs (inhibitors of apoptosis proteins), which are counterbalanced by SMAC and HtrA2/OMI. HIV-encoded proteins also trigger the mitochondrial pathway: gp120 induces the phosphorylation of p53; Tat and gp120 promote BAX insertion into the mitochondrial membrane and subsequent release of cytochrome c; Vpr has a direct effect on the mitochondrial permeability transition pore (PTP); gp120, Tat, and Nef inhibit expression of Bcl-2, whereas protease cleaves it; and Nef inhibits the expression of Bcl-xL. CXCR4 activation by gp120 and Nef results in apoptosis via disruption of mitochondrial membrane potential and MAPK p38 activation. Gp120-mediated stimulation of CCR5 and CD4 triggers apoptosis via activation of caspase-3 and -8.

in a dose-dependent manner.[32] Using human peripheral blood lymphocytes, malignant T cells, and CD4 or CXCR4 transfectants, apoptosis was found to be induced by both cell surface receptors, CD4 and CXCR4.[32] A role for CXCR4 in envelope-mediated CD4+ T cell apoptosis is further confirmed by the fact that a small molecule directed against CXCR4, AMD3100, inhibits cell-surface-expressed HIV-1 envelope-induced apoptosis.[33] Also, HIV-1 envelope-mediated CD4+ T cell apoptosis is inhibited by the cognate ligand of CXCR4, stromal-derived factor-1α (SDF-1α), again suggesting the involvement of CXCR4 in this phenomenon. The CXCR4-mediated CD4+ T cell apoptosis is Fas independent.[32] In addition, HIV-1 virions, with mutant Env that binds CXCR4 but that are defective for CD4 binding or membrane fusion, induce apoptosis, whereas CXCR4-binding-defective mutants do not.[33] Thus, HIV-1 virions can induce apoptosis through a CXCR4- or CCR5-dependent pathway that does not require Env/CD4 signaling or membrane fusion. This suggests that HIV-1 variants with increased envelope/receptor affinity or co-receptor binding site exposure may promote T lymphocyte apoptosis *in vivo* by accelerating bystander cell death. Nevertheless, some reports suggest that HIV-1 envelope-mediated CD4+ T cell death depends mostly on envelope–CD4-receptor interactions, as CCR5-using, as well as CXCR4-using, envelopes elicited this response.[34] These observations show that CD4+ T cell apoptosis can be mediated via CXCR4, and to a lesser extent CCR5, especially by HIV-1 envelope expressed on the surface of neighboring infected MØ.[32,35] The presence of CXCR4-mediated apoptosis, in addition to FasL-, TRAIL-, and TNF-mediated killing, indicates that CD4+ T cell apoptosis in HIV infection is a complex phenomenon that involves both immune activation and direct stimulation of CD4+ T cells by the HIV-1 envelope.

CD8+ T Cell Apoptosis

As well as CD4+ T cell apoptosis, CD8+ T cell apoptosis is observed in peripheral blood isolated from HIV-infected subjects and could account for increased CD8+ T cell turnover during HIV infection. CD8+ T cells isolated from the peripheral blood of HIV-infected subjects are activated cells with an increased cell surface expression of Fas and TNFR2.[13,28] Although Fas–FasL interaction was reported to be involved in CD8+ T cell apoptosis, the apoptosis seen in HIV infection is only partially inhibited by neutralizing anti-Fas and anti-FasL antibodies.[13] The TNF–TNFR death pathway is also altered in HIV-positive individuals. Although an early report found that peripheral T cells from HIV-positive individuals were resistant to apoptosis that was induced by ligation of TNFR,[36] more recent studies have shown that CD8+ T cells from HIV-positive individuals are susceptible to TNFR1- and TNFR2-induced apoptosis.[13,37] The ligand for TNFRs, TNF, is detected at increased levels in the serum of symptomatic individuals, and elevated levels of soluble TNFR2 have been found to be predictive of HIV disease progression.[38] TNF is partially responsible for activation-induced T cell apoptosis in animal models,[39] and TNFR2 stimulation is involved in the apoptosis of mature CD8+ T cells.[39–41] Apoptosis of CD8+ T cells during HIV infection has been shown to result from the interaction between membrane-bound TNF, which is expressed on the surface of activated MØ, and TNFR2 expressed on the surface of activated CD8+ T cells.[13] TNFR2 stimulation of T cells results in decreased intracellular levels of the apoptosis-protective protein Bcl-xL, a member of the Bcl-2 family.[41] Impaired induction of Bcl-xL has also been observed in PBMCs isolated from HIV-infected patients. Thus, TNFR2 stimulation on CD8+ T cells by membrane-bound TNF, expressed on the surface of MØ, decreases the intracellular levels of antiapoptotic proteins belonging to the Bcl-2 family and results in CD8+ T cell death. Susceptibility to TNFR-mediated apoptosis has been reported to be related to Bcl-2 expression, and early recruitment of caspase-3 and caspase-8 is needed to transduce the apoptotic signals. Thus, exacerbated TNFR-mediated cell death of T cells from HIV-infected individuals is associated with both alteration of Bcl-2 expression and activation of caspase-3 and caspase-8.[37] Cross-linking of death receptors leads to the activation of an array of caspases, and both initiator caspase-8 and effector caspase-3 are activated after the ligation of TNFR1 and TNFR2 on T cells from HIV-positive individuals

(Figure 15.2).[37] The active forms of both caspases are expressed *in vivo* by several HIV-encoded proteins, such as Tat, Env, Nef, and Vpr.[34,42–44] The function of CD8[+] T cells killed by MØ via TNF–TNFR2 interaction is still under investigation, but TNFR2 stimulation has been shown to be involved in the depletion of cytotoxic T lymphocytes (CTLs).[40] Also, CD8[+] T cell apoptosis has been reported to be caused by APC, because removal of monocytes or addition of antibodies to CD80 and CD86 reduced apoptosis. A unique CD8[bright]CD28[dim] T cell population died after costimulation by monocytes. Because this population was increased in patients with undetectable viremia, abnormal APCs may contribute to continued CD8[+] T cell exhaustion by inducing apoptosis.[18] It has also been reported that lymphocyte-reactive autoantibodies in HIV-1-infected persons facilitate the deletion of CD8[+] T cells by MØ.[45]

Dendritic Cell-Mediated T Cell Apoptosis

After measles and other viral infections and incubation with dsRNA, DCs become cytotoxic and, consequently, exhibit natural killer function through upregulation of type I interferon secretion that enhances TRAIL expression (Figure 15.1). In HIV infection, such a mechanism might be responsible for massive apoptosis of uninfected lymphocytes and increase specific immunity through cross-presentation of antigens from infected cells killed by DCs.[46] Exposure of normal monocyte-derived DCs to HIV-1 virions leads to the induction of apoptosis in cocultured CD4[+] as well as CD8[+] T cells.[47]

Among several mechanisms of immune evasion used by HIV is the progressive destruction of virus-specific effectors, such as CD4[+] T-helper cells, either through their direct infection during cognate interaction with infected DCs or through a bystander process that involves the upregulation of expression of death receptors and their ligands and the downregulation of expression of Bcl-2 family survival factors, leading to autocrine or paracrine destruction. After their recruitment into infected lymphoid sites, naive CD4[+] T cell precursors might also be killed directly after specific priming and infection by HIV-infected DCs.[48] Accordingly, during acute infection, rapidly proliferating HIV-specific memory CD4[+] T cells are highly susceptible to HIV infection. Also, they have been found to contain higher levels of virus DNA than other memory CD4[+] T cells,[49] suggesting that they are preferentially infected *in vivo* and, consequently, that they are preferentially lost. Finally, destruction of activated HIV-specific CD4[+] T cell effectors might result from fratricide mediated by FasL- or TNF-expressing killers that are generated by the persistent immune activation and the influence of HIV-encoded proteins. Failure to detect HIV-specific CD4[+] T cells *ex vivo* might also be due to their anergy, resulting from interaction with peripheral blood DCs,[50] their *in vivo* inhibition by high levels of viremia,[51] or their suppression by CD4[+]CD25[+] regulatory T cells.[52]

T Cell-Mediated Apoptosis

Both *in vivo* and *in vitro*, HIV infection is associated with an activated T-cell phenotype,[53–57] increased expression of Fas, enhanced susceptibility to Fas-mediated killing,[36,58–63] and increased T cell expressed FasL after T cell receptor stimulation,[19,64] suggesting a role for Fas/FasL in HIV-associated AICD.

As CD4[+] T cells can upregulate FasL, either by direct *in vivo* HIV infection or through the effect of virus proteins such as gp120, Tat, or Nef, they can become possible killers of activated Fas-expressing cells, which are found in high numbers in HIV-positive individuals. This hypothesis is supported by the observation that, *in vitro*, activated CD4[+] T cells that express FasL can kill Fas-expressing CD8[+] T cells independent of antigen recognition.[9,65] Also, HIV-specific CTLs are potential effectors for killing Fas-expressing activated lymphocytes.[10] An MHC class-I-restricted CTL clone specific for Nef, derived from an HIV-positive individual, is able to mediate both perforin- and Fas-expressing cell deaths.[66] The observation that retinoic acid inhibits the expression of FasL and the resultant CD4[+] T cell apoptosis *ex vivo*[67,68] further supports a causal role for

Fas–FasL interactions in the CD4+ T cell death that is induced by HIV. Therefore, some of the HIV-specific effectors might be harmful to the immune system of an HIV-positive individual through the FasL-dependent destruction of uninfected Fas-expressing T cells, which are induced by non-specific, HIV-driven immune stimulation. Altered differentiation of CTLs might be linked to increased apoptosis, as differently differentiated CD8+ T cell subsets have been found to have different susceptibilities to Fas-induced apoptosis.[69] Accordingly, rescue from apoptosis of the Bcl-2[low] CD8+ T cell subset, which is detected in the lymph nodes and blood of individuals that are chronically infected with HIV,[70,71] can be induced by incubation with interleukin (IL)-2 and IL-15.[72,73] IL-15, which is mainly produced by MØ, has many activities in common with IL-2: it is required for the survival of memory CD8+ T cells, natural killer (NK) cells, and NK-T cells, and it modulates immune functions of PBMCs from HIV-positive individuals by increasing NK-cell cytotoxicity and enhancing T cell proliferation in response to mitogens and opportunistic antigens.[73] In addition, IL-15 is a potent survival factor in the prevention of AICD of CD4+ and CD8+ T cells from HIV-positive individuals and functions by upregulating the expression of Bcl-2.[73] Under the same experimental conditions, IL-2, but not IL-10 or -12, has a protective effect on CD4+ and CD8+ T cells.[73] However, IL-15 is not able to prevent Fas-induced apoptosis of T cells *in vitro*,[73] in contrast to its *in vivo* effect in a mouse model.[74] IL-12 and -10 partially inhibit Fas-induced apoptosis of CD4+ T cells and CD8+ T cells from HIV-positive individuals, respectively.[73,75] IL-15 could, therefore, be an effective promoter of innate immune responses by contributing to the development and survival of TH1 cells. Accordingly, recent attention has turned toward IL-15 as a possible alternative immunotherapy for HIV-positive individuals.

AUTOLOGOUS CELL-MEDIATED KILLING VIA THE RELEASE OF SOLUBLE FACTORS

CYTOTOXIC HIV PROTEINS

Env

The influence of HIV on the extrinsic pathways of cell death has been reported both in productively infected cells and in bystander cells. Cross-linking of CD4+ T cells by gp120 activates the Fas–FasL pathway[17] and induces T cell apoptosis (Figure 15.2). Pretreatment with a human polyclonal anti-gp120 antibody that recognizes the CD4 binding region (5145) as well as with an anti-Fas ligand mAb blocked apoptosis in CD4+ T cells but not in CD8+ T cells. A soluble factor was detected that induced apoptosis in CD4+ and CD8+ T cells and B cells.[76] CD8+ T cells express the CD4 receptor after activation, thereby rendering them susceptible to direct infection by the virus.[77,78] In addition, the enhanced expression of CD4 antigen on CD8+ T cells would be expected to render these double-positive cells more susceptible to the effects of gp120 cross-linking and, subsequently, apoptosis. β-Chemokines inhibit AICD of lymphocytes from HIV-infected individuals,[79] and the administration of CXCR4 antagonists blocks the apoptotic response to the HIV envelope.[33,80–82] Activation of the mitogen-activated protein kinase (MAPK) leads to gp120-induced expression of chemokines such as monocyte chemoattractant protein-1 (MCP-1) and macrophage inflammatory protein-1 (MIP-1β) and TNF-α. Dysregulation of MØ function by gp120/chemokine receptor signaling contributes to local inflammation and injury and further recruits additional inflammatory and target cells.[83]

As disease progresses, HIV-1 infected patients show a loss of NK cytotoxic function and a gradual loss of CD4+ T cells and NK cells. A concomitant killing *in vitro* of both gp120-coated CD4+ peripheral blood lymphocytes and NK cells has been observed in the antibody-dependent cell-mediated cytotoxicity (ADCC) system.[84]

The envelope gp120$_{SF2}$ and SDF-1 differ in the cell type on which they stimulate CXCR4 to induce neuronal apoptosis, but both ligands use the p38 MAPK pathway for death signaling. Glycoprotein gp120$_{SF2}$-induced neuronal apoptosis depends predominantly on an indirect pathway via activation of chemokine receptors on MØ/microglia, whereas SDF-1 may act directly on neurons or astrocytes.[85]

Nef, Tat, Vpr

Nef-expressing T cells coexpress FasL,[86] thereby becoming potential killers of uninfected Fas-expressing T cells. Similarly, Tat, which is secreted by infected cells, upregulates Fas and FasL on uninfected cells and enhances their susceptibility to Fas-mediated apoptosis.[63,87] Recently, soluble Nef protein was shown to trigger apoptosis of CD4[+] T cells via stimulation of CXCR4.[88] Synthetic Vpr induces apoptosis via a direct effect on the mitochondrial permeability transition pore, resulting in the mitochondrial release of apoptogenic proteins such as cytochrome c (Figure 15.2).[89–91]

Cytotoxic Factors

Neuronal apoptosis within the central nervous system is a characteristic feature of AIDS dementia, and it represents a common mechanism of neuronal death induced by neurotoxins, e.g., glutamate, released from HIV-infected MØ. The N-methyl-D-aspartate (NMDA) glutamate receptor/Bcl-2-regulated apoptotic pathway contributes significantly to HIV/MØ-induced neuronal apoptosis, and Bcl-2 family proteins protect neurons against the spectrum of primary HIV-1 isolates.[92] An unusual syncytium-inducing HIV-1 primary isolate from the central nervous system that is restricted to CXCR4 replicates efficiently in MØ and induces neuronal apoptosis via the release of soluble factors.[92] Primary HIV-1 (BaL)-infected monocytes produce a 66 kDa protein (FLJ21908) that can be detected in brain and lymph tissue from HIV-1-infected patients who suffer from AIDS dementia. Supernatants from these cells induce apoptosis in PBMC, CD4[+] T cells, CD8[+] T cells, and B cells.[93] Administration of lipopolysaccharide (LPS) to HIV-1-infected huPBMC-NOD-SCID (human PBMC into nonobese diabetic severe combined immunodeficient) mice induces infiltration of HIV-1-infected human cells in the perivascular region of the brain, and neuronal apoptosis is found in MØ-tropic but not T cell-tropic HIV-1-infected brains. The apoptotic neurons were frequently colocalized with the HIV-1-infected MØ that expressed TRAIL. Administration of a neutralizing antibody against TRAIL but not TNF-α or FasL blocked the neuronal apoptosis in the HIV-1-infected brain. These results strongly suggest a critical contribution of TRAIL expressed on HIV-1-infected MØ to neuronal apoptosis.[94] Also, focal inflammation in brain tissue with HIV-1 encephalitis may upregulate neuronal fractalkine levels, which, in turn, may be a neuroimmune modulator recruiting peripheral MØ into the brain.[95]

AUTOLOGOUS CELL-MEDIATED KILLING RESULTS IN IMMUNE SUPPRESSION

Immune Suppression

The degree of apoptosis in lymph nodes obtained from HIV-positive persons is three to four times higher than that observed in the lymph nodes obtained from HIV-negative persons.[96] A significant correlation is observed between intensity of apoptosis and degree of activation of the lymphoid tissue associated with HIV infection.[96] In lymph nodes from HIV-positive individuals, *in vivo* apoptosis is detected predominantly in uninfected bystander cells,[12] arguing for indirect mechanisms of cell destruction, as also indicated by the presence of apoptotic CD8[+] T cells, B cells, and DCs in the lymphoid organs of HIV-positive individuals.[96,97] The destruction of HIV-specific CD4[+] T cell precursors might occur *in vivo* in lymph nodes after HIV binding and ligation of the homing receptor CD62L.[98,99] Large numbers of CD8[+]CD45RO[+] T cells infiltrate both the paracortex and the germinal centers. These cells express low levels of Bcl-2 and correspond to CD8[+] apoptotic cells in the paracortical areas of the HIV-infected lymph nodes. In parallel, the MØ population is expanded and activated, and apoptotic bodies are seen in the cytoplasm of the activated CD68+RFD-7+RFD-1+ MØ. The downregulation of Bcl-2 is detected *in vivo* in the lymph nodes and blood of HIV-positive individuals.[96] Cells that express low levels of Bcl-2 have an activated phenotype *in vivo*,[71,72] indicating that the premature priming for spontaneous apoptosis of patient T cells is mainly the result of immune activation that is induced by repeated virus antigen exposure, thereby tipping the balance in favor of the proapoptotic Bcl-2 family members — a mechanism that is known to contribute to the physiological homeostasis of activated T cells *in vivo*.[100] Spontaneous T cell apoptosis is normally prevented by cytokines and, accordingly, IL-2 and IL-15 can promote the survival of patient T cells *in vitro* by upregulating the expression of Bcl-2.[72,73] Interestingly, nonpathogenic HIV-1 infection in chimpanzees is associated with a lack of immune activation, a low level of spontaneous T cell apoptosis, and normal expression of Bcl-2.[58] In the lymph nodes of HIV-positive individuals, syncytia express markers of early apoptosis, such as tissue transglutaminase.[97] The relevance of fusion-induced apoptosis to AIDS pathogenesis is indicated by the positive correlation between infection with syncytium-inducing HIV or SIV isolates and the decline in CD4+ T cell numbers *in vivo*.[101,102]

Numerous studies show that apoptosis in lymph nodes, rectal mucosa, and peripheral blood lymphocyte (PBL) subsets from patients infected with HIV decreases dramatically in response to protease inhibitor–based HIV treatment.[103-109] This effect is seen for spontaneous apoptosis, apoptosis in response to T cell receptor ligation, apoptosis in response to mitogenic stimulation, and apoptosis in response to Fas receptor ligation.[103,106,108,109] The decrease in apoptosis is rapid and is seen as early as 4 days after protease inhibitor therapy is initiated[103]; it occurs in all patients within 14 days.[106-109] Because the decrease precedes significant changes in viral replication, it has been suggested that protease inhibitors may be antiapoptotic,[110,111] possibly by virtue of inhibiting the activity of effector proteases involved in apoptosis (Figure 15.2). Highly active antiretroviral therapy (HAART) rapidly reduces apoptosis in lymphoid tissue and significantly decreases apoptosis in PBLs.[103]

In animal models, CD4+ and CD8+ T cells from SIVmac251-infected macaques undergo caspase-dependent death after Fas ligation.[112] During the first 2 weeks of infection of rhesus macaques with SIVmac251 isolate, peak numbers of apoptotic cells in the lymph node T cell areas were significantly higher in the future rapid progressors than in the future slow progressors and were correlated with subsequent viremia levels measured 6 months after infection. Thus, extensive apoptosis induction in peripheral lymphoid organs is an early and predictive event that may play a crucial role in impairing the capacity of the immune system to control viral replication and progression toward disease.[113] A recent study using a mouse model showed that chronic immune activation has an important role in T cell immunodeficiency. Transgenic mice that constitutively express CD70, a molecule that is induced by antigen activation of stimulated lymphocytes, develop features that are characteristic of HIV infection—that is, a progressive conversion of naive T cells to effector memory cells, which leads to the depletion of the naive T cell pool and ultimately to death from opportunistic infections.[114] Collectively, these observations provide further evidence that chronic immune activation is a principal cause of T cell dysfunction and apoptosis, contributing markedly to AIDS pathogenesis. Lymphocytic choriomeningitis virus (LCMV)-specific T cells in the peripheral tissue (peritoneal cavity, lung, fat pads) react much less with the apoptotic marker annexin-V than those in the spleen and lymph nodes. In comparison to lymphoid tissue, T cells in the periphery express lower levels of Fas and FasL and are resistant to AICD *in vitro*,[115] suggesting a critical role for lymphoid tissues, and potentially APC, in T cell-mediated apoptosis.

RESERVOIRS

The main obstacle to viral eradication in HIV-infected patients is the presence of chronically infected latent reservoir cells, such as MØ, and latently infected CD4+ T lymphocytes.[116-119] In these cellular populations, HIV infection is not associated with apoptosis but with a chronic productively infected phenotype. CD4+ T cells that contain latent proviral DNA form reservoirs for HIV, as they can survive for a long time with HIV integrated in their genomes.[120] By infecting MØ, HIV can favor the recruitment of T cells, and the viral protein Nef has been shown to be a major player in this scenario. Infection of MØ by HIV-1 bearing a functional Nef gene, or by an adenovirus engineered to express Nef, results in two phenotypic consequences.[121] First, MØ secrete the chemokine MIP-1α and MIP-1β in response to Nef expression. These chemokines promote the chemotaxis of resting T cells, thus increasing the likelihood that CD4+ T cells will encounter HIV virions released from the infected MØ before these are cleared by the reticuloendothelial system. Second, MØ that express Nef or are stimulated through the CD40 receptor release soluble CD23 and soluble intercellular adhesion molecules (ICAM) that render T lymphocytes permissive to HIV-1 infection. Thus, Nef expands the cellular reservoir of HIV-1 by permitting the infection of resting T lymphocytes.[122] This chemotaxis might not be restricted to CD4+ T cells but could also target CD8+ T cells, thereby increasing MØ-mediated CD8+ T cell death and the subsequent immune suppression in HIV-infected subjects. Increased survival of productively infected CD4+ T cells has been reported to require intercellular contacts between MØ and HIV-infected T cells expressing the Nef protein.[123] Furthermore, it has been suggested that Nef might interact with cellular proteins belonging to the TNF–TNFR and Fas–FasL pathways, thereby blocking the apoptotic process. The Nef protein has been reported to assemble with and inhibit the catalytic activity of the apoptosis signal-regulating kinase 1 (ASK1), which is known to be involved in death signaling mediated via both Fas and TNFR1.[124] The resistance to apoptosis in productively infected CD4+ T cells could result from a blockade of the proapoptotic pathway or the activation of antiapoptotic pathways within the infected T cells in contact with MØ. As well as the actions of the Nef protein, the second exon of the regulatory HIV protein Tat stimulates the production of IL-2 by activating a CD28 responsive element, which is located within the promoter of the IL-2 gene.[125,126] This might also allow for the survival and proliferation of activated infected T cells in contact with MØ.

Nonapoptotic infected cells may serve as viral reservoirs, and it is unlikely that phenotypic and functional abnormalities of infected cells will be reversed by merely inhibiting apoptosis. Also, latently infected CD4+ T cells have a markedly prolonged half-life, which limits the probability that viral reservoirs can be eliminated by interference in viral replication alone.[127] Recent estimates based on the half-life of latently infected cells suggest that 60 years of viral suppression would be required to eliminate viral reservoirs.[117] A possible way to achieve viral eradication is to target infected MØ and latently infected CD4+ T cells to undergo apoptosis after infection. It was recently proposed that treatment with a procaspase-3 analogue that contains an HIV protease–specific sequence in its prodomain may cause apoptosis of all infected cells.[128]

CONCLUSION

Previous studies estimated that only 0.00001 to 0.01% of HIV-1 virions are infectious *in vitro* and *in vivo*. Thus, apoptosis of uninfected bystander cells may contribute to HIV-1 pathogenicity *in vivo*. The ability of HIV-1 to induce bystander apoptosis is influenced by several viral and nonviral factors, including activation of APCs and T lymphocytes, and the production of soluble viral factors and cytotoxic cellular factors. Thus, APCs play a pivotal role in the regulation of bystander cell apoptosis following HIV infection. In addition to highly active antiretroviral regimens, new therapeutic approaches that target APCs could result in increased resistance to apoptosis in uninfected T lymphocytes and favor the depletion of productively infected T lymphocytes. This could ultimately enhance immune restoration and limit the formation of viral reservoirs in HIV-infected patients.

ACKNOWLEDGMENTS

We thank Dominique Thomasset and Aurélie Vacheret for technical and secretarial assistance.

REFERENCES

1. Gougeon, M.L., Ledru, E., Naora, H., Bocchino, M., and Lecoeur, H., HIV, cytokines and programmed cell death. A subtle interplay, *Ann. N.Y. Acad. Sci.,* 926, 30, 2000.
2. Selliah, N. and Finkel, T.H., Biochemical mechanisms of HIV induced T cell apoptosis, *Cell Death Differ.,* 8, 127, 2001.
3. Badley, A.D., McElhinny J.A., Leibson, P.J., Lynch, D.H., Alderson, M.R., and Paya, C.V., Upregulation of Fas ligand expression by human immunodeficiency virus in human macrophages mediates apoptosis of uninfected T lymphocytes, *J. Virol.,* 70, 199, 1996.
4. Badley, A.D., Dockrell, D., Simpson, M., Schut, R., Lynch, D.H., Leibson, P., and Paya, C.V., Macrophage-dependent apoptosis of CD4+ T lymphocytes from HIV-infected individuals is mediated by FasL and tumor necrosis factor, *J. Exp. Med.,* 185, 55, 1997.
5. Oyaizu, N., Adachi, Y., Hashimoto, F., McCloskey, T.W., Hosaka, N., Kayagaki, N., Yagita, H., and Pahwa, S., Monocytes express Fas ligand upon CD4 cross-linking and induce CD4+ T cells apoptosis: a possible mechanism of bystander cell death in HIV infection, *J. Immunol.,* 158, 2456, 1997.
6. Cottrez, F., Manca, F., Dalgleish, A.G., Arenzana-Seisdedos, F., Capron, A., and Groux, H., Priming of human CD4+ antigen-specific T cells to undergo apoptosis by HIV-infected monocytes. A two-step mechanism involving the gp120 molecule, *J. Clin. Invest.,* 99, 257, 1997.
7. Orlikowsky, T., Wang, Z.Q., Dudhane, A., Horowitz, H., Riethmuller, G., and Hoffmann, M., Cytotoxic monocytes in the blood of HIV type 1-infected subjects destroy targeted T cells in a CD95-dependent fashion, *AIDS Res. Hum. Retroviruses,* 13, 953, 1997.
8. Nardelli, B., Gonzalez, C.J., Schechter, M., and Valentine, F.T., CD4+ blood lymphocytes are rapidly killed *in vitro* by contact with autologous human immunodeficiency virus-infected cells, *Proc. Natl. Acad. Sci. U.S.A.,* 92, 7312, 1995.
9. Kameoka, M., Suzuki, S., Kimura, T., Fujinaga, K., Auwanit, W., Luftig, R.B., and Ikuta, K., Exposure of resting peripheral blood T cells to HIV 1 particles generates CD25+ killer cells in a small subset, leading to induction of apoptosis in bystander cells, *Int. Immunol.,* 9, 1453, 1997.
10. Kojima, H., Eshima, K., Takayama, H., and Sitkovsky, M.V., Leukocyte function-associated antigen-1-dependent lysis of Fas+ (CD95+/Apo-1+) innocent bystanders by antigen-specific CD8+ CTL, *J. Immunol.,* 159, 2728, 1997.
11. Oyaizu, N., McCloskey, T.W., Coronesi, M., Chirmule, N., Kalyanaraman, V.S., and Pahwa, S., Accelerated apoptosis in peripheral blood mononuclear cells (PBMCs) from human immunodeficiency virus type-1 infected patients and in CD4 cross-linked PBMCs from normal individuals, *Blood,* 82, 3392, 1993.
12. Finkel, T.H., Tudor-Williams, G., Banda, N.K., Cotton, M.F., Curiel, T., Monks, C., Baba, T.W., Ruprecht, R.M., and Kupfer, A., Apoptosis occurs predominantly in bystander cells and not in productively infected cells of HIV- and SIV-infected lymph nodes, *Nat. Med.,* 1, 129, 1995.
13. Herbein, G., Mahlknecht, U., Batliwalla, F., Gregersen, P.K., Pappas, T., Butler, J., O'Brien, W.A., and Verdin, E., Apoptosis of CD8+ T cells is mediated by macrophages through interaction of HIV gp120 with chemokine receptor CXCR4, *Nature,* 395, 189, 1998.
14. Mosier, D.E., Gulizia, R.J., McIsaac, P.D., Torbett, B.E., and Levy, J.A., Rapid loss of CD4+ T cells in human-PBL-SCID mice by noncytopathic HIV isolates, *Science,* 260, 689, 1993.
15. Gendelman, H.E., Ehrlich, G.D., Baca, L.M., Conley, S., Ribas, J., Kalter, D.C., Meltzer, M.S., Poiesz, B.J., and Nara, P., The inability of human immunodeficiency virus to infect chimpanzee monocytes can be overcome by serial viral passage *in vivo, J. Virol.,* 65, 3853, 1991.
16. Schuitemaker, H., Mayaard, L., Koostra, N.A., Dubbes, R., Otto, S.A., Tersmette, M., Heeney, J.L., and Miedema, F., Lack of T cell dysfunction and programmed cell death in human immunodeficiency virus type 1-infected chimpanzees correlates with absence of monocytotropic variants, *J. Infect. Dis.,* 168, 1140, 1993.
17. Banda, N.K., Bernier, J., Kurahara, D.K., Kurrle, R., Haigwood, N., Sekaly, R.P., and Finkel, T.H., Crosslinking CD4 by human immunodeficiency virus gp120 primes T cells for activation-induced apoptosis, *J. Exp. Med.,* 176, 1099, 1992.
18. Lewis, D.E., Ng Tang, D.S., Wang, X., and Kozinetz, C., Costimulatory pathways mediate monocyte-dependent lymphocyte apoptosis in HIV, *Clin. Immunol.,* 90, 302, 1999.

19. Sloand, E.M., Yound, N.S., Kumar, P., Weichold, F.F., Sato, T., and Maciejewski, J.P., Role of Fas ligand and receptor in the mechanism of T-cell depletion in acquired immunodeficiency syndrome: effect on CD4+ lymphocyte depletion and human immunodeficiency virus replication, *Blood,* 89, 1357, 1997.

20. Miller, M.D., Warmerdam, M.T., Page, K.A., Feinberg, M.B., and Greene, W.C., Expression of the human immunodeficiency virus type 1 (HIV-1) Nef gene during HIV-1 production increases progeny particle infectivity independently of gp160 or viral entry, *J. Virol.,* 69, 579, 1995.

21. Mitra, D., Steiner, M., Lynch, D.H., Staiano-Coico, L., and Laurence, J., HIV-1 upregulates Fas ligand expression in CD4+ T cells *in vitro* and *in vivo:* association with Fas-mediated apoptosis and modulation by aurintricarboxylic acid, *Immunology,* 87, 581, 1996.

22. Silvestris, F., Camarda, G., Cafforio, P., and Dammacco, F., Upregulation of Fas ligand secretion in non-lymphopenic stages of HIV-1 infection, *AIDS,* 12, 1103, 1998.

23. Hosaka, N., Oyaizu, N., Kaplan, M.H., Yagita, H., and Pahwa, S., Membrane and soluble forms of Fas (CD95) and Fas ligand in peripheral blood mononuclear cells and in plasma from human immunodeficiency virus-infected persons, *J. Infect. Dis.,* 178, 1030, 1998.

24. Dockrell, D.H., Badley, A.D., Villacian, J.S., Heppelmann, C.J., Algeciras, A., Ziesmer, S., Yagita, H., Lynch, D.H., Roche, P.C., Leibson, P.J., and Paya, C.V., The expression of Fas ligand by macrophages and its upregulation by human immunodeficiency virus infection, *J. Clin. Invest.,* 101, 2394, 1998.

25. Orlikowsky, T.W., Wang, ZQ, Dudhane, A., Dannecker, G.E., Niethammer, D., Wormser, G.P., Hoffmann, M.K., and Horowitz, H.W., Dexamethasone inhibits CD4 T cell deletion mediated by macrophages from human immunodeficiency virus-infected persons, *J. Infect. Dis.,* 184, 1328, 2001.

26. Lum, J.J., Pilon, A.A., Sanchez-Dardon, J., Phenix, B.N., Kim, J.E., Mihowich, J., Jamison, K., Hawley-Foss, N., Lynch, D.H., and Badley, A.D., Induction of cell death in human immunodeficiency virus-infected macrophages and resting memory CD4 T cells by TRAIL/Apo2, *J. Virol.,* 75, 11128, 2001.

27. Zheng, L., Fisher, G., Miller, R.E., Peschon, J., Lynch, D.H., and Lenardo, M.J., Induction of apoptosis in mature T cells by tumour necrosis factor, *Nature,* 377, 348, 1995.

28. Katsikis, P.D., Garcia-Ojeda, M.E., Torres-Roca, J.F., Tijoe, I.M., Smith, C.A., Herzenberg, L.A., and Herzenberg, L.A., Interleukin-1 converting enzyme-like protease involvement in Fas-induced and activation-induced peripheral blood T cell apoptosis in HIV infection. TNF-related apoptosis-inducing ligand can mediate activation-induced T cell death in HIV infection, *J. Exp. Med.,* 186, 1365, 1997.

29. Jeremias, I., Herr, I., Boehler, T., and Debatin, K.M., TRAIL/Apo-2-ligand-induced apoptosis in human T cells, *Eur. J. Immunol.,* 28, 143, 1998.

30. Zhang, M., Li, X., Pang, X., Ding, l., Wood, O., Clouse, K., Hewlett, I., and Dayton, A.I., Identification of a potential HIV-induced source of bystander-mediated apoptosis in T cells: upregulation of TRAIL in primary human macrophages by HIV-1 Tat, *J. Biomed. Sci.,* 8, 290, 2001.

31. Yang, Y., Tikhonov, I., Ruckwardt, T.J., Djavani, M., Zapata, J.C., Pauza, C.D., and Salvato, M.S., Monocytes treated with human immunodeficiency virus Tat kill uninfected CD4(+) cells by a tumor necrosis factor-related apoptosis-induced ligand-mediated mechanism, *J. Virol.,* 77, 6700, 2003.

32. Berndt, C., Mopps, B., Angermuller, S., Gierschik, P., and Krammer, P.H., CXCR4 and CD4 mediate a rapid CD95-independent cell death in CD4(+) T cells, *Proc. Natl. Acad. Sci. U.S.A.,* 95, 12556, 1998.

33. Holm, G.H., Zhang, C., Gorry, P.R., Peden, K., Schols, D., De Clercq, E., and Gabuzda, D., Apoptosis of bystander T cells induced by human immunodeficiency virus type 1 with increased envelope/receptor affinity and coreceptor binding site exposure, *J. Virol.,* 78, 4541, 2004.

34. Cicala, C., Arthos, J., Rubbert, A., Selig, S., Wildt, K., Cohen, O.J., and Fauci, A.S., HIV-1 envelope induces activation of caspase-3 and cleavage of focal adhesion kinase in primary human CD4(+) T cells, *Proc. Natl. Acad. Sci. U.S.A.,* 97, 1178, 2000.

35. Penn, M.L., Grivel, J.C., Schramm, B., Goldsmith, M.A., and Margolis, L., CXCR4 utilization is sufficient to trigger CD4+ T cell depletion in HIV-1-infected human lymphoid tissue, *Proc. Natl. Acad. Sci. U.S.A.,* 96, 663, 1999.

36. Katsikis, P.D., Wunderlich, E.S., Smith, C.A., and Herzenberg, L.A., Fas antigen stimulation induces marked apoptosis of T lymphocytes in human immunodeficiency virus-infected individuals, *J. Exp. Med.,* 181, 2029, 1995.

37. de Oliveira Pinto, L.M., Garcia, S., Lecoeur, H., Rapp, C., and Gougeon, M.L., Increased sensitivity of T lymphocytes to tumor necrosis factor receptor 1 (TNFR1)- and TNFR2-mediated apoptosis in HIV infection: relation to expression of Bcl-2 and active caspase-8 and caspase-3, *Blood,* 99, 1666, 2002.

38. Zangerle, R., Gallati, H., Sarcletti, M., Wachter, H., and Fuchs, D., Tumor necrosis factor and soluble tumor necrosis factor receptors in individuals with human immunodeficiency virus infection, *Immunol. Lett.,* 41, 229, 1994.

39. Zheng, L., Fisher, G., Miller, R.E., Peschon, J., Lynch, D.H., and Lenardo, M.J., Induction of apoptosis in mature T cells by tumour necrosis factor, *Nature,* 377, 348, 1995.

40. Alexander-Miller, M.A., Derby, M.A., Sarin, A., Henkart, P.A., and Berzofsky, J.A., Supraoptimal peptide-major histocompatibility complex causes a decrease in Bc1-2 levels and allows tumor necrosis factor receptor II-mediated apoptosis of cytotoxic T lymphocytes, *J. Exp. Med.,* 188, 1391, 1998.

41. Lin, R.H., Hwang, Y.W., Yang, B.C., and Lin, C.S., TNF receptor-2-triggered apoptosis is associated with the downregulation of Bcl-xL on activated T cells and can be prevented by CD28 costimulation, *J. Immunol.,* 158, 598, 1997.

42. Bartz, S.R. and Emerman, M., Human immunodeficiency virus type 1 Tat induces apoptosis and increases sensitivity to apoptotic signals by up-regulating FLICE/caspase-8, *J. Virol.,* 73, 1956, 1999.

43. Rasola, A., Gramaglia, D., Boccaccio, C., and Comoglio, P.M., Apoptosis enhancement by the HIV-1 Nef protein, *J. Immunol.,* 166, 81, 2001.

44. Stewart, S.A., Poon, B., Song, J.Y., and Chen, I.S., Human immunodeficiency virus type 1 Vpr induces apoptosis through caspase activation, *J. Virol.,* 74, 3105, 2000.

45. Wang, Z.Q., Horowitz, H.W., Orlikowsky, T., Dudhane, A., Weinstein, A., and Hoffmann, M.K., Lymphocyte-reactive autoantibodies in human immunodeficiency virus type 1-infected persons facilitate the deletion of CD8 T cells by macrophages, *J. Infect. Dis.,* 178, 404, 1998.

46. Servet, C., Zitvogel, L., and Hosmalin, A., Dendritic cells in innate immune responses against HIV, *Curr. Mol. Med.,* 2, 739, 2002.

47. Suzuki, S., Tobiume, M., Kameoka, M., Sato, K., Takahashi, T.A., Mukai, T., and Ikuta, K., Exposure of normal monocyte-derived dendritic cells to human immunodeficiency virus type-1 particles leads to the induction of apoptosis in co-cultured CD4+ as well as CD8+ T cells, *Microbiol. Immunol.,* 44, 111, 2000.

48. Patterson, S., Rae, A., Hockey, N., Gilmour, J., and Gotch, F., Plasmacytoid dendritic cells are highly susceptible to human immunodeficiency virus type 1 infection and release infectious virus, *J. Virol.,* 75, 6710, 2001.

49. Douek, D.C., Brenchley, J.M., Betts, M.R., Ambrozak, D.R., Hill, B.J., Okamoto, Y., Casazza, J.P., Kuruppu, J., Kunstman, K., Wolinsky, S., Grossman, Z., Dybul, M., Oxenius, A., Price, D.A., Connors M., and Koup, R.A., HIV preferentially infects HIV-specific CD4+ T cells, *Nature,* 417, 95, 2002.

50. Kuwana, M., Kaburaki J., Wright, T.M., Kawakami, Y., and Ikeda, Y., Induction of antigen-specific human CD4(+) T cell anergy by peripheral blood DC precursors, *Eur. J. Immunol.,* 31, 2547, 2001.

51. McNeil, A.C., Shupert, W.L., Iyasere, C.A., Hallahan, C.W., Mican, J.A., Davey, R.T. Jr, and Connors, M., High-level HIV-1 viremia suppresses viral antigen-specific CD4(+) T cell proliferation, *Proc. Natl. Acad. Sci. U.S.A.,* 98, 13878, 2001.

52. Belkaid, Y., Piccirillo, C.A., Mendez, S., Shevach, E.M., and Sacks, D.L., CD4+CD25+ regulatory T cells control *Leishmania major* persistence and immunity, *Nature,* 420, 502, 2002.

53. Gougeon, M.L., Lecoeur, H., Dulioust, A., Enouf, M.G., Crouvoiser, M., Goujard, C., Debord, T., and Montagnier, L., Programmed cell death in peripheral lymphocytes from HIV-infected persons: increased susceptibility to apoptosis of CD4 and CD8 T cells correlates with lymphocyte activation and with disease progression, *J. Immunol.,* 156, 3509, 1996.

54. Miedema, F., Petit, A.J., Terpstra, F.G., Schattenkerk, J.K., De Wolf, F., Al, B.J., Ross, M., Lange, J.M., Danner, S.A., Goudsmith, J., Immunological abnormalities in human immunodeficiency virus (HIV)-infected asymptomatic homosexual men. HIV affects the immune system before CD4+ T helper cell depletion occurs, *J. Clin. Invest.,* 82, 1908, 1988.

55. Clerici, M., Lucey, D.R., Berzofsky, J.A., Pinto, L.A., Wynn, T.A., Blatt, S.P., Dolan, M.J., Hendrix, C.W., Wolf, S.F., and Shearer, G.M., Restoration of HIV-specific cell-mediated immune responses by interleukin-12 *in vitro, Science,* 262, 1721, 1993.

56. Graziosi, C., Pantaleo, G., Gantt, K.R., Fortin, J.P., Demarest, J.F., Cohen, O.J., Sekaly, R.P., and Fauci, A.S., Lack of evidence for the dichotomy of TH1 and TH2 predominance in HIV-infected individuals, *Science,* 265, 248, 1994.

57. Meyaard, L., Otto, S.A., Jonker, R.R., Mijnster, M.J., Keet, R.P.M., and Miedema, F., Programmed death of T cells in HIV-1 infection, *Science,* 257, 217, 1992.

58. Gougeon, M.L., Lecoeur, H., Boudet, F., Ledru, E., Marzabal, S., Boullier, S., Roue, R., Nagata, S., and Heeney, J., Lack of chronic immune activation in HIV-infected chimpanzees correlates with the resistance of T cells to Fas/Apo-1 (CD95)-induced apoptosis and preservation of a T helper 1 phenotype, *J. Immunol.,* 158, 2964, 1997.

59. Gehri, R., Hahn, S., Rothen, M. Steuerwald, M., Nuesch, R., and Erb, P., The Fas receptor in HIV infection: expression on peripheral blood lymphocytes and role in the depletion of T cells, *AIDS,* 10, 9, 1996.

60. Kobayashi, N., Hamamoto, Y., Yamamoto, N., Ishii, A., Yonehara, M., and Yonehara, S., Anti-Fas monoclonal antibody is cytocidal to human immunodeficiency virus-infected cells without augmenting viral replication, *Proc. Natl. Acad. Sci. U.S.A.,* 87, 9620, 1990.

61. Silvestris, F., Cafforio, P., Frassanito, M.A., Tucci, M., Romito, A., Nagata, S., and Dammacco, F., Overexpression of Fas antigen on T cells in advanced HIV-1 infection: differential ligation constantly induces apoptosis, *AIDS,* 10, 131, 1996.

62. Aries, S.P., Schaaf, B., Muller, C., Dennin, R.H., and Dalhoff, K., Fas (CD95) expression on CD4+ T cells from HIV-infected patients increases with disease progression, *J. Mol. Med.,* 73, 591, 1995.

63. Westendorp, M.O., Frank, R., Ochsenbauer, C., Stricker, K., Dhein, J., Walczak, H., Debatin, K.M., and Krammer, P.H., Sensitization of T cells to CD95-mediated apoptosis by HIV-1 Tat and gp120, *Nature,* 375, 497, 1995.

64. Baumler, C.B., Bohler, T., Herr, I., Benner, A., Krammer, P.H., and Debatin, K.M., Activation of the CD95 (APO-1/Fas) system in T cells from human immunodeficiency virus type-1-infected children, *Blood,* 88, 1741, 1996.

65. Piazza, C., Gilardini Montani, M.S., Moretti, S., Cundari, E., and Piccolella, E., Cutting edge: CD4+ T cells kill CD8+ T cells via Fas/Fas ligand-mediated apoptosis, *J. Immunol.,* 158, 1503, 1997.

66. Garcia, S., Fevrier, M., Dadaglio, G., Lecoeur, H., Riviere, Y., and Gougeon, M.L., Potential deleterious effect of anti-viral cytotoxic lymphocyte through the CD95 (FAS/APO-1)-mediated pathway during chronic HIV infection, *Immunol. Lett.,* 57, 53, 1997.

67. Szondy, Z., Lecoeur, H., Fesus, L., and Gougeon, M.L., All-trans retinoic acid inhibition of anti-CD3-induced T cell apoptosis in human immunodeficiency virus infection mostly concerns CD4 T lymphocytes and is mediated via regulation of CD95 ligand expression, *J. Infect. Dis.,* 178, 1288, 1998.

68. Yang, Y., Mercep, M., Ware, C.F., and Ashwell, J.D., Fas and activation-induced Fas ligand mediate apoptosis of T cell hybridomas: inhibition of Fas ligand expression by retinoic acid and glucocorticoids, *J. Exp. Med.,* 181, 1673, 1995.

69. Mueller, Y.M., de Rosa, S.C., Hutton, J.A., Witek, J., Roederer, M., Altman, J.D., and Katskikis, P.D., Increased CD95/Fas-induced apoptosis of HIV-specific CD8(+) T cells, *Immunity,* 15, 871, 2001.

70. Bofill, M., Gombert, W., Borthwick, N.J., Akbar, A.N., McLaughlin, J.E., Lee, C.A., Johnson, M.A., Pinching, A.J., and Janossy, G., Presence of CD3+CD8+Bcl-2(low) lymphocytes undergoing apoptosis and activated macrophages in lymph nodes of HIV-1+ patients, *Am. J. Pathol.,* 146, 1542, 1995.

71. Boudet, F., Lecoeur, H., and Gougeon, M.L., Apoptosis associated with *ex vivo* down-regulation of Bcl-2 and upregulation of Fas in potential cytotoxic CD8+ T lymphocytes during HIV infection, *J. Immunol.,* 156, 2282, 1996.

72. Adachi, Y., Oyaizu, N., Than, S., McCloskey, T.W., and Pahwa, S., IL-2 rescues *in vitro* lymphocyte apoptosis in patients with HIV infection: correlation with its ability to block culture-induced down-modulation of Bcl-2, *J. Immunol.,* 157, 4184, 1996.

73. Naora, H. and Gougeon, M.L., Interleukin-15 is a potent survival factor in the prevention of spontaneous but not CD95-induced apoptosis in CD4 and CD8 T lymphocytes of HIV-infected individuals. Correlation with its ability to increase Bcl-2 expression, *Cell Death Differ.,* 6, 1002, 1999.

74. Bulfone-Paus, S., Ungureanu, D., Pohl, T., Lindner, G., Paus, R., Ruckert, R., Krause, H., and Kunzendorf, U., Interleukin-15 protects from lethal apoptosis *in vivo, Nat. Med.,* 3, 1124, 1997.

75. Estaquier, J., Tanaka, M., Suda, T., Nagata, S., Golstein, P., and Ameisen, J.C., Fas-mediated apoptosis of CD4+ and CD8+ T cells from human immunodeficiency virus-infected persons: differential *in vitro* preventive effect of cytokines and protease antagonists, *Blood,* 87, 4959, 1996.

76. Chen, H., Yip, Y.K., George, I., Tyorkin, M., Salik, E., and Sperber, K., Chronically HIV-1-infected monocytic cells induce apoptosis in cocultured T cells, *J. Immunol.,* 161, 4257, 1998.

77. Flamand, L., Crowley, R.W., Luso, P., Colombini-Hatch, S., Margolis, D.M., and Gallo, R.C., Activation of CD8+ T lymphocytes through the T cell receptor turns on CD4 gene expression: implications for HIV pathogenesis, *Proc. Natl. Acad. Sci. U.S.A.,* 95, 3111, 1998.

78. Yang, L.P., Riley, J.L., Carroll, R.G., June, C.H., Hoxie, J., Patterson, B.K., Ohshima, Y., Hodes, R.J., and Delespesse, G., Productive infection of neonatal CD8+ T lymphocytes by HIV-1, *J. Exp. Med.,* 187, 1139, 1998.

79. Pinto, L.A., Williams, M.S., Dolan, M.J., Henkart, P.A., and Shearer, G.M., β-Chemokines inhibit activation-induced death of lymphocytes from HIV-infected individuals, *Eur. J. Immunol.,* 30, 2048, 2000.

80. Blanco, J., Barretin, J., Henson, G., Bridger, G., De Clercq, E., Clotet, B., and Este, J.A., The CXCR4 antagonist AMD3100 efficiently inhibits cell-surface-expressed human immunodeficiency virus type 1 envelope-induced apoptosis, *Antimicrob. Agents Chemother.,* 44, 51, 2000.

81. Decrion, A.Z., Varin, A., Estavoyer, J.M., and Herbein, G., CXCR4-mediated T cell apoptosis in HIV infection, *J. Gen. Virol.,* 85, 1471, 2004.

82. Blanco, J., Jacotot, E., Cabrera, C., Cardona, A., Clotet, B., De Clercq, E., and Este, J.A., The implication of the chemokine receptor CXCR4 in HIV-1 envelope protein-induced apoptosis is independent of the G protein-mediated signalling, *AIDS,* 13, 909, 1999.

83. Lee, C., Liu, Q.H., Tomkowicz, B., Yi, Y., Freedman, B.D., and Collman, R.G., Macrophage activation through CCR5- and CXCR4-mediated gp120-elicited signaling pathways, *J. Leukoc. Biol.,* 74, 676, 2003.

84. Jewett, A., Cavalcanti, M., Giorgi, J., and Bonavida, B., Concomitant killing *in vitro* of both gp120-coated CD4+ peripheral T lymphocytes and natural killer cells in the antibody-dependent cellular cytotoxicity (ADCC) system, *J. Immunol.,* 158, 5492, 1997.

85. Kaul, M. and Lipton, S.A., Chemokines and activated macrophages in HIV gp120-induced neuronal apoptosis, *Proc. Natl. Acad. Sci. U.S.A.,* 96, 8212, 1999.

86. Zauli, G., Gibellini, D., Secchiero, P., Dutartre, H., olive, D., Capitani, S., and Collette, Y., Human immunodeficiency virus type 1 Nef protein sensitizes CD4(+) T lymphoid cells to apoptosis via functional upregulation of the CD95/CD95 ligand pathway, *Blood,* 93, 1000, 1999.

87. Li, C.J., Friendman, D.J., Wang, C., Metelev, V., and Parde, A.B., Induction of apoptosis in uninfected lymphocytes by HIV-1 Tat protein, *Science,* 268, 429, 1995.

88. James, C.O., Huang, M.B., Khan, M., Garcia-Barrio, M., Powell, M.D., and Bond, V.C., Extracellular Nef protein targets CD4+ T cells for apoptosis by interacting with CXCR4 surface receptors, *J. Virol.,* 78, 3099, 2004.

89. Jacotot, E., Ravagnan, L., Loeffler, M., Ferri, K.F., Vieira, H.L., Zamzami, N., Costantini, P., Druillennec, S., Hoebeke, J., Briand, J.P., Irinopoulo, T., Daugas, E., Susin, S.A., Cointe, D., Xie, Z.H., Reed, J.C., Rocques, B.P., and Kroemer, G., The HIV-1 viral protein R induces apoptosis via a direct effect on the mitochondrial permeability transition pore, *J. Exp. Med.,* 191, 33, 2000.

90. Jacotot, E.Ferri, K.F., El Hamel, C., Brenner, C., Druillennec, S., Hoebeke, J., Rustin, P., Metivier, D., Lenoir, C., Geuskens, M., Vieira, H.L., Loeffler, M., Belzacqu, A.S., Briand, J.P., Zamzami, N., Edelman, L., Xie, Z.H., Reed, J.C., Roques, B.P., and Kroemer, G., Control of mitochondrial membrane permeabilization by adenine nucleotide translocator interacting with HIV-1 viral protein R and Bcl-2, *J. Exp. Med.,* 193, 509, 2001.

91. Roumier T., Vieira, H.L., Castedo, M, Ferri, K.F., Boya, P., Andreau, K., Druillennec, S., Joza, N., Penninger, J.M., Roques, B., and Kroemer, G., The C-terminal moiety of HIV-1 Vpr induces cell death via caspase-independent mitochondrial pathway, *Cell Death Diff.,* 9, 1212, 2002.

92. Yi, Y., Chen, W., Frank, I., Cutilli, J., Singh, A., Starr-Spires, L., Sulcove, J., Kolson, D.L., and Collman, R.G., An unusual syncytia-inducing human immunodeficiency virus type 1 primary isolate from the central nervous system that is restricted to CXCR4, replicates efficiently in macrophages, and induces neuronal apoptosis, *J. Neurovirol.,* 9, 432, 2003.

93. Sperber, K., Beuria, P., Singha, N., Gelman, I., Cortes, P., Chen, H., and Kraus, T., Induction of apoptosis by HIV-1-infected monocytic cells, *J. Immunol.,* 170, 1566, 2003.

94. Miura, Y., Misawa, N., Kawano, Y., Okada, H., Inagaki, Y., Yamamoto, N., Ito, M., Yagita, H., Okumura, K., and Mizusawa, H., Tumor necrosis factor-related apoptosis-inducing ligand induces neuronal death in a murine model of HIV central nervous system infection, *Proc. Natl. Acad. Sci. U.S.A.,* 100, 2777, 2003.

95. Tong, N., Perry, S.W., Zhang, Q., James, H.J., Guo H., Brooks, A., Bal, H., Kinnear, S.A., Fine, S., Epstein, L.G., Dairaghi, D., Schall, T.J., Gendelman, H.E., Dewhurst, S., Sharer, L.R., and Gelbard, H.A., Neuronal fractalkine expression in HIV-1 encephalitis: roles for macrophage recruitment and neuroprotection in the central nervous system, *J. Immunol.,* 164, 1333, 2000.

96. Muro-Cacho, C.A., Pantaleo, G., and Fauci, A.S., Analysis of apoptosis in lymph nodes of HIV-infected persons. Intensity of apoptosis correlates with the general state of activation of the lymphoid tissue and not with stage of disease or viral burden, *J. Immunol.,* 154, 5555, 1995.

97. Amendola, A., Gougeon, M.L., Poccia, F., Bondurand, A., Fesus, L., and Piacentini, M., Induction of "tissue" transglutaminase in HIV pathogenesis: evidence for high rate of apoptosis of CD4+ T lymphocytes and accessory cells in lymphoid tissues, *Proc. Natl. Acad. Sci. U.S.A.,* 93, 11057, 1996.

98. Chen, J.J., Huang, J.C., Shirtliff, M., Briscoe, E., Ali, S., Cesani, F., Paar, D., and Cloyd, M.W., CD4 lymphocytes in the blood of HIV(+) individuals migrate rapidly to lymph nodes and bone marrow: support for homing theory of CD4 cell depletion, *J. Leukoc. Biol.,* 72, 271, 2002.

99. Wang, L., Chen, J.J., Gelman, B.B., Konig, R., and Cloyd, M.W., A novel mechanism of CD4 lymphocyte depletion involves effects of HIV on resting lymphocytes: induction of lymph node homing and apoptosis upon secondary signaling through homing receptors, *J. Immunol.,* 162, 268, 1999.

100. Hildeman, D.A., Zhu, Y., Mitchell, T.C., Bouillet, P., Strasser, A., Kappler, J., and Marrack, P., Activated T cell death *in vivo* mediated by proapoptotic Bcl-2 family member Bim, *Immunity,* 16, 759, 2002.

101. Etemad-Moghadam, B., Sun, Y., Nicholson, E.K., Fernandes, M., Liou, K., Gomila, R., Lee, J., and Sodroski, J., Envelope glycoprotein determinants of increased fusogenicity in a pathogenic simian-human immunodeficiency virus (SHIV-KB9) passaged *in vivo,* *J. Virol.,* 74, 4433, 2000.

102. Blaak, H., Van't Wout, A.B., Brouwer, M., Hooibrink, B., Hovenkamp, E., and Schuitemaker, H., *In vivo* HIV-1 infection of CD45RA(+)CD4(+) T cells is established primarily by syncytium-inducing variants and correlates with the rate of CD4(+) T cell decline, *Proc. Natl. Acad. Sci. U.S.A.,* 97, 1269, 2000.

103. Badley, A.D., Parato, K., Cameron, D.W., Kravcik, S., Phenix, B.N., Ashby, D., Kumar, A., Lynch, D.H., Tschopp, J., and Angel, J.B., Dynamic correlation of apoptosis and immune activation during treatment of HIV infection, *Cell Death Differ.,* 6, 420, 1999.

104. Chavan, S.J., Tamma, S.L., Kaplan, M., Gersten, M., and Pahwa, S.G., Reduction in T cell apoptosis in patients with HIV disease following antiretroviral therapy, *Clin. Immunol.,* 93, 24, 1999.

105. Kotler, D.P., Shimada, T., Snow, G., Winson, G., Chen, W., Zhao, M., Inada, Y., and Clayton, F., Effect of combination antiretroviral therapy upon rectal mucosal HIV RNA burden and mononuclear cell apoptosis, *AIDS,* 12, 597, 1998.

106. Johnson, N. and Parkin, J.M., Anti-retroviral therapy reverses HIV-associated abnormalities in lymphocyte apoptosis, *Clin. Exp. Immunol.,* 113, 229, 1998.

107. Aries, S.P., Weyrich, K., Schaaf, B., Hansen, F., Dennin, R.H., and Dalhoff, K., Early T-cell apoptosis and Fas expression during antiretroviral therapy in individuals infected with human immunodeficiency virus-1, *Scand. J. Immunol.,* 48, 86, 1998.

108. Badley, A.D., Dockrell, D.H., Algeciras, A., Ziesmer, S., Landay, A., Lederman, M.M., Connick, E., Kessler, H., Kuritzkes, D., Lynch, D.H., Roche, P., Yagita, H., and Paya, C.V., *In vivo* analysis of Fas/FasL interactions in HIV-infected patients, *J. Clin. Invest.,* 102, 79, 1998.

109. Bohler, T., Walcher, J., Holzl-Wenig, G., Geiss, M., Buchholz, B., Linde, R., and Debatin, K.M., Early effects of antiretroviral combination therapy on activation, apoptosis and regeneration of T cells in HIV-1-infected children and adolescents, *AIDS,* 13, 779, 1999.

110. Sloand, E.M., Kumar, P.N., Kim, S., Chaudhuri, A., Weichold, F.F., and Young, N.S., Human immunodeficiency virus type 1 protease inhibitor modulates activation of peripheral blood CD4(+) T cells and decreases their susceptibility to apoptosis *in vitro* and *in vivo,* *Blood,* 94, 1021, 1999.

111. Phenix, B.N., Angel, J.B., Mandy, F., Kravcik, S., Parato, K., Chambers, K.A., Gallicano, K., Hawley-Foss, N., Casol, S., Camerson, D.W., and Badley, A.D., Decreased HIV-associated T cell apoptosis by HIV protease inhibitors, *AIDS Res. Hum. Retroviruses,* 16, 559, 2000.

112. Arnoult, D., Petit, F., Lelievre, J.D., Lecossier, D., Hance, A., Monceaux, V., Hurtrel, B., Ho Tsong Fang R, and Ameisen, J.C., Caspase-dependent and -independent T-cell death pathways in pathogenic

simian immunodeficiency virus infection: relationship to disease progression, *Cell Death Differ.*, 10, 1240, 2003.

113. Monceaux, V., Estaquier, J., Fevrier, M., Cumont, M.C., Riviere, Y., Aubertin, A.M., Ameisen, J.C., and Hurtrel, B., Extensive apoptosis in lymphoid organs during primary SIV infection predicts rapid progression towards AIDS, *AIDS*, 17, 1585, 2003.

114. Tesselaar, K., Arens, R., van Schijndel, G.M., Baars, P.A., van der Valk, M.A., Borst, J., van Oers, M.H., and van Lier, R.A., Lethal T cell immunodeficiency induced by chronic costimulation via CD27-CD70 interactions, *Nat. Immunol.*, 4, 49, 2003.

115. Wang, X.Z., Stepp, S.E., Brehm, M.A., Chen, H.D., Selin, L.K., and Welsh, R.M., Virus-specific CD8 T cells in peripheral tissues are more resistant to apoptosis than those in lymphoid organs, *Immunity*, 18, 631, 2003.

116. Schrager, L.K. and D'Souza, M.P., Cellular and anatomical reservoirs of HIV-1 in patients receiving potent antiretroviral combination therapy, *JAMA*, 280, 67, 1998.

117. Finzi, D., Blankson, J., Siliciano, J.D., Margolick, J.B., Chadwick, K., Pierson, T., Smith, K., Lisziewicz, J., Lori, F., Flexner, C., Quinn, T.C., Chaisson, R.E., Rosenberg, E., Walker, B., Gange, S., Gallant, J., and Siliciano, R.F., Latent infection of CD4+ T cells provides a mechanism for lifelong persistence of HIV-1, even in patients on effective combination therapy, *Nat. Med.*, 5, 512, 1999.

118. Ho, D.D., Toward HIV eradication or remission: the tasks ahead, *Science*, 280, 1866, 1998.

119. Wein, L.M., D'Amato, R.M., and Perelson, A.S., Mathematical analysis of antiretroviral therapy aimed at HIV-1 eradication or maintenance of low viral loads, *J. Theor. Biol.*, 192, 81, 1998.

120. Mahlknecht, U. and Herbein, G., Macrophages and T-cell apoptosis in HIV infection: a leading role for accessory cells? *Trends Immunol.*, 22, 256, 2001.

121. Swingler, S., Mann, A., Jacque, J., Brichacek, B., Sasseville, V.G., Williams, K., Lackner, A.A., Janoff E.N., Wang, R., Fisher, D., and Stevenson, M., HIV-1 Nef mediates lymphocyte chemotaxis and activation by infected macrophages, *Nat. Med.*, 5, 997, 1999.

122. Swingler, S., Brichacek, B., Jacque, J.M., Ulich, C., Zhou, J., and Stevenson, M., HIV-1 Nef intersects the macrophage CD40L signalling pathway to promote resting-cell infection, *Nature*, 424, 213, 2003.

123. Mahlknecht, U., Deng, C., Lu, M.C., Greenough, T.C., Sullivan, J.L., O'Brien, W.A., and Herbein, G., Resistance to apoptosis in HIV-infected CD4+ T lymphocytes is mediated by macrophages: role for Nef and immune activation in viral persistence, *J. Immunol.*, 165, 6437, 2000.

124. Geleziunas, R., Xu, W., Takeda, K., Ichijo, H., and Greene, W.C., HIV-1 Nef inhibits ASK1-dependent death signalling providing a potential mechanism for protecting the infected host cell, *Nature*, 410, 834, 2001.

125. Ott, M., Emiliani, S., Van Lint, C., Herbein, G., Lovett, J., Chirmule, N., McCloskey, T., Pahwa, S., and Verdin E., Immune hyperactivation of HIV-1-infected T cells mediated by Tat and the CD28 pathway, *Science*, 275, 1481, 1997.

126. Westendorp, M.O., Li-Weber, M., Frank, R.W., and Krammer, P.H., Human immunodeficiency virus type 1 Tat upregulates interleukin-2 secretion in activated T cells, *J. Virol.*, 68, 4177, 1994.

127. Zhang, Z., Schuler, T., Zupancic, M., Wietgrefe, S., Staskus, K.A., Reimann, K.A., Reinhart, T.A., Rogan, M., Cavert, W., Miller, C.J., Veazey, R.S., Notermans, D., Little, S., Danner, S.A., Richman, D.d., Havlir, D., Wong, J., Jordan, H.L., Schacker, T.W., Racz, P., Tenner-Racz, K., Letvin N.L., Wolinsky, S., and Haase, A.T., Sexual transmission and propagation of SIV and HIV in resting and activated CD4+ T cells, *Science*, 286, 1353, 1999.

128. Vocero-Akbani, A.M., Heyden, N.V., Lissy, N.A., Ratner, L., and Dowdy, S.F., Killing HIV-infected cells by transduction with an HIV protease-activated caspase-3 protein, *Nat. Med.*, 5, 29, 1999.

16 Molecular Mechanisms of HIV-1 Syncytial Apoptosis

Christophe Lemaire
Jean-Luc Perfettini
Guido Kroemer
Catherine Brenner

CONTENTS

INTRODUCTION

Acquired immunodeficiency syndrome (AIDS) is a pathology caused by the retrovirus human immunodeficiency virus type 1 and type 2 (HIV-1 and HIV-2), which generally leads to the death of patients in the absence of an efficient therapy. During the progression of the disease, HIV infection is responsible for a dramatic loss of CD4+ T lymphocytes, which are key regulators of the immune response.[1–3] This decrease in the number of CD4+ T cells impairs immune system function and decreases its capacity to protect the patient from pathologies such as infection and certain cancers. Typically, HIV-1-infected cells and uninfected CD4+ T lymphocytes, also called bystander cells, die by apoptosis. This applies to the death of infected cells, bystander cells, as well as multinucleated cells, namely, the syncytia.[4] More precisely, bystander cells can die either by encounter with Fas-mediated apoptosis pathways or after interaction with HIV-released proteins (gp120, Vpr, or Tat), whereas syncytia undergo a series of alterations culminating in mitotic catastrophe and cell death by apoptosis.[4]

Recently, the characterization of the mechanism of syncytial death has experienced major advances (for review, see Castedo et al.[5]). Using a cellular model that reproduces the cell-to-cell fusion phenomenon, it was established that the envelope glycoprotein complex (Env)-mediated formation of syncytia involves the activation of kinases, of transcription factors (nuclear factor-B [NF-B] and p53), and of caspases; nuclear fusion; and disruption of mitochondria integrity.[6–9] The main observations were reproduced *in vivo* in HIV-1-infected patients and *ex vivo* in HIV-1-infected CD4+ T lymphocytes. This enhanced knowledge of the molecular events associated with syncytial death may help us to better understand the process of the CD4+ lymphocyte depletion during AIDS progression.

THE FORMATION OF SYNCYTIA

Syncytia are multinucleated giant cells resulting from a cell-to-cell fusion mechanism. Syncytia can be induced by the fusogenic action of cytokines (e.g., IL-4- and IL-13-induced fusion of monocytes–macrophages), as well as by receptor–ligand interactions at the cell surface (e.g., HIV-1-infected T lymphocytes fusion, osteoclast activation during osteoporosis). Thus, syncytia formation can be a physiological process and can contribute to disease progression by accelerating the cell death of infected cells, such as HIV-1-infected cells.[10]

Among HIV-1 proteins, gp120 may be the most important apoptogenic protein. The glycoprotein (gp)120 is expressed in a complex with gp41, namely, the envelope glycoprotein complex Env, on HIV-1 virions and at the surface of HIV-1-infected cells. The gp120 interacts with CD4, a cell surface receptor, and CXCR4 or CCR5, two co-receptors belonging to the chemokine receptor family.[4] The gp41 allows membrane interaction between a fusion domain of the protein and the cell surface. These Env-driven protein interactions trigger an apoptosis signal that activates a cascade of events leading to syncytia formation or single-cell killing.[11] In syncytial death, the interaction of gp120 triggers the exposure of gp41 and favors insertion into the cellular membrane of a hydrophobic fusion peptide. In turn, the structure of gp41 is rearranged into a six-helix hairpin structure linking together cellular membranes and allowing them to fuse.[12] After membrane fusion, death follows in a process involving deregulation of mitosis, karyogamy, caspase activation, inhibition of survival signals, and mitochondrial membrane permeabilization (MMP; for reviews see Castedo et al.[5,13]).

DEREGULATION OF MITOSIS AND KARYOGAMY

Syncytia resulting from the fusion of cells expressing an Env gene with cells expressing the CD4/CXCR4 complex undergo apoptosis through a mitochondrion-controlled pathway that is initiated by the accumulation of cyclin B and the activation of cyclin B-dependent kinase 1 (CDK1; Figure 16.1).[7,8] This step is followed by the phosphorylation of p53 on serine 15 by the mammalian target of rapamycin (mTOR/FRAP). This activates p53-mediated transcription, which results in the expression of genes that induce MMP and cell death (Figure 16.1). This pathway depends on the interaction of Env with co-receptors, because pharmacologic inhibitors targeting gp41 and gp120 delay the formation and the death of syncytia. Moreover, pharmacologic inhibitors such as inhibitors of CDK1 (e.g., roscovitin), mTOR/FRAP (e.g., rapamycin), and p53 (e.g., cyclic pifithrin-α), or overexpression of Bcl-2, a mitochondrial oncoprotein known to inhibit MMP and apoptosis, all independently reduce the syncytial death (Figure 16.1).[7,8]

In syncytia, CDK1 induces a phosphorylation/depolymerization of lamin that culminates in the process of the fusion of nuclei, i.e., karyogamy, before the entry of the cell into an early mitotic prophase (Figure 16.1).[7,8] Thus, experimental manipulations that decrease CDK1 activity, such as microinjection of an antibody against CDK1, pharmacologic inhibition, and transient transfection of syncytia with a dominant–negative gene of CDK1, prevent the karyogamy. As a consequence, these procedures inhibit all the subsequent hallmarks of apoptosis.[7,8]

Subsequent to CDK1 activation, mTOR/FRAP that is located in the cytosol translocates to the nucleus and interacts physically with p53 to phosphorylate it on two serine residues (Figure 16.2).[7,8] The relocation of mTOR/FRAP and the phosphorylation of p53 do not occur in roscovitin-treated syncytia, attesting to the importance of CDK1 in this process. Moreover, inhibition of mTOR/FRAP by rapamycin, its specific pharmacologic inhibitor, reduces apoptosis in several paradigms of syncytium-dependent death, including primary CD4+ lymphoblasts infected by HIV-1. Concomitantly, rapamycin inhibits p53 phosphorylation on serine 15 (p53S15), mitochondrial translocation of Bax, loss of mitochondrial transmembrane potential ($\Delta\Psi$m), mitochondrial release of apoptogenic proteins (e.g., cytochrome c and apoptosis-inducing factor [AIF]), and nuclear chromatin condensation. In addition, transfection with the dominant negative p53 gene has a similar

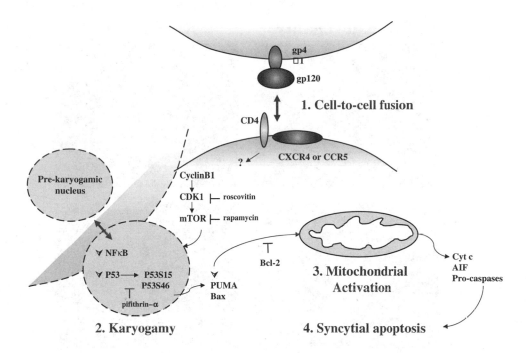

FIGURE 16.1 The syncytial death. Syncytia are multinucleated giant cells arising from the fusion of HIV-1 Env-expressing cells and CD4- and CCR5/CXCR4-expressing cells. After formation, syncytia die by apoptosis after a four-step process. (1) Cell-to-cell fusion: The interaction of cell surface proteins (gp120-GP41 complex or Env, CD4, CXCR4, or CCR5) promotes the fusion of membrane and cytoplasm, the accumulation of cyclin B1, and the activation of CDK1. (2) Karyogamy: This elicits the nucleoplasm fusion, the karyogamy, and the cooperation of two dominant transcription factors, p53, and NF-B. Thus, after nuclear translocation, the kinase mTOR/FRAP phosphorylates p53 on serine 15 and 46, which induces an increased expression of the proapoptotic Puma and Bax. (3) Mitochondrial activation: Mitochondria and caspase activations that manifest by membrane permeabilization, the release of proteins (e.g., cytochrome c and AIF), follow these events into the cytosol and, finally, the chromatin condensation, and (4) syncytial apoptosis. CDK1, mTOR/FRAP, and P53-specific pharmacologic agents as well as Bcl-2 can inhibit syncytial death.

antiapoptotic action as rapamycin, upstream of the Bax upregulation/translocation.[7] Among a panel of kinases known to phosphorylate p53, including ERK, p38 kinase, DNA-PK, ATM, ATR, and mTOR/FRAP, only the mTOR/FRAP expression was enhanced during syncytia formation. The expression of the other kinases was unaffected. In contrast, the protein phosphatase PP2A that inhibits mTOR/FRAP was downregulated at a late stage of syncytia, well after mTOR/FRAP upregulation.[7] Thus, the ability of mTOR/FRAP to initiate phosphorylation of p53 on serine 15 is associated with the initiation of syncytial death.[7]

Interestingly, the p53S15 phosphorylation and mTOR/FRAP upregulation occur also *in vivo*, in HIV-1-infected patients, in which it can be detected in pre-apoptotic and apoptotic syncytia in lymph nodes as well as in peripheral blood mononuclear cells, correlating with viral load.[8] In summary, phosphorylation of p53S15 by mTOR/FRAP plays a critical role in syncytial apoptosis driven by HIV-1 Env.

COOPERATION OF DOMINANT TRANSCRIPTION FACTORS

In the same co-culture model of cells expressing the HIV-1 Env with cells expressing CD4, NF-B, p53, and AP1 are activated in response to the initiation of Env-elicited apoptosis.[9] The repressor

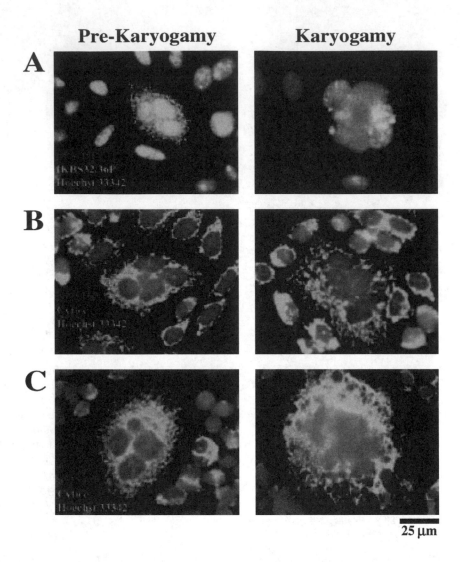

Pre-Karyogamy   **Karyogamy**

A

B

C

25 μm

FIGURE 16.2 p53 phosphorylation in Env-dependent syncytia. (A–C) Representative immunofluorescence microphotographs of prekaryogamic (18 h) and karyogamic syncytia (36 h) labeled with antibodies specific for I-B S32/36P (A), p53S15P (A and B, light area surrounding dark area), p53S46P (A and C, dark dots in the circumference), or cytochrome *c* (B and C, light area). Nuclei are stained with Hoechst 33342 for chromatin condensation.

NF-B exhibits antimitotic and antiapoptotic effects and prevents the Env-elicited phosphorylation of p53 on serine 15 and 46, as well as the activation of AP1. Thus, NF-B inhibition fully prevents the syncytial accumulation of cyclin B1, the p53-dependent transcription and phosphorylation, and karyogamy.[9] Therefore, NF-B exerts its action on the aberrant cyclin B1/Cdk1-mediated entry into mitosis and upstream p53 activation.

Transfection with dominant-negative p53 abolished syncytial apoptosis and AP1 activation and increased the lifetime of syncytia.[9] In addition, 85% of the transcriptional effects of HIV-1 Env were blocked by the p53 pharmacologic inhibitor pifithrin-, as revealed by microarrays.[9] Moreover, macroarrays identified Puma, a proapoptotic "BH3-only" protein from the Bcl-2 family known to activate Bax/Bak,[14] as being transcribed in response to Env, in a p53-dependent manner (Figure 16.1). Downmodulation of Puma by antisense oligonucleotides, as well as RNA interference of Bax and

Bak, prevented Env-induced apoptosis. HIV-1-infected primary lymphoblasts upregulated Puma *in vitro*.[9] Accordingly, circulating CD4+ lymphocytes from untreated HIV-1-infected donors contained enhanced amounts of Puma protein, and these elevated Puma levels dropped with administration of antiretroviral therapy. NF-B and p53 activation were observed in lymph node biopsies of HIV-1-infected patients.[9] Altogether, these results establish that NF-B and p53 are the dominant and competing transcription factors participating in HIV-1 infection, CD4+ depletion, and syncytial death.

MITOCHONDRIAL AND CASPASE ACTIVATION

Depending on the model studied (e.g., the cell type, the apoptotic signal), apoptosis can be promoted via two general pathways: the death receptor pathway (e.g., Fas/CD95, TNF) or the mitochondrial pathway, namely, the intrinsic pathway (Figure 16.3). These pathways occur preferentially in type-I cells and in type-II cells,[15] and also in HIV-1-mediated cell death.[16] The mitochondrial pathway is characterized by a mitochondrial checkpoint (for review see Kroemer and Reed[17]), corresponding to an increase in MMP and a subsequent relocalization of apoptogenic

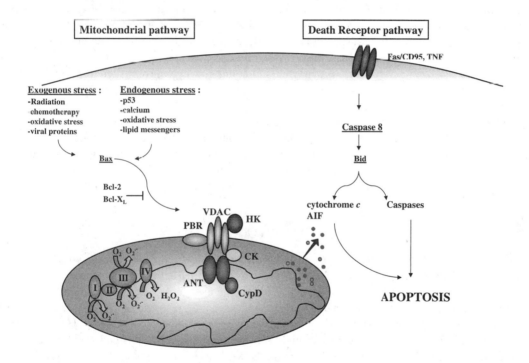

FIGURE 16.3 The signaling pathways of apoptosis. Apoptosis can be initiated via two major pathways: the death receptor pathway and the mitochondrial pathway. The occurrence of one pathway may be dependent on the cell type (type I vs. type II) or the type of proapoptotic trigger. Endogenous stress signal as well as endogenous stimuli that converge to the mitochondrion elicit the mitochondrial pathway. As a consequence, the mitochondrial membrane permeabilizes, intermembrane space proteins (e.g., cytochrome *c*, AIF) are released, and the degradation phase of apoptosis that leads to cell death is initiated. Constitutive mitochondrial proteins from the respiratory chain (complexes I, II, III, and IV, CI, CII, CIII, and CIV) and the permeability transition pore complex (peripheral benzodiazepine receptor [PBR], voltage-dependent anion channel [VDAC], hexokinase [HK], creatine kinase [CK], adenine nucleotide translocase [ANT], cyclophilin D [CypD]) can contribute to the process of membrane permeabilization that is regulated by the Bax/Bcl-2 family proteins. In contrast, the death receptor pathway is triggered by exogenous ligands of the cell surface receptors, such as Fas/CD95 or TNF. The early activation of caspase-8 and the cleavage of the proapoptotic protein Bid characterize this pathway. In some cases, this pathway can also be inhibited by Bcl-2.

proteins from the mitochondria into the cytosol, such as cytochrome c, AIF, second mitochondria-derived activator of caspases (Smac)/direct inhibitors of apoptosis protein-binding protein with low pI (DIABLO), procaspases, and Endo G.[18-20] Although the precise mechanism of release of these various proteins is still unclear, they may cross the outer mitochondrial membrane via either specific or nonspecific pores. These pores are composed of oligomerized members from the Bax/Bcl-2 family (Bax, Bid)[21] and may also contain other proteins, such as the voltage-dependent anion channel (VDAC)[22] or the adenine nucleotide translocase (ANT), two members of the permeability transition pore complex (PTPC).[23]

As described above, as an intermediate stage of syncytial death, phosphorylation of p53 leads to an increased expression of Bax and its translocation to mitochondrial membranes, triggering the activation of mitochondria and caspases. Thus, after a conformational activation of Bax and the exposure of its N-terminus, the permeability of mitochondrial membranes is increased and allows for the release of intermembrane proteins such as cytochrome c and AIF from mitochondria.[6] This phase is independent of caspases, because their inhibition does not suppress the translocation of AIF and cytochrome c, yet it prevents all signs of nuclear apoptosis. In contrast, translocation of Bax to mitochondria leads to MMP, which is inhibited by microinjected Bcl-2 protein or Bcl-2 gene transfection.[6] Bcl-2 also prevents the subsequent nuclear chromatin condensation and DNA fragmentation. The two proteins initiate two different cascades of events downstream of MMP. Thus, microinjection of AIF into syncytia suffices to trigger rapid, caspase-independent cytochrome c release. Neutralization of endogenous AIF by injection of an antibody prevents all signs of spontaneous apoptosis occurring in syncytia, including the cytochrome c release and nuclear apoptosis. In contrast, cytochrome c neutralization prevents only nuclear apoptosis and does not affect AIF release.[6]

HIV-1 protease inhibitors can inhibit apoptosis at the mitochondrial level, which might help to prevent the HIV-1-mediated loss of lymphoid cells.[24] Thus, it seems that the pathogenesis of AIDS and the pharmacologic interventions associated with this disease affect the mitochondrial regulation of apoptosis, which, therefore, largely determines the outcome of HIV-1 infection. In the future, it would be interesting to determine whether HIV-1 protease inhibitors can also inhibit syncytial death.

CONCLUDING REMARKS

The following molecular sequence of events orchestrates the death of Env-induced syncytia *in vitro*: CDK1/mTOR/FRAP/p53/Bax/MMP/AIF and cytochrome c release/caspase activation/nuclear apoptosis (for reviews, see Castedo et al.[5,13]). A CDK1-dependent mitotic catastrophe and the fusion of nucleoplasm accompany this signaling pathway. Pharmacologic agents as well as genetic manipulation can inhibit it. Finally, most of the events described were observed *in vivo* in HIV-1 patients or *ex vivo* in infected CD4+ lymphoblasts, underscoring the relevance of this pathway for the AIDS-associated enhanced death of lymphoid cells, such as CD4+ T lymphocytes.

The HIV-1-mediated pathways leading to cell death are complex and multiple yet sometimes overlap. Cell death can result from specific effects of the virus as well as from effects of viral isolated proteins and indirect effects mediated by the host.[24] For instance, at least *in vitro*, the induction of apoptosis by the HIV-1 proteins Tat, Pr, and Vpr, by syncytia formation and in bystander cells, favors the induction of MMP, which, in turn, coordinates the degradation phase of apoptosis (i.e., nuclear chromatin condensation, fragmentation, and so on). More precisely, Vpr, Pr, and Tat can interact physically with mitochondrial proteins such as ANT, Bcl-2, and the superoxide dismutase (SOD), respectively, and can act directly to induce MMP without the need of a signaling pathway upstream of the mitochondria. Then the CDK1/mTOR/FRAP/p53/Bax activation cascade that is triggered in response to the interaction of Env with CD4 and CXCR4 or CCR5 may be specific to the Env-mediated syncytial death, whereas the activation of mitochondria and caspases that are subsequent to Bax activation is common to the general mitochondrial apoptosis pathway.

In the future, it is important to determine whether this cascade of events occurs within syncytia formed from the fusion of various cell types and how host factors (e.g., cytokine production profile, metabolic activity, and redox state of the fused cells) affect this cascade of events to modulate the fate of syncytia *in vivo*.

ACKNOWLEDGMENTS

Our work was supported by grants funded by ARC (to CB), CNRS (to CB), and the French Ministry of Research (to CB).

REFERENCES

1. Pantaleo, G. and Fauci, A., Apoptosis in HIV infection, *Nat. Med.,* 1, 118–120, 1995.
2. Badley, A.D., Pilon, A.A., Landay, A., and Lynch, D.H., Mechanisms of HIV-associated lymphocyte apoptosis, *Blood,* 96, 2951–2964, 2000.
3. Gougeon, M., Apoptosis as an HIV strategy to escape immune attack, *Nat. Rev. Immunol.,* 3, 392–404, 2003.
4. Alimonti, J.B., Ball, T.B., and Fowke, K.R., Mechanisms of CD4+ T lymphocyte cell death in human immunodeficiency virus infection and AIDS, *J. Gen. Virol.,* 84, 1649–1661, 2003.
5. Castedo, M., Perfettini, J., Andreau, K., Roumier, T., Piacentini, M., and Kroemer, G., Mitochondrial apoptosis induced by the HIV-1 envelope, *Ann. N.Y. Acad. Sci.,* 1010, 19–28, 2003.
6. Ferri, K.F., Jacotot, E., Blanco, J., Este, J.A., Zamzami, N., Susin, S.A., Xie, Z., Brothers, G., Reed, J.C., Penninger, J.M., and Kroemer, G., Apoptosis control in syncytia induced by the HIV type 1-envelope glycoprotein complex. Role of mitochondria and caspases, *J. Exp. Med.,* 192, 1081–1092, 2000.
7. Castedo, M., Ferri, K.F., Blanco, J., Roumier, T., Larochette, N., Barretina, J., Amendola, A., Nardacci, R., Metivier, D., Este, J.A., Piacentini, M., and Kroemer, G., Human immunodeficiency virus 1 envelope glycoprotein complex-induced apoptosis involves mammalian target of rapamycin/FKBP12-rapamycin-associated protein-mediated p53 phosphorylation, *J. Exp. Med.,* 194, 1097–1110, 2001.
8. Castedo, M., Roumier, T., Blanco, J., Ferri, K., Barretina, J., Tintignac, L., Andreau, K., Perfettini, J., Amendola, A., Nardacci, R., Leduc, P., Ingber, D., Druillennec, S., Roques, B., Leibovitch, S., Vilella-Bach, M., Chen, J., Este, J., Modjtahedi, N., Piacentini, M., and Kroemer, G., Sequential involvement of Cdk1, mTOR and p53 in apoptosis induced by the human immunodeficiency virus-1 envelope, *EMBO J.,* 21, 4070–4080, 2002.
9. Perfettini, J., Roumier, T., Castedo M., Larochette, N., Boya, P., Reynal, B., Lazar, V., Ciccosanti, R., Nardacci, R., Penninger, J., Piacentini, M., and Kroemer, G., NF-B and p53 are the dominant apoptosis-inducing transcription factors elicited by the HIV-1 envelope, *J. Exp. Med.,* (201) 279–289, 2004.
10. Anderson, J.M., Multinucleated giant cells, *Curr. Opin. Hematol.,* 7, 40–47, 2000.
11. Blanco, J., Barretina, J., Ferri, K., Jacotot, E., Gutierrez, A., Cabrera, M., Kroemer, G., Clotet, B., and Este, J.A., Cell-surface-expressed HIV-1 envelope induces the death of CD4 T cells during GP41-mediated hemifusion-like events, *Virology,* 305, 356–365, 2003.
12. Munoz-Barroso, I., Durell, S., Sakaguchi, K., Appella, E., and Blumenthal, R., Dilation of the human immunodeficiency virus-1 envelope glycoprotein fusion pore revealed by the inhibitory action of a synthetic peptide from gp41, *J. Cell Biol.,* 140, 315–323, 1998.
13. Castedo, M., Perfettini, J.-L., Roumier, T., and Kroemer, G., Cyclin-dependent kinase-1: linking apoptosis to cell cycle and mitotic catastrophe, *Cell Death Diff.,* (9) 1287–1293, 2003.
14. Nakano, K. and Vousden, K.H., PUMA, a novel proapoptotic gene, is induced by p53, *Mol. Cell,* 7, 683–694, 2001.
15. Scaffidi, C., Fulda, S., Srinivasan, A., Friesen, C., Li, F., Tomaselli, K., Debatin, K., Krammer, P., and Peter, P., Two CD95 (APO-1/Fas) signaling pathways, *EMBO J.,* 17, 1675–1687, 1998.
16. Petit, F., Arnoult, D., Viollet, L., and Estaquier, J., Intrinsic and extrinsic pathways signaling during HIV-1 mediated cell death, *Biochimie,* 85, 795, 2003.
17. Kroemer, G. and Reed, J., Mitochondrial control of cell death, *Nat. Med.,* 6, 513–519, 2000.

18. Liu, X., Kim, C.N., Yang, J., Jemmerson, R., and Wang, X., Induction of apoptotic program in cell-free extracts: requirement for dATP and cytochrome *c*, *Cell,* 86, 147–157, 1996.

19. Li, L.Y., Luo, X., and Wang, X., Endonuclease G is an apoptotic DNase when released from mitochondria, *Nature,* 412, 95–99, 2001.

20. Susin, S.A., Zamzami, N., Castedo, M., Hirsch, T., Marchetti, P., Macho, A., Daugas, E., Geuskens, M., and Kroemer, G., Bcl-2 inhibits the mitochondrial release of an apoptogenic protease, *J. Exp. Med.,* 184, 1331–1341, 1996.

21. Martinou, J.C., Desagher, S., and Antonsson, B., Cytochrome *c* release from mitochondria: all or nothing, *Nat. Cell Biol.,* 2, E41–E43, 2000.

22. Shimizu, S., Matsuoka, Y., Shinohara, Y., Yoneda, Y., and Tsujimoto, Y., Essential role of voltage-dependent anion channel in various forms of apoptosis in mammalian cells, *J. Cell Biol.,* 152, 237–250, 2001.

23. Halestrap, A. and Brenner, C., The adenine nucleotide translocase: a central component of the mitochondrial permeability transition pore and key player in cell death, *Curr. Med. Chem.,* 10, 1507–1525, 2003.

24. Badley, A., Roumier, T., Lum, J.L., and Kroemer, G., Mitochondrion-mediated apoptosis in HIV-1 infection, *Trends Pharmacol. Sci.,* 24, 298–305, 2003.

17 Nonapoptotic HIV-Induced T Cell Death

Keiko Sakai
Matthew Smith-Raska
Michael J. Lenardo

CONTENTS

INTRODUCTION

Infection with human immunodeficiency virus type 1 (HIV-1) causes a decline in CD4[+] T cells leading to acquired immunodeficiency syndrome (AIDS).[1,2] Although mathematical modeling estimates show that the majority of infected cells die with a half-life of about 2 days,[3–5] the number of infected cells is reported to be approximately constant.[4] This implies that there is the production and destruction of large numbers of infected cells each day. Although it is not clear what ultimately leads to the progressive and selective depletion of CD4[+] T cells, it may be inferred that the destruction of these cells exceeds their replacement (discussed in detail in Chapter 19).

Our understanding of the molecular mechanism of HIV-1 replication and viral protein functions has advanced tremendously. However, key elements of the pathogenesis of HIV-1 infection are not completely defined. The mode of viral killing by acute infection has been a focus of numerous studies but remains highly controversial. A strongly favored hypothesis is that HIV-1 induces apoptosis in infected cells as well as bystander cells (discussed in Chapters 13–16, 19, and 21). However, considerable evidence also indicates that HIV cytopathicity does not require the presence or the upregulation of cellular apoptotic factors. Furthermore, the majority of dying infected cells do not display the hallmarks of apoptosis.[6–12] In fact, it was suggested that direct viral killing could occur via a nonapoptotic/necrotic pathway.[11,12]

Another unresolved issue related to the mode of death is whether HIV-1 cytopathicity occurs in uninfected bystander cells. This could occur by syncytia formation, triggering of surface receptors such as CD4, cytokines, or other toxic influences.[13–15] Contradictory data were reported regarding direct versus indirect viral killing mainly because many studies examined a mixed population of infected and uninfected cells, which prevents the precise measurement of virally induced apoptosis. Moreover, cell death *in vivo* may result from normal homeostatic death mechanisms of activated T cells, infected or not.[16,17] This might explain apparent bystander killing *in vivo*[18,19] as discussed in Chapter 13. Accumulating evidence indicates that direct single-cell killing of HIV-1 predominates in peripheral blood mononuclear cells (PBMCs) from most patients and accounts for the majority of cell death in infected cultures.[20–22] In experimental tissue culture infections, bystander death is not evident.[11,12] There are even newly described mechanisms of "programmed" necrosis that can be triggered by cytokines such as tumor necrosis factor (TNF). This chapter focuses on the mode of death in direct single-cell killing induced by HIV-1 rather than indirect death caused by viral infection. We will discuss a nonapoptotic/necrotic mode of cell death during acute HIV-1 infection and present some of the experimental data that argue against viral killing via apoptosis. We will also review possible mechanisms that could contribute to necrotic cell death induced by HIV-1 infection.

CELL DEATH PATHWAY DURING ACUTE HIV-1 INFECTION

Acute HIV-1 infection causes massive death in virus-producing cells. This can be readily observed in highly infected cultures of transformed cell lines or nontransformed peripheral blood CD4+ T cells.[11,12] What is the basic mechanism of direct viral killing? Multiple pathways could lead to pathological cell death, including forms of programmed cell death, typically apoptosis, and traumatic death by injury called necrosis. Signals might be intertwined enough to ultimately lead to multiple death pathways downstream of the signaling cascade, which could result in the hallmarks of both pathways in infected cells. Therefore, data have to be carefully examined in order to draw firm conclusions.

MODES OF CELL DEATH: APOPTOSIS VS. NONAPOPTOTIC CELL DEATH

Apoptosis refers to a programmed cell-death pathway that exhibits distinct morphological changes catalyzed by a family of cysteinyl aspartate-specific proteases called caspases. The activation of caspases leads to characteristic biochemical changes, including chromatin condensation, DNA fragmentation, externalization of phosphatidylserine on the plasma membrane, and the exposure of a specific mitochondrial membrane protein. These changes are detectable by such biochemical techniques as Terminal dUTP Nuclear End Labeling (TUNEL) assays, Annexin V staining, and APO2.7-antibody staining, (detailed discussion can be found in Chapter 3).[23–25] In addition to the biochemical changes, caspase-activated death has a characteristic morphological appearance on electron microscopy. The cell nucleus becomes smaller and electron dense, the cytoplasm condenses, and the cell membrane is intact but may have a ruffled appearance—all of which are easily distinguished from other modes of cell death and preventable by caspase inhibitors.

Another major form of cell death involves necrosis. In contrast to apoptosis, necrosis exhibits loss of plasma membrane integrity and cellular swelling followed by lysis due to traumatic damage. Unlike apoptosis, the contents of the cell are released upon lysis, which could be advantageous to stimulation of immune responses against pathogens. Although condensation of chromatin does not occur in necrosis, DNA degradation is observed at a later stage, resulting in DNA smearing on gel electrophoresis, similar to a DNA laddering effect during apoptosis.[26,27] These two major forms of death are readily distinguished morphologically by electron microscopy. Unlike apoptosis, necrotic cells become enlarged with fragmentation of the plasma membrane, decreased density of the cytoplasmic region, and dissociation of the nucleus and other organelles. Biochemical methods are also frequently used to distinguish apoptotic vs. necrotic changes in dying cells. Perhaps the easiest test is the detection of the fluorescent dye propidium iodide (PI) binding to the genomic DNA, which occurs only when the plasma membrane integrity is lost. Early PI staining will occur in necrosis but not apoptosis. However, many of the commonly used biochemical assays do not clearly distinguish apoptosis from necrosis.

Efforts to understand the mechanism of CD4[+] T cell depletion led to the extensive investigation of the molecular mechanism of cell death during HIV-1 infection, unfortunately resulting in conflicting hypotheses. Large numbers of studies proposed apoptosis to be the major mode of viral-induced cell death based on the appearance of apoptotic features in infected cultures.[28,29] However, significant data challenged the view that apoptosis is the mechanism of viral-mediated CD4[+] T cell destruction. For example, many experiments designed to disrupt the apoptosis pathway failed to abrogate viral-induced cell death. A number of other studies could not establish a requirement for the apoptosis-inducing caspases in HIV-1 killing.[6,7,11,12,30,31] The surface expression of a key apoptosis receptor in T cells, Fas/APO-1/CD95, and its ligand, FasL, was not upregulated by HIV-1 infection in PBMCs.[10] Studies also demonstrated that viral-induced death could bypass the requirement of the Fas death pathway.[10,11] Over expression of antiapoptotic proteins, Bcl-2, Bcl-xL, and adenovirus E1B 19K did not significantly alleviate cell death in infected populations,[11,32,33] casting further doubt on apoptosis as the major death pathway for HIV-1 killing. Moreover, other studies were unable to establish a correlation between the level of apoptosis and disease progression.[16]

These discrepancies also raise questions regarding the validity of *in vitro* systems used to detect apoptosis in some of the experiments and the conclusions made using them. It is well documented that biochemical changes observed in apoptotic cells are not always specific and, therefore, biochemical assays used to investigate apoptosis do not always detect apoptosis exclusively (discussed in Chapter 3). A typical example is the TUNEL assay used to detect nonspecific DNA fragmentation. DNA fragmentation is a feature of cells undergoing apoptosis, but the assay may also detect DNA degradation observed during necrosis in T cells.[9,26,27,34–36] Another widely used assay for apoptosis, Annexin V staining, detects damaged cells by exposing phosphatidylserine, irrespective of the mode of death.[37,38] Due to these complications, individual biochemical assays do not necessarily make a clear distinction between apoptosis and necrosis when used alone. Additionally, the interpretation of many studies in the literature became more difficult when cell death was examined in assay systems that did not distinguish infected from uninfected cells. Apoptotic features as well as cell death observed in those studies do not always correlate in kinetics or magnitude with the degree of viral infection, which makes it difficult to make inferences on virus-induced cell death.[9,39,40] These complications raise the question of whether cell death is due to a specific and direct effect of HIV-1 infection rather than general adverse conditions in cell culture associated with infection.

The analysis by multiple distinct biochemical tests for apoptosis provided persuasive evidence that necrosis rather than apoptosis was the primary mode of death. Importantly, electron micrographs and video imaging of infected cells confirmed the absence of prominent apoptotic features, such as nucleus condensation and fragmentation, in infected PBMCs and T cells from HIV-positive patients (Figure 17.1).[9,12,41] The involvement of apoptosis in provirus-expressing T cells was also investigated by treating cells with an apoptosis inhibitor, zVAD-fmk.[8,10,11] Whereas general Fas-induced apoptosis was blocked in drug-treated cells, virally induced death in infected cells was not impaired by this treatment, indicating that apoptosis is dispensable for direct HIV-1 killing of T cells. This result was confirmed by experiments using T cell lines deficient in primary apoptosis factors, such as caspase-8, FADD, and RIP.[42,43] Experiments in CD4[+] T cells from patients with autoimmune lymphoproliferative syndrome (ALPS).[10] ALPS is a disorder caused by a genetic defect in the apoptosis signaling pathway induced by the death receptor Fas that leads to a specific resistance to apoptotic stimuli.[44] Patients harboring mutations in the Fas gene have significantly reduced susceptibility to Fas-induced apoptosis. As expected, direct viral killing was not prevented in cells lacking the primary

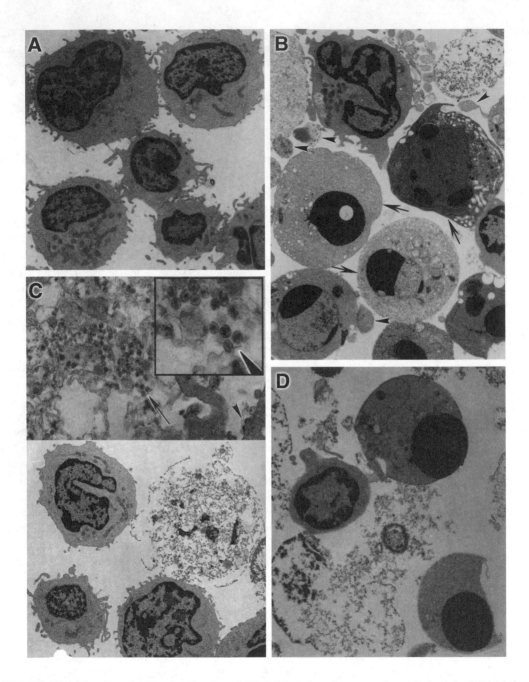

FIGURE 17.1 Electron micrographs of activated CD4+ T lymphocytes comparing cell death by HIV-1 infection with staurosporine-induced apoptosis. Activated CD4+ T lymphocytes were either infected with NL4-3 Env-virus or mock infected for 9 days and then were treated with staurosporine (1 μg/mL) for 4 h. (A) Mock-infected cells. (Original magnification × 4900.) (B) Mock-infected cells with staurosporine added. (Original magnification × 5400.) (C) HIV-1-infected cells. (Lower half, original magnification × 4900; upper half, original magnification × 17,000.) (D) HIV-1-infeceted cells with staurosporine added. (Original magnification × 7200.) In (B), arrows indicate cells with compacted chromatin, a hallmark of apoptosis; arrowheads indicate budding apoptotic bodies. In (C), the arrowhead indicates budding virions; the inset in (C) is a 2.2-fold magnification of the region indicated by the arrow and reveals mature retroviral particles within the debris of a necrotic cell. (From Lenardo, M.J., *J. Virol.*, 76, 5082, 2002. With permission.)

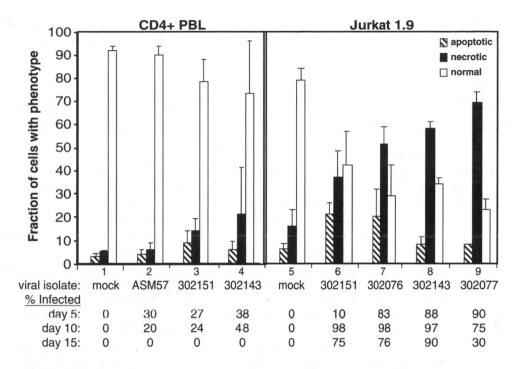

FIGURE 17.2 Quantification of apoptosis and necrosis in peripheral blood CD4⁺ T lymphocytes and Jurkat 1.9 cells infected with HIV-1 primary isolates. Cells were either infected with primary isolates of HIV-1 or mock infected; after 10 days, cells from each sample were analyzed for morphological features of apoptosis or necrosis from transmission electron micrographs of stained microscope slides. Samples 1 to 4 represent peripheral blood CD4⁺ T cells; samples 5 to 9 represent Jurkat cells. The primary isolates used in each infection were as follows: samples 1 and 5, mock-infected controls; sample 2, ASM 57 virus from a long-term nonprogressor; samples 3 and 6, dualtropic virus 302151; samples 4 and 8, dualtropic virus 302143; sample 7, dualtropic virus 302076; and sample 9, dualtropic virus 302077. The bottom of the figure provides the fraction of infected cells, as determined by staining of p24 and flow cytometric analysis for the indicated time points. (From Lenardo, M.J., *J. Virol.*, 76, 5082, 2002. With permission.)

apoptosis factors or in those from ALPS patients.[10,11,42,43] Infection with HIV-1 apparently causes direct cell death in infected CD4⁺ T cells, which is not dependent on the apoptotic pathway.

It should be noted that a small fraction of cells in tissue culture infections is often found to exhibit apoptotic markers *in vitro*.[9–11] One possible explanation is that infected cells show increased vulnerability to apoptosis, possibly due to compromised metabolic conditions within infected cells. Many reports have documented that infected PBMCs and T cell lines become more sensitive to apoptotic stimuli,[29,45–47] which could contribute to the presence of apoptotic features among infected cells. The mechanism by which this increased susceptibility occurs is not well understood. None-theless, the fraction of cells expressing apoptosis markers is minimal, as discussed above, and does not account for massive cell death induced by HIV-1 infection.

To address the issue of direct necrosis vs. bystander death, recent studies have employed a single-cell killing system that allows a clear distinction between infected and uninfected cells. Laboratory strains of HIV-1 were engineered to incorporate a tag, such as placental alkaline phophatase (PLAP), green fluorescent proteins (GFPs), or a mouse heat-stable-antigen/CD24 (HSA) marker, to monitor infection.[10–12,19] These systems also allowed the quantitative cell-by-cell mea-surement of proviral expression by flow cytometry. Careful analyses of provirus/reporter-expressing cells demonstrated that massive cell death occurred in infected cells rather than in bystander cells and that infected cells did not exhibit apoptotic hallmarks such as DNA fragmentation and extracellular flux of phosphatidylserine (Figure 17.1).[11,12] Taken together, evidence strongly indicates that HIV-1 cytopathicity does not involve apoptosis; rather, it predominantly occurs via necrosis. However, this conclusion is based on careful studies carried out *in vitro* and must be further validated by studies of HIV infection *in vivo*.

MOLECULAR MECHANISM OF NONAPOPTOTIC/NECROTIC CELL DEATH

Mitochondrial Dysfunction

A variety of agents that cause lethal cellular damage could result in necrosis or apoptosis via mitochondrial dysfunction. These agents include pathogens and oxidative stress from nitric oxide radicals and reactive oxygen species (ROS). They initiate a death signal by causing drastic changes in the permeability of the mitochondrial inner membrane, which could subsequently trigger the release of proapoptotic molecules such as cytochrome C and Smac/Diablo. This process, called mitochondrial permeability transition (MPT), can lead to either apoptosis or necrosis.[48–51] Studies suggest that the switch between necrosis and apoptosis during mitochondrial dysfunction is modulated by intracellular ATP concentration and caspase activation.[49,52] During MPT, the permeability of the mitochondrial membrane to ions and solutes increases due to the unregulated opening of MPT pores. Subsequent to the MPT pore opening, the mitochondrial transmembrane potential ($\Delta\psi_m$) is perturbed. This is accompanied by the generation of superoxides, uncoupling of oxidative phosphorylation, and ATP depletion, which results in MPT-dependent necrosis.[49–53]

HIV-1 is reported to initiate the loss of $\Delta\psi_m$ upon infection. Three HIV-1 proteins, gp120, Tat, and Vpr, are documented to act as causative agents.[9,14,54–57] The association of soluble gp120 and CXCR4 has been reported to cause opening of the MPT pore followed by the rapid reduction in $\Delta\psi_m$ and, eventually, cell death due to mitochondrial dysfunction.[56,58] In addition to the direct effect, Env-mediated syncytia is also reported to trigger mitochondrial membrane permeabilization and disruption in HeLa cells expressing CD4 and CXCR4.[59] However, it is still controversial whether gp120-induced cell death occurs by apoptosis or necrosis.[21,56,58] Tat is another viral protein proposed to induce MPT and cell death. The constitutive expression of Tat by stable transfection in a T leukemia cell line has been demonstrated to cause a loss of $\Delta\psi_m$ and downregulate mitochondrial superoxide dismutase (SOD2), resulting in the accumulation of ROS and mitochondrial dysfunction.[60,61] Intriguingly, Tat translocates from the nucleus and cytoplasm to mitochondria concomitantly with the $\Delta\psi_m$ disruption,[62] which supports a role of Tat in MPT. During Tat-induced mitochondrial dysfunction, cell death was not abrogated by an inhibitor z-DEVD-cmk for caspase-3, the most important effector caspase.[62] This result suggests that Tat-mediated mitochondrial dysfunction leads to necrotic death. In addition to Env and Tat, Vpr is also suggested to cause death through the mitochondrial pathway. A fraction of Vpr is reported to physically associate with adenine nucleotide translocase (ANT) in mitochondria. This interaction is thought to form large composite ion channels, which facilitates dissipation of $\Delta\psi_m$, leading to cell death.[63,64] A recent study reported an interaction between Vpr and ANT, and the loss of $\Delta\psi_m$ is attributed to the C-terminal region of Vpr (a.a. 52–96), especially R73 and R80 residues,[64] which are also important for the cell-cycle arrest property of Vpr.[65,66] Intriguingly, HIV-1 carrying a mutation at Vpr/R88 (R88A) is reported to be significantly less cytopathic in a Jurkat T cell line,[65,67] suggesting a correlation between Vpr-mediated cytopathicity and mitochondrial dysfunction. Similar to Tat, Vpr-induced MPT and cell death are not inhibited by a caspase inhibitor, zVAD-fmk, or the knockout of mitochondrial apoptosis-inducing factor Apaf-1.[68] These data suggest that Vpr mediates nonapoptotic cell death through the mitochondrial pathway. HIV-induced mitochondrial dysfunction is consistent with the observations that T cells from patients undergo changes in mitochondrial morphology, a reduction of $\Delta\psi_m$, and enhanced ROS generation.[54,55,57] In contrast, long-term nonprogressors have a low incidence of reduced $\Delta\psi_m$,[69] suggesting a correlation between mitochondrial dysfunction and HIV-1 pathogenesis. Further studies are necessary to elucidate the detailed mechanism by which HIV-1 causes the loss of $\Delta\psi_m$.

Cell-Cycle Arrest

HIV-1 infection causes cell-cycle arrest at G2/M phase and Vpr functions as a primary mediator of cell-cycle blockades (Vpr is discussed in Chapter 7). Several groups investigated the correlation between cell-cycle arrest by Vpr and viral-induced cell death. However, the studies failed to reach a consensus (details discussed below).[65,67,70–74] In addition, the mode of death is not yet clearly established in cells arrested by Vpr. Further investigation is necessary to determine whether cell-cycle arrest is a causative factor of cell death.

VIRAL MEDIATORS OF DIRECT, NONAPOPTOTIC CELL DEATH

A major unresolved issue in AIDS pathogenesis is the role of individual viral proteins in HIV-1-induced necrosis. Cell death has been attributed to most HIV-1 genes, including Env, Tat, Nef, Vpr, and Vpu. However, certain studies have also described antiapoptotic properties for nearly all of them, with the exception of Env.[75,76] Functions of HIV-1 proteins in infected cells are well documented in most cases, and cellular factors interacting with them are being uncovered. Nevertheless, a confusing and contradictory picture emerges regarding their roles in death-signal initiation and its mechanism.

ENV

Env is one of the best-studied HIV-1 proteins (discussed in Chapter 6). Various groups have reported that the cross-linking of CD4 by gp120 primes cells for apoptosis and that mutated Env decreases cytopathicity.[6,29,75,76] At the same time, other studies have demonstrated that Env-induced death does not require cellular apoptotic factors and results in necrosis.[6,77,78] In studying the effect of Env on cell death, complications arise from the fact that Env is directly involved in virus entry. Any modification in Env could decrease virus infectivity and indirectly reduce viral killing. To circumvent this problem, recent studies have used VSV-G pseudotyped viruses to investigate the requirement of Env in necrosis.[11,44] The results revealed no difference between Env and Env+ strains of HIV-1$_{NL4-3}$ in their ability to cause direct killing by necrosis in a single round of infection. These data indicate that Env is dispensable for the virally induced necrosis. It should be noted that the deletion study shows only that Env is not required for HIV-1 killing and does not exclude the possibility that Env could participate in cell death.

VIRAL GENOME, REGULATORY PROTEINS

A regulatory protein, Tat, is suggested to be a causative agent of apoptosis; however, antiapoptotic activity has also been documented (details discussed in Chapter 9). Intriguingly, studies with HIV-1 strains carrying mutated Tat also indicate that Tat is not required for direct necrosis during acute infection.[79–81] This conclusion has been underscored by a different experiment using a derivative of an NL4-3 strain, NL-EGFP, that carries mutations in the genome so that it expresses only Tat, Rev, and Vpu as well as GFP inserted in the viral genome.[12] The NL-EGFP virus pseudotyped with VSV-G was highly infectious and replicated in a Jurkat T cell line. However, unlike the wild-type counterpart, it did not cause cytopathicity during an acute single round of infection. Therefore, Tat, Rev, and Vpu, as well as the HIV-1 RNA genome, are not death-inducing factors in this system.

ACCESSORY GENES

Nef

The investigation of viral genetic variations among putative long-term nonprogressors and AIDS patients revealed Nef gene deletions in nonprogressors.[82–84] Moreover, Nef can upregulate an apop-tosis-inducing receptor, Fas, and its ligand, FasL (discussed in Chapter 8),[75,76] although other studies excluded their involvement in CD4+ T cell destruction, as described above. The role of Nef in CD4+ T cell destruction was rendered uncertain by the data that CD4[+] T cell death induced by HIV-1 infection is independent of Fas or apoptosis.[6,7,10–12] Moreover, according to deletion studies, Nef-deficient strains of NL4-3 and HXB2 retain a potent ability to kill infected cells.[11,12,85,86] These data suggest that Nef is also dispensable for HIV-1-induced cell death during an acute single round of infection.

Vpr

Vpr is a well-studied accessory protein that is conserved among primate lentiviruses (details discussed in Chapter 7). It is primarily found in the nucleus. In addition to its effect on mitochondria, Vpr arrests infected cells at the G2/M phase of cell cycle, as discussed above. The cell-cycle arrest by Vpr was suggested to correlate with its ability to kill infected T cells.[65,71,72,74] Vpr is proposed to upregulate provirus expression by causing G2/M cell-cycle arrest, which could indirectly contribute to HIV-1 killing. Studies with Vpr mutants demonstrated that varying degrees of cell-cycle arrest by Vpr induce a proportional amount of cell death relative to arrest,[65] supporting the popular hypothesis that cell-cycle arrest contributes to Vpr killing.[67,72,74] However, other studies demonstrated that the addition of drugs that abrogate the G2/M-phase arrest, including methylxanthines and pentoxifyline, did not prevent Vpr-induced cell death.[65,70,73,87,88] In addition, several Vpr mutants or a GFP fusion abrogate cell-cycle arrest but still maintain their ability to cause cell death,[70,73,89] raising a possibility that cell-cycle arrest may not be the cause of Vpr-mediated cell death. Further clarification is needed to establish a causative relationship between virally induced cell-cycle arrest and HIV-1 killing. It will be especially important to understand how Vpr causes cell cycle arrest and necrotic death from the nucleus, its presumed locus of action.

Vpu

The small accessory protein Vpu is a membrane protein involved in the release of virions from infected cells.[90–92] Its suggested function also includes the induction of CD4 degradation.[93–96] The requirement of Vpu in HIV-1 killing is controversial due to conflicting reports. Vpu is implicated in HIV-1-induced apoptosis in CD4[+] T cells by its ability to inhibit NF-κB activity and, consequently, the expression of antiapoptotic factors, including Bcl-xL. However, several groups reported that viral killing of CD4[+] T cells was not altered with mutant HIV-1 strains, NL4-3ΔVprΔVpuΔNef, HXB-2, and HIV-EGFP, which lack Vpu as well as Vpr and Nef expression.[12,85,86] Therefore, at the present time, it is uncertain as to whether Vpu is necessary for HIV-1 killing.

SUMMARY

Remarkable advances have been made in research on HIV-1 since the discovery of the virus, and the functions of viral proteins are rapidly being uncovered. However, many of the studies on the mode of virally induced cell death have thus far resulted in contradictory conclusions, partially due to technical difficulties and interpretation of data, as discussed above. Moreover, consensus on the roles of viral proteins in HIV-1 killing has not been reached, although numerous reports have assigned molecular functions to HIV-1 proteins during the viral life cycle. Better knowledge of how HIV-1 kills T lymphocytes is likely the key to understanding AIDS pathogenesis and is, therefore, crucial in developing new therapeutic strategies. Further study is necessary to establish a relationship between viral protein functions and cytopathicity, which should lead to an elucidation of the mechanism by which HIV-1 kills CD4[+] T cells.

ACKNOWLEDGMENT

A tremendous amount of work was published in this field, making it impossible to discuss many important studies. The papers cited here make up only a representative and not comprehensive group of publications in this area of research. The authors apologize to the many fine investigators whose work could not be cited for lack of space. We thank Diane Bolton and members of Lenardo Lab for helpful discussions. K.S. is a JSPS research fellow for Japanese biomedical and behavioral research at NIH.

REFERENCES

1. Gallo, R.C., Human retroviruses after 20 years: a perspective from the past and prospects for their future control, *Immunol. Rev.,* 185, 236, 2002.
2. Gallo, R.C., The path to the discoveries of human retroviruses, *J. Hum. Virol.,* 3, 1, 2000.
3. Ho, D.D., Neumann, A.U., Perelson, A.S., Chen, W., Leonard, J.M., and Markowitz, M., Rapid turnover of plasma virions and CD4 lymphocytes in HIV-1 infection, *Nature,* 373, 123, 1995.
4. Perelson, A.S., Neumann, A.U., Markowitz, M., Leonard, J.M., and Ho, D.D., HIV-1 dynamics *in vivo:* virion clearance rate, infected cell life-span, and viral generation time, *Science,* 271, 1582, 1996.
5. Wei, X., Ghosh, S.K., Taylor, M.E., Johnson, V.A., Emini, E.A., Deutsch, P., Lifson, J.D., Bonhoeffer, S., Nowak, M.A., Hahn, B.H., Saag, M.S., and Shaw, G.M., Viral dynamics in human immunodeficiency virus type 1 infection, *Nature,* 373, 117, 1995.
6. Cao, J., Park, I.W., Cooper, A., and Sodroski, J, Molecular determinants of acute single-cell lysis by human immunodeficiency virus type 1, *J. Virol.,* 70, 1340, 1996.
7. Kolesnitchenko, V., King, L., Riva, A., Tani, Y., Korsmeyer, S., and Cohen, D.I., A major human immunodeficiency virus type-1-initiated killing pathway distinct from apoptosis, *J. Virol.,* 71, 9753, 1997.
8. Borthwick, N.J., Wickremasinghe, R.G., Lewin, J., Fairbanks, L.D., and Bofill, M., Activation-associated necrosis in human immunodeficiency virus infection, *J. Infec. Dis.,* 179, 352, 1999.
9. Plymale, D.R., Tang, D.S., Comardelle, A.M., Fermin, C.D., Lewis, D.E., and Garry, R.F., Both necrosis and apoptosis contribute to HIV-1-induced killing of CD4 cells, *AIDS,* 13, 1827, 1999.
10. Gandhi, R.T., Chen, B.K., Straus, S.E., Dale, J.K., Lenardo, M.J., and Baltimore, D., HIV-1 directly kills CD4+ T cells by a Fas-independent mechanism, *J. Exp. Med.,* 187, 1113, 1998.
11. Bolton, D.L., Hahn, B.-I., Park, E.A., Lehnhoff, L.L., Hornung, F., and Lenardo, M.J., Death of CD4+ T-cell lines caused by human immunodeficiency virus type 1 does not depend on caspases or apoptosis, *J. Virol.,* 76, 5094, 2002.
12. Lenardo, M.J., Angelman, S.B., Bounkeua, V., Dimas, J., Duvall, M.G., Graubard, M.B., Hornung, F., Selkirk, M.C., Speirs, C.K., Trageser, C., Orenstein, J.O., and Bolton, D.L., Cytopathic killing of peripheral blood CD4+ T lymphocytes by human immunodeficiency virus type 1 appears necrotic rather than apoptotic and does not require *Env, J. Virol.,* 76, 5082, 2002.
13. Banda, N.K., Bernier, J., Kurahara, D.K., Kurrle, R., Haigwood, N., Sekaly, R.P., and Finkel, T.H., Crosslinking CD4 by human immunodeficiency virus gp120 primes T cells for activation-induced apoptosis, *J. Exp. Med.,* 176, 1099, 1992.
14. Blanco, J., Barretina, J., Ferri, K.F., Jacotot, E., Gutierrez, A., Armand-Ugon, M., Cabrera, C., Kroemer, G., Clotet, B., and Este, J.A., Cell-surface-expressed HIV-1 envelope induces the death of CD4 T cells during GP41-mediated hemifusion-like events, *Virology,* 305, 318, 2003.
15. LaBonte, J.A., Patel, T., Hofmann, W., and Sodroski, J., Importance of membrane fusion mediated by human immunodeficiency virus envelope glycoproteins for lysis of primary CD4-positive T cells, *J. Virol.,* 74, 10690, 2000.
16. Muro-Cacho, C.A., Pantaleo, G., and Fauci, A.S., Analysis of apoptosis in lymph nodes of HIV-infected persons. Intensity of apoptosis correlates with the general state of activation of the lymphoid tissue and not with stage of disease or viral burden, *J. Immunol.,* 154, 5555, 1995.
17. Douek, D.C., Disrupting T-cell homeostasis: how HIV-1 infection causes disease, *AIDS Rev.,* 5, 172, 2003.
18. Finkel, T.H., Tudor-Williams, G., Banda, N.K., Cotton, M.F., Curiel, T., Monks, C., Baba, T.W., Ruprecht, R.M., and Kupfer, A., Apoptosis occurs predominantly in bystander cells and not in productively infected cells of HIV- and SIV-infected lymph nodes, *Nat. Med.,* 1, 129, 1995.
19. Herbein, G., Van Lint, C., Lovett, J.L., and Verdin, E., Distinct mechanisms trigger apoptosis in human immunodeficiency virus type 1-infected and in uninfected bystander T lymphocytes, *J. Virol.,* 72, 660, 1998.
20. Somasundaran, M. and Robinson, H.L., A major mechanism of human immunodeficiency virus-induced cell killing does not involve cell fusion, *J. Virol.,* 61, 3114, 1987.

21. Stevenson, M., Meier, C., Mann, A.M., Chapman, N., and Wasiak, A., Envelope glycoprotein of HIV induces interference and cytolysis resistance in CD4+ cells: mechanism for persistence in AIDS, *Cell,* 53, 483, 1988.

22. Sylwester, A., Murphy, S., Shutt, D., and Soll, D.R., HIV-induced T cell syncytia are self-perpetuating and the primary cause of T cell death in culture, *J. Immunol.,* 158, 3996, 1997.

23. Steensma, D.P., Timm, M., and Witzig, T.E., Flow cytometric methods for detection and quantification of apoptosis, *Methods Mol. Med.,* 85, 323, 2003.

24. Watanabe, M., Hitomi, M., van der Wee, K., Rothenberg, F., Fisher, S.A., Zucker, R., Svoboda, K.K., Goldsmith, E.C., Heiskanen, K.M., and Nieminen, A.L., The pros and cons of apoptosis assays for use in the study of cells, tissues, and organs, *Microsc. Microanal.,* 8, 375, 2002.

25. Allen, R.T., Hunter, W.J. III, and Agrawal, D.K., Morphological and biochemical characterization and analysis of apoptosis, *J. Pharmacol. Toxicol. Methods,* 37, 215, 1997.

26. Chautan, M., Chazal, G., Cecconi, F., Gruss, P., and Golstein, P., Interdigital cell death can occur through a necrotic and caspase-independent pathway, *Curr. Biol.,* 9, 967, 1999.

27. Vercammen, D., Vandenabeele, P., Beyaert, R., Declercq, W., and Fiers, W., Tumour necrosis factor-induced necrosis vs. anti-Fas-induced apoptosis in L929 cells, *Cytokine,* 9, 801, 1997.

28. Jaworowski, A. and Crowe, S.M., Does HIV cause depletion of CD4+ T cells *in vivo* by the induction of apoptosis? *Immunol. Cell Biol.,* 77, 90, 1999.

29. Gougeon, M.L. and Montagnier, L., Programmed cell death as a mechanism of CD4 and CD8 T cell deletion in AIDS. Molecular control and effect of highly active anti-retroviral therapy, *Ann. N.Y. Acad. Sci.,* 887, 199, 1999.

30. Chinnaiyan, A.M., Woffendin, C., Dixit, V.M., and Nabel, G.J., The inhibition of pro-apoptotic ICE-like proteases enhances HIV replication, *Nat. Med.,* 3, 333, 1997.

31. Meyaard, L., Otto, S.A., Keet, I.P., Roos, M.T., and Miedema, F., Programmed death of T cells in human immunodeficiency virus infection. No correlation with progression to disease, *J. Clin. Invest.,* 93, 982, 1994.

32. Antoni, B.A., Sabbatini, P., Rabson, A.B., and White, E., Inhibition of apoptosis in human immuno-deficiency virus-infected cells enhances virus production and facilitates persistent infection, *J. Virol.,* 69, 2384, 1995.

33. Collette, Y., Dutartre, H., Benziane, A., and Olive, D., The role of HIV1 Nef in T-cell activation: Nef impairs induction of Th1 cytokines and interacts with the Src family tyrosine kinase Lck, *Res. Virol.,* 148, 52, 1997.

34. Mangili, F., Cigala, C., and Santambrogio, G., Staining apoptosis in paraffin sections. Advantages and limits, *Anal. Quant. Cytol. Histol.,* 21, 273, 1999.

35. Kerr, J.F., Gobe, G.C., Winterford, C.M., and Harmon, B.V., Anatomical methods in cell death, *Methods Cell Biol.,* 46, 1, 1995.

36. Compton, M.M., A biochemical hallmark of apoptosis: internucleosomal degradation of the genome, *Cancer Metastasis Rev.,* 11, 105, 1992.

37. Vermes, I., Haanen, C., Steffens-Nakken, H., and Reutelingsperger, C., A novel assay for apoptosis. Flow cytometric detection of phosphatidylserine expression on early apoptotic cells using fluorescein labelled Annexin V, *J. Immunol. Methods,* 184, 39, 1995.

38. Koopman, G., Reutelingsperger, C.P., Kuijten, G.A., Keehnen, R.M., Pals, S.T., and van Oers, M.H., Annexin V for flow cytometric detection of phosphatidylserine expression on B cells undergoing apoptosis, *Blood,* 84, 1415, 1994.

39. Park, I.W., Kondo, E., Bergeron, L., Park, J., and Sodroski, J., Effects of human immunodeficiency virus type 1 infection on programmed cell death in the presence or absence of Bcl-2, *J. Acquir. Immune Defic. Syndr. Hum. Retrovirol.,* 12, 321, 1996.

40. Bergeron, L., Sullivan, N., and Sodroski, J., Target cell-specific determinants of membrane fusion within the human immunodeficiency virus type 1 gp120 third variable region and gp41 amino terminus, *J. Virol.,* 66, 2389, 1992.

41. Cotton, M.F., Cassella, C., Rapaport, E.L., Tseng, P.O., Marschner, S., and Finkel, T.H., Apoptosis in HIV-1 infection, *Behring Inst. Mitt.,* 97, 220, 1996.

42. Julius, M., Maroun, C.R., and Haughn, L., Distinct roles for CD4 and CD8 as co-receptors in antigen receptor signalling, *Immunol. Today,* 14, 177, 1993.

43. Juo, P., Woo, M.S., Kuo, C.J., Signorelli, P., Biemann, H.P., Hannun, Y.A., and Blenis, J., FADD is required for multiple signaling events downstream of the receptor Fas, *Cell Growth Differ.,* 10, 797, 1999.

44. Lenardo, M., Chan, K.M., Hornung, F., McFarland, H., Siegel, R., Wang, J., and Zheng, L., Mature T lymphocyte apoptosis—immune regulation in a dynamic and unpredictable antigenic environment, *Annu. Rev. Immunol.,* 17, 221, 1999.

45. Groux, H., Torpier, G., Monte, D., Mouton, Y., Capron, A., and Ameisen, J.C., Activation-induced death by apoptosis in CD4+ T cells from human immunodeficiency virus-infected asymptomatic individuals, *J. Exp. Med.,* 175, 331, 1992.

46. Alimonti, J.B., Ball, T.B., and Fowke, K.R., Mechanisms of CD4+ T lymphocyte cell death in human immunodeficiency virus infection and AIDS, *J. Gen. Virol.,* 84, 1649, 2003.

47. Yang, Y., Dong, B., Mittelstadt, P.R., Xiao, H., and Ashwell, J.D., HIV Tat binds Egr proteins and enhances Egr-dependent transactivation of the Fas ligand promoter, *J. Biol. Chem.,* 277, 19482, 2002.

48. Hunter, D.R., Haworth, R.A., and Southard, J.H., Relationship between configuration, function, and permeability in calcium-treated mitochondria, *J. Biol. Chem.,* 251, 5069, 1976.

49. Kroemer, G., Dallaporta, B., and Resche-Rigon, M., The mitochondrial death/life regulator in apoptosis and necrosis, *Annu. Rev. Physiol.,* 60, 619, 1998.

50. Bernardi, P., Petronilli, V., Di Lisa, F., and Forte, M., A mitochondrial perspective on cell death, *Trends Biochem. Sci.,* 26, 112, 2001.

51. Denecker, G., Vercammen, D., Declercq, W., and Vandenabeele, P., Apoptotic and necrotic cell death induced by death domain receptors, *Cell. Mol. Life Sci.,* 58, 356, 2001.

52. Nicotera, P. and Melino, G., Regulation of the apoptosis-necrosis switch, *Oncogene,* 23, 2757, 2004.

53. Kim, J.S., He, L., and Lemasters, J.J., Mitochondrial permeability transition: a common pathway to necrosis and apoptosis, *Biochem. Biophys. Res. Commun.,* 304, 463, 2003.

54. Carbonari, M., Cibati, M., Pesce, A.M., Sbarigia, D., Grossi, P., D'Offizi, G., Luzi, G., and Fiorilli, M., Frequency of provirus-bearing CD4+ cells in HIV type 1 infection correlates with extent of *in vitro* apoptosis of CD8+ but not of CD4+ cells, *AIDS Res. Hum. Retroviruses,* 11, 789, 1995.

55. Cossarizza, A., Mussini, C., Mongiardo, N., Borghi, V., Sabbatini, A., De Rienzo, B., and Franceschi, C., Mitochondria alterations and dramatic tendency to undergo apoptosis in peripheral blood lymphocytes during acute HIV syndrome, *AIDS,* 11, 19, 1997.

56. Roggero, R., Robert-Hebmann, V., Harrington, S., Roland, J., Vergne, L., Jaleco, S., Devaux, C., and Biard-Piechaczyk, M., Binding of human immunodeficiency virus type 1 gp120 to CXCR4 induces mitochondrial transmembrane depolarization and cytochrome c-mediated apoptosis independently of Fas signaling, *J. Virol.,* 75, 7637, 2001.

57. Macho, A., Castedo, M., Marchetti, P., Aguilar, J.J., Decaudin, D., Zamzami, N., Girard, P.M., Uriel, J., and Kroemer, G., Mitochondrial dysfunctions in circulating T lymphocytes from human immunodeficiency virus-1 carriers, *Blood,* 86, 2481, 1995.

58. Berndt, C., Mopps, B., Angermuller, S., Gierschik, P., and Krammer, P.H., CXCR4 and CD4 mediate a rapid CD95-independent cell death in CD4(+) T cells, *Proc. Natl. Acad. Sci. U.S.A.,* 95, 12556, 1998.

59. Ferri, K.F., Jacotot, E., Blanco, J., Este, J.A., Zamzami, N., Susin, S.A., Xic, Z., Brothers, G., Reed, J.C., Penninger, J.M., and Kroemer, G., Apoptosis control in syncytia induced by the HIV type 1-envelope glycoprotein complex: role of mitochondria and caspases, *J. Exp. Med.,* 192, 1081, 2000.

60. Creaven, M., Hans, F., Mutskov, V., Col, E., Caron, C., Dimitrov, S., and Khochbin, S., Control of the histone-acetyltransferase activity of Tip60 by the HIV-1 transactivator protein, Tat, *Biochem.,* 38, 8826, 1999.

61. Westendorp, M.O., Shatrov, V.A., Schulze-Osthoff, K., Frank, R., Kraft, M., Los, M., Krammer, P.H., Droge, W., and Lehmann, V., HIV-1 Tat potentiates TNF-induced NF-B activation and cytotoxicity by altering the cellular redox state, *EMBO J.,* 14, 546, 1995.

62. Macho, A., Calzado, M.A., Jimenez-Reina, L., Ceballos, E., Leon, J., and Munoz, E., Susceptibility of HIV-1-TAT transfected cells to undergo apoptosis. Biochemical mechanisms, *Oncogene,* 18, 7543, 1999.

63. Jacotot, E., Ravagnan, L., Loeffler, M., Ferri, K.F., Vieira, H.L., Zamzami, N., Costantini, P., Druillennec, S., Hoebeke, J., Briand, J.P., Irinopoulou, T., Daugas, E., Susin, S.A., Cointe, D., Xie, Z.H., Reed, J.C., Roques, B.P., and Kroemer, G., The HIV-1 viral protein R induces apoptosis via a direct effect on the mitochondrial permeability transition pore, *J. Exp. Med.,* 191, 33, 2000.

64. Jacotot, E., Ferri, K.F., El Hamel, C., Brenner, C., Druillennec, S., Hoebeke, J., Rustin, P., Metivier, D., Lenoir, C., Geuskens, M., Vieira, H.L., Loeffler, M., Belzacq, A.S., Briand, J.P., Zamzami, N., Edelman, L., Xie, Z.H., Reed, J.C., Roques, B.P., and Kroemer, G., Control of mitochondrial membrane permeabilization by adenine nucleotide translocator interacting with HIV-1 viral protein rR and Bcl-2, *J. Exp. Med.,* 193, 509, 2001.

65. Stewart, S.A., Poon, B., Jowett, J.B., and Chen, I.S., Human immunodeficiency virus type 1 Vpr induces apoptosis following cell cycle arrest, *J. Virol.,* 71, 5579, 1997.

66. Di Marzio, P., Choe, S., Ebright, M., Knoblauch, R., and Landau, N.R., Mutational analysis of cell cycle arrest, nuclear localization and virion packaging of human immunodeficiency virus type 1 Vpr, *J. Virol.,* 69, 7909, 1995.

67. Yao, X.J., Mouland, A.J., Subbramanian, R.A., Forget, J., Rougeau, N., Bergeron, D., and Cohen, E.A., Vpr stimulates viral expression and induces cell killing in human immunodeficiency virus type 1-infected dividing Jurkat T cells, *J. Virol.,* 72, 4686, 1998.

68. Roumier, T., Vieira, H.L., Castedo, M., Ferri, K.F., Boya, P., Andreau, K., Druillennec, S., Joza, N., Penninger, J.M., Roques, B., and Kroemer, G., The C-terminal moiety of HIV-1 Vpr induces cell death via a caspase-independent mitochondrial pathway, *Cell Death Differ.,* 9, 1212, 2002.

69. Moretti, S., Marcellini, S., Boschini, A., Famularo, G., Santini, G., Alesse, E., Steinberg, S.M., Cifone, M.G., Kroemer, G., and De Simone, C., Apoptosis and apoptosis-associated perturbations of peripheral blood lymphocytes during HIV infection: comparison between AIDS patients and asymptomatic long-term non-progressors, *Clin. Exp. Immunol.,* 122, 364, 2000.

70. Waldhuber, M.G., Bateson, M., Tan, J., Greenway, A.L., and McPhee, D.A., Studies with GFP-Vpr fusion proteins: induction of apoptosis but ablation of cell-cycle arrest despite nuclear membrane or nuclear localization, *Virology,* 313, 91, 2003.

71. Watanabe, N., Yamaguchi, T., Akimoto, Y., Rattner, J.B., Hirano, H., and Nakauchi, H., Induction of M-phase arrest and apoptosis after HIV-1 Vpr expression through uncoupling of nuclear and centrosomal cycle in HeLa cells, *Exp. Cell Res.,* 258, 261, 2000.

72. Zhu, Y., Gelbard, H.A., Roshal, M., Pursell, S., Jamieson, B.D., and Planelles, V., Comparison of cell cycle arrest, transactivation, and apoptosis induced by the simian immunodeficiency virus SIVagm and human immunodeficiency virus type 1 Vpr genes, *J. Virol.,* 75, 3791, 2001.

73. Nishizawa, M., Kamata, M., Mojin, T., Nakai, Y., and Aida, Y., Induction of apoptosis by the Vpr protein of human immunodeficiency virus type 1 occurs independently of G(2) arrest of the cell cycle, *Virol.,* 276, 16, 2000.

74. Fukumori, T., Akari, H., Yoshida, A., Fujita, M., Koyama, A.H., Kagawa, S., and Adachi, A., Regulation of cell cycle and apoptosis by human immunodeficiency virus type 1 Vpr, *Microbes Infect.,* 2, 1011, 2000.

75. Roshal, M., Zhu, Y., and Planelles, V., Apoptosis in AIDS, *Apoptosis,* 6, 103, 2001.

76. Selliah, N. and Finkel, T.H., HIV-1 NL4-3, but not IIIB, inhibits JAK3/STAT5 activation in CD4(+) T cells, *Virology,* 286, 412, 2001.

77. Stevenson, M., Haggerty, S., Lamonica, C., Mann, A.M., Meier, C., and Wasiak, A., Cloning and characterization of human immunodeficiency virus type 1 variants diminished in the ability to induce syncytium-independent cytolysis, *J. Virol.,* 64, 3792, 1990.

78. Koga, Y., Nakamura, K., Sasaki, M., Kimura, G., and Nomoto, K., The difference in gp160 and gp120 of HIV type 1 in the induction of CD4 downregulation preceding single-cell killing, *Virology,* 201, 137, 1994.

79. Dayton, A.I., Sodroski, J.G., Rosen, C.A., Goh, W.C., and Haseltine, W.A., The trans-activator gene of the human T cell lymphotropic virus type III is required for replication, *Cell,* 44, 941, 1986.

80. Sodroski, J., Goh, W.C., Rosen, C., Tartar, A., Portetelle, D., Burny, A., and Haseltine, W., Replicative and cytopathic potential of HTLV-III/LAV with Sor gene deletions, *Science,* 231, 1549, 1986.

81. Rogel, M.E., Wu, L.I., and Emerman, M., The human immunodeficiency virus type 1 Vpr gene prevents cell proliferation during chronic infection, *J. Virol.,* 69, 882, 1995.

82. Premkumar, D.R., Ma, X.Z., Maitra, R.K., Chakrabarti, B.K., Salkowitz, J., Yen-Lieberman, B., Hirsch, M.S., and Kestler, H.W., The Nef gene from a long-term HIV type 1 nonprogressor, *AIDS Res. Hum. Retroviruses,* 12, 337, 1996.

83. Rhodes, D.I., Ashton, L., Solomon, A., Carr, A., Cooper, D., Kaldor, J., and Deacon, N., Characterization of three Nef-defective human immunodeficiency virus type 1 strains associated with long-term nonprogression. Australian Long-Term Nonprogressor Study Group, *J. Virol.,* 74, 10581, 2000.

84. Salvi, R., Garbuglia, A.R., Di Caro, A., Pulciani, S., Montella, F., and Benedetto, A., Grossly defective Nef gene sequences in a human immunodeficiency virus type 1-seropositive long-term nonprogressor, *J. Virol.,* 72, 3646, 1998.

85. Rapaport, E., Casella, C.R., Ikl, Mustafa, F., Isaak, D., and Finkel, T.H., Mapping of HIV-1 determinants of apoptosis in infected T cells, *Virology,* 252, 407, 1998.

86. Gibbs, J.S., Regier, D.A., and Desrosiers, R.C., Construction and *in vitro* properties of SIVmac mutants with deletions in "nonessential" genes, *AIDS Res. Hum. Retroviruses,* 10, 607, 1994.

87. Lum, J.J., Cohen, O.J., Nie, Z., Weaver, J.G., Gomez, T.S., Yao, X.J., Lynch, D., Pilon, A.A., Hawley, N., Kim, J.E., Chen, Z., Montpetit, M., Sanchez-Dardon, J., Cohen, E.A., and Badley, A.D., Vpr R77Q is associated with long-term nonprogressive HIV infection and impaired induction of apoptosis, *J. Clin. Invest.,* 111, 1547, 2003.

88. Somasundaran, M., Sharkey, M., Brichacek, B., Luzuriaga, K., Emerman, M., Sullivan, J.L., and Stevenson, M., Evidence for a cytopathogenicity determinant in HIV-1 Vpr, *Proc. Natl. Acad. Sci. U.S.A.,* 99, 9503, 2002.

89. Jian, H. and Zhao, L.J., Pro-apoptotic activity of HIV-1 auxiliary regulatory protein Vpr is subtype-dependent and potently enhanced by nonconservative changes of the leucine residue at position 64, *J. Biol. Chem.,* 278, 44326, 2003.

90. Klimkait, T., Strebel, K., Hoggan, M.D., Martin, M.A., and Orenstein, J.M., The human immunodeficiency virus type 1-specific protein Vpu is required for efficient virus maturation and release, *J. Virol.,* 64, 621, 1990.

91. Terwilliger, E.F., Cohen, E.A., Lu, Y.C., Sodroski, J.G., and Haseltine, W.A., Functional role of human immunodeficiency virus type 1 Vpu, *Proc. Natl. Acad. Sci. U.S.A.,* 86, 5163, 1989.

92. Strebel, K., Klimkait, T., Maldarelli, F., and Martin, M.A., Molecular and biochemical analyses of human immunodeficiency virus type 1 Vpu protein, *J. Virol.,* 63, 3784, 1989.

93. Tanaka, M., Ueno, T., Nakahara, T., Sasaki, K., Ishimoto, A., and Sakai, H., Downregulation of CD4 is required for maintenance of viral infectivity of HIV-1, *Virology,* 311, 316, 2003.

94. Levesque, K., Finzi, A., Binette, J., and Cohen, E.Al., Role of CD4 receptor down-regulation during HIV-1 infection, *Curr. HIV Res.,* 2, 51, 2004.

95. Lama, J., The physiological relevance of CD4 receptor down-modulation during HIV infection, *Curr. HIV Res.,* 1, 167, 2003.

96. Bour, S. and Strebel, K., The HIV-1 Vpu protein: a multifunctional enhancer of viral particle release, *Microbes Infect.,* 5, 1029, 2003.

18 Apoptosis in Organ Culture and Animal Models of HIV Disease

Shailesh K. Choudhary
David Camerini

CONTENTS

INTRODUCTION

Animal models have great utility in the biomedical research of human diseases. Organ culture also has been used extensively to mimic infection of organs, with complex structure and multiple cell types. Many aspects of complex diseases cannot be duplicated in simpler systems. With regard to acquired immunodeficiency syndrome (AIDS), many of the key questions regarding pathogenesis can only be answered in these complex systems that more closely mimic human immunodeficiency virus (HIV)-1 infection of humans. For example, there is evidence for direct killing of HIV-1 infected cells as well as indirect killing of bystander cells by a variety of mechanisms. Lymph nodes and thymus are major sites of HIV-1 infection *in vivo* and, thus, it is important to study which mechanism or mechanisms of T-cell depletion are operative in each tissue after infection with R5 or X4 HIV-1. Although many questions can be answered with the use of organ culture and animal models, some questions can be addressed only in animal models. This is the case for questions regarding the replacement of T cells. It is not clear whether T cells that are killed by HIV-1 directly or indirectly are replaced primarily by new T cells derived from the thymus or by expansion of those mature T cells that exist. This may also vary depending on a variety of factors, including the age of the infected host and viral phenotype, including co-receptor specificity.

APOPTOSIS IN HIV-1 INFECTED ORGAN CULTURE

The thymus is the primary lymphoid organ where T cells are generated during fetal and neonatal development. In adults, however, peripheral proliferation of T cells is sufficient to maintain normal immune homeostasis. Nevertheless, the thymus functions throughout life, although at reduced levels, and responds to T cell depletion induced by HIV-1.[1,2] Clinical studies of HIV-1-infected children and adults suggest that HIV-1 infection causes thymic dysfunction, increased involution, and thymocyte depletion.[3,4] HIV-1 infection of the thymus, particularly of infants, is documented and is frequently associated with rapid progression to AIDS.[5–8] Reduced thymic volume, impaired delayed-type hypersensitive responses, and spontaneous apoptosis of thymocytes are associated with HIV-1 infection.[9,10] Alteration of thymocyte maturation by HIV-1 could have an impact on the composition and function of peripheral T cell populations and may interfere with immune responses.[11] Recent data on thymic function during highly active antiretroviral therapy (HAART) suggest that thymic recovery may be achieved in some patients on HAART.[5,12,13] Extensive thymic structural damage, however, may hamper immune reconstitution, especially in pediatric patients for whom the thymus is crucial for generating naive T cells. The mechanisms by which HIV-1 causes depletion of thymocytes and destruction of the thymus are not well understood. Thymocytes develop in a specialized microenvironment that is required for their normal maturation and selection. As the thymus is difficult to study in humans, several model systems of human thymopoiesis were developed, including fetal thymic organ culture (FTOC) and the severe combined immune deficient (SCID) mouse implanted with human fetal thymus and liver tissue (SCID-hu Thy/Liv mouse). HIV-1 can infect, replicate, and cause depletion of thymocytes in these experimental models.[14–18]

FETAL THYMIC ORGAN CULTURE (FTOC)

Dissecting human thymus into small pieces that contain several intact lobules creates FTOC. Tissue pieces are infected with HIV-1 and cultured on sterile membrane filters with or without collagen rafts in enriched media. Several groups have used human FTOC as a model to study HIV-1 infection and pathogenesis.[15,19–26] FTOC can be infected with both CCR5-tropic HIV-1 (R5 HIV-1) and CXCR4-tropic HIV-1 (X4 HIV-1) and exhibits pathology similar to that observed in the SCID-hu Thy/Liv model.[15,21,25] Both R5 HIV-1 and X4 HIV-1 deplete thymocytes in a time-dependent manner; however, X4 HIV-1 exhibits greater cytopathic effect (CPE) than R5 HIV-1 in FTOC (Figure 18.1A).[15,25] Many cytokines that are induced in HIV-1 infected individuals[27–30] are also induced by HIV-1 infection of FTOC. These include interleukin (IL)-10, TGF-β, IFN-α, CXCL10, and Oncostatin M.[23,25,31,32] FTOC was shown to be responsive to several antiviral agents[15] and, therefore, is a useful thymus model for the study of HIV-1 replication and its inhibition.

The mechanisms by which HIV-1 infection causes thymic damage are not well understood. Nevertheless, apoptosis is clearly the ultimate cause of CD4+ thymocyte depletion during HIV-1 infection of FTOC.[15] The nature of the signals that initiate apoptosis in the HIV-1-infected thymus, however, is not known. R5 HIV-1 clones preferentially deplete the CCR5+ subset of CD4+ thymocytes, whereas X4 HIV-1 clones spare this thymocyte subset, indicating that direct killing of infected cells is likely the predominant mechanism of thymocyte depletion.[25] HIV-1 can infect thymocytes at several different stages of maturation, which may contribute to direct CPE.[33,34] Alternatively, changes in the thymic milieu due to HIV-1 infection may induce apoptosis of uninfected thymocytes.[14,34] HIV-1 infection of thymocytes impedes thymocyte maturation at early stages, including CD3−, CD4−, CD8−, CD25−, and CD44+.[11] HIV-1 *nef* gene expression in thymic precursor cells and their colonization in FTOC and SCID-hu also lead to impaired T cell development.[32,35] Progenitor cells derived from infants born to HIV-positive mothers have decreased cloning efficiency and generate fewer T cells in FTOC. This impaired progenitor cell function may be responsible, at least in part, for lower naive CD4 counts and reduced thymic output in HIV-negative

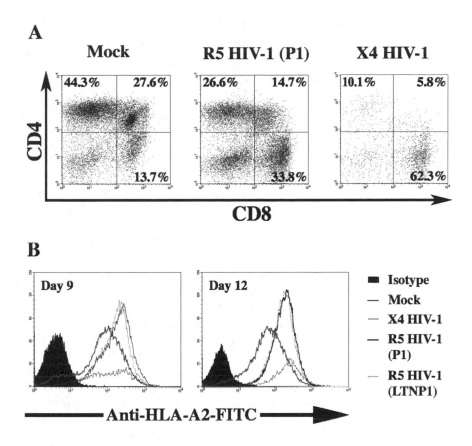

FIGURE 18.1 Pathogenesis of HIV-1 in thymocytes. FTOC was infected with the X4 HIV-1 molecular clone NL4-3, R5 HIV-1 progressor clone P1 (ACH142 *E11), R5 HIV-1 long-term nonprogressor clone LTNP 1 (ACH441 39.2g10) or was mock infected. FTOC was carried in the presence of IL-2 (20 IU/ml), IL-4 (0.5 ng/ml) and IL-7 (1 ng/ml). FTOC was maintained for 12 days with daily media changes. (A) R5 and X4 HIV-1 deplete CD4+ thymocytes in FTOC. Thymocytes were recovered 12 days after infection and stained for surface expression of CD4 and CD8. Representative dot plots for one of three independent experiments are shown. (B) Up-regulation of cell, surface MHC-1 protein in FTOC infected with HIV-1. Thymocytes were recovered 9 and 12 days after infection and stained with monoclonal antibody BB7-2 fluorescein isothiocyanate (FITC) directed at HLA-A2.

infants of HIV-positive mothers.[36] Berkowitz et al. showed that CXCR4 is expressed at various stages of thymocyte development and is highly expressed on immature (CD3−, CD4+, CD8−) intrathymic T progenitor (ITTP) cells.[37] X4 HIV-1 can infect and deplete ITTP, thereby interfering with thymocyte development.[11,14] In HIV-1-infected FTOC, an increased expression of MHC class I (MHC I) was observed on ITTP and double-positive (DP) thymocytes due to increased IFN-α and IL-10 production (Figure 18.1B).[23,31] This increased expression of MHC I interferes with thymocyte development and was shown to be responsible for generation of dysfunctional CD8+ T cells.[22] ITTP cells, however, express undetectable levels of CCR5 and are, therefore, not targets for R5 HIV-1 infection or direct killing.[37] IL-7 exhibits a strong antiapoptotic effect on early lymphoid progenitor cells,[38–40] in part by increasing intracellular Bcl-2 levels.[41,42] Moreover, IL-7 treatment of FTOC increases the proliferation of immature CD3−/CD3low thymocytes and stimulates T cell receptor (TCR)-$\alpha\beta$ gene rearrangements.[42]

Secondary Lymphoid Tissue Organ Culture (LTOC)

Secondary lymphoid organs are the sites where the majority of lymphocytes reside.[43] They are also the sites of antigen presentation and lymphocyte activation and are, therefore, crucial for both viral replication and pathogenesis.[43] Mucosal lymphoid tissue is the dominant site of HIV-1 infection during the initial, acute phase of infection.[44,45] Mucosal CD4+ T cells constitute 50% of all CD4+ T cells and predominantly express CCR5. Mucosal T cells are, therefore, targets of R5 HIV-1, which predominates early in infection.[44–46] The infection of macaques with simian immunodeficiency virus (SIV) is reported to cause severe mucosal CD4+ T cell depletion within a few weeks.[47] It is likely that HIV-1 infection of humans also causes this devastating depletion of mucosal CD4+ thymocytes within a few weeks of infection, thereby undermining peripheral CD4+ T cell counts.[48] HIV-1 also invades and replicates in other lymphoid organs during the acute and chronic phases of infection. An accumulation of HIV-1 has been observed in or on the follicular dendritic cells (FDCs) of lymph node germinal centers, thereby infecting lymphocytes and macrophages, which reside in or pass through the lymph node.[49] FDCs may constitute a major reservoir of HIV-1 infection.[50–54] HIV-1 infection of lymphoid tissue ultimately results in destruction of lymphoid tissue architecture that is important for the peripheral expansion of T cells.[55] Suppression of HIV-1 replication by HAART may restrict lymphoid tissue damage and facilitate subsequent immune reconstitution.[48]

Apoptosis plays a major role in HIV-1-mediated destruction of lymphocytes and lymphoid tissue (LT) architecture. Enhanced apoptosis can be seen in lymphoid tissue of HIV-1-infected individuals.[56,57] In situ labeling of lymph nodes from HIV-1-infected children and SIV-1-infected macaques has suggested that apoptosis occurs predominately in bystander cells, not in productively infected cells.[56,58] A significant correlation was observed between the degree of cellular activation and the intensity of apoptosis in lymphoid tissue associated with HIV-1 infection, which was independent of the rate of progression to disease or viral load.[56] The induced expression of the peripheral lymph node homing receptor (CD62L) by abortive HIV-1 infection of resting memory T cells was also implicated as a cause of apoptosis of these populations in the lymph node.[59–61] More systematic studies of mechanisms of HIV-1-induced apoptosis in LT, however, require an animal or organ culture model. Primary cell culture lacks the complexity of LT and may not represent the true nature of signals that initiate cell death pathways.

Margolis and colleagues first developed human tonsil histoculture as an ex vivo model of HIV-1 infection and pathogenesis.[62,63] Spleen and lymph node histocultures were used later.[62–64] We will collectively refer to them as lymphoid tissue organ culture (LTOC). Secondary lymphoid organs were dissected into 2- to 3-mm-diameter tissue blocks and cultured on collagen gel supports at the liquid/air interface. LTOC was infected ex vivo with HIV, and cultures were maintained up to 4 weeks. Immunohistochemical analysis of tonsil histoculture revealed well-defined tissue architecture typical of normal lymphoid organs.[63] LTOC can be productively infected with both X4 and R5 HIV-1 and does not require stimulation by phytohemagglutinin (PHA) or IL-2 for infection.[62,63] Langerhans cells that play a key role in transmission of HIV-1 in mucosal and lymph nodes can transmit R5 and X4 HIV-1 infection to LTOC ex vivo.[65] Eckstein and colleagues observed that HIV-1 infection, replication, and cell depletion in tonsil histoculture are not dependent on target cell maturation or proliferation status.[66] Moreover, the system accurately reflects viral phenotype, HIV-1 Nef and Vpr mutants replicated to lower levels than the wild-type virus.[67,68] HIV-1 Nef variants, which exhibited greater CD4 downmodulating activity, caused severe depletion of CD4+ T cells in lymphoid tissues.[69] Furthermore, X4 HIV-1 exhibited greater CPEs than R5 HIV-1.[62–64,68,70–72] The difference in CPE, however, could not be correlated with viral replication, as both R5 and X4 HIV-1 replicated to similar levels.[62,64] The replication of R5 HIV-1 in macrophages may have contributed to its high viral load.[67] Grivel and colleagues observed that both R5 and X4 HIV-1 are equally cytopathic for their targets in lymphoid tissue.[73] The CD4+CCR5+ population, however, constitutes a minor fraction of CD4+ cells and, therefore, their depletion by R5 HIV-1 does not have a major impact on the total CD4$^+$ T cell population. R5 HIV-1 does cause AIDS in humans, but the mechanisms by which it depletes CD4$^+$ T cells remain largely unclear.

Both direct and indirect cell killing were reported in HIV-1-infected LTOC. The productive infection of lymphoid tissue *ex vivo* with X4 HIV-1 induced apoptosis in CD4$^+$ T cells but spared the uninfected CD8$^+$ T cells.[73,74] Similarly, R5 HIV-1 infection induced apoptosis in CD4$^+$ cells; however, it was restricted to the CD4$^+$CCR5$^+$ subset, sparing the CD4$^+$CXCR4$^+$CCR5 subset.[74] Moreover, no significant CD8$^+$ T cell apoptosis was observed during HIV-1 infection.[74] These results suggest that apoptosis cannot be attributed to destruction of general tissue architecture or creation of conditions detrimental to the viability of all cells within the same tissue block.[73,74] Nevirapine, a nonnucleoside analogue reverse transcriptase inhibitor (NNRTI), protected uninfected cells residing near productively infected cells, further suggesting that HIV-1 causes cell death primarily by direct killing of infected cells.[75] This is in contrast to observations made by Jekle and colleagues, who found that apoptosis of uninfected bystander CD4$^+$ T cells is a major mechanism of lymphocyte depletion in LTOC by X4 HIV-1 but not R5 HIV-1.[76] Moreover, X4 HIV-1-induced bystander apoptosis was linked to viral envelope glycoprotein gp120. Both AMD 3100 and the anti-CXCR4 monoclonal antibody 12G5, which specifically blocks the binding of X4 HIV-1 to CXCR4, prevented bystander killing. This is consistent with the recent observation that increased co-receptor affinity leads to increased apoptosis for both R5 and X4 HIV-1.[77] NF-κB and p53 are the major apoptosis-inducing factors that are activated by the HIV-1 envelope. Perfettini and colleagues observed enhanced IkB phosphorylation, p53 activation, and higher PUMA levels in lymph nodes of HIV-1 infected people than in those of uninfected controls. The increase in PUMA level was found to be p53 dependent and necessary for syncytial apoptosis.[78]

APOPTOSIS IN ANIMAL MODELS OF HIV-1 DISEASE

Animal models of AIDS are crucial in understanding the complexity of HIV-1 infection of susceptible cells throughout the body and the diverse homeostatic and immune mechanisms that respond to the loss of cells and combat the virus. This is particularly difficult for AIDS, because HIV-1 infects only humans and chimpanzees. In the early years of the AIDS epidemic, studies were carried out in chimpanzees held in primate colonies. Chimpanzees, however, are an endangered species and are too valuable for extensive studies. The lack of another good HIV-1 animal model that permits viral infection, replication, and disease hampered the study of HIV-1 pathogenesis. In the last few years, several promising developments suggest that it may be possible to create a rodent model of AIDS, although this still faces obstacles. Current rodent models of HIV-1 cannot reproduce many aspects of AIDS in humans. Monkey models of AIDS reproduce the human disease in many ways but remain expensive and are limited by the difficulty in working with primates. Monkeys develop an AIDS-like disease only when infected with SIV or a laboratory-made SIV/HIV-1 hybrid called SHIV. SIV and SHIV, however, are quite different from HIV-1. Since the late 1980s, SCID mice have been successfully used to engraft both human fetal tissues and human peripheral blood lymphocytes (PBL) to create SCID-hu Thy/Liv and hu-PBL-SCID models, respectively. SCID-hu Thy/Liv mice are used to study HIV-1 infection of the thymus, whereas the hu-PBL-SCID model is used to assess viral phenotype and vaccine efficacy *in vivo*. It is clear, however, that a single animal model is unlikely to be useful for addressing all the questions in HIV-1 pathogenesis; therefore, the quest for the development of new models of AIDS continues.

HIV-1-Infected SCID-hu Thy/Liv Mouse

The SCID-hu mouse is a small animal model created by implanting human (hu) fetal thymus and liver under the kidney capsule of a SCID mouse, resulting in a conjoint thymus/liver (Thy/Liv) organ. The conjoint Thy/Liv organ has morphology similar to normal human thymus—it contains human hematopoietic progenitor cells and the stromal microenvironment necessary for their differentiation and maturation.[79] All the major populations of thymocytes are seen in Thy/Liv organs, including DN, ITTP, DP, CD4SP, and CD8SP thymocytes in normal proportions. SCID-hu thymocytes are functionally competent with a normal TCR Vβ repertoire and can undergo a selection process that restricts or tolerizes developing thymocytes to self-antigens. The SCID-hu Thy/Liv mouse is a unique model of the thymus because thymopoiesis continues for up to a year, thus allowing long-term experiments that accurately recapitulate infection of the thymus. Infection of Thy/Liv implants with HIV-1 results in depletion of CD4+ thymocytes and destruction of thymic architecture.[16–18,21,80–82] Moreover, the SCID-hu mouse model of HIV-1 infection accurately reflects the effects of diverse viral phenotypes seen in infected people. Unlike many tissue culture models, mutation of accessory genes such as Vpr, Vif, Vpu, or Nef slows the replication and moderates the cytopathic effects of HIV-1 in the SCID-hu mouse.[83,84] Protease inhibitor (PI)-resistant HIV-1 variants have impaired replication capacity in the Thy/Liv organ, which is consistent with the rebound of CD4+ T cell counts in patients showing viral rebound on PI therapy.[26] Both X4 and R5 HIV-1 replicate and deplete thymocytes in the SCID-hu Thy/Liv mouse model.[17,85] X4 HIV-1 and AIDS-associated R5 HIV-1 clones, however, replicate more rapidly and exhibit greater CPE than pre-AIDS R5 HIV-1 clones.[16,17] The increased affinity for CCR5 of AIDS-associated R5 HIV-1 may be responsible for greater CPEs, because an increase in co-receptor affinity can cause an increase in apoptosis as well as an increase in viral replication.[77]

The mechanisms by which HIV-1 causes depletion of thymocytes remain controversial and warrant further systematic study. Direct and indirect killing of thymocytes have been reported in infection of SCID-hu Thy/Liv organs.[14,18,33] Jamieson and colleagues observed a rapid phase of CD4+ thymocyte depletion during the peak of viral replication, suggesting direct killing of HIV-1-infected thymocytes.[33] The integrity of the thymic microenvironment is essential for T cell development; perturbation of this environment by HIV-1 infection of the thymus may cause profound effects on T cell development and function.[34] HIV-1 infection of Thy/Liv organ causes depletion of hematopoietic progenitor cells[86,87] and destruction of thymic epithelial cells.[34,88] An increase in expression of MHC-I on thymic epithelial cells,[22] immature ITTP, and DP thymocytes was associated with HIV-1 infection.[23,31] This elevation of MHC-I levels may interfere with the negative and positive selection of thymocytes, because these processes are tightly regulated by interaction with MHC molecules. This may result in disruption of T cell development or generation of dysfunctional T cells.[22] This is well supported by evidence in transgenic mice, in which overexpression of MHC-I on thymocytes resulted in severe thymocyte depletion.[89]

Apoptosis plays a major role in the depletion of thymocytes from HIV-1-infected Thy/Liv organs. Thymocytes with condensed nuclei, fragmented DNA, and partial chromosomal loss were observed in HIV-1-infected Thy/Liv organs but not in uninfected organs.[14,18,80] Zaitseva and colleagues found overlapping regions of apoptotic thymocytes and immature dendritic cells in HIV-1-infected thymus, implicating a role for DC in clearing apoptotic thymocytes.[90] Immature DP thymocytes, which express higher levels of CXCR4 and CD4 but lower levels of Bcl-2 than mature SP thymocytes, are preferentially infected and depleted by X4 HIV-1 infection of the Thy/Liv organ.[81] Human thymocytes express high levels of Fas on DN and DP thymocytes but little on mature SP thymocytes.[91] Cross-linking Fas with agonist MAb, however, did not cause apoptosis in human thymocytes. This is in contrast to murine thymocytes, for which soluble Fas-ligand-induced apoptosis is seen both *in vitro* and *in vivo*.[91] This species-specific resistance to apoptosis by Fas could be related to the expression of the membrane-bound Fas decoy receptor on human thymocytes.[92] Human thymocytes, however, are sensitive to tumor necrosis factor (TNF)-α-induced apoptosis, indicating that such resistance was specific for Fas, not for all death receptors.[91,92]

Jamieson and colleagues observed few apoptotic cells at the peak of viral replication in Thy/Liv organs, suggesting a role for necrosis in HIV-1-induced thymocyte depletion.[33] Lenardo and colleagues also observed necrotic death rather than apoptotic cell death in peripheral blood mononuclear cells (PBMCs) and several T cell lines.[93,94] These results, however, should be interpreted with caution. Apoptosis is an active process, and mitochondrial respiratory function plays a crucial role in the decision between apoptotic vs. necrotic cell death. Mitochondrial involvement was implicated in HIV-1-mediated cell death. We and others observed an increase in caspase-3 activity in HIV-1 infected thymocytes and monocytes.[95] Electron transport chain complex I was recently shown to be a substrate of caspase-3, and its destruction results in the shutdown of adenosine triphosphate (ATP) production.[96,97] This will lead to intracellular ATP exhaustion and mitochondrial rupture, causing necrotic cell death, even though cell death was originally initiated by apoptosis. We observed that treatment of thymocytes with z-VAD-FMK provides partial protection against cell death induced by HIV-1. Cycloheximide treatment, however, was much more protective against cell death, suggesting that an active apoptotic process rather than a passive process is responsible for cell death in HIV-1-infected thymocytes (Shailesh K. Choudhary and David Camerini, unpublished observations).

HIV-1 Infected hu-PBL-SCID Mice

The hu-PBL-SCID mouse provides another model of HIV-1 infection, which is created by intraperitoneal injection of human PBMCs into SCID mice. The model was created by Mosier and colleagues in 1988, and since then it has been used extensively to study HIV-1 pathogenesis, anti-HIV therapy, and vaccine efficacy *in vivo*.[98–112] Human lymphocytes, including T cells, B cells, and NK cells, survive for an extended period of time, secrete human immunoglobulin, and generate secondary antibody responses in hu-PBL-SCID mice.[98,99,113] Human cells populate not only the site of peritoneal injection but also lymph nodes, spleen, and bone marrow and exhibit activated CD45RO+ memory populations in both CD4+ and CD8+ T cell subsets.[114,115] This activated state of the T cells, therefore, mimics the lymphocyte activation seen in chronic HIV-1 infection and provides an excellent target for HIV-1 infection in hu-PBL-SCID mice.[116] Both R5 and X4 HIV-1 can infect and deplete CD4+ T cells in hu-PBL-SCID mice.[99,100,113,117] The rapid depletion of CD4+ T cells caused by R5 HIV-1 in hu-PBL-SCID mice was striking, because for the first time, it revealed that R5 HIV-1 could be highly cytopathic in a model system. The higher replication and CPE of R5 HIV-1 in this system was supported, in part, by increased expression of CCR5 on CD4+ T cells in hu-PBL-SCID mice. Hu-PBL-SCID mice engrafted with PBL from CCR5Δ32/Δ32 donors were resistant to R5 HIV-1 infection, and dualtropic HIV-1 exhibited slower replication kinetics in CCR5Δ32/Δ32 hu-PBL-SCID mice.[118] Higher expression of CCR5 on CD4+ T cells, along with the high proportion of cells with memory phenotype, makes hu-PBL-SCID mice an excellent choice for R5 HIV-1 pathogenesis studies.[115,117] Moreover, different viral isolates exhibited differential CPEs in hu-PBL-SCID mice, reflecting the biological heterogeneity of HIV-1 observed in infected individuals.[119] For example, Nef-deficient HIV-1 showed lower CD4+ depletion ability than wild-type and exhibited delayed replication kinetics.[120,121]

Many modified hu-PBL-SCID mice models have recently been developed to engraft PBL better. SCID mice possess macrophages and NK cells, which may contribute to destruction of xenotransplanted tissue and, therefore, shorten the duration of HIV-1 replication. Such shortcomings were improved by crossing SCID mice with nonobese diabetic (NOD) mice, which have a functional defect in NK cells.[122,123] Hu-PBL-NOD-SCID mice engraft PBLs better and produce more virus than hu-PBL-SCID mice, and human PBLs invade the brain in response to CNS HIV-1 infection in these mice.[122,124] Hu-PBL-NOD-SCID mice may, therefore, be an excellent choice for studying HIV-1 pathogenesis, especially in the brain.

Although hu-PBL-SCID mice were extensively used to study HIV-1 pathogenesis, few investigators focused their attention on cell death pathways after HIV-1 infection in this system. Miura and colleagues observed many apoptotic cells in the spleens of HIV-1-infected hu-PBL-NOD-SCID mice. The majority of these apoptotic cells, however, were not productively infected, indicating that bystander cell killing was the major mechanism of cell depletion in this model.[125] A combination of terminal deoxynucleotidyltransferase dUTP nick end labeling (TUNEL) assay along with immunoassaying of TNF family members, including TNF-α, FasL, and TNF-related apoptosis-inducing ligand (TRAIL), demonstrated increased expression of TRAIL on apoptotic cells. Furthermore, administration of neutralizing anti-TRAIL antibody but not anti-FasL MAb provided protection against HIV-1-induced apoptosis in hu-PBL-NOD-SCID mice, thus linking TRAIL to depletion of bystander CD4$^+$ T cells.[125] Administration of lipopolysaccharides (LPSs) to HIV-1-infected hu-PBL-NOD-SCID mice induced infiltration of HIV-1-infected human cells into the brain and neuronal apoptosis.[126] Neuronal apoptosis, however, was blocked by antibodies against human TRAIL, implicating a role for TRAIL expressed on HIV-1-infected macrophages in neuronal apoptosis. Another study suggested a role for the antiapoptotic Bcl-2 protein family, because overexpression of Bcl-2 or Bcl-xL in the neuronal cell line NT-2 provided protection against HIV-1-induced apoptosis.[127] del Real and colleagues observed increased expression of TNF-α in serum of HIV-1-infected hu-PBL-SCID mice; however, its relevance in the context of HIV-1-induced apoptosis was not investigated.[128]

HIV-1-INFECTED HUMANIZED MICE AND RATS

The development of an HIV-1 permissive small animal model would greatly facilitate the study of HIV-1-mediated apoptosis and pathogenesis. Attempts to develop a rodent model of AIDS, however, were hampered by the inability of HIV-1 to infect primary rodent cells productively.[129,130] The restrictions occur at various stages of the viral life cycle, including at entry, transcription, and assembly.[131–142] Goldstein and colleagues engineered mice to express human CD4 and CCR5 on their T cells (hu-CD4/CCR5 transgenic mice) in order to remove the block to viral entry.[143] PBMCs and splenocytes from hu-CD4/CCR5 transgenic mice expressed human CD4 and human CCR5 and were susceptible to limited infection with R5 HIV-1. HIV-1-specific Gag DNA sequences were detected in lymphocytes from the spleen and lymph nodes of hu-CD4/CCR5 transgenic mice.[143] Entry of HIV-1 in these transgenic mouse T cells, however, did not result in production of significant numbers of viral particles. The absence of HIV-1 Tat/Rev RNA signals in infected cells indicated a block in, or before, viral transcription in cells from hu-CD4/CCR5 transgenic mice.[143] Similarly, Koito and colleagues developed transgenic mice expressing human CD4 and CXCR4 on their CD4$^+$ T cells.[144] Thymocytes from these mice were susceptible to *in vitro* infection with HIV-1 but to a lower level than human PBMCs. Moreover, lack of *in vivo* data makes the hu-CD4/CXCR4 transgenic mouse a less convincing candidate for an animal model of HIV-1 infection.

The hu-CD4/CCR5 transgenic rat, on the other hand, may be a more promising humanized rodent model for HIV-1 infection. Primary normal rat T cells, macrophages, and microglia expressed considerable levels of early HIV-1 gene products when infected with VSV-G pseudotyped replication-competent HIV-1. Moreover, HIV-1 infection of normal rat macrophages and microglia, but not lymphocytes, produced substantial levels of HIV-1 p24 CA and infectious virions.[145] These encouraging results led the Goldsmith group to develop a transgenic rat that expressed human CD4 and human CCR5 on several diverse rat cell populations.[146] Co-expression of hu-CD4 and hu-CCR5 was detected on rat (r) CD4$^+$ T cells, macrophages, and microglia but not on r-CD8$^+$ T cells or B cells. Moreover, *ex vivo* infection of primary macrophages or microglia from hu-CD4/CCR5 transgenic rats resulted in productive infection and generation of replication-competent virus. In contrast, transgenic expression of human CD4 and CCR5 made rat T cells permissible for viral entry but did not result in productive HIV-1 infection. Hu-CD4/CCR5 transgenic rats, however, can be infected with HIV-1 *in vivo*. Lymphoid organs from HIV-1-infected hu-CD4/CCR5 transgenic rats contained 2-LTR circles, integrated provirus, and early viral gene products, demonstrating susceptibility to HIV-1 *in vivo*. Furthermore, low levels of plasma viremia were detected in the early phase of infection, suggesting that the hu-CD4/CCR5 transgenic rat may be a promising candidate for development into an effective small animal model of HIV-1 infection.[146]

HIV-1 Transgenic Mice

Several investigators have developed transgenic (TG) mice containing the entire HIV-1 genome. Expression of the complete HIV-1 genome in transgenic mice led to the development of an AIDS-like disease and caused early death.[147–152] R5 HIV-1, clone JR-CSF transgenic mice had high levels of plasma viremia, which could be further increased by treatment with the superantigen *Staphylococcus* enterotoxin B (SEB) or by infection with *Mycobacterium tuberculosis*.[152] Both T cells and monocytes from JR-CSF TG mice produced low levels of HIV-1, which could infect human PBMCs.[152] Injection of JR-CSF TG mouse leukocytes into SCID-hu Thy/liv mice resulted in productive infection, indicating that these cells provided an *in vivo* source of replication-competent virus.[150] Moreover, JR-CSF TG mouse leukocytes were a persistent *in vivo* reservoir of HIV-1, despite treatment with HAART.[150] HIV-1 was also detected in microglia of these transgenic mice, which produced more MCP-1 in response to *in vivo* stimulation with LPS than controls.[151] X4 HIV-1, clone NL4-3 (CD4C/HIV) transgenic mice expressed all three size classes of HIV-1 mRNAs (8.8-kb unspliced, 4.3-kb singly spliced, and 2.0-kb multiply spliced) at high levels in the thymus and moderate levels in the spleen and lymph nodes.[148] T cells, dendritic cells (DC), and macrophages all expressed the HIV-1 genome in these transgenic mice. The expression of full-length NL4-3 caused depletion of CD4+ T cells from all lymphoid organs (thymus, spleen, and lymph nodes) and caused their atrophy.[148] Lower expression of CD40 ligands along with reduced IL-4 production by NL4-3 transgenic CD4+ T cells caused impaired germinal center (GC) and follicular dendritic cell (FDC) function[153] similar to the degeneration of GC and the FDC network seen in HIV-1-infected people.[50,154] Jolicoeur and colleagues observed a progressive decrease in the number of DCs in lymph nodes with selective accumulation of CD11b^{HI} MHC class II^{Lo} HIV+ immature DC and a concomitant decrease in antigen presentation capacity.[153] These alterations resulted in serious immune dysfunction and development of an AIDS-like disease in NL4-3 TG mice.

After it became clear that transgenic mice expressing HIV-1 provirus in their CD4+ T cells, DCs, and macrophages developed AIDS-like pathology, attempts were made to delineate the HIV-1 genes responsible for lymphocyte depletion. Transgenic mice containing the HIV-1 clone NL4-3 genome lacking the Gag-Pol genes (HIV$^{\Delta Gag/\Delta Pol}$ transgenic mice) had impaired growth, cachexia, and rapid progression to death, which suggested the involvement of other HIV-1 genes in the development of AIDS-like disease.[155–157] Jay and colleagues created TG mice bearing HIV-1 NL4-3 genomes with deleted Gag, Pol, and most of the Env coding sequence but retaining all the open-reading frames for accessory genes except Vpu. Profound depletion of T cells was observed in all lymphoid compartments, suggesting that neither virion production nor the envelope was required for T cell depletion.[158] Detection of apoptotic cells by TUNEL assay suggested that lymphocyte depletion in the thymus, spleen, and lymph node was the result of accelerated T cell death by apoptosis. Mature T cells in the peripheral lymphoid organs and in the thymic medulla, however, were less sensitive to apoptosis than immature T cells in the thymic cortex.[158] Increased expression of TNF-α was observed in sera of HIV$^{\Delta Gag/\Delta Pol}$ transgenic mice. Administration of anti-TNF-α MAb or human chorionic gonadotropins (hCGs) resulted in reduced TNF-α levels in sera by approximately 75% and prevented mortality, therefore implicating TNF-α in the pathology of HIV$^{\Delta Gag/\Delta Pol}$ transgenic mice.[155]

Jolicoeur and colleagues and others observed that the HIV-1 Nef gene is a major disease determinant in transgenic mice.[148,159–164] A fatal AIDS-like disease occurred in transgenic mice that harbored functional Nef (CD4C/HIV^{Nef}) but not in mice that had all HIV-1 genes except Nef.[159] Transgenic expression of the SIVmac239 *nef* gene under the control of the human CD4 promoter (CD4C/SHIV-Nef^{SIV} transgenic mice) also resulted in AIDS-like disease with low levels of circulating CD8+ and CD4+ T cells, low T cell numbers in the thymus and peripheral lymphoid organs, thymic atrophy, splenomegaly, weight loss, wasting, and premature death.[165] Peripheral TG CD4+ T cells from CD4C/HIV^{Nef} transgenic mice exhibited an activated/memory phenotype with higher expression of CD25 and CD69 on CD44^{high}, CD45RB^{low}, CD62L^{low}, and CD4+ T cells, even in the absence of TCR-mediated stimulation.[162] Furthermore, thymocytes from these transgenic mice showed hypersensitivity to stimulation with CD3 MAb; constitutive increases in tyrosine phosphorylation of several substrates, including LAT and p44/p42 MAPK (Erk-1/Erk-2), were also observed in these transgenic mice.[159] The chronic activation of transgenic CD4[+] T cells and their hypersensitivity to TCR/CD3-mediated signals caused apoptosis after limited cell division.[162] This contributed to exhaustion of the T cell pool, thymic atrophy, and the low numbers of CD4[+] T cells observed in CD4C/HIV[Nef] transgenic mice. The importance of Nef in this model is evident, because mutation in the SH3 ligand-binding domain of Nef resulted in abrogation of hypersensitivity to CD3 stimulation, and mice bearing this mutant Nef failed to develop disease, despite partial downmodulation of CD4 cell surface expression.[161]

HIV[Tat] transgenic mice that express Tat under the control of the astrocyte-specific glial fibrillary acidic protein promoter in a doxycycline (Dox)-dependent manner were developed by He and colleagues to study neuropathological disorders associated with HIV-1 infection.[166] The expression of Tat occurred exclusively in astrocytes and was Dox dependent. Expression of Tat in the brain caused damage to the cerebellum and cortex, astrocytosis, degeneration of neuronal dendrites, neuronal apoptosis, and increased infiltration of activated monocytes and T lymphocytes similar to what is seen in the brains of AIDS patients.[166] HIV[Tat] transgenic mice, therefore, support the hypothesized role of Tat protein in HIV-1 neuropathogenesis and may be helpful in delineating the molecular mechanisms of Tat-induced neurotoxicity and apoptosis.[166]

The HIV[Vpr] transgenic mouse was developed by Yasuda and colleagues to investigate CPE mediated by Vpr *in vivo*.[167] The X4 HIV-1 clone NL4-3 Vpr was expressed under the control of the mouse CD4 enhancer/promoter in transgenic mice. HIV[Vpr] transgenic mice had reduced T cell numbers in both the thymus and peripheral blood due to enhanced apoptosis. The highest-level expression of Vpr was observed in the thymus followed by lymph node (50-fold lower) and spleen (100-fold lower), which was correlated with tissue phenotype; the thymus exhibited atrophy, whereas the spleen and lymph node appeared to be normal in HIV[Vpr] transgenic mice. Crossing the HIV[Vpr] transgenic mice with Gld-mice, which have deficient FasL genes, still caused thymic atrophy, suggesting that FasL is not required for apoptosis induced by Vpr. Increased expression of Bcl-x, Bax, and caspase-1 genes, however, was observed in thymuses of these transgenic mice, suggesting the involvement of mitochondrial pathways in Vpr-mediated apoptosis.

HIV-1-mediated apoptosis was implicated in neuronal damage, resulting in cognitive and motor dysfunction, known as HIV-1-associated dementia.[168] HIV-1 gp120 was demonstrated to promote apoptosis directly on isolated neurons.[169,170] HIV-1 gp120 (HIV[gp120]) transgenic mice exhibited a 25% reduction in CNS dendritic cells compared with wild-type littermates.[171] Administration of caspase inhibitors prevented dendritic cell damage associated with gp120 expression in HIV[gp120] transgenic mice. Moreover, crossing HIV-1 gp120 transgenic mice with Casp-1/C285G transgenic mice, which have reduced production of mature IL-1β in brain, caused dendritic cell numbers to return to the number found in wild-type littermates. This suggests that dendritic cell injury in HIV-1 gp120 transgenic mice is due to inhibition of caspase-1 or blockade of mature IL-1β production.[171] Transgenic mice, therefore, provide a novel *in vivo* model with which to systematically investigate the signals and pathways associated with HIV-1 apoptosis.

HIV-1-INFECTED CHIMPANZEES

Besides humans, chimpanzees are the only species that can be infected with HIV-1.[172–176] Chimpanzees were experimentally infected with various HIV-1 isolates, but infection did not normally result in AIDS or other HIV-1-associated pathogenesis. One exception, however, was animal C499. This animal was part of a superinfection study and was infected with three different HIV isolates: HIV-1$_{SF2}$, HIV-1$_{LAV}$, and HIV$_{NDK}$.[177] The animal developed an AIDS-like syndrome with severe CD4[+] T cell loss with a concomitant increase in viral load, thrombocytopenia, and chronic diarrhea associated with cryptosporidium infection.[178] Blood transfusion from C499 to an HIV-1-negative chimpanzee (C455) resulted in rapid CD4[+] T cell loss in the recipient animal, suggesting that *in vivo* evolution of a chimpanzee-pathogenic HIV-1 strain occurred.[178] HIV-1 isolation from chimps C499 and C455 and subsequent molecular cloning suggested that virulent clones were likely recombinants derived from HIV-1$_{SF2}$ and HIV-1$_{LAV}$.[179] Infection of other chimps with C499 HIV-1 isolates resulted in moderate to high viral load in plasma and peripheral CD4+ T cell depletion.[180–182] None of the viral isolates derived from chimp C499, however, caused AIDS-like disease in experimentally infected chimpanzees, even after years of infection.[183,184] Attention, therefore, focuses on the potential mechanisms for the lack of disease progression in HIV-1-infected chimps. Gougeon and colleagues observed that HIV-1 infection of chimpanzees did not result in chronic activation of T cells, in contrast to HIV-1-infected humans.[185] CD4+ and CD8+ T cells from infected chimpanzees were less susceptible to apoptosis mediated by HIV-1 or TCR ligation.[185–189] Furthermore, Fas ligation by agonistic antibodies or recombinant human Fas ligand failed to induce apoptosis on CD4+ or CD8+ T cells from HIV-1-infected chimpanzees.[185] These observations could be partly explained by the fact that chimpanzee T cells are resistant to FasL upregulation and oxidative stress.[188] T cells from HIV-1-infected chimpanzees exhibited normal responses to antigen and TCR-mediated stimulation.[187] Chimpanzees that exhibited signs of progression to AIDS, however, had increased levels of activated T cells compared to those chimps that did not progress, suggesting that chronic T cell activation is a primary determinant of disease progression in HIV-1-infected chimpanzees.[181,182]

SHIV-INFECTED MACAQUES

Simian–human immunodeficiency viruses (SHIVs) are chimeric viruses in which *env* and the associated auxiliary genes *tat*, *rev*, and *vpu* of SIVmac are replaced with the corresponding regions of HIV-1 to develop a better primate model of HIV-1 infection.[190,191] Some SHIVs, including SHIV89.6P, are pathogenic in infected macaques, causing progressive loss of CD4+ T cells and development of an AIDS-like disease.[192–194] Infection of macaques with pathogenic SHIVs, therefore, may provide insights into the mechanisms of CD4+ T cell depletion in HIV-1-infected people.

Infection of macaques with pathogenic SHIV strains caused apoptosis in both CD4+ and CD8+ T cell populations. In contrast, nonpathogenic SHIV strains were inefficient inducers of apoptosis in rhesus macaque PBMCs.[195–198] Similarly, pathogenic SHIVs caused increased FasL expression in both CD4+ and CD8+ T cells, whereas nonpathogenic SHIVs were conspicuous by their inability to increase FasL expression on PBMCs of infected macaques.[197,199] These results suggest that induction of apoptosis is closely associated with the pathogenic process induced by SHIV in macaques. Apoptotic cells were detected in the lymph nodes and the thymuses of infected macaques.[197,199] An *in situ* TUNEL assay showed increased numbers of apoptotic cells in the paracortical regions of lymph nodes and medulla of thymus, which colocalized to a large extent, with sites of active viral replication and CD4+ T cell depletion.[200] Pathogenic SHIV infection caused increased expression of activation markers on macaque CD4+ T cells, including HLA-DR, CD25, and CD69.[201] Expression of CD95L was higher on HLA-DR+ T cells than HLA-DR T cells.[199] Activation of PBMCs with PHA caused substantial increases in apoptosis of T cells, predominantly among CD4+ T cells.[201] Furthermore, T cells from blood and lymphoid tissue did not proliferate in response to PHA, suggesting the generation of dysfunctional T cells.[201] These data are consistent with previous observation that apoptosis correlated with immune system activation and higher expression of CD95, CD95L, TNF, and IFN-γ.[56,199,202,203] The chronic activation of the immune system in SHIV-infected macaques, therefore, causes an increased rate of T cell destruction and impedes immune system regeneration, which may be responsible for the development of simian AIDS.

SUMMARY AND FUTURE PERSPECTIVES

A wide variety of organ culture and animal models has been developed to study HIV-1 replication and the pathogenesis of AIDS. Apoptosis induced by HIV-1 among infected and bystander cells is an important component of the pathogenic process that leads to AIDS after HIV-1 infection. Animal and organ culture models play an important role in these studies, because they most closely mimic natural infection by HIV-1. Molecular pathways of apoptosis can be studied in cell lines, but their relevance to AIDS must be verified with more realistic models and, if possible, with cells or tissues from HIV-1-infected people.

Animal and organ culture models can also be used to test treatments for AIDS, including immune reconstitution and gene therapy. Thus, it will be interesting to see whether IL-7 treatment of HIV-1-infected FTOCs can protect ITTPs from HIV-1-mediated killing and enhance *de novo* naive T cell generation. Hematopoietic progenitor cells could be transduced with anti-HIV-1 or antiapoptotic genes and used to colonize FTOC or SCID-hu mice in order to check their effects on destructive events associated with HIV-1 infection of the thymus. Previous work has shown that transduction of human CD34[+] cells with anti-HIV-1 genes inhibits HIV-1 replication in monocytes and lymphocytes produced *in vitro*.[204–209] Animal and organ culture models will continue to be vital for the study of HIV-1-mediated apoptosis and its prevention.

REFERENCES

1. Jamieson, B.D., Douek, D.C., Killian, S., Hultin, L.E., Scripture-Adams, D.D., Giorgi, J.V., Marelli, D., Koup, R.A., and Zack, J.A., Generation of functional thymocytes in the human adult, *Immunity,* 10 (5), 569–575, 1999.
2. Poulin, J.F., Viswanathan, M.N., Harris, J.M., Komanduri, K.V., Wieder, E., Ringuette, N., Jenkins, M., McCune, J.M., and Sekaly, R.P., Direct evidence for thymic function in adult humans, *J. Exp. Med.,* 190 (4), 479–486, 1999.
3. Rosenzweig, M., Clark, D.P., and Gaulton, G.N., Selective thymocyte depletion in neonatal HIV-1 thymic infection, *AIDS,* 7 (12), 1601–1605, 1993.
4. Gaulton, G.N., Scobie, J.V., and Rosenzweig, M., HIV-1 and the thymus, *AIDS,* 11 (4), 403–414, 1997.
5. Douek, D.C., Koup, R.A., McFarland, R.D., Sullivan, J.L., and Luzuriaga, K., Effect of HIV on thymic function before and after antiretroviral therapy in children, *J. Infect. Dis.,* 181 (4), 1479–1482, 2000.
6. Douek, D.C., McFarland, R.D., Keiser, P.H., Gage, E.A., Massey, J.M., Haynes, B.F., Polis, M.A., Haase, A.T., Feinberg, M.B., Sullivan, J.L., Jamieson, B.D., Zack, J.A., Picker, L.J., and Koup, R.A., Changes in thymic function with age and during the treatment of HIV infection, *Nature,* 396 (6712), 690–695, 1998.
7. Kourtis, A.P., Ibegbu, C., Nahmias, A.J., Lee, F.K., Clark, W.S., Sawyer, M.K., and Nesheim, S., Early progression of disease in HIV-infected infants with thymus dysfunction, *N. Engl. J. Med.,* 335 (19), 1431–1436, 1996.
8. Nahmias, A.J., Clark, W.S., Kourtis, A.P., Lee, F.K., Cotsonis, G., Ibegbu, C., Thea, D., Palumbo, P., Vink, P., Simonds, R.J., and Nesheim, S.R., Thymic dysfunction and time of infection predict mortality in human immunodeficiency virus-infected infants. CDC Perinatal AIDS Collaborative Transmission Study Group, *J. Infect. Dis.,* 178 (3), 680–685, 1998.
9. Kalayjian, R.C., Landay, A., Pollard, R.B., Taub, D.D., Gross, B.H., Francis, I.R., Sevin, A., Pu, M., Spritzler, J., Chernoff, M., Namkung, A., Fox, L., Martinez, A., Waterman, K., Fiscus, S.A., Sha, B., Johnson, D., Slater, S., Rousseau, F., and Lederman, M.M., Age-related immune dysfunction in health and in human immunodeficiency virus (HIV) disease: association of age and HIV infection with naive CD8+ cell depletion, reduced expression of CD28 on CD8+ cells, and reduced thymic volumes, *J. Infect. Dis.,* 187 (12), 1924–1933, 2003.
10. Vigano, A., Vella, S., Saresella, M., Vanzulli, A., Bricalli, D., Di Fabio, S., Ferrante, P., Andreotti, M., Pirillo, M., Dally, L.G., Clerici, M., and Principi, N., Early immune reconstitution after potent antiretroviral therapy in HIV-infected children correlates with the increase in thymus volume, *AIDS,* 14 (3), 251–261, 2000.
11. Knutsen, A.P., Roodman, S.T., Freeman, J.J., Mueller, K.R., and Bouhasin, J.D., Inhibition of thymopoiesis of CD34+ cell maturation by HIV-1 in an *in vitro* CD34+ cell and thymic epithelial organ culture model, *Stem Cells,* 17 (6), 327–338, 1999.
12. Bohler, T., Walcher, J., Holzl-Wenig, G., Geiss, M., Buchholz, B., Linde, R., and Debatin, K.M., Early effects of antiretroviral combination therapy on activation, apoptosis and regeneration of T cells in HIV-1-infected children and adolescents, *AIDS,* 13 (7), 779–789, 1999.

13. Ye, P., Kourtis, A.P., and Kirschner, D.E., Reconstitution of thymic function in HIV-1 patients treated with highly active antiretroviral therapy, *Clin. Immunol.*, 106 (2), 95–105, 2003.

14. Su, L., Kaneshima, H., Bonyhadi, M., Salimi, S., Kraft, D., Rabin, L., and McCune, J.M., HIV-1-induced thymocyte depletion is associated with indirect cytopathogenicity and infection of progenitor cells *in vivo*, *Immunity*, 2 (1), 25–36, 1995.

15. Bonyhadi, M.L., Su, L., Auten, J., McCune, J.M., and Kaneshima, H., Development of a human thymic organ culture model for the study of HIV pathogenesis, *AIDS Res. Hum. Retroviruses*, 11 (9), 1073–1080, 1995.

16. Camerini, D., Su, H.P., Gamez-Torre, G., Johnson, M.L., Zack, J.A., and Chen, I.S., Human immunodeficiency virus type 1 pathogenesis in SCID-hu mice correlates with syncytium-inducing phenotype and viral replication, *J. Virol.*, 74 (7), 3196–3204, 2000.

17. Scoggins, R.M., Taylor, J.R., Jr., Patrie, J., van't Wout, A.B., Schuitemaker, H., and Camerini, D., Pathogenesis of primary R5 human immunodeficiency virus type 1 clones in SCID-hu mice, *J. Virol.*, 74 (7), 3205–3216, 2000.

18. Bonyhadi, M.L., Rabin, L., Salimi, S., Brown, D.A., Kosek, J., McCune, J.M., and Kaneshima, H., HIV induces thymus depletion *in vivo*, *Nature*, 363 (6431), 728–732, 1993.

19. Rosenzweig, M., Bunting, E.M., and Gaulton, G.N., Neonatal HIV-1 thymic infection, *Leukemia*, 8 (Suppl. 1), S163–S165, 1994.

20. Rosenzweig, M., Bunting, E.M., Damico, R.L., Clark, D.P., and Gaulton, G.N., Human neonatal thymic organ culture: an *ex vivo* model of thymocyte ontogeny and HIV-1 infection, *Pathobiology*, 62 (5–6), 245–251, 1994.

21. Duus, K.M., Miller, E.D., Smith, J.A., Kovalev, G.I., and Su, L., Separation of human immunodeficiency virus type 1 replication from Nef-mediated pathogenesis in the human thymus, *J. Virol.*, 75 (8), 3916–3924, 2001.

22. Keir, M.E., Rosenberg, M.G., Sandberg, J.K., Jordan, K.A., Wiznia, A., Nixon, D.F., Stoddart, C.A., and McCune, J.M., Generation of CD3+CD8low thymocytes in the HIV type 1-infected thymus, *J. Immunol.*, 169 (5), 2788–2796, 2002.

23. Keir, M.E., Stoddart, C.A., Linquist-Stepps, V., Moreno, M.E., and McCune, J.M., IFN- secretion by type 2 predendritic cells up-regulates MHC class I in the HIV-1-infected thymus, *J. Immunol.*, 168 (1), 325–331, 2002.

24. Miller, E.D., Duus, K.M., Townsend, M., Yi, Y., Collman, R., Reitz, M., and Su, L., Human immunodeficiency virus type 1 IIIB selected for replication *in vivo* exhibits increased envelope glycoproteins in virions without alteration in coreceptor usage: separation of *in vivo* replication from macrophage tropism, *J. Virol.*, 75 (18), 8498–8506, 2001.

25. Choudhary, S.K., Choudhary, N.R., Kimbrell, K.C., Colasanti, J., Ziogas, A., Kwa, D., Schuitemaker, H., and Camerini, D., R5 HIV-1 infection of fetal thymic organ culture induces cytokine and CCR5 expression, *J. Virol.*, 79 (1), 458–471, 2005.

26. Stoddart, C.A., Liegler, T.J., Mammano, F., Linquist-Stepps, V.D., Hayden, M.S., Deeks, S.G., Grant, R.M., Clavel, F., and McCune, J.M., Impaired replication of protease inhibitor-resistant HIV-1 in human thymus, *Nat. Med.*, 7 (6), 712–718, 2001.

27. Finnegan, A., Roebuck, K.A., Nakai, B.E., Gu, D.S., Rabbi, M.F., Song, S., and Landay, A.L., IL-10 cooperates with TNF- to activate HIV-1 from latently and acutely infected cells of monocyte/macrophage lineage, *J. Immunol.*, 156 (2), 841–851, 1996.

28. Kekow, J., Wachsman, W., McCutchan, J.A., Cronin, M., Carson, D.A., and Lotz, M., Transforming growth factor and noncytopathic mechanisms of immunodeficiency in human immunodeficiency virus infection, *Proc. Natl. Acad. Sci. U.S.A.*, 87 (21), 8321–8325, 1990.

29. Kolb, S.A., Sporer, B., Lahrtz, F., Koedel, U., Pfister, H.W., and Fontana, A., Identification of a T cell chemotactic factor in the cerebrospinal fluid of HIV-1-infected individuals as interferon- inducible protein 10, *J. Neuroimmunol.*, 93 (1–2), 172–181, 1999.

30. Ostrowski, M.A., Gu, J.X., Kovacs, C., Freedman, J., Luscher, M.A., and MacDonald, K.S., Quantitative and qualitative assessment of human immunodeficiency virus type 1 (HIV-1)-specific CD4+ T cell immunity to Gag in HIV-1-infected individuals with differential disease progression: reciprocal interferon- and interleukin-10 responses, *J. Infect. Dis.*, 184 (10), 1268–1278, 2001.

31. Kovalev, G., Duus, K., Wang, L., Lee, R., Bonyhadi, M., Ho, D., McCune, J.M., Kaneshima, H., and Su, L., Induction of MHC class I expression on immature thymocytes in HIV-1-infected SCID-hu Thy/Liv mice: evidence of indirect mechanisms, *J. Immunol.*, 162 (12), 7555–7562, 1999.

32. Miller, E.D., Smith, J.A., Lichtinger, M., Wang, L., and Su, L., Activation of the signal transducer and activator of transcription 1 signaling pathway in thymocytes from HIV-1-infected human thymus, *AIDS,* 17 (9), 1269–1277, 2003.

33. Jamieson, B.D., Uittenbogaart, C.H., Schmid, I., and Zack, J.A., High viral burden and rapid CD4+ cell depletion in human immunodeficiency virus type 1-infected SCID-hu mice suggest direct viral killing of thymocytes *in vivo, J. Virol.,* 71 (11), 8245–8253, 1997.

34. Stanley, S.K., McCune, J.M., Kaneshima, H., Justement, J.S., Sullivan, M., Boone, E., Baseler, M., Adelsberger, J., Bonyhadi, M., Orenstein, J., Fox, C.H., and Fauci, A.S., Human immunodeficiency virus infection of the human thymus and disruption of the thymic microenvironment in the SCID-hu mouse, *J. Exp. Med.,* 178 (4), 1151–1163, 1993.

35. Verhasselt, B., Naessens, E., Verhofstede, C., De Smedt, M., Schollen, S., Kerre, T., Vanhecke, D., and Plum, J., Human immunodeficiency virus Nef gene expression affects generation and function of human T cells, but not dendritic cells, *Blood,* 94 (8), 2809–2818, 1999.

36. Nielsen, S.D., Jeppesen, D.L., Kolte, L., Clark, D.R., Sorensen, T.U., Dreves, A.M., Ersboll, A.K., Ryder, L.P., Valerius, N.H., and Nielsen, J.O., Impaired progenitor cell function in HIV-negative infants of HIV-positive mothers results in decreased thymic output and low CD4 counts, *Blood,* 98 (2), 398–404, 2001.

37. Berkowitz, R.D., Beckerman, K.P., Schall, T.J., and McCune, J.M., CXCR4 and CCR5 expression delineates targets for HIV-1 disruption of T cell differentiation, *J. Immunol.,* 161 (7), 3702–3710, 1998.

38. von Freeden-Jeffry, U., Solvason, N., Howard, M., and Murray, R., The earliest T lineage-committed cells depend on IL-7 for Bcl-2 expression and normal cell cycle progression, *Immunity,* 7 (1), 147–154, 1997.

39. Murray, R., Suda, T., Wrighton, N., Lee, F., and Zlotnik, A., IL-7 is a growth and maintenance factor for mature and immature thymocyte subsets, *Int. Immunol.,* 1 (5), 526–531, 1989.

40. Peschon, J.J., Morrissey, P.J., Grabstein, K.H., Ramsdell, F.J., Maraskovsky, E., Gliniak, B.C., Park, L.S., Ziegler, S.F., Williams, D.E., Ware, C.B., Meyer, J.D., and Davison, B.L., Early lymphocyte expansion is severely impaired in interleukin 7 receptor-deficient mice, *J. Exp. Med.,* 180 (5), 1955–1960, 1994.

41. Napolitano, L.A., Stoddart, C.A., Hanley, M.B., Wieder, E., and McCune, J.M., Effects of IL-7 on early human thymocyte progenitor cells *in vitro* and in SCID-hu Thy/Liv mice, *J. Immunol.,* 171 (2), 645–654, 2003.

42. Okamoto, Y., Douek, D.C., McFarland, R.D., and Koup, R.A., Effects of exogenous interleukin-7 on human thymus function, *Blood,* 99 (8), 2851–2858, 2002.

43. Levy, J.A., Pathogenesis of human immunodeficiency virus infection, *Microbiol. Rev.,* 57 (1), 183–289, 1993.

44. Stebbing, J., Gazzard, B., and Douek, D.C., Where does HIV live? *N. Engl. J. Med.,* 350 (18), 1872–1880, 2004.

45. Douek, D.C., Picker, L.J., and Koup, R.A., T cell dynamics in HIV-1 infection, *Annu. Rev. Immunol.,* 21, 265–304, 2003.

46. Brenchley, J.M., Schacker, T.W., Ruff, L.E., Price, D.A., Taylor, J.H., Beilman, G.J., Nguyen, P.L., Khoruts, A., Larson, M., Haase, A.T., and Douek, D.C., CD4+ T cell depletion during all stages of HIV disease occurs predominantly in the gastrointestinal tract, *J. Exp. Med.,* 200 (6), 749–759, 2004.

47. Veazey, R.S., DeMaria, M., Chalifoux, L.V., Shvetz, D.E., Pauley, D.R., Knight, H.L., Rosenzweig, M., Johnson, R.P., Desrosiers, R.C., and Lackner, A.A., Gastrointestinal tract as a major site of CD4+ T cell depletion and viral replication in SIV infection, *Science,* 280 (5362), 427–431, 1998.

48. Guadalupe, M., Reay, E., Sankaran, S., Prindiville, T., Flamm, J., McNeil, A., and Dandekar, S., Severe CD4+ T-cell depletion in gut lymphoid tissue during primary human immunodeficiency virus type 1 infection and substantial delay in restoration following highly active antiretroviral therapy, *J. Virol.,* 77 (21), 11708–11717, 2003.

49. Schacker, T., Little, S., Connick, E., Gebhard, K., Zhang, Z.Q., Krieger, J., Pryor, J., Havlir, D., Wong, J.K., Schooley, R.T., Richman, D., Corey, L., and Haase, A.T., Productive infection of T cells in lymphoid tissues during primary and early human immunodeficiency virus infection, *J. Infect. Dis.,* 183 (4), 555–562, 2001.

50. Pantaleo, G., Graziosi, C., Demarest, J.F., Butini, L., Montroni, M., Fox, C.H., Orenstein, J.M., Kotler, D.P., and Fauci, A.S., HIV infection is active and progressive in lymphoid tissue during the clinically latent stage of disease, *Nature,* 362 (6418), 355–358, 1993.

51. Embretson, J., Zupancic, M., Ribas, J.L., Burke, A., Racz, P., Tenner-Racz, K., and Haase, A.T., Massive covert infection of helper T lymphocytes and macrophages by HIV during the incubation period of AIDS, *Nature,* 362 (6418), 359–362, 1993.

52. Haase, A.T., Population biology of HIV-1 infection: viral and CD4+ T cell demographics and dynamics in lymphatic tissues, *Annu. Rev. Immunol.,* 17, 625–656, 1999.

53. Heath, S.L., Tew, J.G., Szakal, A.K., and Burton, G.F., Follicular dendritic cells and human immunodeficiency virus infectivity, *Nature,* 377 (6551), 740–744, 1995.

54. Spiegel, H., Herbst, H., Niedobitek, G., Foss, H.D., and Stein, H., Follicular dendritic cells are a major reservoir for human immunodeficiency virus type 1 in lymphoid tissues facilitating infection of CD4+ T-helper cells, *Am. J. Pathol.,* 140 (1), 15–22, 1992.

55. Schacker, T.W., Nguyen, P.L., Beilman, G.J., Wolinsky, S., Larson, M., Reilly, C., and Haase, A.T., Collagen deposition in HIV-1 infected lymphatic tissues and T cell homeostasis, *J. Clin. Invest.,* 110 (8), 1133–1139, 2002.

56. Muro-Cacho, C.A., Pantaleo, G., and Fauci, A.S., Analysis of apoptosis in lymph nodes of HIV-infected persons. Intensity of apoptosis correlates with the general state of activation of the lymphoid tissue and not with stage of disease or viral burden, *J. Immunol.,* 154 (10), 5555–5566, 1995.

57. Zhang, Z.Q., Notermans, D.W., Sedgewick, G., Cavert, W., Wietgrefe, S., Zupancic, M., Gebhard, K., Henry, K., Boies, L., Chen, Z., Jenkins, M., Mills, R., McDade, H., Goodwin, C., Schuwirth, C.M., Danner, S.A., and Haase, A.T., Kinetics of CD4+ T cell repopulation of lymphoid tissues after treatment of HIV-1 infection, *Proc. Natl. Acad. Sci. U.S.A.,* 95 (3), 1154–1159, 1998.

58. Finkel, T.H., Tudor-Williams, G., Banda, N.K., Cotton, M.F., Curiel, T., Monks, C., Baba, T.W., Ruprecht, R.M., and Kupfer, A., Apoptosis occurs predominantly in bystander cells and not in productively infected cells of HIV- and SIV-infected lymph nodes, *Nat. Med.,* 1 (2), 129–134, 1995.

59. Wang, L., Robb, C.W., and Cloyd, M.W., HIV induces homing of resting T lymphocytes to lymph nodes, *Virology,* 228 (2), 141–152, 1997.

60. Wang, L., Chen, J.J., Gelman, B.B., Konig, R., and Cloyd, M.W., A novel mechanism of CD4 lymphocyte depletion involves effects of HIV on resting lymphocytes: induction of lymph node homing and apoptosis upon secondary signaling through homing receptors, *J. Immunol.,* 162 (1), 268–276, 1999.

61. Chen, J.J., Huang, J.C., Shirtliff, M., Briscoe, E., Ali, S., Cesani, F., Paar, D., and Cloyd, M.W., CD4 lymphocytes in the blood of HIV(+) individuals migrate rapidly to lymph nodes and bone marrow: support for homing theory of CD4 cell depletion, *J. Leukoc. Biol.,* 72 (2), 271–278, 2002.

62. Glushakova, S., Baibakov, B., Margolis, L.B., and Zimmerberg, J., Infection of human tonsil histocultures: a model for HIV pathogenesis, *Nat. Med.,* 1 (12), 1320–1322, 1995.

63. Glushakova, S., Baibakov, B., Zimmerberg, J., and Margolis, L.B., Experimental HIV infection of human lymphoid tissue: correlation of CD4+ T cell depletion and virus syncytium-inducing/non-syncytium-inducing phenotype in histocultures inoculated with laboratory strains and patient isolates of HIV type 1, *AIDS Res. Hum. Retroviruses,* 13 (6), 461–471, 1997.

64. Schramm, B., Penn, M.L., Speck, R.F., Chan, S.Y., De Clercq, E., Schols, D., Connor, R.I., and Goldsmith, M.A., Viral entry through CXCR4 is a pathogenic factor and therapeutic target in human immunodeficiency virus type 1 disease, *J. Virol.,* 74 (1), 184–192, 2000.

65. Blauvelt, A., Glushakova, S., and Margolis, L.B., HIV-infected human Langerhans cells transmit infection to human lymphoid tissue *ex vivo, AIDS,* 14 (6), 647–651, 2000.

66. Eckstein, D.A., Penn, M.L., Korin, Y.D., Scripture-Adams, D.D., Zack, J.A., Kreisberg, J.F., Roederer, M., Sherman, M.P., Chin, P.S., and Goldsmith, M.A., HIV-1 actively replicates in naive CD4(+) T cells residing within human lymphoid tissues, *Immunity,* 15 (4), 671–682, 2001.

67. Eckstein, D.A., Sherman, M.P., Penn, M.L., Chin, P.S., De Noronha, C.M., Greene, W.C., and Goldsmith, M.A., HIV-1 Vpr enhances viral burden by facilitating infection of tissue macrophages but not nondividing CD4+ T cells, *J. Exp. Med.,* 194 (10), 1407–1419, 2001.

68. Glushakova, S., Grivel, J.C., Suryanarayana, K., Meylan, P., Lifson, J.D., Desrosiers, R., and Margolis, L., Nef enhances human immunodeficiency virus replication and responsiveness to interleukin-2 in human lymphoid tissue *ex vivo, J. Virol.,* 73 (5), 3968–3974, 1999.

69. Glushakova, S., Munch, J., Carl, S., Greenough, T.C., Sullivan, J.L., Margolis, L., and Kirchhoff, F., CD4 down-modulation by human immunodeficiency virus type 1 Nef correlates with the efficiency of viral replication and with CD4(+) T-cell depletion in human lymphoid tissue *ex vivo, J. Virol.,* 75 (21), 10113–10117, 2001.

70. Glushakova, S., Grivel, J.C., Fitzgerald, W., Sylwester, A., Zimmerberg, J., and Margolis, L.B., Evidence for the HIV-1 phenotype switch as a causal factor in acquired immunodeficiency, *Nat. Med.,* 4 (3), 346–349, 1998.

71. Kreisberg, J.F., Kwa, D., Schramm, B., Trautner, V., Connor, R., Schuitemaker, H., Mullins, J.I., van't Wout, A.B., and Goldsmith, M.A., Cytopathicity of human immunodeficiency virus type 1 primary isolates depends on coreceptor usage and not patient disease status, *J. Virol.,* 75 (18), 8842–8847, 2001.

72. Fitzgerald, W., Sylwester, A.W., Grivel, J.C., Lifson, J.D., and Margolis, L.B., Noninfectious X4 but not R5 human immunodeficiency virus type 1 virions inhibit humoral immune responses in human lymphoid tissue *ex vivo, J. Virol.,* 78 (13), 7061–7068, 2004.

73. Grivel, J.C. and Margolis, L.B., CCR5- and CXCR4-tropic HIV-1 are equally cytopathic for their T-cell targets in human lymphoid tissue, *Nat. Med.,* 5 (3), 344–346, 1999.

74. Grivel, J.C., Malkevitch, N., and Margolis, L., Human immunodeficiency virus type 1 induces apoptosis in CD4(+) but not in CD8(+) T cells in *ex vivo*-infected human lymphoid tissue, *J. Virol.,* 74 (17), 8077–8084, 2000.

75. Grivel, J.C., Biancotto, A., Ito, Y., Lima, R.G., and Margolis, L.B., Bystander CD4+ T lymphocytes survive in HIV-infected human lymphoid tissue, *AIDS Res. Hum. Retroviruses,* 19 (3), 211–216, 2003.

76. Jekle, A., Keppler, O.T., De Clercq, E., Schols, D., Weinstein, M., and Goldsmith, M.A., *In vivo* evolution of human immunodeficiency virus type 1 toward increased pathogenicity through CXCR4-mediated killing of uninfected CD4 T cells, *J. Virol.,* 77 (10), 5846–5854, 2003.

77. Holm, G.H., Zhang, C., Gorry, P.R., Peden, K., Schols, D., De Clercq, E., and Gabuzda, D., Apoptosis of bystander T cells induced by human immunodeficiency virus type 1 with increased envelope/receptor affinity and coreceptor binding site exposure, *J. Virol.,* 78 (9), 4541–4551, 2004.

78. Perfettini, J.L., Roumier, T., Castedo, M., Larochette, N., Boya, P., Raynal, B., Lazar, V., Ciccosanti, F., Nardacci, R., Penninger, J., Piacentini, M., and Kroemer, G., NF-B and p53 are the dominant apoptosis-inducing transcription factors elicited by the HIV-1 envelope, *J. Exp. Med.,* 199 (5), 629–640, 2004.

79. McCune, J., Kaneshima, H., Krowka, J., Namikawa, R., Outzen, H., Peault, B., Rabin, L., Shih, C.C., Yee, E., Lieberman, M., Weissman, I., and Shultz, L., The SCID-hu mouse: a small animal model for HIV infection and pathogenesis, *Annu. Rev. Immunol.,* 9, 399–429, 1991.

80. Kaneshima, H., Su, L., Bonyhadi, M.L., Connor, R.I., Ho, D.D., and McCune, J.M., Rapid-high, syncytium-inducing isolates of human immunodeficiency virus type 1 induce cytopathicity in the human thymus of the SCID-hu mouse, *J. Virol.,* 68 (12), 8188–8192, 1994.

81. Kitchen, S.G. and Zack, J.A., CXCR4 expression during lymphopoiesis: implications for human immunodeficiency virus type 1 infection of the thymus, *J. Virol.,* 71 (9), 6928–6934, 1997.

82. Aldrovandi, G.M., Feuer, G., Gao, L., Jamieson, B., Kristeva, M., Chen, I.S., and Zack, J.A., The SCID-hu mouse as a model for HIV-1 infection, *Nature,* 363 (6431), 732–736, 1993.

83. Su, L., Kaneshima, H., Bonyhadi, M.L., Lee, R., Auten, J., Wolf, A., Du, B., Rabin, L., Hahn, B.H., Terwilliger, E., and McCune, J.M., Identification of HIV-1 determinants for replication *in vivo, Virology,* 227 (1), 45–52, 1997.

84. Aldrovandi, G.M. and Zack, J.A., Replication and pathogenicity of human immunodeficiency virus type 1 accessory gene mutants in SCID-hu mice, *J. Virol.,* 70 (3), 1505–1511, 1996.

85. Berkowitz, R.D., Alexander, S., Bare, C., Linquist-Stepps, V., Bogan, M., Moreno, M.E., Gibson, L., Wieder, E.D., Kosek, J., Stoddart, C.A., and McCune, J.M., CCR5- and CXCR4-utilizing strains of human immunodeficiency virus type 1 exhibit differential tropism and pathogenesis *in vivo, J. Virol.,* 72 (12), 10108–10117, 1998.

86. Jenkins, M., Hanley, M.B., Moreno, M.B., Wieder, E., and McCune, J.M., Human immunodeficiency virus-1 infection interrupts thymopoiesis and multilineage hematopoiesis *in vivo, Blood,* 91 (8), 2672–2678, 1998.

87. Koka, P.S., Fraser, J.K., Bryson, Y., Bristol, G.C., Aldrovandi, G.M., Daar, E.S., and Zack, J.A., Human immunodeficiency virus inhibits multilineage hematopoiesis *in vivo, J. Virol.,* 72 (6), 5121–5127, 1998.

88. Joshi, V.V. and Oleske, J.M., Pathologic appraisal of the thymus gland in acquired immunodeficiency syndrome in children. A study of four cases and a review of the literature, *Arch. Pathol. Lab. Med.,* 109 (2), 142–146, 1985.

89. Schulz, R. and Mellor, A.L., Self major histocompatibility complex class I antigens expressed solely in lymphoid cells do not induce tolerance in the CD4+ T cell compartment, *J. Exp. Med.,* 184 (4), 1573–1578, 1996.

90. Zaitseva, M., Kawamura, T., Loomis, R., Goldstein, H., Blauvelt, A., and Golding, H., Stromal-derived factor 1 expression in the human thymus, *J. Immunol.,* 168 (6), 2609–2617, 2002.

91. Jenkins, M., Keir, M., and McCune, J.M., Fas is expressed early in human thymocyte development but does not transmit an apoptotic signal, *J. Immunol.,* 163 (3), 1195–1204, 1999.

92. Jenkins, M., Keir, M., and McCune, J.M., A membrane-bound Fas decoy receptor expressed by human thymocytes, *J. Biol. Chem.,* 275 (11), 7988–7993, 2000.

93. Lenardo, M.J., Angleman, S.B., Bounkeua, V., Dimas, J., Duvall, M.G., Graubard, M.B., Hornung, F., Selkirk, M.C., Speirs, C.K., Trageser, C., Orenstein, J.O., and Bolton, D.L., Cytopathic killing of peripheral blood CD4(+) T lymphocytes by human immunodeficiency virus type 1 appears necrotic rather than apoptotic and does not require Env, *J. Virol.,* 76 (10), 5082–5093, 2002.

94. Bolton, D.L., Hahn, B.I., Park, E.A., Lehnhoff, L.L., Hornung, F., and Lenardo, M.J., Death of CD4(+) T-cell lines caused by human immunodeficiency virus type 1 does not depend on caspases or apoptosis, *J. Virol.,* 76 (10), 5094–5107, 2002.

95. Sperber, K., Beuria, P., Singha, N., Gelman, I., Cortes, P., Chen, H., and Kraus, T., Induction of apoptosis by HIV-1-infected monocytic cells, *J. Immunol.,* 170 (3), 1566–1578, 2003.

96. Ricci, J.E., Munoz-Pinedo, C., Fitzgerald, P., Bailly-Maitre, B., Perkins, G.A., Yadava, N., Scheffler, I.E., Ellisman, M.H., and Green, D.R., Disruption of mitochondrial function during apoptosis is mediated by caspase cleavage of the p75 subunit of complex I of the electron transport chain, *Cell,* 117 (6), 773–786, 2004.

97. Ricci, J.E., Gottlieb, R.A., and Green, D.R., Caspase-mediated loss of mitochondrial function and generation of reactive oxygen species during apoptosis, *J. Cell Biol.,* 160 (1), 65–75, 2003.

98. Mosier, D.E., Gulizia, R.J., Baird, S.M., and Wilson, D.B., Transfer of a functional human immune system to mice with severe combined immunodeficiency, *Nature,* 335 (6187), 256–259, 1988.

99. Mosier, D.E., Gulizia, R.J., Baird, S.M., Wilson, D.B., Spector, D.H., and Spector, S.A., Human immunodeficiency virus infection of human-PBL-SCID mice, *Science,* 251 (4995), 791–794, 1991.

100. Mosier, D.E., Gulizia, R.J., MacIsaac, P.D., Torbett, B.E., and Levy, J.A., Rapid loss of CD4+ T cells in human-PBL-SCID mice by noncytopathic HIV isolates, *Science,* 260 (5108), 689–692, 1993.

101. Poignard, P., Sabbe, R., Picchio, G.R., Wang, M., Gulizia, R.J., Katinger, H., Parren, P.W., Mosier, D.E., and Burton, D.R., Neutralizing antibodies have limited effects on the control of established HIV-1 infection *in vivo, Immunity,* 10 (4), 431–438, 1999.

102. Yoshida, A., Tanaka, R., Murakami, T., Takahashi, Y., Koyanagi, Y., Nakamura, M., Ito, M., Yamamoto, N., and Tanaka, Y., Induction of protective immune responses against R5 human immunodeficiency virus type 1 (HIV-1) infection in hu-PBL-SCID mice by intrasplenic immunization with HIV-1-pulsed dendritic cells: possible involvement of a novel factor of human CD4(+) T-cell origin, *J. Virol.,* 77 (16), 8719–8728, 2003.

103. Lapenta, C., Santini, S.M., Logozzi, M., Spada, M., Andreotti, M., Di Pucchio, T., Parlato, S., and Belardelli, F., Potent immune response against HIV-1 and protection from virus challenge in hu-PBL-SCID mice immunized with inactivated virus-pulsed dendritic cells generated in the presence of IFN-, *J. Exp. Med.,* 198 (2), 361–367, 2003.

104. Lapenta, C., Santini, S.M., Proietti, E., Rizza, P., Logozzi, M., Spada, M., Parlato, S., Fais, S., Pitha, P.M., and Belardelli, F., Type I interferon is a powerful inhibitor of *in vivo* HIV-1 infection and preserves human CD4(+) T cells from virus-induced depletion in SCID mice transplanted with human cells, *Virology,* 263 (1), 78–88, 1999.

105. Delhem, N., Hadida, F., Gorochov, G., Carpentier, F., de Cavel, J.P., Andreani, J.F., Autran, B., and Cesbron, J.Y., Primary Th1 cell immunization against HIVgp160 in SCID-hu mice coengrafted with peripheral blood lymphocytes and skin, *J. Immunol.,* 161 (4), 2060–2069, 1998.

106. Boyle, M.J., Connors, M., Flanigan, M.E., Geiger, S.P., Ford, H., Jr., Baseler, M., Adelsberger, J., Davey, R.T., Jr., and Lane, H.C., The human HIV/peripheral blood lymphocyte (PBL)-SCID mouse. A modified human PBL-SCID model for the study of HIV pathogenesis and therapy, *J. Immunol.,* 154 (12), 6612–6623, 1995.

107. Koup, R.A., Safrit, J.T., Weir, R., and Gauduin, M.C., Defining antibody protection against HIV-1 transmission in Hu-PBL-SCID mice, *Semin. Immunol.,* 8 (4), 263–268, 1996.

108. Santini, S.M., Lapenta, C., Logozzi, M., Parlato, S., Spada, M., Di Pucchio, T., and Belardelli, F., Type I interferon as a powerful adjuvant for monocyte-derived dendritic cell development and activity *in vitro* and in Hu-PBL-SCID mice, *J. Exp. Med.,* 191 (10), 1777–1788, 2000.

109. Picchio, G.R., Valdez, H., Sabbe, R., Landay, A.L., Kuritzkes, D.R., Lederman, M.M., and Mosier, D.E., Altered viral fitness of HIV-1 following failure of protease inhibitor-based therapy, *J. Acquir. Immune Defic. Syndr.,* 25 (4), 289–295, 2000.

110. Vieillard, V., Jouveshomme, S., Leflour, N., Jean-Pierre, E., Debre, P., De Maeyer, E., and Autran, B., Transfer of human CD4(+) T lymphocytes producing interferon in Hu-PBL-SCID mice controls human immunodeficiency virus infection, *J. Virol.,* 73 (12), 10281–10288, 1999.

111. Okamoto, Y., Eda, Y., Ogura, A., Shibata, S., Amagai, T., Katsura, Y., Asano, T., Kimachi, K., Makizumi, K., and Honda, M., In SCID-hu mice, passive transfer of a humanized antibody prevents infection and atrophic change of medulla in human thymic implant due to intravenous inoculation of primary HIV-1 isolate, *J. Immunol.,* 160 (1), 69–76, 1998.

112. Gauduin, M.C., Parren, P.W., Weir, R., Barbas, C.F., Burton, D.R., and Koup, R.A., Passive immunization with a human monoclonal antibody protects hu-PBL-SCID mice against challenge by primary isolates of HIV-1, *Nat. Med.,* 3 (12), 1389–1393, 1997.

113. Reinhardt, B., Torbett, B.E., Gulizia, R.J., Reinhardt, P.P., Spector, S.A., and Mosier, D.E., Human immunodeficiency virus type 1 infection of neonatal severe combined immunodeficient mice xenografted with human cord blood cells, *AIDS Res. Hum. Retroviruses,* 10 (2), 131–141, 1994.

114. Mosier, D.E., Viral pathogenesis in hu-PBL-SCID mice, *Semin. Immunol.,* 8 (4), 255–262, 1996.

115. Rizza, P., Santini, S.M., Logozzi, M.A., Lapenta, C., Sestili, P., Gherardi, G., Lande, R., Spada, M., Parlato, S., Belardelli, F., and Fais, S., T-cell dysfunctions in hu-PBL-SCID mice infected with human immunodeficiency virus (HIV) shortly after reconstitution: *in vivo* effects of HIV on highly activated human immune cells, *J. Virol.,* 70 (11), 7958–7964, 1996.

116. Gougeon, M.L., Lecoeur, H., Dulioust, A., Enouf, M.G., Crouvoiser, M., Goujard, C., Debord, T., and Montagnier, L., Programmed cell death in peripheral lymphocytes from HIV-infected persons: increased susceptibility to apoptosis of CD4 and CD8 T cells correlates with lymphocyte activation and with disease progression, *J. Immunol.,* 156 (9), 3509–3520, 1996.

117. Fais, S., Lapenta, C., Santini, S.M., Spada, M., Parlato, S., Logozzi, M., Rizza, P., and Belardelli, F., Human immunodeficiency virus type 1 strains R5 and X4 induce different pathogenic effects in hu-PBL-SCID mice, depending on the state of activation/differentiation of human target cells at the time of primary infection, *J. Virol.,* 73 (8), 6453–6459, 1999.

118. Picchio, G.R., Gulizia, R.J., and Mosier, D.E., Chemokine receptor CCR5 genotype influences the kinetics of human immunodeficiency virus type 1 infection in human PBL-SCID mice, *J. Virol.,* 71 (9), 7124–7127, 1997.

119. Gomez-Roman, V.R., Vazquez, J.A., del Carmen Basualdo, M., Estrada, F.J., Ramos-Kuri, M., and Soler, C., Nef/long terminal repeat quasispecies from HIV type 1-infected Mexican patients with different progression patterns and their pathogenesis in hu-PBL-SCID mice, *AIDS Res. Hum. Retroviruses,* 16 (5), 441–452, 2000.

120. Gulizia, R.J., Collman, R.G., Levy, J.A., Trono, D., and Mosier, D.E., Deletion of Nef slows but does not prevent CD4-positive T-cell depletion in human immunodeficiency virus type 1-infected human-PBL-SCID mice, *J. Virol.,* 71 (5), 4161–4164, 1997.

121. Kawano, Y., Tanaka, Y., Misawa, N., Tanaka, R., Kira, J.I., Kimura, T., Fukushi, M., Sano, K., Goto, T., Nakai, M., Kobayashi, T., Yamamoto, N., and Koyanagi, Y., Mutational analysis of human immunodeficiency virus type 1 (HIV-1) accessory genes: requirement of a site in the Nef gene for HIV-1 replication in activated CD4+ T cells *in vitro* and *in vivo, J. Virol.,* 71 (11), 8456–8466, 1997.

122. Koyanagi, Y., Tanaka, Y., Kira, J., Ito, M., Hioki, K., Misawa, N., Kawano, Y., Yamasaki, K., Tanaka, R., Suzuki, Y., Ueyama, Y., Terada, E., Tanaka, T., Miyasaka, M., Kobayashi, T., Kumazawa, Y., and Yamamoto, N., Primary human immunodeficiency virus type 1 viremia and central nervous system invasion in a novel hu-PBL-immunodeficient mouse strain, *J. Virol.,* 71 (3), 2417–2424, 1997.

123. Kataoka, S., Satoh, J., Fujiya, H., Toyota, T., Suzuki, R., Itoh, K., and Kumagai, K., Immunologic aspects of the nonobese diabetic (NOD) mouse. Abnormalities of cellular immunity, *Diabetes,* 32 (3), 247–253, 1983.

124. Koyanagi, Y., Tanaka, Y., Tanaka, R., Misawa, N., Kawano, Y., Tanaka, T., Miyasaka, M., Ito, M., Ueyama, Y., and Yamamoto, N., High levels of viremia in hu-PBL-NOD-scid mice with HIV-1 infection, *Leukemia,* 11 (Suppl. 3), 109–112, 1997.

125. Miura, Y., Misawa, N., Maeda, N., Inagaki, Y., Tanaka, Y., Ito, M., Kayagaki, N., Yamamoto, N., Yagita, H., Mizusawa, H., and Koyanagi, Y., Critical contribution of tumor necrosis factor-related

apoptosis-inducing ligand (TRAIL) to apoptosis of human CD4+ T cells in HIV-1-infected hu-PBL-NOD-SCID mice, *J. Exp. Med.,* 193 (5), 651–660, 2001.

126. Miura, Y., Misawa, N., Kawano, Y., Okada, H., Inagaki, Y., Yamamoto, N., Ito, M., Yagita, H., Okumura, K., Mizusawa, H., and Koyanagi, Y., Tumor necrosis factor-related apoptosis-inducing ligand induces neuronal death in a murine model of HIV central nervous system infection, *Proc. Natl. Acad. Sci. U.S.A.,* 100 (5), 2777–2782, 2003.

127. Chen, W., Sulcove, J., Frank, I., Jaffer, S., Ozdener, H., and Kolson, D.L., Development of a human neuronal cell model for human immunodeficiency virus (HIV)-infected macrophage-induced neurotoxicity: apoptosis induced by HIV type 1 primary isolates and evidence for involvement of the Bcl-2/Bcl-xL-sensitive intrinsic apoptosis pathway, *J. Virol.,* 76 (18), 9407–9419, 2002.

128. del Real, G., Llorente, M., Bosca, L., Hortelano, S., Serrano, A., Lucas, P., Gomez, L., Toran, J.L., Redondo, C., and Martinez, C., Suppression of HIV-1 infection in linomide-treated SCID-hu-PBL mice, *AIDS,* 12 (8), 865–872, 1998.

129. Maddon, P.J., Dalgleish, A.G., McDougal, J.S., Clapham, P.R., Weiss, R.A., and Axel, R., The T4 gene encodes the AIDS virus receptor and is expressed in the immune system and the brain, *Cell,* 47 (3), 333–348, 1986.

130. Lores, P., Boucher, V., Mackay, C., Pla, M., Von Boehmer, H., Jami, J., Barre-Sinoussi, F., and Weill, J.C., Expression of human CD4 in transgenic mice does not confer sensitivity to human immunodeficiency virus infection, *AIDS Res. Hum. Retroviruses,* 8 (12), 2063–2071, 1992.

131. Mariani, R., Chen, D., Schrofelbauer, B., Navarro, F., Konig, R., Bollman, B., Munk, C., Nymark-McMahon, H., and Landau, N.R., Species-specific exclusion of APOBEC3G from HIV-1 virions by Vif, *Cell,* 114 (1), 21–31, 2003.

132. Mariani, R., Rasala, B.A., Rutter, G., Wiegers, K., Brandt, S.M., Krausslich, H.G., and Landau, N.R., Mouse–human heterokaryons support efficient human immunodeficiency virus type 1 assembly, *J. Virol.,* 75 (7), 3141–3151, 2001.

133. Mariani, R., Rutter, G., Harris, M.E., Hope, T.J., Krausslich, H.G., and Landau, N.R., A block to human immunodeficiency virus type 1 assembly in murine cells, *J. Virol.,* 74 (8), 3859–3870, 2000.

134. Hatziioannou, T., Cowan, S., Von Schwedler, U.K., Sundquist, W.I., and Bieniasz, P.D., Species-specific tropism determinants in the human immunodeficiency virus type 1 capsid, *J. Virol.,* 78 (11), 6005–6012, 2004.

135. Bieniasz, P.D., Grdina, T.A., Bogerd, H.P., and Cullen, B.R., Recruitment of a protein complex containing Tat and cyclin T1 to TAR governs the species specificity of HIV-1 Tat, *EMBO J.,* 17 (23), 7056–7065, 1998.

136. Bieniasz, P.D. and Cullen, B.R., Multiple blocks to human immunodeficiency virus type 1 replication in rodent cells, *J. Virol.,* 74 (21), 9868–9877, 2000.

137. Zimmerman, C., Klein, K.C., Kiser, P.K., Singh, A.R., Firestein, B.L., Riba, S.C., and Lingappa, J.R., Identification of a host protein essential for assembly of immature HIV-1 capsids, *Nature,* 415 (6867), 88–92, 2002.

138. Garber, M.E., Wei, P., KewalRamani, V.N., Mayall, T.P., Herrmann, C.H., Rice, A.P., Littman, D.R., and Jones, K.A., The interaction between HIV-1 Tat and human cyclin T1 requires zinc and a critical cysteine residue that is not conserved in the murine CycT1 protein, *Genes Dev.,* 12 (22), 3512–3527, 1998.

139. Dragic, T., Litwin, V., Allaway, G.P., Martin, S.R., Huang, Y., Nagashima, K.A., Cayanan, C., Maddon, P.J., Koup, R.A., Moore, J.P., and Paxton, W.A., HIV-1 entry into CD4+ cells is mediated by the chemokine receptor CC-CKR-5, *Nature,* 381 (6584), 667–673, 1996.

140. Choe, H., Farzan, M., Sun, Y., Sullivan, N., Rollins, B., Ponath, P.D., Wu, L., Mackay, C.R., LaRosa, G., Newman, W., Gerard, N., Gerard, C., and Sodroski, J., The -chemokine receptors CCR3 and CCR5 facilitate infection by primary HIV-1 isolates, *Cell,* 85 (7), 1135–1148, 1996.

141. Deng, H., Liu, R., Ellmeier, W., Choe, S., Unutmaz, D., Burkhart, M., Di Marzio, P., Marmon, S., Sutton, R.E., Hill, C.M., Davis, C.B., Peiper, S.C., Schall, T.J., Littman, D.R., and Landau, N.R., Identification of a major co-receptor for primary isolates of HIV-1, *Nature,* 381 (6584), 661–666, 1996.

142. Feng, Y., Broder, C.C., Kennedy, P.E., and Berger, E.A., HIV-1 entry cofactor: functional cDNA cloning of a seven-transmembrane, G protein-coupled receptor, *Science,* 272 (5263), 872–877, 1996.

143. Browning, J., Horner, J.W., Pettoello-Mantovani, M., Raker, C., Yurasov, S., DePinho, R.A., and Goldstein, H., Mice transgenic for human CD4 and CCR5 are susceptible to HIV infection, *Proc. Natl. Acad. Sci. U.S.A.*, 94 (26), 14637–14641, 1997.

144. Sawada, S., Gowrishankar, K., Kitamura, R., Suzuki, M., Suzuki, G., Tahara, S., and Koito, A., Disturbed CD4+ T cell homeostasis and *in vitro* HIV-1 susceptibility in transgenic mice expressing T cell line-tropic HIV-1 receptors, *J. Exp. Med.*, 187 (9), 1439–1449, 1998.

145. Keppler, O.T., Yonemoto, W., Welte, F.J., Patton, K.S., Iacovides, D., Atchison, R.E., Ngo, T., Hirschberg, D.L., Speck, R.F., and Goldsmith, M.A., Susceptibility of rat-derived cells to replication by human immunodeficiency virus type 1, *J. Virol.*, 75 (17), 8063–8073, 2001.

146. Keppler, O.T., Welte, F.J., Ngo, T.A., Chin, P.S., Patton, K.S., Tsou, C.L., Abbey, N.W., Sharkey, M.E., Grant, R.M., You, Y., Scarborough, J.D., Ellmeier, W., Littman, D.R., Stevenson, M., Charo, I.F., Herndier, B.G., Speck, R.F., and Goldsmith, M.A., Progress toward a human CD4/CCR5 transgenic rat model for *de novo* infection by human immunodeficiency virus type 1, *J. Exp. Med.*, 195 (6), 719–736, 2002.

147. Leonard, J.M., Abramczuk, J.W., Pezen, D.S., Rutledge, R., Belcher, J.H., Hakim, F., Shearer, G., Lamperth, L., Travis, W., Fredrickson, T., Notkins, A.L., and Martin, M.A., Development of disease and virus recovery in transgenic mice containing HIV proviral DNA, *Science*, 242 (4886), 1665–1670, 1988.

148. Hanna, Z., Kay, D.G., Cool, M., Jothy, S., Rebai, N., and Jolicoeur, P., Transgenic mice expressing human immunodeficiency virus type 1 in immune cells develop a severe AIDS-like disease, *J. Virol.*, 72 (1), 121–132, 1998.

149. Dickie, P., Gazzinelli, R., and Chang, L.J., Models of HIV type 1 proviral gene expression in wild-type HIV and MLV/HIV transgenic mice, *AIDS Res. Hum. Retroviruses*, 12 (12), 1103–1116, 1996.

150. Wang, E.J., Pettoello-Mantovani, M., Anderson, C.M., Osiecki, K., Moskowitz, D., and Goldstein, H., Development of a novel transgenic mouse/SCID-hu mouse system to characterize the *in vivo* behavior of reservoirs of human immunodeficiency virus type 1-infected cells, *J. Infect. Dis.*, 186 (10), 1412–1421, 2002.

151. Wang, E.J., Sun, J., Pettoello-Mantovani, M., Anderson, C.M., Osiecki, K., Zhao, M.L., Lopez, L., Lee, S.C., Berman, J.W., and Goldstein, H., Microglia from mice transgenic for a provirus encoding a monocyte-tropic HIV type 1 isolate produce infectious virus and display *in vitro* and *in vivo* upregulation of lipopolysaccharide-induced chemokine gene expression, *AIDS Res. Hum. Retroviruses*, 19 (9), 755–765, 2003.

152. Browning Paul, J., Wang, E.J., Pettoello-Mantovani, M., Raker, C., Yurasov, S., Goldstein, M.M., Horner, J.W., Chan, J., and Goldstein, H., Mice transgenic for monocyte-tropic HIV type 1 produce infectious virus and display plasma viremia: a new *in vivo* system for studying the postintegration phase of HIV replication, *AIDS Res. Hum. Retroviruses*, 16 (5), 481–492, 2000.

153. Poudrier, J., Weng, X., Kay, D.G., Pare, G., Calvo, E.L., Hanna, Z., Kosco-Vilbois, M.H., and Jolicoeur, P., The AIDS disease of CD4C/HIV transgenic mice shows impaired germinal centers and autoantibodies and develops in the absence of IFN- and IL-6, *Immunity*, 15 (2), 173–185, 2001.

154. Rosenberg, Y.J., Lewis, M.G., Greenhouse, J.J., Cafaro, A., Leon, E.C., Brown, C.R., Bieg, K.E., and Kosco-Vilbois, M.H., Enhanced follicular dendritic cell function in lymph nodes of simian immunodeficiency virus-infected macaques: consequences for pathogenesis, *Eur. J. Immunol.*, 27 (12), 3214–3222, 1997.

155. De, S.K., Devadas, K., and Notkins, A.L., Elevated levels of tumor necrosis factor (TNF-) in human immunodeficiency virus type 1-transgenic mice: prevention of death by antibody to TNF-, *J. Virol.*, 76 (22), 11710–11714, 2002.

156. De, S.K., Wohlenberg, C.R., Marinos, N.J., Doodnauth, D., Bryant, J.L., and Notkins, A.L., Human chorionic gonadotropin hormone prevents wasting syndrome and death in HIV-1 transgenic mice, *J. Clin. Invest.*, 99 (7), 1484–1491, 1997.

157. Santoro, T.J., Bryant, J.L., Pellicoro, J., Klotman, M.E., Kopp, J.B., Bruggeman, L.A., Franks, R.R., Notkins, A.L., and Klotman, P.E., Growth failure and AIDS-like cachexia syndrome in HIV-1 transgenic mice, *Virology*, 201 (1), 147–151, 1994.

158. Tinkle, B.T., Ueda, H., Ngo, L., Luciw, P.A., Shaw, K., Rosen, C.A., and Jay, G., Transgenic dissection of HIV genes involved in lymphoid depletion, *J. Clin. Invest.*, 100 (1), 32–39, 1997.

159. Hanna, Z., Kay, D.G., Rebai, N., Guimond, A., Jothy, S., and Jolicoeur, P., Nef harbors a major determinant of pathogenicity for an AIDS-like disease induced by HIV-1 in transgenic mice, *Cell*, 95 (2), 163–175, 1998.

160. Hanna, Z., Priceputu, E., Kay, D.G., Poudrier, J., Chrobak, P., and Jolicoeur, P., *In vivo* mutational analysis of the N-terminal region of HIV-1 Nef reveals critical motifs for the development of an AIDS-like disease in CD4C/HIV transgenic mice, *Virology*, 327 (2), 273–286, 2004.

161. Hanna, Z., Weng, X., Kay, D.G., Poudrier, J., Lowell, C., and Jolicoeur, P., The pathogenicity of human immunodeficiency virus (HIV) type 1 Nef in CD4C/HIV transgenic mice is abolished by mutation of its SH3-binding domain, and disease development is delayed in the absence of Hck, *J. Virol.*, 75 (19), 9378–9392, 2001.

162. Weng, X., Priceputu, E., Chrobak, P., Poudrier, J., Kay, D.G., Hanna, Z., Mak, T.W., and Jolicoeur, P., CD4+ T cells from CD4C/HIVNef transgenic mice show enhanced activation *in vivo* with impaired proliferation *in vitro* but are dispensable for the development of a severe AIDS-like organ disease, *J. Virol.*, 78 (10), 5244–5257, 2004.

163. Lindemann, D., Wilhelm, R., Renard, P., Althage, A., Zinkernagel, R., and Mous, J., Severe immuno-deficiency associated with a human immunodeficiency virus 1 NEF/3-long terminal repeat transgene, *J. Exp. Med.*, 179 (3), 797–807, 1994.

164. Skowronski, J., Parks, D., and Mariani, R., Altered T cell activation and development in transgenic mice expressing the HIV-1 Nef gene, *EMBO J.*, 12 (2), 703–713, 1993.

165. Simard, M.C., Chrobak, P., Kay, D.G., Hanna, Z., Jothy, S., and Jolicoeur, P., Expression of simian immunodeficiency virus Nef in immune cells of transgenic mice leads to a severe AIDS-like disease, *J. Virol.*, 76 (8), 3981–3995, 2002.

166. Kim, B.O., Liu, Y., Ruan, Y., Xu, Z.C., Schantz, L., and He, J.J., Neuropathologies in transgenic mice expressing human immunodeficiency virus type 1 Tat protein under the regulation of the astrocyte-specific glial fibrillary acidic protein promoter and doxycycline, *Am. J. Pathol.*, 162 (5), 1693–1707, 2003.

167. Yasuda, J., Miyao, T., Kamata, M., Aida, Y., and Iwakura, Y., T cell apoptosis causes peripheral T cell depletion in mice transgenic for the HIV-1 Vpr gene, *Virology*, 285 (2), 181–192, 2001.

168. Adle-Biassette, H., Levy, Y., Colombel, M., Poron, F., Natchev, S., Keohane, C., and Gray, F., Neuronal apoptosis in HIV infection in adults, *Neuropathol. Appl. Neurobiol.*, 21 (3), 218–227, 1995.

169. Hesselgesser, J., Taub, D., Baskar, P., Greenberg, M., Hoxie, J., Kolson, D.L., and Horuk, R., Neuronal apoptosis induced by HIV-1 gp120 and the chemokine SDF-1 is mediated by the chemokine receptor CXCR4, *Curr. Biol.*, 8 (10), 595–598, 1998.

170. Meucci, O., Fatatis, A., Simen, A.A., Bushell, T.J., Gray, P.W., and Miller, R.J., Chemokines regulate hippocampal neuronal signaling and gp120 neurotoxicity, *Proc. Natl. Acad. Sci. U.S.A.*, 95 (24), 14500–14505, 1998.

171. Garden, G.A., Budd, S.L., Tsai, E., Hanson, L., Kaul, M., D'Emilia, D.M., Friedlander, R.M., Yuan, J., Masliah, E., and Lipton, S.A., Caspase cascades in human immunodeficiency virus-associated neurodegeneration, *J. Neurosci.*, 22 (10), 4015–4024, 2002.

172. Alter, H.J., Eichberg, J.W., Masur, H., Saxinger, W.C., Gallo, R., Macher, A.M., Lane, H.C., and Fauci, A.S., Transmission of HTLV-III infection from human plasma to chimpanzees: an animal model for AIDS, *Science*, 226 (4674), 549–552, 1984.

173. Francis, D.P., Feorino, P.M., Broderson, J.R., McClure, H.M., Getchell, J.P., McGrath, C.R., Swenson, B., McDougal, J.S., Palmer, E.L., Harrison, A.K., Barresinoussi, F., Chermann, J.C., Montagnier, L., Curran, J.W., Cabradilla, C.D., and Kalyanaraman, V.S., Infection of chimpanzees with lymphad-enopathy-associated virus, *Lancet*, 2 (8414), 1276–1277, 1984.

174. Ghosh, S.K., Fultz, P.N., Keddie, E., Saag, M.S., Sharp, P.M., Hahn, B.H., and Shaw, G.M., A molecular clone of HIV-1 tropic and cytopathic for human and chimpanzee lymphocytes, *Virology*, 194 (2), 858–864, 1993.

175. Shibata, R., Hoggan, M.D., Broscius, C., Englund, G., Theodore, T.S., Buckler-White, A., Arthur, L.O., Israel, Z., Schultz, A., Lane, H.C., and Martin, M.A., Isolation and characterization of a syncytium-inducing, macrophage/T-cell line-tropic human immunodeficiency virus type 1 isolate that readily infects chimpanzee cells *in vitro* and *in vivo*, *J. Virol.*, 69 (7), 4453–4462, 1995.

176. Ondoa, P., Davis, D., Kestens, L., Vereecken, C., Garcia Ribas, S., Fransen, K., Heeney, J., and van der Groen, G., *In vitro* susceptibility to infection with SIVcpz and HIV-1 is lower in chimpanzee than in human peripheral blood mononuclear cells, *J. Med. Virol.*, 67 (3), 301–311, 2002.

177. Fultz, P.N., Srinivasan, A., Greene, C.R., Butler, D., Swenson, R.B., and McClure, H.M., Superinfection of a chimpanzee with a second strain of human immunodeficiency virus, *J. Virol.*, 61 (12), 4026–4029, 1987.

178. Novembre, F.J., Saucier, M., Anderson, D.C., Klumpp, S.A., O'Neil, S.P., Brown, C.R., II, Hart, C.E., Guenthner, P.C., Swenson, R.B., and McClure, H.M., Development of AIDS in a chimpanzee infected with human immunodeficiency virus type 1, *J. Virol.,* 71 (5), 4086–4091, 1997.

179. Mwaengo, D.M. and Novembre, F.J., Molecular cloning and characterization of viruses isolated from chimpanzees with pathogenic human immunodeficiency virus type 1 infections, *J. Virol.,* 72 (11), 8976–8987, 1998.

180. Novembre, F.J., de Rosayro, J., Nidtha, S., O'Neil, S.P., Gibson, T.R., Evans-Strickfaden, T., Hart, C.E., and McClure, H.M., Rapid CD4(+) T-cell loss induced by human immunodeficiency virus type 1(NC) in uninfected and previously infected chimpanzees, *J. Virol.,* 75 (3), 1533–1539, 2001.

181. O'Neil, S.P., Novembre, F.J., Hill, A.B., Suwyn, C., Hart, C.E., Evans-Strickfaden, T., Anderson, D.C., deRosayro, J., Herndon, J.G., Saucier, M., and McClure, H.M., Progressive infection in a subset of HIV-1-positive chimpanzees, *J. Infect. Dis.,* 182 (4), 1051–1062, 2000.

182. Davis, I.C., Girard, M., and Fultz, P.N., Loss of CD4+ T cells in human immunodeficiency virus type 1-infected chimpanzees is associated with increased lymphocyte apoptosis, *J. Virol.,* 72 (6), 4623–4632, 1998.

183. Prince, A.M., Chimpanzee pathogenic strains of HIV type 1 (Letter), *AIDS Res. Hum. Retroviruses,* 13 (15), 1259, 1997.

184. Prince, A.M., Allan, J., Andrus, L., Brotman, B., Eichber, J., Fouts, R., Goodall, J., Marx, P., Murthy, K.K., McGreal, S., and Noon, C., Virulent HIV strains, chimpanzees, and trial vaccines, *Science,* 283 (5405), 1117–1118, 1999.

185. Gougeon, M.L., Lecoeur, H., Boudet, F., Ledru, E., Marzabal, S., Boullier, S., Roue, R., Nagata, S., and Heeney, J., Lack of chronic immune activation in HIV-infected chimpanzees correlates with the resistance of T cells to Fas/Apo-1 (CD95)-induced apoptosis and preservation of a T helper 1 phenotype, *J. Immunol.,* 158 (6), 2964–2976, 1997.

186. Gougeon, M.L., Garcia, S., Heeney, J., Tschopp, R., Lecoeur, H., Guetard, D., Rame, V., Dauguet, C., and Montagnier, L., Programmed cell death in AIDS-related HIV and SIV infections, *AIDS Res. Hum. Retroviruses,* 9 (6), 553–563, 1993.

187. Schuitemaker, H., Meyaard, L., Kootstra, N.A., Dubbes, R., Otto, S.A., Tersmette, M., Heeney, J.L., and Miedema, F., Lack of T cell dysfunction and programmed cell death in human immunodeficiency virus type 1-infected chimpanzees correlates with absence of monocytotropic variants, *J. Infect. Dis.,* 168 (5), 1140–1147, 1993.

188. Ehret, A., Westendorp, M.O., Herr, I., Debatin, K.M., Heeney, J.L., Frank, R., and Krammer, P.H., Resistance of chimpanzee T cells to human immunodeficiency virus type 1 Tat-enhanced oxidative stress and apoptosis, *J. Virol.,* 70 (9), 6502–6507, 1996.

189. Estaquier, J., Idziorek, T., de Bels, F., Barre-Sinoussi, F., Hurtrel, B., Aubertin, A.M., Venet, A., Mehtali, M., Muchmore, E., Michel, P., Mouton, Y., Girard, M., and Ameisen, J.C., Programmed cell death and AIDS: significance of T-cell apoptosis in pathogenic and nonpathogenic primate lentiviral infections, *Proc. Natl. Acad. Sci. U.S.A.,* 91 (20), 9431–9435, 1994.

190. Shibata, R., Sakai, H., Kiyomasu, T., Ishimoto, A., Hayami, M., and Adachi, A., Generation and characterization of infectious chimeric clones between human immunodeficiency virus type 1 and simian immunodeficiency virus from an African green monkey, *J. Virol.,* 64 (12), 5861–5868, 1990.

191. Shibata, R., Kawamura, M., Sakai, H., Hayami, M., Ishimoto, A., and Adachi, A., Generation of a chimeric human and simian immunodeficiency virus infectious to monkey peripheral blood mononuclear cells, *J. Virol.,* 65 (7), 3514–3520, 1991.

192. Joag, S.V., Li, Z., Foresman, L., Stephens, E.B., Zhao, L.J., Adany, I., Pinson, D.M., McClure, H.M., and Narayan, O., Chimeric simian/human immunodeficiency virus that causes progressive loss of CD4+ T cells and AIDS in pig-tailed macaques, *J. Virol.,* 70 (5), 3189–3197, 1996.

193. Shibata, R., Maldarelli, F., Siemon, C., Matano, T., Parta, M., Miller, G., Fredrickson, T., and Martin, M.A., Infection and pathogenicity of chimeric simian-human immunodeficiency viruses in macaques: determinants of high virus loads and CD4 cell killing, *J. Infect. Dis.,* 176 (2), 362–373, 1997.

194. Reimann, K.A., Li, J.T., Veazey, R., Halloran, M., Park, I.W., Karlsson, G.B., Sodroski, J., and Letvin, N.L., A chimeric simian/human immunodeficiency virus expressing a primary patient human immunodeficiency virus type 1 isolate Env causes an AIDS-like disease after *in vivo* passage in rhesus monkeys, *J. Virol.,* 70 (10), 6922–6928, 1996.

195. Reinberger, S., Spring, M., Nisslein, T., Stahl-Hennig, C., Hunsmann, G., and Dittmer, U., Kinetics of lymphocyte apoptosis in macaques infected with different simian immunodeficiency viruses or simian/human immunodeficiency hybrid viruses, *Clin. Immunol.*, 90 (1), 141–146, 1999.

196. Iida, T., Kita, M., Kuwata, T., Miura, T., Ibuki, K., Ui, M., Hayami, M., and Imanishi, J., Apoptosis induced by *in vitro* infection with simian–human immunodeficiency chimeric virus in macaque and human peripheral blood mononuclear cells, *AIDS Res. Hum. Retroviruses*, 17 (15), 1387–1393, 2001.

197. Iida, T., Ichimura, H., Shimada, T., Ibuki, K., Ui, M., Tamaru, K., Kuwata, T., Yonehara, S., Imanishi, J., and Hayami, M., Role of apoptosis induction in both peripheral lymph nodes and thymus in progressive loss of CD4+ cells in SHIV-infected macaques, *AIDS Res. Hum. Retroviruses*, 16 (1), 9–18, 2000.

198. Yoshino, N., Ryu, T., Sugamata, M., Ihara, T., Ami, Y., Shinohara, K., Tashiro, F., and Honda, M., Direct detection of apoptotic cells in peripheral blood from highly pathogenic SHIV-inoculated monkey, *Biochem. Biophys. Res. Commun.*, 268 (3), 868–874, 2000.

199. Sasaki, Y., Ami, Y., Nakasone, T., Shinohara, K., Takahashi, E., Ando, S., Someya, K., Suzaki, Y., and Honda, M., Induction of CD95 ligand expression on T lymphocytes and B lymphocytes and its contribution to apoptosis of CD95-up-regulated CD4+ T lymphocytes in macaques by infection with a pathogenic simian/human immunodeficiency virus, *Clin. Exp. Immunol.*, 122 (3), 381–389, 2000.

200. Igarashi, T., Brown, C.R., Byrum, R.A., Nishimura, Y., Endo, Y., Plishka, R.J., Buckler, C., Buckler-White, A., Miller, G., Hirsch, V.M., and Martin, M.A., Rapid and irreversible CD4+ T-cell depletion induced by the highly pathogenic simian/human immunodeficiency virus SHIV(DH12R) is systemic and synchronous, *J. Virol.*, 76 (1), 379–391, 2002.

201. Wallace, M., Waterman, P.M., Mitchen, J.L., Djavani, M., Brown, C., Trivedi, P., Horejsh, D., Dykhuizen, M., Kitabwalla, M., and Pauza, C.D., Lymphocyte activation during acute simian/human immunodeficiency virus SHIV(89.6PD) infection in macaques, *J. Virol.*, 73 (12), 10236–10244, 1999.

202. Mitra, D., Steiner, M., Lynch, D.H., Staiano-Coico, L., and Laurence, J., HIV-1 upregulates Fas ligand expression in CD4+ T cells *in vitro* and *in vivo*: association with Fas-mediated apoptosis and modulation by aurintricarboxylic acid, *Immunology*, 87 (4), 581–585, 1996.

203. Badley, A.D., Pilon, A.A., Landay, A., and Lynch, D.H., Mechanisms of HIV-associated lymphocyte apoptosis, *Blood*, 96 (9), 2951–2964, 2000.

204. Bahner, I., Kearns, K., Hao, Q.L., Smogorzewska, E.M., and Kohn, D.B., Transduction of human CD34+ hematopoietic progenitor cells by a retroviral vector expressing an RRE decoy inhibits human immunodeficiency virus type 1 replication in myelomonocytic cells produced in long-term culture, *J. Virol.*, 70 (7), 4352–4360, 1996.

205. Bonyhadi, M.L., Moss, K., Voytovich, A., Auten, J., Kalfoglou, C., Plavec, I., Forestell, S., Su, L., Bohnlein, E., and Kaneshima, H., RevM10-expressing T cells derived *in vivo* from transduced human hematopoietic stem-progenitor cells inhibit human immunodeficiency virus replication, *J. Virol.*, 71 (6), 4707–4716, 1997.

206. Bauer, G., Valdez, P., Kearns, K., Bahner, I., Wen, S.F., Zaia, J.A., and Kohn, D.B., Inhibition of human immunodeficiency virus-1 (HIV-1) replication after transduction of granulocyte colony-stimulating factor-mobilized CD34+ cells from HIV-1-infected donors using retroviral vectors containing anti-HIV-1 genes, *Blood*, 89 (7), 2259–2267, 1997.

207. Gervaix, A., Schwarz, L., Law, P., Ho, A.D., Looney, D., Lane, T., and Wong-Staal, F., Gene therapy targeting peripheral blood CD34+ hematopoietic stem cells of HIV-infected individuals, *Hum. Gene Ther.*, 8 (18), 2229–2238, 1997.

208. Su, L., Lee, R., Bonyhadi, M., Matsuzaki, H., Forestell, S., Escaich, S., Bohnlein, E., and Kaneshima, H., Hematopoietic stem cell-based gene therapy for acquired immunodeficiency syndrome: efficient transduction and expression of RevM10 in myeloid cells *in vivo* and *in vitro*, *Blood*, 89 (7), 2283–2290, 1997.

209. Yu, M., Leavitt, M.C., Maruyama, M., Yamada, O., Young, D., Ho, A.D., and Wong-Staal, F., Intracellular immunization of human fetal cord blood stem/progenitor cells with a ribozyme against human immunodeficiency virus type 1, *Proc. Natl. Acad. Sci. U.S.A.*, 92 (3), 699–703, 1995.

Section 3

Clinical Consequences of HIV-Induced Cell Death

19 T Cell Dynamics and the Role of Apoptosis in HIV Infection

Nienke Vrisekoop
Mette D. Hazenberg
Frank Miedema
José A.M. Borghans

CONTENTS

INTRODUCTION

CD4+ T lymphocyte loss is one of the hallmarks of human immunodeficiency virus (HIV) infection. Because the percentage of T cells undergoing apoptosis is dramatically increased during HIV infection, it may not seem surprising that CD4+ T cells are progressively lost during HIV infection. There is no consensus, however, on the mechanisms behind increased T cell apoptosis in HIV infection and on the reason why HIV induces an instant loss of CD4+ T cells, whereas CD8+ T cell numbers decline only at late stages of infection. Even though our insights into the different effects of HIV on the immune system have improved during the last decades, many questions remain.

Ameisen and Capron[1] were the first to propose that CD4+ T cell loss in HIV infection might be due to inappropriate induction of programmed T cell death. They proposed that HIV primes CD4+ T cells to a state in which they become sensitive to T cell apoptosis upon further stimulation. Factors proposed to be involved in HIV-induced inappropriate apoptosis may include direct cytopathicity of HIV in productively infected CD4+ T cells and apoptosis of CD4+ T cells via CD4 cross-linking by viral proteins. Later, it was argued that CD4+ T cells are eventually lost in HIV infection, because the immune system gets exhausted by continuously trying to replenish the CD4+ T cell pool,[2] whereas others argued that the key problem in HIV infection is interference with T cell renewal, because HIV induces increased apoptosis in the thymus.[3–5] In contrast, we and others proposed that apoptosis in HIV infection is a natural result of increased immune activation induced by the virus and that it is this state of chronic immune activation that leads to the progressive loss of CD4+ T cells in HIV infection.[6,7] In this chapter, the strengths and weaknesses of the different explanations for increased T cell apoptosis and its relation to CD4+ T cell loss during HIV infection are reviewed.

INAPPROPRIATE T CELL APOPTOSIS: DIRECT HIV-INDUCED CYTOPATHICITY AND CD4 CROSS-LINKING

The idea that CD4+ T cell loss in HIV infection is due to direct killing of infected CD4+ T cells by the virus has been fueled by *in vitro* studies in which CD4+ T cells that were infected in culture were shown to undergo apoptosis[8] and by the well-described positive correlation between HIV viral load and the rate of disease progression.[9,10] A number of experimental observations cannot, however, easily be reconciled with this hypothesis.

First, if most CD4+ T cell loss were due to direct killing of HIV-infected cells, one would expect a strong correlation between the number of infected CD4+ T cells and disease progression. Surprisingly, however, the number of infected cells appeared to correlate poorly with plasma viral ribonucleic acid (RNA) load,[11] which is known to be one of the strongest predictors of disease progression. Additionally, the level of *in vitro* cytopathicity of HIV appeared not to correlate with the extent of CD4+ T cell depletion, as no difference in CD4+ T cell loss could be found between severe combined immunodeficient (SCID)-hu mice infected with noncytopathic or cytopathic HIV isolates.[12] Also, in simian immunodeficiency virus (SIV)-infected primates, viral cytopathicity did not correlate with disease progression. *In vitro*, SIV is equally cytopathic to rhesus macaque and sooty mangabey lymphocytes. However, SIV infection in rhesus macaques is associated with CD4+ T cell loss and progression to acquired immunodeficiency syndrome (AIDS), whereas CD4+ T cell numbers in sooty mangabeys are not affected by SIV infection.[13]

Second, several investigators have argued that the number of productively infected CD4+ T cells is too low to account for all the CD4+ T cell loss occurring during HIV infection. Hardly any (<0.01%) plasma virus was found to be infectious,[10] and replication-competent integrated HIV-1 deoxyribonucleic acid (DNA) appeared to be present in only a minor (<0.05%) fraction of susceptible T cells.[11] However, if infected cells are lost rapidly, low numbers of infected cells measured at any moment in time may still add up to a large total number of infected cells. Nevertheless, the T cell subset undergoing apoptosis seems much larger and easier to detect than the subset of T cells that are productively infected, suggesting that T cell apoptosis is not confined to the relatively small number of productively infected T cells. Indeed, when lymph node biopsies from HIV-infected children or SIV-infected macaques were analyzed *in situ*, only a minor fraction of apoptotic lymphocytes was found to be productively infected with HIV, and, vice versa, only a few productively infected cells turned out to undergo apoptosis.[14] Hence, the bulk of CD4+ T cell loss cannot be explained by direct cytopathicity of the virus.

Thus, direct cytopathicity alone does not account for the increased apoptosis levels observed in HIV infection. *In vitro* studies have suggested that the viral protein gp120 can induce cross-linking of CD4 molecules on the T cell surface, which, in the absence of T cell receptor triggering, leads to CD4+ T cell apoptosis.[15–17] Similarly, the viral proteins Tat and Nef have beeen shown to exert apoptotic effects on CD4+ T cells.[18] Indeed, the latter mechanisms of HIV-induced CD4+ T cell apoptosis may play a larger role in CD4+ T cell depletion *in vivo* than that of direct cytopathicity, as they are not confined to the relatively small fraction of T cells that are productively infected by HIV.

Interestingly, however, data from our laboratory and from others have shown that T cell apoptosis in HIV infection is not limited to CD4+ T cells. In fact, the highest percentage of cells undergoing apoptosis during HIV infection is typically observed in the CD8+ T cell population.[17,19–21] Thus, although the different mechanisms described above offer an explanation as to why levels of apoptosis are increased in the CD4+ T cell pool, they fail to explain most of the T cell apoptosis, i.e., that occurring in the CD8+ T cell pool, during HIV infection.

T CELL EXHAUSTION INDUCED BY CONTINUOUS HIGH T CELL TURNOVER

A seminal paper from David Ho's laboratory changed the insights into HIV dynamics altogether by showing that the viral population in infected individuals is highly dynamic, even during the asymptomatic phase of HIV infection, in which the total viral load is known to be relatively stable. Ho and collaborators showed that after initiation of highly active antiretroviral treatment (HAART), the HIV viral load decreases rapidly. The authors argued that before treatment was started, the virus must have turned over just as rapidly but that the rapid loss of viral particles pretreatment remained invisible, because it was balanced by high virus production. When HAART prevented the generation of new virus, the rapid decay of the virus suddenly became evident.[2] In the same paper, Ho et al. reported a rapid increase in CD4+ T cell counts after the initiation of HAART and argued along the same lines that CD4+ T cells were apparently destroyed and replaced at a rapid pace during untreated HIV infection. It was proposed that CD4+ T cells are eventually lost during HIV infection, because the immune system cannot cope with the persistently high CD4+ T cell turnover that is required to keep the T cell pool at its homeostatic level.[2]

We studied if one could find any evidence for such high turnover–induced exhaustion of the CD4+ T lymphocyte pool in HIV infection. To this end, we measured CD4+ T cell telomeres, which are known to shorten upon T cell division, in HIV-infected individuals and age-matched healthy controls. A long history of increased CD4+ T cell proliferation in HIV infection should be apparent by a shortened average telomere length in HIV-infected individuals compared to controls. Surprisingly, however, we found no reduction in the average CD4+ T cell telomere length, but we did find a significant shortening of the CD8+ T cell telomeres in HIV patients compared to healthy controls.[22] Thus, if anything, the CD8+ lymphocyte pool in HIV+ individuals seemed to have a more extensive history of increased T cell division than the CD4+ T cell pool. In addition, studies from our laboratory and other laboratories showed that the initial steep rise in CD4+ T cell counts following the start of HAART is most likely caused by redistribution of lymphocytes from the lymphoid tissues to the peripheral blood rather than by rapid production of new CD4+ T cells. Immune reconstitution in HIV-infected adults on HAART appeared to occur biphasically due to rapid redistribution of memory T cells and a gradual increase of naive T cells, and this seldom leads to normalization of the CD4+ T cell count.[23–25] The rapid increase in CD4+ T cell numbers upon initiation of HAART thus does not imply that CD4+ T cells have a short half-life during untreated HIV infection. We and others estimated CD4+ T cell dynamics during HIV infection by measuring Ki67 expression and using BrdU labeling and came to the conclusion that CD4+ T cell turnover during HIV infection is increased maximally only three- to fivefold.[26–29] Interestingly, CD8+ T cell turnover is also increased severalfold during untreated HIV infection.[30] The reason for this increased T cell turnover remains controversial.

Several researchers have proposed that CD4+ T cell proliferation rates in HIV infection are increased because of homeostatic mechanisms that are compensating for CD4+ T cell loss.[2,31] We refer to this hypothesis as the "pull hypothesis"—the immune system is trying to keep the total T cell number at a fixed homeostatic level, and when many cells are lost due to HIV infection, many new cells have to be generated. In contrast, we and others have proposed that increased T cell proliferation in HIV infection is directly caused by the virus, which is chronically stimulating the immune system, and that T cell loss may be a result of this chronic immune stimulation.[7] The latter hypothesis, referred to as the "push hypothesis," will be later discussed.

If increased T cell turnover rates in HIV infection would reflect a homeostatic response to low numbers of CD4+ T cells, one would expect T cell turnover to remain increased as long as CD4+

T cell numbers are below the normal homeostatic level. We have shown, however, that the increased T cell turnover characteristic of HIV infection is rapidly lost during HAART, despite sustained low CD4+ T cell numbers.[30] This, and the observation that CD8+ T cell division rates are also increased during untreated infection, argues against the idea that increased T cell turnover during HIV infection is due to a homeostatic response.

These data also show that T cell proliferation may be hardly increased during a phase of immune recovery. It is generally assumed that during a phase of T cell depletion, homeostatic responses (e.g., increases in thymic function and peripheral T cell proliferation) drive T cell recovery. However, in a study of 22 lymphopenic patients with hematologic malignancies who received allogeneic human leukocyte antigen (HLA)-matched peripheral stem cell transplantation, we found no evidence for increased thymic function. T cell division was increased only in a subset of lymphopenic patients, the division was specifically related to clinical events, such as graft versus host disease or episodes of infectious diseases, and proportions of dividing cells declined rapidly after transplantation despite still very low T cell numbers.[32] Together, these studies cast serious doubts on the natural homeostatic capacities of the immune system during lymphopenia.

INTERFERENCE WITH THYMIC FUNCTION

Maintenance of a full peripheral CD4+ T cell pool may be hampered not only by HIV-induced peripheral T cell death but also by a lack of *de novo* T cell generation.[33] Because thymopoiesis is generally thought to be the main source of T cell generation, it has been proposed that HIV-induced apoptosis of thymocytes in the presence of slightly increased—or even normal[4]—peripheral T cell death might explain why CD4+ T cell numbers are gradually lost during asymptomatic HIV infection.[3,5,34] Interference with thymic function would also explain why HIV infection is associated with a gradual decline in both naive CD4+ and naive CD8+ T cells.[35,36] However, current data do not unequivocally show that thymic output is severely decreased throughout HIV infection, and even if it was, recent data have brought into question the conclusion that this could explain a severe loss of CD4+ T cells in the periphery.

Experiments in mice and data from humans do suggest that HIV is able to infect the thymus. To analyze the effect of HIV infection on thymopoiesis *in vivo*, McCune and collaborators used the SCID-hu mouse model, in which human fetal liver and thymus are implanted under the kidney capsule of immunodeficient SCID mice.[37,38] These mice show T lymphopoiesis for profound periods of time, and the implanted thymus is permissive for infection with HIV.[39,40] Upon intrathymic HIV injection, infected cells and increased programmed cell death could be seen throughout the thymus, leading to a loss of CD4+CD8+ double-positive thymocytes, a decrease of the CD4+:CD8+ thymocyte ratio, and an overall reduction of thymocyte cellularity.[3] Importantly, HIV infection in the SCID-hu mouse may give rise to more severe effects on the thymus than during natural HIV infection, because HIV in this model system is typically injected directly into the thymus. Evidence for HIV-induced thymocyte loss is, however, not restricted to the SCID-hu mouse model, as it was also observed in some fetuses aborted from HIV-infected mothers,[41] in thymus biopsies from HIV-infected children,[42,43] in thymic tissue studied at autopsy from individuals who died of HIV infection,[44] and in SIV-infected rhesus monkeys.[45] It should be noted, however, that so far, mostly syncytium-inducing (SI) HIV isolates have been shown to interfere with thymopoiesis. When SCID-hu mice were infected with non-SI (NSI) HIV variants, changes in thymus composition occurred much more slowly,[46,47] suggesting that loss of thymic output may only play a role in peripheral CD4+ T cell loss relatively late during disease progression. In a rhesus monkey model of HIV infection, loss of thymic output could be observed only at late stages of disease, just before the animals developed AIDS-like symptoms.[48] However, because of a lack of direct tools to measure ongoing thymic function *in vivo*, it remains to be determined how abrogation of thymic function affects peripheral blood T cell numbers.

To study more directly whether HIV infection impairs thymic output in humans, an *ex vivo* marker of recent thymic emigrants would be helpful. Measuring the number of T cells with a naive phenotype does not suffice to measure thymic output, because naive T cells are relatively long-lived and may retain their naive phenotype upon homeostatic proliferation. Recently, several laboratories have tried to quantify thymic output more directly by measuring T cell receptor excision circles (TRECs), the DNA by-products of T cell receptor gene rearrangements.[49] As TRECs are extrachromosomal DNA circles, they are not copied during T cell division and are, therefore, only produced upon *de novo* T cell generation in the thymus. The fact that the average TREC content (i.e., the average number of TRECs per T cell) of CD4+ and CD8+ T cells decreases with age was taken to reflect the age-related involution of the thymus and thereby was taken as evidence that T cell TREC contents can be used as a reliable measure of thymic output.[5,50] We showed, however, that TREC data need to be interpreted with great caution, because not only thymic output but also T cell death and proliferation influence the average TREC content of a T cell population.[51,52]

In HIV-infected individuals, the average TREC content of the naive CD4+ and CD8+ T cell population is typically decreased as compared to healthy controls. Taking the TREC content of the naive T cell population as a direct measure of thymic output, it has been argued that thymic output is severely reduced, even during the asymptomatic phase of HIV infection. Using a mathematical model, we showed, however, that decreased TREC contents in HIV-infected individuals are much more likely due to increased T cell proliferation than to thymic impairment. In fact, because of the longevity of naive T cells, a decrease in thymic output hardly affects the TREC content of the naive T cell population.[51] The TREC contents of peripheral T cells from thymectomized individuals indeed start to decrease only years after thymectomy.[53] Because there is ample evidence for increased T cell proliferation during HIV infection, decreased TREC contents in HIV-infected individuals should thus not be taken as evidence for decreased thymic output.

It remains to be determined to what extent infection of the thymus contributes to HIV disease progression. In fact, even if HIV infection of the thymus were to decrease the effective thymic output, it is not known to what extent that would affect the peripheral T cell population. In a small group of HIV-infected individuals who were thymectomized for myasthenia gravis before HIV infection, thymectomy did not preclude long-term survival. Increases in CD4+ T cell numbers when those thymectomized individuals were treated with HAART turned out to be similar to those in nonthymectomized HIV-infected individuals on HAART.[44]

Thymic output in adults was estimated to be low, on the order of 10^7 to 10^8 naive T cells per day.[29] Because of this, naive T cell reconstitution after T cell depletion of various causes, such as bone marrow/stem cell transplantation, monoclonal antibody treatment,[54–56] and HIV infection,[25] is very slow, and thymectomy in HIV-negative individuals does not significantly affect peripheral blood T cell numbers.[53] It can thus be expected that in adults, HIV-induced abrogation of thymic function will not significantly change peripheral blood T cell numbers. However, in children, in whom thymic function is optimal, HIV infection of the thymus may contribute significantly to HIV-induced naive T cell loss but only if thymic function plays a major role in daily T cell homeostasis. To study this, we followed changes in total body T cell numbers and TREC numbers with age in a group of healthy children of different ages. We studied total body TREC numbers instead of TREC contents, because total TRECs are not influenced by T cell proliferation and are thus a more direct measure of thymic output than TREC contents.[52] If the major source of T cells in young children were the thymus, one would expect the dynamics of T cell and TREC numbers to be similar. Instead, we found large discrepancies between the two. Whereas total body naive T cell numbers increased during the first years of life, total TREC numbers remained stable during this period, suggesting that the increase in naive T cell numbers was largely due to peripheral T cell proliferation.[57] Thus, even in children, thymic output seems to be less critical in maintaining or increasing peripheral blood naive T cell numbers, and HIV-related impairment of thymic function may, therefore, not have a significant impact on peripheral blood T cell numbers. Even complete abrogation of thymic output in juvenile macaques that were thymectomized was shown to have little impact on the peripheral T cell compartment, both in healthy and in SIV-infected macaques.[58,59]

Together, these data suggest that inappropriate apoptosis of maturing T cells in the HIV-infected thymus may hamper normal thymic function but that its effect on peripheral blood T cell numbers may not be significant, both in children and in adults.

APPROPRIATE T CELL DEATH: CHRONIC IMMUNE ACTIVATION IN HIV PATHOGENESIS

Because the moderate increase in CD4[+] and CD8[+] T cell turnover in HIV infection is not likely to be a homeostatic response to compensate for CD4[+] T cell loss (earlier referred to as the "pull" hypothesis), the alternative "push" hypothesis has been put forward. As already mentioned, we and others have proposed that the increased levels of T cell proliferation and the preferential loss of naive T cells during HIV-1 infection are caused by chronic immune activation in response to the persistently active virus, and hypothesized that CD4[+] T cell loss may be an "appropriate" result of this chronic immune stimulation.[6,7,30,60]

One of the strongest arguments for the idea that HIV induces increased T cell proliferation is that shortly after initiation of HAART, when viral reproduction is already prevented but CD4[+] T cell numbers are still low, T cell proliferation decreases dramatically.[30] Moreover, increases in the expression of the proliferation marker Ki67 during HIV-1 infection are not confined to the CD4[+] T cell pool and are, in fact, even more profound in the CD8[+] T cell pool, which becomes depleted only in the final stages of HIV infection.[26,27,29,30] In line with this observation, HIV-1-induced telomere shortening,[22] T cell apoptosis,[17,20,21,61] and TREC dilution[51] have been found to be most pronounced in the CD8[+] T cell population.

If chronic immune activation is a key player in HIV-1 pathogenesis, one would expect the level of immune activation post-seroconversion to be associated with disease progression. High proportions of proliferating Ki67[+]CD4[+] and Ki67[+]CD8[+] T cells post-seroconversion and high levels of the activation markers CD70, CD38, and HLA-DR expressed by CD4[+] and CD8[+] T cells were found to be associated with rapid disease progression.[62–64] In fact, the level of immune activation turned out to be an even better predictor for disease progression than HIV viral load.[63,64] Interestingly, CD8[+] T cell activation was found to correlate positively with HIV viral load,[63,65,66] which is strongly associated with the rate of disease progression. Because CD8[+] T cells play an important role in controlling viral replication, in principle, such a positive correlation would not be expected unless chronic T cell activation is the key problem in HIV infection.

Interestingly, we found that increased levels of immune activation before seroconversion are also predictive for disease progression after an individual becomes infected with HIV. In a study of homosexual men, we showed that low CD4[+] T cell counts and high CD70 expression by CD4[+] T cells before seroconversion were associated with fast disease progression upon HIV infection.[64] Similarly, HIV-1-infected injecting drug users with low CD4[+] TREC contents pre-seroconversion showed a steeper decline in CD4[+] T cell count throughout infection than those with high pre-seroconversion CD4[+] TREC contents.[67] Thus, the level of immune activation before HIV-1 infection seems to affect disease progression.

Also, during HAART, the level of immune activation was shown to be correlated with changes in CD4[+] T cell counts.[66,68] HAART causes a rapid decrease in plasma virus load, thereby alleviating chronic immune activation, leading to an immediate decrease in T cell proliferation and apoptosis levels. High residual levels of immune activation during HAART turned out to be associated with poor immune reconstitution, again underscoring the central role of immune activation in HIV infection.[66]

Based on these data, the alternative paradigm emerged that hyperactivation of the immune system is the key factor in HIV pathogenesis and that increased levels of T cell apoptosis are a natural result of this hyperactivated state of the immune system.[7,60] High levels of apoptosis were not only described for HIV infection but also, for example, during the acute phase of immune response in mice infected with acute lymphocytic choriomeningitis virus (LCMV)[69] and humans infected with Epstein–Barr virus[70,71] or cytomegalovirus.[72] Meanwhile, many additional data have been found to support this view.

In Vivo Labeling Studies

Recent T cell labeling studies using BrdU, deuterated glucose, and deuterated water have confirmed that T cell turnover is increased in both the CD4+ and CD8+ T cell population of HIV-infected individuals.[28,73–76] Upon half-an-hour *in vivo* pulse-labeling of T cells from HIV-infected individuals with BrdU, Kovacs et al. identified at least two distinct T cell subpopulations, one characterized by a slow and the other by a faster rate of T cell turnover.[75] HIV infection turned out to increase the average turnover of CD4+ and CD8+ T lymphocytes by activating resting cells and driving these cells into the rapidly proliferating subpopulation. Interestingly, however, HIV did not affect the turnover of either of the two subpopulations, and HAART decreased the proportion of labeled cells in the rapidly proliferating pool but did not affect the decay rate of labeled cells.[75]

Ribeiro et al. reported that after a 7-day infusion of deuterated glucose, an increased transition of CD4+ T cells from the resting to the activated compartment could be measured in HIV-infected individuals compared to healthy controls.[76] In contrast, Ribeiro et al. found an HIV-related increase in proliferation and death rates of activated CD4+ T cells.[75] This discrepancy between the two studies may be due to the different labeling techniques. Whereas cessation of label administration leads to an immediate decay of deuterium-labeled DNA, BrdU-labeled cells show much slower decay kinetics. Because BrdU measures the proportion of positive cells rather than label-positive DNA strands (as in the case of deuterium studies), cell division in the absence of label infusion can actually increase the amount of BrdU+ cells, because the label is divided equally over the daughter cells, rendering them BrdU+. Thus, the net decay rate of BrdU+ cells in the presence of continuous cell division will be lower than the actual death rate.[77]

A general problem of *in vivo* labeling techniques such as those described here is that they are biased toward measuring the dynamics of the T cell population with the highest turnover. Whereas the cells that pick up the label are truly representative of the whole T cell population, including rapidly and slowly dividing T cell populations, measuring the rate at which the label is lost upon label retraction is biased because of the dominance of the specific population of T cells with rapid turnover.[60,78] It is, therefore, not surprising that such division-dependent labeling techniques consistently predicted higher T cell death rates than proliferation rates. The longer the label is administered, the more closely the estimated T cell death rate will get to the true average death rate of the whole T cell population.[78] In this respect, the recent introduction of deuterated water labeling is a big step ahead, as it provides a nontoxic alternative for BrdU labeling and can more easily be administered for prolonged periods of time compared to deuterated glucose.

Despite the above-mentioned shortcomings of these labeling techniques, these studies, which are the first to show *in vivo* T cell kinetics, confirmed the hypothesis that T cells that recently divided and, hence, picked up the label tend to have a relatively short life span.[7,60] The fact that recently divided cells in HIV-infected patients are prone to die argues against a homeostatic response to CD4+ T cell depletion and is consistent with the idea that T cell division and death are driven by immune activation.

HIV-Infected Individuals with Discordant Responses to Antiviral Therapy

Another argument in favor of the idea that immune activation is the key player in HIV infection comes from HIV-infected individuals who show discordant responses to antiviral therapy. Most patients treated with antiretroviral therapy show a significant decline in viral load concomitant with an increase in peripheral blood T cell numbers. However, some patients are only virologically responsive to antiviral therapy but do not show immunologic improvement ("virologic responders"), whereas others are responsive only in immunologic terms, showing increases in peripheral blood T cell numbers despite detectable viremia ("immunologic responders"). Immunologic responders have lower levels of T cell activation, proliferation, and apoptosis compared to untreated HIV-infected individuals, despite the presence of detectable virus, and show sustained CD4[+] T cell gains.[68,79,80] Thus, antiviral therapy can have beneficial effects in terms of CD4[+] T cell counts, even in the presence of detectable plasma viremia, most likely because it reduces the level of immune activation. On the other hand, in patients with poor immunologic outcome despite good virologic responses, high *ex vivo* rates of apoptosis have been found.[81] In general, immunologic discordant responders have a favorable clinical outcome compared to virologic discordant responders,[82,83] suggesting that controlling the level of immune activation is more important than controlling HIV viral load.

HIV-2

Differences in the natural course of HIV-1 and HIV-2 infection have also been taken as evidence for a central role of immune activation in HIV infection. In HIV-2-infected patients, the asymptomatic phase takes much longer compared to HIV-1 infection, because CD4[+] T cells decline much more slowly. HIV-2-infected patients have lower levels of T cell activation, proliferation, and apoptosis and a lower plasma viral load than HIV-1-infected individuals. When HIV-1- and HIV-2-infected individuals were stratified on the basis of CD4[+] T cell counts, the levels of T cell activation and proliferation were shown to be similar despite large differences in plasma viral load.[65,84] Sousa et al. interpreted this close association between immune activation and the CD4[+] T cell count as evidence that the level of immune activation, and not viral load, determines disease outcome.[65] One could, however, also reason the other way around, namely, that the *rate* at which CD4[+] T cells are lost, which is known to be higher in HIV-1 compared to HIV-2 infection, should apparently be due to differences in HIV viral load, because it is the only apparent difference between HIV-1- and HIV-2-infected individuals with similar CD4[+] T cell counts.

SIV

If in HIV infection high levels of immune activation are associated with a bad prognosis, one would also expect the reverse, namely, that low levels of immune activation is associated with a good prognosis. The latter situation seems to be the case in some SIV-infected primates. In sooty mangabeys, one of the natural hosts of SIV, CD4[+] T cells are typically not depleted, despite high viral loads. Both immune activation and T cell apoptosis were shown to be unaffected in these animals when compared to noninfected sooty mangabeys. Sooty mangabeys may have been selected in the presence of SIV to give only a limited immune response against the virus. Transmission of SIV to other primate species that are not natural hosts to SIV, such as macaques, is known to induce CD4[+] T cell decline and progression to AIDS. SIV infection of rhesus macaques is characterized by increased levels of immune activation.[13] *In vivo* labeling of dividing T cells using BrdU confirmed these findings by showing that in SIV-infected macaques, the turnover of both CD4[+] and CD8[+] T cells was increased compared with uninfected macaques.[77] These data support the view that low levels of T cell activation are associated with a good prognosis, regardless of the presence of high levels of virus load, again underscoring the central role of chronic immune activation in HIV and SIV pathogenesis.

Collectively, these data changed our viewpoint on HIV infection drastically. Where once it was thought that "inappropriate apoptosis" was the main problem in HIV infection and increased T cell proliferation was a response to cope with this loss of cells, much research is now based on the opposite idea that chronic T cell activation is driving T cell loss through a natural process of "appropriate apoptosis" of activated T cell clones.

IS CHRONIC IMMUNE ACTIVATION ALONE ENOUGH
TO INDUCE CD4+ T CELL LOSS?

To study if chronic immune activation could induce CD4+ T cell loss directly, Tesselaar et al. studied T cell dynamics in other situations of chronic immune stimulation independent of HIV infection.[85] A transgenic mouse model was developed in which CD70 is constitutively expressed on the cell surface of B cells. This leads to continuous stimulation of the T lymphocyte pool via interaction with CD27 on the T cell surface. Remarkably, these mice developed a phenotype reminiscent of HIV infection, in that their peripheral T cell pool was progressively depleted and the mice eventually died of opportunistic infections.[85]

A similar HIV-independent disturbance of the T cell pool reminiscent of HIV infection was observed in a large group of healthy Ethiopians analyzed in the Ethio-Netherlands AIDS Research Project (ENARP). HIV-negative Ethiopians were found to have highly increased levels of immune activation compared to HIV-negative Dutch individuals, probably related to differences in domestic antigen exposure and the high antigenic burden by common parasitic infections in Ethiopia.[86,87] In fact, these healthy individuals have low CD4+ T cell counts, low percentages of CD4+ and CD8+ naive T cells, expansion of CD8+ memory and effector T cell subsets, and decreased TREC contents, comparable to HIV-infected Dutch individuals.[51,86,87] Changes in the peripheral T cell pool of healthy Ethiopians appeared to occur at very early ages. Whereas cord blood samples from healthy Ethiopians and Dutch individuals were virtually identical with respect to TREC contents and T cell subsets, within a few years of childhood, considerable differences were observed (Tsegaye et al., submitted).[88] Thus, also in this situation, chronic antigen exposure and chronically increased immune activation were associated with progressive CD4+ T cell decline.

The fact that a considerable T cell loss can be observed in situations of chronic immune activation independent of HIV infection suggests that CD4+ T cell loss in HIV infection may be a direct result of chronic stimulation induced by HIV. Of note, even in healthy individuals, the total number of T cells, including the number of naive CD4+ and CD8+ T cells, declines with increasing age.[89,90]

WHY ARE ONLY CD4+ AND NOT CD8+ T CELLS
LOST IN HIV INFECTION?

Although many of the data discussed in this chapter underline the importance of chronic immune activation in HIV pathogenesis, one question remains to be answered: If chronic immune activation causes CD4+ T cells to decline, and if both CD4+ and CD8+ T cell turnover are increased during HIV infection, why then is the CD8+ T cell pool not progressively depleted? As already mentioned, increases in T cell proliferation and apoptosis, telomere shortening, and TREC dilution in HIV-infected individuals are even more pronounced in CD8+ T cells than in CD4+ T cells.

To answer this question, it is important to note that although CD4+ T cell depletion is the main characteristic of HIV infection, other changes in T cell subsets typically occur during HIV infection. After an initial increase due to peripheral expansion, the CD4+ memory T cell population is progressively depleted. Because of a massive expansion of the memory CD8+ T cell pool in response to HIV infection, total numbers of CD8+ T cells are typically increased throughout the asymptomatic phase of infection. However, depletion of the naive T cell compartment is not restricted to CD4+ T cells, as naive CD8+ T cells begin to decline shortly after infection as well.[35,36,64] Eventually, shortly preceding AIDS diagnosis, even total CD8+ T cell numbers start to decline.

We argue that naive CD8+ T cell loss might be as important as naive CD4+ T cell loss and is driven by similar mechanisms. Thus, based on data outlined in this chapter, we hypothesized that HIV induces continuous activation of naive CD4+ and naive CD8+ T cells that are recruited to become memory and effector T cells and are thereby lost from the naive T cell pools. Because naive T cells are difficult to replace, continuous naive T cell recruitment will lead to naive T cell depletion.[7] In addition, intrinsic differences in cell survival kinetics between activated CD4[+] and CD8[+] T cells are likely to be the cause of differences between total CD4[+] and CD8[+] T cell numbers in the blood of HIV-infected individuals.

Differences in CD4[+] and CD8[+] T cell dynamics upon antigen-induced T cell activation are not specific for HIV infection. Given the different physiological tasks of CD4[+] T cells (giving help to induce other immune responses) and CD8[+] T cells (creating effector cells to actively kill infected cells), it may not be surprising that the CD8[+] T cell population is more prone to induce and maintain large clones of effector cells than the CD4[+] T-helper population. In mice infected with LCMV, LCMV-specific CD8[+] T cells expanded about 20-fold more than LCMV-specific CD4[+] T cells. Moreover, after the contraction phase of the immune response, when the antigen was cleared, the size of the antigen-specific CD8[+] T cell pool remained stable, whereas LCMV-specific CD4[+] T cells continued to decrease over time.[91] A similar decline in antigen-specific CD4[+] T cell memory cells was found in mice upon infection with *Listeria monocytogenes*.[92] Furthermore, Ahmed et al. recently showed that murine memory CD8[+] T cells are less sensitive to apoptosis than naive CD8[+] T cells.[93] In contrast, when primed T cell receptor transgenic CD4[+] T cells were rechallenged after transfer to normal mice, the transgenic cells initially expanded but progressively declined due to apoptosis after 2 days.[94]

These results indicate that the size of the memory CD4[+] T cell pool diminishes with time, whereas expanded CD8[+] T cell clones show a much longer survival, also during other infections. Interestingly, as indicated above, HIV-negative Ethiopians with chronically increased levels of immune activation also show a selective loss of naive CD4[+] T cells, naive CD8[+] T cells, and memory CD4[+] T cells but an expansion of memory/effector CD8[+] T cells, supporting the idea that the CD4[+] T cell population is generally more prone to be depleted upon chronic immune activation than the CD8[+] T cell population.

In HIV-infected individuals, labeling studies with deuterated glucose confirmed differences in T cell dynamics between CD4[+] and CD8[+] T cells. After a 7-day labeling period, Mohri et al. observed that CD4[+] and CD8[+] T cell proliferation were increased in HIV-1 infected humans as compared with uninfected individuals. Remarkably, however, T cell death rates were increased only in the CD4[+] T cell population, which led them to conclude that CD4[+] T cell depletion in HIV infection is caused by selectively increased CD4[+] T cell destruction. More recently, the same data were analyzed using a two-compartment model, describing a resting T cell pool and an activated pool. When the labeling dynamics of CD4[+] and CD8[+] T cells in HIV-infected individuals were compared to those of healthy controls, it was found that CD4[+] T cells, but not CD8[+] T cells, entered the activated compartment more quickly during HIV infection. CD4[+] T cells in the activated pool had increased proliferation and death rates, but HIV infection did not affect the size of the activated pool. In contrast, the activated CD8[+] T cell compartment became larger during HIV infection, due to diminished reversion of activated cells into the resting pool, whereas proliferation and death rates remained unaffected by HIV infection.[76] The fact that CD8[+] T cells in HIV infection have shorter telomeres and more severely diluted TRECs than CD4[+] T cells also suggests that activated CD4[+] T cells are more prone to die than activated CD8[+] T cells.

CONCLUSION: AN INTEGRATED VIEW

Taken together, HIV infection is associated with an increased level of T cell apoptosis, which, by itself, is not inappropriate, because it is part of the physiological response to immune activation. However, the continuously increased levels of apoptosis characteristic of HIV infection reflect the continuous presence of a very active virus, causing chronic activation of the immune system that, because lost T cells are difficult to replace, leads to depletion of the naive CD4[+] and naive CD8[+] T cell pools as well as the memory CD4[+] T cell pool (Figure 19.1). Thus, by wearing the immune system out, in case of HIV-1 infection, an appropriate immune response ultimately may lead to unsuitable immune deficiency.

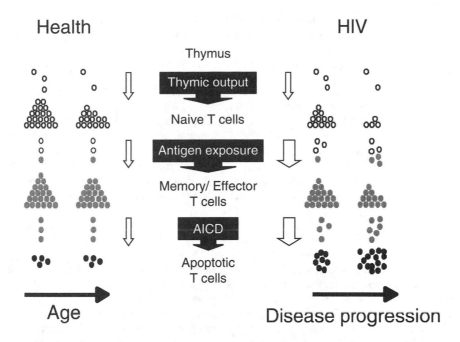

FIGURE 19.1 Increased naive T cell consumption due to immune hyperactivation. Thymic output of T cells to the naive compartment is low and constant in healthy as well as in HIV-infected adults. As one ages, upon encountering infections, naive cells will enter the memory/effector T cell pool, leading to a gradual decrease in the naive T cell compartment. Upon infection, activated and expanded cells will die of activation-induced cell death (AICD). During HIV infection, the same process occurs, only at a faster rate due to chronic immune activation by the virus. (Adapted from Hazenberg, M.D., et al., *Nat. Immunol.*, 1, 285–289, 2000.)

ACKNOWLEDGMENT

We would like to thank Rob de Boer for critical reading of this manuscript. This research has been financially supported by AIDS Fond Netherlands (grants 7010 and 7011) and The Netherlands Organization for Scientific Research (916.36.003).

REFERENCES

1. Ameisen, J.C. and Capron, A., Cell dysfunction and depletion in AIDS: the programmed cell death hypothesis, *Immunol. Today,* 12, 102, 1991.
2. Ho, D.D., Neumann, A.U., Perelson, A.S., Chen, W., Leonard, J.M., and Markowitz, M., Rapid turnover of plasma virions and CD4 lymphocytes in HIV-1 infection, *Nature,* 373, 123, 1995.
3. Bonyhadi, M.L., Rabin, L., Salimi, S., Brown, D.A., Kosek, J., McCune, J.M., and Kaneshima, H., HIV induces thymus depletion *in vivo, Nature,* 363, 728, 1993.
4. Su, L., Kaneshima, H., Bonyhadi, M., Salimi, S., Kraft, D., Rabin, L., and McCune, J.M., HIV induced thymocyte depletion is associated with indirect cytopathogenicity and infection of progenitor cells *in vivo, Immunity,* 2, 25, 1995.
5. Douek, D.C., McFarland, R.D., Keiser, P.H., Gage, E.A., Massey, J.M., Haynes, B.F., Polis, M.A., Haase, A.T., Feinberg, M.B., Sullivan, J.L., Jamieson, B.D., Zack, J.A., Picker, L.J., and Koup, R.A., Changes in thymic function with age and during the treatment of HIV infection, *Nature,* 396, 690, 1998.
6. Miedema, F., Meyaard, L., Koot, M., Klein, M.R., Roos, M.Th.L., Groenink, M., Fouchier, R.A.M., Van't Wout, A.B., Tersmette, M., Schellekens, P.Th.A., and Schuitemaker, H., Changing virus–host interactions in the course of HIV-1 infection, *Immunol. Rev.,* 140, 35, 1994.

7. Hazenberg, M.D., Hamann, D., Schuitemaker, H., and Miedema, F., T cell depletion in HIV-1 infection: how CD4⁺ T cells go out of stock, *Nat. Immunol.,* 1, 285, 2000.

8. Laurent-Crawford, A.G., Krust, B., Muller, S., Rivière, Y., Rey-Cuillé, M.A., Béchet, J.-M., Montagnier, L., and Hovanessian, A.G., The cytopathic effect of HIV is associated with apoptosis, *Virology,* 185, 829, 1991.

9. Terai, C., Kornbluth, R.S., Pauza, C.D., Richman, D.D., and Carson, D.A., Apoptosis as a mechanism of cell death in cultured T lymphoblasts acutely infected with HIV-1, *J. Clin. Invest.,* 87, 1710, 1991.

10. Piatak, M., Jr., Saag, M.S., Yang, L.C., Clark, S.J., Kappes, J.C., Luk, K.-C., Hahn, B.H., Shaw, G.M., and Lifson, J.D., High levels of HIV-1 in plasma during all stages of infection determined by competitive PCR, *Science,* 259, 1749, 1993.

11. Chun, T.-W., Carruth, L., Finzi, D., Shen, X., DiGiuseppe, J.A., Taylor, H., Hermankova, M., Chadwick, K., Margolick, J., Quinn, T.C., Kuo, Y.-H., Brookmeyer, R., Zeiger, M.A., Bardltch-Crovo, P., and Siliciano, R.F., Quantification of latent tissue reservoirs and total body viral load in HIV-1 infection, *Nature,* 387, 183, 1997.

12. Mosier, D.E., Gulizia, R.J., MacIsaac, P.D., Torbett, B.E., and Levy, J.A., Rapid loss of CD4+ T cells in human-PBL-SCID mice by noncytopathic HIV isolates, *Science,* 260, 689, 1993.

13. Silvestri, G., Sodora, D.L., Koup, R.A., Paiardini, M., O'Neil, S.P., McClure, H.M., Staprans, S.I., and Feinberg, M.B., Nonpathogenic SIV infection of sooty mangabeys is characterized by limited bystander immunopathology despite chronic high-level viremia, *Immunity,* 18, 441, 2003.

14. Finkel, T.H., Tudor-Williams, G., Banda, N.K., Cotton, M.F., Curiel, T., Monks, C., Baba, T.W., Ruprecht, R.M., and Kupfer, A., Apoptosis occurs predominantly in bystander cells and not in productively infected cells of HIV- and SIV-infected lymph nodes, *Nat. Med.,* 1, 129, 1995.

15. Newell, M.K., Haughn, L.J., Maroun, C.R., and Julius, M.H., Death of mature T cells by separate ligation of CD4 and the T-cell receptor for antigen, *Nature,* 347, 286, 1990.

16. Banda, N.K., Bernier, J., Kurahara, D.K, Kurrle, R., Haigwood, N., Sekaly, R.-p., and Helman Finkel, T., Crosslinking CD4 by human immunodeficiency virus gp120 primes T cells for activation-induced apoptosis, *J. Exp. Med.,* 176, 1099, 1992.

17. Oyaizu, N., McCloskey, T.W., Coronesi, M., Chirmule, N., Kalyanaraman, V.S., and Pahwa, S., Accelerated apoptosis in peripheral blood mononuclear cells (PBMCs) from human immunodeficiency virus type-1 infected patients and in CD4 cross-linked PBMCs from normal individuals, *Blood,* 82, 3392, 1993.

18. Badley, A.D., Pilon, A.A., Landay, A., and Lynch, D.H., Mechanisms of HIV-associated lymphocyte apoptosis, *Blood,* 96, 2951, 2000.

19. Meyaard, L., Otto, S.A., Jonker, R.R., Mijnster, M.J., Keet, R.P.M. and Miedema, F., Programmed death of T cells in HIV-1 infection, *Science,* 257, 217, 1992.

20. Groux, H., Torpier, G., Monté, D., Mouton, Y., Capron, A., and Ameisen, J.C., Activation-induced death by apoptosis in CD4+ T cells from human immunodeficiency virus-infected asymptomatic individuals, *J. Exp. Med.,* 175, 331, 1992.

21. Gougeon, M., Garcia, S., Heeney, J., Tschopp, R., Lecoeur, H., Guetard, D., Rame, V., Dauguet, R., and Montagnier, L., Programmed cell death in AIDS-related HIV and SIV infections, *AIDS Res. Hum. Retroviruses,* 9, 553, 1993.

22. Wolthers, K.C., Wisman, G.B.A., Otto, S.A., De Roda Husman, A.M., Schaft, N., De Wolf, F., Goudsmit, J., Coutinho, R.A., Van der Zee, A.G.J., Meyaard, L., and Miedema, F., T-cell telomere length in HIV-1 infection: no evidence for increased CD4⁺ T cell turnover, *Science,* 274, 1543, 1996.

23. Pakker, N.G., Notermans, D.W., De Boer, R.J., Roos, M.T.L., Wolf, F., Hill, A., Leonard, J.M., Danner, S.A., Miedema, F., and Schellekens, P.T.A., Biphasic kinetics of peripheral blood T cells after triple combination therapy in HIV-1 infection: a composite of redistribution and proliferation, *Nat. Med.,* 4, 208, 1998.

24. Notermans, D.W., Pakker, N.G., Hamann, D., Foudraine, N.A., Kauffmann, R.H., Meenhorst, P.L., Goudsmit, J., Ross, M.Th.L., Schellekens, P.T.A., Miedema, F., and Danner, S.A., Immune reconstitution after 2 years of successful potent antiretroviral therapy in previously untreated human immunodeficiency virus type 1 infected adults, *J. Infect. Dis.,* 180, 1050, 1999.

25. Hunt, P.W., Deeks, S.G., Rodriguez, B., Valdez, H., Shade, S., Abrams, D., Krone, M., Neilands, T., Lederman, M., and Martin, J.N., Continued CD4⁺ T cell count increases in HIV-infected adults experiencing 4 years of viral suppression on antiretroviral therapy, *AIDS,* 17, 1907, 2003.

26. Fleury, S., De Boer, R.J., Rizzardi, G.P., Wolthers, K.C., Otto, S.A., Welbon, C.C., Graziosi, C., Knabenhans, C., Soudeyns, H., Bart, P.-A., Gallant, S., Corpataux, J.-M. Gillet, M., Meylan, P., Schnyder, P., Meuwly, J.Y., Spreen, W., Glauser, M.P., Miedema, F., and Pantaleo, G., Limited CD4+ T-cell renewal in early HIV-1 infection: effect of highly active antiretroviral therapy, *Nat. Med.,* 4, 794, 1998.

27. Sachsenberg, N., Perelson, A.S., Yerly, S., Schokmel, G.A., Leduc, D., Hirschel, B., and Perrin, L., Turnover of CD4+ and CD8+ T lymphocytes in HIV-1 infection as measured by Ki-67 antigen, *J. Exp. Med.,* 187, 1295, 1998.

28. Hellerstein, M., Hanley, M.B., Cesar, D., Siler, S., Papageorgopoulos, C., Wieder, E., Schmidt, D., Hoh, R., Neese, R., Macallan, D., Deeks, S., and McCune, J.M., Directly measured kinetics of circulating T lymphocytes in normal and HIV-1 infected humans, *Nat. Med.,* 5, 83, 1999.

29. Clark, D.R., De Boer, R.J., Wolthers, K.C., and Miedema, F., T cell dynamics in HIV-1 infection, *Adv. Immunol.,* 73, 301, 1999.

30. Hazenberg, M.D., Cohen Stuart, J.W.T, Otto, S.A., Borleffs, J.C.C., Boucher, C.A., De Boer, R.J., Miedema, F., and Hamann, D., T cell division in human immunodeficiency virus (HIV-1)-infection is mainly due to immune activation: a longitudinal analysis in patients before and during highly active anti-retroviral therapy, *Blood,* 95, 249, 2000.

31. Wei, X., Ghosh, S.K., Taylor, M.E., Johnson, V.A., Emini, E.A., Deutsch, P., Lifson, J.D., Bonhoeffer, S., Nowak, M.A., Hahn, B.H., Saag, M.S., and Shaw, G.M., Viral dynamics in human immunodeficiency virus type 1 infection, *Nature,* 373, 117, 1995.

32. Hazenberg, M.D., Otto, S.A., De Pauw, E.S., Roelofs, H., Fibbe, W.E., Hamann, D., and Miedema, F., T cell receptor excision circle (TREC) and T cell dynamics after allogeneic stem cell transplantation are related to clinical events, *Blood,* 99, 3449, 2002.

33. Wolthers, K.C., Schuitemaker, H., and Miedema, F., Rapid CD4+ T-cell turnover in HIV-1 infection: a paradigm revisited, *Immunol. Today,* 19, 44, 1998.

34. McCune, J.M., The dynamics of CD4+ T-cell depletion in HIV disease, *Nature,* 410, 974, 2001.

35. Rabin, R.L., Roederer, M., Maldonado, Y., Petru, A., and Herzenberg, L.A., Altered representation of naive and memory CD8 T cell subsets in HIV-infected children, *J. Clin. Invest.,* 95, 2054, 1995.

36. Roederer, M., Gregson Dubs, J., Anderson, M.T., Raju, P.A., Herzenberg, L.A., and Herzenberg, L., CD8 naive T cell counts decrease progressively in HIV-infected adults, *J. Clin. Invest.,* 95, 2061, 1995.

37. McCune, J.M., Namikawa, R., Kaneshima, H., Shultz, L.D., Lieberman M., and Weissman, I.L., The SCID-hu mouse: murine model for the analysis of human hematolymphoid differentiation and function, *Science,* 241, 1632, 1988.

38. McCune, J., Kaneshima, H., Krowka, J., Namikawa, R., Outzen, H., Peault, B., Rabin, L., Shih, C.C., Yee, E., Lieberman, M., The SCID-hu mouse: a small animal model for HIV infection and pathogenesis, *Annu. Rev. Immunol.,* 9, 399, 1991.

39. Namikawa, R., Kaneshima, H., Lieberman, M., Weissman, I.L., and McCune, J.M., Infection of the SCID-hu mouse by HIV-1, *Science,* 242, 1684, 1988.

40. Namikawa, R., Weilbaecher, K.N., Kaneshima, H., Yee, E.J., and McCune, J.M., Long-term human hematopoiesis in the SCID-hu mouse, *J. Exp. Med.,* 172, 1055, 1990.

41. Papiernik, M., Brossard, Y., Mulliez, N., Roume, J., Brechot, C., Barin, F., Goudeau, A., Bach, J.F., Griscelli. C., and Henrion, R., Thymic abnormalities in fetuses aborted from human immunodeficiency virus type 1 seropositive women, *Pediatrics,* 89, 297, 1992.

42. Joshi, V.V., Oleska, J.M., Saad, S., Gadol, C., Connor, E., Bobila, R., and Minnefor, A.B., Thymus biopsy in children with acquired immunodeficiency syndrome, *Arch. Path. Lab. Med.,* 110, 837, 1986.

43. Rosenzweig, M., Clark, D.P., and Gaulton, G.N., Selective thymocyte depletion in neonatal HIV-1 infection, *AIDS,* 7, 1601, 1993.

44. Haynes, B.F., Hale, L.P., Weinhold, K.J., Patel, D.D., Liao, H.-X., Bressler, P.B., Jones, D.M., Demarest, J.F., Gebhard-Mitchell, K., Haase, A.T., and Bartlett, J.A., Analysis of the adult thymus in reconstitution of T lymphocytes in HIV-1 infection, *J. Clin. Invest.,* 103, 453, 1999.

45. Baskin, G.B., Murphey-Corb, M., Martin, L.N., Davision-Fairburn, B., Hu, F.S., and Kuebler, D., Thymus in simian immunodeficiency virus-infected rhesus monkeys, *Lab. Invest.,* 65, 400, 1991.

46. Kaneshima, H., SU, L., Bonyhadi, M.L., Connor, R.I., Ho, D.D., and McCune, J.M., Rapid-high, syncytium-inducing isolates of human immunodeficiency virus type 1 induce cytopathicity in the human thymus of SCID-hu mouse, *J. Virol.,* 68, 8188, 1994.

47. Berkowitz, R.D., Alexander, S., Bare, C., Linquist-Stepps, V., Bogan, M., Moreno, M.E., Gibson, L., Wieder, E.D., Kosek, J., Stoddard, C.L., and McCune, J.M., CCR5- and CXCR4-utilizing strains of human immunodeficiency virus type 1 exhibit differential tropism and pathogenesis *in vivo, J. Virol.,* 72, 10108, 1998.

48. Sopper, S., Nierwetberg, D., Halbach, A., Sauer, U., Scheller, C., Stahl-Henning, C., Matz-Rensing, K., Schafer, F., Schneider, T., ter, M., V, and Muller, J.G., Impact of simian immunodeficiency virus (SIV) infection on lymphocyte numbers and T-cell turnover in different organs of rhesus monkeys, *Blood,* 101, 1213, 2003.

49. Kong, F.-K., Chen, C.-L.H., Six, A., Hockett, R.D., and Cooper, M.D., T cell receptor gene deletion circles identify recent thymic emigrants in the peripheral T cell pool, *Proc. Natl. Acad. Sci. U.S.A.,* 96, 1536, 1999.

50. Zhang, L., Lewin, S.R., Markowitz, M., Lin, H.-H., Skulsky, E., Karanicolas, R., He, Y., Jin, X., Tuttleton, S., Vesanen, M., Spiegel, H., Kost, R., Van Lunzen, J., Stellbrink, H.-J., Wolinsky, S., Borkowsky, W., Palumbo, P., Kostrikis, L.G., and Ho, D.D., Measuring recent thymic emigrants in blood of normal and HIV-1-infected individuals before and after effective therapy, *J. Exp. Med.,* 190, 725, 1999.

51. Hazenberg, M.D., Otto, S.A., Cohen Stuart, J.W.T., Verschuren, M.C.M., Borleffs, J.C.C., Boucher, C.A.B., Coutinho, R.A., Lange, J.M.A., Rinke de Wit, T.F., Tsegaye, A., Van Dongen, J.J.M., Hamann, D., De Boer, R.J., and Miedema, F., Increased cell division but not thymic dysfunction rapidly affects the TREC content of the naive T cell population in HIV-1 infection, *Nat. Med.,* 6, 1036, 2000.

52. Hazenberg, M.D., Verschuren, M.C.M., Hamann, D., Miedema, F., and Van Dongen, J.J.M., T cell receptor excision circles as markers for recent thymic emigrants: basic aspects, technical approach, and guidelines for interpretation, *J. Mol. Med.,* 79, 631, 2001.

53. Sempowski, G.D., Thomasch, J.R., Gooding, M.E., Hale, L.P., Edwards, L.J., Ciafaloni, E., Sanders, D.B., Massey, J.M., Douek, D.C., Koup, R.A., and Haynes, B.F., Effect of thymectomy on human peripheral blood T cell pools in myasthenia gravis, *J. Immunol.,* 166, 2808, 2001.

54. Fujimaki, K., Maruta, A., Yoshida, M., Kodama, F., Matsuzaki, M., Fujisawa, S., Kanamori, H., and Ishigatsubo, Y., Immune reconstitution assessed during five years after allogeneic bone marrow transplantation, *Bone Marrow Transplant.,* 27, 1275, 2001.

55. Rep, M., Van Oosten, B.W., Ross, M.T.L., Adèr, H.J., Polman, C.H., and Van Lier, R., Treatment with depleting CD4 monoclonal antibody results in a preferential loss of circulating naive T cells but does not affect IFN-γ secreting TH1 cells in humans, *J. Clin. Invest.,* 99, 2225, 1997.

56. Mackall, C.L., Fleisher, T.A., Brown, M., Andrich, M.P., Chen, C., Feuerstein, I.M., Horowitz, M.E., Magrath, I.T., Shad, A.T., Steinberg, S.M., Wexler, L.H., and Gress, R.E., Age, thymopoiesis, and CD4+ T-lymphocyte regeneration after intensive chemotherapy, *N. Engl. J. Med.,* 332, 143, 1995.

57. Hazenberg, M.D., Otto, S.A., Van Rossum, A.M., Scherpbier, H.J., De Groot, R., Kuijpers, T.W., Lange, J.M., Hamann, D., De Boer, R.J., Borghans, J.A., and Miedema, F., Establishment of the CD4+ T cell pool in healthy and untreated HIV-1 infected children, *Blood,* 104, 3513, 2004.

58. Muthukumar, A., Wozniakowski, A., Matthews, C., Douek, D.C., Johnson, R.P., McClure, H.M., Koup, R.A., and Sodora, D.L., Impact of Thymectomy on SIV Infection in Macaques, presented at XIV International AIDS Conference, Barcelona, Spain, 2004.

59. Arron, S.T., Ribeiro, R.M., Gettie, A., Bohm, R., Blanchard, J., Yu, J., Perelson, A.S., Ho, D.D., and Zhang, L., Impact of Thymectomy on the Peripheral T-Cell Pool in Rhesus Macaques before and after Infection with SIV, presented at Ninth Conference on Retroviruses and Opportunistic Infections, abstract 101, 2002.

60. Grossman, Z., Meier-Schellersheim, M., Sousa, A.E., Victorino, R.M.M., and Paul, W.E., CD4 T-cell depletion in HIV infection: are we closer to understanding the cause? *Nat. Med.,* 8, 319, 2002.

61. Meyaard, L., Otto, S.A., Jonker, R.R., Mijnster, M.J., Keet, R.P.M., and Miedema, F., Programmed death of T cells in HIV-1 infection, *Science,* 257, 217, 1992.

62. Simmonds, P., Beatson, D., Cuthbert, R.J.G., Watson, H., Reynolds, B., Peutherer, J.F., Parry, J.V., Ludlam, C.A., and Steel, C.M., Determinants of HIV disease progression: six-year longitudinal study in the Edinburgh haemophilia/HIV cohort, *Lancet,* 338, 1159, 1991.

63. Giorgi, J.V., Hultin, L.E., McKeating, J.A., Johnson, T.D., Owens, B., Jacobson, L.P., Shih, R., Lewis, J., Wiley, D.J., Phair, J.P., Wolinsky, S.M., and Detels, R., Shorter survival in advanced human immunodeficiency virus type 1 infection is more closely associated with T lymphocyte activation than with plasma virus burden or virus chemokine coreceptor usage, *J. Infect. Dis.,* 179, 859, 1999.

64. Hazenberg, M.D., Otto, S.A., van Benthem, B.H., Roos, M.T., Coutinho, R.A., Lange, J.M., Hamann, D., Prins, M., and Miedema, F., Persistent immune activation in HIV-1 infection is associated with progression to AIDS, *AIDS,* 17, 1881, 2003.

65. Sousa, A.E., Carneiro, J., Meier-Schellersheim, M., Grossman, Z., and Victorino, R.M., CD4 T cell depletion is linked directly to immune activation in the pathogenesis of HIV-1 and HIV-2 but only indirectly to the viral load, *J. Immunol.,* 169, 3400, 2002.

66. Hunt, P.W., Martin, J.N., Sinclair, E., Bredt, B., Hagos, E., Lampiris, H., and Deeks, S.G., T cell activation is associated with lower CD4+ T cell gains in human immunodeficiency virus-infected patients with sustained viral suppression during antiretroviral therapy, *J. Infect. Dis.,* 187, 1534, 2003.

67. Van Asten, L., Danisman, F., Otto, S.A., Borghans, J.A., Hazenberg, M.D., Coutinho, R.A., Prins, M., and Miedema, F., Pre-seroconversion immune status predicts the rate of CD4 T cell decline following HIV infection, *AIDS,* 18, 1885, 2004.

68. Hazenberg, M.D., Otto, S.A., Wit, F.W.N.M., Lange, J.M.A., Hamann, D., and Miedema, F., Discordant responses during antiretroviral therapy: role of immune activation and T cell redistribution rather than true CD4 T cell loss, *AIDS,* 16, 1287, 2002.

69. Razvi, E.S. and Welsh, R.M., Programmed cell death of T lymphocytes during acute viral infection: a mechanism for virus-induced immune deficiency, *J. Virol.,* 67, 5754, 1993.

70. Moss, D.J., Bishop, C.J., Burrows, S.R., and Ryan, J.M., T lymphocytes in infectious mononucleosis. I. T cell death *in vitro, Clin. Exp. Immunol.,* 60, 61, 1985.

71. Uehara, T., Miyawaki, T., Ohta, K., Tamaru, Y., Yokoi, T., Nakamura, S., and Taniguchi, N., Apoptotic cell death of primed CD45RO+ T lymphocytes in Epstein–Barr virus induced infectious mononucleosis, *Blood,* 80, 452, 1992.

72. Van den Berg, A.P., Meyaard, L., Otto, S.A., Van Son W.J., Klompmaker, I.J., Mesander, G., De Leij, L.H.F.M., Miedema, F., and The, T.H., Cytomegalovirus infection associated with a decreased proliferative capacity and increased rate of apoptosis of peripheral blood lymphocytes, *Transplant. Proc.,* 27, 936, 1995.

73. Hellerstein, M.K., Hoh, R.A., Hanley, M.B., Cesar, D., Lee, D., Neese, R.A., and McCune, J.M., Subpopulations of long-lived and short-lived T cells in advanced HIV-1 infection, *J. Clin. Invest.,* 112, 956, 2003.

74. Mohri, H., Perelson, A.S., Tung, K., Ribeiro, R.M., Ramratnam, B., Markowitz, M., Kost, R., Hurley, A., Weinberger, L., Cesar, D., Hellerstein, M., and Ho, D.D., Increased turnover of T lymphocytes in HIV-1 infection and its reduction by anti-retroviral therapy, *J. Exp. Med.,* 194, 1277, 2001.

75. Kovacs, J.A., Lempicki, R.A., Sidorov, I.A., Adelsberger, J.W., Herpin, B., Metcalf, J.A., Sereti, I., Polis, M.A., Davey, R.T., Tavel, J., Falloon, J., Stevens, R., Lambert, L., Dewar, R., Schwartzentruber, D.J., Anver, M.R., Baseler, M.W., Masur, H., Dimitrov, D.S., and Lane, H.C., Identification of dynamically distinct subpopulations of T lymphocytes that are differentially affected by HIV, *J. Exp. Med.,* 194, 1731, 2001.

76. Ribeiro, R.M., Mohri, H., Ho, D.D., and Perelson, A.S., *In vivo* dynamics of T cell activation, proliferation, and death in HIV-1 infection: why are CD4+ but not CD8+ T cells depleted? *Proc. Natl. Acad. Sci. U.S.A.,* 99, 15572, 2002.

77. De Boer, R.J., Mohri, H., Ho, D.D., and Perelson, A.S., Turnover rates of B cells, T cells, and NK cells in simian immunodeficiency virus-infected and uninfected rhesus macaques, *J. Immunol.,* 170, 2479, 2003.

78. Asquith, B., Debacq, C., Macallan, D.C., Willems, L., and Bangham, C.R., Lymphocyte kinetics: the interpretation of labelling data, *Trends Immunol.,* 23, 596, 2002.

79. Deeks, S.G., Hoh, R., Grant, R.M., Wrin, T., Barbour, J.D., Narvaez, A., Cesar, D., Abe, K., Hanley, M.B., Hellmann, N.S., Petropoulos, C.J., McCune, J.M., and Hellerstein, M.K., CD4+ T cell kinetics and activation in human immunodeficiency virus-infected patients who remain viremic despite long-term treatment with protease inhibitor-based therapy, *J. Infect. Dis.,* 185, 315, 2002.

80. Meroni, L., Varchetta, S., Manganaro, D., Gatti, N., Riva, A., Monforte, A., and Galli, M., Reduced levels of CD4 cell spontaneous apoptosis in human immunodeficiency virus-infected patients with discordant response to protease inhibitors, *J. Infect. Dis.*, 186, 143, 2002.

81. Pitrak, D.L., Bolanos, J., Hershow, R., and Novak, R.M., Discordant CD4 T lymphocyte responses to antiretroviral therapy for HIV infection are associated with *ex vivo* rates of apoptosis, *AIDS*, 15, 1317, 2001.

82. Piketty, C., Weiss, L., Thomas, F., Si-Mohammed, A., Bélec, L., and Kazatchkine, M.D., Long-term clinical outcome of human immunodeficiency virus-infected patients with discordant immunologic and virologic responses to a protease inhibitor-containing regimen, *J. Infect. Dis.*, 183, 1328, 2001.

83. Grabar, S., Le Moing, V., Goujard, C., Leport, C., Kazatchkine, M.D., Costagliola, D., and Weiss, L., Clinical outcome of patients with HIV-1 infection according to immunologic and virologic response after 6 months of highly active antiretroviral therapy, *Ann. Int. Med.*, 133, 401, 2000.

84. Michel, P., Toure Balde, A., Roussilhon, C., Aribot, G., Sarthou, J.-L., and Gougeon, M.-L., Reduced immune activation and T cell apoptosis in human immunodeficiency virus type 2 compared with type 1: correlation of T cell apoptosis with β_2 microglobulin concentration and disease evolution, *J. Infect. Dis.*, 181, 64, 2000.

85. Tesselaar, K., Arens, R., van Schijndel, G.M., Baars, P.A., Van der Valk, M.A., Borst, J., van Oers, M.H., and Van Lier, R.A., Lethal T cell immunodeficiency induced by chronic costimulation via CD27–CD70 interactions, *Nat. Immunol.*, 4, 49, 2003.

86. Messele, T., Abdulkadir, M., Fontanet, A.L., Petros, B., Hamann, D., Koot, M., Ross, M.T.L., Schellekens, P.T.A., Miedema, F., and Rinke de Wit, T.F., Reduced naive and increased activated CD4 and CD8 cells in healthy adult Ethiopians compared with their Dutch counterparts, *J. Clin. Exp. Immunol.*, 115, 443, 1999.

87. Kalinkovich, A., Wisman, Z., Greenberg, Z., Nahmias, J., Eitan, S., Stein, M., and Bentwich, Z., Decreased CD4 and increased CD8 counts with T cell activation is associated with chronic helminth infection, *Clin. Exp. Immunol.*, 114, 414, 1998.

88. Tsegaye, A., Wolday, D., Otto, S., Petros, B., Assefa, T., Alebachew, T., Hailu, E., Adugna, F., Measho, W., Dorigo, W., Fontanet, A.L., van Baarle, D., and Miedema, F., Immunophenotyping of blood lymphocytes at birth, during childhood, and during adulthood in HIV-1-uninfected Ethiopians, *Clin. Immunol.*, 109, 338, 2003.

89. Hulstaert, F., Hannet, I., Deneys, V., Munhyeshuli, V., Reichert, T., De Bruyere, M., and Strauss, K., Age-related changes in human blood lymphocyte subpopulations. II. Varying kinetics of percentage and absolute count measurements, *Clin. Immunol. Immunopathol.*, 70, 152, 1994.

90. Fagnoni, F.F., Vescovini, R., Passeri, G., Bologna, G., Pedrazzoni, M., Lavagetto, G., Casti, A., Franceshi, C., Passeri, M., and Sansoni, P., Shortage of circulating naive CD8+ T cells provides new insights on immunodeficiency in aging, *Blood*, 95, 2860, 2000.

91. Homann, D., Teyton, L., and Oldstone, M.B.A., Differential regulation of anti-viral T-cell immunity results in stable CD8+ but declining CD4+ T-cell memory, *Nat. Med.*, 7, 913, 2001.

92. Schiemann, M., Busch, V., Linkemann, K., Huster, K.M., and Busch, D.H., Differences in maintenance of CD8+ and CD4+ bacteria-specific effector-memory T cell populations, *Eur. J. Immunol.*, 33, 2875, 2003.

93. Grayson, J.M., Harrington, L.E., Lanier, J.G., Wherry, E.J., and Ahmed, R., Differential sensitivity of naive and memory CD8+ T cells to apoptosis *in vivo*, *J. Immunol.*, 169, 3760, 2002.

94. Hayashi, N., Liu, D., Min, B., Ben Sasson, S.Z., and Paul, W.E., Antigen challenge leads to *in vivo* activation and elimination of highly polarized TH1 memory T cells, *Proc. Natl. Acad. Sci. U.S.A.*, 99, 6187, 2002.

20 Alteration of the Apoptotic Pathways in the Thymus during HIV Infection

Luchino Y. Cohen
Marie-Lise Dion
Rafick-Pierre Sekaly

CONTENTS

INTRODUCTION

Depletion of peripheral CD4 T lymphocytes is a hallmark of HIV infection and is the result of many pathogenic mechanisms. The very high viral load observed during HIV primary infection leads to high levels of CD4 T cell activation and infection, which results in their elimination by apoptosis. This cell loss must be compensated for by the production of naive T cells by the thymus. Unfortunately, thymic function is altered in the presence of HIV, and thymic output is reduced, which contributes to the reduction in peripheral CD4 lymphocyte pool. The apoptosis rate of thymocytes increases in HIV-infected patients, due to the pathogenic effect of HIV proteins, of HIV replication in thymocytes, and of perturbations in cytokine levels that result from the hyper-activation of the immune system. This review will focus on the alteration of T cell precursor survival in HIV-infected individuals as well as the molecular mechanisms involved in programmed cell death induced by HIV and that lead to the pathogenic effect of HIV on thymic function.

THE APOPTOTIC MACHINERY IN THYMOCYTES

As in most mammalian cells, there are two main pathways that lead to programmed cell death (apoptosis) in thymocytes (Figure 20.1). The extrinsic pathway is triggered by the aggregation of death receptors (DRs), such as Fas/CD95, TNFR1/CD120a, or TNF-related apoptosis-inducing ligand (TRAIL) receptor/DR5, and leads to the activation of a caspase cascade that includes caspase-8, caspase-3, -6, and -7.[1] Apoptosis triggered by DRs is dependent on the function of Fas-associated protein with death domain (FADD), an adaptor protein required for caspase-8 recruitment to DR, and is inhibited by FLICE inhibitory protein (FLIP), a protein homologous to caspase-8 but lacking its proteolytic activity. The other apoptotic pathway is activated after DNA damage and triggers p53 activation, cytochrome c release from the mitochondria, formation of the apoptosome containing

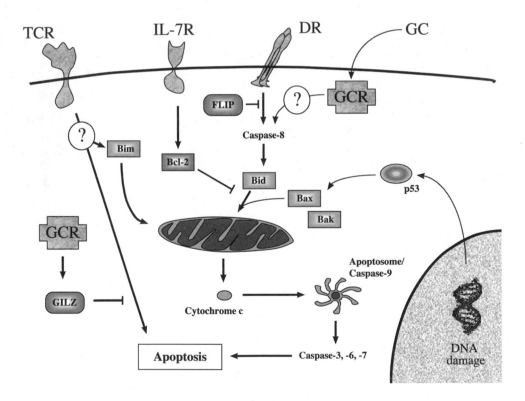

FIGURE 20.1 Molecular mechanisms involved in thymocyte cell death. Thymocyte apoptosis can be triggered through the extrinsic pathway following death receptor (DR) engagement, the intrinsic pathway induced by DNA damage, the presence of glucocorticoids (GC), or T cell receptor (TCR) cross-linking by a high-affinity ligand (negative selection). GC can also antagonize TCR-induced apoptosis by increasing the expression of the GC-induced leucine zipper (GILZ) protein. Most of these apoptotic stimuli lead to cytochrome c release from the mitochondria to the cytosol and the activation of downstream caspases (caspase-3, -6, and -7). IL-7 also protects thymocytes from apoptosis by increasing levels of Bcl-2 expression.

caspase-9, and subsequent activation of effector caspases (caspase-3, -6, and -7). This mechanism, called the intrinsic pathway, is inhibited by the antiapoptotic members of the Bcl-2 family, such as Bcl-2, Bcl-xL, or Bfl-1, whereas the proapoptotic members Bax, Bak, Bid, or Bad are required for cytochrome *c* release, subsequent caspase-9 activation, and cell death.[2] Both pathways are functional in murine thymocytes that can be killed after anti-Fas stimulation[3] or DNA damage[4] *in vitro*. Caspase inhibitors block cell death induced through both pathways.

Glucocorticoids (GCs) also trigger massive cell death among double-positive (DP) thymocytes, through a caspase-dependent pathway.[5,6] Bcl-2 and Bcl-xL overexpression protects DP thymocytes from GC-mediated killing.[7,8] The GC receptor activates caspase-8 through a mechanism that is not well understood, which results in the release of cytochrome *c* from the mitochondria and caspase-9 activation (Figure 20.1).[9] Paradoxically, GCs are also able to antagonize apoptosis after T cell receptor (TCR) cross-linking.[10] The presence of GCs triggers the expression of GC-induced leucine zipper (GILZ),[11] which inhibits TCR-mediated mitogen activated protein kinase (MAPK) activation[12] and AP-1 and NF-κB activity.[13,14] The role of GCs in thymic development is not completely understood, because positive and negative selection are unaffected in GC receptor-deficient mice.[15] Furthermore, although thymocytes can be killed *in vitro* by exposure to any of these stimuli, the signal that will determine whether a T cell progenitor undergoes apoptosis in the thymus is highly dependent on the recombination events that lead to the production of a functional TCR, as described below.

ROLE OF APOPTOSIS IN T CELL DEVELOPMENT

In more than 95% of cases, the fate of a developing T cell in the thymus is death; thus, apoptosis is a routine process within the thymus. Two main options can lead a thymocyte to programmed cell death: either the lack of TCR signaling, termed "death by neglect," or a signal that is too strong, termed "negative selection," both of which contribute to maintaining a functional yet not autoreactive T cell pool. The remaining 5% of thymocytes able to engage a major histocompatibility complex (MHC)–peptide complex with relatively weak affinity are positively selected and survive the maturation process to be released in the periphery first as recent thymic emigrants (RTEs), which later become mature T cells (Figure 20.2).[16]

Death by Neglect

During their maturation, thymocytes rearrange the V(D)J segments of the TCR beta (TCRB) locus followed by the TCR alpha (TCRA) locus, leading to the production of T cells with a diverse repertoire of TCRs. Such a random process evidently will yield a large number of "out of frame" rearrangements as well as TCRs that are unable to bind to the MHC–peptide complex or even remain on the surface of the cell, thus rendering many thymocytes unfit to continue differentiation. These thymocytes undergo apoptosis through death by neglect, due to a lack of signal either from the pre-TCR (pTα–β) or from the TCR (α–β). In mice that cannot rearrange the β chain (Rag1[−/−], Rag2[−/−], and severe combined immunodeficient [SCID] mice) that leads to the absence of pre-TCR signaling, there is a developmental arrest and apoptosis at the CD4[−] CD8[−] CD25[+] CD44[−] pro-T3 stage (where the pre-TCR is usually expressed).[17,18] DR signaling and the extrinsic pathway may play a role in the depletion of these thymocytes. Although Fas deficiency (*lpr* mice) does not increase the survival of thymocytes lacking the pre-TCR to the DP stage, deletion of FADD or expression of a dominant negative mutant of FADD permits their survival and differentiation.[19,20] This suggests that a DR other than Fas is involved in death by neglect. Cell death due to the absence of pre-TCR signaling may also involve the tumor suppressor p53 and the intrinsic pathway, because RAG-deficient or SCID mice lacking p53 contain thymocytes past the pro-T3 stage.[21–23] Moreover, many reports have underlined the importance of the intrinsic pathway, modulated by the Bcl-2 family proteins, in death by neglect of DP thymocytes. Among them, proapoptotic Bax and Bak

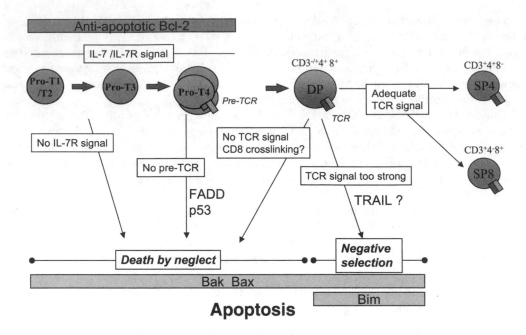

FIGURE 20.2 T cell development and apoptosis in the murine thymus. Double-negative thymocytes (pro-T1 to pro-T4) that do not express a functional TCR undergo apoptosis by a process called "death by neglect." Once they mature into CD4+ CD8+ double-positive (DP) thymocytes and express the TCR, the absence of TCR signaling also results in death by neglect. Engagement of the TCR with a weak affinity triggers DP cell survival and differentiation into single-positive (SP) mature thymocytes (positive selection), whereas a high-affinity interaction with the MHC–peptide complex results in the deletion of the T cell clone by negative selection.

seem to play a role in the induction of cell death by triggering the disruption of mitochondrial transmembrane potential in "neglected" thymocytes. Thymuses reconstituted with Bak$^{-/-}$ Bax$^{-/-}$ hematopoietic cells display an accumulation of CD4$^-$ CD8$^-$ thymocytes that do not differentiate into DP cells. These thymocytes do not die by neglect, as they maintain a constant viable cell count for 7 days as *in vitro* culture in medium alone.[24] Therefore, signals leading to the mitochondria are required for the induction of death by neglect. After TCR rearrangement produces a functional receptor, TCR engagement on DP thymocytes can provide either a rescue signal, which counteracts death by neglect (positive selection), or a death signal (negative selection).

NEGATIVE SELECTION

Negative selection is triggered after high-affinity TCR engagement by MHC–self-peptide complexes. The role of DR in this process has been extensively studied, and most studies suggest that DRs are not required for negative selection. Mice expressing a FADD dominant negative form, which inhibits signaling though Fas, TNFR1, DR3, and TRAILR2, exhibit normal or even increased negative selection compared with their wild-type counterparts, again suggesting a possible prosurvival role of FADD.[25] However, the importance of TRAIL in negative selection recently became a matter of debate. An impaired deletion of thymocytes by TCR ligation was initially observed in TRAIL-deficient mice.[26] Contrarily, using various "thymic selection" models such as antibody-mediated TCR cross-linking, stimulation with endogenous superantigens, or exposure to exogenous superantigen *in vitro*, another group could not identify a defect in negative selection in thymocytes from TRAIL$^{-/-}$ mice.[27] Thus, the role of TRAIL in negative selection remains to be confirmed.

In contrast, the intrinsic pathway plays a critical role during negative selection, although the antiapoptotic protein Bcl-2 is not involved. When overexpressed, Bcl-2 protects immature thymocytes from selective death stimuli *in vitro*, including glucocorticoids, irradiation, and TCR cross-linking. However, *in vivo*, Bcl-2 overexpression only modestly reduces negative selection mediated by histocompatibility (HY) antigen[28] or endogenous superantigens.[7] On the other hand, the proapoptotic BH3-only Bcl-2 family member Bim is essential for negative selection, as Bim-deficient transgenic thymocytes expressing a HY-specific TCR are resistant to apoptosis in the presence of the HY negative-selecting peptide.[29] Moreover, ligation of the TCR induces the accumulation of Bim and its association with Bcl-xL,[29] confirming its involvement during negative selection. Bax and Bak are also important players in thymocyte deletion, as negative selection is impaired in thymocytes from Bax and Bak double knockout mice.[24] These observations support a critical role of mitochondria in the control of apoptosis during negative selection, although Bcl-2 does not seem to be involved. The downstream events leading to cell death are not so clear. Mice deficient in either apoptotic activating factor 1 (Apaf-1) or caspase-9 do not show abnormalities in thymocyte selection,[30,31] suggesting that formation of the apoptosome is not essential during negative selection.

In summary, a large number of studies have focused on the molecular pathways involved in the deletion of immature thymocytes *in vivo* by death by neglect or negative selection. Although the survival of T cell progenitors is without doubt controlled by the mitochondria, the exact mechanisms involved in thymocyte deletion in the thymus still need to be clarified.

IMPACT OF HIV INFECTION ON THYMIC FUNCTION

ALTERATION OF THYMIC OUTPUT AFTER HIV INFECTION

During HIV infection, *de novo* T cell production by the thymus is critical for immune reconstitution, as it ensures the replenishment of the T cell pool with a functionally diverse TCR repertoire. Patients infected by protease inhibitor (PI)-resistant HIV strains with diminished capacity to infect thymocytes maintain normal CD4 T cell counts despite high viremia.[32] Many studies have demonstrated the deleterious effect of HIV on thymopoiesis. Thymic specimens from HIV-infected individuals exhibit a profound disorganization of the thymic stroma associated with depletion of thymocytes.[33,34] Immuno-histochemical studies in these tissues as well as in the SCID-hu Thy/Liv mice (SCID mice with a human thymus grafted on the renal capsule) further demonstrated that HIV-infected thymic tissue contained RNA and viral proteins within the resident thymocytes, confirming their susceptibility to infection.[35–37]

Early reports indirectly suggested a decrease in thymic output, as the frequency of naive T cell (CD45RA+) in the peripheral blood was found to drop with HIV disease progression.[38–40] However, this decrease in naive T cell numbers could be the result of increased peripheral T cell activation, a hallmark of HIV infection. HIV-infected individuals consistently show the upregulation of several activation markers, such as CD38, HLA-DR, and Ki67, reflecting their heightened immune activation and the increased recruitment of naive cells to the effector memory compartment.[41–44] Recently, quantification of the frequency of T cell receptor excision circles (TRECs), which are markers for intrathymic TCR rearrangements, has led to much insight on thymic function in HIV-infected patients. Many studies demonstrate the reduced frequency of TRECs in peripheral blood mononuclear cells (PBMCs) and purified naive T cells in untreated, HIV-infected individuals.[45–48] A large-scale study (297 HIV-infected patients who ranged in age from 20 to 59 years) using sorted CD4 and CD8 T cells demonstrated that TREC frequency increases in CD4 T cells early during HIV infection and slowly declines, whereas CD8 T cells harbor much lower TRECs throughout disease progression.[49] Longitudinal analysis of SIV-infected macaques also showed that TREC levels decline in CD4 and CD8 T cells at 16 to 30 weeks after infection, which correlates with an increase in CD8 T cell infiltrates and apoptosis in the thymus, linking affected thymic function and reduced TREC frequencies.[50] Finally, in contrast to lower TREC levels in rapid-progressing HIV-infected children, TREC levels in long-term asymptomatic, vertically HIV-infected children are similar to TREC levels in uninfected individuals.[51] Again, this stresses the impact of thymic output in maintaining CD4 T cell count after HIV infection.

However, such observations must be interpreted cautiously given the higher proliferation rate of PBMCs within HIV-infected patients, which could also result in the decrease in TREC numbers. The exact contribution of thymic output or peripheral proliferation to the reduced TREC frequency remains to be assessed.[52,53] Interestingly, highly active antiretroviral therapy (HAART) was shown to trigger an increase in TREC frequency to levels comparable to age-matched controls in HIV-infected children and adults.[54–56] This was, in most cases, accompanied by an increase in thymic size (as determined by computerized tomography [CT] scan),[57] suggesting an improved thymic function when HIV viral load was controlled. But again, the increase in TREC levels is also accompanied by a decrease in peripheral T cell activation and, thus, a diminished TREC dilution effect.[43,58] Still, the observed increase in TCRβ variable region diversity within CD45RA+ CD8+ T cells in HIV-infected children after HAART suggests a replenishment of the T cell repertoire with new T cells from the thymus and, thus, an increase in thymic function.[59] Taken together, these data suggest that thymic function seems to be reestablished following suppression of viral load, which correlates with observations demonstrating the cytopathic effect of thymocyte infection by HIV.

SEVERAL THYMOCYTE SUBSETS ARE SUSCEPTIBLE TO HIV INFECTION

Most thymocytes, including the CD3– CD4+ CD8– cells (immature single-positive [ISP]), CD3+ CD4+ CD8+ (DP cells), and CD3+ CD4+ CD8– single-positive (SP) cells, carry the CD4 molecule and are thus a potential breeding ground for the HIV virus. However, various thymocyte subsets exhibit a different sensitivity to HIV infection, and this mainly depends on their levels of HIV co-receptor expression (Figure 20.3). Although CXCR4 expression is high in DN, ISP, and DP

CD4	-	±	+	+	+	-
CXCR4	-	±	++	+	-	-
CCR5	-	-	-	±	-	-
HIV infection			+	+	+	
HIV replication			low		+	

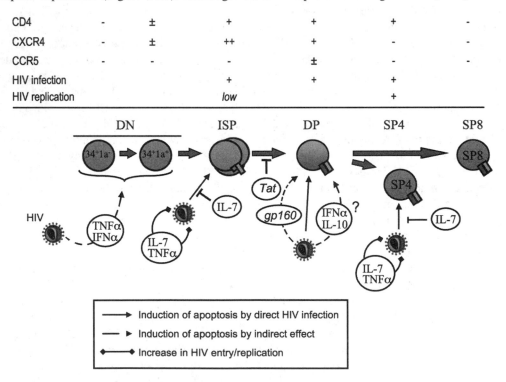

FIGURE 20.3 Influence of HIV on human thymic development. HIV infection directly causes an increased apoptotic rate of immature single-positive (ISP), DP, or CD4+ SP thymocytes (plain arrows). Thymocyte maturation is also indirectly altered by the increase in cytokine levels or the presence of HIV proteins (dotted arrows). An increase in TNF-α or IL-7 levels enhances HIV replication in thymocytes and the pathogenic effect of HIV on thymic function. The levels of CD4 or HIV co-receptor expression are indicated for each subset, as well as the susceptibility to infection and the capacity to sustain HIV replication.

thymocytes, it decreases considerably in mature SP thymocytes.[60–62] On the contrary, CCR5 expression is very low throughout the differentiation process and is restricted to 1 to 3% of thymocytes.[61–63] Correlating with these observations, CXCR4-using viruses (X4) have a stronger cytopathic effect on thymocytes than the CCR5-using strains (R5) in the SCID-hu mouse model[61] or in SIV-infected infant macaques.[64] In the SCID-hu model, thymic dysfunction was mainly attributed to a preferential infection of ISP by HIV.[65] Furthermore, SIV infection seems to inhibit the developmental potential of ISP; when exposed to SIV, these cells differentiate less in a xenogeneic monkey–mouse fetal thymus organ culture (FTOC) system than ISP from uninfected monkeys.[66]

Experiments using mixed thymocyte–stromal cells co-cultures have demonstrated that even though HIV can efficiently enter ISP and DP cells, virus production occurs only in SP CD4+ cells.[67] Although this subpopulation possesses a low level of CXCR4 expression, exposure to interleukin (IL)-7, a cytokine secreted by thymic epithelial cells (TECs), upregulates the CXCR4 molecule to the cell surface, rendering them susceptible to X4 HIV-1 entry.[68] Furthermore, it has been shown that IL-7 regulates TNF-α expression, which, in turn, induces the activation of the NF-κB transcription factor, required for high replication of HIV.[67,69] Finally, although CD8 SP thymocytes are not directly susceptible to HIV infection, HIV provirus has been detected in this subset. Infection of DP thymocyte has been shown to produce CD4− CD8+ SP cells that contain HIV provirus, which could contribute to the latent HIV reservoir, because after stimulation in the periphery, these cells can sustain HIV replication.[70,71] In summary, several thymocyte subsets are susceptible to HIV infection (ISP and DP) and are able to sustain replication (SP). The evidence that HIV infection results in the apoptosis of some of these subsets will now be described.

MODULATION OF APOPTOSIS IN THE THYMUSES OF INFECTED INDIVIDUALS

INFLUENCE OF HIV ON THYMOCYTE APOPTOSIS RATE

Because apoptosis plays such a critical role during thymocyte maturation, it is not surprising that modifications in thymocytes' apoptosis rates were observed in the presence of HIV at different steps of maturation. An early study by the group of McCune et al. showed that thymocyte depletion resulted from apoptotic cell death in human thymuses implanted in SCID-hu mice and infected with various HIV strains.[65] Experiments performed in FTOC confirmed that in the presence of HIV, thymocyte loss occurred by programmed cell death and resulted in the preferential destruction of CD4+ cells.[72] Another study confirmed the rapid CD4+ cell loss after HIV infection of thymic implants in SCID-hu mice, although very low levels of apoptosis were reported *in vivo* by using either the TdT-mediated dUTP nicked-end labeling (TUNEL) assay or Annexin-V staining.[73] Increases in the apoptotic rate and the number of apoptotic cells were observed in the thymus 7 to 14 days after infection of macaques by SIV[74] and 14 to 28 days after infection by the chimeric virus SHIV 89.6P.[75] Increases in DNA fragmentation and caspase activity were detected in CD4+ thymocytes 14 to 21 days after infection just after peak viremia.[76] Therefore, the rate of thymocyte apoptosis is elevated during HIV or SIV infection.

Results obtained using different experimental models confirm that HIV or SIV infection causes the apoptosis of several thymocyte subsets (Figure 20.3). During SIV infection, early thymocyte progenitors (CD34+ cells) were shown to undergo apoptosis 7 to 14 days after infection, leading to a decrease in CD34+ cells in SIV-infected thymus at day 21.[74] In a case of HIV-1 neonatal infection, a marked depletion of DP and CD4+ SP thymocytes was also demonstrated.[77] HIV infection of human thymocytes seems to be directly responsible for the destruction of CD4+ CD−3 ISP thymocytes.[65] This subset, which can sustain *in vitro* a low HIV replication in the presence of TEC, is even more sensitive to HIV-induced apoptosis than DP thymocytes.[78] In macaques, SIVmac239 infection first results in a slight and transient decrease in the percentage of DP thymocytes 7 to 14 days after infection. Then, a gradual loss of CD4+ SP cells occurs 21 days after infection.[74,76] Recent evidence in humans also suggests that thymocyte apoptosis is the direct consequence of infection *in vivo*, because patients infected with protease inhibitor–resistant HIV strains that do not replicate in thymocytes have an abundant thymic mass, as shown by CT.[32] Overall, these observations indicate that HIV infection is directly responsible for an increase in the apoptotic rate of ISP, DP, and CD4[+] SP thymocytes.

APOPTOSIS IN THE THYMIC STROMA

HIV infection was also associated with the disruption of the thymic microenvironment.[79] TEC function is affected in the presence of HIV, although conflicting observations were published regarding their susceptibility to HIV infection. Several studies described HIV replication in TEC cultures *in vitro* using HIV1-IIIB and HIV1-NDK but not HIV1-LAI.[80–82] Another group reported that TECs were not infected by HIV1/IIIB or HIV1/Ba-L.[83] In the SCID-hu mice model, TECs were shown to endocytose large amounts of HIV/JR-CSF, but productive infection of this cell type could not be confirmed.[84] However in the same study, uninfected TECs undergoing apoptosis were observed, suggesting that TEC survival is altered in the presence of HIV.

Thymic dendritic cells (DCs) were also shown to be susceptible to HIV infection by macrophage-tropic strains[85] but not HIV-1/IIIB,[86] although DC survival decreased in the presence of HIV1/IIIB *in vitro*.[86] Therefore, survival of thymic epithelial and dendritic cells may be reduced in the presence of HIV.

MECHANISMS INVOLVED IN THE MODULATION OF APOPTOSIS BY HIV IN THE THYMUS

ROLE OF CYTOKINES IN HIV-MEDIATED PERTURBATION OF THYMOCYTE APOPTOSIS

Influence of IL-7 Levels

An overall increase in plasma of IL-7 levels was demonstrated during HIV infection. Moreover, a negative correlation between circulating IL-7 levels and CD4[+] T cell count was reported,[87,88] which likely reflects a compensatory response. In HIV-infected children, the drop in CD4[+] T cells correlates with a marked increase in IL-7 plasma levels; also, a drastic decrease in viral load after HAART treatment leads to a recovery of CD4 levels, with a concomitant decrease in IL-7 levels.[89] These observations suggest that the drop in CD4[+] T cells in HIV-infected children induces IL-7 production by stromal cells in lymph nodes as part of a homeostatic mechanism aimed at maintaining normal levels of peripheral CD4[+] cells. However, this increase in IL-7 production can also increase HIV replication (Figure 20.3). IL-7 was shown to increase the infectivity of HIV-1 in thymocyte cultures,[90] and administration of IL-7 to Thy/Liv SCID-hu mice infected with HIV results in increased viral load.[91]

Because HIV replication usually leads to the destruction of T cells, two mechanisms could explain the contradictory effects of IL-7, which increases CD4 cell numbers despite the fact that this cytokine also enhances HIV replication. First, IL-7 exerts an antiapoptotic effect on various thymocyte subsets by increasing Bcl-2 expression.[92] A culture of CD4[+] SP thymocytes in the absence of IL-7 results in a decrease in Bcl-2 levels, which is compensated for by the addition of IL-7 in the medium. IL-7R-deficient mice exhibit a severe block in thymocyte maturation, which is rescued by Bcl-2 overexpression.[93–96] Interestingly, HIV p24 protein was detected exclusively in Bcl-2-positive CD4[+] thymocytes infected *in vitro*, suggesting that Bcl-2 expression protects infected thymocytes from apoptosis and allows HIV replication.[78] Therefore, the high levels of IL-7 maintain infected thymocyte viability and increase HIV replication, which is detrimental for thymic function. The second mechanism involved in IL-7's positive effect on HIV replication is the enhancement of HIV co-receptor expression. The interaction of TECs with mature CD4[+] thymocytes *in vitro* results in higher levels of CXCR4 expression by thymocytes, through a mechanism dependent on IL-7 production by the stromal cells.[68] Pretreatment of naive T cells with IL-7 also increases surface expression of CXCR4 but not CCR5. The presence of IL-7 even triggers the productive infection of T cell–adapted strains and primary isolates of HIV in resting, naive T cells.[87,97] Therefore, elevated levels of IL-7 enhance HIV replication in thymocytes by increasing thymocyte sensitivity to infection and by protecting infected cells from apoptosis through the intrinsic pathway.

Role of TNF-α

Thymocytes can be killed by TNF-α *in vitro* and *in vivo*,[98,99] although this cytokine is not absolutely required for negative selection, as shown by a normal clonal deletion in the presence of anti-TNF neutralizing antibody.[100] However, TNFR-deficient mice exhibit thymic hypertrophy, with a 60% increase in total thymocytes but no alteration in the CD4/CD8 subset proportions.[101] Therefore, thymic output is affected by TNF-α, which reduces thymic cellularity by triggering the extrinsic pathway. During the course of HIV infection, TNF-α plasma levels are elevated[102–104] and affect thymocyte function in three different ways. First, TNF-α produced in the bone marrow during HIV infection stimulates apoptosis in hematopoietic progenitors,[105] reducing the thymic input of lymphoid precursors. Second, early T cell precursors such as DN thymocytes are sensitive to TNF-α-mediated apoptosis,[101] and increased levels of TNF-α in HIV-infected patients are likely to affect the survival of this subset. Finally, TNF-α-mediated NF-κB activation in CD4[+] ISP and CD4[+] SP thymocytes increases HIV replication.[67,106] Furthermore, IL-7 produced by TEC may amplify the negative effect of TNF-α by sustaining the expression of the p75 TNF receptor in thymocytes.[67] Therefore, the increase in systemic TNF-α levels after HIV infection leads to the deletion of several thymocyte subsets and enhances HIV replication in the thymus.

Perturbation of Thymic Function by IFN-α

As part of the natural innate immune response, triggering of the Toll-like receptors (TLR) by viruses such as HIV induces the secretion of type I interferons, including interferon-α (IFN-α) and interferon-β (IFN-β). Increased levels of serum IFN-α are observed during primary HIV infection.[107] Moreover, IFN-α levels increase in patients' serum in parallel with disease progression and are associated with p24 antigen levels.[107] It was shown that the human thymus contains IFN-α-producing CD11c-negative plasmacytoid dendritic cells.[108] In an FTOC model as well as the SCID-hu Thy/Liv model, IFN-α is secreted in the thymus after HIV infection.[109] This secretion of IFN-α within the thymus may have detrimental consequences for thymopoiesis, as the administration of IFN-α to a mouse causes a severe thymic atrophy, a decrease in thymic cellularity, and an impairment in progenitor T cell development.[110] As for its effects on thymic stroma, the addition of recombinant IFN-α, IFN-β, or IFN-γ to human thymic epithelial cell cultures is sufficient to induce differentiation and apoptosis of these cells.[111]

Furthermore, HIV-induced IFN-α upregulates the expression of MHC class I (MHC-I) molecules on thymocytes as well as on thymic epithelium in the SCID-hu Thy/Liv model,[109] which may have important consequences on the threshold of negative and positive selection and, hence, on DP thymocyte apoptotic rate. HIV infection causes the generation of CD3[+]CD8[low] thymocytes in SCID-hu Thy/Liv mice,[112] suggesting that the IFN-α-induced increase in MHC-I expression enhances the avidity of the TCR/MHC interaction and decreases the requirement for CD8 in thymocytes expressing MHC-I restricted TCR. IFN-α may also directly influence apoptosis by activating proapoptotic molecules involved in thymocyte depletion. In a tumor cell line, IFN-α-induced apoptosis involved the activation of the proapoptotic proteins Bak and Bax,[113] two major players in death by neglect and negative selection of thymocytes, as discussed above. Therefore, IFN-α seems to disrupt the thymic microenvironment and to affect both the quality and number of the exported thymocytes.

Influence of IL-10

Finally, IL-10 plasma levels also increase during the course of HIV infection and progression to AIDS.[114] PBMCs from infected patients produce higher amounts of IL-10 following stimulation.[115,116] This response is thought to contribute to the Th1/Th2 imbalance triggered by HIV infection. IL-10 expression is induced in HIV-1-infected thymus organs and may result in increased MHC-I expression on DP thymocytes by HIV.[117] Overexpression of MHC-I molecules on DP thymocytes could interfere with thymocyte maturation and may contribute to HIV-1-induced thymocyte depletion.

ROLE OF GLUCOCORTICOIDS

Adrenal function is elevated following HIV infection. Compared with control asymptomatic patients, individuals suffering from lymphadenopathy have higher levels of adrenocorticotropic hormone (ACTH) and basal cortisol.[118] In HIV-infected children, an increased cortisol secretion may be associated with specific central nervous system damage.[119] Elevated GC levels in HIV-infected patients could hypothetically cause an increase in thymocyte death during HIV infection. However, GCs can also antagonize TCR-mediated apoptosis in thymocytes and may protect HIV-infected T cells from apoptosis. HIV may mimic this mechanism through the action of the Vpr protein on the GC receptor. After the entry of HIV into T cells, Vpr is able to trigger GC receptor translocation to the nucleus, through its interaction with a GC receptor–binding protein Rip-1,[120] and is able to act as a cofactor for GC-mediated gene expression.[121,122] Expression of Vpr in T cells antagonizes TCR-induced apoptosis by inhibition of NF-κB activity through a mechanism dependent on the function of the GC receptor.[123] Vpr-mediated GC receptor activation also increases HIV replication, because the HIV long terminal repeat (LTR) contains a functional GC response element,[124] and GCs were also shown to synergize with TNF-α for the induction of HIV expression.[125] Although these results were obtained in peripheral T cells, the presence of Vpr in infected thymocytes may allow a higher level of HIV replication, particularly in CD4+ SP cells, by antagonizing negative selection through a GC receptor–dependent mechanism.

ROLE OF THE CD95/FAS PATHWAY

Peripheral CD4 T cell depletion during HIV infection occurs through several mechanisms, including sensitization of CD4+ T cells to Fas-mediated apoptosis through the extrinsic pathway. CD4 cross-linking by the HIV-envelope protein stimulates Fas expression and increases CD4+ T cell apoptosis after Fas cross-linking.[126–128] The Tat protein also causes CD4+ T cell death by upregulating the expression of Fas ligand in T cells.[129] Finally, HIV infection induces the production and release of Fas ligand by monocytic cells, which can kill peripheral T cells.[130] During the course of SIV infection in macaques, the expression of Fas and Fas ligand increases slightly in the thymus,[76] although there is no evidence that this response plays a role in thymocyte depletion. Moreover, some studies have revealed discrepancies between the sensitivity of human and murine thymocytes to Fas-mediated apoptosis. Murine DP thymocytes are sensitive to apoptosis induced by anti-Fas antibody *in vitro*[3] and in FTOC,[131,132] as well as by Fas ligand *in vivo*.[133] However, human thymocytes exhibit a different expression pattern for Fas, which is found in DN thymocytes that are not susceptible to apoptosis induced by Fas cross-linking using antibodies or Fas ligand.[133] Some human thymocytes even express a Fas decoy receptor, lacking a death domain critical for its proapoptotic function and able to antagonize Fas-ligand function.[134] This lack of sensitivity of human thymocytes to Fas-mediated cell death certainly explains the lack of evidence so far that the Fas pathway plays a role in thymocyte depletion after HIV infection, whereas it is certainly involved in the deletion of peripheral T cells.

ROLE OF HIV PROTEINS IN THYMIC APOPTOSIS MODIFICATIONS

We described above how the HIV protein Vpr affects thymocyte development by activating the GC receptor and interfering with the apoptotic process. Other HIV proteins are involved in thymocyte depletion during HIV infection. Several studies have demonstrated the negative impact of Nef on thymocyte development. A strong downregulation in the level of CD4 on the surface of DP thymocytes and a decrease in the number of CD4+ T cells in the thymus were observed in mice expressing Nef under the control of CD2 regulatory elements. Activation of transgenic thymocytes by anti-CD3 antibody is reduced in these mice.[135] Thymocytes expressing the HIV_{NL4-3} Nef also show altered activation responses in correlation with a decrease in CD4 antigen expression.[136] Confirming these results, retroviral gene transfer of Nef impairs human thymic function. Thymocytes are generated in reduced numbers and exhibit lower CD4 and CD8β cell surface expression. T cells grown from Nef-expressing thymocytes are hyperproliferative *in vitro* upon T cell receptor triggering.[137] However, using CD34+ cells retrovirally transduced with various Nef mutants, it was shown that Nef-mediated impaired thymopoiesis is not due to altered surface marker trafficking. On the contrary, Nef domains critical for interaction of Nef with PACS-1, SH3 domains, and PAK2 are essential for Nef negative effect on T cell development.[138] These studies suggest that Nef expression by infected thymocytes is likely to affect their normal maturation and decrease thymic cellularity.

The presence of HIV-infected cells could also trigger thymocyte cell death in a bystander mechanism. T cell clones expressing the HIV envelope protein gp160 induce normal DP thymocyte apoptosis in co-cultures by a CD4-dependent mechanism.[139] Moreover, addition of recombinant gp120 to FTOC results in an impairment of T cell maturation[66] (Figure 20.3). Another HIV protein could also exert a negative influence through stromal cells. Tat-expressing TECs have a decreased ability to drive the generation of DP thymocytes from DN precursors, as shown by gene transfer-mediated Tat expression in thymic stroma.[140] Finally, it was shown that the HIV protease can cleave and activate the proapoptotic caspase-8, which results in the cleavage of Bid, cytochrome *c* release from the mitochondria, and T cell death.[141] Although this mechanism remains to be confirmed in thymocytes, it is possible that it contributes to thymocyte depletion during HIV replication. In summary, many HIV proteins, including Vpr, Nef, gp120, Tat, and the HIV protease, contribute to thymocyte apoptosis during HIV infection.

CONCLUDING REMARKS

As evidenced by the reduced thymic cellularity, increase in apoptosis, and altered T cell development, HIV infection strongly affects thymic function. The viability of many thymocyte subsets is impaired either as a direct consequence of HIV infection or through indirect mechanisms. Variation in cytokine levels or the simple presence of HIV proteins can alter thymocyte susceptibility to apoptosis. Furthermore, HIV affects thymic stromal cell functions, modifying the thymic microenvironment and leading to a decrease in thymocyte survival. Controlling the viral load after HAART eliminates many of these detrimental effects. However, better comprehension of the mechanisms involved in thymopoiesis will be essential to enhance the immune reconstitution after HAART and may help toward the stimulation of a protective immunity.

ACKNOWLEDGMENTS

The work from our laboratory supporting this review was funded by the CIHR, the NIH, CANVAC, and Genome Canada. MLD is a recipient of a doctoral research award from the CIHR. RPS holds the Canada Research Chair in Human Immunology and is a senior scientist of the CIHR.

REFERENCES

1. Green, D.R., Overview: apoptotic signaling pathways in the immune system, *Immunol. Rev.*, 193 (1), 5–9, 2003.
2. Opferman, J.T. and Korsmeyer, S.J., Apoptosis in the development and maintenance of the immune system, *Nat. Immunol.*, 4 (5), 410–415, 2003.
3. Ogasawara, J., Suda, T., and Nagata, S., Selective apoptosis of CD4+CD8+ thymocytes by the anti-Fas antibody, *J. Exp. Med.*, 181 (2), 485–491, 1995.
4. Clarke, A.R., Purdie, C.A., Harrison, D.J., Morris, R.G., Bird, C.C., Hooper, M.L., and Wyllie, A.H., Thymocyte apoptosis induced by p53-dependent and independent pathways, *Nature*, 362 (6423), 849–852, 1993.
5. Clayton, L.K., Ghendler, Y., Mizoguchi, E., Patch, R.J., Ocain, T.D., Orth, K., Bhan, A.K., Dixit, V.M., and Reinherz, E.L., T-cell receptor ligation by peptide/MHC induces activation of a caspase in immature thymocytes: the molecular basis of negative selection, *EMBO J.*, 16 (9), 2282–2293, 1997.
6. Alam, A., Braun, M.Y., Hartgers, F., Lesage, S., Cohen, L., Hugo, P., Denis, F., and Sekaly, R.P., Specific activation of the cysteine protease CPP32 during the negative selection of T cells in the thymus, *J. Exp. Med.*, 186 (9), 1503–1512, 1997.
7. Sentman, C.L., Shutter, J.R., Hockenbery, D., Kanagawa, O., and Korsmeyer, S.J., Bcl-2 inhibits multiple forms of apoptosis but not negative selection in thymocytes, *Cell*, 67 (5), 879–888, 1991.
8. Grillot, D.A., Merino, R., and Nunez, G., Bcl-XL displays restricted distribution during T cell development and inhibits multiple forms of apoptosis but not clonal deletion in transgenic mice, *J. Exp. Med.*, 182 (6), 1973–1983, 1995.
9. Marchetti, M.C., Di Marco, B., Cifone, G., Migliorati, G., and Riccardi, C., Dexamethasone-induced apoptosis of thymocytes: role of glucocorticoid receptor-associated Src kinase and caspase-8 activation, *Blood*, 101 (2), 585–593, 2003.
10. Vacchio, M.S., Lee, J.Y., and Ashwell, J.D., Thymus-derived glucocorticoids set the thresholds for thymocyte selection by inhibiting TCR-mediated thymocyte activation, *J. Immunol.*, 163 (3), 1327–1333, 1999.
11. D'Adamio, F., Zollo, O., Moraca, R., Ayroldi, E., Bruscoli, S., Bartoli, A., Cannarile, L., Migliorati, G., and Riccardi, C., A new dexamethasone-induced gene of the leucine zipper family protects T lymphocytes from TCR/CD3-activated cell death, *Immunity*, 7 (6), 803–812, 1997.
12. Ayroldi, E., Zollo, O., Macchiarulo, A., Di Marco, B., Marchetti, C., and Riccardi, C., Glucocorticoid-induced leucine zipper inhibits the Raf-extracellular signal-regulated kinase pathway by binding to Raf-1, *Mol. Cell. Biol.*, 22 (22), 7929–7941, 2002.
13. Mittelstadt, P.R. and Ashwell, J.D., Inhibition of AP-1 by the glucocorticoid-inducible protein GILZ, *J. Biol. Chem.*, 276 (31), 29603–29610, 2001.
14. Ayroldi, E., Migliorati, G., Bruscoli, S., Marchetti, C., Zollo, O., Cannarile, L., D'Adamio, F., and Riccardi, C., Modulation of T-cell activation by the glucocorticoid-induced leucine zipper factor via inhibition of nuclear factor B, *Blood*, 98 (3), 743–753, 2001.
15. Purton, J.F., Boyd, R.L., Cole, T.J., and Godfrey, D.I., Intrathymic T cell development and selection proceeds normally in the absence of glucocorticoid receptor signaling, *Immunity*, 13 (2), 179–186, 2000.
16. Palmer, E., Negative selection — clearing out the bad apples from the T-cell repertoire, *Nat. Rev. Immunol.*, 3 (5), 383–391, 2003.
17. Shinkai, Y., Rathbun, G., Lam, K.P., Oltz, E.M., Stewart, V., Mendelsohn, M., Charron, J., Datta, M., Young, F., Stall, A.M., and Alt, F.W., RAG-2-deficient mice lack mature lymphocytes owing to inability to initiate V(D)J rearrangement, *Cell*, 68 (5), 855–867, 1992.
18. Mombaerts, P., Clarke, A.R., Rudnicki, M.A., Iacomini, J., Itohara, S., Lafaille, J.J., Wang, L., Ichikawa, Y., Jaenisch, R., Hooper, M.L., and Tonegawa, S., Mutations in T-cell antigen receptor genes alpha and beta block thymocyte development at different stages, *Nature*, 360 (6401), 225–231, 1992.
19. Kabra, N.H., Kang, C., Hsing, L.C., Zhang, J., and Winoto, A., T cell-specific FADD-deficient mice: FADD is required for early T cell development, *Proc. Natl. Acad. Sci. U.S.A.*, 98 (11), 6307–6312, 2001.
20. Newton, K., Harris, A.W., and Strasser, A., FADD/MORT1 regulates the pre-TCR checkpoint and can function as a tumour suppressor, *EMBO J.*, 19 (5), 931–941, 2000.

21. Guidos, C.J., Williams, C.J., Grandal, I., Knowles, G., Huang, M.T., and Danska, J.S., V(D)J recombination activates a p53-dependent DNA damage checkpoint in Scid lymphocyte precursors, *Genes Dev.,* 10 (16), 2038–2054, 1996.

22. Bogue, M.A., Zhu, C., Aguilar-Cordova, E., Donehower, L.A., and Roth, D.B., p53 is required for both radiation-induced differentiation and rescue of V(D)J rearrangement in Scid mouse thymocytes, *Genes Dev.,* 10 (5), 553–565, 1996.

23. Jiang, D., Lenardo, M.J., and Zuniga-Pflucker, J.C., p53 prevents maturation to the CD4+CD8+ stage of thymocyte differentiation in the absence of T cell receptor rearrangement, *J. Exp. Med.,* 183 (4), 1923–1928, 1996.

24. Rathmell, J.C., Lindsten, T., Zong, W.X., Cinalli, R.M., and Thompson, C.B., Deficiency in Bak and Bax perturbs thymic selection and lymphoid homeostasis, *Nat. Immunol.,* 3 (10), 932–939, 2002.

25. Newton, K., Harris, A.W., Bath, M.L., Smith, K.G., and Strasser, A., A dominant interfering mutant of FADD/MORT1 enhances deletion of autoreactive thymocytes and inhibits proliferation of mature T lymphocytes, *EMBO J.,* 17 (3), 706–718, 1998.

26. Lamhamedi-Cherradi, S.E., Zheng, S.J., Maguschak, K.A., Peschon, J., and Chen, Y.H., Defective thymocyte apoptosis and accelerated autoimmune diseases in TRAIL' mice, *Nat. Immunol.,* 4 (3), 255–260, 2003.

27. Cretney, E., Uldrich, A.P., Berzins, S.P., Strasser, A., Godfrey, D.I., and Smyth, M.J., Normal thymocyte negative selection in TRAIL-deficient mice, *J. Exp. Med.,* 198 (3), 491–496, 2003.

28. Strasser, A., Harris, A.W., von Boehmer, H., and Cory, S., Positive and negative selection of T cells in T cell receptor transgenic mice expressing a Bcl-2 transgene, *Proc. Natl. Acad. Sci. U.S.A.,* 91 (4), 1376–1380, 1994.

29. Bouillet, P., Purton, J.F., Godfrey, D.I., Zhang, L.C., Coultas, L., Puthalakath, H., Pellegrini, M., Cory, S., Adams, J.M., and Strasser, A., BH3-only Bcl-2 family member Bim is required for apoptosis of autoreactive thymocytes, *Nature,* 415 (6874), 922–926, 2002.

30. Hara, H., Takeda, A., Takeuchi, M., Wakeham, A.C., Itie, A., Sasaki, M., Mak, T.W., Yoshimura, A., Nomoto, K., and Yoshida, H., The apoptotic protease-activating factor 1-mediated pathway of apoptosis is dispensable for negative selection of thymocytes, *J. Immunol.,* 168 (5), 2288–2295, 2002.

31. Marsden, V.S., O'Connor, L., O'Reilly, L.A., Silke, J., Metcalf, D., Ekert, P.G., Huang, D.C., Cecconi, F., Kuida, K., Tomaselli, K.J., Roy, S., Nicholson, D.W., Vaux, D.L., Bouillet, P., Adams, J.M., and Strasser, A., Apoptosis initiated by Bcl-2-regulated caspase activation independently of the cytochrome c/Apaf-1/caspase-9 apoptosome, *Nature,* 419 (6907), 634–637, 2002.

32. Stoddart, C.A., Liegler, T.J., Mammano, F., Linquis, T., Stepps, V.D., Hayden, M.S., Deeks, S.G., Grant, R.M., Clavel, F., and McCune, J.M., Impaired replication of protease inhibitor-resistant HIV-1 in human thymus, *Nat. Med.,* 7 (6), 712–718, 2001.

33. Papiernik, M., Brossard, Y., Mulliez, N., Roume, J., Brechot, C., Barin, F., Goudeau, A., Bach, J.F., Griscelli, C., and Henrion, R., Thymic abnormalities in fetuses aborted from human immunodeficiency virus type 1 seropositive women, *Pediatrics,* 89 (2), 297–301, 1992.

34. Prevot, S., Audouin, J., Andre-Bougaran, J., Griffais, R., Le Tourneau, A., Fournier, J.G., and Diebold, J., Thymic pseudotumorous enlargement due to follicular hyperplasia in a human immunodeficiency virus sero-positive patient. Immunohistochemical and molecular biological study of viral infected cells, *Am. J. Clin. Pathol.,* 97 (3), 420–425, 1992.

35. Pekovic, D.D., Gornitsky, M., Ajdukovic, D., Dupuy, J.M., Chausseau, J.P., Michaud, J., Lapointe, N., Gilmore, N., Tsoukas, C., Zwadlo, G., and Popovic, M., Pathogenicity of HIV in lymphatic organs of patients with AIDS, *J. Pathol.,* 152 (1), 31–35, 1987.

36. Namikawa, R., Kaneshima, H., Lieberman, M., Weissman, I.L., and McCune, J.M., Infection of the SCID-hu mouse by HIV-1, *Science,* 242 (4886), 1684–1686, 1988.

37. Burke, A.P., Anderson, D., Benson, W., Turnicky, R., Mannan, P., Liang, Y.H., Smialek, J., and Virmani, R., Localization of human immunodeficiency virus 1 RNA in thymic tissues from asymptomatic drug addicts, *Arch. Pathol. Lab. Med.,* 119 (1), 36–41, 1995.

38. Gruters, R.A., Terpstra, F.G., De Goede, R.E., Mulder, J.W., De Wolf, F., Schellekens, P.T., Van Lier, R.A., Tersmette, M., and Miedema, F., Immunological and virological markers in individuals progressing from seroconversion to AIDS, *AIDS,* 5 (7), 837–844, 1991.

39. Jaleco, A.C., Covas, M.J., Pinto, L.A., and Victorino, R.M., Distinct alterations in the distribution of CD45RO+ T cell subsets in HIV-2 compared with HIV-1 infection, *AIDS,* 8 (12), 1663–1668, 1994.

40. Plaeger-Marshall, S., Isacescu, V., O'Rourke, S., Bertolli, J., Bryson, Y.J., and Stiehm, E.R., T cell activation in pediatric AIDS pathogenesis: three-color immunophenotyping, *Clin. Immunol. Immunopathol.,* 71 (1), 19–26, 1994.

41. Giorgi, J.V., Liu, Z., Hultin, L.E., Cumberland, W.G., Hennessey, K., and Detels, R., Elevated levels of CD38+ CD8+ T cells in HIV infection add to the prognostic value of low CD4+ T cell levels: results of 6 years of follow-up. The Los Angeles Center, Multicenter AIDS Cohort Study, *J. Acquir. Immune Defic. Syndr.,* 6 (8), 904–912, 1993.

42. Liu, Z., Cumberland, W.G., Hultin, L.E., Prince, H.E., Detels, R., and Giorgi, J.V., Elevated CD38 antigen expression on CD8+ T cells is a stronger marker for the risk of chronic HIV disease progression to AIDS and death in the Multicenter AIDS Cohort Study than CD4+ cell count, soluble immune activation markers, or combinations of HLA-DR and CD38 expression, *J. Acquir. Immune Defic. Syndr. Hum. Retrovirol.,* 16 (2), 83–92, 1997.

43. Hazenberg, M.D., Stuart, J.W., Otto, S.A., Borleffs, J.C., Boucher, C.A., de Boer, R.J., Miedema, F., and Hamann, D., T cell division in human immunodeficiency virus (HIV)-1 infection is mainly due to immune activation: a longitudinal analysis in patients before and during highly active antiretroviral therapy (HAART), *Blood,* 95 (1), 249–255, 2000.

44. Hazenberg, M.D., Otto, S.A., van Benthem, B.H., Roos, M.T., Coutinho, R.A., Lange, J.M., Hamann, D., Prins, M., and Miedema, F., Persistent immune activation in HIV-1 infection is associated with progression to AIDS, *AIDS,* 17 (13), 1881–1888, 2003.

45. Douek, D.C., McFarland, R.D., Keiser, P.H., Gage, E.A., Massey, J.M., Haynes, B.F., Polis, M.A., Haase, A.T., Feinberg, M.B., Sullivan, J.L., Jamieson, B.D., Zack, J.A., Picker, L.J., and Koup, R.A., Changes in thymic function with age and during the treatment of HIV infection, *Nature,* 396 (6712), 690–695, 1998.

46. Douek, D.C., Betts, M.R., Hill, B.J., Little, S.J., Lempicki, R., Metcalf, J.A., Casazza, J., Yoder, C., Adelsberger, J.W., Stevens, R.A., Baseler, M.W., Keiser, P., Richman, D.D., Davey, R.T., and Koup, R.A., Evidence for increased T cell turnover and decreased thymic output in HIV infection, *J. Immunol.,* 167 (11), 6663–6668, 2001.

47. Lewin, S.R., Ribeiro, R.M., Kaufmann, G.R., Smith, D., Zaunders, J., Law, M., Solomon, A., Cameron, P.U., Cooper, D., and Perelson, A.S., Dynamics of T cells and TCR excision circles differ after treatment of acute and chronic HIV infection, *J. Immunol.,* 169 (8), 4657–4666, 2002.

48. Ye, P. and Kirschner, D.E., Reevaluation of T cell receptor excision circles as a measure of human recent thymic emigrants, *J. Immunol.,* 168 (10), 4968–4979, 2002.

49. Nobile, M., Correa, R., Borghans, J.A., D'Agostino, C., Schneider, P., De Boer, R.J., and Pantaleo, G., *De novo* T cell generation in patients at different ages and stages of HIV-1 disease, *Blood,* 104 (2), 470–477, 2004.

50. Sodora, D.L., Milush, J.M., Ware, F., Wozniakowski, A., Montgomery, L., McClure, H.M., Lackner, A.A., Marthas, M., Hirsch, V., Johnson, R.P., Douek, D.C., and Koup, R.A., Decreased levels of recent thymic emigrants in peripheral blood of simian immunodeficiency virus-infected macaques correlate with alterations within the thymus, *J. Virol.,* 76 (19), 9981–9990, 2002.

51. Resino, S., Correa, R., Bellon, J.M., and Munoz-Fernandez, M.A., Preserved immune system in long-term asymptomatic vertically HIV-1 infected children, *Clin. Exp. Immunol.,* 132 (1), 105–112, 2003.

52. Hazenberg, M.D., Otto, S.A., Cohen Stuart, J.W., Verschuren, M.C., Borleffs, J.C., Boucher, C.A., Coutinho, R.A., Lange, J.M., Rinke de Wit, T.F., Tsegaye, A., van Dongen, J.J., Hamann, D., de Boer, R.J., and Miedema, F., Increased cell division but not thymic dysfunction rapidly affects the T cell receptor excision circle content of the naive T cell population in HIV-1 infection, *Nat. Med.,* 6 (9), 1036–1042, 2000.

53. Hazenberg, M.D., Borghans, J.A., de Boer, R.J., and Miedema, F., Thymic output: a bad TREC record, *Nat. Immunol.,* 4 (2), 97–99, 2003.

54. Johnston, A.M., Valentine, M.E., Ottinger, J., Baydo, R., Gryszowka, V., Vavro, C., Weinhold, K., St Clair, M., and McKinney, R.E., Immune reconstitution in human immunodeficiency virus-infected children receiving highly active antiretroviral therapy: a cohort study, *Pediatr. Infect. Dis. J.,* 20 (10), 941–946, 2001.

55. Steffens, C.M., Smith, K.Y., Landay, A., Shott, S., Truckenbrod, A., Russert, M., and Al-Harthi, L., T cell receptor excision circle (TREC) content following maximum HIV suppression is equivalent in HIV-infected and HIV-uninfected individuals, *AIDS,* 15 (14), 1757–1764, 2001.

56. Delgado, J., Leal, M., Ruiz-Mateos, E., Martinez-Moya, M., Rubio, A., Merchante, E., de la Rosa, R., Sanchez-Quijano, A., and Lissen, E., Evidence of thymic function in heavily antiretroviral-treated human immunodeficiency virus type 1-infected adults with long-term virologic treatment failure, *J. Infect. Dis.*, 186 (3), 410–414, 2002.

57. Kolte, L., Dreves, A.M., Ersboll, A.K., Strandberg, C., Jeppesen, D.L., Nielsen, J.O., Ryder, L.P., and Nielsen, S.D., Association between larger thymic size and higher thymic output in human immuno-deficiency virus-infected patients receiving highly active antiretroviral therapy, *J. Infect. Dis.*, 185 (11), 1578–1585, 2002.

58. Wellons, M.F., Ottinger, J.S., Weinhold, K.J., Gryszowka, V., Sanders, L.L., Edwards, L.J., Gooding, M.E., Thomasch, J.R., and Bartlett, J.A., Immunologic profile of human immunodeficiency virus-infected patients during viral remission and relapse on antiretroviral therapy, *J. Infect. Dis.*, 183 (10), 1522–1525, 2001.

59. Kou, Z.C., Puhr, J.S., Wu, S.S., Goodenow, M.M., and Sleasman, J.W., Combination antiretroviral therapy results in a rapid increase in T cell receptor variable region repertoire diversity within CD45RA CD8 T cells in human immunodeficiency virus-infected children, *J. Infect. Dis.*, 187 (3), 385–397, 2003.

60. Schabath, R., Muller, G., Schubel, A., Kremmer, E., Lipp, M., and Forster, R., The murine chemokine receptor CXCR4 is tightly regulated during T cell development and activation, *J. Leukoc. Biol.*, 66 (6), 996–1004, 1999.

61. Berkowitz, R.D., Beckerman, K.P., Schall, T.J., and McCune, J.M., CXCR4 and CCR5 expression delineates targets for HIV-1 disruption of T cell differentiation, *J. Immunol.*, 161 (7), 3702–3710, 1998.

62. Taylor, J.R., Jr., Kimbrell, K.C., Scoggins, R., Delaney, M., Wu, L., and Camerini, D., Expression and function of chemokine receptors on human thymocytes: implications for infection by human immunodeficiency virus type 1, *J. Virol.*, 75 (18), 8752–8760, 2001.

63. Dairaghi, D.J., Franz-Bacon, K., Callas, E., Cupp, J., Schall, T.J., Tamraz, S.A., Boehme, S.A., Taylor, N., and Bacon, K.B., Macrophage inflammatory protein-1 induces migration and activation of human thymocytes, *Blood*, 91 (8), 2905–2913, 1998.

64. Reyes, R.A., Canfield, D.R., Esser, U., Adamson, L.A., Brown, C.R., Cheng-Mayer, C., Gardner, M.B., Harouse, J.M., and Luciw, P.A., Induction of simian AIDS in infant rhesus macaques infected with CCR5- or CXCR4-utilizing simian-human immunodeficiency viruses is associated with distinct lesions of the thymus, *J. Virol.*, 78 (4), 2121–2130, 2004.

65. Su, L., Kaneshima, H., Bonyhadi, M., Salimi, S., Kraft, D., Rabin, L., and McCune, J.M., HIV-1-induced thymocyte depletion is associated with indirect cytopathogenicity and infection of progenitor cells *in vivo*, *Immunity*, 2 (1), 25–36, 1995.

66. Neben, K., Heidbreder, M., Muller, J., Marxer, A., Petry, H., Didier, A., Schimpl, A., Hunig, T., and Kerkau, T., Impaired thymopoietic potential of immature CD3()CD4(+)CD8() T cell precursors from SIV-infected rhesus monkeys, *Int. Immunol.*, 11 (9), 1509–1518, 1999.

67. Chene, L., Nugeyre, M.T., Guillemard, E., Moulian, N., Barre-Sinoussi, F., and Israel, N., Thymocyte-thymic epithelial cell interaction leads to high-level replication of human immunodeficiency virus exclusively in mature CD4(+) CD8(.) CD3(+) thymocytes: a critical role for tumor necrosis factor and interleukin-7, *J. Virol.*, 73 (9), 7533–7542, 1999.

68. Schmitt, N., Chene, L., Boutolleau, D., Nugeyre, M.T., Guillemard, E., Versmisse, P., Jacquemot, C., Barre-Sinoussi, F., and Israel, N., Positive regulation of CXCR4 expression and signaling by inter-leukin-7 in CD4+ mature thymocytes correlates with their capacity to favor human immunodeficiency X4 virus replication, *J. Virol.*, 77 (10), 5784–5793, 2003.

69. Chene, L., Nugeyre, M.T., Barre-Sinoussi, F., and Israel, N., High-level replication of human immuno-deficiency virus in thymocytes requires NF-B activation through interaction with thymic epithelial cells, *J. Virol.*, 73 (3), 2064–2073, 1999.

70. Lee, S., Goldstein, H., Baseler, M., Adelsberger, J., and Golding, H., Human immunodeficiency virus type 1 infection of mature CD3hiCD8+ thymocytes, *J. Virol.*, 71 (9), 6671–6676, 1997.

71. Kitchen, S.G., Uittenbogaart, C.H., and Zack, J.A., Mechanism of human immunodeficiency virus type 1 localization in CD4-negative thymocytes: differentiation from a CD4-positive precursor allows productive infection, *J. Virol.*, 71 (8), 5713–5722, 1997.

72. Bonyhadi, M.L., Su, L., Auten, J., McCune, J.M., and Kaneshima, H., Development of a human thymic organ culture model for the study of HIV pathogenesis, *AIDS Res. Hum. Retroviruses*, 11 (9), 1073–1080, 1995.

73. Jamieson, B.D., Uittenbogaart, C.H., Schmid, I., and Zack, J.A., High viral burden and rapid CD4+ cell depletion in human immunodeficiency virus type 1-infected SCID-hu mice suggest direct viral killing of thymocytes *in vivo*, *J. Virol.*, 71 (11), 8245–8253, 1997.

74. Wykrzykowska, J.J., Rosenzweig, M., Veazey, R.S., Simon, M.A., Halvorsen, K., Desrosiers, R.C., Johnson, R.P., and Lackner, A.A., Early regeneration of thymic progenitors in rhesus macaques infected with simian immunodeficiency virus, *J. Exp. Med.*, 187 (11), 1767–1778, 1998.

75. Iida, T., Ichimura, H., Shimada, T., Ibuki, K., Ui, M., Tamaru, K., Kuwata, T., Yonehara, S., Imanishi, J., and Hayami, M., Role of apoptosis induction in both peripheral lymph nodes and thymus in progressive loss of CD4+ cells in SHIV-infected macaques, *AIDS Res. Hum. Retroviruses*, 16 (1), 9–18, 2000.

76. Rosenzweig, M., Connole, M., Forand-Barabasz, A., Tremblay, M.P., Johnson, R.P., and Lackner, A.A., Mechanisms associated with thymocyte apoptosis induced by simian immunodeficiency virus, *J. Immunol.*, 165 (6), 3461–3468, 2000.

77. Rosenzweig, M., Clark, D.P., and Gaulton, G.N., Selective thymocyte depletion in neonatal HIV-1 thymic infection, *AIDS*, 7 (12), 1601–1605, 1993.

78. Guillemard, E., Nugeyre, M.T., Chene, L., Schmitt, N., Jacquemot, C., Barre-Sinoussi, F., and Israel, N., Interleukin-7 and infection itself by human immunodeficiency virus 1 favor virus persistence in mature CD4(+)CD8(−)CD3(+) thymocytes through sustained induction of Bcl-2, *Blood*, 98 (7), 2166–2174, 2001.

79. Autran, B., Guiet, P., Raphael, M., Grandadam, M., Agut, H., Candotti, D., Grenot, P., Puech, F., Debre, P., and Cesbron, J.Y., Thymocyte and thymic microenvironment alterations during a systemic HIV infection in a severe combined immunodeficient mouse model, *AIDS*, 10 (7), 717–727, 1996.

80. Numazaki, K., Bai, X.Q., Goldman, H., Wong, I., Spira, B., and Wainberg, M.A., Infection of cultured human thymic epithelial cells by human immunodeficiency virus, *Clin. Immunol. Immunopathol.*, 51 (2), 185–195, 1989.

81. Braun, J., Valentin, H., Nugeyre, M.T., Ohayon, H., Gounon, P., and Barre-Sinoussi, F., Productive and persistent infection of human thymic epithelial cells *in vitro* with HIV-1, *Virology*, 225 (2), 413–418, 1996.

82. Kourtis, A.P., Ibegbu, C.C., Scinicarrielo, F., and Lee, F.K., Effects of different HIV strains on thymic epithelial cultures, *Pathobiology*, 69 (6), 329–332, 2001.

83. Sandborg, C.I., Imfeld, K.L., Zaldivar, F., Jr., and Berman, M.A., HIV type 1 induction of interleukin 1 and 6 production by human thymic cells, *AIDS Res. Hum. Retroviruses*, 10 (10), 1221–1229, 1994.

84. Stanley, S.K., McCune, J.M., Kaneshima, H., Justement, J.S., Sullivan, M., Boone, E., Baseler, M., Adelsberger, J., Bonyhadi, M., and Orenstein, J., Human immunodeficiency virus infection of the human thymus and disruption of the thymic microenvironment in the SCID-hu mouse, *J. Exp. Med.*, 178 (4), 1151–1163, 1993.

85. Cameron, P.U., Lowe, M.G., Sotzik, F., Coughlan, A.F., Crowe, S.M., and Shortman, K., The interaction of macrophage and non-macrophage tropic isolates of HIV-1 with thymic and tonsillar dendritic cells *in vitro*, *J. Exp. Med.*, 183 (4), 1851–1856, 1996.

86. Beaulieu, S., Kessous, A., Landry, D., Montplaisir, S., Bergeron, D., and Cohen, E.A., *In vitro* characterization of purified human thymic dendritic cells infected with human immunodeficiency virus type 1, *Virology*, 222 (1), 214–226, 1996.

87. Llano, A., Barretina, J., Gutierrez, A., Blanco, J., Cabrera, C., Clotet, B., and Este, J.A., Interleukin-7 in plasma correlates with CD4 T cell depletion and may be associated with emergence of syncytium-inducing variants in human immunodeficiency virus type 1-positive individuals, *J. Virol.*, 75 (21), 10319–10325, 2001.

88. Napolitano, L.A., Grant, R.M., Deeks, S.G., Schmidt, D., De Rosa, S.C., Herzenberg, L.A., Herndier, B.G., Andersson, J., and McCune, J.M., Increased production of IL-7 accompanies HIV-1-mediated T cell depletion: implications for T cell homeostasis, *Nat. Med.*, 7 (1), 73–79, 2001.

89. Correa, R., Resino, S., and Munoz-Fernandez, M.A., Increased interleukin-7 plasma levels are associated with recovery of CD4+ T cells in HIV-infected children, *J. Clin. Immunol.*, 23 (5), 401–406, 2003.

90. Pedroza-Martins, L., Gurney, K.B., Torbett, B.E., and Uittenbogaart, C.H., Differential tropism and replication kinetics of human immunodeficiency virus type 1 isolates in thymocytes: coreceptor expression allows viral entry, but productive infection of distinct subsets is determined at the postentry level, *J. Virol.*, 72 (12), 9441–9452, 1998.

91. Uittenbogaart, C.H., Boscardin, W.J., Anisman-Posner, D.J., Koka, P.S., Bristol, G., and Zack, J.A., Effect of cytokines on HIV-induced depletion of thymocytes *in vivo, AIDS,* 14 (10), 1317–1325, 2000.

92. von Freeden-Jeffry, U., Solvason, N., Howard, M., and Murray, R., The earliest T lineage-committed cells depend on IL-7 for Bcl-2 expression and normal cell cycle progression, *Immunity,* 7 (1), 147–154, 1997.

93. Peschon, J.J., Morrissey, P.J., Grabstein, K.H., Ramsdell, F.J., Maraskovsky, E., Gliniak, B.C., Park, L.S., Ziegler, S.F., Williams, D.E., and Ware, C.B., Early lymphocyte expansion is severely impaired in interleukin 7 receptor-deficient mice, *J. Exp. Med.,* 180 (5), 1955–1960, 1994.

94. von Freeden-Jeffry, U., Vieira, P., Lucian, L.A., McNeil, T., Burdach, S.E., and Murray, R., Lymphopenia in interleukin (IL)-7 gene-deleted mice identifies IL-7 as a nonredundant cytokine, *J. Exp. Med.,* 181 (4), 1519–1526, 1995.

95. Maraskovsky, E., O'Reilly, L.A., Teepe, M., Corcoran, L.M., Peschon, J.J., and Strasser, A., Bcl-2 can rescue T lymphocyte development in interleukin-7 receptor-deficient mice but not in mutant rag-1$^{/}$ mice, *Cell,* 89 (7), 1011–1019, 1997.

96. Akashi, K., Kondo, M., von Freeden-Jeffry, U., Murray, R., and Weissman, I.L., Bcl-2 rescues T lymphopoiesis in interleukin-7 receptor-deficient mice, *Cell,* 89 (7), 1033–1041, 1997.

97. Steffens, C.M., Managlia, E.Z., Landay, A., and Al-Harthi, L., Interleukin-7-treated naive T cells can be productively infected by T cell-adapted and primary isolates of human immunodeficiency virus 1, *Blood,* 99 (9), 3310–3318, 2002.

98. Wang, S.D., Huang, K.J., Lin, Y.S., and Lei, H.Y., Sepsis-induced apoptosis of the thymocytes in mice, *J. Immunol.,* 152 (10), 5014–5021, 1994.

99. Hernandez-Caselles, T. and Stutman, O., Immune functions of tumor necrosis factor. I. Tumor necrosis factor induces apoptosis of mouse thymocytes and can also stimulate or inhibit IL-6-induced proliferation depending on the concentration of mitogenic costimulation, *J. Immunol.,* 151 (8), 3999–4012, 1993.

100. Sytwu, H.K., Liblau, R.S., and McDevitt, H.O., The roles of Fas/APO-1 (CD95) and TNF in antigen-induced programmed cell death in T cell receptor transgenic mice, *Immunity,* 5 (1), 17–30, 1996.

101. Baseta, J.G. and Stutman, O., TNF regulates thymocyte production by apoptosis and proliferation of the triple negative (CD3CD4CD8) subset, *J. Immunol.,* 165 (10), 5621–5630, 2000.

102. Ellaurie, M. and Rubinstein, A., Tumor necrosis factor- in pediatric HIV-1 infection, *AIDS,* 6 (11), 1265–1268, 1992.

103. Lahdevirta, J., Maury, C.P., Teppo, A.M., and Repo, H., Elevated levels of circulating cachectin/tumor necrosis factor in patients with acquired immunodeficiency syndrome, *Am. J. Med.,* 85 (3), 289–291, 1988.

104. Reddy, M.M., Sorrell, S.J., Lange, M., and Grieco, M.H., Tumor necrosis factor and HIV P24 antigen levels in serum of HIV-infected populations, *J. Acquir. Immune Defic. Syndr.,* 1 (5), 436–440, 1988.

105. Yurasov, S.V., Pettoello-Mantovani, M., Raker, C.A., and Goldstein, H., HIV type 1 infection of human fetal bone marrow cells induces apoptotic changes in hematopoietic precursor cells and suppresses their *in vitro* differentiation and capacity to engraft SCID mice, *AIDS Res. Hum. Retroviruses,* 15 (18), 1639–1652, 1999.

106. Rothe, M., Chene, L., Nugeyre, M.T., Braun, J., Barre-Sinoussi, F., and Israel, N., Contact with thymic epithelial cells as a prerequisite for cytokine-enhanced human immunodeficiency virus type 1 replication in thymocytes, *J. Virol.,* 72 (7), 5852–61, 1998.

107. von Sydow, M., Sonnerborg, A., Gaines, H., and Strannegard, O., Interferon- and tumor necrosis factor- in serum of patients in various stages of HIV-1 infection, *AIDS Res. Hum. Retroviruses,* 7 (4), 375–380, 1991.

108. Bendriss-Vermare, N., Barthelemy, C., Durand, I., Bruand, C., Dezutter-Dambuyant, C., Moulian, N., Berrih-Aknin, S., Caux, C., Trinchieri, G., and Briere, F., Human thymus contains IFN--producing CD11c(), myeloid CD11c(+), and mature interdigitating dendritic cells, *J. Clin. Invest.,* 107 (7), 835–844, 2001.

109. Keir, M.E., Stoddart, C.A., Linquis, T., Stepps, V., Moreno, M.E., and McCune, J.M., IFN- secretion by type 2 predendritic cells up-regulates MHC class I in the HIV-1-infected thymus, *J. Immunol.,* 168 (1), 325–331, 2002.

110. Lin, Q., Dong, C., and Cooper, M.D., Impairment of T and B cell development by treatment with a type I interferon, *J. Exp. Med.,* 187 (1), 79–87, 1998.

111. Vidalain, P.O., Laine, D., Zaffran, Y., Azocar, O., Serve, T., Delprat, C., Wild, T.F., Rabourdin-Combe, C., and Valentin, H., Interferons mediate terminal differentiation of human cortical thymic epithelial cells, *J. Virol.,* 76 (13), 6415–6424, 2002.

112. Keir, M.E., Rosenberg, M.G., Sandberg, J.K., Jordan, K.A., Wiznia, A., Nixon, D.F., Stoddart, C.A., and McCune, J.M., Generation of CD3+CD8low thymocytes in the HIV type 1-infected thymus, *J. Immunol.,* 169 (5), 2788–2796, 2002.

113. Panaretakis, T., Pokrovskaja, K., Shoshan, M.C., and Grander, D., Interferon--induced apoptosis in U266 cells is associated with activation of the proapoptotic Bcl-2 family members Bak and Bax, *Oncogene,* 22 (29), 4543–4556, 2003.

114. Stylianou, E., Aukrust, P., Kvale, D., Muller, F., and Froland, S.S., IL-10 in HIV infection: increasing serum IL-10 levels with disease progression — down-regulatory effect of potent anti-retroviral therapy, *Clin. Exp. Immunol.,* 116 (1), 115–120, 1999.

115. Barcellini, W., Rizzardi, G.P., Borghi, M.O., Fain, C., Lazzarin, A., and Meroni, P.L., TH1 and TH2 cytokine production by peripheral blood mononuclear cells from HIV-infected patients, *AIDS,* 8 (6), 757–762, 1994.

116. Clerici, M., Wynn, T.A., Berzofsky, J.A., Blatt, S.P., Hendrix, C.W., Sher, A., Coffman, R.L., and Shearer, G.M., Role of interleukin-10 in T helper cell dysfunction in asymptomatic individuals infected with the human immunodeficiency virus, *J. Clin. Invest.,* 93 (2), 768–775, 1994.

117. Kovalev, G., Duus, K., Wang, L., Lee, R., Bonyhadi, M., Ho, D., McCune, J.M., Kaneshima, H., and Su, L., Induction of MHC class I expression on immature thymocytes in HIV-1-infected SCID-hu Thy/Liv mice: evidence of indirect mechanisms, *J. Immunol.,* 162 (12), 7555–7562, 1999.

118. Verges, B., Chavanet, P., Desgres, J., Vaillant, G., Waldner, A., Brun, J.M., and Putelat, R., Adrenal function in HIV infected patients, *Acta Endocrinol. (Copenh.),* 121 (5), 633–637, 1989.

119. Oberfield, S.E., Cowan, L., Levine, L.S., George, A., David, R., Litt, A., Rojas, V., and Kairam, R., Altered cortisol response and hippocampal atrophy in pediatric HIV disease, *J. Acquir. Immune Defic. Syndr.,* 7 (1), 57–62, 1994.

120. Refaeli, Y., Levy, D.N., and Weiner, D.B., The glucocorticoid receptor type II complex is a target of the HIV-1 Vpr gene product, *Proc. Natl. Acad. Sci. U.S.A.,* 92 (8), 3621–3625, 1995.

121. Sherman, M.P., de Noronha, C.M., Pearce, D., and Greene, W.C., Human immunodeficiency virus type 1 Vpr contains two leucine-rich helices that mediate glucocorticoid receptor coactivation independently of its effects on G(2) cell cycle arrest, *J. Virol.,* 74 (17), 8159–8165, 2000.

122. Kino, T., Gragerov, A., Slobodskaya, O., Tsopanomichalou, M., Chrousos, G.P., and Pavlakis, G.N., Human immunodeficiency virus type 1 (HIV-1) accessory protein Vpr induces transcription of the HIV-1 and glucocorticoid-responsive promoters by binding directly to p300/CBP coactivators, *J. Virol.,* 76 (19), 9724–9734, 2002.

123. Ayyavoo, V., Mahboubi, A., Mahalingam, S., Ramalingam, R., Kudchodkar, S., Williams, W.V., Green, D.R., and Weiner, D.B., HIV-1 Vpr suppresses immune activation and apoptosis through regulation of nuclear factor B, *Nat. Med.,* 3 (10), 1117–1123, 1997.

124. Vanitharani, R., Mahalingam, S., Rafaeli, Y., Singh, S.P., Srinivasan, A., Weiner, D.B., and Ayyavoo, V., HIV-1 Vpr transactivates LTR-directed expression through sequences present within -278 to -176 and increases virus replication *in vitro, Virology,* 289 (2), 334–342, 2001.

125. Bressler, P., Poli, G., Justement, J.S., Biswas, P., and Fauci, A.S., Glucocorticoids synergize with tumor necrosis factor in the induction of HIV expression from a chronically infected promonocytic cell line, *AIDS Res. Hum. Retroviruses,* 9 (6), 547–551, 1993.

126. Katsikis, P.D., Wunderlich, E.S., Smith, C.A., and Herzenberg, L.A., Fas antigen stimulation induces marked apoptosis of T lymphocytes in human immunodeficiency virus-infected individuals, *J. Exp. Med.,* 181 (6), 2029–2036, 1995.

127. Oyaizu, N., McCloskey, T.W., Than, S., Hu, R., Kalyanaraman, V.S., and Pahwa, S., Cross-linking of CD4 molecules upregulates Fas antigen expression in lymphocytes by inducing interferon- and tumor necrosis factor- secretion, *Blood,* 84 (8), 2622–2631, 1994.

128. Desbarats, J., Freed, J.H., Campbell, P.A., and Newell, M.K., Fas (CD95) expression and death-mediating function are induced by CD4 cross-linking on CD4+ T cells, *Proc. Natl. Acad. Sci. U.S.A.,* 93 (20), 11014–11018, 1996.

129. Westendorp, M.O., Frank, R., Ochsenbauer, C., Stricker, K., Dhein, J., Walczak, H., Debatin, K.M., and Krammer, P.H., Sensitization of T cells to CD95-mediated apoptosis by HIV-1 Tat and gp120, *Nature,* 375 (6531), 497–500, 1995.

130. Badley, A.D., McElhinny, J.A., Leibson, P.J., Lynch, D.H., Alderson, M.R., and Paya, C.V., Upregulation of Fas ligand expression by human immunodeficiency virus in human macrophages mediates apoptosis of uninfected T lymphocytes, *J. Virol.*, 70 (1), 199–206, 1996.

131. Anderson, K.L., Anderson, G., Michell, R.H., Jenkinson, E.J., and Owen, J.J., Intracellular signaling pathways involved in the induction of apoptosis in immature thymic T lymphocytes, *J. Immunol.*, 156 (11), 4083–4091, 1996.

132. Zhou, T., Fleck, M., Mueller-Ladner, U., Yang, P., Wang, Z., Gay, S., Matsumoto, S., and Mountz, J.D., Kinetics of Fas-induced apoptosis in thymic organ culture, *J. Clin. Immunol.*, 17 (1), 74–84, 1997.

133. Jenkins, M., Keir, M., and McCune, J.M., Fas is expressed early in human thymocyte development but does not transmit an apoptotic signal, *J. Immunol.*, 163 (3), 1195–1204, 1999.

134. Jenkins, M., Keir, M., and McCune, J.M., A membrane-bound Fas decoy receptor expressed by human thymocytes, *J. Biol. Chem.*, 275 (11), 7988–7993, 2000.

135. Brady, H.J., Pennington, D.J., Miles, C.G., and Dzierzak, E.A., CD4 cell surface downregulation in HIV-1 Nef transgenic mice is a consequence of intracellular sequestration, *EMBO J.*, 12 (13), 4923–4932, 1993.

136. Skowronski, J., Parks, D., and Mariani, R., Altered T cell activation and development in transgenic mice expressing the HIV-1 Nef gene, *EMBO J.*, 12 (2), 703–713, 1993.

137. Verhasselt, B., Naessens, E., Verhofstede, C., De Smedt, M., Schollen, S., Kerre, T., Vanhecke, D., and Plum, J., Human immunodeficiency virus Nef gene expression affects generation and function of human T cells, but not dendritic cells, *Blood*, 94 (8), 2809–2818, 1999.

138. Stove, V., Naessens, E., Stove, C., Swigut, T., Plum, J., and Verhasselt, B., Signaling but not trafficking function of HIV-1 protein Nef is essential for Nef-induced defects in human intrathymic T cell development, *Blood*, 102 (8), 2925–2932, 2003.

139. Minowada, J., Nishizaki, C., Toki, M., Kinzl, P., Otani, T., and Matsuo, Y., *In vitro* study using leukemia cell line panel to demonstrate rapid thymic T cell death due to contact with HIV-1 carrier cell clones, *Leukemia*, 11 (5), 714–722, 1997.

140. Maroder, M., Scarpa, S., Screpanti, I., Stigliano, A., Meco, D., Vacca, A., Stuppia, L., Frati, L., Modesti, A., and Gulino, A., Human immunodeficiency virus type 1 Tat protein modulates fibronectin expression in thymic epithelial cells and impairs *in vitro* thymocyte development, *Cell. Immunol.*, 168 (1), 49–58, 1996.

141. Nie, Z., Phenix, B.N., Lum, J.J., Alam, A., Lynch, D.H., Beckett, B., Krammer, P.H., Sekaly, R.P., and Badley, A.D., HIV-1 protease processes procaspase 8 to cause mitochondrial release of cytochrome c, caspase cleavage and nuclear fragmentation, *Cell Death Differ.*, 9 (11), 1172–1184, 2002.

21 Correlations between Apoptosis and HIV Disease Progression

David H. Dockrell
Anne J. Tunbridge

CONTENTS

INTRODUCTION

Multiple studies have examined apoptosis in CD4 T cells infected with human immunodeficiency virus (HIV) *in vitro* or in CD4 T lymphocytes isolated from HIV-seropositive patients and cultured *ex vivo*.[1] Although these studies documented multiple mechanisms by which HIV can engage signal transduction pathways that initiate apoptosis, evidence that these contribute to CD4 T lymphocyte depletion *in vivo* is more limited. In this regard, a central issue is whether the induction of apoptosis observed represents an essential cause of the selective CD4 T lymphocyte depletion that characterizes HIV infection or whether apoptosis represents an epiphenomenon in association with chronic viral infection and immune activation, as was observed in other viral infections.[2] Attempts to correlate levels of apoptosis with rates of HIV disease progression have produced mixed results—some demonstrate an association,[3,4] and others fail to do so.[5]

RATES OF HIV DISEASE PROGRESSION

Marked individual variation in the rate of HIV disease progression occurs, and to date, a number of individuals with minimal or no disease progression over prolonged periods have been described.[6,7] Multiple definitions were used in the studies cited below, but in general, the term long-term nonprogressors (LTNPs) was applied to individuals who maintain a stable CD4 T lymphocyte count >500/mm^3 for prolonged periods after infection (>10 years in most series) in the absence of antiretroviral therapy.[6] These individuals remain asymptomatic, and the majority have low HIV-1 viral loads.[8–10] In many cases, these individuals have stronger HIV-specific CD8 T lymphocyte responses and higher titers of neutralizing antibodies against HIV-1 than do individuals with progressive HIV disease. More vigorous HIV-specific CD4 T lymphocyte responses are a feature of individuals who control HIV replication without antiretroviral therapy[11] and may be a feature associated with LTNPs.[11,12] Lymph-node architecture is less affected than in those with progressive disease. Many LTNPs do, however, demonstrate lower CD4 T lymphocyte counts than HIV-seronegative controls,[6] and CD4 T lymphocyte counts may decline over longer periods of follow-up in association with a gradual failure in HIV-specific CD8 T lymphocyte responses.[13] Thus, some LTNPs will develop progressive disease on more extended follow-up, even in the presence of attenuated virus, and are more accurately long-term survivors (LTSs) or slow progressors (SPs), not LTNPs.[14] It, therefore, becomes apparent that so-called LTNPs are a heterogeneous group that represents part of a continuum of disease progression with individuals with progressive disease.[15] At the other extreme, individuals who progressed to acquired immunodeficiency syndrome (AIDS) over relatively short periods (<3 to 5 years), or whose CD4 T lymphocyte count declined by at least 20% per year, were labeled rapid progressors (RPs).[16-18]

With longer follow-up, steady attrition of LTNP numbers occurs, as many LTNPs are, in fact, SPs. These considerations are important when interpreting studies that attempt to compare features of apoptosis in LTNPs to those with progressive disease. Studies with small numbers of LTNPs, defined with shorter periods of follow-up, may be underpowered to detect associations between apoptosis and disease progression, as they contain a significant number of individuals who would fail to meet criteria as LTNPs with longer follow-up.

VIRAL AND HOST FACTORS THAT INFLUENCE DISEASE PROGRESSION

HIV disease progression is influenced by a variety of host and virus variables (Table 21.1). Viral factors associated with LTNPs include deletions in the HIV-1 Nef/LTR gene, as illustrated by the Sydney Blood Bank Cohort of LTNPs infected by contaminated blood products.[19] The R77Q mutation in HIV-1 Vpr is also overrepresented in LTNPs.[20] In addition, analysis of the temporal evolution of HIV-1 quasispecies in LTNPs reveals that there is more likely to be development of additional amino acids in the V2 region of HIV-1 Env, which favors CCR5 chemokine receptor usage.[21] The influence of HIV Env on disease progression is also shown by more rapid disease progression in individuals infected with dualtropic syncytium-inducing (SI) strains of HIV than in those infected with non-syncytium-inducing (NSI) strains.[22] Longitudinal analysis of the V3 loop of gp120 in LTNPs also demonstrated acquired mutations that reflect selective pressure of a strong anti-HIV immunologic response,[23] which coincidentally are associated with attenuated cell entry capability.

Host factors that influence disease progression include genetic variability in chemokine co-receptors required for HIV entry. Heterozygosity for the CCR5-Δ32 allele is overrepresented in LTNPs in contrast to HIV-seropositive individuals with progressive disease.[24] In a San Francisco cohort, CCR5-Δ32 heterozygosity occurred in 19.5% of LTNPs but only 10.7% of progressors.[25] Polymorphisms in the CCR5 promoter are also associated with disease progression; CCR5 59029A and 59353C are more common in some cohorts of progressors,[26] and 59353C is associated with slower disease progression in other cohorts.[27] The CCR2B 64I mutant allele was also associated with slower rates of progression in some cohorts.[28] It may exert its effect early in infection,[29] but its impact has not been assessed in all cohorts.[30] It was estimated that the survival rate of 28 to 29% of LTNPs is attributable to heterozygosity for CCR5-Δ32 or CCR2B 64I.[28] Homozygosity for a polymorphism of the 3′ untranslated region of the SDF-1 chemokine gene was associated with slower rates of progression, particularly in the subset of individuals with longer periods of infection.[31] Although less extensively studied, cytokine gene polymorphisms may also affect disease progression; for example, heterozygosity for the IFN-γ gene promoter polymorphism −179T in Afro-Americans is associated with more rapid disease progression.[32]

Immunologic responsiveness may explain the association of certain HLA haplotypes (HLA B27, B51, B57) with slower disease progression. These may enhance the efficiency of presentation

TABLE 21.1
Factors Associated with LTNPs or Slower Rates of Progression

Category	Factor	Results	Ref.
Viral	Nef	Deletions in the HIV-1 Nef/LTR gene more common in LTNPs	19
	Vpr	↑ incidence R77Q mutation in HIV-1 Vpr in LTNPs	20
	Env	HIV-1 Env V2 region more likely to have additional amino acids that favor CCR5 receptor usage in LTNPs	21
	Env	R5 (NSI) Env associated with less rapid progression	22
	GPX	Env gene encodes a selenoprotein homologous to GPX; gene intact in most LTNPs, frequent loss of function mutations in progressors	74
	HIV-2	Associated with ↓ rate of disease progression compared to HIV-1	71
Genetic	CCR5-Δ32	↑ incidence of heterozygosity for CCR5-Δ32 in LTNPs compared to progressors	24, 25
	CCR2B	CCR2B 64I mutant allele associated with ↓ rates of progression in some cohorts	28
	HLA B35/Cw04 haplotype	Heterozygosity at HLA class I loci or lack of alleles B35 and Cw04 associated with 28 to 40% Caucasian LTNPs	34
	KIR3DSI	In combination with Bw4-801le associated with ↓ disease progression	37
	HLA-Bw4	Homozygosity associated with ↓ risk of HIV disease progression	36
Immunologic	CD4 T lymphocyte	Vigorous HIV-specific responses ↑ in LTNPs	11
	Gag-specific CTL	Presence of Gag-specific CTL response associated with slower disease progression	35
	p53	Upregulated in progressors relative to LTNPs or HIV −ve controls	100

Key: ↑, increase; ↓, decrease; HIV+ve, HIV-seropositive; HIV −ve, HIV-seronegative; LTNP, long-term nonprogressor; RP, rapid progressor; CTL, cytotoxic T lymphocyte; Vpr, viral protein r; GPX, human glutathione peroxidase; NSI, nonsyncytial inducing.

of immunodominant epitopes.[33] Conversely, heterozygosity at the HLA class I loci or the lack of alleles B35 and Cw04 is associated with 28 to 40% of Caucasian LTNPs.[34] The B35/Cw04 haplotype in Caucasians LTNPs is associated with decreased natural killer (NK) cell function, and loss of this haplotype is associated with LTNPs.[35] This helps to identify another element in the optimal immunologic response to control HIV replication.

HLA-B alleles are grouped according to the presence of the Bw4 or Bw6 epitope, defined by amino acid residues 79 to 83 of the a1 helix. HLA-Bw4 serves as a ligand for the natural killer cell inhibitory receptor (KIR), and homozygosity of Bw4 is associated with a lower risk of HIV disease progression,[36] whereas polymorphisms of Bw4 are associated with more rapid HIV disease progression.[37] The KIR3DS1 polymorphisms of the KIR, when combined with HLA-B Bw4-801le, are also associated with delayed disease progression.[37] Because cytotoxic T lymphocytes (CTL) also have KIRs, this effect may reflect either NK or CTL function. A major immunologic determinant of rate of progression, however, remains the HIV-specific CTL response. HIV Nef downregulates HLA class I molecules, providing a link between virological and host factors in control of HIV replication.[38] Progression of disease has been linked to loss of Gag-specific CTL responses.[35] The frequency of CD8 T lymphocytes specific for the A2Gag epitope, as detected by tetramer staining, is directly related to CTL responses and is inversely correlated with HIV viral load.[39]

Multiple factors may combine to affect the slow rate of disease progression. For example, in the Sydney Blood Bank Cohort, the individual who maintained LTNP status for the longest period after infection with the Nef-mutant virus was also heterozygous for CCR5-Δ32 and possessed the HLA B57 class I allele.[14] In addition, there is a failure of CCR5-Δ32 heterozygotes to demonstrate a distinct phenotype in comparison to individuals homozygous for the CCR5 wild-type allele.[40]

EVIDENCE FOR AN INFLUENCE OF APOPTOSIS IN DISEASE PROGRESSION IN LENTIVIRUS INFECTION

Primate models provide important insights into the essential features of HIV immunopathogenesis. Chimpanzees infected with HIV-1 have detectable virus in blood and lymph nodes, yet the majority remain asymptomatic, with stable CD4 T lymphocyte counts, and, therefore, resemble LTNPs.[41,42] In infected chimpanzees, there is a lack of chronic lymphocyte activation;[43] however, longitudinal follow-up demonstrates that some chimpanzees can develop a progressive decline in CD4 T lymphocyte numbers and, subsequently, opportunistic infections.[44] Coincident with the delayed disease progression, infected chimpanzees develop higher HIV viral loads and greater expression of lymphocyte activation markers such as CD38 on CD8 T lymphocytes.[45] An additional feature of slow or nonprogressive primate models of retroviral infection is the absence of macrophage infection.[46–48] In contrast to most cases of chimpanzee infection, pathogenic simian immunodeficiency virus (SIV) infections in macaques are characterized by CD4 T lymphocyte depletion and generalized activation of lymphocytes and infection of macrophages.[49]

Apoptosis induction was compared in models of progressive and nonprogressive lentivirus disease. Elevated CD4 T lymphocyte apoptosis is not a feature of HIV infection in chimpanzees that in most cases demonstrate maintenance of CD4 T lymphocyte numbers.[50] Increased levels of CD4 T lymphocyte activation-induced apoptosis are observed in pathogenic models of SIV infection in macaques, however, and are associated with CD4 T cell depletion.[51] *In vivo* apoptosis occurs in bystander uninfected T lymphocytes in pathogenic macaque models.[52] In the first 2 weeks of infection with SIVmac, levels of apoptotic T lymphocytes are predictive of subsequent clinical course and allow for identification of RPs and SPs.[53] Transient apoptosis in thymic progenitors is also demonstrable in juvenile macaques infected with SIV.[54] In a further apathogenic primate model, SIV$_{sm}$ infection of sooty mangabeys results in high levels of viral replication, despite also being characterized by an absence of immunodeficiency.[55] In this model, levels of spontaneous and activation-induced CD4 T lymphocyte apoptosis are unaltered compared with SIV uninfected controls, and limited SIV-specific CD8 T lymphocyte responses are observed. Although immunodeficiency does not occur, CD4 T lymphocyte numbers vary between animals and are inversely correlated with CD8 T lymphocyte activation markers. Altogether, these models suggest that the critical determinants of rate of disease progression are the level of immune activation and CD4 T lymphocyte apoptosis rather than the level of viral replication and lentivirus-specific CTL response.

With primate models, associations between specific molecular pathways of apoptosis induction and patterns of disease progression were identified. The Fas/Fas ligand (FasL) pathway of apoptosis induction was suggested to contribute to CD4 T lymphocyte depletion in some studies but not in others.[56] In a SIV/HIV hybrid virus (SHIV)-infected macaque model, Fas upregulation on T lymphocytes occurred at the same time as the initial burst of viral replication.[57] The initial transient drop in CD4 T lymphocytes coinciding with this was associated with increased spontaneous apoptosis of both CD4 and CD8 T lymphocytes. A second phase of gradual CD4 T lymphocyte decline was associated with enhanced apoptosis in CD4 T lymphocytes, increased Fas expression on CD4 T lymphocytes, and increased FasL in activated CD8 T lymphocytes and B lymphocytes. However, in another macaque model, the transient development of lymphopenia 1 week after infection preceded apoptosis and was not believed to be related to apoptosis.[58] In contrast, infection with a nonpathogenic Nef-deletion derivative of SIV induced lower levels of apoptosis during the more chronic phase of infection, suggesting that CD4 T lymphocyte apoptosis might be more important in the chronic phase of CD4 T lymphocyte depletion than in the initial transient dip that characterizes acute infection.

HIV-infected chimpanzees do not have enhanced levels of CD4 or CD8 T lymphocyte activation or apoptosis after T cell receptor (TCR) activation and are resistant to Fas-mediated apoptosis.[43,59] In the SIV-macaque model of progressive disease, Nef binds to the TCR -chain and stimulates FasL upregulation and cell activation via immunoreceptor tyrosine activation motifs (ITAMs).[60] Nef, which exerts its effect via the transcription factor NFAT, is sufficient for FasL upregulation in a highly pathogenic model that involves the infection of pigtailed macaques with $SIV_{smmPBj14}$.[62] Infection with an SIV strain containing a four-amino-acid deletion in Nef fails to induce the enhanced FasL expression and results in a nonprogressive form of disease in which apoptosis is not obsrved.[61] Importantly, these elegant studies suggest that Nef not only contributes to CD4 T lymphocyte depletion but may also provide a mechanism by which directly infected cells can kill virus-responsive CTLs via FasL, facilitating immune evasion. HIV Tat also contributes to FasL upregulation, but chimpanzee CD4 T cells are resistant to Tat-induced oxidative stress and fail to downregulate manganese superoxide dismutase, unlike lymphocytes in models of progressive lentivirus infection.[63] HIV Vpr is another viral factor that can influence disease progression and induces apoptosis via mitochondrial membrane permeabilization.[64,65] Of interest, there exist marked virus and species-specific differences in the ability of Vpr to induce apoptosis.[66] Chimpanzee cells are resistant to the apoptotic effects of Vpr, and SIV_{agm} Vpr does not induce apoptosis in African green monkey cells. These studies emphasize that the ability of viral factors to influence disease progression correlates to their ability to induce apoptosis in primate models.

LINKS BETWEEN FACTORS IMPLICATED IN DISEASE PROGRESSION AND APOPTOSIS

A number of the factors demonstrated to impact disease progression in cohorts of HIV-seropositive individuals also influence apoptosis (Table 21.2). Viral burden is clearly linked to disease progression.[67] Some studies have documented a correlation between viral load and level of activation-induced or Fas-mediated apoptosis,[68,69] whereas other studies, including one that measured apoptosis in lymphoid tissue, have failed to demonstrate a relationship.[70] These conflicting results may be explained by differences in methodologies and by the measurement of levels of activation-induced as opposed to spontaneous apoptosis. HIV-2 infection is associated with a slower rate of disease progression than HIV-1.[71] A study from Senegal suggests that HIV-2 is also associated with a lower level of CD4 T lymphocyte activation or apoptosis than is HIV-1 in the same population.[72] As mentioned above, all the viral factors demonstrated to affect disease progression (Nef, Tat, Vpr, and Env) also influence apoptosis.[1] Nef and Tat enhance Fas-mediated apoptosis.[61,73] A deletion mutant of Nef has been associated with decreased apoptosis and disease progression in macaques, and Nef-deletion mutants are recognized in cohorts of LTNPs.

Some of the strongest direct evidence for a direct relationship between apoptosis and a viral factor implicated in the rate of disease progression has been provided for Vpr. A specific Vpr R77Q mutation is overrepresented in LTNPs.[20] Mutant Vpr resulted in less mitochondrial membrane permeabilization (MMP) and less caspase activation than did wild-type protein. Importantly, the mutant Vpr was associated with significantly less CD4 and CD8 T lymphocyte depletion when injected into Balb/c mice than was the wild-type protein. In contrast, HIV Env encodes a seleno-protein that is homologous to human glutathione peroxidase (GPX) and that inhibits apoptosis induced by mitochondrial reactive oxygen species (ROS).[74] Of interest is the observation that this gene is intact in isolates from most LTNPs but frequently demonstrates loss-of-function mutations

TABLE 21.2
Evidence for Linkage between Apoptosis and Disease Progression

Category	Factor	Results	Ref.
Viral	Vpr	Vpr R77Q mutation overrepresented in LTNPs; mutant Vpr causes ↓ mitochondrial membrane permeabilization and ↓ caspase activation of wild-type protein; mutant Vpr causes less CD4 and CD8 T lymphocyte depletion in Balb/c mice than in wild-type	20
	Nef	Deletion mutant associated with ↓ apoptosis and ↑ progression in macaques	61
	HIV-1 GPX	Inhibits mitochondrial ROS-induced apoptosis; gene intact in LTNPs, loss of function mutations in progressors	74
Immunologic	Anti-HIV CTL	Decreased HIV-specific CTL response associated with disease progression and HIV-specific CTL have enhanced Fas susceptibility	78, 79
Genetic	CCR5-Δ32	↓ R5 Env-mediated apoptosis in CCR5-Δ32 homozygotes	75, 76
	CXCR4	Used by T-tropic (SI or X4) virus, causes ↑ apoptosis induction and ↑ bystander cell apoptosis in lymph nodes, associated with ↑ progression	75, 77
Animal models	HIV chimpanzee	Levels of apoptosis not increased after HIV exposure in nonpathogenic chimpanzee model	50
	HIV chimpanzee	No increase in CD4 or CD8 T lymphocyte activation or apoptosis after T cell receptor (TCR) activation, resistance to Fas-mediated apoptosis in nonpathogenic chimpanzee model	43, 59
	SIV macaque	Nef deletion strain of SIV fails to induce enhanced FasL expression or apoptosis, causing nonpathogenic disease	61
	SIV macaque	Initial levels of apoptosis in lymphocytes are predictive of rate of disease progression	53
	SIV sooty mangabey	Nonpathogenic model characterized by absence of immunodeficiency but high levels of viral replication; no difference in levels of CD4 cell apoptosis compared to uninfected controls	55
Cohort studies		Blood mononuclear cell apoptosis levels comparable in LTNPs and HIV−ve, but ↑ in HIV+ve progressors	80
		↓ spontaneous apoptosis in PBMCs, ↓ caspase-3 activation in LTNPs compared to RP	82
		↑ levels of apoptosis in CD4 T lymphocytes predicted HIV-related complications (when CD4 count <200 cells/mL)	83
		↑ Fas expression with more advanced disease; ↓ spontaneous apoptosis in individuals with higher CD4 T lymphocyte counts	3
		↓ lymphocyte susceptibility to Fas-mediated apoptosis in 50% LTNPs	88
		LTNPs showed ↓ loss of $\Delta\psi_m$, ↓ superoxide generation, ↓ caspase-1 activation; loss of $\Delta\psi_m$ linked with Fas expression	91

Key: ↑, increase; ↓, decrease; HIV+ve, HIV-seropositive; HIV−ve, HIV-seronegative; SIV, simian immunodeficiency virus; LTNP, long-term nonprogressor; RP, rapid progressor; CTL, cytotoxic T lymphocyte; Vpr, viral protein r; $\Delta\psi_m$, inner mitochondrial transmembrane potential; ROS, reactive oxygen species.

in individuals with progressive disease. Hence, mutation in Vpr can reduce MMP, whereas maintenance of HIV-1 GPX can help prevent ROS-mediated MMP in LTNPs.

HIV Env also contributes to apoptosis induction, which might explain the relationship between changes in gp120 sequence or genetic variation in chemokine receptors and rates of disease progression. Experiments with lymphocytes derived from CCR5-Δ32 homozygous individuals confirm that CCR5 contributes to R5 Env-mediated apoptosis.[75,76] Furthermore, a T-tropic (SI or X4) virus that uses CXCR4 and that is associated with more rapid disease progression is also associated with greater levels of apoptosis induction[75] and enhanced bystander cell apoptosis in lymph nodes.[77]

Many of the other host factors linked to disease progression have a potential direct or indirect link to CTL responses against HIV. Advanced stages of HIV infection may be associated with greater levels of CD8 T lymphocyte apoptosis and a lack of HIV-specific CTL responses.[78] HIV-specific CD8 T lymphocytes have enhanced Fas susceptibility, as compared with CMV-specific T lymphocytes from the same individuals.[79] Hence, levels of CTL apoptosis may have an indirect effect on disease progression.

ANALYSIS OF APOPTOSIS IN PATIENT COHORTS WITH DEFINED RATES OF DISEASE PROGRESSION

Despite the indirect links between levels of apoptosis and markers of disease progression, there is much less information available from well-defined clinical cohorts. A study comparing a small number of LTNPs to individuals with progressive disease or HIV-seronegative controls found that spontaneous, activation-induced, and Fas-mediated apoptosis in unsorted peripheral blood mononuclear cells (PBMCs) was increased in HIV-seropositive individuals but comparable in LTNPs and HIV-seronegative individuals.[80] Levels of spontaneous apoptosis correlated directly with T lymphocyte activation markers and inversely with the percentage of CD4 T lymphocytes. Glutathione was reduced in lymphocytes from LTNPs compared with seronegative controls, suggesting that the LTNP lymphocytes were subject to greater oxidative stress, but the reduction in glutathione was even greater in lymphocytes from individuals with progressive disease. A larger cohort of HIV-seropositive individuals, some of whom were receiving antiretroviral therapy, was studied by Gougeon and colleagues.[81] Culture *ex vivo* resulted in enhanced spontaneous and activation-induced apoptosis in CD4 and CD8 T lymphocytes from HIV-seropositive individuals. LTNPs had similar levels of apoptosis, and RPs had higher levels of apoptosis than in seronegative controls. Apoptosis correlated with activation markers and was inversely correlated with percentage of CD4 T lymphocytes.

Another group analyzing a well-defined group of LTNPs from San Francisco documented decreased spontaneous apoptosis in PBMCs from LTNPs and decreased rates of caspase-3 activation compared with progressors.[82] However, in this study, rates of activation-induced cell death using two separate stimuli (pokeweed mitogen and staphylococcal enterotoxin B) were no different between LTNPs and progressors, although by some, but not all, analyses performed, there was a significant increase in rates of activation-induced apoptosis between LTNPs and seronegative individuals. This study examined only individuals who were not receiving antiretroviral therapy. It should be noted that this study measured apoptosis in PBMCs and used specific comparator groups. Comparator groups for LTNPs were either LTNPs who subsequently developed disease progression (SPs with a comparable duration of infection to LTNPs) or recent seroconverters with comparable CD4 counts to the LTNPs. This allowed the authors to conclude that duration of infection was not associated with level of apoptosis. Increased levels of apoptosis were seen at an early stage of HIV infection in those with progressive disease. Viral load, which was significant in many of this cohort of LTNPs, was not correlated with level of apoptosis in the LTNP but was correlated with level of apoptosis in the recent seroconverters. Furthermore, it was suggested that when the group with long-term infection was considered as a whole (LTNPs and SPs), spontaneous, but not activation-induced, apoptosis correlated with the likelihood of a decline in CD4 T lymphocyte count.

Another study also noted increased spontaneous apoptosis in lymphocytes in asymptomatic seropositive individuals in early stages of disease but lower levels of spontaneous apoptosis in LTNPs.[4] In a longitudinal follow-up of asymptomatic individuals, the rate of CD4 T lymphocyte decline was directly correlated with the level of spontaneous lymphocyte apoptosis. An additional study suggested that spontaneous *ex vivo* apoptosis correlated with complications during short-term follow-up. In this case, high levels of spontaneous apoptosis in CD4 T lymphocytes, as measured by Annexin-V binding, were predictive of HIV-related complications in the subset of individuals with CD4 T lymphocyte counts <200 cells/mL, whereas none of the other virologic or immunologic parameters analyzed displayed this association.[83] A further study analyzed rates of apoptosis in different lymphocyte subsets and found that rates of CD4 T lymphocyte spontaneous apoptosis but not B lymphocyte apoptosis were higher in the subgroup with progressive disease.[84] Importantly, however, the majority of those with progressive disease were on antiretroviral therapy (mainly with dual nucleoside therapy, including AZT [zidovudine]), whereas the slow progressors were not. Antiretroviral therapy may modulate lymphocyte apoptosis,[1] which is an important effect to consider when reviewing the above studies, some of which include patients on antiretroviral therapy and others that do not mention whether or not patients were receiving therapy.

Specific pathways of apoptosis were also investigated in LTNPs. Fas expression was shown to increase with more advanced disease.[3] Once again, spontaneous apoptosis was observed to be lower in individuals with higher CD4 T lymphocyte counts, but the majority of these individuals were also on AZT, with or without a second nucleoside analogue. Although longitudinal follow-up was not included and progression status was not stated, this study suggested a possible link between Fas expression and disease progression. In HIV-infected infants followed for 13 months after delivery, levels of soluble Fas (sFas) and soluble FasL (sFasL) correlated inversely with CD4 T lymphocyte percentages and varied directly with HIV viral loads.[85] However, the opposite was seen in a study of SP hemophiliacs who were followed longitudinally and reported increases in sFas in association with antiretroviral-induced increases in CD4 T lymphocyte count and a negative correlation between viral load and sFas level.[86] In that study, sFasL correlated directly with viral load and inversely with total lymphocyte count. Again, the confounding effects of antiretroviral therapy are apparent, and these studies emphasize the discordance between sFas and sFasL and disease progression. As the function of membrane-bound FasL is better characterized than sFasL, the significance of findings relating levels of sFas and sFasL remains unclear.[87] Intriguingly, decreased susceptibility of lymphocytes to Fas-mediated apoptosis was noted in 9 of 18 (50%) LTNPs in another study.[88] In this study, parents of two of the LTNPs were analyzed, and in each case, one parent showed decreased Fas susceptibility. Creation of a hybrid cell from Fas-susceptible cells and these Fas-resistant cells suggested that the Fas-resistant cells contained a molecule involved in Fas signal transduction that exerted a dominant negative effect. Fas ligation is associated with sphingomyelin breakdown and accumulation of intracellular ceramide.[89] Analysis of a defined Italian cohort of LTNPs (infected by intravenous drug abuse and not receiving antiretroviral therapy) demonstrated lower levels of intracellular ceramide in PBMCs than in a group of individuals with AIDS, although the levels were higher in the LTNPs than in seronegative controls.[90] In addition, LTNPs demonstrated lower percentages of apoptotic CD4 or CD8 T lymphocytes than did individuals with AIDS. In a separate study, the same cohort of LTNPs demonstrated a lower percentage of Fas or FasL positive cells than did the group with AIDS.[91] Fas-mediated apoptosis was also observed to be associated with impairment of CD45 tyrosine phosphatase activity, and CD45 tyrosine phosphatase was not impaired in LTNPs.[92] The impairment of CD45 tyrosine phosphatase activity may be related to oxidative stress and depletion of the intracellular antioxidant glutathione (GSH),[93] although others failed to show restoration of CD45 phosphatase activity after antioxidant treatment.[80]

Evidence was also presented for involvement of the mitochondrial pathway of apoptosis induction in HIV-associated lymphocyte apoptosis.[1] In this regard, the same cohort of Italian LTNPs was studied for evidence of loss of inner mitochondrial transmembrane potential ($\Delta\psi_m$) and superoxide generation.[91] In this study, LTNPs had less loss of $\Delta\psi_m$ and less superoxide generation as well as less caspase-1 activation. Loss of $\Delta\psi_m$ was associated with Fas expression, and lower levels of both

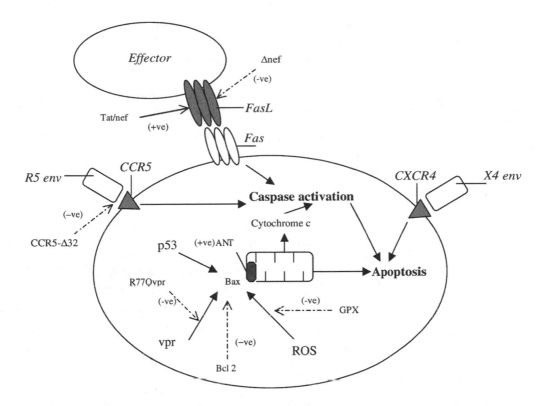

FIGURE 21.1 Factors influencing apoptosis and rate of progression. Outlined are factors implicated in the induction of lymphocyte apoptosis. Factors that result in slower disease progression are represented by dashed arrows, (-ve). Mitochondrial permeabilization is shown in a simplified schematic, as many of the factors interact with each other. (Key: Vpr, viral protein r; ROS, reactive oxygen species; FasL, Fas ligand; GPX, human glutathione peroxidase; ANT, adenine nucleotide translocator.)

parameters were apparent in LTNPs. Furthermore, Ledru and colleagues demonstrated that susceptibility to apoptosis in different lymphocyte subsets was related to the level of Bcl-2 expressed and that this correlated with disease progression.[94] As expression of Bcl-2 family members regulates induction of MMP, this provides further insight into how molecular pathways that regulate apoptosis may influence rates of disease progression.[95] The tumor suppressor p53 also plays a role in induction of mitochondrial pathways of apoptosis in certain settings, such as DNA damage pathways, via its effects on Bcl-2 family members.[96–99] p53 is upregulated in HIV-seropositive progressors relative to LTNPs or seronegative controls.[100] Figure 21.1 summarizes some of the main factors implicated in both modulating apoptosis and retarding disease progression.

IMPACT OF ANTIRETROVIRAL THERAPY ON LEVELS OF LYMPHOCYTE APOPTOSIS

The benefits of combination antiretroviral therapy were demonstrated by improved survival and fewer opportunistic infections.[101] These outcomes were associated with improvement in immunologic dysfunction, such as reduced markers of activation and improved CD4 T lymphocyte response to recall antigens.[102] Despite this, several markers of defective cellular immune responses associated with HIV infection persist, despite the use of highly active antiretroviral therapy (HAART).[103] A large number of studies have suggested that levels of apoptosis are decreased in association with HAART, and these are summarized in Table 21.3.

TABLE 21.3
Relationship of Lymphocyte Apoptosis Levels to Antiretroviral Therapy

Parameter Analyzed	Principal Findings	Follow-up	HAART	Ref.
Mononuclear cell apoptosis in tonsils; spontaneous and Fas-mediated apoptosis in CD4 and CD8 T lymphocytes *ex vivo*	↓ apoptosis in all settings with HAART; ↑ apoptosis during transient infections; no ↓ in elevated FasL expression	48 wk	AZT/3TC/RIT	104
Mononuclear cell apoptosis in rectal mucosa	↓ mononuclear cell apoptosis *in situ*	7 to 10 d	Mixed 2 NA ± IND; NA mostly d4T,3TC	105
Lymphocytes *ex vivo* isolated from tonsils	↓ apoptosis in CD4 and CD8 T cells; apoptotic CD4 T cells mainly memory phenotype and FasL+	48 wk	AZT, 3TC, IND	106
Lymphocyte apoptosis in tonsillar biopsies *in situ*	Normalization of CD4 T cell apoptosis; ↓ in CD8 T cell apoptosis; ↓ in CD4 T cell FasL expression	48 wk	AZT, 3TC, IND	107
Lymphocyte apoptosis in tonsillar biopsies *in situ*	No ↓ in CD4 T cell apoptosis, but ↓ in proliferation	60 wk	AZT, 3TC, RIT	108
D-glucose labeling and analysis of spontaneous apoptosis *ex vivo*	CD4 and CD8 T cell proliferation and apoptosis ↓	5 to 11 wk	ABC, 3TC, EFV, NFV	109
Spontaneous apoptosis	CD4 and CD8 T cell apoptosis ↓ with HAART; CD4 T cell ↑ inversely related to level of CD4 T cell apoptosis	48 wk	ART with 1 or 2 NA (AZT, ddI, ddC, or 3Tc)	149
Spontaneous apoptosis; Fas expression	↓ CD4 T cell apoptosis at wk 26 not wk 8, but Fas expression ↓ at wk 8	26 wk	AZT, 3TC or AZT, ddI ± IND/SAQ	115
Spontaneous apoptosis; not a longitudinal study	Individuals receiving ART had ↓ CD4 T cell apoptosis; apoptosis lower if PI-containing regimen	4 wk	Mixed 2NA ± PI	150
Spontaneous apoptosis	↓ total lymphocyte apoptosis with ART correlated with ↓ in viral load and CD8 T cell activation and inversely correlated with absolute CD4 T cell count	48 wk	d4T or NFV or d4T+NFV × 24 wk then d4T, 3TC, NFV × 24 wk	151
Spontaneous apoptosis in memory and naive subsets of CD4 and CD8 T cells	↓ apoptosis with each parameter associated with increases in numbers of subset and ↓ in activation and viral load	8 wk	d4T, ddI, HD	110
Spontaneous apoptosis	↓ apoptosis associated with a reduction of cells with ↓ Bcl-2	9 d	ART	122
Spontaneous apoptosis	Significant ↓ in CD4 T cell but not CD8 T cell or PBMC apoptosis	8 wk	HAART	152

TABLE 21.3 (Continued)
Relationship of Lymphocyte Apoptosis Levels to Antiretroviral Therapy

Parameter Analyzed	Principal Findings	Follow-up	HAART	Ref.
Spontaneous apoptosis	↓ in apoptosis independent of antiviral effect	8 d	PI	144
Spontaneous and Fas-mediated PBMC apoptosis	↓ spontaneous and Fas-mediated apoptosis early after treatment; ↓ in Fas but not Bcl-2 expression	24 wk	HAART	116
Spontaneous apoptosis in memory and naive subsets of CD4 and CD8 T cells	↓ apoptosis in CD4 and CD8 T cells and in CD4+ CD45RO+ cells in particular; ↓ apoptosis inversely correlated with ↓ activation markers and ↑ CD4 T cell counts	6 mo	PI containing HAART	111
Bcl-2 expression in memory and naive CD4 or CD8 T cells	HAART with undetectable viral load associated with ↑ Bcl-2 as compared to those with detectable viral load or seronegative controls	>3 mo	PI containing HAART	123
Spontaneous apoptosis	↓ apoptosis associated with rise in naive CD4 T cells	12 wk	HAART	153
Spontaneous PBMC apoptosis	↑ antioxidant activity	60 to 72 mo	HAART	154
Spontaneous PBMC apoptosis	In those with virologic response to HAART, increase in CD4 T cells; inversely correlated with PBMC apoptosis	6 mo	HAART	145
Spontaneous apoptosis in lymphocyte subsets	↓ apoptosis in CD4, CD8, and CD4+CD45RO+ cells; lymphocyte apoptosis ↓ at 3 mo but only normalized vs. HIV at 52 wk	52 wk	HAART	112
Spontaneous apoptosis in CD4 or CD8 T lymphocytes; direct comparison in different groups	↓ apoptosis in CD4 and CD8 T lymphocytes in individuals receiving HAART; no difference between PI and NNRTI containing HAART	6 mo	PI or NNRTI containing HAART	155
Spontaneous apoptosis in lymphocyte subsets	↑ apoptosis after 2 wk NA only treatment; apoptosis and Fas expression ↓ on PI; poor immunologic response associated with early increase in CD4 T cell apoptosis and Fas expression	12 wk	2 NA (AZT or d4T+ 3TC) × 2 wk then PI (IND/NFV) added	113
Spontaneous or activation-induced apoptosis in PBMCs	No ↓ in apoptosis with HAART, even after virological failures excluded; ↓ TREC levels after HAART	18 mo	AZT or d4T, 3TC or ddI, IND/RIT/ RIT-SQV	125

(continued)

TABLE 21.3 (Continued)
Relationship of Lymphocyte Apoptosis Levels to Antiretroviral Therapy

Parameter Analyzed	Principal Findings	Follow-up	HAART	Ref.
Spontaneous PBMC apoptosis	↓ PBMC apoptosis with HAART; no difference in ↓ of apoptosis by treatment except a ↓ in apoptosis with PI-containing vs. PI-sparing regimen at wk 1	48 wk	ABC/d4T/3TC or LOP-RIT/NEV	114
Spontaneous or activation-induced apoptosis in PBMCs	After a median period of 5 mo on HAART, activation-induced but not spontaneous apoptosis lower if VL<200 copies/ml than if >200 copies/ml	On HAART 5 mo	2NA + PI	156
Spontaneous, Fas-mediated or activation-induced apoptosis in PBMC subsets from individuals with discordant response to HAART	In individuals with virological treatment failure, no relationship between apoptosis parameters and CD4 T cell ↑; correlation between CD4 T cell recovery and TREC levels	On HAART >1 yr	2NA + PI	126
Fas-mediated CD4 or CD8 T cell apoptosis	↓ Fas-mediated apoptosis after HAART in CD4 and CD8 T cells; ↓ apoptosis in CD4 memory but not naive cells	9 mo	2NA + IND	117
Spontaneous, Fas-mediated, and activation-induced CD4 or CD8 T cell apoptosis	↓ in Fas-mediated and activation-induced but not spontaneous apoptosis in CD4 and CD8 T cells, even without complete virological suppression	10 to 12 wk	Mixed NA, with NEV or NFV or IDV or NFV-SAQ	118
Spontaneous and activation-induced CD4 T cell apoptosis	↓ spontaneous and activation-induced apoptosis	12 wk	2NA + PI (various combinations)	69
Activation-induced apoptosis	↓ CD4 T cell apoptosis on HAART; ↑ apoptosis if HAART stopped	2 to 3 mo (up to 21 mo)	1 or 2 NA (AZT, ddI, ddC, or 3TC)	157
Activation-induced apoptosis	↓ in CD4 T cell apoptosis inversely correlated with percentages of CD4 T cells synthesizing TNF-α	18 mo	AZT or d4T, 3TC or ddI or ddC, IND	158
Spontaneous, Fas-mediated, and activation-induced CD4 or CD8 T cell apoptosis	↓ apoptosis with HAART; all forms of apoptosis inversely correlated with CD4 T cell counts but significant levels of apoptosis persist despite HAART; persistent CD4 T cell apoptosis associated with 3TC; interrupting HAART associated with increased apoptosis	2 yr	2NA + PI	121
Spontaneous or activation-induced apoptosis of CD4 and CD8 T cells	Both treatments ↓ CD4 and CD8 T cell apoptosis	96 wk	HAART or IL-2	120

TABLE 21.3 (Continued)
Relationship of Lymphocyte Apoptosis Levels to Antiretroviral Therapy

Parameter Analyzed	Principal Findings	Follow-up	HAART	Ref.
Spontaneous apoptosis in PBMCs	↓ apoptosis and Fas expression in each treatment arm	12 wk	3TC, d4T, IND ± IL-2 or IL-2 + GM-CSF	119
Spontaneous apoptosis in PBMCs	Spontaneous apoptosis ↓ in both arms but greater decrease with IL-2	28 wk	HAART ± IL-2	159
Spontaneous CD4 and CD8 T cell apoptosis	Spontaneous apoptosis ↑ in both subsets during IL-2 therapy	5 d	HAART + IL-2	160
Spontaneous CD4 and CD8 T cell apoptosis in PBMCs and apoptosis in LN	Transient ↑ in apoptosis with IL-2 in each subset	14 d PBMC 24 mo LN	AZT, ddI, SQV + IL-2	161
Spontaneous apoptosis	Increased after STI	96 wk	STI	162
Spontaneous apoptosis in CD4 and CD8 T cells	↓ apoptosis inversely correlated with a rise in CD4 T cells; ceramide levels fell	5 mo	L-carnitine	163

Key: ↑, increase; ↓, decrease; PBMC, peripheral blood mononuclear cell; D-glucose, deuterated glucose; FasL, Fas ligand; IL-2, interleukin-2; GM-CSF, granulocyte–macrophage colony stimulating factor; LN, lymph node; TREC, T cell receptor excision circle; d, day; wk, week; mo, month; STI, structured treatment interruption; HAART, highly active antiretroviral therapy; ART, antiretroviral therapy (unspecified); NA, nucleoside analogue (various or unspecified); PI, protease inhibitor (various or unspecified); NNRTI, nonnucleoside reverse transcriptase inhibitor; AZT, zidovudine; d4T, stavudine; ddI, didanosine; ddC, zalcitabine; 3TC, lamivudine; ABC, abacavir; EFV, efavirenz; NEV, nevirapine; RIT, ritonavir; SAQ, saquinavir; IND, indinavir; NFV, nelfinavir.

An important source of information included analysis of apoptosis *in vivo*. Analysis of tonsillar biopsies from a subset of individuals enrolled in ACTG 315 demonstrated decreased numbers of Terminal deoxynucleotidyltransferase dUTP nick and labeling (TUNEL) -positive mononuclear cells after initiation of antiretroviral therapy. This was associated with increased CD4 T lymphocyte counts and decreased viral loads, although the exact nature of the TUNEL positive cells was not determined.[104] Kotler and colleagues demonstrated a decline in mononuclear cell apoptosis in the lamina propria of rectal mucosa in the first 7 days after initiation of antiretroviral therapy[105] in association with increases in CD4 T cell numbers and declines in HIV viral burden. Others demonstrated declines in both CD4 and CD8 T lymphocyte apoptosis *ex vivo*, in response to HAART in lymphocytes isolated from tonsillar biopsies.[106] The same group demonstrated decreased percentages of both CD4 and CD8 T lymphocyte apoptosis *in situ* in tonsillar biopsies with HAART. The levels of CD4 T lymphocyte apoptosis normalized, whereas the level of CD8 T lymphocyte apoptosis decreased with 48 weeks of HAART, in association with a decline in HIV viral load.[107] However, another important study did not reach the same conclusion. Zhang and colleagues, analyzing tonsillar biopsies or lymph node biopsies from a group of individuals receiving HAART or seronegative controls, failed to demonstrate decreased CD4 T lymphocyte apoptosis, even though they noted decreased cell proliferation after HAART.[108] In this study, levels of CD4 T lymphocyte apoptosis were enhanced in HIV-seropositive individuals compared with seronegative controls at baseline. The authors interpreted the continued elevation in TUNEL positive cells as reflecting a failure in clearance of early apoptotic cells.

The majority of studies have, however, examined apoptosis *ex vivo* in PBMC or in lymphocyte subsets. The parameter measured was most often spontaneous apoptosis, but Fas-mediated apoptosis or activation-induced apoptosis was also measured. In elegant studies using deuterated glucose to label DNA in proliferating cells and *ex vivo* analysis of PBMC, followed by mathematical modeling, Mohri and colleagues demonstrated that antiretroviral therapy reduced both proliferation and apoptosis in CD4 T lymphocytes in association with increases in CD4 T lymphocyte numbers.[109] This finding is in agreement with the majority of the 40 studies outlined in Table 21.2. Most of these studies show decreased levels of apoptosis either in PBMCs or, more specifically, CD4 and CD8 T lymphocytes. In most cases, this has occurred in response to protease inhibitor (PI)-based therapy. Usually, decreased apoptosis correlated with the immunologic response to therapy, as defined by the increase in CD4 T lymphocyte counts. CD45RO+ memory cells constitute the subset most susceptible to HAART-induced inhibition of apoptosis,[106,110–112] and markers of cell activation are decreased.[106,110,113,114] Apoptosis reduction is associated with decreased Fas expression and decreased Fas-mediated apoptosis,[69,113,115–121] but in a few studies, it is also associated with a rise in Bcl-2 expression.[122,123] Recently, it was suggested that HIV Env can induce p53-mediated apoptosis with upregulation of the proapoptotic BH-3-only Bcl-2 family member Puma. In keeping with this observation, Puma is upregulated in antiretroviral naive, HIV-infected individuals, but this is reversed by initiation of HAART.[124]

Not all studies confirm a beneficial effect of HAART on apoptosis. A number show no reduction in the level of apoptosis by HAART or persistent elevation of apoptosis for long periods after HAART initiation.[115,121,125,126] Elevated sFas levels may not be corrected by up to a year of HAART.[127] The reason for these differences is not immediately obvious, as similar parameters of apoptosis were measured in many studies. One source of variation is that intercurrent illnesses may cause transient elevations in apoptosis, just as they may cause viral blips.[104] The effect of these may be quite marked in series containing small numbers of individuals. Another potentially confounding variable is the effect of antiretroviral therapy, as discussed elsewhere in this book. AZT or 3TC may enhance apoptosis,[121,128] although a further study has suggested that AZT and 3TC may be associated with a more marked decrease in apoptosis than AZT and ddI.[129] The nonnucleoside reverse transcriptase inhibitor (NNRTI) efavirenz may enhance caspase activation and mitochondrial membrane permeabilization *in vitro*.[130] Conversely, the protease inhibitors ritonavir, saquinavir, and nelfinavir were shown to decrease apoptosis in lymphocytes and bone marrow progenitors *in vitro*.[131,132] PIs may prevent activation-associated mitochondrial hyperpolarization.[132,133] Differences may exist between PIs, as evidenced by reports that Indinavir does not inhibit apoptosis.[134,135] In one of these studies, low doses of Indinavir helped promote lymphoproliferation *in vitro*,[135] whereas in a second study, using higher doses, lymphoproliferation was blocked.[134] Although these findings could be interpreted as being due to differences between PIs of different structures, one of these studies also tested Saquinavir and the other Nelfinavir with similar results.[134,135] One of these studies used short-term culture, and this might translate into a later inhibition of apoptosis.[134] The other study involved up to 14 days of *in vitro* culture with PIs before apoptosis assessment and only detected reduced apoptosis with the highest dose of each PI, administered for 14 days to PBMCs from HIV-seropositive but not from HIV-seronegative individuals.[135] The exact reasons for the different findings are unclear.[136] However, a review of individuals receiving postexposure prophylaxis was unable to detect any alteration in apoptosis with AZT; AZT plus 3TC; or AZT, 3TC, plus Indinavir,[137] whereas a different study of postexposure prophylaxis showed reduced susceptibility to apoptosis after Nelfinavir, AZT, and 3TC.[138] This study does not exclude the possibility of an indirect effect on lymphocyte apoptosis in an activation-specific context, such as by HIV infection *in vivo* or by *ex vivo* culture of lymphocytes from HIV-seronegative individuals. Nevertheless, a role for viral independent effects, which include apoptosis inhibition, is supported by reports that PI-based therapy may result in equivalent immunologic responses, despite inferior virological responses compared with regimens not containing a PI.[139,140] Also, the observation that NNRTI-based therapy may be associated with lower levels of CD4 T lymphocyte recovery than PI-based regimens is consistent with this possibility.[141] These findings may be partly explained by differing effects on reduction of apoptosis. The fact that the majority of regimens used in the studies in Table 21.2 contain PIs is in keeping with this observation.

However, some of these studies do not involve PI-based therapy,[110] and a study that compared PI- and NNRTI-based regimens found no differences in rates of CD4 T lymphocyte recovery or apoptosis.[114] One interpretation for these apparently disparate findings is that direct induction of apoptosis, which kills viral-infected cells or depletes activated cells capable of becoming infected with HIV, might aid the virological response.[142] Hence, this helps reduce apoptosis in bystander cells, whereas direct inhibition of bystander apoptosis, particularly in the context of suboptimal virological response by PI, might help boost CD4 T lymphocyte counts.

DISCORDANT RESPONSES TO ANTIRETROVIRAL THERAPY

In many cases, decreased apoptosis correlates with decreased viral load (Table 21.2). However, the response to antiretroviral therapy is often discordant, and either a virological response (as measured by decline in viral load) or immunologic response (as determined by an increase in CD4 T lymphocyte) does not occur.[143] Clinical outcomes in those cases having an immunologic, but not a virological, response are not significantly different from those individuals with a complete response (both virological and immunologic) during 18 months of follow-up.[143] These cases give an ideal opportunity to dissect whether the reduction of apoptosis during HAART is dissociated from control of viral replication or an increase in CD4 T lymphocyte count. Discordant responses are often seen in series involving PI-based regimens, which could be explained by PI effects independent of control of viral replication.[136] The effects of PI monotherapy in the first 8 days of administration include reduced apoptosis, with increases in CD4 and CD8 T lymphocyte counts, but these effects occur in the absence of significant control of viral replication.[144] Conversely, in individuals with a good virological but a poor immunologic response, rates of *ex vivo* apoptosis correlate with CD4 T lymphocyte counts.[145] Under similar circumstances, failure of immunologic response is associated with decreased Bcl-2 expression with enhanced rates of spontaneous apoptosis, compared with individuals demonstrating both immunologic and virological responses to HAART.[146] Interleukin-2 (IL-2) therapy in this setting both increases Bcl-2 expression and enhances CD4 T lymphocyte recovery.

Another insight into discordance is provided by studies looking at the response of HIV-seropositive individuals to glucocorticoids. These agents inhibit apoptosis but have no effect on viral replication, yet they are associated with sustained increases in CD4 T lymphocyte responses.[147,148] However, not all studies analyzing discordant responses identified a relationship between immunologic response and apoptosis induction.[126] Lecossier and colleagues found no link between spontaneous apoptosis and the degree of immune response.[126] In this study, most individuals were receiving ritonavir or indinavir as the PI, along with two unspecified nucleoside analogues. However, PBMCs were cultured without the removal of monocytes, and apoptosis was measured after 48 hours of incubation, which is longer than in many other studies. Total cell loss was measured by comparing initial and postincubation numbers, and although cells may have been in a late stage of apoptosis, quantification mainly involved Annexin-V binding. Interestingly, although activation-induced apoptosis and Fas-mediated apoptosis were increased, there was no increase in levels of spontaneous apoptosis in the HIV-seropositive group compared with seronegative controls. No differences were noted between any apoptosis parameter, and no correlation was apparent between levels of apoptosis and CD4 or CD8 T lymphocyte count, when comparing individuals not on treatment to those with a poor immunologic response. However, the authors noted a significant relationship between T cell receptor excision circles (TRECs) in CD4+CD45RA+ (naive) T lymphocytes, a marker of thymic output, and CD4 T lymphocyte recovery, but no relationship between TRECs and viral load. The authors argued that this supports a correlation between thymic output and CD4 T lymphocyte recovery, which they speculated might be influenced by levels of viral resistance affecting thymic replicative capacity.

CONCLUSIONS

Studies attempting to correlate apoptosis with clinical parameters, such as rate of disease progression or response to therapy, have produced a variety of results. In part, differences can be reconciled by the variety of methods by which apoptosis was induced or measured and by the various criteria used to define groups of differing rates of clinical progression or the variety of antiretroviral agents used. Although a number of studies illustrate no association between specific clinical parameters and apoptosis, arguing for the existence of other contributory factors in the pathogenesis of CD4 T lymphocyte decline during HIV infection, a large number of studies now support a contributory role for apoptosis in clinical outcomes in HIV infection. These, however, are largely associations. The challenge for researchers now is to identify specific models with which to address causality.

ACKNOWLEDGMENTS

This work was supported by a Wellcome Trust Advanced Clinical Fellowship to DHD (#065054).

REFERENCES

1. Badley, A.D., Pilon, A.A., Landay, A., and Lynch, D.H., Mechanisms of HIV-associated lymphocyte apoptosis, *Blood,* 96 (9), 2951–2964, 2000.
2. Akbar, A.N., Borthwick, N., Salmon, M., Gombert, W., Bofill, M., Shamsadeen, N., Pilling, D., Pett, S., Grundy, J.E., and Janossy, G., The significance of low Bcl-2 expression by CD45RO T cells in normal individuals and patients with acute viral infections. The role of apoptosis in T cell memory, *J. Exp. Med.,* 178 (2), 427–438, 1993.
3. Patki, A.H., Georges, D.L., and Lederman, M.M., CD4+-T-cell counts, spontaneous apoptosis, and Fas expression in peripheral blood mononuclear cells obtained from human immunodeficiency virus type 1-infected subjects, *Clin. Diagn. Lab. Immunol.,* 4 (6), 736–741, 1997.
4. Prati, E., Gorla, R., Malacarne, F., Airo, P., Brugnoni, D., Gargiulo, F., Tebaldi, A., Castelli, F., Carosi, G., and Cattaneo, R., Study of spontaneous apoptosis in HIV+ patients: correlation with clinical progression and T cell loss, *AIDS Res. Hum. Retroviruses,* 13 (17), 1501–1508, 1997.
5. Meyaard, L., Otto, S.A., Keet, I.P., Roos, M.T., and Miedema, F., Programmed death of T cells in human immunodeficiency virus infection. No correlation with progression to disease, *J. Clin. Invest.,* 93 (3), 982–988, 1994.
6. Buchbinder, S.P., Katz, M.H., Hessol, N.A., O'Malley, P.M., and Holmberg, S.D., Long-term HIV-1 infection without immunologic progression, *AIDS,* 8 (8), 1123–1128, 1994.
7. Easterbrook, P.J. and Schrager, L.K., Long-term nonprogression in HIV infection: methodological issues and scientific priorities. Report of an international European community—National Institutes of Health Workshop, The Royal Society, London, England, November 27–29, 1995. Scientific Coordinating Committee, *AIDS Res. Hum. Retroviruses,* 14 (14), 1211–1228, 1998.
8. Lifson, A.R., Buchbinder, S.P., Sheppard, H.W., Mawle, A.C., Wilber, J.C., Stanley, M., Hart, C.E., Hessol, N.A., and Holmberg, S.D., Long-term human immunodeficiency virus infection in asymptomatic homosexual and bisexual men with normal CD4+ lymphocyte counts: immunologic and virologic characteristics, *J. Infect. Dis.,* 163 (5), 959–965, 1991.
9. Pantaleo, G., Menzo, S., Vaccarezza, M., Graziosi, C., Cohen, O.J., Demarest, J.F., Montefiori, D., Orenstein, J.M., Fox, C., Schrager, and L.K., Studies in subjects with long-term nonprogressive human immunodeficiency virus infection, *N. Engl. J. Med.,* 332 (4), 209–216, 1995.
10. Cao, Y., Qin, L., Zhang, L., Safrit, J., and Ho, D.D., Virologic and immunologic characterization of long-term survivors of human immunodeficiency virus type 1 infection, *N. Engl. J. Med.,* 332 (4), 201–208, 1995.
11. Rosenberg, E.S., Billingsley, J.M., Caliendo, A.M., Boswell, S.L., Sax, P.E., Kalams, S.A., and Walker, B.D., Vigorous HIV-1-specific CD4+ T cell responses associated with control of viremia, *Science,* 278 (5342), 1447–1450, 1997.
12. Paroli, M., Propato, A., Accapezzato, D., Francavilla, V., Schiaffella, E., and Barnaba, V., The immunology of HIV-infected long-term non-progressors—a current view, *Immunol. Lett.,* 79 (1–2), 127–129, 2001.
13. Feinberg, M.B. and McLean, A.R., AIDS: decline and fall of immune surveillance? *Curr. Biol.,* 7 (3), R136–R140, 1997.

14. Birch, M.R., Learmont, J.C., Dyer, W.B., Deacon, N.J., Zaunders, J.J., Saksena, N., Cunningham, A. L., Mills, J., and Sullivan, J.S., An examination of signs of disease progression in survivors of the Sydney Blood Bank Cohort (SBBC), *J. Clin. Virol.,* 22 (3), 263–270, 2001.

15. Sheppard, H.W., Lang, W., Ascher, M.S., Vittinghoff, E., and Winkelstein, W., The characterization of non-progressors: long-term HIV-1 infection with stable CD4+ T-cell levels, *AIDS,* 7 (9), 1159–1166, 1993.

16. Keet, I.P., Krijnen, P., Koot, M., Lange, J.M., Miedema, F., Goudsmit, J., and Coutinho, R.A., Predictors of rapid progression to AIDS in HIV-1 seroconverters, *AIDS,* 7 (1), 51–57, 1993.

17. Klein, M.R., van Baalen, C.A., Holwerda, A.M., Kerkhof Garde, S.R., Bende, R.J., Keet, I. P., Eeftinck-Schattenkerk, J.K., Osterhaus, A.D., Schuitemaker, H., and Miedema, F., Kinetics of Gag-specific cytotoxic T lymphocyte responses during the clinical course of HIV-1 infection: a longitudinal analysis of rapid progressors and long-term asymptomatics, *J. Exp. Med.,* 181 (4), 1365–1372, 1995.

18. Salerno-Goncalves, R., Lu, W., and Andrieu, J.M., Quantitative analysis of the antiviral activity of CD8(+) T cells from human immunodeficiency virus-positive asymptomatic patients with different rates of CD4(+) T-cell decrease, *J. Virol.,* 74 (14), 6648–6651, 2000.

19. Deacon, N.J., Tsykin, A., Solomon, A., Smith, K., Ludford-Menting, M., Hooker, D.J., McPhee, D.A., Greenway, A.L., Ellett, A., and Chatfield, C., Genomic structure of an attenuated quasi species of HIV-1 from a blood transfusion donor and recipients, *Science,* 270 (5238), 988–991, 1995.

20. Lum, J.J., Cohen, O.J., Nie, Z., Weaver, J.G., Gomez, T.S., Yao, X.J., Lynch, D., Pilon, A.A., Hawley, N., Kim, J.E., Chen, Z., Montpetit, M., Sanchez-Dardon, J., Cohen, E.A., and Badley, A.D., Vpr R77Q is associated with long-term nonprogressive HIV infection and impaired induction of apoptosis, *J. Clin. Invest.,* 111 (10), 1547–1554, 2003.

21. Masciotra, S., Owen, S.M., Rudolph, D., Yang, C., Wang, B., Saksena, N., Spira, T., Dhawan, S., and Lal, R.B., Temporal relationship between V1V2 variation, macrophage replication, and coreceptor adaptation during HIV-1 disease progression, *AIDS,* 16 (14), 1887–1898, 2002.

22. Yu, X.F., Wang, Z., Vlahov, D., Markham, R.B., Farzadegan, H., and Margolick, J.B., Infection with dual-tropic human immunodeficiency virus type 1 variants associated with rapid total T cell decline and disease progression in injection drug users, *J. Infect. Dis.,* 178 (2), 388–396, 1998.

23. Menzo, S., Sampaolesi, R., Vicenzi, E., Santagostino, E., Liuzzi, G., Chirianni, A., Piazza, M., Cohen, O.J., Bagnarelli, P., and Clementi, M., Rare mutations in a domain crucial for V3-loop structure prevail in replicating HIV from long-term non-progressors, *AIDS,* 12 (9), 985–997, 1998.

24. Dean, M., Carrington, M., Winkler, C., Huttley, G.A., Smith, M.W., Allikmets, R., Goedert, J.J., Buchbinder, S.P., Vittinghoff, E., Gomperts, E., Donfield, S., Vlahov, D., Kaslow, R., Saah, A., Rinaldo, C., Detels, R., and O'Brien, S.J., Genetic restriction of HIV-1 infection and progression to AIDS by a deletion allele of the CKR5 structural gene. Hemophilia Growth and Development Study, Multicenter AIDS Cohort Study, Multicenter Hemophilia Cohort Study, San Francisco City Cohort, ALIVE Study, *Science,* 273 (5283), 1856–1862, 1996.

25. Michael, N.L., Chang, G., Louie, L.G., Mascola, J.R., Dondero, D., Birx, D.L., and Sheppard, H.W., The role of viral phenotype and CCR-5 gene defects in HIV-1 transmission and disease progression, *Nat. Med.,* 3 (3), 338–340, 1997.

26. Clegg, A.O., Ashton, L.J., Biti, R.A., Badhwar, P., Williamson, P., Kaldor, J.M., and Stewart, G.J., CCR5 promoter polymorphisms, CCR5 59029A and CCR5 59353C, are under represented in HIV-1-infected long-term non-progressors. The Australian Long-Term Non-Progressor Study Group, *AIDS,* 14 (2), 103–108, 2000.

27. Easterbrook, P.J., Rostron, T., Ives, N., Troop, M., Gazzard, B.G., and Rowland-Jones, S.L., Chemokine receptor polymorphisms and human immunodeficiency virus disease progression, *J. Infect. Dis.,* 180 (4), 1096–1105, 1999.

28. Smith, M.W., Dean, M., Carrington, M., Winkler, C., Huttley, G.A., Lomb, D.A., Goedert, J.J., O'Brien, T.R., Jacobson, L.P., Kaslow, R., Buchbinder, S., Vittinghoff, E., Vlahov, D., Hoots, K., Hilgartner, M.W., and O'Brien, S.J., Contrasting genetic influence of CCR2 and CCR5 variants on HIV-1 infection and disease progression. Hemophilia Growth and Development Study (HGDS), Multicenter AIDS Cohort Study (MACS), Multicenter Hemophilia Cohort Study (MHCS), San Francisco City Cohort (SFCC), ALIVE Study, *Science,* 277 (5328), 959–965, 1997.

29. Mulherin, S.A., O'Brien, T.R., Ioannidis, J.P., Goedert, J.J., Buchbinder, S.P., Coutinho, R.A., Jamieson, B.D., Meyer, L., Michael, N.L., Pantaleo, G., Rizzardi, G.P., Schuitemaker, H., Sheppard, H.W., Theodorou, I.D., Vlahov, D., and Rosenberg, P.S., Effects of CCR5-32 and CCR2-64I alleles on HIV-1 disease progression: the protection varies with duration of infection, *AIDS,* 17 (3), 377–387, 2003.

30. Michael, N.L., Louie, L.G., Rohrbaugh, A.L., Schultz, K.A., Dayhoff, D.E., Wang, C.E., and Sheppard, H.W., The role of CCR5 and CCR2 polymorphisms in HIV-1 transmission and disease progression, *Nat. Med.,* 3 (10), 1160–1162, 1997.

31. Winkler, C., Modi, W., Smith, M.W., Nelson, G.W., Wu, X., Carrington, M., Dean, M., Honjo, T., Tashiro, K., Yabe, D., Buchbinder, S., Vittinghoff, E., Goedert, J.J., O'Brien, T.R., Jacobson, L.P., Detels, R., Donfield, S., Willoughby, A., Gomperts, E., Vlahov, D., Phair, J., and O'Brien, S.J., Genetic restriction of AIDS pathogenesis by an SDF-1 chemokine gene variant. ALIVE Study, Hemophilia Growth and Development Study (HGDS), Multicenter AIDS Cohort Study (MACS), Multicenter Hemophilia Cohort Study (MHCS), San Francisco City Cohort (SFCC), *Science,* 279 (5349), 389–393, 1998.

32. An, P., Vlahov, D., Margolick, J.B., Phair, J., O'Brien, T.R., Lautenberger, J., O'Brien, S.J., and Winkler, C.A., A tumor necrosis factor–inducible promoter variant of interferon- accelerates CD4+ T cell depletion in human immunodeficiency virus-1-infected individuals, *J. Infect. Dis.,* 188 (2), 228–231, 2003.

33. Kaslow, R.A., Carrington, M., Apple, R., Park, L., Munoz, A., Saah, A.J., Goedert, J.J., Winkler, C., O'Brien, S.J., Rinaldo, C., Detels, R., Blattner, W., Phair, J., Erlich, H., and Mann, D.L., Influence of combinations of human major histocompatibility complex genes on the course of HIV-1 infection, *Nat. Med.,* 2 (4), 405–411, 1996.

34. Carrington, M., Nelson, G.W., Martin, M.P., Kissner, T., Vlahov, D., Goedert, J.J., Kaslow, R., Buchbinder, S., Hoots, K., and ⊝'Brien, S.J., HLA and HIV-1: heterozygote advantage and B*35-Cw*04 disadvantage, *Science,* 283 (5408), 1748–1752, 1999.

35. Dubey, D.P., Alper, C.A., Mirza, N.M., Awdeh, Z., and Yunis, E.J., Polymorphic Hh genes in the HLA-B(C) region control natural killer cell frequency and activity, *J. Exp. Med.,* 179 (4), 1193–1203, 1994.

36. Flores-Villanueva, P.O., Yunis, E.J., Delgado, J.C., Vittinghoff, E., Buchbinder, S., Leung, J.Y., Uglialoro, A.M., Clavijo, O.P., Rosenberg, E.S., Kalams, S.A., Braun, J.D., Boswell, S.L., Walker, B.D., and Goldfeld, A.E., Control of HIV-1 viremia and protection from AIDS are associated with HLA-Bw4 homozygosity, *Proc. Natl. Acad. Sci. U.S.A.,* 98 (9), 5140–5145, 2001.

37. Martin, M.P., Gao, X., Lee, J.H., Nelson, G.W., Detels, R., Goedert, J.J., Buchbinder, S., Hoots, K., Vlahov, D., Trowsdale, J., Wilson, M., O'Brien, S.J., and Carrington, M., Epistatic interaction between KIR3DS1 and HLA-B delays the progression to AIDS, *Nat. Genet.,* 31 (4), 429–434, 2002.

38. Schwartz, O., Marechal, V., Le Gall, S., Lemonnier, F., and Heard, J.M., Endocytosis of major histocompatibility complex class I molecules is induced by the HIV-1 Nef protein, *Nat. Med.,* 2 (3), 338–342, 1996.

39. Ogg, G.S., Jin, X., Bonhoeffer, S., Dunbar, P.R., Nowak, M.A., Monard, S., Segal, J.P., Cao, Y., Rowland-Jones, S.L., Cerundolo, V., Hurley, A., Markowitz, M., Ho, D.D., Nixon, D.F., and McMichael, A.J., Quantitation of HIV-1-specific cytotoxic T lymphocytes and plasma load of viral RNA, *Science,* 279 (5359), 2103–2106, 1998.

40. Cohen, O.J., Vaccarezza, M., Lam, G.K., Baird, B.F., Wildt, K., Murphy, P.M., Zimmerman, P.A., Nutman, T.B., Fox, C.H., Hoover, S., Adelsberger, J., Baseler, M., Arthos, J., Davey, R.T., Jr., Dewar, R.L., Metcalf, J., Schwartzentruber, D.J., Orenstein, J.M., Buchbinder, S., Saah, A.J., Detels, R., Phair, J., Rinaldo, C., Margolick, J.B., and Fauci, A.S., Heterozygosity for a defective gene for CC chemokine receptor 5 is not the sole determinant for the immunologic and virologic phenotype of HIV-infected long-term nonprogressors, *J. Clin. Invest.,* 100 (6), 1581–1589, 1997.

41. Watanabe, M., Ringler, D.J., Fultz, P.N., MacKey, J.J., Boyson, J.E., Levine, C.G., and Letvin, N.L, A chimpanzee-passaged human immunodeficiency virus isolate is cytopathic for chimpanzee cells but does not induce disease, *J. Virol.,* 65 (6), 3344–3348, 1991.

42. Johnson, B.K., Stone, G.A., Godec, M.S., Asher, D.M., Gajdusek, D.C., and Gibbs, C.J., Jr., Long-term observations of human immunodeficiency virus-infected chimpanzees, *AIDS Res. Hum. Retroviruses,* 9 (4), 375–378, 1993.

43. Gougeon, M.L., Lecoeur, H., Boudet, F., Ledru, E., Marzabal, S., Boullier, S., Roue, R., Nagata, S., and Heeney, J., Lack of chronic immune activation in HIV-infected chimpanzees correlates with the

resistance of T cells to Fas/Apo-1 (CD95)-induced apoptosis and preservation of a T helper 1 phenotype, *J. Immunol.,* 158 (6), 2964–2976, 1997.

44. Novembre, F.J., de Rosayro, J., Nidtha, S., O'Neil, S.P., Gibson, T.R., Evans-Strickfaden, T., Hart, C.E., and McClure, H.M., Rapid CD4(+) T-cell loss induced by human immunodeficiency virus type 1(NC) in uninfected and previously infected chimpanzees, *J. Virol.,* 75 (3), 1533–1539, 2001.

45. O'Neil, S.P., Novembre, F.J., Hill, A.B., Suwyn, C., Hart, C.E., Evans-Strickfaden, T., Anderson, D.C., deRosayro, J., Herndon, J.G., Saucier, M., and McClure, H.M., Progressive infection in a subset of HIV-1-positive chimpanzees, *J. Infect. Dis.,* 182 (4), 1051–1062, 2000.

46. Desrosiers, R.C., Hansen-Moosa, A., Mori, K., Bouvier, D.P., King, N.W., Daniel, M.D., and Ringler, D.J., Macrophage-tropic variants of SIV are associated with specific AIDS-related lesions but are not essential for the development of AIDS, *Am. J. Pathol.,* 139 (1), 29–35, 1991.

47. Schuitemaker, H., Meyaard, L., Kootstra, N.A., Dubbes, R., Otto, S.A., Tersmette, M., Heeney, J.L., and Miedema, F., Lack of T cell dysfunction and programmed cell death in human immunodeficiency virus type 1-infected chimpanzees correlates with absence of monocytotropic variants, *J. Infect. Dis.,* 168 (5), 1140–1147, 1993.

48. Veazey, R.S., DeMaria, M., Chalifoux, L.V., Shvetz, D.E., Pauley, D.R., Knight, H.L., Rosenzweig, M., Johnson, R.P., Desrosiers, R.C., and Lackner, A.A., Gastrointestinal tract as a major site of CD4+ T cell depletion and viral replication in SIV infection, *Science,* 280 (5362), 427–431, 1998.

49. Whetter, L.E., Ojukwu, I.C., Novembre, F.J., and Dewhurst, S., Pathogenesis of simian immunodeficiency virus infection, *J. Gen. Virol.,* 80 (Pt. 7), 1557–1568, 1999.

50. Heeney, J., Jonker, R., Koornstra, W., Dubbes, R., Niphuis, H., Di Rienzo, A.M., Gougeon, M.L., and Montagnier, L., The resistance of HIV-infected chimpanzees to progression to AIDS correlates with absence of HIV-related T-cell dysfunction, *J. Med. Primatol.,* 22 (2–3), 194–200, 1993.

51. Estaquier, J., Idziorek, T., de Bels, F., Barre-Sinoussi, F., Hurtrel, B., Aubertin, A.M., Venet, A., Mehtali, M., Muchmore, E., and Michel, P., Programmed cell death and AIDS: significance of T-cell apoptosis in pathogenic and nonpathogenic primate lentiviral infections, *Proc. Natl. Acad. Sci. U.S.A.,* 91 (20), 9431–9435, 1994.

52. Finkel, T.H., Tudor-Williams, G., Banda, N.K., Cotton, M.F., Curiel, T., Monks, C., Baba, T.W., Ruprecht, R.M., and Kupfer, A., Apoptosis occurs predominantly in bystander cells and not in productively infected cells of HIV- and SIV-infected lymph nodes, *Nat. Med.,* 1 (2), 129–134, 1995.

53. Monceaux, V., Estaquier, J., Fevrier, M., Cumont, M.C., Riviere, Y., Aubertin, A.M., Ameisen, J.C., and Hurtrel, B., Extensive apoptosis in lymphoid organs during primary SIV infection predicts rapid progression towards AIDS, *AIDS,* 17 (11), 1585–1596, 2003.

54. Wykrzykowska, J.J., Rosenzweig, M., Veazey, R.S., Simon, M.A., Halvorsen, K., Desrosiers, R.C., Johnson, R.P., and Lackner, A.A., Early regeneration of thymic progenitors in rhesus macaques infected with simian immunodeficiency virus, *J. Exp. Med.,* 187 (11), 1767–1778, 1998.

55. Silvestri, G., Sodora, D.L., Koup, R.A., Paiardini, M., O'Neil, S.P., McClure, H.M., Staprans, S.I., and Feinberg, M.B., Nonpathogenic SIV infection of sooty mangabeys is characterized by limited bystander immunopathology despite chronic high-level viremia, *Immunity,* 18 (3), 441–452, 2003.

56. Dockrell, D.H., The multiple roles of Fas ligand in the pathogenesis of infectious diseases, *Clin. Microbiol. Infect.,* 9 (8), 766–779, 2003.

57. Sasaki, Y., Ami, Y., Nakasone, T., Shinohara, K., Takahashi, E., Ando, S., Someya, K., Suzaki, Y., and Honda, M., Induction of CD95 ligand expression on T lymphocytes and B lymphocytes and its contribution to apoptosis of CD95-up-regulated CD4+ T lymphocytes in macaques by infection with a pathogenic simian/human immunodeficiency virus, *Clin. Exp. Immunol.,* 122 (3), 381–389, 2000.

58. Iida, T., Ichimura, H., Ui, M., Shimada, T., Akahata, W., Igarashi, T., Kuwata, T., Ido, E., Yonehara, S., Imanishi, J., and Hayami, M., Sequential analysis of apoptosis induction in peripheral blood mononuclear cells and lymph nodes in the early phase of pathogenic and nonpathogenic SIVmac infection, *AIDS Res. Hum. Retroviruses,* 15 (8), 721–729, 1999.

59. Gougeon, M.L., Lecoeur, H., Heeney, J., and Boudet, F., Comparative analysis of apoptosis in HIV-infected humans and chimpanzees: relation with lymphocyte activation, *Immunol. Lett.,* 51 (1–2), 75–81, 1996.

60. Xu, X.N., Laffert, B., Screaton, G.R., Kraft, M., Wolf, D., Kolanus, W., Mongkolsapay, J., McMichael, A.J., and Baur, A.S., Induction of Fas ligand expression by HIV involves the interaction of Nef with the T cell receptor chain, *J. Exp. Med.,* 189 (9), 1489–1496, 1999.

61. Xu, X.N., Screaton, G.R., Gotch, F.M., Dong, T., Tan, R., Almond, N., Walker, B., Stebbings, R., Kent, K., Nagata, S., Stott, J.E., and McMichael, A.J., Evasion of cytotoxic T lymphocyte (CTL) responses by Nef-dependent induction of Fas ligand (CD95L) expression on simian immunodeficiency virus-infected cells, *J. Exp. Med.,* 186 (1), 7–16, 1997.

62. Hodge, S., Novembre, F.J., Whetter, L., Gelbard, H.A., and Dewhurst, S., Induction of Fas ligand expression by an acutely lethal simian immunodeficiency virus, SIVsmmPBj14, *Virology,* 252 (2), 354–363, 1998.

63. Ehret, A., Westendorp, M.O., Herr, I., Debatin, K.M., Heeney, J.L., Frank, R., and Krammer, P.H., Resistance of chimpanzee T cells to human immunodeficiency virus type 1 Tat-enhanced oxidative stress and apoptosis, *J. Virol.,* 70 (9), 6502–6507, 1996.

64. Ayyavoo, V., Mahboubi, A., Mahalingam, S., Ramalingam, R., Kudchodkar, S., Williams, W.V., Green, D.R., and Weiner, D.B., HIV-1 Vpr suppresses immune activation and apoptosis through regulation of nuclear factor B, *Nat. Med.,* 3 (10), 1117–1123, 1997.

65. Jacotot, E., Ferri, K.F., El Hamel, C., Brenner, C., Druillennec, S., Hoebeke, J., Rustin, P., Metivier, D., Lenoir, C., Geuskens, M., Vieira, H.L., Loeffler, M., Belzacq, A.S., Briand, J.P., Zamzami, N., Edelman, L., Xie, Z.H., Reed, J.C., Roques, B.P., and Kroemer, G., Control of mitochondrial membrane permeabilization by adenine nucleotide translocator interacting with HIV-1 viral protein R and Bcl-2, *J. Exp. Med.,* 193 (4), 509–519, 2001.

66. Chang, L.J., Chen, C.H., Urlacher, V., and Lee, T.Z., Differential apoptosis effects of primate lentiviral Vpr and Vpx in mammalian cells, *J. Biomed. Sci.,* 7 (4), 322–333, 2000.

67. Mellors, J.W., Rinaldo, C.R., Jr., Gupta, P., White, R.M., Todd, J.A., and Kingsley, L.A., Prognosis in HIV-1 infection predicted by the quantity of virus in plasma, *Science,* 272 (5265), 1167–1170, 1996.

68. Karmochkine, M., Parizot, C., Calvez, V., Coutellier, A., Herson, S., Debre, P., and Gorochov, G., Susceptibility of peripheral blood mononuclear cells to apoptosis is correlated to plasma HIV load, *J. Acquir. Immune Defic. Syndr. Hum. Retrovirol.,* 17 (5), 419–423, 1998.

69. Dockrell, D.H., Badley, A.D., Algeciras-Schimnich, A., Simpson, M., Schut, R., Lynch, D.H., and Paya, C.V., Activation-induced CD4+ T cell death in HIV-positive individuals correlates with Fas susceptibility, CD4+ T cell count, and HIV plasma viral copy number, *AIDS Res. Hum. Retroviruses,* 15 (17), 1509–1518, 1999.

70. Muro-Cacho, C.A., Pantaleo, G., and Fauci, A.S., Analysis of apoptosis in lymph nodes of HIV-infected persons. Intensity of apoptosis correlates with the general state of activation of the lymphoid tissue and not with stage of disease or viral burden, *J. Immunol.,* 154 (10), 5555–5566, 1995.

71. Marlink, R., Kanki, P., Thior, I., Travers, K., Eisen, G., Siby, T., Traore, I., Hsieh, C.C., Dia, M.C., Gueye, E.H., and et al., Reduced rate of disease development after HIV-2 infection as compared to HIV-1, *Science,* 265 (5178), 1587–1590, 1994.

72. Michel, P., Balde, A.T., Roussilhon, C., Aribot, G., Sarthou, J.L., and Gougeon, M.L., Reduced immune activation and T cell apoptosis in human immunodeficiency virus type 2 compared with type 1: correlation of T cell apoptosis with 2 microglobulin concentration and disease evolution, *J. Infect. Dis.,* 181 (1), 64–75, 2000.

73. Westendorp, M.O., Frank, R., Ochsenbauer, C., Stricker, K., Dhein, J., Walczak, H., Debatin, K.M., and Krammer, P.H., Sensitization of T cells to CD95-mediated apoptosis by HIV-1 Tat and gp120, *Nature,* 375 (6531), 497–500, 1995.

74. Cohen, I., Boya, P., Zhao, L., Metivier, D., Andreau, K., Perfettini, J.L., Weaver, J.G., Badley, A., Taylor, E.W., and Kroemer, G., Anti-apoptotic activity of the glutathione peroxidase homologue encoded by HIV-1, *Apoptosis,* 9 (2), 181–192, 2004.

75. Arthos, J., Cicala, C., Selig, S.M., White, A.A., Ravindranath, H.M., Van Ryk, D., Steenbeke, T.D., Machado, E., Khazanie, P., Hanback, M.S., Hanback, D.B., Rabin, R.L., and Fauci, A.S., The role of the CD4 receptor vs. HIV coreceptors in envelope-mediated apoptosis in peripheral blood mononuclear cells, *Virology,* 292 (1), 98–106, 2002.

76. Vlahakis, S.R., Algeciras-Schimnich, A., Bou, G., Heppelmann, C.J., Villasis-Keever, A., Collman, R.C., and Paya, C.V., Chemokine-receptor activation by Env determines the mechanism of death in HIV-infected and uninfected T lymphocytes, *J. Clin. Invest.,* 107 (2), 207–215, 2001.

77. Jekle, A., Keppler, O.T., De Clercq, E., Schols, D., Weinstein, M., and Goldsmith, M.A., *In vivo* evolution of human immunodeficiency virus type 1 toward increased pathogenicity through CXCR4-mediated killing of uninfected CD4 T cells, *J. Virol.,* 77 (10), 5846–5854, 2003.

78. Chia, W.K., Freedman, J., Li, X., Salit, I., Kardish, M., and Read, S.E., Programmed cell death induced by HIV type 1 antigen stimulation is associated with a decrease in cytotoxic T lymphocyte activity in advanced HIV type 1 infection, *AIDS Res. Hum. Retroviruses,* 11 (2), 249–256, 1995.

79. Mueller, Y.M., De Rosa, S.C., Hutton, J.A., Witek, J., Roederer, M., Altman, J.D., and Katsikis, P.D., Increased CD95/Fas-induced apoptosis of HIV-specific CD8(+) T cells, *Immunity,* 15 (6), 871–882, 2001.

80. Franceschi, C., Franceschini, M.G., Boschini, A., Trenti, T., Nuzzo, C., Castellani, G., C., S., De Rienzo, B., Roncaglia, R., Portolani, M., Pietrosemoli, P., Meacci, M., Pecorari, M., Sabbatini, A., and Malorni, W., Phenotypic characteristics and tendency to apoptosis of peripheral blood mononuclear cells from HIV+ long term non progressors, *Cell Death Differ.,* 4 (8), 815–823, 1997.

81. Gougeon, M.L., Lecoeur, H., Dulioust, A., Enouf, M.G., Crouvoiser, M., Goujard, C., Debord, T., and Montagnier, L., Programmed cell death in peripheral lymphocytes from HIV-infected persons: increased susceptibility to apoptosis of CD4 and CD8 T cells correlates with lymphocyte activation and with disease progression, *J. Immunol.,* 156 (9), 3509–3520, 1996.

82. Liegler, T.J., Yonemoto, W., Elbeik, T., Vittinghoff, E., Buchbinder, S.P., and Greene, W.C., Diminished spontaneous apoptosis in lymphocytes from human immunodeficiency virus-infected long-term non-progressors, *J. Infect. Dis.,* 178 (3), 669–679, 1998.

83. Wasmuth, J.C., Klein, K.H., Hackbarth, F., Rockstroh, J.K., Sauerbruch, T., and Spengler, U., Prediction of imminent complications in HIV-1-infected patients by markers of lymphocyte apoptosis, *J. Acquir. Immune Defic. Syndr.,* 23 (1), 44–51, 2000.

84. Samuelsson, A., Brostrom, C., van Dijk, N., Sonnerborg, A., and Chiodi, F., Apoptosis of CD4+ and CD19+ cells during human immunodeficiency virus type 1 infection — correlation with clinical progression, viral load, and loss of humoral immunity, *Virology,* 238 (2), 180–188, 1997.

85. Hosaka, N., Oyaizu, N., Than, S., and Pahwa, S., Correlation of loss of CD4 T cells with plasma levels of both soluble form Fas (CD95) Fas ligand (FasL) in HIV-infected infants, *Clin. Immunol.,* 95 (1, Pt. 1), 20–25, 2000.

86. Daniel, V., Susal, C., Weimer, R., Zimmermann, R., Huth-Kuhne, A., and Opelz, G., Increased soluble Fas in HIV-infected hemophilia patients with CD4+ and CD8+ cell count increases and viral load and immune complex decreases, *AIDS Res. Hum. Retroviruses,* 17 (4), 329–335, 2001.

87. Krammer, P.H., CD95's deadly mission in the immune system, *Nature,* 407 (6805), 789–795, 2000.

88. Bottarel, F., Bonissoni, S., Lucia, M.B., Bragardo, M., Bensi, T., Buonfiglio, D., Mezzatesta, C., DiFranco, D., Balotta, C., Capobianchi, M.R., Dianzani, I., Cauda, R., and Dianzani, U., Decreased function of Fas in patients displaying delayed progression of HIV-induced immune deficiency, *Hematol. J.,* 2 (4), 220–227, 2001.

89. Cifone, M.G., De Maria, R., Roncaioli, P., Rippo, M.R., Azuma, M., Lanier, L.L., Santoni, A., and Testi, R., Apoptotic signaling through CD95 (Fas/Apo-1) activates an acidic sphingomyelinase, *J. Exp. Med.,* 180 (4), 1547–1552, 1994.

90. De Simone, C., Cifone, M.G., Alesse, E., Steinberg, S.M., Di Marzio, L., Moretti, S., Famularo, G., Boschini, A., and Testi, R., Cell associated ceramide in HIV-1-infected subjects, *AIDS,* 10 (6), 675–676, 1996.

91. Moretti, S., Marcellini, S., Boschini, A., Famularo, G., Santini, G., Alesse, E., Steinberg, S.M., Cifone, M.G., Kroemer, G., and De Simone, C., Apoptosis and apoptosis-associated perturbations of peripheral blood lymphocytes during HIV infection: comparison between AIDS patients and asymptomatic long-term non-progressors, *Clin. Exp. Immunol.,* 122 (3), 364–373, 2000.

92. Giovannetti, A., Pierdominici, M., Mazzetta, F., Mazzone, A.M., Ricci, G., Prozzo, A., Pandolfi, F., Paganelli, R., and Aiuti, F., HIV type 1-induced inhibition of CD45 tyrosine phosphatase activity correlates with disease progression and apoptosis, but not with anti-CD3-induced T cell proliferation, *AIDS Res. Hum. Retroviruses,* 16 (3), 211–219, 2000.

93. Cayota, A., Vuillier, F., Gonzalez, G., and Dighiero, G, CD4+ lymphocytes from HIV-infected patients display impaired CD45-associated tyrosine phosphatase activity which is enhanced by anti-oxidants, *Clin. Exp. Immunol.,* 104 (1), 11–17, 1996.

94. Ledru, E., Lecoeur, H., Garcia, S., Debord, T., and Gougeon, M.L., Differential susceptibility to activation-induced apoptosis among peripheral Th1 subsets: correlation with Bcl-2 expression and consequences for AIDS pathogenesis, *J. Immunol.,* 160 (7), 3194–3206, 1998.

95. Adams, J.M. and Cory, S., The Bcl-2 protein family: arbiters of cell survival, *Science,* 281 (5381), 1322–1326, 1998.

96. Miyashita, T. and Reed, J.C., Tumor suppressor p53 is a direct transcriptional activator of the human Bax gene, *Cell,* 80 (2), 293–299, 1995.

97. Oda, E., Ohki, R., Murasawa, H., Nemoto, J., Shibue, T., Yamashita, T., Tokino, T., Taniguchi, T., and Tanaka, N., Noxa, a BH3-only member of the Bcl-2 family and candidate mediator of p53-induced apoptosis, *Science,* 288 (5468), 1053–1058, 2000.

98. Cuconati, A., Mukherjee, C., Perez, D., and White, E., DNA damage response and MCL-1 destruction initiate apoptosis in adenovirus-infected cells, *Genes Dev.,* 17 (23), 2922–2932, 2003.

99. Chipuk, J.E., Kuwana, T., Bouchier-Hayes, L., Droin, N.M., Newmeyer, D.D., Schuler, M., and Green, D.R., Direct activation of Bax by p53 mediates mitochondrial membrane permeabilization and apoptosis, *Science,* 303 (5660), 1010–1014, 2004.

100. McLinden, R.J., Lewis, B., Michael, N.L., Redfield, R.R., Birx, D.L., and Kim, J.H., Correlation of tumor suppressor P53 RNA expression with human immunodeficiency virus disease in rapid and slow progressors, *J. Hum. Virol.,* 1 (1), 30–36, 1997.

101. Palella, F.J., Jr., Delaney, K.M., Moorman, A.C., Loveless, M.O., Fuhrer, J., Satten, G.A., Aschman, D.J., and Holmberg, S.D., Declining morbidity and mortality among patients with advanced human immunodeficiency virus infection. HIV Outpatient Study Investigators, *N. Engl. J. Med.,* 338 (13), 853–860, 1998.

102. Autran, B., Carcelain, G., Li, T.S., Blanc, C., Mathez, D., Tubiana, R., Katlama, C., Debre, P., and Leibowitch, J., Positive effects of combined antiretroviral therapy on CD4+ T cell homeostasis and function in advanced HIV disease, *Science,* 277 (5322), 112–116, 1997.

103. Chougnet, C., Fowke, K.R., Mueller, B.U., Smith, S., Zuckerman, J., Jankelevitch, S., Steinberg, S.M., Luban, N., Pizzo, P.A., and Shearer, G.M., Protease inhibitor and triple-drug therapy: cellular immune parameters are not restored in pediatric AIDS patients after 6 months of treatment, *AIDS,* 12 (18), 2397–2406, 1998.

104. Badley, A.D., Dockrell, D.H., Algeciras, A., Ziesmer, S., Landay, A., Lederman, M.M., Connick, E., Kessler, H., Kuritzkes, D., Lynch, D.H., Roche, P., Yagita, H., and Paya, C.V., *In vivo* analysis of Fas/FasL interactions in HIV-infected patients, *J. Clin. Invest.,* 102 (1), 79–87, 1998.

105. Kotler, D.P., Shimada, T., Snow, G., Winson, G., Chen, W., Zhao, M., Inada, Y., and Clayton, F., Effect of combination antiretroviral therapy upon rectal mucosal HIV RNA burden and mononuclear cell apoptosis, *AIDS,* 12 (6), 597–604, 1998.

106. Dyrhol-Riise, A.M., Stent, G., Rosok, B.I., Voltersvik, P., Olofsson, J., and Asjo, B., The Fas/FasL system and T cell apoptosis in HIV-1-infected lymphoid tissue during highly active antiretroviral therapy, *Clin. Immunol.,* 101 (2), 169–179, 2001.

107. Dyrhol-Riise, A.M., Ohlsson, M., Skarstein, K., Nygaard, S.J., Olofsson, J., Jonsson, R., and Asjo, B., T cell proliferation and apoptosis in HIV-1-infected lymphoid tissue: impact of highly active antiretroviral therapy, *Clin. Immunol.,* 101 (2), 180–191, 2001.

108. Zhang, Z.Q., Notermans, D.W., Sedgewick, G., Cavert, W., Wietgrefe, S., Zupancic, M., Gebhard, K., Henry, K., Boies, L., Chen, Z., Jenkins, M., Mills, R., McDade, H., Goodwin, C., Schuwirth, C.M., Danner, S.A., and Haase, A.T., Kinetics of CD4+ T cell repopulation of lymphoid tissues after treatment of HIV-1 infection, *Proc. Natl. Acad. Sci. U.S.A.,* 95 (3), 1154–1159, 1998.

109. Mohri, H., Perelson, A.S., Tung, K., Ribeiro, R.M., Ramratnam, B., Markowitz, M., Kost, R., Hurley, A., Weinberger, L., Cesar, D., Hellerstein, M.K., and Ho, D.D., Increased turnover of T lymphocytes in HIV-1 infection and its reduction by antiretroviral therapy, *J. Exp. Med.,* 194 (9), 1277–1287, 2001.

110. Nokta, M., Rossero, R., Nichols, J., Rosenbaum, M., and Pollard, R.B., Effect of didanosine, stavudine, and hydroxyurea therapy on apoptosis in CD45RA+ and CD45RO+ T lymphocyte subpopulations, *AIDS Res. Hum. Retroviruses,* 15 (3), 255–264, 1999.

111. Ensoli, F., Fiorelli, V., Alario, C., De Cristofaro, M., Santini Muratori, D., Novi, A., Cunsolo, M.G., Mazzetta, F., Giovannetti, A., Mollicone, B., Pinter, E., and Aiuti, F., Decreased T cell apoptosis and T cell recovery during highly active antiretroviral therapy (HAART), *Clin. Immunol.,* 97 (1), 9–20, 2000.

112. Ensoli, F., Fiorelli, V., De Cristofaro, M., Collacchi, B., Santini Muratori, D., Alario, C., Sacco, G., Iebba, F., and Aiuti, F., Endogenous cytokine production protects T cells from spontaneous apoptosis during highly active antiretroviral therapy, *HIV Med.,* 3 (2), 105–117, 2002.

113. Wasmuth, J.C., Hackbarth, F., Rockstroh, J.K., Sauerbruch, T., and Spengler, U., Changes of lympho-cyte apoptosis associated with sequential introduction of highly active antiretroviral therapy, *HIV Med.,* 4 (2), 111–119, 2003.

114. Landay, A.L., Spritzler, J., Kessler, H., Mildvan, D., Pu, M., Fox, L., O'Neil, D., Schock, B., Kuritzkes, D., and Lederman, M.M., Immune reconstitution is comparable in antiretroviral-naive subjects after 1 year of successful therapy with a nucleoside reverse-transcriptase inhibitor- or protease inhibitor-containing antiretroviral regimen, *J. Infect. Dis.,* 188 (10), 1444–1454, 2003.

115. Aries, S.P., Weyrich, K., Schaaf, B., Hansen, F., Dennin, R.H., and Dalhoff, K., Early T-cell apoptosis and Fas expression during antiretroviral therapy in individuals infected with human immunodeficiency virus-1, *Scand. J. Immunol.,* 48 (1), 86–91, 1998.

116. Grelli, S., Campagna, S., Lichtner, M., Ricci, G., Vella, S., Vullo, V., Montella, F., Di Fabio, S., Favalli, C., Mastino, A., and Macchi, B., Spontaneous and anti-Fas-induced apoptosis in lymphocytes from HIV-infected patients undergoing highly active anti-retroviral therapy, *AIDS,* 14 (8), 939–949, 2000.

117. Gougeon, M.L., Lecoeur, H. and Sasaki, Y., Apoptosis and the CD95 system in HIV disease: impact of highly active anti-retroviral therapy (HAART), *Immunol. Lett.,* 66 (1–3), 97–103, 1999.

118. Bohler, T., Walcher, J., Holzl-Wenig, G., Geiss, M., Buchholz, B., Linde, R., and Debatin, K.M., Early effects of antiretroviral combination therapy on activation, apoptosis and regeneration of T cells in HIV-1-infected children and adolescents, *AIDS,* 13 (7), 779–789, 1999.

119. Amendola, A., Poccia, F., Martini, F., Gioia, C., Galati, V., Pierdominici, M., Marziali, M., Pandolfi, F., Colizzi, V., Piacentini, M., Girardi, E., and D'Offizi, G., Decreased CD95 expression on naive T cells from HIV-infected persons undergoing highly active anti-retroviral therapy (HAART) and the influence of IL-2 low dose administration. Irhan Study Group, *Clin. Exp. Immunol.,* 120 (2), 324–332, 2000.

120. Caggiari, L., Zanussi, S., Crepaldi, C., Bortolin, M.T., Caffau, C., D'Andrea, M., and De Paoli, P., Different rates of CD4+ and CD8+ T-cell proliferation in interleukin-2-treated human immunodeficiency virus-positive subjects, *Cytometry,* 46 (4), 233–237, 2001.

121. de Oliveira Pinto, L.M., Lecoeur, H., Ledru, E., Rapp, C., Patey, O., and Gougeon, M.L., Lack of control of T cell apoptosis under HAART. Influence of therapy regimen *in vivo* and *in vitro, AIDS,* 16 (3), 329–339, 2002.

122. Regamey, N., Harr, T., Battegay, M., and Erb, P., Downregulation of Bcl-2, but not of Bax or Bcl-x, is associated with T lymphocyte apoptosis in HIV infection and restored by antiretroviral therapy or by interleukin 2, *AIDS Res. Hum. Retroviruses,* 15 (9), 803–810, 1999.

123. Airo, P., Torti, C., Uccelli, M.C., Malacarne, F., Palvarini, L., Carosi, G., and Castelli, F., Naive CD4+ T lymphocytes express high levels of Bcl-2 after highly active antiretroviral therapy for HIV infection, *AIDS Res. Hum. Retroviruses,* 16 (17), 1805–1807, 2000.

124. Perfettini, J.L., Roumier, T., Castedo, M., Larochette, N., Boya, P., Raynal, B., Lazar, V., Ciccosanti, F., Nardacci, R., Penninger, J., Piacentini, M., and Kroemer, G., NF-κB and p53 are the dominant apoptosis-inducing transcription factors elicited by the HIV-1 envelope, *J. Exp. Med.,* 199 (5), 629–640, 2004.

125. Aladdin, H., Katzenstein, T., Dreves, A.M., Ryder, L., Gerstoft, J., Skinhoj, P., Pedersen, B.K., and Ullum, H., T-cell receptor excisional circles, telomere length, proliferation and apoptosis in peripheral blood mononuclear cells of human immunodeficiency virus-infected individuals after 18 months of treatment induced viral suppression, *Scand. J. Immunol.,* 57 (5), 485–492, 2003.

126. Lecossier, D., Bouchonnet, F., Schneider, P., Clavel, F., and Hance, A.J., Discordant increases in CD4+ T cells in human immunodeficiency virus-infected patients experiencing virologic treatment failure: role of changes in thymic output and T cell death, *J. Infect. Dis.,* 183 (7), 1009–1016, 2001.

127. De Milito, A., Hejdeman, B., Albert, J., Aleman, S., Sonnerborg, A., Zazzi, M., and Chiodi, F., High plasma levels of soluble Fas in HIV type 1-infected subjects are not normalized during highly active antiretroviral therapy, *AIDS Res. Hum. Retroviruses,* 16 (14), 1379–1384, 2000.

128. Benveniste, O., Estaquier, J., Lelievre, J.D., Vilde, J.L., Ameisen, J.C., and Leport, C., Possible mechanism of toxicity of zidovudine by induction of apoptosis of CD4+ and CD8+ T-cells *in vivo, Eur. J. Clin. Microbiol. Infect. Dis.,* 20 (12), 896–897, 2001.

129. Piconi, S., Trabattoni, D., Fusi, M.L., Milazzo, F., Dix, L.P., Rizzardini, G., Colombo, F., Bray, D., and Clerici, M., Effect of two different combinations of antiretrovirals (AZT+ddI and AZT+3TC) on

cytokine production and apoptosis in asymptomatic HIV infection, *Antiviral. Res.,* 46 (3), 171–179, 2000.

130. Pilon, A.A., Lum, J.J., Sanchez-Dardon, J., Phenix, B.N., Douglas, R., and Badley, A.D., Induction of apoptosis by a nonnucleoside human immunodeficiency virus type 1 reverse transcriptase inhibitor, *Antimicrob. Agents Chemother.,* 46 (8), 2687–2691, 2002.

131. Sloand, E.M., Maciejewski, J., Kumar, P., Kim, S., Chaudhuri, A., and Young, N., Protease inhibitors stimulate hematopoiesis and decrease apoptosis and ICE expression in CD34(+) cells, *Blood,* 96 (8), 2735–2739, 2000.

132. Phenix, B.N., Lum, J.J., Nie, Z., Sanchez-Dardon, J., and Badley, A.D., Antiapoptotic mechanism of HIV protease inhibitors: preventing mitochondrial transmembrane potential loss, *Blood,* 98 (4), 1078–1085, 2001.

133. Matarrese, P., Gambardella, L., Cassone, A., Vella, S., Cauda, R., and Malorni, W., Mitochondrial membrane hyperpolarization hijacks activated T lymphocytes toward the apoptotic-prone phenotype: homeostatic mechanisms of HIV protease inhibitors, *J. Immunol.,* 170 (12), 6006–6015, 2003.

134. Chavan, S., Kodoth, S., Pahwa, R., and Pahwa, S., The HIV protease inhibitor Indinavir inhibits cell-cycle progression *in vitro* in lymphocytes of HIV-infected and uninfected individuals, *Blood,* 98 (2), 383–389, 2001.

135. Lu, W. and Andrieu, J.M., HIV protease inhibitors restore impaired T-cell proliferative response *in vivo* and *in vitro*: a viral-suppression-independent mechanism, *Blood,* 96 (1), 250–258, 2000.

136. Phenix, B.N., Cooper, C., Owen, C., and Badley, A.D., Modulation of apoptosis by HIV protease inhibitors, *Apoptosis,* 7 (4), 295–312, 2002.

137. Puro, V. and Ippolito, G., Brief report: effect of antiretroviral agents on T-lymphocyte subset counts in healthy HIV-negative individuals. The Italian Registry on Antiretroviral Postexposure Prophylaxis, *J. Acquir. Immune Defic. Syndr.,* 24 (5), 440–443, 2000.

138 Cooper, C.L., Phenix, B., Mbisa, G., Lum, J.J., Parato, K., Hawley, N., Angel, J.B., and Badley, A. D., Antiretroviral therapy influences cellular susceptibility to apoptosis in vivo, *Front Biosci* 9, 338–41, 2004.

139. Collier, A.C., Coombs, R.W., Schoenfeld, D.A., Bassett, R.L., Timpone, J., Baruch, A., Jones, M., Facey, K., Whitacre, C., McAuliffe, V.J., Friedman, H.M., Merigan, T.C., Reichman, R.C., Hooper, C., and Corey, L., Treatment of human immunodeficiency virus infection with saquinavir, zidovudine, and zalcitabine. AIDS Clinical Trials Group, *N. Engl. J. Med.,* 334 (16), 1011–1017, 1996.

140. Albrecht, M.A., Bosch, R.J., Hammer, S.M., Liou, S.H., Kessler, H., Para, M.F., Eron, J., Valdez, H., Dehlinger, M., and Katzenstein, D.A., Nelfinavir, efavirenz, or both after the failure of nucleoside treatment of HIV infection, *N. Engl. J. Med.,* 345 (6), 398–407, 2001.

141. van Leth, F., Wit, F.W., Reiss, P., Schattenkerk, J.K., van der Ende, M.E., Schneider, M.M., Mulder, J.W., Frissen, P.H., de Wolf, F., and Lange, J.M., Differential CD4 T-cell response in HIV-1-infected patients using protease inhibitor-based or nevirapine-based highly active antiretroviral therapy, *HIV Med.,* 5 (2), 74–81, 2004.

142. Chapuis, A.G., Paolo Rizzardi, G., D'Agostino, C., Attinger, A., Knabenhans, C., Fleury, S., Acha-Orbea, H., and Pantaleo, G., Effects of mycophenolic acid on human immunodeficiency virus infection *in vitro* and *in vivo, Nat. Med.,* 6 (7), 762–768, 2000.

143. Grabar, S., Le Moing, V., Goujard, C., Leport, C., Kazatchkine, M.D., Costagliola, D., and Weiss, L., Clinical outcome of patients with HIV-1 infection according to immunologic and virologic response after 6 months of highly active antiretroviral therapy, *Ann. Intern. Med.,* 133 (6), 401–410, 2000.

144. Phenix, B.N., Angel, J.B., Mandy, F., Kravcik, S., Parato, K., Chambers, K.A., Gallicano, K., Hawley-Foss, N., Cassol, S., Cameron, D.W., and Badley, A.D., Decreased HIV-associated T cell apoptosis by HIV protease inhibitors, *AIDS Res. Hum. Retroviruses,* 16 (6), 559–567, 2000.

145. Pitrak, D.L., Bolanos, J., Hershow, R., and Novak, R.M., Discordant CD4 T lymphocyte responses to antiretroviral therapy for HIV infection are associated with *ex vivo* rates of apoptosis, *AIDS,* 15 (10), 1317–1319, 2001.

146. David, D., Keller, H., Nait-Ighil, L., Treilhou, M.P., Joussemet, M., Dupont, B., Gachot, B., Maral, J., and Theze, J., Involvement of Bcl-2 and IL-2R in HIV-positive patients whose CD4 cell counts fail to increase rapidly with highly active antiretroviral therapy, *AIDS,* 16 (8), 1093–1101, 2002.

147. Andrieu, J.M., Lu, W., and Levy, R., Sustained increases in CD4 cell counts in asymptomatic human immunodeficiency virus type 1-seropositive patients treated with prednisolone for 1 year, *J. Infect. Dis.*, 171 (3), 523–530, 1995.

148. Wallis, R.S., Kalayjian, R., Jacobson, J.M., Fox, L., Purdue, L., Shikuma, C.M., Arakaki, R., Snyder, S., Coombs, R.W., Bosch, R.J., Spritzler, J., Chernoff, M., Aga, E., Myers, L., Schock, B., and Lederman, M.M., A study of the immunology, virology, and safety of prednisone in HIV-1-infected subjects with CD4 cell counts of 200 to 700 mm(-3), *J. Acquir. Immune Defic. Syndr.*, 32 (3), 281–286, 2003.

149. Cotton, M.F., Ikle, D.N., Rapaport, E.L., Marschner, S., Tseng, P.O., Kurrle, R., and Finkel, T.H., Apoptosis of CD4+ and CD8+ T cells isolated immediately *ex vivo* correlates with disease severity in human immunodeficiency virus type 1 infection, *Pediatr. Res.*, 42 (5), 656–664, 1997.

150. Roger, P.M., Breittmayer, J.P., Arlotto, C., Pugliese, P., Pradier, C., Bernard-Pomier, G., Dellamonica, P., and Bernard, A., Highly active anti-retroviral therapy (HAART) is associated with a lower level of CD4+ T cell apoptosis in HIV-infected patients, *Clin. Exp. Immunol.*, 118 (3), 412–416, 1999.

151. Chavan, S.J., Tamma, S.L., Kaplan, M., Gersten, M., and Pahwa, S.G., Reduction in T cell apoptosis in patients with HIV disease following antiretroviral therapy, *Clin. Immunol.*, 93 (1), 24–33, 1999.

152. Dieye, T.N., Van Vooren, J.P., Delforge, M.L., Liesnard, C., Spontaneous apoptosis and highly active antiretroviral therapy (HAART), *Biomed. Pharmacother.*, 54 (1), 16–20, 2000.

153. Nielsen, S.D., Sorensen, T.U., Ersboll, A.K., Ngo, N., Mathiesen, L., Nielsen, J.O., and Hansen, J.E., Decrease in immune activation in HIV-infected patients treated with highly active antiretroviral therapy correlates with the function of hematopoietic progenitor cells and the number of naive CD4+ cells, *Scand. J. Infect. Dis.*, 32 (6), 597–603, 2000.

154. Losa, G.A. and Graber, R., Spontaneous apoptosis, oxidative status and immunophenotype markers in blood lymphocytes of AIDS patients, *Anal. Cell Pathol.*, 21 (1), 11–20, 2000.

155. Benito, J.M., Lopez, M., Martin, J.C., Lozano, S., Martinez, P., Gonzalez-Lahoz, J., and Soriano, V., Differences in cellular activation and apoptosis in HIV-infected patients receiving protease inhibitors or nonnucleoside reverse transcriptase inhibitors, *AIDS Res. Hum. Retroviruses*, 18 (18), 1379–1388, 2002.

156. Roger, P.M., Breittmayer, J.P., Cottalorda, J., Ticchioni, M., Mondain, V., Dellamonica, P., and Bernard, A., Persistent viral load in patients on HAART is associated with a higher level of activation-induced apoptosis of CD4(+) T cells, *J. Antimicrob. Chemother.*, 47 (1), 109–112, 2001.

157. Johnson, N. and Parkin, J.M., Anti-retroviral therapy reverses HIV-associated abnormalities in lymphocyte apoptosis, *Clin. Exp. Immunol.*, 113 (2), 229–234, 1998.

158. Ledru, E., Christeff, N., Patey, O., de Truchis, P., Melchior, J.C., and Gougeon, M.L., Alteration of tumor necrosis factor- T-cell homeostasis following potent antiretroviral therapy: contribution to the development of human immunodeficiency virus-associated lipodystrophy syndrome, *Blood*, 95 (10), 3191–3198, 2000.

159. Pandolfi, F., Pierdominici, M., Marziali, M., Livia Bernardi, M., Antonelli, G., Galati, V., D'Offizi, G., and Aiuti, F., Low-dose IL-2 reduces lymphocyte apoptosis and increases naive CD4 cells in HIV-1 patients treated with HAART, *Clin. Immunol.*, 94 (3), 153–159, 2000.

160. Sereti, I., Herpin, B., Metcalf, J.A., Stevens, R., Baseler, M.W., Hallahan, C.W., Kovacs, J.A., Davey, R.T., and Lane, H.C., CD4 T cell expansions are associated with increased apoptosis rates of T lymphocytes during IL-2 cycles in HIV infected patients, *AIDS*, 15 (14), 1765–1775, 2001.

161. Hengge, U.R., Borchard, C., Esser, S., Schroder, M., Mirmohammadsadegh, A., and Goos, M., Lymphocytes proliferate in blood and lymph nodes following interleukin-2 therapy in addition to highly active antiretroviral therapy, *AIDS*, 16 (2), 151–160, 2002.

162. Grelli, S., Di Traglia, L., Matteucci, C., Lichtner, M., Vullo, V., Di Sora, F., Lauria, F., Montella, F., Favalli, C., Vella, S., Macchi, B., and Mastino, A., Changes in apoptosis after interruption of potent antiretroviral therapy in patients with maximal HIV-1-RNA suppression, *AIDS*, 15 (9), 1178–1181, 2001.

163. Moretti, S., Alesse, E., Di Marzio, L., Zazzeroni, F., Ruggeri, B., Marcellini, S., Famularo, G., Steinberg, S.M., Boschini, A., Cifone, M.G., and De Simone, C., Effect of L-carnitine on human immunodeficiency virus-1 infection-associated apoptosis: a pilot study, *Blood*, 91 (10), 3817–3824, 1998.

22 HIV-1 Infection and Cell Death in the Nervous System

Gareth Jones
Christopher Power

CONTENTS

NEUROLOGICAL DYSFUNCTION ASSOCIATED WITH HIV-1 INFECTION

Infection by the lentivirus human immunodeficiency virus type 1 (HIV-1) results in a variety of syndromes involving both the central and the peripheral nervous systems. Although HIV-1 enters the central nervous system (CNS) early in the course of infection, the chief burden of neurological illness occurs in the later stages of systemic disease concurrent with the development of immunodeficiency.[1] The neuropathological hallmark of HIV-1 infection of the brain is HIV-encephalitis (HIVE), characterized by astrogliosis, microglial nodules, macrophage infiltration, multinucleated giant cells, white matter pallor, reduced synaptic density, and neuronal loss in the cortex and basal ganglia.[2–10] These neuropathological features are associated with a subcortical clinical disorder termed HIV-associated dementia (HAD), which is defined by cognitive, behavioral, and motor dysfunction. Although HAD increasingly develops before profound immunosuppression, it is generally a neurological syndrome associated with advanced stages of HIV-1 infection. Before the introduction of highly active antiretroviral therapy (HAART), the prevalence of HAD in the asymptomatic phase of HIV-1 infection was estimated to be only 0.4%,[11] compared with 16% observed in patients who displayed symptomatic infection.[12] Among children with HIV-1 infection, delays in neurological development occur, although frank encephalopathy may be evident only during advanced infection.[13] In addition to HAD, a less severe form of cognitive impairment termed "minor cognitive/motor disorder" (MCMD) exists in at least 30% of symptomatic HIV-1-infected individuals.[14,15] Manifestations of MCMD seems to indicate a worse prognosis for acquired immunodeficiency syndrome (AIDS)[16] and a high predictive value for the subsequent detection of HIVE at autopsy.[17] The introduction of HAART regimens in the mid-1990s led to a dramatic decline in the death rate due to AIDS, decreased rates of maternal-to-infant transmission, and reductions in the rates of opportunistic infections. In addition, introduction of HAART almost halved the incidence rates of HAD to that now of approximately 10%,[18,19] although the prevalence of HAD is on the rise.[20] In addition, improvements in neurocognitive performance were observed in patients after 6 months of HAART therapy.[21,22] However, HAART does not provide complete protection from or complete reversal of HAD.[23,24] The incidence of HAD as an AIDS-defining illness has actually increased in recent years,[23] and the prevalence of HAD in patients with higher CD4 T cell counts (greater than 200 cells/μl) is also on the rise.[19] Moreover, HAART has had little effect on the prevalence of MCMD, with current rates maintained at approximately 37%,[15] and the frequency of HIVE detected in postmortem tissue has remained constant.[25] These findings support the hypothesis that HAART does not provide complete protection from developing HAD, possibly as a result of the poor penetration of these drugs into the CNS.[26] Although HAART has impacted on the incidence of HAD, it still remains a serious concern during AIDS. This is highlighted in a recent report, in which more than 90% of HIV-1 infected patients who died between 1996 and 2001 were newly diagnosed with HAD within 12 months of death.[27] It has been suggested that HAD is now the most common form of dementia worldwide in people aged 40 years or younger and is a significant independent risk factor for death due to AIDS.[28]

In addition to CNS disorders, HIV-1 infection gives rise to a variety of disorders of the peripheral nervous system (PNS). In fact, peripheral neuropathy has become the chief neurological complication observed among persons infected with HIV-1 in the developed world.[29] Of these peripheral neuropathies, HIV-associated sensory neuropathy (HIV-SN) is the most common form recognized among patients with HIV-1 infection, affecting approximately 30% of both adults and children with AIDS.[30–32] HIV-SN includes neuropathy directly related to HIV disease *per se* (e.g., distal symmetrical polyneuropathy [DSP]) and neuropathy associated with the use of the nucleoside analogue reverse transcriptase inhibitors (NRTIs) including zalcitabine (ddC), stavudine (d4T), and didanosine (ddI) (e.g., antiretroviral toxic neuropathy [ATN]). In both cases, HIV-SN is predominantly defined by pain; patients exhibit hyperalgesia (lowered pain threshold) and allodynia (pain induced by otherwise nonnoxious stimuli), although, in addition, patients may suffer from paresthesia, gait instability, and autonomic dysfunction.[33] The neuropathology of HIV-SN is characterized by the distal degeneration of long axons after a "dying back pattern."[33] Prominent loss of the small, unmyelinated nociceptive sensory neurons of the dorsal root ganglia (DRG) occurs together with inflammation within the nerve, although reduced density of small and large myelinated fibers has also been observed.[30,34] In addition, reduced density of DRG neurons[35] and degeneration of the centrally directed extensions of sensory neurons in gracile tracts of the cervical and upper thoracic spinal cord[36] have been demonstrated in HIV-SN patients. Very recent studies from our group indicate that cats infected with the lentivirus feline immunodeficiency virus (FIV) develop a distal sensory neuropathy that is also defined by axonal injury and inflammation.[37]

A burgeoning field of interest is the interaction between intravenous drug abuse, HIV-1 infection, and neurological impairment. Abuse of opiate drugs, such as heroin and morphine, not only promote HIV-1 infection and the progression to AIDS[38–40] but also seem to increase the frequency and severity of HIVE.[41] In addition, morphine potentiates the toxic effects of HIV-derived proteins.[42,43] In a recent study, morphine enhanced the HIV-1 Tat-induced cell death of both glial precursor cells and type I astrocytes via a μ-opioid receptor-mediated mechanism involving caspase-3 activation.[44] In addition to the opiates, one of the most common drugs of abuse in HIV-1 infected individuals is methamphetamine (a drug chemically similar to the CNS stimulant amphetamine), and abuse of either drug was associated with neurological impairment.[45,46] In a recent study, Rippeth et al. demonstrated that although HIV-1 infection and methamphetamine dependence were both individually associated with neuropsychological defects, the additive deleterious effects of the two were greater than either risk factor alone.[47] Given that intravenous drug abuse remains a driving force behind the HIV-1 epidemic in certain regions (e.g., Eastern Europe and Central Asia), its potential role in the neurological dysfunction observed in HAD requires further attention.

HIV-1 GENE EXPRESSION IN THE NERVOUS SYSTEM

Productive HIV-1 infection of the CNS is chiefly detectable in perivascular macrophages and microglia.[48,49] Although HIV-1 infection of astrocytes has also been documented,[50–53] this seems to be restricted to the expression of mRNA transcripts and protein of the HIV-1 early genes Tat, Nef, and Rev, without production of progeny virus. Several groups have demonstrated the presence of HIV-1-encoded transcripts and proteins in the PNS.[54–61] However, HIV-1 replication in peripheral nerves seems to be sparse and restricted to the macrophage cell population. Productive HIV-1 infection of neurons has not been convincingly demonstrated in either the CNS or the PNS. Although latent proviral HIV-1 infection of neurons has been demonstrated in brain tissue from AIDS patients[62,63] and in DRG tissue from HIV-SN patients,[54] other groups have failed to confirm these findings.[58,59,61,64]

NEURONAL CELL DEATH IN HIV-1 INFECTION

Despite the absence of productive infection of neurons, HIV-1 infection has been associated with neuronal loss in distinct regions of the brain, including the frontal cortex,[65–67] the substantia nigra,[68] the cerebellum,[69] and the putamen.[70] Neuronal loss in the frontal cortex and in other regions of the neocortex has been estimated between 18 and 38%.[65–67] Select neuronal subpopulations, including larger pyramidal cells within the cortex, are at greater risk of cell death and, similarly, neuronal populations that express certain neurotransmitters (e.g., GABA) and proteins (e.g., parvalbumin and calbindin) are more likely to be diminished in the HIV-1-infected brain.[71]

Cellular injury and loss in the brain during HIV-1 infection are also evident from magnetic resonance imaging and spectroscopy studies that revealed brain atrophy in HIV-1-infected patients occurring within cortical, central, brain stem, and cerebellar regions.[72–76] In some studies, the severity of atrophy was correlated with advanced stages of HIV-1 infection[72,73] or with clinical signs of cognitive impairment.[72,73,75] Although necrosis of neurons was demonstrated in HAD patients,[77] it was hypothesized as early as 1990 that the neuronal loss observed in HIV-1-infected patients may also be due to apoptosis.[65,66] Apoptosis is an active process of cell death, characterized by cell shrinkage, chromatin aggregation with genomic fragmentation, and nuclear pyknosis.[78,79] *In vivo*, phagocytic cells normally sequester antigenically modified apoptotic cells, preventing inflammation and damage to the surrounding tissue.[80] In contrast, necrosis is characterized by passive cell swelling and disruption of internal homeostasis, resulting in membrane lysis and release of intracellular constituents that induces a local inflammatory reaction with subsequent injury to the surrounding tissue.[81] If apoptotic neurons are not phagocytosed in a timely fashion, however, secondary necrosis may occur, producing ultrastructural phenotypes characteristic of both apoptosis and necrosis.[82–84] It has been suggested that the intensity of the initial insult dictates whether neurons undergo apoptosis or necrosis; a severe initial challenge will result in necrosis, whereas a milder challenge triggers apoptotic signals via p38 MAPK.[82]

With the development of TUNEL (terminal deoxynucleotidyl dUTP nick end labeling), a technique that identifies the free 3-OH ends of newly cleaved DNA *in situ*,[85] several groups in the mid-1990s demonstrated that neuronal damage in both adult and pediatric AIDS was, in part, due to apoptosis.[86–89] However, neuronal apoptosis seems to be a feature of AIDS, as only rare apoptotic neurons were demonstrated in a few pre-AIDS cases.[87,90] Whether a direct correlation exists between the degree of neuronal apoptosis and the severity of neurological cognitive disorders observed in HIV-1 infection remains unclear. In one study, Adle-Biassette et al. characterized apoptotic neurons in several regions of the brain (including the frontal and temporal cortex, basal ganglia, and brain stem) taken postmortem from 20 HIV/AIDS patients.[91] In this study, no global correlation was observed between neuronal apoptosis and cognitive dysfunction, the presence of HIVE, or microglial activation. However, within the cerebral cortex, neuronal apoptosis correlated with cerebral atrophy. Moreover, within the basal ganglia, apoptotic neurons seemed to be more abundant in the vicinity of activated microglia, which contained HIV-1 p24 immunoreactivity, suggesting that some topographical associations may exist. Furthermore, Gelbard et al. demonstrated that apoptotic neurons in the cerebral cortex and basal ganglia of children with HIVE were present in the vicinity of perivascular inflammatory cell infiltrates containing HIV-1 infected macrophages and multinucleated giant cells.[88]

TUNEL analysis also identified DNA strand breaks in the DRG of HIV-1-infected individuals.[92] TUNEL-positive neurons were observed only in the DRG of patients with AIDS and were more abundant in patients with peripheral neuropathy. However, TUNEL-positive neurons did not correlate with neuronal loss or demonstrate morphological features of apoptosis, suggesting that these neurons might be "primed" to apoptosis rather than fully committed to apoptosis.

In addition to neuronal cell death, apoptosis of other CNS cell types in the brains of HIV/AIDS patients was reported, including astrocytes[86,89,93] and, more rarely, multinucleated giant cells.[86] In addition, astrocyte apoptosis was demonstrated in primary human brain cultures following exposure to HIV-1.[89] Significantly greater numbers of apoptotic astrocytes were detected in the brains of HIV/AIDS patients with rapidly progressing dementia compared with slow progressors,[93] and detection of apoptotic astrocytes seemed to be more common in HAD patients compared with nondemented HIV/AIDS patients,[89] suggesting a role for astrocyte cell loss in the neuropathogenesis of HAD. Given the trophic role of astrocytes in the CNS, HIV-1-mediated astrocyte cell death may result in several deleterious events, including alterations in the composition of the extracellular environment (e.g., concentrations of excitotoxic neurotransmitters such as glutamate), reduced production of neurotropic factors, and impaired maintenance of the blood–brain barrier. Several studies in AIDS patients have demonstrated CNS microvascular abnormalities and increased blood–brain barrier permeability.[94–96] HIV-1 Tat was shown to induce apoptosis of human brain microvascular endothelial cells *in vitro* via a nitric oxide–mediated mechanism,[97] and apoptotic endothelial cells were identified in the brain of HIV/AIDS patients.[89] Thus, HIV-1-induced endothelial cell apoptosis may contribute to the microvascular pathologies observed in HIV/AIDS patients. Animal models of HAD also revealed cell death via apoptosis in the CNS after lentivirus infection, when infection of macaques with a neurovirulent strain of another lentivirus, simian immunodeficiency virus (SIV), resulted in apoptosis of neurons, endothelial cells, and glial cells.[98]

Although there is no clear consensus as to the underlying mechanism of HIV-induced neuropathogenesis, two main theories dominate (Figure 22.1). First, although HIV-1 does not infect neurons productively, if at all, HIV-1-encoded proteins may injure neurons directly without requiring the intermediary functions of nonneuronal cells. Alternatively, neuronal apoptosis may result indirectly from the secretion of neurotoxic factors by resident brain macrophages or microglia in response to HIV-1 infection, stimulation by viral proteins, or immune activation. These two theories are in no way mutually exclusive, and current data exist to support each hypothesis.

DIRECT EFFECT OF VIRAL PROTEINS

HIV-1 gp120 envelope protein was shown to be neurotoxic in serum-free primary neuronal cultures (serum-free neuronal cultures contain few nonneuronal cells), in neuroblastoma cell lines, and in mature neuronal differentiated human NT2 cells.[99–102] Soluble HIV-1 gp120 can bind CXCR4, and perhaps CCR5, on neurons in the absence of CD4[100] to induce neuronal signaling and apoptosis,[103] and blocking this interaction was shown to prevent gp120-induced neuronal apoptosis.[104] Recombinant gp120 from the macrophage tropic HIV-1 BaL isolate also induced apoptosis in cultured human neuronal NT2 cells,[101] suggesting that HIV-1-induced neuronal cell death may be mediated through the direct interaction of gp120 with CCR5 expressed on neurons. In addition to gp120, it was suggested that HIV-1 gp41, the transmembrane region of the envelope protein, is neurotoxic.[105] Similarly, full-length envelope-derived proteins from FIV are directly cytotoxic to neurons.[106,107] Recent work from our group indicates that both replication-competent and replication-incompetent virions are highly cytotoxic to neuronal cultures in an envelope-sequence-dependent manner.[108]

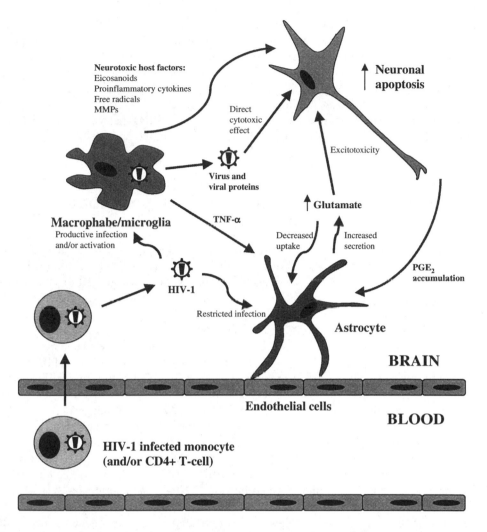

FIGURE 22.1 Neuronal apoptosis results from the direct cytotoxicity of HIV-1 proteins, the production of neurotoxic factors from HIV-1-infected or -activated macrophages/microglia, and the disruption of glutamate homeostasis, resulting in neuronal cell death through an excitotoxic pathway involving hyperstimulation of the NMDA receptor. MMPs, matrix metalloproteinases; TNF-α, tumor necrosis factor-α; PGE_2, prostaglandin E_2.

Although *in vitro* studies demonstrated that CXCR4-dependent HIV-1 isolates induced the highest level of neuronal death,[109] the majority of brain-derived viruses are CCR5 dependent.[110,111] Some CCR5/CXCR4 dualtropic viruses were identified in the brains of HIV-1-infected individuals,[111] but detection of virus that is solely CXCR4 dependent is rare.

Experiments using neuroblastoma cell lines have supported the hypothesis that several other HIV-1 viral proteins can directly injure neurons without the intermediary role of nonneuronal cells. HIV-1 Tat can be taken up into rat PC12 neuronal cells by a receptor-mediated mechanism.[112] Compared with BSA-treated control cells, HIV-1 Tat (at a concentration as low as 1 nM) induced significantly greater apoptosis of PC12 cells, 72 h after stimulation.[113] In addition, HIV-1 Tat was shown to induce apoptosis in human neuroblastoma SK-N-MC cells in a dose-dependent manner.[114] In this study, the specificity for Tat-induced apoptosis was confirmed using Tat-neutralizing anti-serum. These findings are underscored by the early observation that the Tat protein from the animal lentivirus maedi-visna virus is also neurotoxic.[115] Experiments using human NT2 cells (stably differentiated postmitotic neurons) demonstrated that extracellular HIV-1 Vpr induced apoptosis in a dose-dependent fashion.[116] In addition, this latter group showed that when delivered intracellularly, HIV-1 Vpr was capable of inducing apoptosis of mature neuronal differentiated human NT2 cells.[117] Trillo-Pazos et al. demonstrated that 100 ng/ml of recombinant HIV-1 Nef caused a 30% decrease in cell number in the human SK-N-SH neuroblastoma cell line after a single exposure over 3 days.[118] At this time, fragmented nuclei were detected in the cell line, suggesting that the Nef-induced toxicity might also be mediated via apoptosis. However, these studies are complicated by the observation that HIV-1-infected cells minimally release the Nef protein.

SOLUBLE FACTORS FROM MACROPHAGES/MICROGLIA EXPOSED TO HIV-1

It has been reported that neuronal apoptosis may result indirectly due to the secretion of neurotoxic factors from brain resident macrophages or microglia in response to HIV-1 infection, stimulation by viral proteins, or immune activation. Apoptotic neurons in the basal ganglia of HIV-1-infected patients were observed to be more abundant in the vicinity of mutinucleated giant cells and HIV-1 p24 immunoreactive cells.[91] Although HIV-1 infects both CD4$^+$ T cells and cells of the macrophage/microglia lineage, the results of a recent study underscore the importance of only the latter cell type in the secretion of neurotoxic host factors upon HIV-1 infection.[101] Cell culture media from HIV-1-infected T cells induced high levels of neuronal apoptosis; however, depletion of HIV-1 virions from the media reversed this effect. Although virion-containing cell culture media from HIV-1-infected macrophages induced greater neuronal apoptosis than virion-depleted macrophage supernatants, significant cell death was still observed in the latter. These results suggest that upon HIV-1 infection, both macrophages and T cells secrete virus factors that mediate neuronal apoptosis. In contrast, neurotoxic host factors were secreted from HIV-1-infected macrophages but not T cells.

HIV-1-infected brain mononuclear phagocytes are capable of producing a long list of potential neurotoxins, which include eicosanoids (e.g., arachidonic acid and platelet-activating factor [PAF]), proinflammatory cytokines (e.g., tumor necrosis factor-α [TNF-α] and interleukin-1β [IL-1β]), free radicals (e.g., nitric oxide and superoxide anion), and neurotoxic amines (e.g., NTox).[119–121] In addition, immune activation of macrophages by HIV-1 gp120 induces the production of arachidonic acid, TNF-α and IL-1β,[122] and the glutamate-like agonist L-cysteine,[123] an endogenous neurotoxin that acts via excessive N-methyl-D-aspartate (NMDA) receptor activation.[124,125] L-cysteine production by gp120-stimulated macrophages was reduced when the cells were pretreated with a TNF-α neutralizing antibody or an IL-1 receptor antagonist, demonstrating that production of this neurotoxin was partially mediated by the cytokines TNF-α and IL-1β.[123] Importantly, TNF-α and IL-1β are elevated in the brain, spinal cord, and cerebrospinal fluid of AIDS patients.[126,127] Similar to gp120, HIV-1 Tat was also shown to induce the production of neurotoxins by macrophages. Injection of synthetic Tat peptide into mouse brain induced lesions in areas that were associated with expression of TNF-α, IL-1β, IL-6, and inducible nitric oxide synthase (iNOS) in macrophages/microglia and astrocytes.[128] Blockade of TNF-α resulted in decreased IL-1β and iNOS expression, along with a reduction in the volume of the brain lesions, underlying a central role of TNF-α in Tat-mediated neurotoxicity. TNF-α was also demonstrated to inhibit glutamate uptake by astrocytes.[129] Thus, Tat-induced TNF-α expression may also result in increased levels of extracellular glutamate, leading to hyperactivation of NMDA receptors and subsequent neuronal cell death. When expressed in primary human macrophages, HIV-1 Tat from HAD patients induced elevated release and activation of matrix metalloproteinase (MMP)-2 and MMP-7.[130] Conditioned media from these cells induced neuronal death, which was inhibited by anti-MMP-2 or anti-MMP-7 antibodies but not by antibodies against MMP-9 or Tat. Interestingly, this phenomenon was not observed for Tat from nondemented HIV-1-infected patients, indicating that HIV-1 Tat also causes neuronal death through an indirect mechanism that is Tat sequence dependent and involves the induction of specific MMPs. HIV-1 Tat was also demonstrated to mediate neuronal cell death through enhancing the activity of glycogen synthase kinase (GSK)-3β,[131] an enzyme shown to play a direct role in the regulation of neuronal apoptosis.[132] Although Tat was shown to associate with GSK-3β, direct addition of Tat to purified GSK-3β had no effect on enzyme activity. Instead, it was suggested that HIV-1 Tat influences GSK-3β by an indirect mechanism, involving TNF-α-mediated induction of PAF, which, in turn, induces GSK-3β activity.[131]

The above list of neurotoxins is by no means exhaustive, and the identities of potentially neurotoxic substances released by HIV-1-infected or HIV-1-activated macrophages are currently under intense investigation. Recent findings from our laboratory identified a novel mechanism by which HIV-1 infection of the CNS may result in neuronal apoptosis, and this highlights the selective pathological proteolysis of tissue proteins by proteases in the brain.[133] Although increased MMP expression and activity were associated with multiple neurological diseases,[134] including those caused by HIV-1 and other related lentiviruses,[135–137] the causality and neurotoxic mechanism remain unclear. We recently demonstrated that HIV-1 infection of human macrophages resulted in elevated levels of secreted pro-MMP-2, although macrophages infected with an HIV-1 Env-deleted luciferase vector pseudotyped by a vesicular stomatis virus envelope did not show increased MMP-2 activity. Thus, MMP-2 induction was specific to HIV-1 Env-mediated infection and not a nonspecific cellular response to viral envelope. Recombinant pro-MMP-2, but not a catalytically inactive mutant of MMP-2, induced cell death in LAN-2 neuronal cells in a dose-dependent manner. This neurotoxicity of pro-MMP-2 was blocked by treatment with the synthetic MMP inhibitor Prinomastat, confirming the neurotoxic effects of MMP-2 proteolytic activity. MMPs are now recognized as pivotal processing enzymes of a wide variety of bioactive mediators that control many cellular processes.[138] Chemokines, such as stromal-derived factor (SDF)-1, are abundantly expressed in the CNS[139] and constitute a recently recognized class of MMP substrates.[140,141] MMP-2 efficiently cleaves SDF-1 to yield a truncated molecule that lacks CXCR4 agonist activity in chemotaxis assays and has a greater than 100-fold reduction in CXCR4 binding affinity.[140] Unexpectedly, TUNEL analysis indicated that LAN-2 neuronal cells (but not other nonneuronal cell lines expressing CXCR4) treated with SDF-1(5-67), a synthetic analogue of the cleaved SDF-1, showed higher levels of neuronal apoptosis than in those treated with full-length SDF-1.[133] A marked reduction in neurotoxicity was found after pretreatment of the cells with an SDF-1-neutralizing antibody or with pertussis toxin, demonstrating the specificity of cleaved SDF-1(5-67) as a novel cytotoxin selective for neurons acting through an as-yet-unidentified $G_{\alpha i}$-coupled receptor. Cleaved SDF-1(5-67) and MMP-2 also proved to be neurotoxic in a rodent model that recapitulated several aspects of HIV-1 infection of the brain, including the nature and site of neuropathy with concurrent motor abnormalities.[133] These findings emphasize the role of proteolysis of constitutively expressed proteins to yield novel molecules in the pathogenesis of neurodegenerative diseases, as also illustrated in Alzheimer's and Huntington's diseases.

The proapoptotic transcription factor p53 was proposed as a candidate mediator of HIV-induced neuronal injury. HeLa cells expressing CD4 undergo p53-dependent apoptosis upon exposure to HIV-1 gp120,[142,143] and there is evidence of p53 accumulation in CNS tissue from monkeys with SIV encephalitis (SIVE)[144] and in neurons of HAD patients.[145] In addition, p53 mediates neuronal apoptosis induced by excitotoxins.[146,147] Using murine cerebrocortical cultures, Garden et al. recently demonstrated that neurotoxic concentrations of HIV-1 gp120 induced p53 accumulation in both neurons and microglia.[148] Neurons from mice deficient in p53 were resistant to gp120-induced apoptosis, whereas microglia from p53-deficient mice lost the ability to mediate gp120-induced neuronal apoptosis, suggesting that p53-dependent signaling is required in both neurons and microglia for gp120-induced neuronal apoptosis.[148]

In addition to cells of the macrophage/microglia lineage, recent data from our group also demonstrate that neurotoxic factors are released by astrocytes upon exposure to HIV-1.[149] Expression of the envelope protein of the brain-derived HIV-1 strain JRFL in primary astrocytes resulted in the production of soluble factors that were highly toxic to LAN-2 neuronal cells. As no detectable

HIV-1 gp120 was present in the conditioned media from these astrocytes, other molecules released in response to expression of the HIV-1 envelope were responsible for the cytotoxicity of the astrocyte-conditioned media. Moreover, the inability to block this neurotoxic effect with antagonists for the NMDA and AMPA receptors indicated that these neurotoxins do not act via the glutamate receptors. Data from our group also demonstrated upregulation of proteinase-activated receptor 1 (PAR-1) expression in astrocytes during HIVE.[150] The fact that mRNA expression levels for PAR-1 and its agonist, prothrombin, were greater in HIV-1-infected patients than in those individuals with multiple sclerosis suggests that this increase is not solely an effect of inflammation but is possibly related to a virus-induced event. *In vitro* experiments revealed that supernatants from astrocytic cells stimulated with selective agonists for PAR-1 induced neurotoxicity in human fetal neurons via an NMDA receptor-mediated mechanism.

DISTINCT HIV-1 STRAINS AND NEUROLOGICAL DISEASE

Although the role of specific lentivirus strains that do or do not induce neurological disease remains controversial, evidence from animal models suggests that distinct SIV and FIV strains are responsible for development of brain disease.[151–153] In particular, specific sequences within the Env gene were shown to influence the development of animal lentivirus neurological disease.[154,155] Specific mutations were also identified within the V1 and V3 domains of brain-derived gp120 sequences that differed between AIDS patients with and without HAD.[156,157] However, it is clear that brain-derived HIV-1 sequences from AIDS patients with or without HAD exhibit significantly greater molecular diversity among HAD patients, depending on the individual viral gene, which may impact on neuropathogenesis.[156–159]

MECHANISMS OF NEURONAL APOPTOSIS

Apoptosis may occur via two major pathways, the "extrinsic pathway" induced by the ligation of certain transmembrane proteins with cytoplasmic death domains and the "intrinsic pathway" resulting from depolarization of mitochondrial membranes and the release of cytochrome *c* from the mitochondria. Both pathways involve the activation of cellular caspases, and both are in no way mutually exclusive. These pathways are well defined in neurons and seem to be perturbed in several neurodegenerative diseases, including Alzheimer's disease, amyotrophic lateral sclerosis, and stroke (reviewed in Friedlander[160]).

EXTRINSIC PATHWAY

The binding of TNF-α to TNF-α receptor-1 (TNFR1) activates TNFR1-associated death domain protein (TRADD), which, in turn, interacts with Fas-associated death domain protein (FADD) to induce apoptosis.[161] The finding that TNFR1 is expressed on some neurons, coupled with the fact that HIV-1-infected and gp120-stimulated microglia release TNF-α, suggests a role for TNF-α-mediated apoptosis of neurons in HIV-1 infection. Expression of TNF-α and its receptor is elevated in brains from patients with HAD,[162] and TNF-α is capable of stimulating apoptosis in human neurons.[163] TNF-α was also implicated in mediating gp120-induced apoptosis of sensory neurons.[164] In this study, gp120 ligation of CXCR4 on Schwann cells induced their production of RANTES, which, in turn, induced TNF-α production by DRG neurons to mediate neurotoxicity in an autocrine manner.

In contrast to the evidence presented above, TNF-α was shown to have a neuroprotective role. Pretreatment of hippocampal, cortical, or septal neurons with TNF-α was shown to protect from glucose-deprivation neurotoxicity by maintaining Ca^{2+} homeostasis.[165] The neuroprotection afforded by TNF-α seems to be mediated through its activation of the transcription factor NF-κB and the subsequent upregulation of antiapoptotic proteins (reviewed in Saha and Pahan[166]). In particular,

TNF-α was shown to induce the expression of the chemokine CX3CL1 (fractalkine) in astrocytes,[167] which, in turn, protects neurons from HIV-1 gp120-mediated toxicity.[168] It is thought that the timing and the duration of exposure to TNF-α are important factors in determining its neurotoxic or neuroprotective role. For example, TNF-α was shown to be neuroprotective when applied before ischemic stress, but it becomes neurotoxic when applied after the same insult.[169] In addition, the role of TNF-α may be affected by the expression of other elements within the diseased brain. For example, in combination with IL-1β, TNF-α induces expression of iNOS in glial cells.[170] In contrast, when given alone, TNF-α has limited effect on the expression of the same gene.[171]

Two recent studies suggest the contribution of TNF-related apoptosis-inducing ligand (TRAIL) expressed on HIV-1-infected macrophages to neuronal apoptosis.[172,173] TRAIL is a member of the TNF superfamily of proteins, showing homology to TNF and Fas ligand (CD95L). TRAIL is capable of inducing apoptosis in target cells by binding to receptors that contain conserved death domains (e.g., TRAIL-R1 and TRAIL-R2). Although originally thought to induce apoptosis only in tumor or transformed cells, TRAIL was later shown to induce apoptosis in virally infected cells. TRAIL mediates activation-induced cell death in peripheral blood mononuclear cells (PBMCs) from HIV-1-infected individuals,[174] and lymphocytes from HIV-1-infected patients seem to be more sensitive to TRAIL-induced apoptosis compared to uninfected controls.[175,176] Furthermore, large numbers of HIV-1-uninfected CD4+ T cells undergo TRAIL-mediated apoptosis in HIV-infected lymphoid organs.[177] In addition to secreting neurotoxic agents, HIV-1-exposed macrophages were recently shown to induce neuronal apoptosis via a TRAIL-mediated process.[172,173] TRAIL-R1 and TRAIL-R2 were detected in both human fetal neuron cultures and on neurons in brain tissue from HIVE patients.[172] In agreement with earlier studies,[178,179] Ryan et al. demonstrated that exposure of human macrophages to HIV-1 resulted in their increased expression of TRAIL.[172] Although not previously detected in normal brain tissue,[180] TRAIL-expressing macrophages were detected in brain tissue from HIVE patients and were spatially associated with caspase-3-expressing neurons. In a series of *in vitro* experiments using human fetal neuron cultures, Ryan et al. demonstrated that recombinant human TRAIL protein-induced apoptosis could be prevented by inhibition of either caspase-8 or caspase-9, suggesting that both the extrinsic pathway and amplification of this by the intrinsic pathway were required for this process.[172] Similar observations were seen in a murine model in which human PBMCs were transplanted into nonobese diabetic severe combined immunodeficiency mice.[173] HIV-1 infection in combination with LPS administration induced infiltration of infected human cells into the perivascular region of the brain and neuronal apoptosis, which frequently colocalized with HIV-1-infected macrophages expressing TRAIL. In both studies, the specificity of TRAIL-induced apoptosis was demonstrated using a neutralizing antibody against human TRAIL.

A role for the Fas/Fas ligand system in mediating neuronal apoptosis in HIV-1 infection remains to be determined. Aquaro et al. demonstrated that anti-Fas antibody could reverse the programmed cell death observed in astrocytes exposed to supernatants from HIV-infected macrophages,[181] and elevated levels of Fas and Fas ligand were reported in the brains of HAD patients.[182] However, further studies are required to elucidate the extent to which this system may play a role in HAD.

INTRINSIC PATHWAY

Dysregulation of neuronal cell Ca^{2+} homeostasis may play a central role in both HIV-1 gp120 and Tat-induced apoptosis of neurons through the intrinsic pathway. As discussed previously, HIV-1 gp120-induced neurotoxicity may occur through a direct action of neurons or through an indirect mechanism involving toxic products released from gp120-stimulated microglia and macrophages. One such toxic product is arachidonic acid, which facilitates NMDA-evoked currents in neurons[183] and stimulates glutamate release from astrocytes[184] while simultaneously inhibiting its uptake by glial cells.[184–186] Thus, the net result is a hyperactivation of NMDA receptors, resulting in potentially lethal levels of Ca^{2+} influx. In support, blockade of NMDA receptors but not non-NMDA receptors was shown to inhibit HIV-1 gp120-induced neurotoxicity.[187] HIV-1 gp120 also stimulates Na^+/H^+ exchange in astrocytes, resulting in increased intracellular pH.[188] Intracellular alkalinization can, in turn, activate ion-motive transporters, resulting in increased extracellular K^+ levels, inhibition of glutamate uptake, enhanced glutamate release, and depolarization of neuronal membranes,[189] thus again priming NMDA receptors on neurons for hyperactivation by an increased pool of extracellular glutamate. An additional mechanism of HIV-1 gp120-induced neurotoxicity is the direct binding of the viral protein to CXCR4 expressed on neurons. Recently, Bachis et al. demonstrated that HIV-1 gp120-mediated apoptosis of cerebellar granule cells could be prevented by the CXCR4 inhibitor AMD3100 but not by the glutamate receptor antagonist MK801, suggesting that in their system, the toxic mechanism of gp120 primarily involves activation of the CXCR4 receptor.[190] CXCR4 is a G protein–coupled receptor that, upon ligand activation, triggers a transient increase in cytosolic free Ca^{2+},[104] suggesting again that dysregulation of Ca^{2+} homeostasis may be a key event in HIV-1 gp120-induced neurotoxicity. HIV-1 Tat protein interacts directly with neurons, inducing phosphatidylinositol 3-kinase activity,[191] increased IP_3 levels, and subsequent release of Ca^{2+} from IP_3-sensitive ER stores.[192] This ability of HIV-1 Tat to induce IP_3-sensitive release of Ca^{2+} in neurons led to the speculation that exogenous HIV-1 Tat interacts with CXCR4, a G protein–linked receptor that is coupled to inositol hydrolysis and IP_3-mediated Ca^{2+} release. Indeed, neurons express CXCR4, and interaction of HIV-1 Tat with CXCR4 was demonstrated in human PBMCs,[193] though this remains to be demonstrated in neurons. Inhibition of IP_3-mediated Ca^{2+} release from the ER was shown to inhibit Tat-induced neurotoxicity, underscoring the importance of inositol signaling in this process.[192] In addition, HIV-1 Tat induces Ca^{2+} influx from the extracellular environment through at least four types of neuronal calcium channels, including L- and N-type voltage-gated channels, and the NMDA and non-NMDA ionotropic glutamate receptors.[194–198] The elevation of cytosolic Ca^{2+} is an early event in the neuronal apoptosis cascade and seems to trigger mitochondrial calcium uptake, production of reactive oxygen species, and mitochondrial stress, resulting in mitochondrial membrane depolarization and subsequent release of apoptotic factors.[199,200]

Experimental data from both *in vitro* and *in vivo* studies from the Bagetta laboratory led to the suggestion that an underlying mechanism of neuronal cell death triggered by HIV-1 gp120 involves the chemokine receptor–mediated enhancement of brain IL-1β and subsequent enhancement of cyclooxygenase (COX-2) expression.[201] Intracerebroventricular (i.c.v.) injection of recombinant HIV-1 gp120 IIIB protein in adult rats induced apoptosis of neurons in the brain neocortex, which was preceded by enhanced IL-1β expression.[202] Double-labeling immunofluorescence experiments established that the main sources of gp120-enhanced IL-1β expression in the rat neocortex were neurons and microglia.[203] Subchronic i.c.v. administration of recombinant IL-1β induced apoptosis in the neocortex of rat brain, whereas combined treatment with HIV-1 gp120 and an inhibitor of caspase-1 (the enzyme that converts pro-IL-1β into biologically active IL-1β) resulted in a reduction in neuronal cell death compared with animals treated with HIV-1 gp120 alone, implicating a role for IL-1β in neocortical cell death. Pretreatment with a nonneurotoxic dose of SDF-1 reduced both gp120-induced IL-1β production and neuronal apoptosis, implicating the role of CXCR4 receptor stimulation in mediating gp120 neurotoxicity.[202] The role of IL-1β in mediating neurotoxicity induced by HIV-1 gp120 is also supported by *in vitro* data, when exposure of human CHP100 neuroblastoma cells to the viral protein enhanced IL-1β secretion and cell death.[204] Again, inhibition of IL-1β production and blockade of the IL-1β receptor both protected the neuroblastoma cells from HIV-1 gp120-induced cell death. In addition to IL-1β, i.c.v. injection of HIV-1 gp120 caused early enhancement of functionally active COX-2 expression in the neocortex of rat, resulting in increased prostaglandin E_2 (PGE_2) production.[205,206] HIV-1 gp120 induced accumulation of PGE_2 in human CHP100 neuroblastoma cells by enhancing the COX activity of the arachidonate-metabolizing enzyme prostaglandin H synthase, whereas inhibitors of this process provided protection against the cell death triggered by HIV-1 gp120.[207] Interestingly, inhibitors of IL-1β production also prevented the enhanced COX-2 expression otherwise induced by HIV-1 gp120, suggesting that IL-1β may be the signal by which the viral protein enhanced COX-2 expression.[204] Products of the arachidonic cascade, especially PGE_2, stimulate the release of glutamate from astrocytes in a Ca^{2+}-dependent manner.[208] Thus, it was suggested[201] that HIV-1 gp120, through an IL-1β-dependent mechanism, induces COX-2 expression in neuronal cells to convert arachidonic acid into PGE_2, which accumulates. Elevated PGE_2 levels, in turn, increase synaptic glutamate levels by stimulating its release from astrocyes, triggering neuronal cell oxidative stress and apoptotic death through an excitotoxic pathway involving stimulation of the NMDA receptor.

FUTURE PERSPECTIVES: IMPLICATIONS FOR THERAPY

As discussed, neuronal apoptosis in HIV-1 infection is, in part, mediated through an excitotoxic pathway involving hyperstimulation of the NMDA receptor. Therefore, antagonists of the NMDA receptors may represent potential therapeutic agents (Table 22.1). One such agent, memantine, was shown to block the neurotoxicity of HIV-1 gp120 and Tat *in vitro*.[187,209] The results from clinical trials of memantine treatment in Alzheimer's disease (AD) have been encouraging, and memantine provides clinical benefits in the functional abilities of patients with moderate to severe AD.[210–212] A double-blind, placebo-controlled phase II trial of memantine for HAD is currently under way, and the results are keenly awaited. Dysregulation of neuronal cell Ca^{2+} homeostasis seems to play a central role in many of the pathways mediating neuronal apoptosis. Thus, calcium channel blockers, non-NMDA receptor antagonists, intracellular calcium chelators, and agents that block mitochondrial membrane permeability may also be beneficial. Given that caspases mediate the apoptotic pathway, caspase inhibitors may also be useful in preventing the neuronal apoptosis seen in HAD. However, due to the risk of promoting oncogenic processes, extreme care will need to be taken to make sure that normal physiological responses are not interrupted.

As previously discussed, the proapoptotic transcription factor p53 was proposed as a candidate mediator of HIV-induced neuronal injury. Recent findings from our group demonstrated that human growth hormone (GH) prevents HIV-1 Tat-induced neuronal death by inhibiting p53 expression in neurons.[145] In addition, *in vivo* treatment with human GH reduced Tat-related neurobehavioral and

---

TABLE 22.1
Potential Therapeutic Interventions

Mechanism	Potential Therapeutics
Maintenance of neuronal cell Ca^{2+} homeostasis	NMDA receptor antagonists (e.g., memantine); non-NMDA receptor antagonists; Ca^{2+} channel blockers (e.g., nimodipine); intracellular Ca^{2+} chelators
Inhibition of apoptotic pathway	Blockers of mitochondrial membrane permeability (e.g., minocycline); caspase-3, caspase-8, and caspase-9 inhibitors; inhibitors of proapoptotic factors (e.g., growth hormone)
Inhibition of neurotoxic factors from macrophages/microglia	Inhibitors of macrophage/microglia activation (e.g., minocycline); inhibitors of PAF (e.g., lexipafant); inhibitors of MMP activity (e.g., prinomastat); inhibitors of TNF-α (e.g., CN-1189); antioxidants (e.g., OPC-14117, selegiline)
Neuroprotection	Neurotrophins (e.g., NGF)
Improved delivery of HAART regimens to the CNS	

Note: PAF, platelet-activating factor; MMP, matrix metalloproteinase; TNF-α, tumor necrosis factor-α; NGF, nerve growth factor.

neuropathological changes in an animal model. Finally, human GH was well tolerated among patients with HIV-related cognitive impairment and, more importantly, within this small group of patients, GH improved their neurocognitive performance without adverse effects on systemic immune parameters.[213] Hence, these findings suggest that GH may be protective in a neurodegenerative disease. These observations reflect previous studies indicating that GH protects neurons during hypoxic ischemia[213] and enhances rescue of neurons after spinal cord injury.[214] Results from previous clinical trials have demonstrated that recombinant human GH treatment is tolerated and increases body weight, physical function, and quality of life in HIV-1-infected individuals with HIV-associated weight loss.[215,216] In addition, concomitant administration of recombinant human GH with HAART has been shown to enhance T cell maturation, differentiation, and function in HIV-1-infected patients.[217] Thus, recombinant human GH treatment seems to be an exciting approach, benefiting not only neurological disorders associated with HIV-1 infection but also metabolic and immunologic concerns.

Another potential therapy for HAD is the tetracycline derivative minocycline, a broad-spectrum antimicrobicide with anti-inflammatory properties. In the CNS, minocycline was shown to exert many effects, including the prevention of microglial activation, inhibiting IL-1β activation by interfering with the induction of IL-1-converting enzyme mRNA, decreasing induction of iNOS mRNA and preventing NOS protein expression, inhibiting NO-induced neurotoxicity, and reducing COX-2 expression and PGE2 production.[218–221] Minocycline also inhibits cytochrome c release from mitochondria,[222] which is a potent stimulus for activation of caspase-9 and caspase-3 and subsequent apoptosis. Thus, minocycline may interfere with many of the potential pathways that were suggested to play a role in mediating neuronal cell death in HAD. Of importance, minocycline treatment is safe and well tolerated and was approved by the U.S. Food and Drug Administration as an antibiotic that may be used for a wide variety of infections. In a small randomized clinical trial, the antioxidants selegiline and OPC-14117 showed beneficial trends toward improvement in HIV-related neurocognitive impairments.[223] Similarly, the PAF antagonist lexipafant also demonstrated beneficial trends in a small randomized clinical trial.[224]

The recent findings from our laboratory highlighting the selective pathological proteolysis of tissue proteins by proteases in the brain have revealed a new therapeutic target that could potentially be used to treat HAD. Reducing the cleavage of SDF-1 with synthetic MMP inhibitor drugs such as Prinomastat, which are already in phase 3 clinical trials for the treatment of cancer,[225,226] would have a twofold benefit, both in stabilizing full-length SDF-1, which protects CD4+CXCR4+ cells from further infection and subsequent signal transduction via CXCR4 on neurons, and in protecting neurons by reducing the levels of the toxic cleaved SDF-1 product.

Although the introduction of HAART regimens has led to a drastic decline in the death rate due to AIDS, they do not provide complete protection from developing HAD, possibly as a result of the poor penetration of these drugs into the CNS.[26] A potential neuroprotective role for HIV-1 protease inhibitors was demonstrated, in which the HIV-1 protease inhibitor ritonavir protected hippocampal neurons against oxidative stress-induced apoptosis.[227] Thus, with the development of improved systems to deliver these drugs to the CNS, HAART regimens may also play a more significant role in the treatment of HAD.

REFERENCES

1. Johnson, R.T., McArthur, J.C., and Narayan, O., The neurobiology of human immunodeficiency virus infections, *FASEB J.,* 2, 2970, 1988.
2. Sharer, L.R. and Kapila, R., Neuropathologic observations in acquired immunodeficiency syndrome (AIDS), *Acta Neuropathol. (Berl.),* 66, 188, 1985.
3. Budka, H., Costanzi, G., Cristina, S., Lechi, A., Parravicini, C., Trabattoni, R., and Vago, L., Brain pathology induced by infection with the human immunodeficiency virus (HIV). A histological, immunocytochemical, and electron microscopical study of 100 autopsy cases, *Acta Neuropathol. (Berl.),* 75, 185, 1987.

4. Wiley, C.A., Masliah, E., Morey, M., Lemere, C., DeTeresa, R., Grafe, M., Hansen, L., and Terry, R., Neocortical damage during HIV infection, *Ann. Neurol.,* 29, 651, 1991.

5. Weis, S., Haug, H., and Budka, H., Neuronal damage in the cerebral cortex of AIDS brains: a morphometric study, *Acta Neuropathol. (Berl.),* 85, 185, 1993.

6. Aylward, E.H., Henderer, J.D., McArthur, J.C., Brettschneider, P.D., Harris, G.J., Barta, P.E., and Pearlson, G.D., Reduced basal ganglia volume in HIV-1-associated dementia: results from quantitative neuroimaging, *Neurology,* 43, 2099, 1993.

7. Nielsen, S.L., Petito, C.K., Urmacher, C.D., and Posner, J.B., Subacute encephalitis in acquired immune deficiency syndrome: a postmortem study, *Am. J. Clin. Pathol.,* 82, 678, 1984.

8. Aylward, E.H., Brettschneider, P.D., McArthur, J.C., Harris, G.J., Schlaepfer, T.E., Henderer, J.D., Barta, P.E., Tien, A.Y., and Pearlson, G.D., Magnetic resonance imaging measurement of gray matter volume reductions in HIV dementia, *Am. J. Psychiatry,* 152, 987, 1995.

9. Kozlowski, P.B., Brudkowska, J., Kraszpulski, M., Sersen, E.A., Wrzolek, M.A., Anzil, A.P., Rao, C., and Wisniewski, H.M., Microencephaly in children congenitally infected with human immunodeficiency virus — a gross-anatomical morphometric study, *Acta Neuropathol. (Berl.),* 93, 136, 1997.

10. Stout, J.C., Ellis, R.J., Jernigan, T.L., Archibald, S.L., Abramson, I., Wolfson, T., McCutchan, J.A., Wallace, M.R., Atkinson, J.H., and Grant, I., Progressive cerebral volume loss in human immunodeficiency virus infection: a longitudinal volumetric magnetic resonance imaging study. HIV Neurobehavioral Research Center Group, *Arch. Neurol.,* 55, 161, 1998.

11. Miller, E.N., Selnes, O.A., McArthur, J.C., Satz, P., Becker, J.T., Cohen, B.A., Sheridan, K., Machado, A.M., Van Gorp, W.G., and Visscher, B., Neuropsychological performance in HIV-1-infected homosexual men: The Multicenter AIDS Cohort Study (MACS), *Neurology,* 40, 197, 1990.

12. McArthur, J.C., Neurologic manifestations of AIDS, *Medicine (Baltimore),* 66, 407, 1987.

13. Laverda, A.M., Cogo, P., Condini, A., Cattelan, C., Giaquinto, C., Cozzani, S., Ruga, E., Viero, F., De Rossi, A., and Del Mistro, A., How frequent and how early does the neurological involvement in HIV-positive children occur? Preliminary results of a prospective study, *Childs Nerv. Syst.,* 6, 406, 1990.

14. Janssen, R.S., Nwanyanwu, O.C., Selik, R.M., and Stehr-Green, J.K., Epidemiology of human immunodeficiency virus encephalopathy in the United States, *Neurology,* 42, 1472, 1992.

15. Sacktor, N., McDermott, M.P., Marder, K., Schifitto, G., Selnes, O.A., McArthur, J.C., Stern, Y., Albert, S., Palumbo, D., Kieburtz, K., De Marcaida, J.A., Cohen, B., and Epstein, L., HIV-associated cognitive impairment before and after the advent of combination therapy, *J. Neurovirol.,* 8, 136, 2002.

16. Sacktor, N.C., Bacellar, H., Hoover, D.R., Nance-Sproson, T.E., Selnes, O.A., Miller, E.N., Dal Pan, G.J., Kleeberger, C., Brown, A., Saah, A., and McArthur, J.C., Psychomotor slowing in HIV infection: a predictor of dementia, AIDS and death, *J. Neurovirol.,* 2, 404, 1996.

17. Cherner, M., Masliah, E., Ellis, R.J., Marcotte, T.D., Moore, D.J., Grant, I., and Heaton, R.K., Neurocognitive dysfunction predicts postmortem findings of HIV encephalitis, *Neurology,* 59, 1563, 2002.

18. Brodt, H.R., Kamps, B.S., Gute, P., Knupp, B., Staszewski, S., and Helm, E.B., Changing incidence of AIDS-defining illnesses in the era of antiretroviral combination therapy, *AIDS,* 11, 1731, 1997.

19. Sacktor, N., Lyles, R.H., Skolasky, R., Kleeberger, C., Selnes, O.A., Miller, E.N., Becker, J.T., Cohen, B., and McArthur, J.C., HIV-associated neurologic disease incidence changes: Multicenter AIDS Cohort Study, 1990–1998, *Neurology,* 56, 257, 2001.

20. Sacktor, N., The epidemiology of human immunodeficiency virus-associated neurological disease in the era of highly active antiretroviral therapy, *J. Neurovirol.,* 8 (Suppl. 2), 115, 2002.

21. Ferrando, S., van Gorp, W., McElhiney, M., Goggin, K., Sewell, M., and Rabkin, J., Highly active antiretroviral treatment in HIV infection: benefits for neuropsychological function, *AIDS,* 12, F65, 1998.

22. Tozzi, V., Balestra, P., Galgani, S., Narciso, P., Ferri, F., Sebastiani, G., D'Amato, C., Affricano, C., Pigorini, F., Pau, F.M., De Felici, A., and Benedetto, A., Positive and sustained effects of highly active antiretroviral therapy on HIV-1-associated neurocognitive impairment, *AIDS,* 13, 1889, 1999.

23. Dore, G.J., Correll, P.K., Li, Y., Kaldor, J.M., Cooper, D.A., and Brew, B.J., Changes to AIDS dementia complex in the era of highly active antiretroviral therapy, *AIDS,* 13, 1249, 1999.

24. Major, E.O., Rausch, D., Marra, C., and Clifford, D., HIV-associated dementia, *Science,* 288, 440, 2000.

25. Masliah, E., DeTeresa, R.M., Mallory, M.E., and Hansen, L.A., Changes in pathological findings at autopsy in AIDS cases for the last 15 years, *AIDS,* 14, 69, 2000.

26. Enting, R.H., Hoetelmans, R.M., Lange, J.M., Burger, D.M., Beijnen, J.H., and Portegies, P., Anti-retroviral drugs and the central nervous system, *AIDS,* 12, 1941, 1998.

27. Welch, K. and Morse, A., The clinical profile of end-stage AIDS in the era of highly active antiretroviral therapy, *AIDS Patient Care STDS,* 16, 75, 2002.

28. Ellis, R.J., Deutsch, R., Heaton, R.K., Marcotte, T.D., McCutchan, J.A., Nelson, J.A., Abramson, I., Thal, L.J., Atkinson, J.H., Wallace, M.R., and Grant, I., Neurocognitive impairment is an independent risk factor for death in HIV infection. San Diego HIV Neurobehavioral Research Center Group, *Arch. Neurol.,* 54, 416, 1997.

29. Brinley, F.J., Jr., Pardo, C.A., and Verma, A., Human immunodeficiency virus and the peripheral nervous system workshop, *Arch. Neurol.,* 58, 1561, 2001.

30. Araujo, A.P., Nascimento, O.J., and Garcia, O.S., Distal sensory polyneuropathy in a cohort of HIV-infected children over five years of age, *Pediatrics,* 106, E35, 2000.

31. Tagliati, M., Grinnell, J., Godbold, J., and Simpson, D.M., Peripheral nerve function in HIV infection: clinical, electrophysiologic, and laboratory findings, *Arch. Neurol.,* 56, 84, 1999.

32. Keswani, S.C., Pardo, C.A., Cherry, C.L., Hoke, A., and McArthur, J.C., HIV-associated sensory neuropathies, *AIDS,* 16, 2105, 2002.

33. Pardo, C.A., McArthur, J.C., and Griffin, J.W., HIV neuropathy: insights in the pathology of HIV peripheral nerve disease, *J. Peripher. Nerv. Syst.,* 6, 21, 2001.

34. Griffin, J.W. and McArthur, J.C. Peripheral neuropathies associated with HIV infection, in *The Neurology of AIDS,* Gendelman, H.E., Epstein, L., and Swindells, S., Eds., Chapman & Hall, New York, 1998, p. 275.

35. Bradley, W.G., Shapshak, P., Delgado, S., Nagano, I., Stewart, R., and Rocha, B., Morphometric analysis of the peripheral neuropathy of AIDS, *Muscle Nerve,* 21, 1188, 1998.

36. Rance, N.E., McArthur, J.C., Cornblath, D.R., Landstrom, D.L., Griffin, J.W., and Price, D.L., Gracile tract degeneration in patients with sensory neuropathy and AIDS, *Neurology,* 38, 265, 1988.

37. Kennedy, J.M., Hoke, A., Zhu, Y., Johnston, J.B., van Marle, G., Silva, C., Zochodne, D.W., and Power, C., Peripheral neuropathy in lentivirus infection: evidence of inflammation and axonal injury, *AIDS,* 18, 1241, 2004.

38. Rouveix, B., Opiates and immune function. Consequences on infectious diseases with special reference to AIDS, *Therapie,* 47, 503, 1992.

39. Peterson, P.K., Sharp, B.M., Gekker, G., Jackson, B., and Balfour, H.H., Jr., Opiates, human peripheral blood mononuclear cells, and HIV, *Adv. Exp. Med. Biol.,* 288, 171, 1991.

40. Arora, P.K., Morphine-induced immune modulation: does it predispose to HIV infection? *NIDA Res. Monogr.,* 96, 150, 1990.

41. Bell, J.E., Brettle, R.P., Chiswick, A., and Simmonds, P., HIV encephalitis, proviral load and dementia in drug users and homosexuals with AIDS. Effect of neocortical involvement, *Brain,* 121 (Pt. 11), 2043, 1998.

42. Nath, A., Human immunodeficiency virus (HIV) proteins in neuropathogenesis of HIV dementia, *J. Infect. Dis.,* 186 (Suppl. 2), S193, 2002.

43. Gurwell, J.A., Nath, A., Sun, Q., Zhang, J., Martin, K.M., Chen, Y., and Hauser, K.F., Synergistic neurotoxicity of opioids and human immunodeficiency virus-1 Tat protein in striatal neurons *in vitro, Neuroscience,* 102, 555, 2001.

44. Khurdayan, V.K., Buch, S., El-Hage, N., Lutz, S.E., Goebel, S.M., Singh, I.N., Knapp, P.E., Turchan-Cholewo, J., Nath, A., and Hauser, K.F., Preferential vulnerability of astroglia and glial precursors to combined opioid and HIV-1 Tat exposure *in vitro, Eur. J. Neurosci.,* 19, 3171, 2004.

45. Simon, S.L., Domier, C.P., Sim, T., Richardson, K., Rawson, R.A., and Ling, W., Cognitive performance of current methamphetamine and cocaine abusers, *J. Addict. Dis.,* 21, 61, 2002.

46. McKetin, R. and Mattick, R.P., Attention and memory in illicit amphetamine users: comparison with non-drug-using controls, *Drug Alcohol Depend.,* 50, 181, 1998.

47. Rippeth, J.D., Heaton, R.K., Carey, C.L., Marcotte, T.D., Moore, D.J., Gonzalez, R., Wolfson, T., and Grant, I., Methamphetamine dependence increases risk of neuropsychological impairment in HIV infected persons, *J. Int. Neuropsychol. Soc.,* 10, 1, 2004.

48. Wiley, C.A., Schrier, R.D., Nelson, J.A., Lampert, P.W., and Oldstone, M.B., Cellular localization of human immunodeficiency virus infection within the brains of acquired immune deficiency syndrome patients, *Proc. Natl. Acad. Sci. U.S.A.,* 83, 7089, 1986.

49. Sharer, L.R., Pathology of HIV-1 infection of the central nervous system. A review, *J. Neuropathol. Exp. Neurol.,* 51, 3, 1992.

50. Bagasra, O., Lavi, E., Bobroski, L., Khalili, K., Pestaner, J.P., Tawadros, R., and Pomerantz, R.J., Cellular reservoirs of HIV-1 in the central nervous system of infected individuals: identification by the combination of *in situ* polymerase chain reaction and immunohistochemistry, *AIDS,* 10, 573, 1996.

51. Ranki, A., Nyberg, M., Ovod, V., Haltia, M., Elovaara, I., Raininko, R., Haapasalo, H., and Krohn, K., Abundant expression of HIV Nef and Rev proteins in brain astrocytes *in vivo* is associated with dementia, *AIDS,* 9, 1001, 1995.

52. Saito, Y., Sharer, L.R., Epstein, L.G., Michaels, J., Mintz, M., Louder, M., Golding, K., Cvetkovich, T.A., and Blumberg, B.M., Overexpression of Nef as a marker for restricted HIV-1 infection of astrocytes in postmortem pediatric central nervous tissues, *Neurology,* 44, 474, 1994.

53. Tornatore, C., Chandra, R., Berger, J.R., and Major, E.O., HIV-1 infection of subcortical astrocytes in the pediatric central nervous system, *Neurology,* 44, 481, 1994.

54. Brannagan, T.H., III, Nuovo, G.J., Hays, A.P., and Latov, N., Human immunodeficiency virus infection of dorsal root ganglion neurons detected by polymerase chain reaction *in situ* hybridization, *Ann. Neurol.,* 42, 368, 1997.

55. Cornford, M.E., Ho, H.W., and Vinters, H.V., Correlation of neuromuscular pathology in acquired immune deficiency syndrome patients with cytomegalovirus infection and zidovudine treatment, *Acta Neuropathol. (Berl.),* 84, 516, 1992.

56. Gherardi, R., Lebargy, F., Gaulard, P., Mhiri, C., Bernaudin, J.F., and Gray, F., Necrotizing vasculitis and HIV replication in peripheral nerves, *N. Engl. J. Med.,* 321, 685, 1989.

57. Grafe, M.R. and Wiley, C.A., Spinal cord and peripheral nerve pathology in AIDS: the roles of cytomegalovirus and human immunodeficiency virus, *Ann. Neurol.,* 25, 561, 1989.

58. Nagano, I., Shapshak, P., Yoshioka, M., Xin, K., Nakamura, S., and Bradley, W.G., Increased NADPH-diaphorase reactivity and cytokine expression in dorsal root ganglia in acquired immunodeficiency syndrome, *J. Neurol. Sci.,* 136, 117, 1996.

59. Rizzuto, N., Cavallaro, T., Monaco, S., Morbin, M., Bonetti, B., Ferrari, S., Galiazzo-Rizzuto, S., Zanette, G., and Bertolasi, L., Role of HIV in the pathogenesis of distal symmetrical peripheral neuropathy, *Acta Neuropathol. (Berl.),* 90, 244, 1995.

60. Vital, A., Beylot, M., Vital, C., Delors, B., Bloch, B., and Julien, J., Morphological findings on peripheral nerve biopsies in 15 patients with human immunodeficiency virus infection, *Acta Neuropathol. (Berl.),* 83, 618, 1992.

61. Yoshioka, M., Shapshak, P., Srivastava, A.K., Stewart, R.V., Nelson, S.J., Bradley, W.G., Berger, J.R., Rhodes, R.H., Sun, N.C., and Nakamura, S., Expression of HIV-1 and interleukin-6 in lumbosacral dorsal root ganglia of patients with AIDS, *Neurology,* 44, 1120, 1994.

62. Nuovo, G.J., Gallery, F., MacConnell, P., and Braun, A., *In situ* detection of polymerase chain reaction-amplified HIV-1 nucleic acids and tumor necrosis factor- RNA in the central nervous system, *Am. J. Pathol.,* 144, 659, 1994.

63. Torres-Munoz, J., Stockton, P., Tacoronte, N., Roberts, B., Maronpot, R.R., and Petito, C.K., Detection of HIV-1 gene sequences in hippocampal neurons isolated from postmortem AIDS brains by laser capture microdissection, *J. Neuropathol. Exp. Neurol.,* 60, 885, 2001.

64. Takahashi, K., Wesselingh, S.L., Griffin, D.E., McArthur, J.C., Johnson, R.T., and Glass, J.D., Localization of HIV-1 in human brain using polymerase chain reaction/*in situ* hybridization and immuno-cytochemistry, *Ann. Neurol.,* 39, 705, 1996.

65. Ketzler, S., Weis, S., Haug, H., and Budka, H., Loss of neurons in the frontal cortex in AIDS brains, *Acta Neuropathol. (Berl.),* 80, 92, 1990.

66. Everall, I.P., Luthert, P.J., and Lantos, P.L., Neuronal loss in the frontal cortex in HIV infection, *Lancet,* 337, 1119, 1991.

67. Everall, I.P., Luthert, P.J., and Lantos, P.L., Neuronal number and volume alterations in the neocortex of HIV infected individuals, *J. Neurol. Neurosurg. Psychiatry,* 56, 481, 1993.

68. Reyes, M.G., Faraldi, F., Senseng, C.S., Flowers, C., and Fariello, R., Nigral degeneration in acquired immune deficiency syndrome (AIDS), *Acta Neuropathol. (Berl.),* 82, 39, 1991.

69. Graus, F., Ribalta, T., Abos, J., Alom, J., Cruz-Sanchez, F., Mallolas, J., Miro, J.M., Cardesa, A., and Tolosa, E., Subacute cerebellar syndrome as the first manifestation of AIDS dementia complex, *Acta Neurol. Scand.,* 81, 118, 1990.

70. Everall, I., Luthert, P., and Lantos, P., A review of neuronal damage in human immunodeficiency virus infection: its assessment, possible mechanism and relationship to dementia, *J. Neuropathol. Exp. Neurol.*, 52, 561, 1993.

71. Masliah, E., Ge, N., Achim, C.L., Hansen, L.A., and Wiley, C.A., Selective neuronal vulnerability in HIV encephalitis, *J. Neuropathol. Exp. Neurol.*, 51, 585, 1992.

72. Elovaara, I., Poutiainen, E., Raininko, R., Valanne, L., Virta, A., Valle, S.L., Lahdevirta, J., and Iivanainen, M., Mild brain atrophy in early HIV infection: the lack of association with cognitive deficits and HIV-specific intrathecal immune response, *J. Neurol. Sci.*, 99, 121, 1990.

73. Dal Pan, G.J., McArthur, J.H., Aylward, E., Selnes, O.A., Nance-Sproson, T.E., Kumar, A.J., Mellits, E.D., and McArthur, J.C., Patterns of cerebral atrophy in HIV-1-infected individuals: results of a quantitative MRI analysis, *Neurology*, 42, 2125, 1992.

74. Manji, H., Connolly, S., McAllister, R., Valentine, A.R., Kendall, B.E., Fell, M., Durrance, P., Thompson, A.J., Newman, S., and Weller, I.V., Serial MRI of the brain in asymptomatic patients infected with HIV: results from the UCMSM/Medical Research Council neurology cohort, *J. Neurol. Neurosurg. Psychiatry*, 57, 144, 1994.

75. Hall, M., Whaley, R., Robertson, K., Hamby, S., Wilkins, J., and Hall, C., The correlation between neuropsychological and neuroanatomic changes over time in asymptomatic and symptomatic HIV-1-infected individuals, *Neurology*, 46, 1697, 1996.

76. Chang, L., Ernst, T., Leonido-Yee, M., Witt, M., Speck, O., Walot, I., and Miller, E.N., Highly active antiretroviral therapy reverses brain metabolite abnormalities in mild HIV dementia, *Neurology*, 53, 782, 1999.

77. Ho, D.D., Bredesen, D.E., Vinters, H.V., and Daar, E.S., The acquired immunodeficiency syndrome (AIDS) dementia complex, *Ann. Intern. Med.*, 111, 400, 1989.

78. Wyllie, A.H., Kerr, J.F., and Currie, A.R., Cell death: the significance of apoptosis, *Int. Rev. Cytol.*, 68, 251, 1980.

79. Kerr, J.F., Wyllie, A.H., and Currie, A.R., Apoptosis: a basic biological phenomenon with wide-ranging implications in tissue kinetics, *Br. J. Cancer*, 26, 239, 1972.

80. Savill, J., Fadok, V., Henson, P., and Haslett, C., Phagocyte recognition of cells undergoing apoptosis, *Immunol. Today*, 14, 131, 1993.

81. Schwartz, L.M., Smith, S.W., Jones, M.E., and Osborne, B.A., Do all programmed cell deaths occur via apoptosis? *Proc. Natl. Acad. Sci. U.S.A.*, 90, 980, 1993.

82. Bonfoco, E., Krainc, D., Ankarcrona, M., Nicotera, P., and Lipton, S.A., Apoptosis and necrosis: two distinct events induced, respectively, by mild and intense insults with N-methyl-D-aspartate or nitric oxide/superoxide in cortical cell cultures, *Proc. Natl. Acad. Sci. U.S.A.*, 92, 7162, 1995.

83. Ankarcrona, M., Dypbukt, J.M., Bonfoco, E., Zhivotovsky, B., Orrenius, S., Lipton, S.A., and Nicotera, P., Glutamate-induced neuronal death: a succession of necrosis or apoptosis depending on mitochondrial function, *Neuron*, 15, 961, 1995.

84. Dreyer, E.B., Zhang, D., and Lipton, S.A., Transcriptional or translational inhibition blocks low dose NMDA-mediated cell death, *Neuroreport*, 6, 942, 1995.

85. Gavrieli, Y., Sherman, Y., and Ben-Sasson, S.A., Identification of programmed cell death *in situ* via specific labeling of nuclear DNA fragmentation, *J. Cell Biol.*, 119, 493, 1992.

86. Petito, C.K. and Roberts, B., Evidence of apoptotic cell death in HIV encephalitis, *Am. J. Pathol.*, 146, 1121, 1995.

87. Adle-Biassette, H., Levy, Y., Colombel, M., Poron, F., Natchev, S., Keohane, C., and Gray, F., Neuronal apoptosis in HIV infection in adults, *Neuropathol. Appl. Neurobiol.*, 21, 218, 1995.

88. Gelbard, H.A., James, H.J., Sharer, L.R., Perry, S.W., Saito, Y., Kazee, A.M., Blumberg, B.M., and Epstein, L.G., Apoptotic neurons in brains from paediatric patients with HIV-1 encephalitis and progressive encephalopathy, *Neuropathol. Appl. Neurobiol.*, 21, 208, 1995.

89. Shi, B., De Girolami, U., He, J., Wang, S., Lorenzo, A., Busciglio, J., and Gabuzda, D., Apoptosis induced by HIV-1 infection of the central nervous system, *J. Clin. Invest.*, 98, 1979, 1996.

90. An, S.F., Giometto, B., Scaravilli, T., Tavolato, B., Gray, F., and Scaravilli, F., Programmed cell death in brains of HIV-1-positive AIDS and pre-AIDS patients, *Acta Neuropathol. (Berl.)*, 91, 169, 1996.

91. Adle-Biassette, H., Chretien, F., Wingertsmann, L., Hery, C., Ereau, T., Scaravilli, F., Tardieu, M., and Gray, F., Neuronal apoptosis does not correlate with dementia in HIV infection but is related to microglial activation and axonal damage, *Neuropathol. Appl. Neurobiol.*, 25, 123, 1999.

92. Adle-Biassette, H., Bell, J.E., Creange, A., Sazdovitch, V., Authier, F.J., Gray, F., Hauw, J.J., and Gherardi, R., DNA breaks detected by *in situ* end-labelling in dorsal root ganglia of patients with AIDS, *Neuropathol. Appl. Neurobiol.*, 24, 373, 1998.

93. Thompson, K.A., McArthur, J.C., and Wesselingh, S.L., Correlation between neurological progression and astrocyte apoptosis in HIV-associated dementia, *Ann. Neurol.*, 49, 745, 2001.

94. Power, C., Kong, P.A., Crawford, T.O., Wesselingh, S., Glass, J.D., McArthur, J.C., and Trapp, B.D., Cerebral white matter changes in acquired immunodeficiency syndrome dementia: alterations of the blood–brain barrier, *Ann. Neurol.*, 34, 339, 1993.

95. Smith, T.W., DeGirolami, U., Henin, D., Bolgert, F., and Hauw, J.J., Human immunodeficiency virus (HIV) leukoencephalopathy and the microcirculation, *J. Neuropathol. Exp. Neurol.*, 49, 357, 1990.

96. Petito, C.K. and Cash, K.S., Blood–brain barrier abnormalities in the acquired immunodeficiency syndrome: immunohistochemical localization of serum proteins in postmortem brain, *Ann. Neurol.*, 32, 658, 1992.

97. Kim, T.A., Avraham, H.K., Koh, Y.H., Jiang, S., Park, I.W., and Avraham, S., HIV-1 Tat-mediated apoptosis in human brain microvascular endothelial cells, *J. Immunol.*, 170, 2629, 2003.

98. Adamson, D.C., Dawson, T.M., Zink, M.C., Clements, J.E., and Dawson, V.L., Neurovirulent simian immunodeficiency virus infection induces neuronal, endothelial, and glial apoptosis, *Mol. Med.*, 2, 417, 1996.

99. Meucci, O., Fatatis, A., Simen, A.A., Bushell, T.J., Gray, P.W., and Miller, R.J., Chemokines regulate hippocampal neuronal signaling and gp120 neurotoxicity, *Proc. Natl. Acad. Sci. U.S.A.*, 95, 14500, 1998.

100. Hesselgesser, J., Halks-Miller, M., DelVecchio, V., Peiper, S.C., Hoxie, J., Kolson, D.L., Taub, D., and Horuk, R., CD4-independent association between HIV-1 gp120 and CXCR4: functional chemokine receptors are expressed in human neurons, *Curr. Biol.*, 7, 112, 1997.

101. Xu, Y., Kulkosky, J., Acheampong, E., Nunnari, G., Sullivan, J., and Pomerantz, R.J., HIV-1-mediated apoptosis of neuronal cells: proximal molecular mechanisms of HIV-1-induced encephalopathy, *Proc. Natl. Acad. Sci. U.S.A.*, 101, 7070, 2004.

102. Dreyer, E.B., Kaiser, P.K., Offermann, J.T., and Lipton, S.A., HIV-1 coat protein neurotoxicity prevented by calcium channel antagonists, *Science*, 248, 364, 1990.

103. Hesselgesser, J., Taub, D., Baskar, P., Greenberg, M., Hoxie, J., Kolson, D.L., and Horuk, R., Neuronal apoptosis induced by HIV-1 gp120 and the chemokine SDF-1 is mediated by the chemokine receptor CXCR4, *Curr. Biol.*, 8, 595, 1998.

104. Zheng, J., Thylin, M.R., Ghorpade, A., Xiong, H., Persidsky, Y., Cotter, R., Niemann, D., Che, M., Zeng, Y.C., Gelbard, H.A., Shepard, R.B., Swartz, J.M., and Gendelman, H.E., Intracellular CXCR4 signaling, neuronal apoptosis and neuropathogenic mechanisms of HIV-1-associated dementia, *J. Neuroimmunol.*, 98, 185, 1999.

105. Adamson, D.C., Kopnisky, K.L., Dawson, T.M., and Dawson, V.L., Mechanisms and structural determinants of HIV-1 coat protein, gp41-induced neurotoxicity, *J. Neurosci.*, 19, 64, 1999.

106. Bragg, D.C., Meeker, R.B., Duff, B.A., English, R.V., and Tompkins, M.B., Neurotoxicity of FIV and FIV envelope protein in feline cortical cultures, *Brain Res.*, 816, 431, 1999.

107. Gruol, D.L., Yu, N., Parsons, K.L., Billaud, J.N., Elder, J.H., and Phillips, T.R., Neurotoxic effects of feline immunodeficiency virus, FIV-PPR, *J. Neurovirol.*, 4, 415, 1998.

108. Zhang, K., Rana, F., Silva, C., Ethier, J., Wehrly, K., Chesebro, B., and Power, C., Human immunodeficiency virus type 1 envelope-mediated neuronal death: uncoupling of viral replication and neurotoxicity, *J. Virol.*, 77, 6899, 2003.

109. Ohagen, A., Ghosh, S., He, J., Huang, K., Chen, Y., Yuan, M., Osathanondh, R., Gartner, S., Shi, B., Shaw, G., and Gabuzda, D., Apoptosis induced by infection of primary brain cultures with diverse human immunodeficiency virus type 1 isolates: evidence for a role of the envelope, *J. Virol.*, 73, 897, 1999.

110. Chan, S.Y., Speck, R.F., Power, C., Gaffen, S.L., Chesebro, B., and Goldsmith, M.A., V3 recombinants indicate a central role for CCR5 as a coreceptor in tissue infection by human immunodeficiency virus type 1, *J. Virol.*, 73, 2350, 1999.

111. Ohagen, A., Devitt, A., Kunstman, K.J., Gorry, P.R., Rose, P.P., Korber, B., Taylor, J., Levy, R., Murphy, R.L., Wolinsky, S.M., and Gabuzda, D., Genetic and functional analysis of full-length human immunodeficiency virus type 1 Env genes derived from brain and blood of patients with AIDS, *J. Virol.*, 77, 12336, 2003.

112. Liu, Y., Jones, M., Hingtgen, C.M., Bu, G., Laribee, N., Tanzi, R.E., Moir, R.D., Nath, A., and He, J.J., Uptake of HIV-1 Tat protein mediated by low-density lipoprotein receptor-related protein disrupts the neuronal metabolic balance of the receptor ligands, *Nat. Med.*, 6, 1380, 2000.

113. Zauli, G., Milani, D., Mirandola, P., Mazzoni, M., Secchiero, P., Miscia, S., and Capitani, S., HIV-1 Tat protein down-regulates CREB transcription factor expression in PC12 neuronal cells through a phosphatidylinositol 3-kinase/AKT/cyclic nucleoside phosphodiesterase pathway, *FASEB J.*, 15, 483, 2001.

114. Ramirez, S.H., Sanchez, J.F., Dimitri, C.A., Gelbard, H.A., Dewhurst, S., and Maggirwar, S.B., Neurotrophins prevent HIV Tat-induced neuronal apoptosis via a nuclear factor-B (NF-B)-dependent mechanism, *J. Neurochem.*, 78, 874, 2001.

115. Strijbos, P.J., Zamani, M.R., Rothwell, N.J., Arbuthnott, G., and Harkiss, G., Neurotoxic mechanisms of transactivating protein Tat of Maedi-Visna virus, *Neurosci. Lett.*, 197, 215, 1995.

116. Patel, C.A., Mukhtar, M., and Pomerantz, R.J., Human immunodeficiency virus type 1 Vpr induces apoptosis in human neuronal cells, *J. Virol.*, 74, 9717, 2000.

117. Patel, C.A., Mukhtar, M., Harley, S., Kulkosky, J., and Pomerantz, R.J., Lentiviral expression of HIV-1 Vpr induces apoptosis in human neurons, *J. Neurovirol.*, 8, 86, 2002.

118. Trillo-Pazos, G., McFarlane-Abdulla, E., Campbell, I.C., Pilkington, G.J., and Everall, I.P., Recombinant nef HIV-IIIB protein is toxic to human neurons in culture, *Brain Res.*, 864, 315, 2000.

119. Genis, P., Jett, M., Bernton, E.W., Boyle, T., Gelbard, H.A., Dzenko, K., Keane, R.W., Resnick, L., Mizrachi, Y., and Volsky, D.J., Cytokines and arachidonic metabolites produced during human immunodeficiency virus (HIV)-infected macrophage-astroglia interactions: implications for the neuropathogenesis of HIV disease, *J. Exp. Med.*, 176, 1703, 1992.

120. Bukrinsky, M.I., Nottet, H.S., Schmidtmayerova, H., Dubrovsky, L., Flanagan, C.R., Mullins, M.E., Lipton, S.A., and Gendelman, H.E., Regulation of nitric oxide synthase activity in human immunodeficiency virus type 1 (HIV-1)-infected monocytes: implications for HIV-associated neurological disease, *J. Exp. Med.*, 181, 735, 1995.

121. Giulian, D., Yu, J., Li, X., Tom, D., Li, J., Wendt, E., Lin, S.N., Schwarcz, R., and Noonan, C., Study of receptor-mediated neurotoxins released by HIV-1-infected mononuclear phagocytes found in human brain, *J. Neurosci.*, 16, 3139, 1996.

122. Wahl, L.M., Corcoran, M.L., Pyle, S.W., Arthur, L.O., Harel-Bellan, A., and Farrar, W.L., Human immunodeficiency virus glycoprotein (gp120) induction of monocyte arachidonic acid metabolites and interleukin 1, *Proc. Natl. Acad. Sci. U.S.A.*, 86, 621, 1989.

123. Yeh, M.W., Kaul, M., Zheng, J., Nottet, H.S., Thylin, M., Gendelman, H.E., and Lipton, S.A., Cytokine-stimulated, but not HIV-infected, human monocyte-derived macrophages produce neurotoxic levels of l -cysteine, *J. Immunol.*, 164, 4265, 2000.

124. Olney, J.W., Zorumski, C., Price, M.T., and Labruyere, J., L-cysteine, a bicarbonate-sensitive endogenous excitotoxin, *Science*, 248, 596, 1990.

125. Lipton, S.A., Choi, Y.B., Pan, Z.H., Lei, S.Z., Chen, H.S., Sucher, N.J., Loscalzo, J., Singel, D.J., and Stamler, J.S., A redox-based mechanism for the neuroprotective and neurodestructive effects of nitric oxide and related nitroso-compounds, *Nature*, 364, 626, 1993.

126. Tyor, W.R., Glass, J.D., Baumrind, N., McArthur, J.C., Griffin, J.W., Becker, P.S., and Griffin, D.E., Cytokine expression of macrophages in HIV-1-associated vacuolar myelopathy, *Neurology*, 43, 1002, 1993.

127. Tyor, W.R., Glass, J.D., Griffin, J.W., Becker, P.S., McArthur, J.C., Bezman, L., and Griffin, D.E., Cytokine expression in the brain during the acquired immunodeficiency syndrome, *Ann. Neurol.*, 31, 349, 1992.

128. Philippon, V., Vellutini, C., Gambarelli, D., Harkiss, G., Arbuthnott, G., Metzger, D., Roubin, R., and Filippi, P., The basic domain of the lentiviral Tat protein is responsible for damages in mouse brain: involvement of cytokines, *Virology*, 205, 519, 1994.

129. Fine, S.M., Angel, R.A., Perry, S.W., Epstein, L.G., Rothstein, J.D., Dewhurst, S., and Gelbard, H.A., Tumor necrosis factor inhibits glutamate uptake by primary human astrocytes. Implications for pathogenesis of HIV-1 dementia, *J. Biol. Chem.*, 271, 15303, 1996.

130. Johnston, J.B., Zhang, K., Silva, C., Shalinsky, D.R., Conant, K., Ni, W., Corbett, D., Yong, V.W., and Power, C., HIV-1 Tat neurotoxicity is prevented by matrix metalloproteinase inhibitors, *Ann. Neurol.*, 49, 230, 2001.

131. Maggirwar, S.B., Tong, N., Ramirez, S., Gelbard, H.A., and Dewhurst, S., HIV-1 Tat-mediated activation of glycogen synthase kinase-3 contributes to Tat-mediated neurotoxicity, *J. Neurochem.*, 73, 578, 1999.

132. Pap, M. and Cooper, G.M., Role of glycogen synthase kinase-3 in the phosphatidylinositol 3-kinase/Akt cell survival pathway, *J. Biol. Chem.*, 273, 19929, 1998.

133. Zhang, K., McQuibban, G.A., Silva, C., Butler, G.S., Johnston, J.B., Holden, J., Clark-Lewis, I., Overall, C.M., and Power, C., HIV-induced metalloproteinase processing of the chemokine stromal cell derived factor-1 causes neurodegeneration, *Nat. Neurosci.*, 6, 1064, 2003.

134. Yong, V.W., Power, C., Forsyth, P., and Edwards, D.R., Metalloproteinases in biology and pathology of the nervous system, *Nat. Rev. Neurosci.*, 2, 502, 2001.

135. Conant, K., McArthur, J.C., Griffin, D.E., Sjulson, L., Wahl, L.M., and Irani, D.N., Cerebrospinal fluid levels of MMP-2, 7, and 9 are elevated in association with human immunodeficiency virus dementia, *Ann. Neurol.*, 46, 391, 1999.

136. Johnston, J.B., Silva, C., and Power, C., Envelope gene-mediated neurovirulence in feline immunodeficiency virus infection: induction of matrix metalloproteinases and neuronal injury, *J. Virol.*, 76, 2622, 2002.

137. Berman, N.E., Marcario, J.K., Yong, C., Raghavan, R., Raymond, L.A., Joag, S.V., Narayan, O., and Cheney, P.D., Microglial activation and neurological symptoms in the SIV model of NeuroAIDS: association of MHC-II and MMP-9 expression with behavioral deficits and evoked potential changes, *Neurobiol. Dis.*, 6, 486, 1999.

138. McCawley, L.J. and Matrisian, L.M., Matrix metalloproteinases: they're not just for matrix anymore! *Curr. Opin. Cell Biol.*, 13, 534, 2001.

139. Ohtani, Y., Minami, M., Kawaguchi, N., Nishiyori, A., Yamamoto, J., Takami, S., and Satoh, M., Expression of stromal cell-derived factor-1 and CXCR4 chemokine receptor mRNAs in cultured rat glial and neuronal cells, *Neurosci. Lett.*, 249, 163, 1998.

140. McQuibban, G.A., Gong, J.H., Tam, E.M., McCulloch, C.A., Clark-Lewis, I., and Overall, C.M., Inflammation dampened by gelatinase A cleavage of monocyte chemoattractant protein-3, *Science*, 289, 1202, 2000.

141. McQuibban, G.A., Butler, G.S., Gong, J.H., Bendall, L., Power, C., Clark-Lewis, I., and Overall, C.M., Matrix metalloproteinase activity inactivates the CXC chemokine stromal cell-derived factor-1, *J. Biol. Chem.*, 276, 43503, 2001.

142. Castedo, M., Ferri, K.F., Blanco, J., Roumier, T., Larochette, N., Barretina, J., Amendola, A., Nardacci, R., Metivier, D., Este, J.A., Piacentini, M., and Kroemer, G., Human immunodeficiency virus 1 envelope glycoprotein complex-induced apoptosis involves mammalian target of rapamycin/FKBP12-rapamycin-associated protein-mediated p53 phosphorylation, *J. Exp. Med.*, 194, 1097, 2001.

143. Ferri, K.F., Jacotot, E., Blanco, J., Este, J.A., Zamzami, N., Susin, S.A., Xie, Z., Brothers, G., Reed, J.C., Penninger, J.M., and Kroemer, G., Apoptosis control in syncytia induced by the HIV type 1-envelope glycoprotein complex: role of mitochondria and caspases, *J. Exp. Med.*, 192, 1081, 2000.

144. Jordan-Sciutto, K.L., Wang, G., Murphy-Corb, M., and Wiley, C.A., Induction of cell-cycle regulators in simian immunodeficiency virus encephalitis, *Am. J. Pathol.*, 157, 497, 2000.

145. Silva, C., Zhang, K., Tsutsui, S., Holden, J.K., Gill, M.J., and Power, C., Growth hormone prevents human immunodeficiency virus-induced neuronal p53 expression, *Ann. Neurol.*, 54, 605, 2003.

146. Hughes, P.E., Alexi, T., Yoshida, T., Schreiber, S.S., and Knusel, B., Excitotoxic lesion of rat brain with quinolinic acid induces expression of p53 messenger RNA and protein and p53-inducible genes Bax and Gadd-45 in brain areas showing DNA fragmentation, *Neuroscience*, 74, 1143, 1996.

147. Morrison, R.S., Wenzel, H.J., Kinoshita, Y., Robbins, C.A., Donehower, L.A., and Schwartzkroin, P.A., Loss of the p53 tumor suppressor gene protects neurons from kainate-induced cell death, *J. Neurosci.*, 16, 1337, 1996.

148. Garden, G.A., Guo, W., Jayadev, S., Tun, C., Balcaitis, S., Choi, J., Montine, T.J., Moller, T., and Morrison, R.S., HIV associated neurodegeneration requires p53 in neurons and microglia, *FASEB J.*, 18, 1141, 2004.

149. van Marle, G., Ethier, J., Silva, C., Mac Vicar, B.A., and Power, C., Human immunodeficiency virus type 1 envelope-mediated neuropathogenesis: targeted gene delivery by a Sindbis virus expression vector, *Virology*, 309, 61, 2003.

150. Boven, L.A., Vergnolle, N., Henry, S.D., Silva, C., Imai, Y., Holden, J., Warren, K., Hollenberg, M.D., and Power, C., Up-regulation of proteinase-activated receptor 1 expression in astrocytes during HIV encephalitis, *J. Immunol.*, 170, 2638, 2003.

151. Power, C., Buist, R., Johnston, J.B., Del Bigio, M.R., Ni, W., Dawood, M.R., and Peeling, J., Neurovirulence in feline immunodeficiency virus-infected neonatal cats is viral strain specific and dependent on systemic immune suppression, *J. Virol.*, 72, 9109, 1998.

152. Phillips, T.R., Prospero-Garcia, O., Puaoi, D.L., Lerner, D.L., Fox, H.S., Olmsted, R.A., Bloom, F.E., Henriksen, S.J., and Elder, J.H., Neurological abnormalities associated with feline immunodeficiency virus infection, *J. Gen. Virol.*, 75 (Pt. 5), 979, 1994.

153. Mankowski, J.L., Flaherty, M.T., Spelman, J.P., Hauer, D.A., Didier, P.J., Amedee, A.M., Murphey-Corb, M., Kirstein, L.M., Munoz, A., Clements, J.E., and Zink, M.C., Pathogenesis of simian immunodeficiency virus encephalitis: viral determinants of neurovirulence, *J. Virol.*, 71, 6055, 1997.

154. Mankowski, J.L., Spelman, J.P., Ressetar, H.G., Strandberg, J.D., Laterra, J., Carter, D.L., Clements, J.E., and Zink, M.C., Neurovirulent simian immunodeficiency virus replicates productively in endothelial cells of the central nervous system *in vivo* and *in vitro, J. Virol.*, 68, 8202, 1994.

155. Johnston, J.B., Jiang, Y., van Marle, G., Mayne, M.B., Ni, W., Holden, J., McArthur, J.C., and Power, C., Lentivirus infection in the brain induces matrix metalloproteinase expression: role of envelope diversity, *J. Virol.*, 74, 7211, 2000.

156. Power, C., McArthur, J.C., Johnson, R.T., Griffin, D.E., Glass, J.D., Perryman, S., and Chesebro, B., Demented and nondemented patients with AIDS differ in brain-derived human immunodeficiency virus type 1 envelope sequences, *J. Virol.*, 68, 4643, 1994.

157. Power, C., McArthur, J.C., Nath, A., Wehrly, K., Mayne, M., Nishio, J., Langelier, T., Johnson, R.T., and Chesebro, B., Neuronal death induced by brain-derived human immunodeficiency virus type 1 envelope genes differs between demented and nondemented AIDS patients, *J. Virol.*, 72, 9045, 1998.

158. Van Marle, G., Rourke, S.B., Zhang, K., Silva, C., Ethier, J., Gill, M.J., and Power, C., HIV dementia patients exhibit reduced viral neutralization and increased envelope sequence diversity in blood and brain, *AIDS,* 16, 1905, 2002.

159. Bratanich, A.C., Liu, C., McArthur, J.C., Fudyk, T., Glass, J.D., Mittoo, S., Klassen, G.A., and Power, C., Brain-derived HIV-1 Tat sequences from AIDS patients with dementia show increased molecular heterogeneity, *J. Neurovirol.*, 4, 387, 1998.

160. Friedlander, R.M., Apoptosis and caspases in neurodegenerative diseases, *N. Engl. J. Med.*, 348, 1365, 2003.

161. Hsu, H., Xiong, J., and Goeddel, D.V., The TNF receptor 1-associated protein TRADD signals cell death and NF-B activation, *Cell,* 81, 495, 1995.

162. Wesselingh, S.L., Takahashi, K., Glass, J.D., McArthur, J.C., Griffin, J.W., and Griffin, D.E., Cellular localization of tumor necrosis factor mRNA in neurological tissue from HIV-infected patients by combined reverse transcriptase/polymerase chain reaction *in situ* hybridization and immunohistochemistry, *J. Neuroimmunol.*, 74, 1, 1997.

163. Pulliam, L., Zhou, M., Stubblebine, M., and Bitler, C.M., Differential modulation of cell death proteins in human brain cells by tumor necrosis factor and platelet activating factor, *J. Neurosci. Res.*, 54, 530, 1998.

164. Keswani, S.C., Polley, M., Pardo, C.A., Griffin, J.W., McArthur, J.C., and Hoke, A., Schwann cell chemokine receptors mediate HIV-1 gp120 toxicity to sensory neurons, *Ann. Neurol.*, 54, 287, 2003.

165. Cheng, B., Christakos, S., and Mattson, M.P., Tumor necrosis factors protect neurons against metabolic-excitotoxic insults and promote maintenance of calcium homeostasis, *Neuron,* 12, 139, 1994.

166. Saha, R.N. and Pahan, K., Tumor necrosis factor- at the crossroads of neuronal life and death during HIV-associated dementia, *J. Neurochem.*, 86, 1057, 2003.

167. Yoshida, H., Imaizumi, T., Fujimoto, K., Matsuo, N., Kimura, K., Cui, X., Matsumiya, T., Tanji, K., Shibata, T., Tamo, W., Kumagai, M., and Satoh, K., Synergistic stimulation, by tumor necrosis factor- and interferon-, of fractalkine expression in human astrocytes, *Neurosci. Lett.*, 303, 132, 2001.

168. Cotter, R., Williams, C., Ryan, L., Erichsen, D., Lopez, A., Peng, H., and Zheng, J., Fractalkine (CX3CL1) and brain inflammation: implications for HIV-1-associated dementia, *J. Neurovirol.*, 8, 585, 2002.

169. Venters, H.D., Tang, Q., Liu, Q., VanHoy, R.W., Dantzer, R., and Kelley, K.W., A new mechanism of neurodegeneration: a proinflammatory cytokine inhibits receptor signaling by a survival peptide, *Proc. Natl. Acad. Sci. U.S.A.*, 96, 9879, 1999.

170. Pahan, K., Liu, X., McKinney, M.J., Wood, C., Sheikh, F.G., and Raymond, J.R., Expression of a dominant-negative mutant of p21(ras) inhibits induction of nitric oxide synthase and activation of nuclear factor-B in primary astrocytes, *J. Neurochem.*, 74, 2288, 2000.

171. Pahan, K., Sheikh, F.G., Liu, X., Hilger, S., McKinney, M., and Petro, T.M., Induction of nitric-oxide synthase and activation of NF-B by interleukin-12 p40 in microglial cells, *J. Biol. Chem.*, 276, 7899, 2001.

172. Ryan, L.A., Peng, H., Erichsen, D.A., Huang, Y., Persidsky, Y., Zhou, Y., Gendelman, H.E., and Zheng, J., TNF-related apoptosis-inducing ligand mediates human neuronal apoptosis: links to HIV-1-associated dementia, *J. Neuroimmunol.*, 148, 127, 2004.

173. Miura, Y., Koyanagi, Y., and Mizusawa, H., TNF-related apoptosis-inducing ligand (TRAIL) induces neuronal apoptosis in HIV-encephalopathy, *J. Med. Dent. Sci.*, 50, 17, 2003.

174. Katsikis, P.D., Garcia-Ojeda, M.E., Torres-Roca, J.F., Tijoe, I.M., Smith, C.A., and Herzenberg, L.A., Interleukin-1 -converting enzyme-like protease involvement in Fas-induced and activation-induced peripheral blood T cell apoptosis in HIV infection. TNF-related apoptosis-inducing ligand can mediate activation-induced T cell death in HIV infection, *J. Exp. Med.*, 186, 1365, 1997.

175. Jeremias, I., Herr, I., Boehler, T., and Debatin, K.M., TRAIL/Apo-2-ligand-induced apoptosis in human T cells, *Eur. J. Immunol.*, 28, 143, 1998.

176. Lum, J.J., Pilon, A.A., Sanchez-Dardon, J., Phenix, B.N., Kim, J.E., Mihowich, J., Jamison, K., Hawley-Foss, N., Lynch, D.H., and Badley, A.D., Induction of cell death in human immunodeficiency virus-infected macrophages and resting memory CD4 T cells by TRAIL/Apo2l, *J. Virol.*, 75, 11128, 2001.

177. Miura, Y., Misawa, N., Maeda, N., Inagaki, Y., Tanaka, Y., Ito, M., Kayagaki, N., Yamamoto, N., Yagita, H., Mizusawa, H., and Koyanagi, Y., Critical contribution of tumor necrosis factor-related apoptosis-inducing ligand (TRAIL) to apoptosis of human CD4+ T cells in HIV-1-infected hu-PBL-NOD-SCID mice, *J. Exp. Med.*, 193, 651, 2001.

178. Zhang, M., Li, X., Pang, X., Ding, L., Wood, O., Clouse, K., Hewlett, I., and Dayton, A.I., Identification of a potential HIV-induced source of bystander-mediated apoptosis in T cells: upregulation of TRAIL in primary human macrophages by HIV-1 Tat, *J. Biomed. Sci.*, 8, 290, 2001.

179. Yang, Y., Tikhonov, I., Ruckwardt, T.J., Djavani, M., Zapata, J.C., Pauza, C.D., and Salvato, M.S., Monocytes treated with human immunodeficiency virus Tat kill uninfected CD4(+) cells by a tumor necrosis factor-related apoptosis-induced ligand-mediated mechanism, *J. Virol.*, 77, 6700, 2003.

180. Dorr, J., Waiczies, S., Wendling, U., Seeger, B., and Zipp, F., Induction of TRAIL-mediated glioma cell death by human T cells, *J. Neuroimmunol.*, 122, 117, 2002.

181. Aquaro, S., Panti, S., Caroleo, M.C., Balestra, E., Cenci, A., Forbici, F., Ippolito, G., Mastino, A., Testi, R., Mollace, V., Calio, R., and Perno, C.F., Primary macrophages infected by human immunodeficiency virus trigger CD95-mediated apoptosis of uninfected astrocytes, *J. Leukoc. Biol.*, 68, 429, 2000.

182. Vargas, D.L., Guo, L., Skolasky, R., McArthur, J.C., and Pardo, C.A. The Fas and FasL Pathways in the Pathogenesis of HIV Dementia, in 10th Conference on Retroviruses and Opportunistic Infections, Session 83, Boston, MA, 2003.

183. Miller, B., Sarantis, M., Traynelis, S.F., and Attwell, D., Potentiation of NMDA receptor currents by arachidonic acid, *Nature*, 355, 722, 1992.

184. Vesce, S., Bezzi, P., Rossi, D., Meldolesi, J., and Volterra, A., HIV-1 gp120 glycoprotein affects the astrocyte control of extracellular glutamate by both inhibiting the uptake and stimulating the release of the amino acid, *FEBS Lett.*, 411, 107, 1997.

185. Barbour, B., Szatkowski, M., Ingledew, N., and Attwell, D., Arachidonic acid induces a prolonged inhibition of glutamate uptake into glial cells, *Nature*, 342, 918, 1989.

186. Dreyer, E.B. and Lipton, S.A., The coat protein gp120 of HIV-1 inhibits astrocyte uptake of excitatory amino acids via macrophage arachidonic acid, *Eur. J. Neurosci.*, 7, 2502, 1995.

187. Nath, A., Haughey, N.J., Jones, M., Anderson, C., Bell, J.E., and Geiger, J.D., Synergistic neurotoxicity by human immunodeficiency virus proteins Tat and gp120: protection by memantine, *Ann. Neurol.*, 47, 186, 2000.

188. Benos, D.J., McPherson, S., Hahn, B.H., Chaikin, M.A., and Benveniste, E.N., Cytokines and HIV envelope glycoprotein gp120 stimulate Na+/H+ exchange in astrocytes, *J. Biol. Chem.*, 269, 13811, 1994.

189. Barres, B.A., Chun, L.L., and Corey, D.P., Ion channels in vertebrate glia, *Annu. Rev. Neurosci.*, 13, 441, 1990.

190. Bachis, A. and Mocchetti, I., The chemokine receptor CXCR4 and not the N-methyl-D-aspartate receptor mediates gp120 neurotoxicity in cerebellar granule cells, *J. Neurosci. Res.,* 75, 75, 2004.

191. Milani, D., Mazzoni, M., Borgatti, P., Zauli, G., Cantley, L., and Capitani, S., Extracellular human immunodeficiency virus type-1 Tat protein activates phosphatidylinositol 3-kinase in PC12 neuronal cells, *J. Biol. Chem.,* 271, 22961, 1996.

192. Haughey, N.J., Holden, C.P., Nath, A., and Geiger, J.D., Involvement of inositol 1,4,5-trisphosphate-regulated stores of intracellular calcium in calcium dysregulation and neuron cell death caused by HIV-1 protein tat, *J. Neurochem.,* 73, 1363, 1999.

193. Xiao, H., Neuveut, C., Tiffany, H.L., Benkirane, M., Rich, E.A., Murphy, P.M., and Jeang, K.T., Selective CXCR4 antagonism by Tat: implications for *in vivo* expansion of coreceptor use by HIV-1, *Proc. Natl. Acad. Sci. U.S.A.,* 97, 11466, 2000.

194. Bonavia, R., Bajetto, A., Barbero, S., Albini, A., Noonan, D.M., and Schettini, G., HIV-1 Tat causes apoptotic death and calcium homeostasis alterations in rat neurons, *Biochem. Biophys. Res. Commun.,* 288, 301, 2001.

195. Perez, A., Probert, A.W., Wang, K.K., and Sharmeen, L., Evaluation of HIV-1 Tat induced neurotoxicity in rat cortical cell culture, *J. Neurovirol.,* 7, 1, 2001.

196. Nath, A., Psooy, K., Martin, C., Knudsen, B., Magnuson, D.S., Haughey, N., and Geiger, J.D., Identification of a human immunodeficiency virus type 1 Tat epitope that is neuroexcitatory and neurotoxic, *J. Virol.,* 70, 1475, 1996.

197. Cheng, J., Nath, A., Knudsen, B., Hochman, S., Geiger, J.D., Ma, M., and Magnuson, D.S., Neuronal excitatory properties of human immunodeficiency virus type 1 Tat protein, *Neuroscience,* 82, 97, 1998.

198. Haughey, N.J., Nath, A., Mattson, M.P., Slevin, J.T., and Geiger, J.D., HIV-1 Tat through phosphorylation of NMDA receptors potentiates glutamate excitotoxicity, *J. Neurochem.,* 78, 457, 2001.

199. Kruman, II, Nath, A., and Mattson, M.P., HIV-1 protein Tat induces apoptosis of hippocampal neurons by a mechanism involving caspase activation, calcium overload, and oxidative stress, *Exp. Neurol.,* 154, 276, 1998.

200. Kruman, I., Guo, Q., and Mattson, M.P., Calcium and reactive oxygen species mediate staurosporine-induced mitochondrial dysfunction and apoptosis in PC12 cells, *J. Neurosci. Res.,* 51, 293, 1998.

201. Corasaniti, M.T., Bellizzi, C., Russo, R., Colica, C., Amantea, D., and Di Renzo, G., Caspase-1 inhibitors abolish deleterious enhancement of COX-2 expression induced by HIV-1 gp120 in human neuroblastoma cells, *Toxicol. Lett.,* 139, 213, 2003.

202. Corasaniti, M.T., Piccirilli, S., Paoletti, A., Nistico, R., Stringaro, A., Malorni, W., Finazzi-Agro, A., and Bagetta, G., Evidence that the HIV-1 coat protein gp120 causes neuronal apoptosis in the neocortex of rat via a mechanism involving CXCR4 chemokine receptor, *Neurosci. Lett.,* 312, 67, 2001.

203. Bagetta, G., Corasaniti, M.T., Berliocchi, L., Nistico, R., Giammarioli, A.M., Malorni, W., Aloe, L., and Finazzi-Agro, A., Involvement of interleukin-1 in the mechanism of human immunodeficiency virus type 1 (HIV-1) recombinant protein gp120-induced apoptosis in the neocortex of rat, *Neuroscience,* 89, 1051, 1999.

204. Corasaniti, M.T., Bilotta, A., Strongoli, M.C., Navarra, M., Bagetta, G., and Di Renzo, G., HIV-1 coat protein gp120 stimulates interleukin-1 secretion from human neuroblastoma cells: evidence for a role in the mechanism of cell death, *Br. J. Pharmacol.,* 134, 1344, 2001.

205. Bagetta, G., Corasaniti, M.T., Paoletti, A.M., Berliocchi, L., Nistico, R., Giammarioli, A.M., Malorni, W., and Finazzi-Agro, A., HIV-1 gp120-induced apoptosis in the rat neocortex involves enhanced expression of cyclo-oxygenase type 2 (COX-2), *Biochem. Biophys. Res. Commun.,* 244, 819, 1998.

206. Corasaniti, M.T., Strongoli, M.C., Piccirilli, S., Nistico, R., Costa, A., Bilotta, A., Turano, P., Finazzi-Agro, A., and Bagetta, G., Apoptosis induced by gp120 in the neocortex of rat involves enhanced expression of cyclooxygenase type 2 and is prevented by NMDA receptor antagonists and by the 21-aminosteroid U-74389G, *Biochem. Biophys. Res. Commun.,* 274, 664, 2000.

207. Maccarrone, M., Bari, M., Corasaniti, M.T., Nistico, R., Bagetta, G., and Finazzi-Agro, A., HIV-1 coat glycoprotein gp120 induces apoptosis in rat brain neocortex by deranging the arachidonate cascade in favor of prostanoids, *J. Neurochem.,* 75, 196, 2000.

208. Bezzi, P., Carmignoto, G., Pasti, L., Vesce, S., Rossi, D., Rizzini, B.L., Pozzan, T., and Volterra, A., Prostaglandins stimulate calcium-dependent glutamate release in astrocytes, *Nature,* 391, 281, 1998.

209. Muller, W.E., Schroder, H.C., Ushijima, H., Dapper, J., and Bormann, J., gp120 of HIV-1 induces apoptosis in rat cortical cell cultures: prevention by memantine, *Eur. J. Pharmacol.,* 226, 209, 1992.

210. Doody, R., Wirth, Y., Schmitt, F., and Mobius, H.J., Specific functional effects of memantine treatment in patients with moderate to severe Alzheimer's disease, *Dement. Geriatr. Cogn. Disord.,* 18, 227, 2004.

211. Reisberg, B., Doody, R., Stoffler, A., Schmitt, F., Ferris, S., and Mobius, H.J., Memantine in moderate-to-severe Alzheimer's disease, *N. Engl. J. Med.,* 348, 1333, 2003.

212. Winblad, B. and Poritis, N., Memantine in severe dementia: results of the 9M-Best study (benefit and efficacy in severely demented patients during treatment with memantine), *Int. J. Geriatr. Psychiatry,* 14, 135, 1999.

213. Scheepens, A., Sirimanne, E.S., Breier, B.H., Clark, R.G., Gluckman, P.D., and Williams, C.E., Growth hormone as a neuronal rescue factor during recovery from CNS injury, *Neuroscience,* 104, 677, 2001.

214. Nyberg, F., Growth hormone in the brain: characteristics of specific brain targets for the hormone and their functional significance, *Front Neuroendocrinol.,* 21, 330, 2000.

215. Schambelan, M., Mulligan, K., Grunfeld, C., Daar, E.S., LaMarca, A., Kotler, D.P., Wang, J., Bozzette, S.A., and Breitmeyer, J.B., Recombinant human growth hormone in patients with HIV-associated wasting. A randomized, placebo-controlled trial. Serostim Study Group, *Ann. Intern. Med.,* 125, 873, 1996.

216. Serano 9037 Study Team, Growth hormone improves lean body mass, physical performance, and quality of life in subjects with HIV-associated weight loss or wasting on highly active antiretroviral therapy, *J. Acquir. Immune Defic. Syndr.,* 35, 367, 2004.

217. Pires, A., Pido-Lopez, J., Moyle, G., Gazzard, B., Gotch, F., and Imami, N., Enhanced T-cell maturation, differentiation and function in HIV-1-infected individuals after growth hormone and highly active antiretroviral therapy, *Antivir. Ther.,* 9, 67, 2004.

218. Lin, S., Zhang, Y., Dodel, R., Farlow, M.R., Paul, S.M., and Du, Y., Minocycline blocks nitric oxide-induced neurotoxicity by inhibition p38 MAP kinase in rat cerebellar granule neurons, *Neurosci. Lett.,* 315, 61, 2001.

219. Yrjanheikki, J., Tikka, T., Keinanen, R., Goldsteins, G., Chan, P.H., and Koistinaho, J., A tetracycline derivative, minocycline, reduces inflammation and protects against focal cerebral ischemia with a wide therapeutic window, *Proc. Natl. Acad. Sci. U.S.A.,* 96, 13496, 1999.

220. Tikka, T., Fiebich, B.L., Goldsteins, G., Keinanen, R., and Koistinaho, J., Minocycline, a tetracycline derivative, is neuroprotective against excitotoxicity by inhibiting activation and proliferation of microglia, *J. Neurosci.,* 21, 2580, 2001.

221. Tikka, T.M. and Koistinaho, J.E., Minocycline provides neuroprotection against *N*-methyl-D-aspartate neurotoxicity by inhibiting microglia, *J. Immunol.,* 166, 7527, 2001.

222. Zhu, S., Stavrovskaya, I.G., Drozda, M., Kim, B.Y., Ona, V., Li, M., Sarang, S., Liu, A.S., Hartley, D.M., Wu du, C., Gullans, S., Ferrante, R.J., Przedborski, S., Kristal, B.S., and Friedlander, R.M., Minocycline inhibits cytochrome *c* release and delays progression of amyotrophic lateral sclerosis in mice, *Nature,* 417, 74, 2002.

223. Dana Consortium on the Therapy of HIV Dementia and Related Cognitive Disorders, A randomized, double-blind, placebo-controlled trial of deprenyl and thioctic acid in human immunodeficiency virus-associated cognitive impairment. Dana Consortium on the Therapy of HIV Dementia and Related Cognitive Disorders, *Neurology,* 50, 645, 1998.

224. Schifitto, G., Sacktor, N., Marder, K., McDermott, M.P., McArthur, J.C., Kieburtz, K., Small, S., and Epstein, L.G., Randomized trial of the platelet-activating factor antagonist lexipafant in HIV-associated cognitive impairment. Neurological AIDS Research Consortium, *Neurology,* 53, 391, 1999.

225. Overall, C.M. and Lopez-Otin, C., Strategies for MMP inhibition in cancer: innovations for the post-trial era, *Nat. Rev. Cancer,* 2, 657, 2002.

226. Coussens, L.M., Fingleton, B., and Matrisian, L.M., Matrix metalloproteinase inhibitors and cancer: trials and tribulations, *Science,* 295, 2387, 2002.

227. Wan, W. and DePetrillo, P.B., Ritonavir protects hippocampal neurons against oxidative stress-induced apoptosis, *Neurotoxicology,* 23, 301, 2002.

23 Involvement of Apoptosis in Complications of HIV and Its Treatment

David Nolan

CONTENTS

INTRODUCTION: OVERVIEW OF COMPLICATIONS ASSOCIATED WITH HIV THERAPY

Over the past decade there has been a dramatic transformation in the management of human immunodeficiency virus (HIV) infection — one that has brought about an increased focus on the adverse effects of antireroviral therapy. Although antiretroviral drugs were available in clinical practice from approximately 1987 onward with the introduction of zidovudine (ZDV, also known as azidothymidine or AZT), HIV therapy was initially characterized by the rapid development of drug resistance in patients receiving treatment. This meant that after a period of adequate response to treatment generally lasting several months, the inevitable development of drug resistance meant that patients again experienced an inexorable decline in immunologic status due to a high burden of productive viral infection, becoming susceptible to opportunistic infections and other acquired immunodeficiency syndrome (AIDS)-defining illnesses that were associated with considerable morbidity and, eventually, mortality. However, this era came to an end in approximately 1996, with the introduction of combination treatment regimens (commonly referred to as "highly active anti-retroviral therapy" [HAART]) that were capable of profoundly suppressing HIV replication for prolonged periods, thereby reducing the risk of drug resistance and also halting and subsequently reversing the effects of the virus on immune function in infected patients. Accordingly, there are profound prognostic benefits associated with HAART that are now well documented in observational cohort studies.[1-3]

However, although the use of HAART regimens for HIV-infected patients has fundamentally altered the prognosis of this disease, this relatively new paradigm in which HIV infection can be viewed as a chronic, manageable disease has also increased awareness that long-term drug toxicities have the potential to cause significant morbidity and even mortality in this patient population. Recent analyses suggest that the burden of serious drug toxicities (that is, grade 4 reactions) remains greater than the risk of disease-associated complications, even in patients with advanced immune deficiency and AIDS.[4,5] To some extent, this simply provides further testimony to the beneficial disease-modifying effects of HAART, although these results also indicate that clinicians and patients alike need to be conscious of drug toxicity as a possible cause of any symptoms during (lifelong) therapy.

There are many potential complications associated with antiretroviral drug use, reflecting the great variety of HIV drugs that are available and widely used in clinical practice. The potential role of apoptosis in the pathogenesis of these adverse events will be the subject of this chapter, which will also develop the theme that each individual drug has a specific toxicity profile (even with drug classes, as will be discussed), so that both clinical management and scientific research need to be guided by this principle. The discussion will commence with the drug class in longest use for the treatment of HIV infection, the nucleoside analogue reverse transcriptase inhibitors (NRTIs), taking examples of toxicity syndromes that highlight underlying mechanisms, and will then move on to examine the complications associated more strongly with HIV protease inhibitor (PI) drug therapy.

COMPLICATIONS OF THERAPY WITH NUCLEOSIDE ANALOGUE REVERSE TRANSCRIPTASE INHIBITORS

NRTI drugs were the first to demonstrate clinical efficacy against HIV infection, and they continue to be used in HAART regimens in combination with HIV protease inhibitors (PIs) or nonnucleoside reverse transcriptase inhibitors (NNRTIs). The long-term use of NRTI drugs was associated with a number of clinically relevant toxicities, including myopathy, hematological toxicities, hyperlactatemia and lactic acidosis, neuropathy, pancreatitis, and, more recently, a syndrome of pathological loss of subcutaneous fat tissue (lipoatrophy). Importantly, the toxicity profile of each NRTI drug within this class is unique in terms of the overall risk of long-term complications as well as the tissue specificity of its toxic effects (Figure 23.1). It is also notable that NRTI drugs entered clinical practice at different times (see Figure 23.2), so that some drugs have spanned the "pre-HAART" and "HAART" eras (e.g., zidovudine, didanosine), whereas others were introduced as a component of HAART regimens (e.g., stavudine, lamivudine). There are also newer agents for which clinical experience is limited at present (e.g., abacavir, tenofovir). Hence, the history of NRTI-associated toxicities reflects the availability of specific drugs at the time as well as the influence of other factors that may have been different in the HAART era compared with the period that preceded it.

MECHANISMS OF ACTION: THERAPEUTIC AND TOXIC EFFECTS OF NRTI THERAPY

The therapeutic activity of NRTI drugs is conferred by their ability to inhibit HIV reverse transcriptase, the viral ribonucleic acid (RNA)-dependent deoxyribonucleic acid (DNA) polymerase that allows the virus to create a DNA sequence based on its own RNA template (Figure 23.3). Following intracellular phosphorylation, NRTI drugs compete with naturally occurring nucleotides for utilization by HIV reverse transcriptase. Incorporation of the "false" (NRTI) substrate then causes DNA chain termination, as these nucleotides lack a hydroxyl group that is critical for chain elongation. The effectiveness of NRTI drugs as inhibitors of HIV replication is, therefore, determined primarily by their ability to inhibit HIV reverse transcriptase. However, the efficiency with

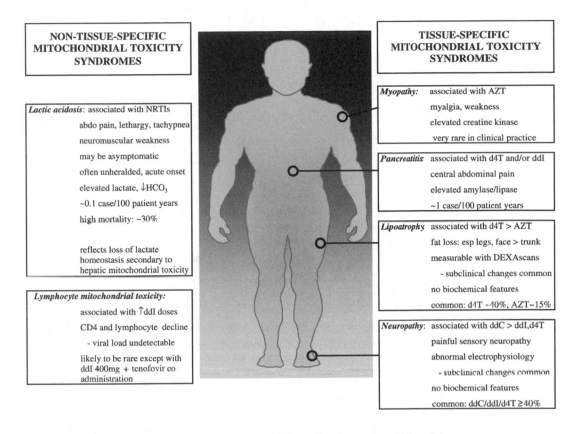

FIGURE 23.1 Clinical syndromes attributed to NRTI-associated mitochondrial toxicity.

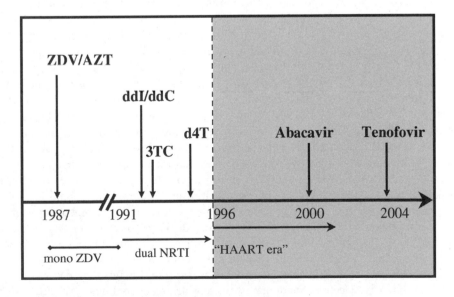

FIGURE 23.2 Time line for the introduction of NRTI drugs into clinical practice in the "pre-HAART" and "HAART" eras.

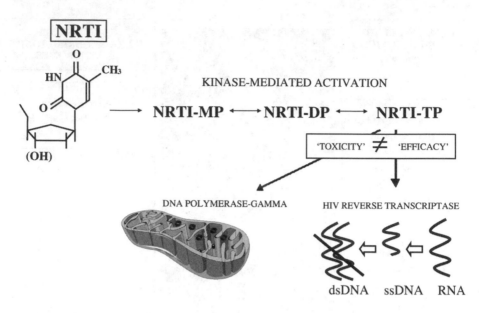

FIGURE 23.3 The basis of NRTI-associated mitochondrial toxicity, according to the "pol-" model. NRTI drugs differ from natural nucleoside compounds in that the 5-hydroxyl group (OH) is modified or removed, so that incorporation into a nascent DNA chain leads to chain termination. Note that the affinity of a given NRTI drug for DNA polymerase-γ (a determinant of its ability to cause mitochondrial DNA depletion) is not related to its affinity for HIV reverse transcriptase (a determinant of antiretroviral efficacy).

which NRTI drugs are delivered to the appropriate cell population and intracellular compartment—determined by factors such as oral bioavailability, cellular transport mechanisms, and the efficiency of cellular kinases responsible for activating NRTI drugs to their triphosphate form—is also an important factor in determining drug efficacy.

The study of NRTI-associated toxicity syndromes, now more than a decade old, was also informed by their capacity to inhibit a specific host DNA polymerase (mitochondrial DNA polymerase-γ) through a mechanism that is similar to its therapeutic activity as an HIV reverse transcriptase inhibitor.[6–8] However, it is worth stating at the outset that the toxic and therapeutic profiles of individual NRTI drugs are not correlated, so that increased efficacy is not necessarily accompanied by increased toxicity.[9] Over the past 15 years, there has been an increasing trend toward the development and clinical use of NRTI drugs (e.g., abacavir, tenofovir) that combine high levels of antiviral efficacy with benign toxicity profiles both *in vitro* and (thus far) *in vivo*,[8,10] providing cause for optimism that the prevalence of NRTI-associated toxicities will decline over time.

The fact that mitochondria appear to be specifically targeted by NRTI therapy forms the basis for tissue-specific "mitochondrial toxicity" syndromes that may also involve apoptotic cell death as a factor in pathogenesis. As summarized in earlier chapters, mitochondria are both energy powerhouses of the cell and critical regulators of cell death. These organelles are also unique in containing their own genome that is extrachromosomal and replicates independently of nuclear DNA. Hence, mitochondrial DNA can be synthesized in postmitotic and resting cells, using a mitochondrial DNA polymerase (polymerase-γ), with evolutionary origins[11] and enzymatic activities[12] that are distinct from the multiple DNA polymerases involved in nuclear DNA synthesis. Importantly, DNA pol- is also unable to discriminate strongly in favor of naturally occurring nucleotides (i.e., deoxynucleotides) and against nucleoside analogue drugs,[12] so that NRTI drugs are capable of inhibiting cellular mitochondrial DNA synthesis via their effects on DNA pol- (Figure 23.3). The basic premise of the pol- hypothesis is that NRTI-induced inhibition of DNA pol- leads to cellular mitochondrial DNA depletion (i.e., reduced numbers of mitochondrial DNA copies per

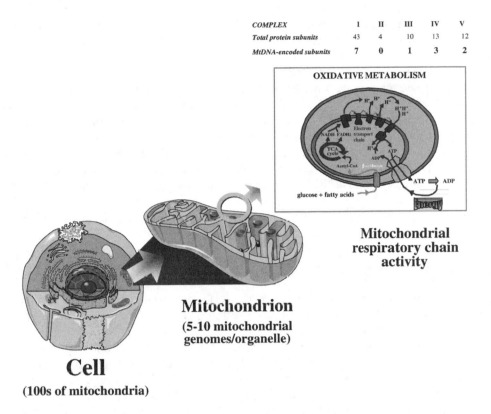

COMPLEX	I	II	III	IV	V
Total protein subunits	43	4	10	13	12
MtDNA-encoded subunits	7	0	1	3	2

FIGURE 23.4 The mitochondrial organelle and its role in oxidative phosphorylation. Mitochondria are ubiquitous organelles responsible for cellular energy production through oxidative phosphorylation. This requires the concerted action of mitochondrial DNA-encoded subunits (13) and nuclear DNA-encoded subunits (~70) of the five respiratory chain complexes. Note also that all proteins involved in determining mitochondrial organellar number and structure, as well as in maintaining biochemical pathways within the mitochondrial matrix (e.g., oxidation, Krebs' cycle), are nuclear DNA encoded.

cell). Cellular, and ultimately organ, toxicity is the consequence of loss of mitochondrial bioenergetic function, when mitochondrial DNA levels fall beyond a critical level at which the production of the 13 mitochondrial DNA-encoded protein subunits of the mitochondrial respiratory chain is insufficient to meet energy requirements (see Figure 23.4).

In vitro studies have provided important information regarding the potential of specific NRTI drugs to cause DNA polymerase-γ inhibition, mitochondrial DNA depletion, and mitochondrial dysfunction.[8,9,13] A consistent ranking of the potency of mitochondrial DNA synthesis inhibition has emerged from these studies, with decreasing potency associated with ddC (zalcitabine) >> ddI (didanosine) > d4T (stavudine) > AZT (zidovudine). These studies also suggest that no significant mitochondrial polymerase inhibition occurs at pharmacologic concentrations of lamivudine, abacavir, and tenofovir. One caveat to these findings is that NRTI-associated mitochondrial toxicity is tissue specific, so that *in vitro* studies should ideally use human cell cultures that are relevant to the clinical toxicity profile of the drug.

These studies can also shed light on the intracellular activation pathways of specific drugs, which also may influence their ability to affect mitochondria. For example, the thymidine analogues zidovudine and stavudine share a common activation pathway that involves thymidine kinase (TK) in the formation of the monophosphate form of the drug. TK comprises two isoforms: TK1, which is cytosolic and active in mitotically active cells only; and TK2, which is mitochondria specific and active in postmitotic cells. Of interest, stavudine—but not zidovudine—retains anti-HIV activity within the pharmacologic dose range in TK1-deficient cells (IC_{50} for HIV = 0.27 μM with TK, 2.5 μM in TK-deficient CEM cells)[14] and is efficiently phosphorylated to its active triphosphate derivative within isolated mitochondria.[15] This suggests that this thymidine analogue is more likely to target mitochondrial DNA, particularly in postmitotic tissues (e.g., nerves) in which mitochondrial DNA synthesis accounts for a greater proportion of nucleoside utilization while nuclear DNA synthesis is latent.

The ability of NRTI drugs to induce cellular mitochondrial toxicity may, therefore, be considered in two parts. A drug's affinity for DNA polymerase-γ may be considered a primary determinant, but the relevance of this effect to the development of tissue toxicity is modulated by other factors, such as that drug's ability to access the mitochondrial compartment within the cell cytosol in an activated form. This is also highlighted by the discovery of a mitochondrial inner membrane transporter that is able to facilitate the entry of active NRTI diphosphate and triphosphate derivatives into mitochondria.[16]

We now move to specific syndromes that have been linked to NRTI treatment in the clinical setting, with a particular emphasis on the distinct toxicity profiles of individual drugs, and the likely role of mitochondrial toxicity and apoptosis in pathogenesis.

ZIDOVUDINE MYOPATHY: A TALE OF TWO TREATMENT ERAS

Myopathy associated with zidovudine therapy represents the most well-known mitochondrial toxicity syndrome linked to NRTI therapy, as it was the subject of seminal studies and publications at a time when mitochondrial disorders in general were beginning to be appreciated in the late 1980s and early 1990s.[17–20] Apoptosis was not directly investigated in these studies—although cell death and tissue destruction were certainly suggested by marked loss of muscle mass in affected areas[21]—but it was possible to demonstrate abnormal mitochondrial morphology in clinical muscle biopsy specimens, along with significant mitochondrial DNA depletion. Muscle morphology appeared typical of genetically determined mitochondrial myopathy syndromes that were newly described at the time, with evidence of enlarged mitochondria with distorted crystal architecture and the presence of "ragged red" fibers.[17–19,21]

However, an interesting aspect of this clinical syndrome is its rarity in the current treatment era (~4% in a recent study involving a cohort of 1000 patients)[22] compared with previous estimates of ~25%,[23] despite the continuous and frequent use of zidovudine throughout the past 17 years. The most obvious explanation for this dramatic decline is that zidovudine dosing was reduced from 1200 mg (monotherapy) to 500 mg per day (within HAART regimens), although the metabolic activation of zidovudine also appears to be upregulated in poorly controlled and advanced HIV disease, presumably due to increased activity of the rate-limiting TK1 enzyme.[24] It also is notable that prolonged and advanced HIV disease is associated with myopathic syndromes and that disease- and treatment-associated pathology may have converged during this period.[18,19,21,25] As discussed below, this synergy between HIV and treatment also is observed in the case of neuropathy, although this toxicity syndrome remains prevalent in contemporary practice.

LIPOATROPHY: PATHOLOGICAL FAT LOSS AS A MITOCHONDRIAL TOXICITY

There is now compelling clinical and pathological evidence that the choice and duration of NRTI drugs are the dominant risk factors for developing pathological loss of fat tissue, referred to as subcutaneous fat wasting or lipoatrophy. Moreover, pathogenesis studies have directly implicated apoptotic cell death within adipocyte cell populations as a consequence of NRTI-induced mitochondrial toxicity.

Clinical studies have demonstrated that NRTI therapy alone provides sufficient conditions for the development of lipoatrophy[26–28] and that NRTI therapy is an independent risk factor for its occurrence in HAART-treated individuals.[29,30] Clinical trials data have now confirmed the findings of observational cohort studies that stavudine therapy is associated with an approximately twofold increased risk of lipoatrophy compared with zidovudine, so that the risk of clinically apparent lipoatrophy over 30 months

Lipodystrophy model

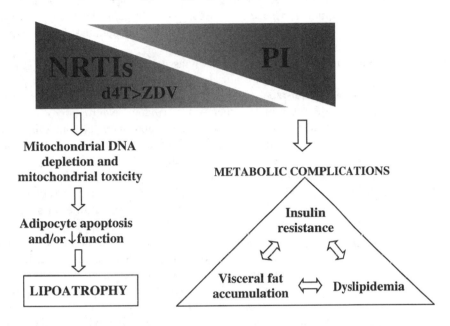

FIGURE 23.5 Model of lipodystrophy pathogenesis, incorporating contributions of NRTI and HIV protease inhibitor therapy.

is approximately 10 to 20% in zidovudine-treated individuals and 40 to 50% among those treated with stavudine.[30–32] Switching NRTI therapy from stavudine to abacavir in two clinical trials[33,34] was also associated with significant improvements in fat wasting. In contrast, more than 30 trials investigating HIV protease inhibitor discontinuation as a therapeutic strategy (reviewed in Dreschler and Powderly[35]) failed to demonstrate beneficial effects on fat wasting, although metabolic abnormalities such as dyslipidemia and insulin resistance improved (see Figure 23.5).

The histopathological correlates of this syndrome also indicate that adipose tissue–specific mitochondrial toxicity and widespread adipocyte cell death are central to lipoatrophy pathogenesis. At the ultrastructural level, abnormal mitochondrial architecture and organellar proliferation are prominent features.[36–38] Here, the involvement of apoptotic cell death also was characterized, so that increased adipocyte apoptosis associated with lipoatrophy was shown to improve after discontinuing stavudine therapy in favor of abacavir[39] but not after switching HIV protease inhibitor therapy.[40] In keeping with a role for mitochondrial toxicity in lipoatrophy pathogenesis, studies using precise real-time polymerase chain reaction (PCR) quantitation methods demonstrated significant mitochondrial DNA depletion particularly associated with stavudine compared with zidovudine,[41,42] and associations were found between NRTI therapy, adipocyte mitochondrial DNA depletion, adipose tissue toxicity, and the eventual development of clinical fat loss.[42,43]

NEUROPATHY: CONTRIBUTIONS OF BOTH HIV DISEASE AND NRTI THERAPY

There are a number of issues to be considered in addressing potential associations between NRTI therapy, mitochondrial toxicity, and neuropathy. First, routine clinical assessment of potential neuropathic changes—relying on the assessment of neurological function rather than more visible changes such as lipoatrophy—may underestimate the prevalence and potential long-term impact of neuropathy in HIV-infected patients. Second, there is a definite contribution of HIV disease *per se* to the pathogenesis of a form of distal sensory neuropathy that is clinically and electrophysiologically indistinguishable from so-called "toxic neuropathy from antiretroviral drugs" (reviewed in Keswani et al.[44]). It is, therefore, difficult to dissect the relative contribution of disease-associated and drug-associated factors in the syndrome, as these effects are likely to be synergistic. The clinical syndrome common to both HIV-associated and toxic sensory polyneuropathy is dominated by peripheral pain and dysesthesia, with rare motor involvement.

In general, the relative risk of toxic polyneuropathy associated with specific NRTI drugs correlates well with their ability to inhibit mitochondrial DNA polymerase-γ *in vitro*, with greatest risk associated with zalcitabine, with equivalent affinity for mitochondrial DNA polymerase-γ and HIV reverse transcriptase that provides for a narrow therapeutic window.[9] Neuropathy was a consistent and dose-limiting toxicity in trials of zalcitabine, occurring in approximately one-third of patients receiving low-dose therapy (2.25 mg/day) for less than a year,[45,46] whereas higher doses were associated with almost universal development of neuropathy (100% with >0.04 mg/kg/day, 80% with 0.04 mg/kg/day).[45] Consistent with the pol- hypothesis, Dalakas and colleagues[47] provided *in vivo* evidence of significant mitochondrial DNA depletion in peripheral nerves associated with this complication of zalcitabine therapy, although studies investigating potential roles of apoptosis were not published to date.

Stavudine and didanosine also were associated with increased risk of neuropathy in the large Johns Hopkins AIDS Service cohort study ($n = 1116$). Compared with a crude incidence rate of 6.8 cases per 100 person years for didanosine, stavudine use was associated with a relative risk of 1.4, and concurrent use of these drugs further increased risk (3.5-fold relative risk for stavudine/ didanosine compared with didanosine alone). Concurrent hydroxyurea therapy had an additional effect (7.8-fold relative risk of stavudine/didanosine/hydroxyurea compared with didanosine alone, giving a crude incidence of 28.6 cases per 100 person years).[48]

As already mentioned, markers of progressive HIV disease correlate with the development of sensory polyneuropathy, with evidence for associations between reduced CD4+ T cell counts and increased risk of neuropathy (reviewed in Weissman[49]), and increased pain scores in patients with higher HIV viral load.[50] However, in a cohort of 144 ambulatory patients with longer duration of exposure to stavudine or didanosine and average CD4+ T cell counts $>400 \times 10^6$/L, the prevalence of sensory neuropathy was found to be 44%,[51] suggesting that among patients with prolonged survival and effective immunologic and virological outcomes, choice and duration of NRTI therapy will ultimately become the dominant risk factors for neuropathy. In this analysis, use of stavudine, didanosine, or zalcitabine compared with other NRTI drugs was associated with an odds ratio of 26 of developing clinical neuropathy, whereas disease-associated factors were not found to be significant in multivariate analysis.

"LYMPHOCYTE MITOCHONDRIAL TOXICITY" AND DIDANOSINE THERAPY: AN INTERSECTION OF THERAPEUTIC AND TOXIC EFFECTS

Intriguing results from a recent study has suggested that a suboptimal immunologic response to antiretroviral therapy (i.e., loss of CD4+ T cell counts on treatment) may represent a complication of NRTI therapy in selected circumstances.[52] These data emerged from a Spanish cohort study involving 302 patients, in which regimens containing standard doses of didanosine (ddI; 400 mg) and tenofovir were associated with significant decreases in CD4+ and CD8+ T cells as well as total lymphocyte counts (with a decline of >100 CD4+ T cells at 48 weeks in more than 50%). Plasma levels of ddI appeared to be pharmacologically "boosted" in all patients receiving coadministered ddI 400 mg and tenofovir and returned to expected levels when the ddI dose was reduced to 250 mg.[52] More recently, a case series was reported of four patients who developed an identical clinical syndrome while receiving zidovudine and ddI with a mean reduction of CD4+ lymphocyte count from 442 to 220 cells/μL,[53] when switching from ddI to lamivudine proved to be an effective strategy (mean CD4+ lymphocytes increased to 385 cells/μL).[53]

This is not to suggest that immunologic responses to ddI are generally inferior to other NRTI drugs, as this drug has demonstrated efficacy in large clinical trials.[54,55] However, this emerging field of investigation provides an interesting insight into the tissue-specific nature of NRTI toxicity and the potential for drug toxicity to directly impinge on the efficacy of HIV treatment. Associations between ddI therapy and peripheral blood mitochondrial DNA depletion have been noted by a number of groups,[41,56–58] whereas this drug has not been associated with any significant effect on mitochondrial DNA in subcutaneous fat samples obtained concurrently with blood samples.[41] Hence, there is a sound basis for considering a potential role for ddI toxicity when "discordant" responses to treatment are observed. Other chapters addressed the critical role of lymphocyte apoptosis in determining T cell dynamics both before and after treatment, placing these data in a broader context.

LACTIC ACIDOSIS AND HYPERLACTATEMIA

Lactic acidosis is probably the most recognizable feature of mitochondrial toxicity in clinical disease, in which loss of mitochondrial oxidative function leads to increased reliance on "anaerobic" metabolism and the inevitable accumulation of lactate (and thus, of acid). In the setting of NRTI therapy, there is now an appreciation of a spectrum of clinical phenotypes associated with elevated systemic lactate levels.[59] At one end of this spectrum is a relatively common syndrome of mild, asymptomatic, nonprogressive hyperlactatemia (generally < 2.5 mmol/L), which appears to represent a "compensated" homeostatic system in which elevated lactate production is balanced by effective mechanisms of lactate clearance. Although the degree of hyperlactatemia seems to be greater in the presence of stavudine or ddI therapy than in the presence of zidovudine or abacavir, this syndrome appears to be benign, irrespective of the choice of NRTI therapy.[60–64]

In contrast, lactic acidosis and hepatic steatosis are uncommon (approximately 1 to 2 per 1000 person years) but carry a high mortality rate (~50%),[65] representing a syndrome of "decompensated" lactate metabolism that allows the rapid, progressive accumulation of lactate and development of acidosis in affected patients. A critical aspect of lactic acidosis is its unpredictability, as it typically occurs in patients who were on stable NRTI regimens for months or even years, and it is not heralded by increased lactate levels before the development of the fulminant syndrome. Host risk factors have been identified for NRTI-associated lactic acidosis, including concurrent liver disease, female gender, and obesity. Although the majority of reported cases in recent years have involved stavudine-based HAART, it is important to recognize that cases involving zidovudine therapy also occur (reviewed in John and Mallal[59]). A more recently identified clinical syndrome accompanying lactic acidosis is progressive, severe neuromuscular weakness mimicking the Guillain–Barré syndrome.[66] Initial reports by the Food and Drug Administration identified 25 cases, including seven fatal cases (notably, NRTI therapy was continued after the diagnosis was made in all but one of these individuals). A more recent retrospective analysis identified 55 cases (44 associated with stavudine therapy), with a 25% mortality rate among the 36 patients with documented follow-up.[67]

With regard to the pathogenesis of hyperlactatemia syndromes, evidence has been provided from studies using exogenous lactate loading that show that increased lactate production is the fundamental defect in chronic asymptomatic hyperlactatemia rather than decreased clearance of lactate (which primarily involves the liver).[68] Lactate clearance seems to be upregulated as a compensatory response to an elevated systemic lactate load.[68] In contrast, lactic acidosis represents a complete failure of lactate homeostasis, in which lactic acid accumulation progresses inexorably while NRTI therapy is maintained. In this scenario, the prominent role of liver mitochondrial dysfunction (indicated by the presence of marked microvesicular hepatic steatosis),[69,70] leading to a shift from hepatic lactate consumption to lactate production, would be predicted to have dramatic consequences for systemic lactate metabolism. The capacity for selected NRTI drugs (zalcitabine, ddI, and stavudine) to induce significant mitochondrial DNA depletion in liver biopsy samples[71] also lends support to the possibility that functional hepatic reserve may be impaired and predisposition to hepatic steatosis (reflecting mitochondrial dysfunction and impaired hepatic β-oxidation) may be increased.

COMPLICATIONS ASSOCIATED WITH THE USE OF HIV PROTEASE INHIBITORS

There has been intense research interest in the adverse effects of HIV protease inhibitors and the influences of these drugs on cellular biology that may be related to their therapeutic effects (e.g., inhibition of the proapoptotic activities of HIV protease) and that may also be independent of their antiretroviral properties (e.g., mitochondrial membrane–stabilizing effects of nelfinavir,[72] inhibition of caspase-dependent apoptosis associated with ritonavir[73,74]). As reviewed in Chapter 25 and elsewhere,[75] HIV protease inhibitors, therefore, exhibit pleiotropic activities that generally promote antiapoptotic rather than proapoptotic activities in nonmalignant cell populations, a view that is supported by the fact that complications of these drugs do not involve tissue destruction (in contrast to the effects of NRTI drugs on adipose tissue, for example) but, rather, represent alterations in whole-body metabolism. The most notable complications associated with drugs from the HIV protease inhibitor group include dyslipidemia and insulin resistance, as reviewed elsewhere.[76] However, the following examples are provided that may illustrate specific associations between HIV protease inhbitors and the modulation of apoptosis. It is an interesting conundrum that in these cases, the same cellular mechanisms that provide protection from apoptotic cell death may also be implicated in the pathogenesis of adverse events.

DIARRHEA ASSOCIATED WITH NELFINAVIR: A COMMON PATHWAY WITH ANTIAPOPTOTIC PROPERTIES?

Although elucidating the pathogenesis of diarrhea associated with the HIV protease inhibitor nelfinavir has not generated the same level of enthusiasm within the research community as other topics, this issue is, nevertheless, of considerable clinical importance and has produced a seminal paper. Approximately 40 to 50% of patients treated with nelfinavir develop chronic diarrhea, which requires drug treatment in one third of cases and also is a relatively frequent reason for drug discontinuation.[77] This effect is specific to nelfinavir within the HIV protease inhibitor drug class, although less severe diarrhea also was observed with lopinavir therapy.[77] In their impressive study on this subject, Paul Rufo, Patricia Lin, and their colleagues at Harvard University and Johns Hopkins University characterized this as a secretory form of diarrhea (i.e., involving excessive intestinal fluid secretion) among eight patients admitted for investigation and also went on to demonstrate that the underlying cellular defect involves altered cellular calcium metabolism in studies using intestinal (T84) cell lines.[78] This mechanism involves the potentiation of muscarinic signaling that triggers increased chloride conductance (and thus water and sodium secretion into the intestinal lumen), so that this physiological response is not appropriately downregulated. At a more fundamental level, this defect may be described as a prolongation of cellular calcium entry in response to the release of intracellular calcium stores ("store-operated" calcium entry).[78]

This topic will be addressed further, after examining another effect of nelfinavir treatment that is also of biological and clinical significance and one that is directly interested in the regulation of apoptosis by this drug. In this study, Andrew Badley and his colleagues from the University of Ottawa demonstrated that nelfinavir stabilized the mitochondrial membrane potential in T lymphocyte cell lines (Jurkat and H9), inhibiting apoptotic signaling by preventing the opening of the permeability transition pore and the subsequent release of mitochondrial apoptotic mediators such as cytochrome c and apoptosis-inducing factor (AIF)[79] (see also Chapter 25). These data, therefore, describe an important antiapoptotic function of nelfinavir that is clinically relevant,[80] that is independent of antiretroviral activity, and that targets immune cells intrinsically involved in HIV-1 infection.

Can these disparate nelfinavir effects—one potentially therapeutic, the other clinically problematic—be attributed to a common mechanism? Any answer is speculative, but it is notable that mitochondria play an important role in buffering intracellular calcium flux and, therefore, contribute to the regulation of store-operated calcium uptake in response to hormonal signals mediated via phosphoinositide pathways (e.g., muscarinic acetylcholine activation).[81,82] Hence, the ability of nelfinavir to maintain the mitochondrial transmembrane potential in a more energized state may also prolong store-operated calcium signaling and inhibit its physiological downregulation.[83] The obvious therapeutic benefits associated with antiapoptotic HIV drugs may, therefore, require consideration of potential adverse effects associated with reduced flexibility of intracellular calcium signaling, although it must be stressed again that such associations are entirely unproven at this point.

RITONAVIR, PROTEASOMAL INHIBITION, AND METABOLIC COMPLICATIONS

A second example in which an intrinsic effect of HIV protease inhibitor therapy may contribute to multiple outcomes involves ritonavir. This drug selectively inhibits the chymotrypsin-like activity of the 20S proteasome at therapeutic concentrations (although enhancing its tryptic activity), thus potentially affecting antigen presentation and cytotoxic T lymphocyte responses.[84,85] This attribute may also contribute to the antiapoptotic properties of ritonavir, in keeping with complex effects of proteasome inhibitors on apoptosis,[86] although there is more substantial evidence that this drug inhibits the activation of selected caspases.[73,74,87]

Proteasomal inhibition caused by ritonavir is likely to provide an explanation for the specific effect of this drug to cause elevated triglyceride levels after short-term drug exposure, even in HIV-uninfected individuals.[88] This seems to reflect increased hepatic triglyceride synthesis and secretion, which, in turn, is related to reduced proteasomal degradation of a key transcriptional factor (sterol regulatory binding protein [SREBP-1]) involved in regulating lipid metabolism in response to insulin and fatty acid stimuli.[89–93] This means that hepatic triglyceride production is constitutively activated due to the nuclear retention of SREBP-1.[90–92] It is uncertain at present whether the propensity of ritonavir therapy to cause hypertriglyceridemia is associated with increased risk of cardiovascular disease, but this example highlights how tissue-specific effects of HIV protease inhibitor may have diverse consequences mediated by a single mechanism.

CONCLUSIONS

One of the messages that is becoming increasingly clear from the large body of research devoted to antiretroviral drug toxicity is that each drug has a unique toxicity profile, implying that although considering drug "class effects" can be useful to some extent (guiding research into NRTI-associated mitochondrial toxicity, for example), important differences remain between individual drugs. These may be apparent in terms of overall risk of toxicity, tissue specificity (bearing in mind that one mechanism can give rise to disparate outcomes in different tissues), and phenotypic expression. With regard to the role of apoptosis in these toxicity syndromes, it is fair to say that there are limited data that directly relate apoptosis to clinical or pathological outcomes, compared with the larger body of evidence relating to the involvement of apoptosis in HIV disease pathogenesis and progression reviewed elsewhere in this book. Nevertheless, the mechanisms underlying these drug toxicities are becoming better defined, revealing complex and intriguing insights into the regulation of whole-body metabolism as well as cellular integrity and function.

REFERENCES

1. Jensen-Fangel, S., Pedersen, L., Pedersen, C., Larsen, C.S., Tauris, P., Moller, A., Sorensen, H.T., and Obel, N., Low mortality in HIV-infected patients starting highly active antiretroviral therapy: a comparison with the general population, *AIDS*, 18, 89, 2004.

2. van Sighem, A.I., van de Wiel, M.A., Ghani, A.C., Jambroes, M., Reiss, P., Gyssens, I.C., Brinkman, K., Lange, J.M., and de Wolf, F., Mortality and progression to AIDS after starting highly active antiretroviral therapy, *AIDS,* 17, 2227, 2003.

3. Palella, F.J., Jr., Delaney, K.M., Moorman, A.C., Loveless, M.O., Fuhrer, J., Satten, G.A., Aschman, D.J., and Holmberg, S.D., Declining morbidity and mortality among patients with advanced human immunodeficiency virus infection. HIV Outpatient Study Investigators, *N. Engl. J. Med.,* 338, 853, 1998.

4. Reisler, R.B., Han, C., Burman, W.J., Tedaldi, E.M., and Neaton, J.D., Grade 4 events are as important as AIDS events in the era of HAART, *J. Acquir. Immune Defic. Syndr.,* 34, 379, 2003.

5. Brown, S.T., The Effects of Non-AIDS Serious Adverse Events Equals or Exceeds that of AIDS Outcomes in Patients with Advanced Multi-Drug Resistant HIV Disease, presented at 11th Conference on Retroviruses and Opportunistic Infections, February 8–11, San Francisco, 2004, Abstract 874.

6. Kakuda, T.N., Pharmacology of nucleoside and nucleotide reverse transcriptase inhibitor-induced mitochondrial toxicity, *Clin. Ther.,* 22, 685, 2000.

7. Lewis, W., Day, B.J., and Copeland, W.C., Mitochondrial toxicity of NRTI antiviral drugs: an integrated cellular perspective, *Nat. Rev. Drug Discov.,* 2, 812, 2003.

8. Lee, H., Hanes, J., and Johnson, K.A., Toxicity of nucleoside analogues used to treat AIDS and the selectivity of the mitochondrial DNA polymerase, *Biochemistry,* 42, 14711, 2003.

9. Martin, J.L., Brown, C.E., Matthews-Davis, N., and Reardon JE., Effects of antiviral nucleoside analogs on human DNA polymerases and mitochondrial DNA synthesis, *Antimicrob. Agents Chemother.,* 38, 2743, 1994.

10. Nolan, D. and Mallal, S., Thymidine analogue-sparing highly active antiretroviral therapy (HAART), *J. HIV Ther.,* 8, 2, 2003.

11. Burgers, P.M., Koonin, E.V., Bruford, E., Blanco, L., Burtis, K.C., Christman, M.F., Copeland, W.C., Friedberg, E.C., Hanaoka, F., Hinkle, D.C., Lawrence, C.W., Nakanishi, M., Ohmori, H., Prakash, L., Prakash, S., Reynaud, C.A., Sugino, A., Todo, T., Wang, Z., Weill, J.C., and Woodgate, R., Eukaryotic DNA polymerases: proposal for a revised nomenclature, *J. Biol. Chem.,* 276, 43487, 2001.

12. Longley, M.J., Ropp, P.A., Lim, S.E., and Copeland, W.C. Characterization of the native and recombinant catalytic subunit of human DNA polymerase- Identification of residues critical for exonuclease activity and dideoxynucleotide sensitivity, *Biochemistry,* 37, 10529, 1998.

13. Birkus, G., Hitchcock, M.J., and Cihlar, T., Assessment of mitochondrial toxicity in human cells treated with tenofovir: comparison with other nucleoside reverse transcriptase inhibitors, *Antimicrob. Agents Chemother.,* 46, 716, 2002.

14. Turriziani, O., Simeoni, E., Dianzani, F., and Antonelli, G., Anti-HIV antiviral activity of stavudine in a thymidine kinase-deficient cellular line, *Antiviral. Ther.,* 3, 191, 1998.

15. Cui, L., Locatelli, L., Xie, M.Y., and Sommadossi, J.P., Effect of nucleoside analogs on neurite regeneration and mitochondrial DNA synthesis in PC-12 cells, *J. Pharmacol. Exp. Ther.,* 280, 1228, 1997.

16. Dolce, V., Fiermonte, G., Runswick, M.J., Palmieri, F., and Walker, J.E., The human mitochondrial deoxynucleotide carrier and its role in the toxicity of nucleoside antivirals, *Proc. Natl. Acad. Sci. U.S.A.,* 98, 2284, 2001.

17. Arnaudo, E., Dalakas, M., Shanske, S., Moraes, C.T., DiMauro, S., and Schon, E.A., Depletion of muscle mitochondrial DNA in AIDS patients with zidovudine-induced myopathy, *Lancet,* 337, 508, 1991.

18. Dalakas, M.C., Illa, I., Pezeshkpour, G.H., Laukaitis, J.P., Cohen, B., and Griffin, J.L., Mitochondrial myopathy caused by long-term zidovudine therapy, *N. Engl. J. Med.,* 322, 1098, 1990.

19. Pezeshkpour, G., Illa, I., and Dalakas, M.C., Ultrastructural characteristics and DNA immunocytochemistry in human immunodeficiency virus and zidovudine-associated myopathies, *Hum. Pathol.,* 22, 1281, 1991.

20. Weissman, J.D., Constantinitis, I., Hudgins, P., and Wallace, D.C., 31P magnetic resonance spectroscopy suggests impaired mitochondrial function in AZT-treated HIV-infected patients, *Neurology,* 42, 619, 1992.

21. Mhiri, C., Baudrimont, M., Bonne, G., Geny, C., Degoul, F., Marsac, C., Roullet, E., and Gherardi, R., Zidovudine myopathy: a distinctive disorder associated with mitochondrial dysfunction, *Ann. Neurol.,* 29, 606, 1991.

22. Manfredi, R., Motta, R., Patrono, D., Calza, L., Chiodo, F., and Boni, P., A prospective case-control survey of laboratory markers of skeletal muscle damage during HIV disease and antiretroviral therapy, *AIDS,* 16, 1969, 2002.

23. Grau, J.M., Masanes, F., Pedrol, E., Casademont, J., Fernandez-Sola, J., and Urbano-Marquez, A., Human immunodeficiency virus type 1 infection and myopathy: clinical relevance of zidovudine therapy, *Ann. Neurol.,* 34, 206, 1993.

24. Anderson, P.L., Kakuda, T.N., and Lichtenstein, K.A., The cellular pharmacology of nucleoside- and nucleotide-analogue reverse-transcriptase inhibitors and its relationship to clinical toxicities, *Clin. Infect. Dis.,* 38, 743, 2004.

25. Morgello, S., Wolfe, D., Godfrey, E., Feinstein, R., Tagliati, M., and Simpson, D.M., Mitochondrial abnormalities in human immunodeficiency virus-associated myopathy, *Acta Neuropathol. (Berl.),* 90, 366, 1995.

26. Carr, A., Miller, J., Law, M., and Cooper, D.A., A syndrome of lipoatrophy, lactic acidemia and liver dysfunction associated with HIV nucleoside analogue therapy: contribution to protease inhibitor-related lipodystrophy syndrome, *AIDS,* 14, F25, 2000.

27. Mallal, S.A., John, M., Moore, C.B., James, I.R., and McKinnon, E.J., Contribution of nucleoside analogue reverse transcriptase inhibitors to subcutaneous fat wasting in patients with HIV infection, *AIDS,* 14, 1309, 2000.

28. Saint-Marc, T., Partisani, M., Poizot-Martin, I., Rouviere, O., Bruno, F., Avellaneda, R., Lang, J.M., Gastaut, J.A., and Touraine, J.L., Fat distribution evaluated by computed tomography and metabolic abnormalities in patients undergoing antiretroviral therapy: preliminary results of the LIPOCO study, *AIDS,* 14, 37, 2000.

29. van der Valk, M., Gisolf, E.H., Reiss, P., Wit, F.W., Japour, A., Weverling, G.J., and Danner, S.A., Increased risk of lipodystrophy when nucleoside analogue reverse transcriptase inhibitors are included with protease inhibitors in the treatment of HIV-1 infection, *AIDS,* 15, 847, 2001.

30. Chene, G., Angelini, E., Cotte, L., Lang, J.M., Morlat, P., Rancinan, C., May, T., Journot, V., Raffi, F., Jarrousse, B., Grappin, M., Lepeu, G., and Molina, J.M., Role of long-term nucleoside-analogue therapy in lipodystrophy and metabolic disorders in human immunodeficiency virus-infected patients, *Clin. Infect. Dis.,* 34, 649, 2002.

31. Amin, J., Moore, A., Carr, A., French, M.A., Law, M., Emery, S., and Cooper, D.A., Combined analysis of two-year follow-up from two open-label randomized trials comparing efficacy of three nucleoside reverse transcriptase inhibitor backbones for previously untreated HIV-1 infection: Ozcombo 1 and 2, *HIV Clin. Trials,* 4, 252, 2003.

32. Joly, V., Flandre, P., Meiffredy, V., Leturque, N., Harel, M., Aboulker, J.P., and Yeni P., Increased risk of lipoatrophy under stavudine in HIV-1-infected patients: results of a substudy from a comparative trial, *AIDS,* 16, 2447, 2002.

33. Martin, A., Reversibility of lipoatrophy in HIV-infected patients 2 years after switching from a thymidine analogue to abacavir: the MITOX Extension Study, *AIDS,* 18, 1029, 2004.

34. McComsey, G.A., Ward, D.J., Hessenthaler, S.M., Sension, M.G., Shalit, P., Lonergan, J.T., Fisher, R.L., Williams, V.C., and Hernandez, J.E., Improvement in lipoatrophy associated with highly active antiretroviral therapy in human immunodeficiency virus-infected patients switched from stavudine to abacavir or zidovudine: the results of the TARHEEL study, *Clin. Infect. Dis.,* 38, 263, 2004.

35. Dreschler, H. and Powderly, W.G., Switching effective antiretroviral therapy: a review, *Clin. Infect. Dis.,* 35, 1219, 2002.

36. Nolan, D., Hammond, E., Martin, A., Taylor, L., Herrmann, S., McKinnon, E., Metcalf, C., Latham, B., and Mallal, S., Mitochondrial DNA depletion and morphologic changes in adipocytes associated with nucleoside reverse transcriptase inhibitor therapy, *AIDS,* 17, 1329, 2003.

37. Lloreta, J., Domingo, P., Pujol, R.M., Arroyo, J.A., Baixeras, N., Matias-Guiu, X., Gilaberte, M., Sambeat, M.A., and Serrano, S., Ultrastructural features of highly active antiretroviral therapy-associated partial lipodystrophy, *Virchows Arch.,* 441, 599, 2002.

38. Walker, U.A., Bickel, M., Lutke Volksbeck, S.I., Ketelsen, U.P., Schofer, H., Setzer, B., Venhoff, N., Rickerts, V., and Staszewski, S., Evidence of nucleoside analogue reverse transcriptase inhibitor-associated genetic and structural defects of mitochondria in adipose tissue of HIV-infected patients, *J. Acquir. Immune Defic. Syndr.,* 29, 117, 2002.

39. Thompson, K., Lal, L., Thompson, K.A., McLean, C.A., Ross, L.L., Hernandez, J., Wesselingh, S.L., and McComsey, G., Improvements in Body Fat and Mitochondrial DNA Levels Are Accompanied by Decreased Adipose Tissue Apoptosis after Replacement of Stavudine Therapy with Either Abacavir

or Stavudine, presented at 10th Conference on Retroviruses and Opportunistic Infections, February 10–14, Boston, MA, 2003, Abstract 728.

40. Domingo, P., Matias-Guiu, X., Pujol, R.M., Domingo, J.C., Arroyo, J.A., Sambeat, M.A., and Vazquez, G., Switching to nevirapine decreases insulin levels but does not improve subcutaneous adipocyte apoptosis in patients with highly active antiretroviral therapy-associated lipodystrophy, *J. Infect. Dis.,* 184, 1197, 2001.

41. Cherry, C., Nolan, D., James I, Mallal S., McKinnon, E., French, M., Hammond, E., Gahan, M., McArthur, J., and Wesselingh, S., Longitudinal Associations between Antiretroviral Treatments and Quantification of Tissue Mitochondrial DNA from Ambulatory Subjects with HIV Infection, presented at 10th Conference on Retroviruses and Opportunistic Infections, February 10–14, Boston, MA, 2003, Abstract 133.

42. Nolan, D.., Hammond, E., James, I., McKinnon, E., and Mallal, S, Contribution of nucleoside-analogue reverse transcriptase inhibitor therapy to lipoatrophy from the population to the cellular level, *Antivir. Ther.,* 8, 617, 2003.

43. Hammond, E., Nolan, D., James, I., Metcalf, C., and Mallal, S., Reduction of mitochondrial DNA content and respiratory chain activity occurs in adipocytes within 6–12 months of commencing nucleoside reverse transcriptase inhibitor therapy, *AIDS,* 18, 563, 2004.

44. Keswani, S.C., Pardo, C.A., Cherry, C.L., Hoke, A., and McArthur, J.C., HIV-associated sensory neuropathies, *AIDS,* 16, 2105, 2002.

45. Berger, A.R., Arezzo, J.C., Schaumburg, H.H., Skowron, G., Merigan, T., Bozzette, S., Richman, D., and Soo, W., 2,3-dideoxycytidine (ddC) toxic neuropathy: a study of 52 patients, *Neurology,* 43, 358, 1993.

46. Blum, A.S., Dal Pan, G.J., Feinberg, J., Raines, C., Mayjo, K., Cornblath, D.R., and McArthur, J.C., Low-dose zalcitabine-related toxic neuropathy: frequency, natural history, and risk factors, *Neurology,* 46, 999, 1996.

47. Dalakas, M.C., Semino-Mora, C., and Leon-Monzon, M., Mitochondrial alterations with mitochondrial DNA depletion in the nerves of AIDS patients with peripheral neuropathy induced by 23-dideoxycytidine (ddC), *Lab. Invest.,* 81, 1537, 2001.

48. Moore, R.D., Wong, W.M., Keruly, J.C., and McArthur, J.C., Incidence of neuropathy in HIV-infected patients on monotherapy vs. those on combination therapy with didanosine, stavudine and hydroxyurea, *AIDS,* 14, 273, 2000.

49. Simpson, D.M., Selected peripheral neuropathies associated with human immunodeficiency virus infection and antiretroviral therapy, *J. Neurovirol.,* 8 (Suppl. 2), 33, 2002.

50. Simpson, D.M., Haidich, A.B., Schifitto, G., Yiannoutsos, C.T., Geraci, A.P., McArthur, J.C., and Katzenstein, D.A., Severity of HIV-associated neuropathy is associated with plasma HIV-1 RNA levels, *AIDS,* 16, 407, 2002.

51. Cherry, C., McArthur J, Costello K, Mijch, A., and Wesselingh, S., Increasing Prevalence of Neuropathy in the Era of Highly Active Antiretroviral Therapy (HAART), presented at 9th Conference on Retroviruses and Opportunistic Infections, February 24–28, Seattle, WA, 2002, Abstract 69.

52. Negredo, E., Molto, J., Burger, D., Viciana, P., Ribera, E., Paredes, R., Juan, M., Ruiz, L., Puig, J., Pruvost, A., Grassi, J., Masmitja, E., and Clotet, B. Unexpected CD4 cell count decline in patients receiving didanosine and tenofovir-based regimens despite undetectable viral load, *AIDS,* 18, 459, 2004.

53. Zucman, D., Lymphocyte Toxicity of Nucleoside Analog Therapy, presented at XV International AIDS Conference, July 11–16, Bangkok, Thailand, 2004, Abstract WePeB5954.

54. Fisher, M., Nucleoside combinations for antiretroviral therapy: efficacy of stavudine in combination with either didanosine or lamivudine, *AIDS,* 12 (Suppl. 3), S9, 1998.

55. Lichterfeld, M., Nischalke, H.D., Bergmann, F., Wiesel, W., Rieke, A., Theisen, A., Fatkenheuer, G., Oette, M., Carls, H., Fenske, S., Nadler, M., Knechten, H., Wasmuth, J.C., and Rockstroh, J.K. Long-term efficacy and safety of ritonavir/indinavir at 400/400 mg twice a day in combination with two nucleoside reverse transcriptase inhibitors as first line antiretroviral therapy, *HIV Med.,* 3, 37, 2002.

56. Reiss, P., Casula, M., de Ronde, A., Weverling, G.J., Goudsmit, J., and Lange, J.M., Greater and more rapid depletion of mitochondrial DNA in blood of patients treated with dual (zidovudine + didanosine or zidovudine + zalcitabine) vs. single (zidovudine) nucleoside reverse transcriptase inhibitors, *HIV Med.,* 5, 11, 2004.

57. Petit, C., Mathez, D., Barthelemy, C., Leste-Lasserre, T., Naviaux, R.K., Sonigo, P., and Leibowitch, J., Quantitation of blood lymphocyte mitochondrial DNA for the monitoring of antiretroviral drug-induced mitochondrial DNA depletion, *Acquir. Immune Defic. Syndr.*, 33, 461, 2003.

58. Miura, T., Goto, M., Hosoya, N., Odawara, T., Kitamura, Y., Nakamura, T., and Iwamoto, A., Depletion of mitochondrial DNA in HIV-1-infected patients and its amelioration by antiretroviral therapy, *J. Med. Virol.*, 70, 497, 2003.

59. John, M. and Mallal, S., Hyperlactatemia syndromes in people with HIV infection, *Curr. Opin. Infect. Dis.*, 15, 23, 2002.

60. John, M., Moore, C.B., James, I.R., Nolan, D., Upton, R.P., McKinnon, E.J., and Mallal, S.A., Chronic hyperlactatemia in HIV-infected patients taking antiretroviral therapy, *AIDS*, 15, 717, 2001.

61. Boubaker, K., Flepp, M., Sudre, P., Furrer, H., Haensel, A., Hirschel, B., Boggian, K., Chave, J.P., Bernasconi, E., Egger, M., Opravil, M., Rickenbach, M., Francioli, P., and Telenti, A., Hyperlactatemia and antiretroviral therapy: the Swiss HIV Cohort Study, *Clin. Infect. Dis.*, 33, 1931, 2001.

62. Moyle, G.J., Datta, D., Mandalia, S., Morlese, J., Asboe, D., and Gazzard, B.G., Hyperlactataemia and lactic acidosis during antiretroviral therapy: relevance, reproducibility and possible risk factors, *AIDS*, 16, 1341, 2002.

63. Vrouenraets, S.M., Treskes, M., Regez, R.M., Troost, N., Smulders, Y.M., Weigel, H.M., Frissen, P.H., and Brinkman, K., Hyperlactataemia in HIV-infected patients: the role of NRTI-treatment, *Antivir. Ther.*, 7, 239, 2002.

64. Hocqueloux, L., Alberti, C., Feugeas, J.P., Lafaurie, M., Lukasiewicz, E., Bagnard, G., Carel, O., Erlich, D., and Molina, J.M., Prevalence, risk factors and outcome of hyperlactataemia in HIV-infected patients, *HIV Med.*, 4, 18, 2003.

65. Falco, V., Rodriguez, D., Ribera, E., Martinez, E., Miro, J.M., Domingo, P., Diazaraque, R., Arribas, J.R., Gonzalez-Garcia, J.J., Montero, F., Sanchez, L., and Pahissa, A., Severe nucleoside-associated lactic acidosis in human immunodeficiency virus-infected patients: report of 12 cases and review of the literature, *Clin. Infect. Dis.*, 34, 838, 2002.

66. Wooltorton, E., HIV drug stavudine (Zerit, d4T) and symptoms mimicking Guillain–Barre syndrome, *CMAJ*, 166, 1067, 2002.

67. Simpson, D., Estanislao, L., Marcus, K., Truffa, K., McArthur J., Lucey, B., Dorfman, D., Naismith, R., Guerrero, I., Mendez, J., Lonergan, T., Landau, Y., Harris, M., Montaner, J., and Clifford, D., HIV-Associated Neuromuscular Weakness Syndrome, presented at 10th Conference on Retroviruses and Opportunistic Infections, February 10–14, Boston, MA, 2003.

68. Leclercq, P., Roth, H., Bosseray, A., and Leverve, X., Investigating lactate metabolism to estimate mitochondrial status, *Antiviral. Ther.*, 6 (Suppl. 4), 16, 2001.

69. Arenas-Pinto, A., Grant, A.D., Edwards, S., and Weller, I.V., Lactic acidosis in HIV infected patients: a systematic review of published cases, *Sex Transm. Infect.*, 79, 340, 2003.

70. Gerard, Y., Maulin, L., Yazdanpanah, Y., De La Tribonniere, X., Amiel, C., Maurage, C.A., Robin, S., Sablonniere, B., Dhennain, C., and Mouton, Y., Symptomatic hyperlactataemia: an emerging complication of antiretroviral therapy, *AIDS*, 14, 2723, 2000.

71. Walker, U.A., Depletion of mitochondrial DNA in liver under antiretroviral therapy with didanosine, stavudine, or zalcitabine, *Hepatology*, 39, 311, 2004.

72. Phenix, B.N., Lum, J.J., Nie, Z., Sanchez-Dardon, J., and Badley, A.D., Antiapoptotic mechanism of HIV protease inhibitors: preventing mitochondrial transmembrane potential loss, *Blood*, 98, 1078, 2001.

73. Sloand, E.M., Kumar, P.N., Kim, S., Chaudhuri, A., Weichold, F.F., and Young, N.S., Human immunodeficiency virus type 1 protease inhibitor modulates the activation of peripheral blood CD4+ T cells and decreases their susceptibility to apoptosis *in vitro* and *in vivo*, *Blood*, 94, 1021, 1999.

74. Weichold, F.F., Bryant, J.L., Pati, S., Barabitskaya, O., Gallo, R.C., and Reitz, M.S. Jr., HIV-1 protease inhibitor ritonavir modulates susceptibility to apoptosis of uninfected T cells, *J. Hum. Virol.*, 2, 261, 1999.

75. Phenix, B.N., Cooper, C., Owen, C., and Badley, A.D., Modulation of apoptosis by HIV protease inhibitors, *Apoptosis*, 7, 295, 2002.

76. Nolan, D., Metabolic complications associated with HIV protease inhibitor therapy, *Drugs*, 63, 2555, 2003.

77. Guest, J.L., Differences in rates of diarrhea in patients with human immunodeficiency virus receiving lopinavir-ritonavir or nelfinavir, *Pharmacotherapy*, 24, 727, 2004.

78. Rufo, P.A., Diarrhea-associated HIV-1 APIs potentiate muscarinic activation of Cl⁻ secretion by T84 cells via prolongation of cytosolic calcium signaling, *Am. J. Physiol. Cell Physiol.,* 286, C998, 2004.

79. Badley, A.D., *In vitro* and *in vivo* effects of HIV protease inhibitors in apoptosis, *Cell Death Differ,* 2005 Mar 11; doi. 1038.1038/sj cdd4401580 [Epub ahead of print].

80. Cooper, C.L., Phenix, B., Mbisa, G., Lum, J.J., Parato, K., Hawley, N., Angel, J.B., and Badley, A.D., Antiretroviral therapy influences cellular susceptibility to apoptosis *in vivo, Front Biosci.,* 9, 338, 2004.

81. Glitsch, M.D., Bakowski, D., and Parekh, A.B., Store-operated Ca^{2+} entry depends on mitochondrial Ca^{2+} uptake, *EMBO J.,* 21, 6744, 2002.

82. Parekh, A.B., Store-operated Ca^{2+} entry: dynamic interplay between endoplasmic reticulum, mitochondria, and plasma membrane, *J. Physiol.,* 547, 333, 2003.

83. Gilabert, J.A., Bakowski, D., and Parekh, A.B., Energised mitochondria increase the dynamic range over which inositol 1,4,5-triphosphate activates store-operated calcium influx, *EMBO J.,* 20, 2672, 2001

84. Andre, P., Groettrup, M., Klenerman, P., de Giuli, R., Booth, B.L. Jr, Cerundolo, V., Bonneville, M., Jotereau, F., Zinkernagel, R.M., and Lotteau, V., An inhibitor of HIV-1 protease modulates proteasome activity, antigen presentation, and T cell responses, *Proc. Natl. Acad. Sci. U.S.A.,* 95, 13120, 1998.

85. Schmidtke, G., Holzhutter, H.G., Bogyo, M., Kairies, N., Groll, M., de Giuli, R., Emch, S., and Groettrup, M., How an inhibitor of the HIV-I protease modulates proteasome activity, *J. Biol. Chem.,* 274, 35734, 1999.

86. Wojcik, C., Regulation of apoptosis by the ubiquitin and proteasome pathway, *J. Cell Mol. Med.,* 6, 25, 2002.

87. Sloand, E.M., Maciejewski, J., Kumar, P., Kim, S., Chaudhuri, A., and Young, N., Protease inhibitors stimulate hematopoiesis and decrease apoptosis and ICE expression in CD34+ cells, *Blood,* 96, 2735, 2000.

88. Purnell, J.Q., Zambon, A., Knopp, R.H., Pizzuti, D.J., Achari, R., Leonard, J.M., Locke, C., and Brunzell, J.D., Effect of ritonavir on lipids and post-heparin lipase activities in normal subjects, *AIDS,* 14, 51, 2000.

89. Lenhard, J.M., Croom, D.K., Weiel, J.E., and Winegar, D.A., HIV protease inhibitors stimulate hepatic triglyceride synthesis, *Arterioscler. Thromb. Vasc. Biol.,* 20, 2625, 2000.

90. Nguyen, A.T., Gagnon, A., Angel, J.B., and Sorisky, A., Ritonavir increases the level of active ADD-1/SREBP-1 protein during adipogenesis, *AIDS,* 14, 2467, 2000.

91. Riddle, T.M., Kuhel, D.G., Woollett, L.A., Fichtenbaum, C.J., and Hui, D.Y., HIV protease inhibitor induces fatty acid and sterol biosynthesis in liver and adipose tissues due to the accumulation of activated sterol regulatory element-binding proteins in the nucleus, *J. Biol. Chem.,* 276, 37514, 2001.

92. Riddle, T.M., Schildmeyer, N.M., Phan, C., Fichtenbaum, C.J., and Hui, D.Y., The HIV protease inhibitor ritonavir increases lipoprotein production and has no effect on lipoprotein clearance in mice, *Lipid Res.,* 43, 1458, 2002.

93. Liang, J.S., Distler, O., Cooper, D.A., Jamil, H., Deckelbaum, R.J., Ginsberg, H.N., and Sturley, S.L., HIV protease inhibitors protect apolipoprotein B from degradation by the proteasome: a potential mechanism for protease inhibitor-induced hyperlipidemia, *Nat. Med.,* 7, 1327, 2001.

24 Apoptosis as a Pathogenic Mechanism of HIV-Associated Opportunistic Infections

Sara Warren
Xian-Ming Chen
Nicholas F. LaRusso
Andrew D. Badley

CONTENTS

INTRODUCTION

A balance of apoptosis is necessary to prevent abnormal cellular accumulations or, conversely, heightened cell turnover. A small shift in either direction may have serious clinical and pathophysiological repercussions. In immunosuppressed patients, the development of opportunistic infections shifts the apoptotic balance. Such infections enhance immune activation, which promotes immune cell turnover and apoptosis. Moreover, many such pathogens directly alter the apoptotic balance in order to favor their own pathogenesis.

In general, infectious agents (bacteria, viruses, and protozoa) and their toxins subvert normal apoptotic pathways in order to achieve one or more of four specific goals: invasion, immune escape, tissue damage, or cellular transformation.

Invasion: Cryptosporidium parvum causes a self-limited diarrheal disease in immunocompetent individuals but life-threatening biliary tract disease in those with immunodeficiency. *In vitro, C. parvum* directly induces apoptosis of human biliary epithelial cells through a Fas/Fas ligand-mediated pathway in which both are upregulated after contact with *C. parvum*. The result is caspase-dependent biliary cell apoptosis (see below).[1] A similar mechanism is employed by *Pseudomonas aeruginosa*, which enhances endogenous Fas/Fas ligand production and, consequently, induces apoptosis of pulmonary epithelia.[2,3] Other bacterial pathogens, including *Legionella pneumophila*,[4,5] *Clostridium* difficile toxin B,[6] and group A streptococcus,[7] also induce epithelial apoptosis as part of their pathogenesis, and although the phenomena are well described, the molecular mechanisms are undefined. In contrast, intestinal pathogens, including *Salmonella* species and enteroinvasive *Escherichia coli*, induce intestinal epithelial cell apoptosis that is only partially inhibited by administration of anti-tumor necrosis factor (TNF)- antibodies or nitric oxide scavengers,[8] suggesting that this form of apoptosis is mediated through a non-Fas-dependent pathway.

Immune escape: The host response to extracellular bacteria typically involves activation of the humoral immune response in order to generate antibodies, which interfere with bacterial attachment to host cells, neutralize bacterial toxins, facilitate phagocytosis, or directly activate the complement system. Intracellular bacteria typically induce cell-mediated immune responses, most commonly a delayed-type hypersensitivity (DTH) response. CD4 T helper cells (TH) are essential to the generation of both the humoral and DTH responses. TH cells are activated by encountering an antigen. Depending on the prevailing cytokine environment, TH cells can develop into discrete subsets. The predominance of interleukin (IL)-4 promotes a TH_2 cytokine profile, whereas a predominance of IL-12 promotes a TH_1 cytokine profile. Most commonly, naive TH cells, mast cells, and natural killer (NK) or NK-T cells produce IL-4 after exposure to antigen. Conversely, IL-12 production occurs after macrophage exposure to foreign antigens, most often, intracellular pathogens or bacterial products. Specifically, CD4 T cells skewed toward the TH_2 response activate B cells, which results in a humoral response. Conversely, the generation of a TH_1 skewed response mostly produces IL-2 and interferon (IFN)-, which supports inflammation and activates cytotoxic T cells and macrophages. Together, these result in enhanced T cell killing and macrophage-mediated lysis of invading pathogens.

It has recently become apparent that different bacterial pathogens possess mechanisms of inducing apoptosis to inhibit the normal host immune response. The following bacteria were described to induce apoptosis of antigen-presenting cells: *Shigella flexneri, Bordetella pertussis, Listeria monocytogenes, Salmonella typhimurium, Salmonella enteritidis, E. coli, Actinobacillus actinomycetemcomitans, P. aeruginosa, Yersinia enterocolitica, Staphylococcus aureus, L. pneumophila, Helicobacter pylori, Mycobacterium tuberculosis,* and *Mycobacterium bovis*.[9] In the case of salmonella, the invasive toxin SipB is both necessary and sufficient for macrophage apoptosis. SipB enters target-cell macrophages and directly cleaves and activates procaspase-1, which results in caspase-1 activity and consequent cytotoxicity.[10] Similarly, the secreted IpaB toxin of *S. flexneri* induces macrophage apoptosis both *in vitro* and *in vivo* by directly cleaving and activating procaspase-1; the resultant apoptotic cascade is caspase-1 dependent, as caspase-1 inhibitors block the process.[11–14] Furthermore, *Streptococcus pyogenes* infection can cause life-threatening streptococcal diseases, including toxic-shock syndrome and necrotizing fasciitis with high mortality (40 to 60%).[15] The streptococcal pyrogenic exotoxin B (SPE B) toxin of *S. pyogenes* was postulated to contribute to the unchecked spread of bacteria by virtue of its ability to induce the apoptotic death of polymorphonuclear leukocytes, monocytes, and macrophages. It was, therefore, postulated that this effect results in the death of macrophages before they are able to ingest and destroy the invading bacteria.[16]

Bacteria may also alter mitochondrial regulation of apoptosis. Mitochondria are involved in the initiation, decision, and degradation phases of apoptosis.[17] *Neisseria meningitidis* and *Neisseria gonorrhoeae* are Gram-negative bacteria that encode pore proteins that localize to the outer Gram-negative cell wall. Porin B (PorB) from *N. meningitidis* (PorBm) or from *N. gonorrhoeae* (PorBg) can translocate from the bacterial cell wall into the mitochondria of mammalian cells. Although both proteins associate with the voltage-dependent anion channel (VDAC) component of the mitochondrial permeability transition pore complex (PTPC), PorBg inhibits $\delta\psi_m$,[18] whereas PorBm induces loss of $\delta\psi_m$ and, consequently, initiates apoptosis.[19] The vacuolating cytotoxin of *H. pylori* (VacA) is essential for its cytotoxic effects on gastric mucosa. The N-terminal fragment of VacA, p34, localizes to mitochondria, resulting in loss of $\delta\psi_m$ and release of apoptotic mediators. This is inhibited by Bcl-2, which suggests a mechanism that involves the PTPC,[20] although the protein to which p34 binds is not yet defined.

Tissue damage: Many infections cause tissue damage by enhanced apoptosis. Such may be the case for the ocular pathology of toxoplasmosis. In mice inoculated with *Toxoplasma gondii*, the Fas/Fas ligand pathway is activated and destroys ocular tissue.[21] Similarly, the induction of apoptosis was implicated in the effects of *Clostridium difficile* toxin B on the intestinal surface, which ultimately results in a profound, watery, life-threatening diarrhea.[6] Bartonella infections may also represent an example of altered apoptotic regulation by bacteria; however, in this context, bartonella seems to inhibit endothelial cell apoptosis, resulting in endothelial cell accumulation and angioproliferation.[22]

Transformation: A number of viral pathogens may promote the oncogenic transformation of target cells teleologically in an effort to enhance their survival and increase production of progeny viruses. Viral transformation is often associated with the expression of gene products that inhibit apoptosis, thereby promoting survival of infected target cells. Examples include activation of the PI3 kinase-Akt survival pathways by Epstein–Barr virus (EBV), papillomavirus, Kaposi's sarcoma-associated herpesvirus (KSHV), polyomavirus, hepatitis B virus (HBV), and hepatitis C virus (HCV).[23] Many of the viruses that more efficiently induce transformation may have multiple mechanisms of inhibiting apoptosis, as exemplified by KSHV. KSHV not only activates Akt survival pathways but also directly enhances NFκB activity as well as inhibits caspase-8 activation through viral FLICE-inhibitory protein (FLIP) expression.[24]

This chapter will illustrate one disease of patients with human immunodeficiency virus (HIV) that is characterized by excessive apoptosis (cryptosporidium) and one disease characterized by insufficient apoptosis (EBV).

CRYPTOSPORIDIUM

Cryptosporidium is a genus of enteric protozoans within the protist phylum Apicomplexa.[25,26] At least 10 species within the genus (*C. parvum, C. muris, C. wrairi, C. felis, C. meleagridis, C. baileyi, C. serpentis, C. nasorum, C. saurophilum,* and *C. andersoni*) are currently recognized on the basis of differences in host specificity and oocyst morphology.[27,28] Immunocompromised individuals, such as patients with acquired immunodeficiency syndrome (AIDS), can be infected with most of the species, including *C. parvum, C. felis, C. muris,* and *C. meleagridis.*[27–29] *C. parvum,* the most common species that infects AIDS patients, contains eight chromosomes with a 10.4-Mb haploid genome, of which a complete map was recently completed and two distinct genotypes—human type 1 and bovine type 2—were identified to infect humans.[30,31] Humans are infected by ingesting cryptosporidium oocysts. After being ingested, oocysts excyst in the gastrointestinal tract, releasing infective sporozoites that, in turn, invade intestinal epithelial cells. Soon after attachment, cryptosporidium induces protrusion of the host cell membrane to cover the parasite and resides in an intracellular but extracytoplasmic parasitophorous vacuole. In this parasitophorous vacuole, the parasite undergoes further intracellular development through asexual as well as sexual cycles, which eventually leads to formation of new oocysts.[27,32,33] Although most oocysts formed are thick-walled and are excreted from the host in fecal material, some oocysts are thin-walled and can excyst within the same host, which leads to a new cycle of development.[27]

EPIDEMIOLOGY OF CRYPTOSPORIDIAL INFECTION IN HUMANS

Human cryptosporidium infection is distributed in urban and rural populations in both developed and developing countries and on all six continents. Oocyst excretion rates for the general population vary from 1 to 3% in industrialized countries and can reach 10% in developing nations.[33] Cryptosporidial infection accounts for 2.2% (0.26 to 22%) and 6.1% (1.4 to 40.9%) of diarrhea in immunocompetent persons in industrialized and developing countries, respectively.[34] Infection is more common in immunocompromised subjects, especially those with AIDS.[33] In the pre-HAART (highly active antiretroviral therapy) era, 3 to 4% of AIDS patients in developed countries and as many as 50% of AIDS patients in the developing world were infected. Since the introduction of HAART, the incidence of cryptosporidial infection has fallen in the developed world.[32,35–37] However, it is still a clinically significant problem in AIDS patients who do not have access to HAART.[32]

CRYPTOSPORIDIAL CLINICAL MANIFESTATIONS

The intestinal tract is the primary site of cryptosporidiosis. The duration and severity of clinical symptoms of intestinal cryptosporidiosis depend largely on the immune status of the infected individual. Cryptosporidiosis in otherwise healthy individuals is usually a self-limiting illness with a median duration of 9 to 15 days.[38] Although infection can be asymptomatic, the most common clinical manifestation is profuse watery diarrhea containing mucus but rarely blood or leukocytes. Other symptoms include nausea, vomiting, cramp-like abdominal pain, and mild fever.[38] In AIDS patients, a wide spectrum of disease symptoms are present, from asymptomatic shedding of cryptosporidial oocysts to a fulminant cholera-like illness, which result in significant morbidity and mortality. The diarrhea in AIDS patients is usually watery, and stool frequency can be as much as 10 times a day with a mean volume of 1 liter; these patients can experience a 10% drop in body weight and usually develop severe malabsorption. Most never clear the infection and ultimately have a shorter survival life than AIDS patients without cryptosporidiosis.[37,39,40] Several studies have shown a significant relationship between the immune status of AIDS patients and the severity of the diarrheal disease. Four clinical patterns of disease have been identified: asymptomatic carriage, transient infection (diarrhea lasts for less than 2 months followed by a complete remission of symptoms and loss of cryptosporidium from fecal specimens), chronic infection (diarrhea lasts for 2 months or more with persistence of the parasite in the stool and biopsy specimens), and fulminant infection (patients pass at least 2 l of watery stool daily).[38] Chronic infection is the most common presentation (59.7%), followed by transient (28.7%), fulminant (7.8%), and asymptomatic (3.9%) infection. Fulminant infection occurs only in patients with CD4+ T cell counts of <50 cells/mm³. Transient or asymptomatic infection is more likely associated with a higher CD4+ T cell count. Extraintestinal cryptosporidiosis has been principally reported in AIDS patients and may involve the lungs, middle ear, biliary tract, pancreas, and stomach.[38] Biliary cryptosporidiosis is the most common extraintestinal manifestation of cryptosporidiosis.

CRYPTOSPORIDIOSIS AND EPITHELIAL CELL APOPTOSIS

C. parvum–Associated Epithelial Cell Apoptosis in Patients

Cryptosporidium does not infect intestinal tissue beyond the most superficial surface epithelia, producing a range of abnormalities in crypt and villous architecture, including villous atrophy and crypt hyperplasia. The organism attaches intimately to the microvillous membrane surface and, consequently, causes a loss of microvilli and effacement and, thus, results in malabsorption. These changes are usually accompanied by a mixed inflammatory cell infiltrate within the lamina propria.[38,41] In AIDS patients with intestinal cryptosporidiosis, an increase of intestinal epithelial cell apoptosis has been reported, and a correlation between cell apoptosis and intensity of C. parvum infection has been noted.[42] The histopathologic findings on ampullary biopsy of biliary cryptosporidiosis in AIDS patients include acute and chronic submucosal infiltrates and are characteristic of a nonspecific inflammatory response. Typically, there is periductal inflammation with interstitial edema, neutrophilic infiltrate, and hyperplasia and dilatation of the periductal glands.[43,44] Apoptosis in cholangiocytes adjacent to C. parvum-infected biliary epithelia was found in the gallbladder of an AIDS patient with biliary cryptopsporidiosis.[45] The pathological changes that develop around the portal tracts of AIDS patients with chronic cryptosporidial infections can mimic the histological changes seen in primary sclerosing cholangitis (PSC). Obviously, cryptosporidium-induced epithelial cell apoptosis can further damage the intestinal and biliary epithelial barrier, resulting in an absorptive dysfunction, and contribute to the pathogenesis of intestinal and biliary cryptosporidiosis.[46,47]

C. parvum–Associated Epithelial Cell Apoptosis In Vitro

In an in vitro model of biliary cryptosporidiosis, we first reported direct cytopathic effects in cultured monolayers of biliary epithelial cells (i.e., cholangiocytes) infected with C. parvum. Widespread apoptosis of cholangiocytes within hours after exposure to C. parvum was detected. C. parvum-induced apoptosis was later reported in vitro in human ileocecal carcinoma (HCT-8) and Caco-2 cells and in vivo in a mouse model.[48–50] Furthermore, it was found that C. parvum infection attenuated epithelial apoptosis induced by strongly proapoptotic agents such as staurosporine (an inhibitor of protein kinase C), etoposide (a DNA topoisomerase II inhibitor), and 5-fluorouracil (a thymidine monophosphate synthesis inhibitor), suggesting that C. parvum developed strategies to limit apoptosis in order to facilitate its growth and maturation in the early period after epithelial cell infection.[45] A possible explanation is that C. parvum prevents induction of high levels of epithelial cell apoptosis early after infection, when the parasite depends on the host cell for growth and development. Additionally, the induction of moderate levels of epithelial cell apoptosis, rather than necrosis, soon after infection may limit the host inflammatory response, which could be detrimental to the survival of the parasite. Alternatively, deletion of infected epithelial cells by apoptosis may benefit the host, because it allows for maintenance of epithelial barrier integrity.[49,51]

Mechanisms of C. parvum–Associated Epithelial Cell Apoptosis

C. parvum–induced apoptosis of human biliary epithelial cells could be inhibited by Fas-receptor- or Fas-ligand-neutralizing antibodies, and C. parvum infection induced Fas-ligand membrane surface translocation and stimulated Fas-receptor or Fas-ligand protein expression. Uninfected biliary epithelia or Fas-sensitive Jurkat cells undergo apoptosis via a Fas receptor/Fas ligand–dependent pathway when co-cultured with infected biliary epithelia. Additionally, caspases, a family of intracellular cysteine proteases and downstream effectors of the Fas receptor,[52] were implicated in the molecular machinery responsible for apoptosis.[1] Furthermore, blocking apoptosis with the caspase inhibitor increased the percentage of infected cells, suggesting that infected cells may attempt to eliminate the parasite through apoptosis.[47]

Nuclear factor-κ B (NF-κB) activation was shown to prevent apoptosis of TNF-α-stimulated cells in several experimental models.[53] Early studies demonstrated that *C. parvum* infection of intestinal epithelial cells is paralleled by the activation of NF-κB target genes.[54–56] More recently, McCole et al.[49] indicated that *C. parvum*–induced apoptosis was limited by the concomitant activation of NF-κB. Using an *in vitro* model of biliary cryptosporidiosis, we found that nuclear translocation of the NF-κB family of proteins (p65 and p50) was observed in cells exposed to *C. parvum*, and activation of NF-κB was found in infected cells but not in bystander uninfected cells. In addition, inhibition of NF-κB activation resulted in apoptosis in infected cells and significantly enhanced *C. parvum*–induced apoptosis in bystander uninfected cells. These observations support the concept that although *C. parvum* induces host cell apoptosis in bystander uninfected host cells, which may limit spread of the infection, it directly activates the NF-κB/I(κ)B system in infected biliary epithelia, thus protecting infected cells from death and facilitating parasite survival and propagation.

Synergistic Effects of *C. parvum* Infection with HIV-1-Associated Soluble Factors to Induce Epithelial Cell Apoptosis

There is, as yet, no evidence to support a role for direct infection of the biliary epithelium by the HIV virus. Previous studies have shown that cholangiocytes do not express CD4, CXCR4/5, and galactosylceramide, the receptors for HIV direct infection or injury.[57] No direct HIV-1 infection of cholangiocytes in humans has ever been reported. As mentioned above, AIDS-cholangiopathy is now recognized as a result of opportunistic infections within the biliary tree, and in most series, cryptosporidium is the single most common identifiable pathogen. Moreover, immunocompromised patients, because of profound immunosuppressive therapy, do not appear to be affected by biliary cryptosporidiosis; thus, HIV-infected patients appear to be uniquely susceptible. However, in our lab and in others, cholangiocytes were successfully infected with *C. parvum in vitro*,[58,59] suggesting that HIV infection is not a prerequisite for *C. parvum* infection of cholangiocytes. Therefore, synergistic pathologic effects in response to dual infection may exist.

Recent studies indicate that HIV infection of T cells can promote release of a number of HIV-1-related biologically active peptides (e.g., TAT, gp120, Nef, and Vpr).[60–62] Of particular importance is the discovery that HIV-1 TAT protein, a 16 kDa peptide essential for viral replication, is released in a biologically active soluble form and can be taken up by many cell types by an unknown mechanism.[63] Furthermore, TAT was reported to induce cytokine release and to mediate apoptosis in T cells, monocytes, and endothelial cells.[64–68] TAT was reported to induce apoptosis in many cell types. It was reported that TAT stimulates matrix metalloprotease activity in monocytes *in vitro*.[69] TAT does not seem to upregulate FasL directly but rather synergies with other signaling pathways (e.g., CD4 dependent) to regulate FasL expression.[69] Most recent studies indicate that TAT upregulates caspase-8 in epithelial cells, a downstream enzyme system responsible for Fas-associated apoptosis,[70] and thus makes cells more susceptible to Fas/FasL-associated apoptosis. Therefore, TAT could serve two functions to influence Fas/FasL-mediated apoptosis: to increase susceptibility via caspase-8 upregulation and to enhance FasL production at the transcriptional level. We found that when cholangiocytes were exposed to *C. parvum* in the presence of recombinant biologically active HIV-1 TAT protein at a concentration (100 ng/ml) comparable to serum levels in AIDS patients,[71] a significant increase of cholangiocyte apoptosis was found.[58] In contrast, no change of the infection rate of *C. parvum* to cholangiocytes was found in the presence of TAT. Thus, it seems that HIV-1 will influence *C. parvum*–induced cholangiocyte apoptosis with *C. parvum* via TAT by modulation of the Fas/FasL apoptotic pathway and thus will act synergistically to cause AIDS cholangiopathy.

DIAGNOSIS

Definitive diagnosis of cryptosporidium infection requires microscopic detection of the parasite in tissue or in body fluids, because cryptosporidiosis does not produce a specific clinical syndrome. However, clinical, endoscopic, immunologic, and molecular techniques all have a place in the diagnosis and clinical assessment of human cryptosporidiosis.[38] The identification of oocysts in stool with the aid of special stains, such as the modified acid-fast stain, or with the assistance of immunofluorescent labeled antibodies is the approach most commonly used in diagnostic laboratories. Polymerase chain reaction (PCR)-based techniques are available as research tests. Serology is of limited value, because many healthy individuals already have antibodies to *C. parvum*. Endoscopic appearances in the intestines are usually normal, although in cases of severe partial villous atrophy, there may be a diminution in the fold pattern in the duodenum.

When biliary disease is suspected, ultrasonography is the best initial diagnostic test. However, the most sensitive test to use to diagnose biliary tract disease in HIV-infected patients is endoscopic retrograde cholangiopancreatography (ERCP).

THERAPY

The treatment of cryptosporidiosis is far from satisfactory. There is no antimicrobial chemotherapeutic agent that will reliably eradicate the organism. A number of agents seem to have some activity against the organism, although they are usually suppressive and not curative. Paromomycin, azithromycin, and most recently, nitazoxanide are commonly used.[72,73] However, as these antiparasitic drugs are, at best, modestly effective, preliminary clinical results using these agents are not uniform and well-designed placebo-controlled studies, including assessment of drug combinations, are needed to confirm which, if any, agents are effective.[73–76]

Because the clinical course of cryptosporidiosis in immunocompromised individuals depends largely upon the immune status of the host, treatment options for these patients should aim to resolve the immune system function and to lead to resolution of clinical symptoms. The best treatment for cryptosporidiosis in AIDS patients is HAART.[32,35-37,74] It can partially restore host immune function and resolve cryptosporidium infection. Some nucleoside antivirals have a direct effect on growth of *C. parvum in vitro*.[77] Therapy of biliary cryptosporidiosis in AIDS patients is dependent on HAART, whereas endoscopic sphincterotomy may provide symptomatic relief for patients with abdominal pain or cholangitis associated with papillary stenosis, as it facilitates drainage and decompression of the biliary system and thus improves quality of life.[36,78] Patients with diffuse intra- and extrahepatic sclerosing cholangitis alone have few specific treatment options.

EPSTEIN–BARR VIRUS

Epstein–Barr virus (EBV) is a human Gammaherpesvirinae that infects B lymphocytes. EBV is ubiquitous, and up to 95% of the adult population has been infected with it. Most infections are acquired through the orthopharynx, usually during childhood, and do not seem to cause any disease. However, when immunocompromised, EBV can favor the initiation of a number of malignancies. The EBV genome is linear and composed of double-stranded DNA of more than 170,000 base pairs. The DNA is contained in an icosahedral capsid, surrounded by a loosely packed tegument and finally by an outer envelope coated with glycoprotein spikes. When it is introduced to a host, infection occurs when the EBV gp350/220 binds to the B lymphocyte surface receptor CD 20.[79]

After infection, either the viral genome is incorporated into the host cell's chromosomal DNA or the viral genome is circularized and exists as an episome.[80]

EBV CLINICAL MANIFESTATIONS

The majority of human infections with EBV are asymptomatic. However, in different contexts, EBV has been implicated in the pathogenesis of Hodgkin's lymphoma (HL), Burkitt's lymphoma (BL), posttransplant proliferative disorder (PTLD), primary effusion lymphoma (PEL), and nasopharyngeal carcinoma (NPC). HIV patients and other immunocompromised individuals have a much greater susceptibility to EBV-related infections, because the T lymphocytes are not able to restrict the virus to a latent pool of B lymphocytes.[79] In most circumstances, lymphomas that result from EBV infection are restricted to the B lymphocytes, but there are cases in which the EBV-associated malignancies occur in T lymphocytes or NK cells.[81]

EBV infection can either exist in the lytic form or develop into one of three latency phases. Most frequently, the latency seen *in vitro* is type III, which is that characterized by expression of nuclear antigens EBNA-1, EBNA-2, EBNA-3A, EBNA-3B, EBNA-3C, and EBNA-LP and membrane proteins LMP1, LMP2A, and LMP2B.[79] In type-I latency, most frequently seen in BL, only EBNA1 and LMP2A are observed, and in type-II latency (nasopharyngeal cancers), EBNA1, LMP1, and LMP2 are expressed.[79] It is commonly believed that the different types of latency observed can be attributed to different viral/host interactions and are explained by the utilization of different promoters (Table 24.1).

Burkitt's Lymphoma

BL accounts for 30% of all EBV-related lymphomas. It was the first characterized and most frequently continues to be observed as lymphoid malignancies located on the jaw and neck.[82] Almost all cases of endemic BL are infected with EBV, but only 20% of sporadic BL cases are positive for EBV.[79] Both types of BL feature a chromosomal translocation of either t(8,14), t(2,8), or t(8,22), which causes the *c-myc* oncogene to relocate near the immunoglobin gene of chromosome 2, 14, or 22.[83] The effect of this translocation is to abrogate the biological functions of *c-myc*, which is implicated in the regulation of cell proliferation, differentiation, and apoptosis.

Lymphoproliferative Disorders

The family of lymphoproliferative disorders (LPDs) begins with polyclonal B cell hyperplasia, which may progress to monoclonal lymphoma.[84] The LPDs are observed most frequently in primary immunodeficient patients[83] but more frequently are being observed in posttransplant patients as immunosuppressant drugs become more effective.[85] A risk factor for LPDs may be the development of EBV viremia, which has led to recommendations to monitor EBV viremia and reduce immunosuppressant

Table 24.1

Latency	Proteins Expressed	Associated Disorders
I	LMP2A, EBNA1	Burkitt's lymphoma, primary effusion lymphoma, NK-cell leukemia
II	LMP1, LMP2A, EBNA1	EBV-positive Hodgkin's disease, nasopharyngeal carcinoma
III	EBNA-1, EBNA-2, EBNA-3, EBNA-4, EBNA-5, EBNA-6, EBNA-LP, LMP1, LMP2A, LMP2B	LCLs, infectious mononucleosis, posttransplant proliferative disorder

treatments if plasma viral copy numbers begin increasing rapidly.[86] Primary effusion lymphoma (PEL) is a type of LPD that occurs in AIDS patients and is characterized by liquid growth in the body cavity, without a solid-mass tumor.[87] It is almost always seen in patients with an EBV infection of type-I latency who are usually coinfected with Karposi's sarcoma–associated herpesvirus. EBV may also be associated with T cell lymphomas[88] and NK-cell leukemia.

EBV Proteins That Modulate Apoptosis

LMP1

Latent membrane protein 1 (LMP1) has a dramatic influence upon infected cells, because it radically affects the molecular signaling that occurs through its interactions with TRAFs, TRADD, JAK3, activation markers, and antiapoptotic molecules and has a cumulative effect of reducing the susceptibility of a cell to apoptosis. LMP1 is an integral membrane protein of 386 amino acids that is encoded by the BNLF1 gene. LMP1 is composed of three regions, each with differing functions. The short amino terminus (17 amino acids) is responsible for tethering the protein to the membrane in the proper orientation, and it is also the site of ubiquination. Six hydrophobic transmembrane domains (200 amino acids total) promote oligomerization centered around the protein rafts in the membrane.[89] The carboxy terminus varies in length depending on the number of repeats of an 11 amino acid segment but averages approximately 200 amino acids.[90] Within the carboxy terminus are three C-terminal activation regions (CTAR) regions that are associated with the protein's antiapoptotic effects (detailed below). LMP1 has a typical half-life of 2 to 5 h within the cell, dependent on cell type, after which it is cleaved at Leu 242 and releases the carboxy cytosolic tail from the transmembrane region. Additionally, LMP1 is subject to phosphorylation within the C-terminal domain at Ser 313 and Thr 324.[91]

Wang et al. first identified LMP1 as possessing oncogenic (and antiapoptotic) properties when they transfected it into NIH3T3 and Rat-1 cell lines and observed a resulting increase in contact inhibition and anchorage independent growth.[92] Furthermore, after LMP1-expressing cells were introduced into nude mice, it induced the growth of lymphoid tumors.[92] LMP1 has four functions that favor its oncogenic properties: transfection with LMP1 can induce DNA synthesis, similar to the DNA synthesis seen during B cell activation; LMP1 has the ability to upregulate the production of antiapoptotic molecules; LMP1 interferes with the cell cycle of progression; and evidence of LMP1 control of matrix metalloproteins and IL-8 suggests that it may play a role in angiogenesis and metastasis.[89]

In addition to its transforming and toxic properties, LMP1 disrupts the signal transduction pathway of B and T cells. LMP1 expression induces an upregulation in the cell adhesion molecules ICAM-1, LFA1, and LFA3. It also increases activity of the CD71 transferrin receptor; activation markers CD21, CD23, and CD40; and antiapoptotic molecules Bcl-2 and A20.[93–95] However, this upregulation is seen only in the latent infections of EBV. The lytic form of LMP1, which lacks the 128 amino acids on the amino terminus, does not display the transactivating or antiapoptotic activity[94] and it does not impact DNA synthesis or the cell size:volume ratio.[96]

Through the presence of two NF-κB response elements located on the A20 promoter, LMP1 is able to enhance expression and activity of A20 (a zinc finger protein that confers resistance to the TNF). Transient transfection of Jurkat T cells with LMP1 increases CAT activity of an A20 promoter CAT construct, whereas mutating the B-binding sites of the A20 promoter negates any increase in CAT activity, indicating that the activation of A20 is mediated by NF-κB.[97] Similarly, the ability of LMP1 to transactivate HIV replication is eliminated when the HIV LTR NF-κB binding site is deleted.[98] It has been suggested that LMP1 influences NF-κB-dependent activation by phosphorylating and degrading the IB inhibitor of NF-κB, which then allows the NF-κB to translocate to the nucleus.[99] Additionally, the ability of LMP1 to activate NF-κB may account for the property of LMP-1 to upregulate ICAM1, a cellular adhesion molecule.

LMP1 is similar in function, although not in amino acid homology, to members of the TNF receptors, especially TNFR1 and CD40,[100] because it is able to bind to many of the same proteins as do TNFRs. These interactions occur primarily through the CTARs located on the carboxy terminus of LMP1.

CTAR1

LMP1 is able to interact with the TNF receptor–associated factor (TRAF) proteins. These six proteins (TRAF1-6) are involved with signal transduction and interact with the cytoplasmic tail of TNF receptors. LMP1 was shown to interact with TRAF1, TRAF2, and TRAF3, and the first 44 amino acids of the carboxy terminus are mandatory for that binding to occur. The interaction of TRAF1/TRAF2 heterodimers with the LMP1 TRAF domain results in cellular activation, ultimately resulting in B phosphorylation and nuclear translocation of NF-B.

Stimulation of the TRAFs initiates a signal cascade that results in the activation of not only the NF-B pathway but also the mitogen-activated protein kinase (MAPK) pathway, the c-Jun N-terminal kinase (JNK) pathway, and the PI3K pathway, all of which repress antiapoptotic signaling within the cell.[100] The decreased rates of apoptosis promote an environment in which tumor development may be favorable. For example, TRAF2 binds to CTAR1 and becomes activated, after which it, in turn, activates NIK. NIK activation leads to the activation of the NF-B pathway, which initiates transcription for a number of antiapoptotic molecules, including A20, Bcl-2, Bfl-1, and cIAP2.[100] A similar schema and result occur for the p38/MAPK pathway,[101] and although both pathways were proven to operate independently of each other, both are initiated through TRAF2.

CTAR2

The NF-B and MAPK pathways are capable of being activated through not only CTAR1 but also CTAR2. For that to occur, TRADD interacts via its N-terminus with the C-terminus of CTAR2.[100] The TRADD activating site has proven crucial for the LMP1 transformation of B lymphocytes, because LMP mutants lacking the last three amino acids are incapable of transforming B cells.[102] The CTAR2 recruitment of TRADD is also responsible for the activation of the JNK pathway.[103] Stimulation of the JNK pathway upregulates phosphorylation and transcriptional activity of c-Jun (a mitogenic transcription factor). This leads to the binding of TRAF2 and TRADD to CTAR2 and the formation of AP-1 complexes, which increase transcription associated with certain promoters.[103] However, the requirement for CTAR2 has proven to be cell-type specific and could happen through CTAR1 if TRAF1 were present.[101] TRADD can induce apoptosis by means of TRAF2-mediated NF-B activation or promote apoptosis through association with Fas-associated death domain (FADD) protein, which initiates caspase-8 clustering activation and ultimately results in apopto-sis.[104] Because LMP1 lacks TNFR death domains yet interacts with TRADD, the TNFR-mediated antiapoptotic signals of NF-B are favored after LMP1 activation of TRADD.

Additionally, there is some evidence that LMP1 can act as a tumor suppressor as well. Floett-mann et al. transfected LMP1 into an EBV-negative cell and observed an inhibition of the cell cycle progression.[105] More telling, EBV-negative BL transfected into LMP-1 decreases the tumor-ogenicity and clonality from control cells.[106] In total, this suggests that LMP1 plays a complicated role within the progression of an EBV-infected cell to a cancerous state; however, the overall effect seems to be antiapoptotic.

EBNA1

EBNA1 is a DNA-binding protein encoded by the BKRF1 gene of EBV. It is 641 amino acids long and is the only EBV latent protein to be expressed in all three forms of viral latency.[107] EBNA1 functions in the virus by dimerizing and binding DNA, where it acts as a transcriptional enhancer to assist in maintaining and replicating the viral genome,[108] and EBNA1 is restrained to the nucleus by *c-myc*, the nuclear matrix attachment factor. When the *c-myc* gene is silenced or translocated, as often occurs in BL, the EBNA1 invades the cytoplasm and is not able to maintain the EBV genome.[109] Within the infected B cell, EBNA1 prevents cytotoxic T cell responses through a glycine-alanine rich region that interferes with the major histocompatibility complex (MHC), thus inhibiting the EBNA1 peptide antigen processing that would normally occur within an infected cell. Additionally, EBNA1 has been observed to have oncogenic properties, as 20% of transgenic mice that expressed EBNA1 developed B cell lymphoma.[110] Moreover, EBNA1 is expressed in more than 95% of endemic BL cases, but in sporadic BL, the instance of EBNA1 is less than 20%.

EBNA2

Similar to EBNA1, EBNA2 is a transcription factor that is coded for by the BYRF1 gene. Interestingly, it is the first protein to be expressed after infection of B lymphocytes[111] and is critical to the immortalization of B cells.[112] It exists in two different forms (EBNA2A and EBNA2B), and both forms are required to immortalize cells.[113] Rickinson et al. showed, by comparisons of complementary deficient strains, that EBNA2B is much less efficient at transforming B cells than EBNA2A.[114] The genetic variation between the two strains occurs in the middle of the DNA and most significantly features a 42 base pair deletion in EBNA2B.[113] Within the viral genome, EBNA2 is a transactivator of LMP1 and LMP2A, the Cp promoter (that regulates latent EBNA1) and HIV-LTR. In addition, EBNA2 is responsible for some of the reprogramming of cell gene expression by activating a variety of promoters.[115] Within the cellular genome, it transactivates CD21 and CD23 along with *c fgr*, another proto-oncogene.[111]

EBNA3, EBNA4, EBNA5, EBNA6

EBNA3, EBNA4, and EBNA6 (also called EBNA3A, EBNA3B, and EBNA3C) have been shown to act to arrest the infected B lymphocyte's growth and division cycle by disrupting the G2/M checkpoint.[116] After a cell becomes infected, EBNA2 and EBNA5 are expressed within the first 32 h, which corresponds with the progression from G0 to G1. As the cell moves into S phase, an increase in the levels of EBNA3, EBNA4, and EBNA6 can be observed.[116] EBV modifies the activities of the tumor suppressor gene pRb by modulating the signaling pathways that affect it and thus obliterating its checkpoint function.[117] EBNA5, in combination with EBNA2, has been shown to upregulate the activity of cyclin D2 and stimulate the resting B cell into G1 phase. The G1 checkpoint is then circumvented by EBNA6 as it disrupts the cyclin/CDK-pRB-E2F signaling pathway,[118] and EBNA6 is also thought to disrupt the G2/M progression.

EBNA-LP

When a B lymphoctye becomes infected with EBV, EBNA2 and EBNA-leader protein (LP) are the first two viral proteins to be expressed.[119] EBNA-LP is a phosphoprotein that is the first protein encoded on the multicistronic mRNA that codes for all the EBNAs. It is located in both the cytoplasm and nucleus (where it associates with the nuclear matrix) and has been shown to stimulate EBNA2-mediated transcriptional activation of genes such as LMP1.[120] It is not essential for viral survival, but it has been shown to bind to a number of molecules involved in viral life cycle, cellular mechanics, antiapoptotic molecules Bcl-2, p53, and pRb, and HS1-associated protein HAX-1.[119] HAX-1 has been implicated in the regulation of apoptosis. The antiapoptotic characteristic of HAX-1 has been attributed to the Bcl-2 homologous domains BH1 and BH2 that are known to be the apoptosis regulatory sites in Bcl-2.[121] Consequently, EBNA-LP modifies the net cellular susceptibility toward apoptosis.[120,121] These results imply that as soon as a cell becomes infected with EBV, the viral proteins are enhancing antiapoptotic and protransformation signals.

EBER

EBERs (Epstein–Barr encoded RNAs) are viral protein transcripts that are found in almost all infected cells. They are usually approximately 250 nt nonpolydenylated RNA polymerase III transcripts and are usually found in the nucleus.[122] EBERs are thought to inhibit the activity of double-stranded RNA-activated protein kinase and thus provide resistance to the host antiviral defense of interferon-α induced apoptosis. However, they are not required for viral maintenance or replication, and they are not required for the immortalization of B cells. There is also some speculation that EBERs enhance tumorigenicity.[122]

DIAGNOSIS AND THERAPY

Diagnosis of EBV-associated malignancies requires tissue diagnosis, often with molecular techniques, to confirm the presence of EBV-encoded transcripts.[123] After being diagnosed, EBV-associated malignancies are often amenable to standard anti-neuroplastic therapies of chemotherapy and radiation. However, in such cases, reversal of underlying immunosuppression is an important adjunct therapy. When applied to the therapy of HIV-associated EBV malignancies, immune reformation associated with HAART can independently reduce EBV viral replication[124] as well as improve survival from such malignancies.[125]

CONCLUSIONS

A pivotal role of apoptosis is emerging in the pathogenesis of HIV-induced immunosuppression. This immunosuppression involves an increased susceptibility toward opportunistic infections and malignancies that drive their own pathogenesis, in part, by subverting apoptotic signaling pathways. With the emerging interest in developing therapeutic agents that directly modify apoptotic responses, both HIV infection and complications of HIV are, therefore, candidates for testing the abilities of such agents to alter disease outcomes.

REFERENCES

1. Chen, X.M., Gores, G.J., Paya, C.V., and LaRusso, N.F., Cryptosporidium parvum induces apoptosis in biliary epithelia by a Fas/Fas ligand-dependent mechanism, *Am. J. Physiol.*, 277, G599–G608, 1999.
2. Grassme, H., Kirschnek, S., Riethmueller, J., Riehle, A., von Kurthy, G., Lang, F., Weller, M., and Gulbins, E., CD95/CD95 ligand interactions on epithelial cells in host defense to *Pseudomonas aeruginosa*, *Science,* 290, 527–530, 2000.
3. Hauser, A.R. and Engel, J.N., *Pseudomonas aeruginosa* induces type-III-secretion-mediated apoptosis of macrophages and epithelial cells, *Infect. Immun.,* 67, 5530–5537, 1999.
4. Gao, L.Y. and Abu Kwaik, Y., Activation of caspase 3 during *Legionella pneumophila*-induced apoptosis, *Infect. Immun.,* 67, 4886–4894, 1999.
5. Gao, L.Y. and Abu Kwaik, Y., Apoptosis in macrophages and alveolar epithelial cells during early stages of infection by *Legionella pneumophila* and its role in cytopathogenicity, *Infect. Immun.,* 67, 862–870, 1999.
6. Fiorentini, C., Fabbri, A., Falzano, L., Fattorossi, A., Matarrese, P., Rivabene, R., and Donelli, G., *Clostridium* difficile toxin B induces apoptosis in intestinal cultured cells, *Infect. Immun.,* 66, 2660–2665, 1998.
7. Tsai, P.J., Lin, Y.S., Kuo, C.F., Lei, H.Y., and Wu, J.J., Group A streptococcus induces apoptosis in human epithelial cells, *Infect. Immun.,* 67, 4334–4339, 1999.
8. Kim, J.M., Eckmann, L., Savidge, T.C., Lowe, D.C., Witthoft, T., and Kagnoff, M.F., Apoptosis of human intestinal epithelial cells after bacterial invasion, *J. Clin. Invest.,* 102, 1815–1823, 1998.
9. Kornfeld, H., Mancino, G., and Colizzi, V., The role of macrophage cell death in tuberculosis, *Cell Death Differ.,* 6, 71–78, 1999.

10. Hersh, D., Monack, D.M., Smith, M.R., Ghori, N., Falkow, S., and Zychlinsky, A., The salmonella invasin SipB induces macrophage apoptosis by binding to caspase-1, *Proc. Natl. Acad. Sci. U.S.A.,* 96, 2396–2401, 1999.

11. Alnemri, E.S., Livingston, D.J., Nicholson, D.W., Salvesen, G., Thornberry, N.A., Wong, W.W., and Yuan, J., Human ICE/CED-3 protease nomenclature, *Cell,* 87, 171, 1996.

12. Chen, Y., Smith, M.R., Thirumalai, K., and Zychlinsky, A., A bacterial invasin induces macrophage apoptosis by binding directly to ICE, *EMBO J.,* 15, 3853–3860, 1996.

13. McSorley, S.J., Xu, D., and Liew, F.Y., Vaccine efficacy of salmonella strains expressing glycoprotein 63 with different promoters, *Infect. Immun.,* 65, 171–178, 1997.

14. Zychlinsky, A., Prevost, M.C., and Sansonetti, P.J., *Shigella flexneri* induces apoptosis in infected macrophages, *Nature,* 358, 167–169, 1992.

15. Eriksson, B.K., Andersson, J., Holm, S.E., and Norgren, M., Epidemiological and clinical aspects of invasive group A streptococcal infections and the streptococcal toxic shock syndrome, *Clin. Infect. Dis.,* 27, 1428–1436, 1998.

16. Kuo, C.F., Wu, J.J., Tsai, P.J., Kao, F.J., Lei, H.Y., Lin, M.T., and Lin, Y.S., Streptococcal pyrogenic exotoxin B induces apoptosis and reduces phagocytic activity in U937 cells, *Infect. Immun.,* 67, 126–130, 1999.

17. Kroemer, G. and Reed, J.C., Mitochondrial control of cell death, *Nat. Med.,* 6, 513–519, 2000.

18. Muller, A., Gunther, D., Brinkmann, V., Hurwitz, R., Meyer, T.F., and Rudel, T., Targeting of the pro-apoptotic VDAC-like porin (PorB) of *Neisseria gonorrhoeae* to mitochondria of infected cells, *EMBO J.,* 19, 5332–5343, 2000.

19. Massari, P., Ho, Y., Wetzler, L.M., *Neisseria meningitidis* porin PorB interacts with mitochondria and protects cells from apoptosis, *Proc. Natl. Acad. Sci. U.S.A.,* 97, 9070–9075, 2000.

20. Galmiche, A., Rassow, J., Doye, A., Cagnol, S., Chambard, J.C., Contamin, S., de Thillot, V., Just, I., Ricci, V., Solcia, E., Van Obberghen, E., and Boquet, P., The N-terminal 34 kDa fragment of *Helicobacter pylori* vacuolating cytotoxin targets mitochondria and induces cytochrome *c* release, *EMBO J.,* 19, 6361–6370, 2000.

21. Hu, S., Schwartzman, J.D., and Kasper, L.H., Apoptosis within mouse eye induced by *Toxoplasma gondii, Chin. Med. J. (Engl.),* 114, 640–644, 2001.

22. Kirby, J.E. and Nekorchuk, D.M., Bartonella-associated endothelial proliferation depends on inhibition of apoptosis, *Proc. Natl. Acad. Sci. U.S.A.,* 99, 4656–4661, 2002.

23. Coorey, S., The pivotal role of phosphatidylinositol 3-kinase-Akt signal transduction in virus survival, *J. Gen. Virol.,* 85, 1065–1076, 2004.

24. Guasparri, I., Keller, S.A., and Cesarman, E., KSHV vFLIP is essential for the survival of infected lymphoma cells, *J. Exp. Med.,* 199, 993–1003, 2004.

25. Zhu, G., Keithly, J.S., and Philippe, H., What is the phylogenetic position of cryptosporidium? *Intl. J. Sys. Evol. Microbiol.,* 50, 1673–1681, 2000.

26. Tzipori, S. and Griffiths, J.K., Natural history and biology of *Cryptosporidium parvum, Adv. Parasitol.,* 40, 5–36, 1998.

27. O'Donoghue, P.J., Cryptosporidium and cryptosporidiosis in man and animals, *Int. J. Parasitol.,* 25, 139–195, 1995.

28. Xiao, L., Morgan, U.M., Fayer, R., Thompson, R.C., and Lal, A.A., Cryptosporidium systematics and implications for public health, *Parasitol. Today,* 16, 287–292, 2000.

29. Hoepelman, I.M., Human cryptosporidiosis, *Int. J. STD and AIDS,* 7, 28–33, 1996.

30. Blunt, D.S., Khramtsov, N.V., Upton, S.J., and Montelone, B.A., Molecular karyotype analysis of *Cryptosporidium parvum:* evidence for eight chromosomes and a low-molecular-size molecule. *Clin. Diagn. Lab. Immunol.,* 4, 11–13, 1997.

31. Okhuysen, P.C., Chappell, C.L., Crabb, J.H., Sterling, C.R., and DuPont, H.L., Virulence of three distinct *Cryptosporidium parvum* isolates for healthy adults, *J. Infect. Dis.,* 180, 1275–1281, 1999.

32. Chen, X.M., Keithly, J.S., Paya, C.V., and LaRusso, N.F., Cryptosporidiosis, *N. Engl. J. Med.,* 346, 1723–1731, 2002.

33. Meinhardt, P.L., Casemore, D.P., and Miller, K.B., Epidemiologic aspects of human cryptosporidiosis and the role of waterborne transmission, *Epidemiol. Rev.,* 18, 118–136, 1996.

34. Guerrant, R.L., Cryptosporidiosis: an emerging, highly infectious threat, *Emerg. Infect. Dis.,* 3, 51–57, 1997.

35. Miller, J.R., Decreasing cryptosporidiosis among HIV-infected persons in New York City, 1995–1997, *J. Urban Health,* 75, 601–602, 1998.

36. Carr, A., Marriott, D., Field, A., Vasak, E., and Cooper, D.A., Treatment of HIV-1-associated microsporidiosis and cryptosporidiosis with combination antiretroviral therapy, *Lancet,* 351, 256–261, 1998.

37. Clark, D.P., New insights in human cryptosporidiosis, *Clin. Microbiol. Rev.,* 12, 554–563, 1999.

38. Farthing, M.J., Clinical aspects of human cryptosporidiosis, *Contrib. Microbiol.,* 6, 50–74, 2000.

39. Hunter, P.R. and Nichols, G., Epidemiology and clinical features of cryptosporidium infection in immunocompromised patients, *Clin. Microbiol. Rev.,* 15, 145–154, 2002.

40. Manabe, Y.C., Clark, D.P., Moore, R.D., Lumadue, J.A., Dahlman, H.R., Belitsos, P.C., Chaisson, R.E., and Sears, C.L., Cryptosporidiosis in patients with AIDS: correlates of disease and survival, *Clin. Infect. Dis.,* 27, 536–542, 1998.

41. Argenzio, R.A., Liacos, J.A., Levy, M.L., Meuton, D.J., Lecce, J.G., and Powell, D.W., Villous atrophy, crypt hyperplasia, cellular infiltration, and impaired glucose-Na+ absorption in enteric cryptosporidiosis of pigs, *Gastroenterology,* 98, 1129–1140, 1990.

42. Lumadue, J.A., Manabe, Y.C., Moore, R.D., Belitsos, P.C., Sears, C.L., and Clark, D.P., A clinicopathologic analysis of AIDS-related cryptosporidiosis, *AIDS,* 12, 2459–2566, 1998.

43. Teixidor, H.S., Godwin, T.A., and Ramirez, E.A., Cryptosporidiosis of the biliary tract in AIDS, *Radiology,* 180, 51–56, 1991.

44. Hasan, F.A., Jeffers, L.J., Dickinson, G., Otrakji, C.L., Greer, P.J., Reddy, K.R., and Schiff, E.R., Hepatobiliary cryptosporidiosis and cytomegalovirus infection mimicking metastatic cancer to the liver, *Gastroenterology,* 100, 1743–1748, 1991.

45. Chen, X.M., Levine, S.A., Tietz, P., Krueger, E., McNiven, M.A., Jefferson, D.M., Mahle, M., and LaRusso, N.F., *Cryptosporidium parvum* is cytopathic for cultured human biliary epithelia via an apoptotic mechanism, *Hepatology,* 28, 906–913, 1998.

46. McCole, D.F., Eckmann, L., Laurent, F., and Kagnoff, M.F., Intestinal epithelial cell apoptosis following *Cryptosporidium parvum* infection, *Infect. Immun.,* 68, 1710–1713, 2000.

47. Ojcius, D.M., Perfettini, J.L., Bonnin, A., and Laurent, F., Caspase-dependent apoptosis during infection with *Cryptosporidium parvum*, *Microbes Infect.,* 1, 1163–1168, 1999.

48. Widmer, G., Corey, E.A., Stein, B., Griffiths, J.K., and Tzipori, S., Host cell apoptosis impairs *Cryptosporidium parvum* development *in vitro, J. Parasitol.,* 86, 922–928, 2000.

49. McCole, D.F., Eckmann, L., Laurent, F., and Kagnoff, M.F., Intestinal epithelial cell apoptosis following *Cryptosporidium parvum* infection, *Infect. Immun.,* 68, 1710–1713, 2000.

50. Motta, I., Gissot, M., Kanellopoulos, J.M., and Ojcius, D.M., Absence of weight loss during cryptosporidium infection in susceptible mice deficient in Fas-mediated apoptosis, *Microbes Infect.,* 4, 821–827, 2002.

51. Deng, M., Rutherford, M.S., and Abrahamsen, M.S., Host intestinal epithelial response to *Cryptosporidium parvum*, *Adv. Drug Deliv. Rev.,* 56, 869–884, 2004.

52. Thornburn, A., Death receptor-induced cell killing, *Cell. Signalling,* 16, 139–144, 2004.

53. Beg, A.A. and Baltimore, D., An essential role for NF-B in preventing TNF--induced cell death, *Science,* 274, 782–784, 1996.

54. Laurent, F., Eckmann, L., Savidge, T.C., Morgan, G., Theodos, C., Naciri, M., and Kagnoff, M.F., *Cryptosporidium parvum* infection of human intestinal epithelial cells induces the polarized secretion of C-X-C chemokines, *Infect. Immun.,* 65, 5067–5073, 1997.

55. Laurent, F., Kagnoff, M.F., Savidge, T.C., Naciri, M., and Eckmann, L., Human intestinal epithelial cells respond to *Cryptosporidium parvum* infection with increased prostaglandin H synthase 2 expression and prostaglandin E_2 and F_2 production, *Infect. Immun.,* 66, 1787–1790, 1998.

56. Seydel, K.B., Zhang, T., Champion, G.A., Fichtenbaum, C., Swanson, P.E., Tzipori, S., Griffiths, J.K., and Stanley, S.L., Jr., *Cryptosporidium parvum* infection of human intestinal xenografts in SCID mice induces production of human tumor necrosis factor and interleukin-8, *Infect. Immun.,* 66, 2379–2382, 1998.

57. Vollenweider, J.M., Chen, X.M., Gores, S.J., Paya, C.V., and LaRusso, N.F., HIV-1 Tat protein enhances *Cryptosporidium parvum*-induced apoptosis in cultured human biliary epithelia, *Hepatology,* 34, A268, 2001.

58. Chen, X.M. and LaRusso, N.F., Mechanisms of attachment and internalization of *Cryptosporidium parvum* to biliary and intestinal epithelial cells, *Gastroenterology,* 118, 368–379, 2000.

59. Verdon, R., Keusch, G.T., Tzipori, S., Grubman, S.A., Jefferson, D.M., and Ward, H.D., An *in vitro* model of infection of human biliary epithelial cells by *Cryptosporidium parvum*, *J. Infect. Dis.,* 175, 1268–1272, 1997.

60. Siliciano, R.F., The role of CD4 in HIV envelope-mediated pathogenesis, *Curr. Top. Microbiol. Immunol.,* 205, 159–179, 1996.

61. Townsley-Fuchs, J., Neshat, M.S., Margolin, D.H., Braun, J., and Goodglick, L., HIV-1 gp120: a novel viral B cell superantigen, *Int. Rev. Immunol.,* 14, 325–338, 1997.

62. Jeang, K.T., Xiao, H., and Rich, E.A., Multifaceted activities of the HIV-1 transactivator of transcription, Tat, *J. Biol. Chem.,* 274, 28837–28840, 1999.

63. Schwarze, S.R., Ho, A., Vocero-Akbani, A., and Dowdy, S.F., *In vivo* protein transduction: delivery of a biologically active protein into the mouse, *Science,* 285, 1569–1572, 1999.

64. Rappaport, J., Joseph, J., Croul, S., Alexander, G., Del Valle, L., Amini, S., and Khalili, K., Molecular pathway involved in HIV-1-induced CNS pathology: role of viral regulatory protein, Tat, *J. Leukocyte Biol.,* 65, 458–465, 1999.

65. Nath, A., Haughey, N.J., Jones, M., Anderson, C., Bell, J.E., and Geiger, J.D., Synergistic neurotoxicity by human immunodeficiency virus proteins Tat and gp120: protection by memantine, *Ann. Neurol.,* 47, 186–194, 2000.

66. Macho, A., Calzado, M.A., Jimenez-Reina, L., Ceballos, E., Leon, J., and Munoz, E., Susceptibility of HIV-1-TAT transfected cells to undergo apoptosis, *Oncogene,* 18, 7543–7551, 1999.

67. Bartz, S.R. and Emerman, M., Human immunodeficiency virus type 1 Tat induces apoptosis and increases sensitivity to apoptotic signals by up-regulating FLICE/caspase-8, *J. Virol.,* 73, 1956–1963, 1999.

68. Seve, M., Favier, A., Osman, M., Hernandez, D., Vaitaitis, G., Flores, N.C., McCord, J.M., and Flores, S.C., The human immunodeficiency virus-1 Tat protein increases cell proliferation, alters sensitivity to zinc chelator-induced apoptosis, and changes Sp1 DNA binding in HeLa cells, *Arch. Biochem. Biophys.,* 361, 165–172, 1999.

69. Kumar, A., Dhawan, S., Mukhopadhyay, A., and Aggarwal, B.B., Human immunodeficiency virus-1-Tat induces matrix metalloproteinase-9 in monocytes through protein tyrosine phosphatase-mediated activation of nuclear transcription factor NF B, *FEBS Lett.,* 462, 140–144, 1999.

70. Bartz, S.R. and Emerman, M., Human immunodeficiency virus type 1 Tat induces apoptosis and increases sensitivity to apoptotic signals by up-regulating FLICE/caspase-8, *J. Virol.,* 73, 1956–1963, 1999.

71. Re, M.C., Furlini, G., Vignoli, M., Ramazzotti, E., Zauli, G., and La Placa, M., Antibody against human immunodeficiency virus type 1 (HIV-1) Tat protein may have influenced the progression of AIDS in HIV-1-infected hemophiliac patients, *Clin. Diagn. Lab. Immunol.,* 3, 230–232, 1996.

72. White, A.C., Jr., Chappell, C.L., Hayat, C.S. et al., Paromomycin for cryptosporidiosis in AIDS: a prospective, double-blind trial, *J. Infect. Dis.,* 170, 419–424, 1994.

73. Rossignol, J.F., Ayoub, A., and Ayers, M.S., Treatment of diarrhea caused by *Cryptosporidium parvum*: a prospective randomized, double-blind, placebo-controlled study of nitazoxanide, *J. Infect. Dis.,* 184, 103–106, 2001.

74. Griffiths, J.K., Treatment for AIDS-associated cryptosporidiosis, *J. Infect. Dis.,* 178, 915–916, 1998.

75. White, A.C., Jr., Cron, S.G., and Chappell, C.L., Paromomycin in cryptosporidiosis, *Clin. Infect. Dis.,* 32, 1516–1517, 2001.

76. Hewitt, R.G., Yiannoutsos, C.T., Higgs, E.S., Carey, J.T., Geisler, P.J., Soave, R., Rosenber, R., Vazquez, G.J., Wheat, L.J., Fass, R.J., Antoninievic, Z., Walawander, A.L., Flanigan, T.P., and Bender, J.F., Paromomycin: no more effective than placebo for treatment of cryptosporidiosis in patients with advanced human immunodeficiency virus infection, *Clin. Infect. Dis.,* 3, 1084–1092, 2000.

77. Woods, K.M. and Upton, S.J., Efficacy of select antivirals against *Cryptosporidium parvum in vitro,* *FEMS Microbiol. Lett.,* 168, 59–63, 1998.

78. Wilcox, C.M. and Monkemuller, K.E., Hepatobiliary diseases in patients with AIDS: focus on AIDS cholangiopathy and gallbladder disease, *Dig. Dis.,* 16, 205–213, 1998.

79. Kanegane, H., Nomura, K., Miyawaki, T., and Tosato, G., Biological aspects of Epstein-Barr virus (EBV) infected lymphocytes in chronic active EBV infection and associated malignancies, *Crit. Rev. Oncol.,* 44, 239–249, 2002.

80. Knecht, H., Berger, C., Al-Homsi, A.S., McQuain, C., and Brouseet, P., Epstein-Barr virus oncogenesis, *Crit. Rev. Oncol./Hemat.,* 26, 117–135, 1997.

81. Jones, J.F., Shurin, S.J., Abramowsky, C., Tubbs, R.R., Sciotto, C.G., Wahl, R., Sands, J., Gottman, D., Katz, B.Z., and Sklar, J., T-cell lymphomas containing Epstein-Barr virus DNA in patients with chronic Epstein-Barr virus infection, *N. Engl. J. Med.,* 318, 733–741, 1988.

82. Burkitt, D., A sarcoma involving the jaws in African children, *Br. J. Surg.,* 45, 218–223, 1958.

83. Magrath, I.T. and Bhatia, K., Pathogenesis of small noncleaved cell lymphomas (Burkitt's lymphoma), in *The Non-Hodgkin's Lymphomas,* 2nd ed., Magrath, I.T., Ed., Arnold, New York, 1997, pp. 385–409.

84. Knowles, D.M., Immunodeficiencies-associated lymphoproliferative disorders, *Mod. Pathol.,* 12, 200–217, 1999.

85. Okano, M., Nakanishi, M., Taguchi, Y., Sakiyama, Y., and Matsumoto, S., Primary immunodeficiency diseases and Epstein-Barr virus-induced lymphoproliferative disorders, *Act. Paediatr. Jpn.,* 34, 385–392, 1992.

86. Hoshino, Y., Kimura, H., Kuzushima, K., Tsurumi, T., Nemoto, K., Kikuta, A., Nishiyama, Y., Kojima, S., Matsuyama, T., and Morishima, T., Early intervention in post-transplant lymphoproliferative disorders based on Epstein-Barr viral load, *Bone Marrow Transplant.,* 26, 199–201, 2000.

87. Aoki, Y., Joknes, K.D., and Tosato, G., Kaposi's sarcoma—associated herpesvirus-encoded interleukin-6, *J. Hematother. Stem Cell Res.,* 9, 137–145, 2000.

88. Harabuchi, Y., Yamanaka, N., Kataura, A., Imai, S., Kinoshita, T., Mizuno, F., and Osato, T., Epstein-Barr virus in nasal T-cell lymphomas in patients with lethal midline granuloma, *Lancet,* 335, 128–130, 1990.

89. Eliopoulos, A.G. and Young, L.S., LMP1 structure and signal transduction, *Sem. in Cancer Biol.,* 11, 435–444, 2001.

90. Liebowitz, D., Wang, D., and Kieff, E., Orientation and patching of the latent infection membrane protein encoded by Epstein-Barr virus, *J. Virol.,* 58, 233–237, 1986.

91. Moorthy, R.K. and Thorley-Lawson, D.A., Biochemical, genetic and functional analyses of the phosphorylation sites of the Epstein-Barr virus-encoded oncogenic latent membrane protein LMP-1, *J. Virol.,* 67, 2637–2645, 1993.

92. Wang, D., Liebowitz, D., and Kieff, E., An ENV membrane protein expressed in immortalized lymphocytes transforms established rodent cells, *Cell,* 43, 831–840, 1985.

93. Peng, M. and Lundgren, E., Transient expression of the Epstein-Barr virus LMP1 gene in human primary B cells induces cellular activation and DNA synthesis, *Oncogene,* 7, 1775–1782, 1992.

94. Huen, D.S., Henderson, S.A., Croom-Carter, D., and Rowe, M., The Epstein-Barr virus latent membrane protein-1 (LMP1) mediates activation of NF-B and cell surface phenotype via two effector regions in its carboxy-terminal cytoplasmic domain, *Oncogene,* 10, 549–560, 1995.

95. Kieff, E., Epstein-Barr virus and its replication, in *Fundamental Virology,* Fileds, B.N., Knipe, D.M., Howley, P.M., Knipe, D.M., and Howley, P.M., Eds., Lippincott-Raven, Philadelphia, 1996, pp. 1109–1163.

96. Wang, D., Liebowitz, D., Wang, F., Gregory, C., Rickinson, A., Larson, R., Springer, T., and Kieff, E., Epstein-Barr latent infection membrane protein alters the human B-lymphocyte phenotype: deletion of the amino terminus abolishes activity, *J. Virol.,* 62, 4173–4184, 1988.

97. Laherty, C.D., Hu, H.M., Opipari, A.W., Wang, F., and Dixit, V.M., The Epstein-Barr virus LMP1 gene product induces A20 zinc finger protein expression by activatin nuclear factor B, *J. Biol. Chem.,* 34, 24157–24160, 1992.

98. Hammarskjold, M.L. and Simurada, M.C., Epstein-Barr virus latent membrane protein transactivates the human immunodeficiency virus type 1 long terminal repeat through induction of NF-B activity, *J. Virol.,* 66, 6496–6501, 1992.

99. Herrero, J.A., Mathew, P., and Paya, C.V., LMP1 activates NF-B by targeting the inhibitory molecule iKBα, *J. Virol.,* 69, 2168–2174, 1995.

100. Li, H.P. and Chang, Y.S., Epstein-Barr virus latent membrane protein 1: structure and function, *J. Biomed. Sci.,* 10, 490–504, 2003.

101. Eliopoulos, A.G., Gallagher, N.J., Blake, S.M., Dawson, C.W., and Young, L.S., Activation of the p38 mitogen-activated protein kinase pathway by Epstein-Barr virus encoded latent membrane protein 1 coregulates interleukin 6 and interleukin 8 production, *J. Biol. Chem.,* 274, 16085–16096, 1999.

102. Izumi, K.M., Kaye, K.M., and Kieff, E.D., The Epstein-Barr virus LMP1 amino acid sequence that engages tumor necrosis factor receptor associated factors is critical for primary B lymphocyte growth transformation, *Proc. Natl. Acad. Sci. U.S.A.,* 94, 1447–1452, 1997.

103. Eliopoulos, A.G., Blake, S.M., Floettmann, J.E., Rowe, M., and Young, L.S., Epstein-Barr virus encoded latent membrane protein 1 activates the JNK pathway through its extreme C terminus via a mechanism involving TRADD and TRAF2, *J. Virol.,* 73, 1023–1035, 1999.

104. Baker, S.J. and Reddy, E.P., Modulation of life and death by the TNF receptor superfamily, *Oncogene,* 17, 3261–3270, 1998.
105. Floettmann, J.E., Ward, K., Rickinson, A.B., and Rowe, M., Cytostatic effect of Epstein-Barr virus latent membrane protein analyzed using tetracycline-regulated expression in B cell lines, *Virology,* 223, 29–40, 1996.
106. Cuomo, L., Ramquist, T., Trivedi, P., Wang, F., Klein, G., and Masucci, M.G., Expression of the Epstein Barr virus (EBV) encoded membrane protein LMP1 impairs the *in vitro* growth, clonability, and tumorigenicity of an EBV-negative Burkitt lymphoma line, *Int. J. Cancer,* 51, 949–955, 1992.
107. Rowe, M., Lear, A.L., Croom-Carter, D., Davies, A.H., and Rickinson, A.B., Three pathways of Epstein-Barr virus gene activation from EBNA1-positive latency in B lymphocytes, *J. Virol.,* 66, 122–131, 1992.
108. Yates, J.L., Warren, N., and Sugden, B., Stable replication of plasmids derived from Epstein-Barr virus genome, *Nature,* 310, 812–815, 1984.
109. Niller, H.H., Salamon, D., Ilg, K., Koroknai, A., Banati, F., Schwarzmann, F., Wolf, H., and Minarovits, J., EBV-associated neoplasms: alternative pathogenetic pathways, *Med. Hypoth.,* 62, 387–391, 2004.
110. Wilson, J.B. and Levine, A.J., The oncogenic potential of Epstein-Barr virus nuclear antigen 1 in transgenic mice, *Curr. Top. Microbiol. Immunol.,* 182, 375–384, 1992.
111. Rooney, C., Howe, J.G., Speck, S.H., and Miller, G., Influences of Burkitt's lymphoma and primary B cells on latent gene expression by the nonimmortalizing PJ3-HR-1 strain of Epstein-Barr virus, *J. Virol.,* 63, 1531–1539, 1989.
112. Yalamanchili, R., Tong, X., Grossman, S., Johannsen, E., Mosialos, G., and Kieff, E., Genetic and biochemical evidence that EBNA2 interaction with a 63-kDa cellular GTG-binding protein is essential for B lymphocyte growth transformation by EBV, *Virology,* 204, 634–641, 1994.
113. Dambaugh, T., Hennessy, K., Chamnankit, L., and Kieff, E., U2 region of Epstein-Barr virus DNA may encode Epstein-Barr nuclear antigen-2, *Proc. Natl. Acad. Sci. U.S.A.,* 81, 7632–7636, 1984.
114. Rickinson, A.B., Young, L.S., and Rowe, M., Influence of the Epstein-Barr virus nuclear antigen EBNA2 on the growth phenotype of virus transformed B cells, *J. Virol.,* 61, 1310–1317, 1987.
115. Farrell, C.J., Lee, J.M., Shin, E.-C., Cebrat, M., Cole, P.A., and Hayward, S.D., Inhibition of Epstein-Barr virus induced growth proliferation by a nuclear antigen EBNA2-TAT peptide, *Proc. Nat. Acad. Sci. U.S.A.,* 101, 4625–4630, 2004.
116. Krauer, K.G., Burgess, A., Buck, M., Flanagan, J., Sculley, T.B., and Gabrielli, B., The EBNA-3 gene family proteins disrupt the G2/M checkpoint, *Oncogene,* 23, 1342–1353, 2004.
117. Cannell, E.J., Farrell, P.J., and Sinclair, A.J., Epstein-Barr virus disrupts the normal cell pathway to regulate Rb activity during immortalisation of primary B cells, *Oncogene,* 13, 1413–1421, 1996.
118. Parker, G.A., Touitou, R., and Allday, M.J., Epstein-Barr virus EBNA3C can disrupt multiple cell cycle checkpoints and induce nuclear division divorced from cytokinesis, *Oncogene,* 19, 700–709, 2000.
119. Alfieri, C., Birkenbach, M., and Kieff, E., Early events in Epstein-Barr virus infection of human B lymphocytes, *Virology,* 181, 595–608, 1991.
120. Matsuda, G., Nakajima, K., Kawaguchi, Y., Yamanashi, Y., and Hiria, K., Epstein-Barr virus (EBV) nuclear antigen leader protein (EBNA-LP) forms complexes with a cellular anti-apoptosis protein Bcl2 or its ENV counterpart BHRF1 through hHS1-associated protein X-1, *Microbiol. Immunol.,* 47, 91–99, 2003.
121. Yin, X.M., Oltvai, A.N., and Korsmeyer, S.J., BH1 and BH2 domains of Bcl-2 are required for inhibition of apoptosis and heterodimerization with Bax, *Nature,* 369, 321–323, 1994.
122. Swaminathan, S., Molecular biology of Epstein-Barr virus and Kaposi's sarcoma-associated herpesvirus, *Semin. Hematol.,* 40, 107–114, 2003.
123. Loughrey, M., Trivett, M., Lade, S., Murray, W., Turner, H., and Waring, P., Diagnostic application of Epstein-Barr virus-encoded RNA *in situ* hybridization, *Pathology,* 26(4), 301–308, 2004.
124. Ling, P.D., Vilchez, R.A., Keitel, W.A., Poston, D.G., Peng, R.S., White, Z.S., Visnegarwala, F., Lewis, D.E., and Butel, J.S., Epstein-Barr virus DNA loads in adult human immunodeficiency virus type 1-infected patients receiving highly active antiretroviral therapy, *Clin. Infect. Dis.,* 37, 1244–1249, 2004.
125. Glaser, S.L., Clarke, C.A., Gulley, M.L., Craig, F.E., DiGiuseppe, J.A., Dorfman, R.F., Mann, R.B., and Ambinder, R.F., Population-based patterns of human immunodeficiency virus-related Hodgkin lymphoma in the Greater San Francisco Bay Area, 1988–1998, *Cancer,* 98, 300–309, 2003.

Section 4

Therapeutic Issues

25 Direct Effects of Anti-HIV Therapeutics on Apoptosis

David Schnepple
Andrew D. Badley

CONTENTS

INTRODUCTION

The most prevalent pharmacotherapeutic paradigm for treatment and management of human immunodeficiency virus (HIV)-related disease remains HAART (highly active antiretroviral therapy). This concurrent administration of multiple antiviral drugs simultaneously targets multiple aspects of the viral life cycle in order to both suppress viral replication and reduce the likelihood of generating viable escape mutants in treated individuals. The success of this approach has been well documented[1] and is attested to by the thousands of HIV-positive individuals with suppressed levels of viral infection on such regimens, although novel compounds are continuously sought to counter escape mutants and reduce the side effects of medication.

Current classes of antiretroviral drugs available for the treatment and management of HIV infection include nucleoside and nonnucleoside reverse transcriptase inhibitors, HIV protease inhibitors (PIs), and fusion inhibitors. The two predominant viral targets for anti-HIV therapeutics have been reverse transcriptase (RT) and HIV protease (PR). Both are essential enzymes in the viral life cycle, and inhibition of either or both have proven to be an effective strategy to suppress viral replication and extend the lives of those infected.

Compounds targeting reverse transcriptase interfere with viral deoxyribonucleic acid (DNA) production, preventing all downstream events, including viral genome integration and messenger ribonucleic acid (mRNA) transcript production. This occurs either through the inhibition of catalytic activity (nonnucleoside reverse transcriptase inhibitors) or through induction of DNA chain termination by incorporation of nonnatural nucleotides that lack further linkage potential (nucleoside reverse transcriptase inhibitors).

Protease inhibitors exert their primary effect through the inhibition of HIV polyprotein cleavage, preventing virion maturation at the end of the life cycle. This results in production of noninfectious particles as well as interferes in the initial polyprotein cleavages necessary for production of individual viral proteins.

Entry inhibitors, broadly defined to include all drugs that interfere with virion attachment and fusion, have only recently come to market, and the first fusion inhibitor T-20 (fuzeon) received

441

U.S. Food and Drug Administration (FDA) approval in 2003. This compound acts to competitively inhibit HIV gp41 from self-associating and prevents the mechanical formation necessary to enable virion–cell fusion. Results to date show this class of drugs to be quite effective, with reduced side effect profiles[2,3] in many respects compared with traditional classes of antiretrovirals, and thus this class of therapeutics is expected to receive increased attention and grow significantly in coming years.

Finally, there are several "miscellaneous" drugs currently prescribed for the treatment of HIV disease. These include hydroxyurea, which acts to reduce the available nucleotide pool for use in DNA synthesis, thereby serving to antagonize viral replication; interleukin (IL)-2, currently used to stimulate lymphocyte replication and activate remaining lymphocytes; and thalidomide, an anti-tumor necrosis factor (TNF) medication that prevents TNF-induced inflammatory processes and TNF-induced cell death, thereby sparing additional lymphocytes and increasing T cell numbers.

Although an ideal pharmacological drug has high selectivity and specificity, the reality is that most compounds in use have multiple targets for interaction. Anti-HIV medications are no different, and in addition to their predominant effects on viral targets, also act on a variety of cellular targets. In this chapter, we will address the current understanding of each of the four major classes of HIV therapeutics with regard to their apoptotic impact. We will discuss likely mechanisms for the impacts of these drugs, and highlight specific activities of compounds, when known.

NUCLEOTIDE REVERSE TRANSCRIPTASE INHIBITORS (NRTIs)

Each of the currently approved therapeutics in this class — abacavir, didanosine, emtricitabine, lamivudine, stavudine, zalcitabine, and zidovudine — exerts its activity after intracellular conversion to an active triphosphate form. These active compounds reduce viral replication by competitively substituting for natural deoxynucleoside triphosphates, causing premature DNA-chain termination when incorporated. Human DNA polymerases generally remain unaffected by this class of compounds, although mitochondrial DNA polymerase-γ is an exception.[4,5] Most long-term toxicity seen with this class of therapeutics is thought to result from mitochondrial DNA inhibition/depletion,[6] possibly coupled with decreases in adenosine triphosphate (ATP) production or increases in reactive oxygen species generation.[7] Supporting a connection between such toxicity and apoptosis, it has been shown that such disturbances are efficient to induce apoptosis in a variety of cells and cell lines upon exposure.

Research into the potential cytotoxic properties of NRTIs began when (in an effort to explore the roots of muscular toxicity and mitochondrial myopathy seen in HIV patients) it was noted that histological evidence of mitochondrial damage correlated with long-term or heavy cumulative doses of zidovudine.[8,9] Further studies by Jay et al. hinted at a dose-dependent phenomenon through the demonstration that low-dose-AZT-receiving patients lacked typical features of zidovudine myopathy.[10] Depletion of muscle mitochondrial DNA was first noted in zidovudine (AZT) monotherapy patients,[11] and this was coupled with results showing impaired cytochrome c oxidase activity, altered morphology,[12] and altered oxidative phosphorylation enzymes in treated mitochondria.[13] Shortly thereafter, it was demonstrated that these were not phenomena related simply to HIV infection, as such deficiencies were absent in HIV-positive patients not receiving AZT.[14] Also, although this research did not directly demonstrate or assess drug-induced apoptosis, it provided a theoretical basis to investigate mitochondrial-dependent apoptosis modulation under these therapies.

Initial studies assessing apoptosis in short-term cultures of HIV-infected peripheral blood mono-nuclear cells (PBMCs) failed to suggest significant differences between AZT-treated and nontreated PBMCs,[15] but confounders related to infection may have obscured examination of direct pharmacologic effects. AZT was later shown to induce biochemical and morphological changes characteristic of apoptosis *in vitro* using the mouse myeloma cell line Sp2/O.[16] Likewise, the treatment of noninfected, organotypic cocultures of spinal ganglia, spinal cord, and skeletal muscle from fetal rats with various concentrations of AZT also demonstrated duration-dependent effects on their morphology characteristic of apoptosis.[17] Further studies by Viora et al. helped explain any previous conflict of research results by indicating that even though stimulated PBMCs seemed to be resistant to apoptosis in short-term culture, both zidovudine and zalcitabine (ddC) inhibited cell growth and cell cycling in CEM and phytohaemagglutinin (PHA)-stimulated PBMCs, and both compounds likewise induced apoptosis (in addition to growth and cycling inhibition) in CEM cell line cells.[18] Of these two NTRIs examined, ddC seemed to be more efficient at reducing cell growth and inducing apoptosis, although a combination of the two seemed to display additive effects. This differential toxicity also matched the relative affinities of the two compounds for mitochondrial DNA polymerase-γ. Interestingly, limited research suggested that chronically HIV-infected T cell lines (MOLT-4/IIIB and CEM/ROD) are selectively more sensitive to apoptosis induced by stavudine than are their parental lines.[19] Because this increased sensitivity to treatment is still presumed to be based on drug interference with DNA polymerase-γ, it hints at additional or enhanced vulnerabilities to mitochondrial damage potentially exploitable in infected cells.

What was more conclusively shown, in multiple large, prospective, open label, and randomized switch studies, is that stavudine therapy (and to a lesser extent, zidovudine therapy) is associated with greater patient risk for developing lipoatrophy.[20,21] Lipoatrophy, which is a fat-wasting syndrome observed in some HIV patients, is recognized to occur most frequently in those patients taking certain NRTIs,[22,23] although individuals receiving particular protease inhibitor therapies are also at risk for developing this syndrome.[24,25] T. Saint-Marc et al. have reported that the use of stavudine significantly correlates with fat wasting in both NRTI and protease inhibitor treatment groups, with odds ratios of 4.13 and 2.08, respectively, compared with zidovudine usage.[26] Additionally, the severity of fat wasting seems to correlate well with the duration of NRTI therapy, which is further confirmed by direct observation of the correlation between mitochondrial DNA depletion and increased apoptotic cells in fat biopsies taken from treated patients.[27]

Therefore, the majority of data available on the direct effects of NRTIs point to some limited ability of NRTIs to induce apoptosis in cells via their mitochondrial toxicity, at high concentrations or over long-term administration. This hypothesis can be examined by comparing the relative mitochondrial polymerase-γ inhibition versus apoptotic potential of currently available NRTIs. Ranking compounds for their ability to interfere with/inhibit DNA polymerase-γ was determined as (shown in decreasing order): zalcitabine, didanosine, stavudine, lamivudine, zidovudine, abacavir, and emtricitabine.[28–32] What remains to be examined is a full account of the relative apoptotic potential across this class of compounds. To date, results have been suggestive that this correlation is tight.[18,19,32] Additional evidence that mitochondrial toxicity may be the apoptotic mechanism functioning in NRTI therapy is suggested by noting common side effects of long-term therapy, which include peripheral neuropathy, myopathies, pancreatitis, and bone marrow suppression, that closely resemble certain heritable mitochondrial disease phenotypes.[33,34] Although the exact mechanisms connecting NRTI-induced mitochondrial toxicity with resultant apoptosis remain unclear, several pathways are possible (see Figure 25.1) and suggested in the models below:

Mitochondrial DNA polymerase-γ interference: The interference of NRTIs with mitochondrial DNA (mtDNA) replication may result in increased mtDNA mutations, deletions, or simply competitive inhibition and continued replication prevention. All these could engender both enhanced vulnerability to and accumulation of additional mtDNA damage as well as directly impact the expression and function of enzymatic subunits encoded by the small mitochondrial genome. Indirect impairment of mitochondrial enzymes through DNA damage (production, replacement problems) would lead to lowered ATP production[35] and uncoupling of oxidative phosphorylation,[36] finally triggering apoptosis.

Electron transport chain or ATP production disruption: Alternately, NRTIs may directly impact the enzymes and processes of energy formation in mitochondria. This is suggested by one study that found that AZT can bind to and inhibit the enzyme adenylate kinase in vitro, resulting in a 30% decrease in ATP production at concentrations of 10 μM zidovudine.[37] Although this concentration is unlikely to be physiological, other cell-culture and cell-free

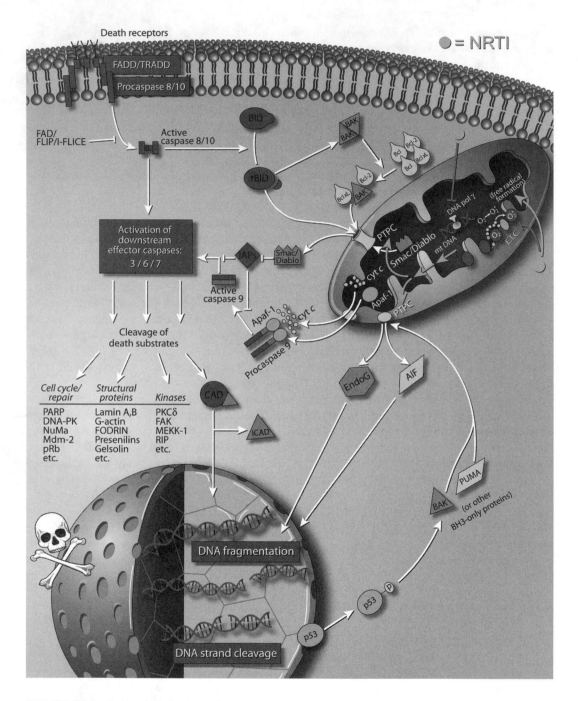

FIGURE 25.1 Apoptotic influences of Nucleoside Reverse Transcriptase Inhibitors (NRTIs). As described in the text, NRTIs likely impact cellular apoptosis through a number of potential mitochondrial interactions. Interference with mitochondrial DNA polymerase gamma, disruption of the electron transport chain (and ATP production), and enhanced reactive radical formation by this class of HIV therapeutics are all proposed to feed forward toward mitochondrial deregulation, which results in the release of pro-apoptotic factors.

studies have shown that AZT also inhibits the adenosine diphosphate (ADP)/adenosine triphosphate (ATP) translocator[38] and demonstrates a dose-dependent inhibition of both cytochrome *c* reductase and NADH-linked respiration.[39]

Reactive radical formation: Finally, NRTIs may exert their toxic effects through the generation of oxygen-free radicals that attack mtDNA and induce mutation.[40] This hypothesis was supported by two studies in which zidovudine-treated mice were killed and their liver levels of oxidized guanosine were determined as a measure of oxidant damage. In both studies, greater levels of oxidized guanosine were found in the AZT-treated mice than in controls, and the administration of antioxidants inhibited this effect.[40,41] Although the pathway of NRTI-promoted radical formation is lacking, in this scenario damage to the mtDNA results from oxidation-induced mutation rather than from substitution-induced termination or misincorporation.

In the interest of preserving perspective, it deserves mention that although there are studies indicating the proapoptotic effects of NRTIs, this is due in part to the success of NRTIs as therapeutics, and the careful clinical examinations of NRTI toxicity limits. When used in practice for the management of HIV infection, multiple drugs are often prescribed together, or therapies are alternated, in a manner that intentionally reduces the incidence or severities of drug toxicities. The result of administration within the therapeutic index for NRTIs has been resoundingly beneficial for patients, and it is certain that in the context of HIV infection, wherein large numbers of uninfected "bystander" T cells are likely to be undergoing apoptosis, administration of NRTIs to these patients reduces their spontaneous lymphocytic apoptosis levels.[42] This reduction in apoptosis generally mirrors a concurrent suppression of viral load and is, therefore, believed to be an indirect effect mediated through a reduction in the quantity of proapoptotic viral proteins and virions. However, there was a recent report suggesting a potential antiapoptotic effect attributable to a model system for AZT, observed in treatment of maedi-visna-virus-infected sheep choroid plexus cells.[43] Researchers observed the inhibition of morphological characteristics, as well as inhibition of caspase-3, -8, and -9 cascades, DNA laddering, and mitochondrial apoptosis-inducing factor (AIF) leakage, although whether this represents a direct protective effect of AZT or, more likely, an indirect result of the reduction in viral products and their downstream effects remains to be seen. Further research is needed, but the blockade of both caspase-dependent and caspase-independent pathways suggested in this report is intriguing.

There is another structurally distinct class of NRTIs that differs from nucleoside inhibitors in that they possess a phosphonyl moiety that makes these compounds resistant to enzymatic hydrolysis. Prototypes for this class of drug include tenofovir disoproxil fumarate and adefovir dipivoxil, and these compounds are able to circumvent the obligatory intracellular phosphorylation step of nucleoside inhibitors before being pharmacologically active. Unfortunately, they also demonstrate some inhibitory activity for DNA polymerase-α, -δ, and -ϵ, which makes them less specific for viral inhibition and reciprocally more toxic.[44] Current research on this class of compounds demonstrates that they can be very efficient at inducing differentiation of *in vitro* tumor-cell models, possess antitumor activity *in vivo*, and cause apoptosis in a wide variety of cell lines.[44–46]

Although clear experiments are lacking, it is likely that such compounds induce apoptosis through the same mechanism(s) as NRTIs. In both a nasopharyngeal carcinoma tumor xenograft mouse model (in which EBV is present) and in MTT assays of the EBV-positive epithelial cell line NPC-KT, to apoptosis as indicated by DNA fragmentation assays and evidence of poly-ADP ribose polymerase (PARP) cleavage, was the result of treatment with micromolar amounts of several phosphonated nucleotide analogues.[47] The observed apoptotic effects in these experiments are believed to be independent of EBV-polymerase inhibition, as that protein is not expressed in latently infected NPCs.[48] The basis for apoptotic activity in this class of compounds is their broader effects on various cellular DNA polymerases. Impacting and decreasing nuclear DNA as well as mitochondrial DNA reasonably exacerbate the toxicities described above for mitochondrial DNA polymerase interference. Although only two nucleotide compounds are currently approved for HIV therapy, there are likely to be further refinements and advances in these phosphonated compounds, and with them will come further elucidation of their abilities to induce apoptosis.

NONNUCLEOSIDE REVERSE TRANSCRIPTASE INHIBITORS (NNRTIs)

NNRTIs are structurally distinct from NRTIs, and rather than inducing DNA-chain termination through the substitution of natural nucleotides, NNRTIs bind to a nonsubstrate-specific site on the HIV reverse transcriptase enzyme. These binding sites, situated physically close to the enzyme active site, and induce on ligation an approximately 2 Å shift in the position of the active site of the enzyme that results in a dramatic reduction in the catalytic activity.[49–52] There is a wide variety of chemical structures and classes that display activity against HIV-1 reverse transcriptase, and because NNRTIs do not necessarily resemble nucleosides in structure, they do not generally present the same issues of toxicity resulting from interactions with human DNA polymerases as do NRTIs. Although there are more than 30 structurally distinct classes of RT inhibitors, they all seem to function through the same mechanism. This remains a disadvantage during treatment with NNRTIs, because only a minor resistance mutation needs to emerge to effectively neutralize all NNRTIs as a treatment option (until treatment substitution or interruption can allow for reversion to wild type). There are currently three FDA-approved NNRTIs: delavirdine, efavirenz, and nevirapine, and although there is an abundance of data from clinical trials of these and other NNRTIs, there is unfortunately a paucity of published literature specifically examining the apoptotic effects of this anti-HIV drug class.[53]

One of the few studies to address this indicated that at least one of the NNRTIs, efavirenz, possesses the ability to modulate apoptosis.[54] Jurkat T cells cultured with physiologically relevant micromolar concentrations of efavirenz (EFV) showed dramatic (70%) inhibition of growth rates compared with control cultures as well as a large increase in cell death. Additionally, by several parameters, EFV-treated cells appeared apoptotic in contrast to control cells (mitochondrial membrane potential changes, Annexin-V staining, histone–DNA complex changes, high caspase-3 activity, and inhibition of the observed cell death by administration of caspase inhibitors). The EFV-induced cell death also was found to be enhanced by coadministration of AZT, and these findings were extended to include PBMCs from HIV-negative donors upon investigation.[54] We reason that such *in vitro* results may simply be masked *in vivo* by the overall rise in CD4 counts after HAART initiation, when viral suppression alleviates the major cause of CD4 apoptosis, and, in fact, some studies have begun to suggest that NNRTIs may induce measurable apoptosis *in vivo*. For example, replacement of the protease inhibitor with an NNRTI in stable individuals on HAART regimens consisting initially of two NRTIs and a protease inhibitor was demonstrated to lead to a decline in CD4 lymphocyte numbers.[55,56]

The available evidence suggests that at least one member of this class possesses apoptotic modulatory activity; however, because this class of therapeutics is so diverse in its molecular structure, generalizations between members of this class are premature. The secondary targets of EFV are unclear, as is whether the other NNRTIs likewise impact apoptosis. Additional studies ideally will address this lack of understanding as well as strive to elucidate the target for this proapoptotic effect of EFV.

PROTEASE INHIBITORS (PIs)

PIs target the aspartyl proteinase of HIV, inhibiting the cleavage of viral polyproteins and the subsequent generation of individual viral proteins. This prevents the formation of functional virions, although it does not prevent upstream infection events of parental virions, including viral DNA integration and replication. When applied as monotherapy, PIs induce dramatic suppression of viral-load levels, although resistance mutations develop that permit viral escape from such inhibition. Such PI-resistance mutations are generally compound specific, however, and thus allow for therapeutic substitution within the class. Additionally, it seems a general rule that the selection of resistance mutations in the protease gene that allow for escape also produce a virus with notably less fitness than wild type, and there is generally rapid reversion to wild-type strains during therapeutic interruptions. Combined in regimens with reverse transcriptase inhibitors or other PIs, this class of drugs has been clearly successful in the suppression of viral load and responsible for large increases in patient survival periods, and with proper stewardship and application of these combination therapies, the potential seems to exist to delay the emergence of resistant/escape HIV strains indefinitely.

At present, there are eight approved PIs (saquinavir, ritonavir, indinavir, nelfinavir, amprenavir, lopinavir, atazanavir, and fosamprenavir), and several more are under development. Although each is associated with particular side effects to a greater or lesser degree, as a class, their toxicity is relatively homogencous. Of note, there is a common characteristic of inhibition and/or reduction in cytochrome P-450 enzymes with PI usage, which provides the basis for potentially harmful drug interactions with PIs. However, aside from side effects and potential interactions, there is an accumulating body of evidence, both direct and indirect, to suggest that protease inhibitors may possess antiapoptotic properties apart from their antiviral activity.[42,57-59] Protcase inhibitors have been shown to affect a variety of cellular processes, including inhibition of the proteasome,[60] alteration of antigen presentation and cytotoxic T lymphocyte (CTL) activity,[61] and enhancement of all-trans retinoic acid signaling.[62] Alteration of lipoprotein metabolism,[63] preadipocyte differentiation,[64] and expression of CD36 have also been noted.[65] Additionally, nelfinavir has been suggested to induce DNA strand cleavage in fully differentiated 3T3-L1 adipocytes,[66] and apoptotic adipocytes have been observed in fat biopsies taken from protease inhibitor therapy receipients.[67] Together, this suggests that some combination of these effects may be partially responsible for the lipid abnormalities seen in some patients on PI-based HAART.[56]

Apart from these studies, there have been few specific examinations of protease inhibitors that suggest static or proapoptotic effects. Two of these came from Ikezoe et al., who initially demonstrated that ritonavir, saquinavir, and indinavir induced the growth arrest of the myelocytic leukemia cell lines HL-60, NB4, and UF-1 and suggested that this effect was due to enhancement of all-trans retinoic acid signaling.[68] Later research suggested that other multiple myeloma cells (U266, RPMI8226, and ARH77, as well as fresh multiple myeloma (MM) cells from patients) were subject not only to growth arrest but also to apoptosis induction by treatment with ritonavir, saquinavir, and nelfinavir (but not indinavir). Treatment with these PIs (except indinavir) was found to block IL-6-stimulated STAT-3 and ERK-1/2 phosphorylation, whereas ritonavir additionally was found to inhibit vascular endothelial growth factor. Notably, treatment did not affect survival of normal B cells and colony formation of myeloid committed cells.[69] Conversely, although Sgardari et al. demonstrated a similar capacity for PIs to prevent or treat proliferative lesions in mice, they found that this effect belonged to both saquinavir and indinavir and that the effect occurred downstream of vascular endothelial growth factor (VEGF, itself a target of STAT-3), through the activation of matrix metalloproteinase-2 and subsequent inhibition of invasion and not as a result of any direct effect on cell growth or viability.[70]

Additional antineoplastic effects were noted by ritonavir, which decreased endothelial production of TNF-α, IL-6, IL-8, VEGF, and adhesion molecules and inhibited NF-κB. After treatment with physiologically relevant concentrations, this resulted in apoptosis of Kaposi's sarcoma derived cell lines *in vitro* and in tumor progression in mouse models.[71] Finally, it is worth noting that the reported static or apoptotic effects of protease inhibitors seem to affect cancerous cells and, particularly, invasive cells, harder. This allows speculation that such effects, if applicable *in vivo*, may provide an additional benefit to HIV+ individuals with compromised immune systems, in that PI therapy may boost their lymphocyte levels as well as directly prevent or reduce their chances of developing certain types of cancers. Also, even though there are reports that indinavir induces blockade of stimulation-induced lymphoproliferation in PBMCs,[72] or that there is actually a reduction in total and percent CD4 and CD8 T cell counts after indinavir monotherapy,[73] the majority of data supports that indinavir (and saquinavir) treatment enhances survival of patients' peripheral blood T cells while concurrently restoring T cell proliferative responses to immune stimuli.[74]

The majority of reports indicate that PIs may directly modulate apoptosis; apart from their antiviral effect that indirectly reduces host cell death, several PIs were shown to possess antiapoptotic biologic activity. Ritonavir can reduce spontaneous apoptosis induced by a variety of stimuli,[75,76] and it was suggested initially that this may be mediated through either a reduction in Fas-L expression by stimulated CD4+ cells or a reduction in ICE (caspase-1) expression in both stimulated and unstimulated CD4+ cells, irregardless of infection.[77] There was initially some controversy as to whether the antiapoptotic effects of protease inhibitors were limited to HIV-infected cells[59,78] or whether these effects occurred in both infected and uninfected cells.[76,79–81] However, recent experiments in various sepsis models (another condition involving large amounts of undesired apoptotic cell death) have conclusively demonstrated the ability of protease inhibitors to modulate these processes and protect against cell death.[82]

The mechanism of this activity is likely to occur at the level of the mitochondria. The inhibition of apoptotic death by nelfinavir was not found to relate to inhibition of caspase-1, -3, -6, -7, or -8 activity, and it was not coincident with changes in the mRNA levels of either pro- or antiapoptotic factors, including caspases, death receptors-4/5, BID, Bfl-1, TRAIL, TNF receptor, XIAP, FAP, cIAP-1/2, Bcl-2, and others.[76] Instead, protease inhibitors seem to prevent loss of mitochondrial potential even after apoptotic stimuli, suggesting that they have their effect at the level of the permeability transition pore complex (PTPC) (see Figure 25.2). Alternately, there was a role suggested for saquinavir and indinavir in ameliorating oxidative stress; these PIs inhibited apoptosis in normal and transgenic mice expressing HIV genes under basal conditions and under morphine-stimulated states of oxidative stress, in dose-dependent fashion.[83] The inhibition of superoxide production conceivably suggests that oxidative stress mitigation is another function of PIs.

FUSION INHIBITORS

The most recent class of anti-HIV medications to reach the market is fusion inhibitors, so called because they interfere with the last stage of virion binding and entry processes. There are three basic steps in HIV virion entry, including CD4 binding, co-receptor binding, and hairpin formation/membrane fusion.[84] Each of these steps is currently being heavily targeted for drug development, and there are multiple compounds in clinical trials for each of the three stages. Currently, however, only enfuvirtide (T-20, fuzeon) has received FDA approval.

HIV cell entry requires active conformational changes by gp41, triggered by co-receptor binding, that lead to so-called prehairpin intermediate formations (coiled–coil helical arrangements) among trimers of gp41. These prehairpin structures then drive into and anchor in the cell membrane before folding back at their middle into the hairpin structure. This action of gp41—its amino and carboxyl termini self-attracting- are what draw cell and virion membranes into contact. Although the gp41 trimers are arranged in the "prehairpin intermediate formation," they are vulnerable to interference.[85,86] Enfuvirtide, a 36-amino acid peptide homologous to particular repeat regions of HIV-1 gp41,[87] was designed to competitively bind to the hydrophobic grooves of the gp41 carboxy terminus, preventing association with the gp41 amino-terminal region.

This type of interference, at the last stages of entry, might not have been expected to affect cellular apoptosis, but interestingly, Barretina et al. reported that not only was *ex vivo* treatment of CD4 lymphocytes with T-20 able to completely prevent syncytium formation but T-20 treatment also fully abrogated death of bystander cells with an IC50 of only 40 ng/ml.[88] As mutant virus resistant to T-20 demonstrated the ability to induce bystander cell death in the presence of the drug,

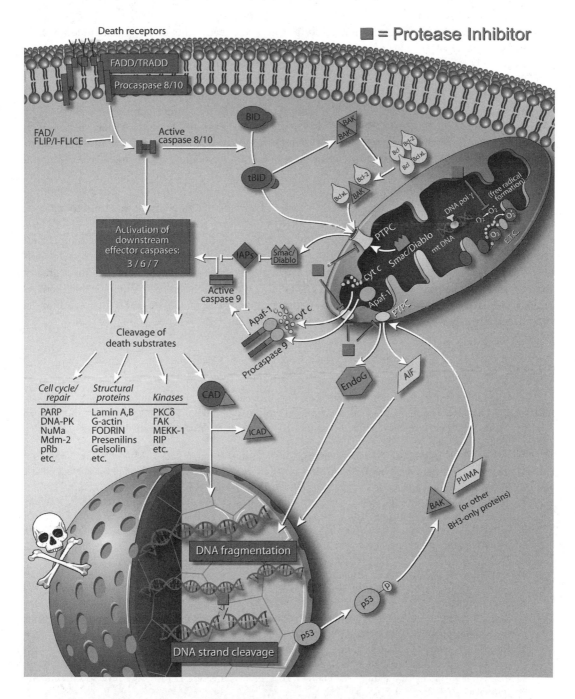

FIGURE 25.2 Apoptotic influences of Protease Inhibitors (PIs). Although there is some suggestion that certain PIs may contribute to DNA strand cleavage (mechanism unknown), most evidence now suggests that PIs may possess anti-apoptotic properties. Such anti-apoptotic effects may be mediated through amelioration of oxidative stress or alternately through a protection of mitochondrial membrane potential (prevention of mitochondrial membrane pore formation and/or release of pro-apoptotic factors).

this suggests that gp41 is involved in mediating single-cell death. This also suggests that any anti-apoptotic property of T-20 may be indirect, through interference with gp120 signaling.

Conversely, others have reported that interfering with HIV binding to either CD4 or co-receptors CXCR4 and CCR5 successfully inhibits bystander apoptosis, whereas T-20 fusion inhibitor displayed no effect.[89] In this system, Holm et al. produced molecularly cloned viruses that differed only in specific *Env* amino acids and tested them for their ability to induce apoptosis in herpesvirus saimiri-immortalized primary CD4 T cells. They observed that amino acid changes resulting in increased *Env* affinity for either CD4 or one of the co-receptors lead to increased abilities to induce bystander apoptosis. Furthermore, the enhanced apoptosis in bystander cells by these mutants could be inhibited by antibodies to CD4, CXCR4, or CCR5, or by the CXCR4 inhibitor AMD3100, but not by T-20.[89] Such results argue that bystander apoptosis is due largely to co-receptor cross-linking; however, further investigations are needed. It should be noted that any compound able to enhance bystander lymphocyte survival in the context of HIV infection will be welcomed.

Future generation fusion inhibitors already are being tested. T-1249 is a current example of a compound that, although apparently more potent, seems to remain—as its forerunner—well tolerated.[2] Two-week monotherapy studies in patients with advanced disease indicate that doses greater than 25 mg once per day significantly raise CD4 cell counts from baseline.[90] Whether this increase results simply from decreased viral load or whether further research indicates these fusion inhibitors can directly prevent bystander apoptosis remains to be seen.

REFERENCES

1. Palella, F.J., Jr., Delaney, K.M., Moorman, A.C., Loveless, M.O., Fuhrer, J., Satten, G.A., Aschman, D.J., and Holmberg, S.D. (1998). Declining morbidity and mortality among patients with advanced human immunodeficiency virus infection. HIV Outpatient Study Investigators. *N Engl J Med* 338(13):853–60.
2. Cohen, C.J., Dusek, A., Green, J., Johns, E.L., Nelson, E., and Recny, M.A. (2002). Long-term treatment with subcutaneous T-20, a fusion inhibitor, in HIV-infected patients: patient satisfaction and impact on activities of daily living. *AIDS Patient Care STDS* 16(7):327–35.
3. Kilby, J.M., Lalezari, J.P., Eron, J.J., Carlson, Cohen, C., Arduino, R.C., Goodgame, J.C., Gallant, J.E., Volberding, P., Murphy, R.L., Valentine, F., Saag, M.S., Nelson, E.L., Sista, P.R., and Dusek, A. (2002). The safety, plasma pharmacokinetics, and antiviral activity of subcutaneous enfuirtide (T-20), a peptide inhibitor of gp41-mediated virus fusion, in HIV-infected adults. *AIDS Res Hum Retroviruses* 18(10):685–93.
4. Pezeshkpour, G., Illa, I., and Dalakas, M.C. (1991). Ultra structural characteristics and DNA immunocytochemisty in human immunodeficiency virus and zidovudine-associated myopathies. *Hum Pathol* 22(12):1281–8.
5. Styrt, B.A., Piazza-Hepp, T.D., and Chikami, G.K. (1996). Clinical toxicity of antiretroviral nucleoside analogs. *Antiviral Res* 31(3):121–35.
6. Reiss, P., Casula, M., de Ronde, A., Weverling, G.J., Goudsmit, J., and Lange, J.M. (2004). Greater and more rapid depletion of mitochondrial DNA in blood of patients treated with dual (zidovudine+didanosine or zidovudine+zalcitabine) vs. single (zidovudine) nucleoside reverse transcriptase inhibitors. *HIV Med* 5(1):11–4.
7. Ozawa, T. (1997). Oxidative damage and fragmentation of mitochondrial DNA in cellular apoptosis. *Biosci Rep* 17(3)237–50.
8. Dalakas, M.C., Illa, I., Pezechkpour, G.H., Laukaitis, J.P., Cohen, B., and Griffin, J.L. (1990). Mitochondrial myopathy caused by long-term zidovudine therapy. *N Engl J Med* 322(16):1098–1105.
9. Chariot, P. and R. Gheradi. (1991). Partial cytochrome c oxidase deficiency and cytoplasmic bodies in patients with zidovudine myopathy. *Neuromuscl Disord* 1(5):357–63.
10. Jay, C., M. Ropka, K. Hench, C. Grady, and M. Dalakas. (1992). Prospective study of myopathy during prolonged low-dose AZT: clinical correlates of AZT mitochondrial myopathy and HIV-associated inflammatory myopathy. *Neurology* 42(Suppl 3):145.

11. Arnaudo, E., Dalakas, M., Shanske, S., Moraes, C.T., DiMauro, S., and Schon E.A. (1991). Depletion of muscle mitochondrial DNA in AIDS patients with zidovudine-induced myopathy. *Lancet* 337(8740):508–10.

12. Mhiri, C., Baudrimont, M., Bonne, G., Geny, C., Degoul, F., Marsac, C., Roullet, E., and Gherardi, R. (1991). Zidovudine myopathy : a distinctive disorder associated with mitochondrial dysfunction. *Ann Neurol* 29(6):606–14.

13. Tomelleri, G., Tonin, P., Spadaro, M., Tilia, G., Orrico, D., Barelli, A., Bonetti, B., Monaco, S., Salviati, A., and Morocutti, C. (1992). AZT-induced itochondrial myopathy. *Ital J Neurol Sci* 13(9):723–8.

14. Chariot, P., Monnet, I., and Gherardi, R. (1993). Cytochrome c oxidase reaction improves histopathological assessment of zidovudine myopathy. *Ann Neurol* 34(4):561–5.

15. Cordiali, Fei, P., Siolmone, M., Viora, M., Vanacore, P., Pugliese, O., Giglio, A., Caprilli, F., and Ameglio, F. (1994). Apoptosis in HIV infection ; protective role of IL-2. *J Biol Regul Homeost Agents* 8(2):60–4.

16. Sailaja, G., Nayak, R., and Antony, A. (1996). Azidothymidine induces apoptosis in mouse myeloma cell line Sp2/0. *Biochem Pharmacol* 52(6):857–62.

17. Schroder, J.M., Kaldenbach, T., and Piroth, W. (1996). Nuclear and mitochondrial changes of co-cultivated spinal cord, spinal ganglia and muscle fibers following treatment with various doses of zidovudine. *Acta Neuropathol (Berl)* 92(2):138–49.

18. Viora, M., DiGienova, G., Rivabene, R., Malorni, W., and Fattorossi, A. (1997). Interference with cell cycle progression and induction of apoptosis by dideoxynucleoside analogs. *Int J Immunopharmacol* 19(6):311–21.

19. Hashimoto, K.I., Tsunoda, R., Okamoto, M., Shigeta, S., and Baba, M. (1997). Stavudine selectively induces apoptosis in HIV type 1-infected cells. *AIDS Res Hum Retroviruses* 13(2):193–9.

20. Martin, A., Smith, D.E., Carr, A., Ringland, C., Amin, J., Emery, S., Hoy, J., Workman, C., Doong, N., Freund, J., Cooper, D.A. (Mitochondrial Toxicity Group). (2004). Reversibility of lipoatrophy in HIV-infected patients 2 years after switching from a thymidine analogue to abacavir: the MITOX Extension Study. *AIDS* 18(7):1029–36.

21. McComsey, G.A., Ward, D.J., Hessenthaler, S.M., Sension, M.G., Shalit, P., Longergan, J.T., Fisher, R.L., Williams, V.C., Hernandez, J.E., Trial to Assess the Regression of Hyperlactatemia and to Evaluate the Regression of Established Lipodystrophy in HIV-1-Positive Subjects (TARHEEL; ESS40010) Study Team. (2004). Improvement in lipoatrophy associated with highly active antiretroviral therapy in human immunodeficiency virus-infected patients switched from stavudine to abacavir or zidovudine: the results of the TARHEEL study. *Clin Infect Dis* 38(2):263–70.

22. Saint-Marc, T. and J.L. Touraine. (1998). The effects of discontinuing stavudine therapy on clinical and metabolic abnormalities in patients suffering from lipodystrophy. *AIDS* 13(15):2188–9.

23. Mallal, S.A., John, M., Moore, C.B., James, I.R., and McKinnon, E.J. (2000). Contribution of nucleoside analogue reverse transcriptase inhibitors to subcutaneous fat wasting in patients with HIV infection. *AIDS* 14(10):1309–16.

24. Lo, J.C., Mulligan, K., Tai, V.W., Algren, H., and Schambelan, M. (1998) Body shape changes in HIV-infected patients. *J Acquir Immune Defic Syndr Hum Retrovirol* 19(3):307–8.

25. Bonfanti, P., Gulisano, C., Ricci, E., Timillero, L., Valsecchi, L., Carradori, S., Pusterla, L., Fortuna, P., Miccolis, S., Magnani, C., Gabbuti, A., Parazzini, F., Martinelli, C., Faggion, I., Landonio, S., Quirino, T., Vigevani, G., Coordinamento Italiano Studio Allergia e Infezione da HIV (CISAI) Group. (2003) Risk factors for lipodystrophy in the CISAI cohort. *Biomed Pharmacother* 57(9):422–7.

26. Saint-Marc, T., Partisani, M., Poizot-Martin, I., Rouviere, O., Bruno, F., Avellandra, R., Lang, J.M., Gastaut, J.A., and Touraine, J.L. (2000). Fat distribution evaluated by computed tomography and metabolic abnormalities in patients undergoing antiretroviral therapy: preliminary results of the LIPOCO study. *AIDS* 14(1):37–49.

27. Nolan D., Hammond, E., James, I., McKinnon, E., and Mallal, S. (2003). Contribution of nucleoside-analogue reverse transcriptase inhibitor therapy to lipoatrophy from the population to the cellular level. *Antivir Ther* 8(6):617–26.

28. Martin, J.L., Brown, C.E., Matthews-Davis, N., and Readon, J.E. (1994). Effects of antiviral nucleoside analogs on human DNA polymerase and mitochondrial DNA synthesis. *Antimicrob Agents Chemother* 38(12):2743–9.

29. Feng, J.Y., Murakami, E., Zorca, S.M., Johnson, A.A., Johnson, K.A., Schinarzi, R.F., Furman, P.A., and Anderson, K.S. (2004). Relationship between antiviral activity and host toxicity: comparison of the incorporation efficiencies of 2',3'-dideoxy-5-fluoro-3'thiacytidine-triphosphate analogs by human immunodeficiency virus type 1 reverse transcriptase and human mitochondrial DNA polymerase. *Antimicrob Agents Chemother* 48(4):1300–6.

30. Lee, H., Hanes, J., and Johnson, K.A. (2003). Toxicity of nucleoside analogues used to treat AIDS and the selectivity of the mitochondrial DNA polymerase. *Biochemistry* 42(50):14711–9.

31. Johnson, A.A., Ray, A.S., Hanes, J., Suo, Z., Colacino, J.M., Anderson, K.S., and Johnson, K.A. (2001). Toxicity of antiviral nucleoside analogs and the human mitochondrial DNA polymerase. *J Biol Chem* 276(44):40847–57.

32. Kakunda, T. N. (2000). Pharmacology of nucleoside and nucleotide reverse transcriptase inhibitor-induced mitochondrial toxicity. *Clin Ther* 22(6):685–708.

33. Lewis, W., Copeland, W.C., and Day, B.J. (2001). Mitochondrial dna depletion, oxidative stress, and mutation : mechanisms of dysfunction from nucleoside reverse transcriptase inhibitors. *Lab Invest* 81(6):777–90.

34. Brinkman, K. and T.N. Kakuda. (2000). Mitochondrial toxicity of nucelsoside analogue reverse transciptase inhibitors: a looming obstacle for long-term antiretroviral therapy? *Curr Opin Infect Dis* 13(1):5–11.

35. Chen, C.H., Vazquez-Padua, M., and Cheng, Y.C. (1991). Effect of anit-human immunodeficiency virus nucleoside analogs on mitochondrial DNA an dits implication for delayed toxicity. *Mol Pharmacol* 39(5):625–8.

36. Hobbs, G.A., Keilbaugh, S.A., Rief, P.M., and Simpson, M.V. (1995) Cellular targets of 3'azido-3'-deoxythymidine: an early (non-delayed) effect on oxidative phosphorylation. *Biochem Pharmacol* 50(3):381–90.

37. Barile, M., Valenti, D., Hobbs, G.A., Abruzzese, M.F., Keilbaugh, S.A., Passarella, S., Quagliariello, E., and Simpson, M.V. (1994). Mechanisms of toxicity of 3'-azido-3'-deoxythymidine. Its interaction with adenylate kinase. *Biochem Pharmacol* 48(7):1405–12.

38. Barile, M., Valenti, D., Passarella, S., and Quagliariello, E. (1997). 3'-Azido-3'deoxythmidine uptake into isolated rat liver mitochondria and impairment of ADP/ATP translocator. *Biochem Pharmacol* 53(7):913–20.

39. Modica-Napolitano, J.S. (1993). AZT causes tissue-specific inhibition of mitochondrial bioenergetic function. *Biochem Biophys Res Commun* 194(1):170–7.

40. Hayakawa, M., Ogawa, T., Sugiyama, S., Tanaka, M., and Ozawa, T. (1991). Massive conversion of guanosine to 8-hydroxy-guanosine in mouse liver mitochondrial DNA by administration of azidothymidine. *Biochem Biophys Res Commun* 176(1):87–93.

41. de la Asuncion, J.G., del Olmo, M.L., Sastre, J., Pallardo, F.V., and Vina, J. (1999). Zidovudine (AZT) causes an oxidation of mitochondrial DNA in mouse liver. *Hepatology* 29(3):985–7.

42. Johnson, N. and J.M. Parkin. (1998). Anti-retroviral therapy reverse HIV-associated abnormalities in lymphocyte apoptosis. *Clin Exp Immunol* 113(2):229–34.

43. Bellet, V., Duval, R., Delebassee, S., Cook-Moreau, J., and Bosgiraud, C. (2004) AZT inhibits Visna/maedi virus-induced apoptosis. *Arch Virol* 149(3):583–601.

44. Hatse, S., Schols, D., De Clercq, E., and Balzarini, J. (1999). 9-(2-Phosphonylmethoyxethyl)adenine induces tumor cell differentiation or cell death by blocking cell cycle progression through the S phase. *Cell Growth Differ* 10(6):435–46.

45. Valerianove, M., Otova, B., Bila, V., Hanzalova, J., Votruba, I., Holy, A., Eckschlager, T., Krejci, O., and Trka, J. (2003). PMEDAP and its N6-substituted derivatives: genotoxic effect and apoptosis in vitro conditions. *Anticancer Res* 23 (6C):4933–9.

46. De Clercq, E., Andrei, G., Balzarini, J., Hatse, S., Liekens, S., Naesens, L., Neyts, J., and Snoeck, R. (1999). Antitumor potential of acyclic nucleoside phosphates. *Nucleosides Nucleotides* 18(4-5): 759–71.

47. Murono, S., Raab-Traub, N., and Pagano, J.S. (2001). Prevention and inhibition of nasopharyngeal carcinoma growth by antiviral phosphonated nucleoside analogs. *Cancer Res* 61(21):7875–7.

48. Hitt, M.M., Allday, M.J., Hara, T., Karran, L., Jones, M.D., Busson, P., Tursz, T., Ernberg, I., and Griffin, B.E. (1989). EBV gene expression in an NPC-related tumour. *Embo J* 8(9):2639–51.

49. Ren, J., Esnouf, R., Hopkins, A., Ross, C., Jones, Y., Stammers, D., and Stuart, D. (1995). The structure of HIV-1 reverse transciptase complexed with 9-chloro-TIBO: lessons for inhibitor design. *Structure* 3(9):915–26.

50. Ren, J., Esnouf, R., Garman, E. Somers, D., Ross, C., Kirby, I., Keeling, J., Darby, G., Jones, and Stuart, D. (1995). High resolution structures of HIV-1 RT from four RT-inhibitor complexes. *Nat Struct Biol* 2(4):293–302.

51. Spense, R.A., Kati, W.M., Anderson, K.S., and Johnson, K.A. (1995). Mechanism of inhibition of HIV-1 reverse transcriptase by nonnucleoside inhibitors. *Science* 267(5200):988–93.

52. Esnouf, R., Ren, J., Ross, C., Jones, Y. Stammers, D., and Stuart, D. (1995). Mechanism of inhibition of HIV-1 reverse transcriptase by non-nucleoside inhibitors. *Nat Struct Biol* 2(4):303–8.

53. De Clercq, E. (1999). Perspectives of non-nucleoside reverse transcriptase inhibitors (NNRTIs) in the therapy of HIV-1 infection. *Farmaco* 54(1-2):26–45.

54. Pilon, A.A., Lum, J.J., Sanchez-Dardon, J., Phenix, B.N., Douglas, R., and Badley, A.D. (2002). Induction of apoptosis by a nonnucleoside human immunodeficiency virus type 1 reverse transcriptase inhibitor. *Antimicrob Agents Chemother* 46(8):2687–91.

55. Barreiro, P., Soriano, V., Blanco, F., Casimiro, C., DelaCruz, J.J., and Gonzalez-Lahoz, J. (2000). Risks and benefits of replacing protease inhibitors by nevirapine in HIV-infected subjects under long-term successful triple combination therapy. *AIDS* 14(7):807–12.

56. Martinez, E., Conget, I., Lozano, L., Casamitjana, R., and Gatell, J.M. (1999). Reversion of metabolic abnormalities after switching from HIV-1 protease inhibitors to nevirapine. *AIDS* 13(7):805–10.

57. Collier, A.C., R.W. Coombs, D.A. Schoenfeld, R.L. Bassett, J. Timpone, A. Baruch, M. Jones, K. Facey, C. Whitacre, V.J. McAuliffe, H.M. Friedman, T.C. Merigan, R.C. Reichman, C. Hooper, and L. Corey. (1996) Treatment of human immunodeficiency virus infection with saquinavir, zidovudine, and zalcitabine. AIDS Clinical Trials Group. *N Engl J Med* 334(16):1011–7.

58. Piketty, C., Castiel, P., Belec, L., Batisse, D., Si Mohamad, A., Gilquin, J., Gonzalez-Canali, G., Jayle, D., Karmochkine, M., Weiss, L., Aboulker, J.P., and Kazatchkine, M.D. (1998). Discrepant responses to triple combination antiretroviral therapy in advanced HIV disease. *AIDS* 12(7):745–50.

59. Roger, P.M., Breittmayer, J.P., Arlotto, C., Pugliese, P., Pradier, C., Bernard-Pomier, G., Dellamonica, P., and Bernard, A. (1999). Highly active anti-retroviral therapy (HAART) is associated with a lower level of CD4+ T cell apoptosis in HIV-infected patients. *Clin Exp Immunol* 118(3):412–6.

60. Pajonk, F., Himmelsbach, J., Riess, K., Sommer, A., and McBride, W.H. (2002). The human immuno-deficiency virus (HIV)-1 protease inhibitor saquinavir inhibits proteasome function and causes apoptosis and radiosensitization in non-HIV-associated human cancer cells. *Cancer Res* 62(18):5230–5.

61. Andre, P., Groettrup, M., Klenerman, P., deGuili R., Booth, B.L., Jr., Cerundolo, V., Bonneville, M., Jotereau, F., Zinkernagel, R.M., and Lotteau, V. (1998). An inhibitor of HIV-1 protease modulates proteasome activity, antigen presentation, and T cell responses. *Proc Natl Acad Sci USA* 95(22):13120–4.

62. Lenhard, J.M., Croom, D.K., Weiel, J.E., and Winegar, D.A. (2000) HIV protease inhibitors stimulate hepatic triglyceride synthesis. *Arterioscler Thromb Vasc Biol* 20(12):2625–9.

63. Berthold, H.K., Parhofer, K.G., Ritter, M.M., Addo, M., Wasmuth, J.C., Schliefer, K., Sspengler, U., and Rockstroh, J.K. (1999). Influence of protease inhibitor therapy on lipoprotein metabolism. *J Intern Med* 246(6):567–75.

64. Zhang, B., MacNaul., K., Szalkowski, D., Li, Z., Berger, J., and Moller, D.E. (1999). Inhibition of adipocyte differentiation by HIV protease inhibitors. *J Clin Endocrinol Metab* 84(11):4274–7.

65. Serghides, L., Nathoo, S., Walmsley, S., and Kain, K.C. (2002). CD36 deficiency induced by antiret-roviral therapy. *AIDS* 16(3):353–8.

66. Dowell, P., Flexner, C., Kwiterovich, P.O., and Lane, M.D. (2000). Suppression of preadipocyte differentiation and promotion of adipocyte death by HIV protease inhibitors. *J Biol Chem* 275(52):41325–32.

67. Domingo, P., Matias-Guiu, X., Pujol, R.M., Francia, E., Lagarda, E., Sambeat, M.A., and Vazques, G. (1999). Subcutaneous adipocyte apoptosis in HIV-1 protease inhibitor-associated lipodystrophy. *AIDS* 13(16):2261–7.

68. Ikezoe, T., Daar, E.S., Hisatake, J., Taguchi, H., and Koeffler, H.P. (2000). Hiv-1 protease inhibitors decrease proliferation and induce differentiation of human myelocytic leukemia cells. *Blood* 96(10):3553–9.

69. Ikezoe, T., Saito, T., Bandobashi, K., Yang, Y., Koeffler, H.P., and Taguchi, H. (2004). HIV-1 protease inhibitor induces growth arrest and apoptosis of human multiple myeloma cells via inactivation of signal transducer and activator of transcription 3 and extracellular signal-regulated kinase 1/2. *Mol Cancer Ther* 3(4):473–9.

70. Sgadari, C., Barillari, G., Toschi, E., Carlei, D., Becigalupo, I., Baccarini, S., Palladino, C., Leone, P., Bugarini, R., Malavasi, L., Cafaro, A., Falchi, M., Valdembri, D., Rezza, G., Bussolino, F., Monini, P., and Ensoli, B. (2002). HIV protease inhibitors are potent anti-angiogenic molecules and promote regressin of Kaposi sarcoma. *Nat Med* 8(3):225–32.

71. Pati, S., Pelser, C.B., Dufraine, J., Bryant, J.L., Reitz, M.S., Jr., and Weichold, F.F. (2002). Antitumorigenic effects of HIV protease inhibitor ritonavir : inhibition of Kaposi sarcoma. *Blood* 99(10): 3771–9.

72. Chavan, S., Kodoth, S., Pahwa, R., and Sahwa, S. (2001). The HIV protease inhibitor Indinavir inhibits cell-cycle progression in vitro in lymphocytes of HIV-infected and uninfected individuals. *Blood* 98(2):383–9.

73. Barker, E., Kahn, J., Fujimaura, S., and Levy, S.A. (1998). Protease inhibitors do not increase the CD4+ cell count in HIV-uninfected individuals. *AIDS* 12(9):1117–8.

74. Lu, W. and Andrieu, J.M. (2000). HIV protease inhibitors restore impaired T-cell proliferative response in vivo and in vitro: a viral-suppression-independent mechanism. *Blood* 96(1):250–8.

75. Weichold, F.F., Bryant, J.L., Pati, S., Barabitskaya, O., Gallo, R.C., and Reitz, M.S., Jr. (1999) HIV-1 protease inhibitor ritonavir modulates susceptibility to apoptosis of uninfected T cells. *J Hum Virol* 2(5):261–9.

76. Phenix, B.N., Lum, J.J., Nie, Z., Sanchez-Dardon, J., and Badley, A.D. (2001). Antiapoptotic mechanism of HIV protease inhibitors : preventing mitochondrial transmembrane potential loss. *Blood* 98(4):1078–85.

77. Sloand, E.M., Kumar, P.N., Kim, S., Chaudhuri, A., Weichold, F.F., and Young, N.S. (1999). Human immunodeficiency virus type 1 protease inhibitor modulates activation of peripheral blood CD4(+) T cells and decreases their susceptibility to apoptosis in vitro and in vivo. *Blood* 94(3):1021–7.

78. Phenix, B.N., Angel, J.B., Mandy, F., Kravcik, S., Parato, K., Chambers, K.A., Gallicano, K., Hawley-Foss, N., Cassol, S., Cameron, D.W., and Badley, A.D. (2000). Decreased HIV-associated T cell apoptosis by HIV protease inhibitors. *AIDS Res Hum Retroviruses* 16(6):559–67.

79. Sloand, E.M., Maciejewski, J.P., Sato, T., Bruny, J., Kumar, P., Kim, S., Weichold, F.F., and Young, N.S. (1998). The role of interleukin-converting enzyme in Fas-mediated apoptosis in HIV-1 infection. *J Clin Invest* 101(1):195–201.

80. Sloand, E.M., Maciejewski, J.P., Kumar, P., Kim, S., Chaudhuri, A., and Young, N.S. (2000). Protease inhibitors stimulate hematopoiesis and decrease apoptosis and ICE expression in CD34(+) cells. *Blood* 96(8):2735–9.

81. Mastroianni, C.M., Mengoni, F., Lichtner, M., D'Agostino, C., d'Ettorre, G., Forcina, G., Marzi, M., Russo, G., Massetti, A.P., and Vullo, V. (2000). Ex vivo and in vitro effect of human immunodeficiency virus protease inhibitors on neutrophil apoptosis. *J Infect Dis* 182(5):1536–9.

82. Weaver, J.G., Rouse, M.S., Steckelberg, J.M., and Badley, A.D. (2004) Improved survival in experimental sepsis with an orally administered inhibitor of apoptosis. *Faseb J* 18(11):1185–91.

83. Mongia, A., Bhaskaran, M., Reddy, K., Manjappa, N., Baqi, N., and Singhal, P.C.. (2004). Protease inhibitors modulate apoptosis in mesangial cells derived from a mouse model of HIVAN. *Kidney Int* 65(3):860–70.

84. Cooley, L.A. and Lewin, S.R. (2003). Hiv-1 cell entry and advances in viral entry inhibitor therapy. *J Clin Virol* 26(2):121–32.

85. Eckert, D.M. and Kim, P.S. (2001). Design of potent inhibitors of HIV-1 entry from the gp41 N-peptide region. *Proc Natl Acad Sci USA* 98(20):11187–92.

86. Root, M.J., Kay, M.S., and Kim, P.S. (2001). Protein design of an HIV-1 entry inhibitor. *Science* 291(5505):884–8.

87. Rimsky, L.T., Shugars, D.C., and Matthews, T.J. (1998). Determinants of human immunodeficiency virus type 1 resistance to gp41-derived inhibitory peptides. *J Virol* 72(2):986–93.

88. Barretina, J., Blanco, J., Armand-Ugon, M., Gutierrez, A., Clotet, B., and Este, J.A. (2003). Anti-HIV-1 activity of enfuvirtide (T-20) by inhibition of bystander cell death. *Antivir Ther* 8(2):155–61.

89. Holm, G.H., Zhang, C., Gorry, P.R., Peden, K., Schols, D., De Clercq, E., and Gabuzda, D. (2004). Apoptosis of bystander T cells induced by human immunodeficiency virus type 1 with increased envelope/receptor affinity and coreceptor binding site exposure. *J Virol* 78(9):4541–4551.

90. Eron, J.J., Gulick, R.M., Bartlet, J.A., Merigan, T., Arduino, R., Kilby, J.M., Yangco, B., Diers, A., Drobnes, C., DeMasi, R., Greenberg, M., Melby, T., Raskino, C., Rusnak, P., Zhang, Y., Spence, R., and Miralles, G.D. (2004) Short-term safety and antiretroviral activity of T-1249, a second-generation fusion inhibitor of HIV. *J Infect Dis* 189(6):1075–83.

26 HIV-1 Reservoirs and Residual Viral Replication during Highly Active Antiretroviral Therapy

Roger J. Pomerantz
Giuseppe Nunnari

CONTENTS

INTRODUCTION

Human immunodeficiency virus type I (HIV-1) replicates in most untreated infected individuals at high levels throughout the infection. This includes the clinical quiescent phase, and levels of this active viral replication directly correlate with disease progression rates and survival.[1-4] Combination therapeutics for HIV-1, or highly active antiretroviral therapy (HAART), has led to dramatic decreases in viral replication *in vivo* to below the clinical limits of detection (i.e., plasma HIV-1 ribonucleic acid [RNA] levels below 50 to 400 copies/ml, depending on the assay system used) and a drastic reduction in morbidity/mortality, at least in the developed world.[5-7] These therapeutic modalities have completely altered the epidemic in many regions. The era of HAART now allows for clearer investigations of classical questions in human retrovirology as well as for the generation of new clinical problems. In fact, mechanisms of viral latency and hidden or "cryptic" viral replication can now be addressed without the "noise" of active virally producing cells and high levels of cell-free virions,[8] as virally suppressive HAART unveils viral persistence.

After an individual is infected with HIV-1, replication of the virus occurs at a rapid rate. However, the half-life of virions in blood plasma seems to be within a time frame of hours.[9] Initially, the great majority of viral replication occurs in activated, productively infected CD4+ T lymphocytes in peripheral blood as well as lymphoid tissue. After the initiation of HAART, there seems to be an initial phase of rapid decay of plasma viral RNA involving productively infected cells, followed by a second phase of decay in what is likely persistence of virus within more long-lived cells, perhaps tissue-bound macrophages.[9,10] The final phase includes some loss of persistently infected cells, such as resting CD4+ T lymphocytes, but this decay period is prolonged and may require many decades of HAART-mediated viral suppression to accomplish.[11,12]

Effective HAART blocks the virus from infecting healthy CD4+ T cells, and the initial drop in HIV-1 levels in the peripheral blood reflects the life spans of cells that were infected before treatment was initiated. Based on the complex viral and cellular kinetics with slow decay of viral-infected cells in several phases, most HIV-1-infected individuals would require complete suppression of viral replication for years if one is even going to consider viral eradication. It is critical to note that true viral eradication may be extremely difficult for many reasons (see below), including the finding that cells other than those in the immune system (e.g., kidney, heart) also seem to be infected at low levels with HIV-1 *in vivo*.[1,13,14] Furthermore, this goal is hindered by several factors, including less than optimal treatment, HIV-1 infection in immunologically privileged sites, partial drug sanctuary compartments, *in vivo* "spikes" or "blips" of plasma viremia occurring in certain patients on virally suppressive HAART, and potential ongoing *de novo* cellular infections, secondary to local bursts of viral replication.[15]

New therapeutic strategies toward possible viral eradication or at least long-term remission need to be fully investigated to rationally design novel antiretroviral approaches.

HIV-1 LATENCY AND PERSISTENCE

Interest in retroviral latency or, more generally, persistence preceded the acquired immunodeficiency syndrome (AIDS) epidemic. Although retroviral latency was initially studied in avian retroviruses, the understanding of these processes *in vivo* has significantly increased in recent years, utilizing HIV-1 as a model.[1] It is critical, at the outset, to delineate the concepts of cellular latency and persistent replication for retroviruses *in vivo*. Viral latency is a general property of many viruses, including herpes viruses. Retroviral latency, however, can be defined as integrated provirus with no active transcription. Low-level chronic yet productive viral expression may best be characterized as persistent or cryptic replication, as HAART seems to not fully ablate all replication from previously infected cells. This can take place after initially successful viral suppressive therapy or antiviral immune responses in the infected host. Such cryptic viral replication may also continue in immunologically privileged sites, such as the brain and testes. As well, transcriptionally active but nonproductive viral infections (i.e., certain viral mRNA are expressed but not intact virions) may also occur (see below).

The HIV-1 replicative cycle contains many possible stages for latency, both pre- and postintegration into the human cell genome.[1] It involves binding to cellular receptors and co-receptors (i.e., CD4 receptor and chemokine co-receptors, CCR5 and CXCR4), with subsequent viral core internalization and reverse transcription of the viral RNA template into a double-stranded DNA intermediate. The viral DNA is then integrated into the host cell genome as proviral DNA. Transcriptional activation of the integrated provirus is controlled by a complex series of interactions between HIV-1 regulatory proteins (e.g., Tat and Rev) and cellular transcription factors, controlled by the state of activation of the host cell. The activated provirus produces various viral messenger RNA species and, in some cases, new virions (reviewed in O'Brien and Pomerantz[1]).

Many studies demonstrated that "clinical" HIV-1 latency, when defined as no viral expression in an entire untreated infected individual, does not exist at the organismal level in any stage of disease.[16–18] In the great majority of HIV-1-infected individuals, some cultivable virus may be

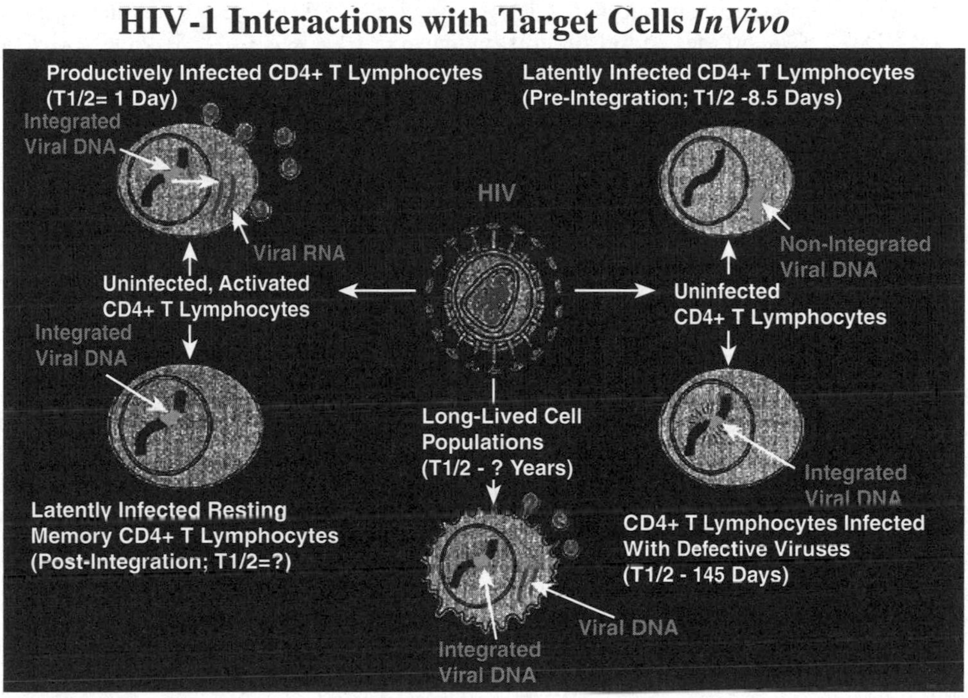

HIV-1 Interactions with Target Cells *In Vivo*

FIGURE 26.1 The in vivo dynamics and cellular reservoirs for HIV-1. This figure illustrates the different cell-types infected with HIV-1 *in vivo* and the possible sites for long-lived infected cell populations and latently infected cells. Red viral genomic RNA; Green, integrated or unintegrated viral DNA; Blue, host cell chromosome. Upper left, productively infected cell; Upper right, pre-integration latently infected; Lower left, integrated latently infected; Lower middle, long-lived population; Lower right, cells with defective proviruses.

recovered at all stages of disease.[16–18] Nevertheless, data demonstrate that in the HIV-1-infected individual, some cells contain proviral DNA but express little or no viral RNA and produce few or no virions.[16,19,20] In addition to postintegration latency, various states of preintegration HIV-1 latency were described both *in vitro* and *in vivo*.[21–25] As such, latency at a cellular level exists *in vivo*, and the number of latently infected cells may vary based on the stage of disease. As HIV-1 infects, *in vivo*, CD4+ T cells, monocyte/macrophages, and other nonimmune-based cells,[14,26,27] the virus may maintain cellular latency by different mechanisms in differing cell types (see Figure 26.1). This possibly represents important research opportunities in further elucidating mechanisms of HIV-1 latency *in vivo*.

CELL-ASSOCIATED AND CELL-FREE VIRUS DURING HAART

RESIDUAL HIV-1 PROVIRUS *IN VIVO* DURING HAART

Persistently infected, nonactivated CD4+ T cells have been demonstrated in the peripheral blood of HIV-1-infected individuals, after treatment effectively suppresses most productive viral infection.[28] Replication-competent viruses can be recovered from these proviral-positive cells after CD8+ T cell depletion *in vitro*.[29–31] The vast majority of these viruses are CCR5-tropic.[32] In addition, this cell reservoir is established soon after primary HIV-1 infection[33] and can be activated by proinflammatory cytokines *in vitro* and potentially *in vivo*.[34] Suppressive HAART initiated before primary HIV-1 seroconversion and during perinatal HIV-1 infection has also been shown to be unable to halt the development of this replication-competent virus reservoir in CD4+ T cells.[35,36]

This replication-competent but latent HIV-1 proviral DNA resides in resting memory or CD45RO CD4[+] T lymphocytes and also in resting naive or CD45RA CD4[+] T lymphocytes, albeit at lower levels.[32] It has been suggested that naive CD4[+] T cells are rarely directly infected but that the provirally harboring naive CD4[+] T cells are generated via reversion from a memory phenotype.[32] Given that infected resting memory and naive CD4[+] T lymphocytes harbor replication-competent proviral DNA, it is not surprising that viral rebound could occur for most, if not every, patient at some time after discontinuation of HAART, as these cells turn over slowly.

Most viruses isolated from resting CD4[+] T cells from the peripheral blood of patients with undetectable levels of viral RNA in plasma have few mutations that confer antiretroviral drug resistance.[29] As resistance mutations in the reverse transcriptase (RT) and protease (PR) genes of HIV-1 are correlated with ongoing viral replication despite treatment,[37] this finding suggests that these viral strains may represent "archival" species from a time before treatment. Notably, however, a recent study showed that in some patients on virally suppressive HAART, with transient spikes of virus of more than 50 copies/ml in plasma, resistance mutations may develop.[38] Data from the above studies demonstrated that although defective proviruses accumulate in CD4[+] T cells *in vivo* (i.e., "viral graveyard sequences"),[39] replication-competent provirus still exists in resting CD4[+] T cells, which may hinder attempts at reducing the viral reservoir and reseed the body with virus if HAART is discontinued. These possibilities, therefore, confound our attempts to obtain pharmacologically mediated viral eradication. In addition, replication-competent HIV-1 was demonstrated in peripheral blood monocytes,[40] natural killer (NK) cells, and CD8[+] T lymphocytes[41,42] isolated from patients on virally suppressive HAART. Also, co-cultures of B cells isolated from HAART patients, with peripheral blood mononuclear cells (PBMCs) isolated from HIV-1-seronegative donors, showed that B cells can lead to infection of PBMCs. Interestingly, in this type of cell, HIV-1 was demonstrated to be bound to the cellular surface through the CD21 cellular receptor.[43]

CELL-ASSOCIATED VIRUS

It was hypothesized that persistently infected CD4[+] T cells containing nondefective but quiescent HIV-1 proviral DNA and low levels of viral replication could account for the replication-competent virus within the peripheral blood CD4[+] T cells and seminal cells of infected individuals on HAART.[8] Viral replication could occur at such low levels that the virus would not be detectable in peripheral body fluids with standard clinical assays. Low-level productively infected cells might infect small numbers of target cells in the surrounding cellular microenvironment. One study demonstrated that, in a minority of patients on suppressive HAART, modest evolution of viral envelope sequences occurred over time.[11] Other studies demonstrated that during what seems to be full suppression of HIV-1 in the plasma, as determined by clinical RNA assays, there is ongoing viral replication in a majority of patients in some cohorts, as shown by evolution of viral sequences in cellular reservoirs.[44,45] These "archival" viruses did not demonstrate antiretroviral resistance mutations, even in those strains that could undergo low-level replication. Nonetheless, some selection bias of the viral genetic variants may confound analyses in these studies.

In studies by Furtado et al.[46] and recently confirmed by another group,[47] sensitive measures of the "footprints" of persistent viral replication were used to evaluate HIV-1 mRNA species and HIV-1 long-terminal repeat (LTR) DNA circles, which are formed by self-ligation of proviral DNA by cellular nuclear ligases after transport of the viral preintegration complex to the nucleus. LTR-DNA circles were demonstrated in the CD4[+] T cells of most patients on suppressive HAART. The LTR circles along with quasi-steady state levels of HIV-1 mRNA demonstrate the existence of low-level viral replication at some time in the recent past. Although a short *in vivo* half-life for HIV-1 2-LTR DNA circles was suggested,[47] further studies are necessary to confirm this finding.

In one study,[44] a further complexity toward characterizing HIV-1 persistence during HAART was also demonstrated, as certain patients had cells with multiply spliced viral RNA out of proportion to unspliced viral RNA, in agreement with *in vitro* models.[48–50] As multiply spliced viral RNA encodes regulatory proteins, whereas unspliced RNA encodes structural proteins, this pattern suggests transcriptionally active but nonproductive infection. These patterns of viral RNA species expression may be based on the state of host cell activation. Thus, several forms of cryptic replication might "reset the virological clock" by infecting previously uninfected cells in localized microenvironments.

Recent data have demonstrated that certain unstimulated, HLA-DR-negative CD4[+] T cells in lymphoid tissue sampled from HIV-1-infected people treated with virally suppressive HAART were positive for low levels of viral RNA.[51] HIV-1 replication was demonstrated previously to occur in activated memory CD4[+] T cells in both peripheral blood and lymphoid tissues.[1,52,53] These data suggest that there may actually be a spectrum of cell types in a relatively inactive state that can still express low levels of viral RNA, a finding supported by others.[50]

Most,[54,55] but possibly not all,[56] patients rebound rapidly with high levels of plasma viral RNA when standard suppressive HAART is discontinued, even when fewer than 50 copies/ml of plasma viral RNA are demonstrated for significant time periods. This observation suggests not only the presence of persistently infected cells but also ongoing viral replication. One study showed that rebounding virus in certain patients, after discontinuation of HAART, does not seem to be from outgrowth of latent provirus in resting peripheral blood CD4[+] T cells, but the cellular reservoir that produces the rebound virus remains unclear.[57] Nevertheless, another recent study demonstrated that rebound virus in some cases can arise from the latent virus in resting CD4[+] T cells.[58]

CELL-FREE VIRUS

Using a laboratory-based reverse transcriptase-polymerase chain reaction (RT-PCR) assay that can quantify at least 5 copies/ml of plasma viral RNA,[59] low but detectable levels of virus were demonstrated in all HIV-1-infected patients studied with clinically undetectable levels of plasma HIV-1 RNA (i.e., less than 50 copies/ml).[60] Of importance, this ongoing viral replication may infect cells in local sites as well as cells at a distance within the body. Of note, however, the quantities of defective virions[39] in residual plasma HIV-1 load are still unknown and remain important for understanding possible viral spread within an infected host.

The viral decay characteristics in cells are different from decay characteristics in plasma.[2] In PBMCs, the first phase represents decay of productively infected and long-lived cells, whereas the second phase is decay of cell-associated viral DNA and mRNA. Viral reservoir decay characteristics in one analysis suggested that 60 years will be necessary for patients treated with suppressive HAART to continue therapy for potential viral eradication.[61] Unfortunately, even this estimate may not represent a best-case scenario for patients, as this analysis did not fully take into account the low-level viral replication that occurs in many patients on HAART.

Some patients with fewer than 50 copies/ml of blood plasma viral RNA on HAART still have low-level bursts of viral replication leading to more than 50 copies/ml in transient spikes,[62] which carry risks of viral resistance evolution.[38] The mean decay half-life of latent replication-competent proviral reservoirs is longer in patients with transient plasma viral RNA spikes than in those patients who consistently maintain plasma HIV-1 RNA levels with fewer than 50 copies/ml.[62] A recent small study suggested that low-level spikes of plasma HIV-1 RNA in patients on previously suppressive HAART do not seem to associate with higher rates of viral rebound, at least in the short term,[63] but other preliminary studies showed lower CD4[+] T cell counts in patients with plasma HIV-1 RNA spikes.[64,65] It will be important to determine whether these viral spikes are due to insufficient plasma drug levels and depressed drug penetration into unique reservoir sites or just continued HIV-1 replication in the lymphoid tissue and peripheral blood cells.

The findings that 60 to 75 years of suppressive HAART would be necessary to possibly allow latent reservoir eradication were suggested to be due more to the slow decay of a truly latent proviral state rather than to ongoing viral replication.[61] The other study mentioned above might lead to some optimism, as a mean half-life of approximately 6 months of the latent viral reservoir in patients who have no spikes and more than 50 copies/ml of plasma viral RNA on HAART may allow viral eradication in this select group of patients in a shorter time period.[62] Nonetheless, this possibility assumes that even in these patients, there exist no other viral reservoirs in peripheral blood, lymphoid tissue, or other solid organs[14] and that subclinical viral replication did not occur.

HIV-1 SANCTUARIES AND THE ROLE OF BLOOD–TISSUE BARRIERS: CENTRAL NERVOUS SYSTEM, TESTES, AND RETINA

Tissues that maintain blood–tissue barriers, secondary to microvascular endothelial cell tight junctions, may limit penetration of certain antiretroviral agents and act as partial "drug sanctuaries." These compartments would potentially include the central nervous system (CNS), the retina, and the testes.[1,8,66] A plasma-membrane-localized drug transporter, the P-glycoprotein, was shown to decrease the penetration of protease inhibitors across certain blood–tissue barriers. As such, substances that block this drug transporter will increase protease inhibitor concentrations across the blood–tissue barriers. This finding may represent a new pharmacologic approach to target relative viral sanctuary sites with higher levels of antiretroviral agents.[67] Nevertheless, most analyses show parallel decay of HIV-1 in tissues, compared with peripheral blood, in patients on virally suppressive HAART.[51] In addition, free-drug levels, rather than protein-bound drug concentrations, were rarely analyzed in body sites outside the peripheral blood.

MALE AND FEMALE GENITAL TRACTS AND HIV-1 COMPARTMENTALIZATION IN THE HAART ERA

The drastic reduction, up to undetectability, of blood plasma HIV-1 RNA viral load obtained in most patients after initiation of HAART is accompanied by a parallel reduction of genital fluids' HIV-1 viral load.[8,68–71] A previous study performed in our laboratories showed evidence of HIV-1 replication-competent virus in the semen of patients on HAART, demonstrating its relevance not only in the potential transmission of the disease but also as a viral reservoir.[8] Moreover, seminal viral load was demonstrated, although at low levels, in the majority of patients on HAART. This might represent an *in vivo* reservoir or sanctuary site that may cause reinfection in the peripheral bloodstream and lymphoid tissue, especially when HAART is discontinued, interfering with attempts at viral remission or eradication.

Semen cells are a mixture of spermatozoa, precursors of germ cells, T lymphocytes, macrophages, and epithelial cells. HIV-1 proviral DNA was detected in several of these cell types,[8,72,73] and cell-free virus can still be detected at very low levels in seminal fluids.[8,69] Of note, there are often significant differences in viral load and viral sequences between semen and peripheral blood, suggesting compartmentalization of HIV-1 replication.[74,75] Recently, subcompartmentalization of HIV-1 quasi-species between seminal plasma and seminal cells was also reported, indicating that HIV-1 in seminal plasma is derived from the prostrate, whereas HIV-1-infected cells in semen originate mostly from the rete testis and epididymis.[76] These data are further supported by a study by Kiessling and colleagues[77] that displayed an increase of HIV-1 seminal plasma viral load of 100-fold over that in blood during an episode of prostatitis. On the other hand, another recent study concluded that seminal fluid mononuclear cells are possible sources for the cell-free virus found in semen, confirming that this remains a controversial issue.[78] Importantly, these studies were performed on subjects not on HAART. As such, they could not address the possible implications of this compartment as a viral reservoir. The principal types of HIV-1-infected cells in semen are monocytes/macrophages and lymphocytes,[73] but the specific individual roles of these two cell types during virally suppressive HAART are not known. The development of viral evolution has to be considered as well as the possibility of reseeding this persistent reservoir either by cell-free virus or by trafficking T cell or monocyte/macrophages. Although the low number of cells and the low seminal viral load make this a very difficult task, characterization of cell-free as well as cell-associated viral strains and tropism is of critical importance in understanding the molecular mechanisms involved in cryptic HIV-1 replication in sanctuaries, where the presence of blood–tissue barriers may reduce the efficacy of HAART.

HIV-1 can also be recovered from vaginal fluids, vaginal cells, and cervical cells.[71,79,80] Most genital virus arises from the cervix and possibly the upper genital tract. The genital-associated lymphoid tissue, composed of endo- and exocervical stromal lymphocytes, monocytes, and dendritic cells, is considered a potential source of HIV-1 replication.[7] Furthermore, epithelial cells have been found to be susceptible to HIV-1 infection and replication and to either release HIV particles or transmit virions by cell-to-cell contact.[81,82]

A recent study analyzed HIV-1 viral load levels in the cervical–vaginal secretions in 122 HIV-1-infected women using a sensitive technique with a lower detection limit of 80 copies/ml.[83] The authors reported that in 40% of women on HAART, cervical vaginal lavages (CVL) samples were positive for the presence of HIV-1 RNA. Moreover, in 25% of women with undetectable virus in plasma, cervical–vaginal shedding was demonstrated, suggesting not only that blood plasma viral load may fail to predict infectivity of genital secretions but also that there is a possibility of compartmentalization.

The latter finding is in agreement with previous studies in which the higher HIV-1 RNA levels in genital fluid than in blood plasma[70,71,84] and the presence of different drug-selected mutations in distinct compartments, such as blood plasma and endocervix and cervicovaginal secretions, supported this hypothesis.[85,86]

The presence of cell-free HIV-1 RNA in cervicovaginal secretions emphasizes the importance of continuing to practice protected sex, even in the era of HAART. Furthermore, the persistence of HIV-1 shedding, even in the presence of HAART, strongly recommends the need for new preventive strategies, such as virucides, in order to block or at least reduce the transmission of HIV-1. The use of microbicides would be paramount, especially in developing countries where sexual transmission plays a major role in the spread of the disease.

TREATMENT OF RESIDUAL HIV-1 DISEASE: PURGING LATENT HIV-1 CELLULAR RESERVOIRS AND THE REQUIREMENT FOR INDUCING DISTINCT VIRAL ARCHIVES FROM LATENCY

STIMULATORY THERAPY

The treatment options for attempting to rid the body of cells infected with HIV-1, which can produce replication-competent virus in patients on effective HAART, are now beginning to be explored based on our increased understanding of molecular lentiviral pathogenesis. One could develop approaches to activate persistently infected cells, which might lead to virus-induced cell death and purging of the viral reservoir. In theory, activation approaches for persistently infected cells would not lead to increases in newly infected cells, because this would be prevented by HAART. This approach was attempted recently, solely utilizing interleukin 2 (IL-2),[87] but it might also include infusions of anti-CD3 antibodies and even interferon- for activating monocyte/macrophage scavenger cells. An initial study using anti-CD3 monoclonal antibodies plus IL-2, in trying to eliminate HIV-1 in patients on suppressive HAART, was unsuccessful; patients had increased viral replication and experienced serious side effects from the high doses of drugs.[88] No intensification therapy (i.e., additional antiretroviral drugs) was added in this clinical trial, in attempts to halt low-level viral replication. Therefore, persistently infected cells were stimulated, but the ongoing cryptic viral replication was probably not affected in a substantial manner.

Some patients on suppressive HAART treated with intermittent IL-2 had decreased or undetectable levels of resting $CD4^+$ T cells containing replication-competent virus.[87] When treatment was interrupted in two patients in whom replication-competent virus could not be found in resting CD4[+] T cells, virus was readily isolated *in vitro*, and high levels of viral RNA in plasma were observed within weeks of stopping therapy.[89] Another larger group of 18 patients with undetectable levels of virus for more than 1 year had therapy interrupted.[90] All patients relapsed to relatively high levels of plasma HIV-1 RNA in 2 to 3 weeks after stopping HAART. Neither relatively long-term suppressive HAART alone nor it combined with IL-2 led to the elimination of HIV-1 infection. These studies suggest that "rebound" viremia after stopping HAART with IL-2 may be produced from yet-unidentified reservoir sites.

Other cytokines, such as IL-7 and -15, were proposed not only to activate latently infected T cells and increase the turnover rate of the latent viral reservoir to promote viral clearance but also as an immune-adjunctive therapy to HAART.

IL-7 and -15 have pleiotropic effects on T cell homeostasis. IL-7 enhances mature T cell survival,[91-93] expansion of both peripheral CD4[+] and CD8[+] T cells in T cell-depleted hosts,[94] and mobilization of pluripotent hematopoietic stem cells from the bone marrow to the peripheral circulation.[95] Importantly, IL-7 improves both HIV-1-specific CD8[+] cytotoxic cellular activity *in vitro* in humans[96] and CD4[+] Th cell-dependent humoral responses and CD8 cytotoxic T cell activity in mice immunized with HIV-1 envelope protein.[97] It was reported that in patients with AIDS, the plasma levels of IL-7 and -15 were higher than in healthy donors and that increased circulating IL-7 levels in HIV-1-infected subjects were inversely correlated with CD4[+] T cell loss and positively correlated with plasma HIV-1 viral load.[98] This probably represents a homeostatic response to lymphopenia, because the high levels of IL-7 decrease when the T cell count increases after a HAART regimen is initiated.[98] The authors hypothesized that IL-7 might play an important role in HIV-1 disease progression.

Of note, IL-7 was shown to be able to induce HIV-1 replication *in vitro* from PBMCs of HIV-1-infected individuals who were not on HAART[99-101] and to increase the expression of HIV-1 Tat mRNA in CD8-depleted PBMCs from HIV-1 chronically infected patients.[99] Moreover, recent *in vitro* studies have indicated that IL-7 has minimal effects on T cell phenotype, whereas it induces substantial expression of latent HIV-1,[102-104] which suggests that IL-7 may be used as an additive therapy to HAART, not only for its immunologic effects but also to purge HIV-1 latent reservoirs.

INTENSIFICATION THERAPY

Some current clinical strategies approach the challenge of eliminating low levels of ongoing replication-competent virus through an "intensification" protocol that relies on administration of additional viral inhibitors apart from the combination of drugs that typically constitute HAART. Candidate drugs used in intensification with HAART include hydroxyurea (HU), although this should only be used in intensification research protocols based on its toxicity, especially in combination with the reverse transcriptase (RT) inhibitor didanosine (ddI).[105] HU inhibits the cellular enzyme ribonucleotide reductase, thereby decreasing intracellular deoxyribonucleotide pools, which indirectly impedes viral RT activity.[106] As such, HU inhibits HIV-1 replication, notably in nonactivated CD4[+] T cells, by indirectly blocking reverse transcriptase, which is dependent on intracellular deoxyribonucleotide triphosphates (dNTPs) as substrates.[106] Recently, it was demonstrated that mycophenolic acid, together with the RT inhibitor abacavir, may be similarly effective at blocking residual HIV-1 replication as ddI and hydroxyurea.[107] Hydroxyurea was used in several HAART regimens, especially combined with ddI, with which it has synergistic effects in inhibiting HIV-1 replication.[106] At least one patient had a dramatically low proviral reservoir in the peripheral blood when treated with a ddI and hydroxyurea-containing regimen and maintained undetectable plasma viral RNA after HAART was discontinued.[56] In addition, a small cohort of patients on ddI and hydroxyurea did not manifest viral rebound when the antiretroviral therapy was discontinued.[108] Unfortunately, another study showed that a quick and profound rebound of plasma HIV-1 developed after treatment with HAART, plus or minus hydroxyurea therapy during primary HIV-1 infection. One patient in this study, however, remained with fewer than 50 copies/ml of plasma HIV-1 RNA off HAART for 46 weeks.[109] However, this is an anecdotal case, and there remains a lack of compelling data from randomized, controlled studies. Also, certain safety issues exist regarding the use of HU in inducing "remissions" in HIV-1 infection.

Cytoreduction therapy may also be useful in removing persistently infected cells and available uninfected target cells. Whether cyclophosphamide as a low-level cytotoxic agent, or even HU at high levels, at which point it is cytotoxic,[110] could be used in this manner will require further study. This approach may be based on the "predator and prey" relationship of HIV-1 with CD4+ T cells.[111] Targeting specific immunotoxins to deplete the HIV-1 proviral reservoir, after stimulation of viral transcription from the proviruses, and potentially ongoing viral replication should now also be readdressed in the era of HAART.[112]

As low levels of ongoing viral replication occur in many HIV-1 infected individuals on effective HAART, "intensification" therapeutics must be added to initial combination therapy to ablate this viral replication if eradication is a goal. Using approaches to purge only the latent viral reservoirs likely will not be successful. Additional studies on the penetration of different pharmacologic agents across blood–tissue barriers (e.g., blood–brain and blood–testes) will be of importance in attacking potential sanctuary sites that may contain both cryptically replicating as well as latently infected cells. Certainly, combinations of each of these approaches listed above may be required for elimination of the HIV-1 reservoirs *in vivo*. As such, one might now begin to think of the treatment of HIV-1 in somewhat of an "oncological" paradigm. This would include effective HAART as "induction" therapy and then additional approaches against HIV-1 latency, cryptic replication, and sanctuary sites for the removal of "residual disease." This approach is not induction followed by "maintenance" therapy but induction followed by aggressive therapies to approach residual viral disease, with the induction HAART continued unaltered. Nonetheless, HAART alone decreases mortality and morbidity in the vast majority of patients maintaining a low viral load, despite residual HIV-1 infection, even in subjects with virological "failure" and drug resistance.[113]

Immune-Based Approaches and Structured Treatment Interruptions (STIs)

As the potential for viral eradication remains an unproven concept and a difficult goal, studies aimed at disease remission have been developed. Gradual improvements in the immune system with effective HAART are tied to increased thymic and possibly extrathymic output of cells.[114,115] Increased anti-HIV-1-specific immune responses have been detected in certain patients who were treated with virally suppressive HAART and then had it discontinued after various periods of time. There is a report of two patients who maintained undetectable levels of plasma viral RNA by clinical assays (fewer than 50 copies/ml).[116] It is now being tested whether some patients may be placed into clinical remission and can be maintained off HAART for significant periods of time (see below). HIV-1-specific CD8+ cytotoxic T lymphocytes (CTLs), which usually become undetectable with fully suppressive HAART, increase when a significant level of viral RNA in plasma redevelops off HAART in patients who were virologically suppressed previously.[117] In addition, HIV-1-specific CD4+ T lymphocyte proliferation, which is usually not found in patients first treated with suppressive HAART months to years after primary infection, remained detectable in infected individuals who were treated at the time of primary infection with fully suppressive HAART,[118] although in one study, early but chronic infection treated with HAART demonstrated preserved anti-HIV-1 CD4+ T lymphocyte responses.[119] These studies suggest that residual, antigen-specific CD8+ T lymphocytes may expand during cyclical periods of starting and stopping HAART. Nevertheless, these findings require additional controlled studies to validate the clinical relevance of the alterations of specific anti-HIV-1 immunologic parameters.

Structured cyclic antiretroviral therapy interruptions are beginning to be studied in attempts to augment anti-HIV-1 immune responses. A few patients show a substantial decrease in plasma HIV-1

RNA set points after STIs. Although augmentation of select immune parameters may occur, there are many cases in these initial studies in which little or no virological effects were demonstrated.[120–125] Diverse antiretroviral immune responses seem to be well-preserved if suppressive HAART is initiated soon after primary HIV-1 infection *in vivo*.[118,126] A recent study showed exciting new data on remissions induced in a small group of patients treated with HAART during primary HIV-1 seroconversion, with subsequent treatment interruptions.[127] Although HIV-1 was not eradicated in these patients, quite low plasma viral RNA levels were induced and maintained off all antiretroviral therapy. Additional studies and long-term follow-up of these patients will be required to determine the durability of this effect and whether any patients initially treated with HAART during chronic HIV-1 infection will benefit from this approach. Unfortunately, STIs in chronically HIV-1-infected individuals treated after primary seroconversion, unlike STIs in patients treated during primary infection, have not been effective.[123] Selected therapeutic vaccines should also be used in future studies. Recent data in simian immunodeficiency virus (SIV)-infected macaques demonstrated viral control (i.e., with low plasma viral set points) when specific therapeutic vaccines were combined with STIs.[128]

The potential use of selected immunotoxins, in synergy with augmentation of antiretroviral immune responses, toward maintaining HIV-1 remission was also proposed and awaits evaluation.[112] Exploring the hypothesis that the immune system may keep the virus in check (i.e., remission) and induce long-term nonprogression without leading to full viral eradication is an intriguing idea that clearly merits further clinical analyses. Converting HIV-1 infections into a manageable long-term disease may be based on the selection of less fit (i.e., lower replicative capacity) or less virulent (i.e., lower CD4+ T cell depletion) viral variants.

HAART has led to dramatic and positive changes in the treatment of HIV-1 infection. Nonetheless, in its present form, HAART is unlikely to lead to viral eradication in the majority of HIV-1-infected individuals within clinically reasonable time periods. Nonetheless, the progress obtained with therapy against HIV-1 within the last decade suggests cautious optimism regarding approaches for continued suppression of residual HIV-1 infection.

Conclusion

In summary, HAART had led to dramatic, positive changes in the clinical management of HIV-1 infection. However, it is not likely that these regimens will lead to elimination of virus in their present forms for the majority of HIV-1-infected individuals, at least within a clinically reasonable time frame. The presence of multiple persistent viral species sequestered in multiple cell types makes this a difficult task. Many questions remain open regarding the nature of HAART-persistent viruses and the cell and tissue types in which they may be sequestered.

The activation of specific subsets of patients' cells *in vitro* with a variety of agents, apart from phytohemagglutinin (PHA) and IL-2 (used in many T cell activation and proliferation studies in cell cultures), is warranted. Compounds such as phorbol myristic acid (PMA) prostratin, IL-7, or IL-15 likely induce viral expression though a pathway distinct from PHA/IL-2. This is important, as latent virus may reside within multiple cell types, and each could have a heterogeneous response in terms of viral expression induced by individual activating agents. This possibility supports the use of a more diverse panel of stimulatory agents *in vitro* and, importantly, in *in vivo* studies. Certain mitogens, such as PHA, are restricted to use *in vitro* and are not viable candidates for *in vivo* clinical application.

Thus, further studies are necessary to try to identify not only which compounds would be able to induce viral expression from latent cellular reservoirs but also the molecular mechanisms involved in maintaining these viral reservoirs constant over time.

Finally, it is important to determine whether currently devised and clinically applied viral intensification and immune-stimulatory protocols are definitively aiding in the elimination of the latent viral pool or whether modifications to current protocols would enhance their efficacy. Such modifications could include use of additional stimulatory compounds or, alternatively, hybrid immunotoxins that are specific recombinant proteins that exhibit preferential targeting and killing of HIV-1-infected cells.[112] Additional stimulatory molecules would broaden the spectrum of cell activation, which is linked to latent viral expression. These compounds, perhaps combined with novel immune-based therapies, could accelerate elimination of the virus so that complete withdrawal from HAART is realistically attainable for at least some HIV-1-infected individuals.

ACKNOWLEDGMENTS

The author wishes to thank Ms. Rita M. Victor and Ms. Brenda O. Gordon for excellent secretarial assistance.

REFERENCES

1. O'Brien, W. and Pomerantz, R.J., AIDS and other diseases due to HIV infection, in *Viral Pathogenesis,* Nathanson, N., Ed., Raven Press, New York, 1997, pp. 813–837.
2. Ho, D.D., Neumann, A.U., Perelson, A.S., Chen, W., Leonard, J.M., and Markowitz, M., Rapid turnover of plasma virions and CD4 lymphocytes in HIV-1 infection, *Nature,* 373, 123–126, 1995.
3. Mellors, J., Rinaldo, C., Gupta, P., White, R.M., Todd, J.A., and Kingsley, L.A., Prognosis in HIV infection predicted by the quantity of virus in plasma, *Science,* 272, 1167–1170, 1996.
4. Shaw, G.M., and Lifson, J.D., High levels of HIV-1 in plasma during all stages of infection determined by competitive PCR, *Science,* 259, 1749–1754, 1993.
5. Gulick, R.M., Mellors, J.W., Havlir, D., Eron, J.J., Gonzalez, C., McMahon, D., Richman, D.D., Valentine, F.T., Jonas, L., Meibohm, A., Emini, E.A., and Chodakewitz, J.A., Treatment with indinavir, zidovudine, and lamivudine in adults with human immunodeficiency virus infection and prior antiretroviral therapy, *N. Engl. J. Med.,* 337, 734–739, 1997.
6. Palella, F.J., Delaney, K.M., Jr., and Moorman, A.C., Declining morbidity and mortality among patients with advanced human immunodeficiency virus infection. HIV Outpatient Study Investigators, *N. Engl. J. Med.,* 338, 853–860, 1998.
7. Hammer, S.M., Squires, K.E., Hughes, M.D., Grimes, J.M., Demeter, L.M., Currier, J.S., Eron, J.J. Feinberg, J.E. Jr., Balfour, H.H., Deyton, L.R. Jr., Chodakewitz, J.A., and Fischl, M.A., A controlled trial of two nucleoside analogues plus indinavir in persons with human immunodeficiency virus infection and CD4 cell counts of 200 per cubic millimeter or less. AIDS Clin. Trial Grp. 320 Study Team, *N. Engl. J. Med.,* 337, 725–733, 1997.
8. Zhang, H., Dornadula, G., Beumont, M., Livornese, L., Van Uitert, B., Henning, K., and Pomerantz, R.J., HIV-1 in the semen of men receiving highly active anti-retroviral therapy, *N. Engl. J. Med.,* 339, 1803–1809, 1998.
9. Perelson, A.S., Essunger, P., Cao, Y., Vesanen, M., Hurley, A., Saksela, K., Markowitz, M., and Ho, D.D., Decay characteristics of HIV-1 infected compartments during combination therapy. *Nature,* 387, 88–92, 1997.
10. Cavert, W., Notermans, D.W., Staskus, K., Wietgrefe, S.W., Zupancic, M., Febhard, K., Henry, K., Zhang, Z-Q., Mills, R., McDade, H., Goudsmith, J., Danner, S. A., and Haase, A.T., Kinetics of response in lymphoid tissues to antiretroviral therapy of HIV-1 infection, *Science,* 276, 960–964, 1997.
11. Zhang, L., Ramratnam, B., Tenner-Racz, K., He, Y., Vesanen, M., Lewin, S., Talal, A., Racz, P., Perelson, A.S., Korber, B.T., Markowitz, M., and Ho, D.D., Quantifying residual HIV-1 replication in patients receiving combination antiretroviral therapy, *N. Engl. J. Med.,* 340, 1605–1613, 1999.
12. Finzi, D., Blankson, J., Siliciano, J.S., Margolick, J.B., Chadwick, K., Person, T., Smith, K., Lisziewicz, J., Lori, F., Flexner, C., Quinn, T.C., Chaisson, R.E., Rosenberg, E., Walker, B., Gange, S., Gallant, J., and Siliciano, R.F., Latent infection of CD4+ T cells provides a mechanism for lifelong persistence of HIV-1, even in patients on effective combination therapy, *Nat. Med.,* 5, 512–517, 1997.
13. Winston, J.A., Bruggeman, L.A., Ross, M.D., Jacobson, J., Ross, L., D'agati, V., Klotman, P.E., Klotman, M.E. Nephropathy and establishment of a renal reservoir of HIV type 1 during primary infection, *N. Engl. J. Med.,* 344, 1979–1984, 2001.

14. Levy, J.A., *HIV-1 and the pathogenesis of AIDS,* ASM Press, Washington, 1994, pp. 29–34.

15. Grossman, Z., Polis, M., Feinberg, M.B., Grossman, Z., Levi, I., Jankelevich, S., Yarchoan, R., Boon, J., DeWolf, F., Lange, J.M.A., Goudsmit, J., Dimitrov, D.S., and Paul, W.E., Ongoing HIV dissemination during HAART, *Nat. Med.,* 5, 1099–1104, 1999.

16. Embretson, J., Zupancic, M., Ribas, J.L., Burke, A., Racz, P., Tenner-Racz, K., and Haase, A.T., Massive covert infection of helper T lymphocytes and macrophages by HIV during the incubation period of AIDS, *Nature,* 362, 357–362, 1993.

17. Pantaleo, G., Graziosi, C., Demarest, J.F., Butini, L., Montroni, M., Fox, C., Orenstein, J., Kotler, D., and Fauci, A.S., HIV infection is active and progressive in lymphoid tissue during the clinically latent stage of disease, *Nature,* 362, 355–358, 1993.

18. Piatak, M., Saag, M., Yang, L., Clark, S., Kappes, J., Luc, K., Hahn, B., Shaw, G., and Lifson, J., High levels of HIV-1 in plasma during all stages of infection by competitive PCR, *Science,* 259, 1749–1754, 1993.

19. Patterson, B., Till, M., Otto, P., Goolsby, C., Furtado, M., McBride, L., and Wolinsky, S., Detection of HIV-1 DNA and RNA in individual cells by PCR-driven *in situ* hybridization and flow cytometry, *Science,* 260, 976–979, 1993.

20. Peng, H., Reinhart, T.A., Retzel, E.F., Staskus, K.A., Zupancic, M., and Haase, A.T., Single cell transcript analysis of human immunodeficiency virus gene expression in the transition from latent to production infection, *Virology,* 206, 16–27, 1995.

21. Zack, J.A., Arrigo, S.J., Weitsman, S.R., Go, A.S., Haislip, A., and Chen, I.S.Y., HIV-1 entry into quiescent primary lymphocytes: molecular analysis reveals a labile, latent viral structure, *Cell,* 61, 213–222, 1990.

22. Zack, J.A., Haislip, A.M., Krogstad, P., and Chen, I.S. Incompletely reverse-transcribed human immunodeficiency virus type 1 genomes in quiescent cells can function as intermediates in the retroviral life-cycle, *J. Virol.,* 66, 1717–1725, 1992.

23. Bukrinsky, M.I., Stanwick, T.L., Dempsey, M.P., and Stevenson, M., Quiescent T lymphocytes as an inducible virus reservoir in HIV-1 infection, *Science,* 254, 423–427, 1991.

24. Stevenson, M., Stanwick, T.L., Dempsey, M.P., and Lamonica, C.A. HIV-1 replication is controlled at the level of T cell activation and proviral integration, *EMBO J.,* 9, 1551–1560, 1990.

25. Spina, C.A., Guatelli, J.C., and Richman, D.D., Establishment of a stable, inducible form of human immunodeficiency virus type 1 DNA in quiescent CD4 lymphocytes *in vitro, J. Virol.,* 69, 2977–2988, 1995.

26. Schnittman, S.M., Psallidopoulos, M.C., Lane, H.C., Thompson, L., Baseler, M., Massari, F., Fox, C.H., Salzman, N.P., and Fauci, A.S., The reservoir for HIV-1 in human peripheral blood is a T cell that maintains expression of CD4, *Science,* 245, 305–308, 1989.

27. Koenig, S., Gendelman, H.E., Orenstein, J.M., Dal Canto, M.C., Pezeshkpour, G.H., Yungbluth, M., Janotta, F., Aksamit, A., Martin, M.A., and Fauci, A.S., Detection of AIDS virus in macrophages in brain tissue from AIDS patients, *Science,* 233, 1089–1093, 1986.

28. Chun, T.W.L., Carruth, D., Finzi, D., Shen, X., DiFiuseppe, J.A., Taylor, H., Hermankova, M., Chadwick, K., Margolick, J., Quinn, T.C., Kuo, Y.H., Brookmeyer, R., Zeiger, M.A., Barditch-Crovo, P., and Siliciano, R.F. Quantification of latent tissue reservoirs and total body viral load in HIV-1 infection, *Nature,* 387, 183–188, 1997.

29. Wong, J.K., Hezareh, M., Gunthard, H.F., Havlir, D.V., Ignacio, C.C., Spina, C.A., and Richman, D.D., Recovery of replication-competent HIV despite prolonged suppression of plasma viremia, *Science,* 278, 1291–1295, 1997.

30. Finzi, D., Hermankova, M., Pierson, T., Carruth, L.M., Buck, C., Chaisson, R.E., Quinn, T.C., Chadwick, K., Margolick, J., Brookmeyer, R., Gallant, J., Markowitz, M., Ho, D.D., Richman, D.D., and Siliciano, R.F., Identification of a reservoir for HIV-1 in patients on highly active anti-retroviral therapy, *Science,* 278, 1295–1300, 1997.

31. Chun, T.W., Stuyver, L., Mizell, S.B., Ehler, L.A., Mican, J.A., Baseler, M., Lloyd, A.L., Nowak, M.A., and Fauci, A.S., Presence of an inducible HIV-1 latent reservoir during highly active anti-retroviral therapy, *Proc. Natl. Acad. Sci. U.S.A.,* 94, 13193–13197, 1997.

32. Pierson, T., Hoffman, T.L., Blankson, J., Finzi, D., Chadwick, K., Margolick, J.B., Buck, C., Siliciano, J.D., Doms, R.W., and Siliciano, R.F., Characterization of chemokine receptor utilization of viruses in the latent reservoir for human immunodeficiency virus type 1, *J. Virol.,* 74, 7824–7833, 2000.

33. Chun, T.W., Engle, D., Berrey, M.M., Shea, T., Corey, L., and Fauci, A.S., Early establishment of a pool of latently infected, resting CD4+ T cells during primary HIV-1 infection, *Proc. Natl. Acad. Sci. U.S.A.,* 95, 8869–8873, 1998.

34. Chun, T.W., Engle, D., Mizeh, S.B., Ehler, L.A., and Fauci, A.S., Induction of HIV-1 replication in latently infected CD4+ T-cells using a combination of cytokines, *J. Exp. Med.,* 188, 83–91, 1998.

35. Persaud, D., Pierson, T., Ruff, C., Finzi, D., Chadwick, K.R., Margolick, J., Margolick, J.B., Ruff, A., Hutton, N., Ray, S., and Siliciano, R.F., A stable latent reservoir for HIV-1 in resting CD4+ T-lymphocytes in infected children, *J. Clin. Invest.,* 105, 995–1003, 2000.

36. Poggi, C., Profizi, N., Djediouane, A., Chollet, L., Hittinger, G., and Lafeuillade, A. Long-term evaluation of triple nucleoside therapy administered from primary HIV-1 infection, *AIDS,* 13, 1213–1220, 1999.

37. Gunthard, H.F., Wong, J.K., Ignacio, C.C., Guatelli, J.C., Riggs, N.L., Havlir, D.V., and Richman, D.D., Human immunodeficiency virus replication and genotypic resistance in blood and lymph nodes after a year of potent anti-retroviral therapy, *J. Virol.,* 72, 2422–2428, 1998.

38. Martinez, M.A., Cababa, M., Ibanez, A., Clotet, B., Arno, A., and Ruiz, L., Antiretroviral resistance during successful therapy of HIV type 1 infection, *Proc. Natl. Acad. Sci. U.S.A.,* 97, 10948–10953, 2000.

39. Sanchez, G., Xu, X., Chermann, J.C., and Hirsch, I., Accumulation of defective viral genomes in peripheral blood mononuclear cells of human immunodeficiency virus type 1-infected individuals, *J. Virol.,* 71, 2233–2240, 1997.

40. Sonza, S., Mutimer, H.P., Oelrichs, R., Jardine, D., Harvey, K., Dunne, A., Purcell, D.F., Birch, C., and Crowe, S.M., Monocytes harbour replication-competent, non-latent HIV-1 in patients on highly active antiretroviral therapy, *AIDS,* 15, 17–22, 2001.

41. Valentin, A., Rosati, M., Patenaude, D.J., Hatzakis, A., Kostrikis, L.G., Lazanas, M., Wyvill, K.M., Yarchoan, R., and Pavlakis, G.N., Persistent HIV-1 infection of natural killer cells in patients receiving highly active antiretroviral therapy, *Proc. Natl. Acad. Sci. U.S.A.,* 99, 7015–7020, 2002.

42. Saha, K., Zhang, J., Gupta, A., Dave, R., Yimen, M., and Zerhouni, B., Isolation of primary HIV-1 that target CD8+ T lymphocytes using CD8 as a receptor, *Nat. Med.,* 7, 65–72, 2001.

43. Moir, S., Malaspina, A., Li, Y., Chun, T.W., Lowe, T., Adelsberger, J., Baseler, M., Ehler, L.A., Liu, S., Davey, R.T., Mican, J.A., Fauci, A.S., B cells of HIV-1 infected patients bind virions through CD21-complement interactions and transmit infectious virus to activated T cells, *J. Exp. Med.,* 195, 637–646, 2000.

44. Gunthard, H.F., Frost, S.D.W., Leigh-Brown, A.J., Ignacio, C.C., Kee, K., Perelson, A.S., Spina, C.A., Havlir, D.V., Hezareh, M., Looney, D.J., Richman, D.D., and Wong, J.K., Evolution of envelope sequences of human immunodeficiency virus type 1 in cellular reservoirs in the setting of potent antiviral therapy, *J. Virol.,* 73, 9404–9412, 1999.

45. Martinez, M.A., Cabana, M., Ibanez, A., Clotet, B., Arno, A., and Ruiz, L.. Human immunodeficiency virus type 1 genetic evolution in patients with prolonged depression of plasma viremia, *Virology,* 256, 180–187, 1999.

46. Furtado, M.R., Callaway, D.S., Phair, J.P., Kunstman, K.J., Stanton, J.L., Macken, C.A., Perelson, A.S., and Wolinsky, S.M., Persistence of HIV-1 transcription in peripheral blood mononuclear cells in patients receiving potent antiretroviral therapy, *N. Engl. J. Med.,* 340, 1614–1622, 1999.

47. Sharkey, M.E.,Teo., I., Greenough, T., Sharova, N., Luzuriaga, K., Sullivan, J.L., Bucy, R.P., Kostrikis, L.G., Haase, A., Veryard, C., Davaro, R.E., Cheeseman, S.H., Daly, J.S., Bova, C., Ellison, R.T., III, Mady, B., Lai, K.K., Moyle, G., Nelson, M., Gazzard, B., Shaunak, S., and Stevenson, M., Persistence of episomal HIV-1 infection intermediates in patients on highly active antiretroviral therapy, *Nat. Med.,* 6, 76–81, 1999.

48. Pomerantz, R.J., Trono, D., Feinberg, M.B., and Baltimore, D., Cells non-productively infected with HIV-1 exhibit an aberrant pattern of viral RNA expression: a molecular model for latency, *Cell,* 61, 1271–1276, 1990.

49. Butera, S.T., Robers, B.D., Lam, L., Hodge, T., and Folks, T.M., Human immunodeficiency virus type 1 RNA expression by four chronically infected cell lines indicates multiple mechanisms of latency, *J. Virol.,* 68, 2726–2730, 1994.

50. Michael, N.L., Morrow, P., Mosca, J., Vahey, M., Burke, D.S., and Redfield, R.R., Induction of human immunodeficiency virus type 1 expression in chronically infected cells is associated primarily with a shift in RNA splicing patterns, *J. Virol.,* 65, 1291–1303, 1991.

51. Zhang, Z.-Q., Schuler, T., Zupancic, M., Wietgrefe, S., Staskus, K.A., Reimann, K.A.,Reinhart, T.A., Rogan, M., Cavert, W., Miller, C.J., Veazey, R.S., Notermans, D., Little, S., Danner, S.A., Richman, D.D., Havlir, D., Wong, J., Jordan, H.L., Schacker, T.W., Racz, P., Tenner-Racz, K., Letvin, N.L., Wolinsky, S., and Haase, A.T., Sexual transmission and propagation of SIV and HIV in resting and activated CD4+ T cells, *Science,* 286, 1353–1357, 1999.

52. Cavert, W., Notermans, D.W., Staskus, K., Wietgrefe, S.W., Zupancic, M., Febhard, K., Henry, K., Zhang, Z.-Q., Mills, R., McDade, H., Goudsmith, J., Danner, S. A., and Haase, A.T., Kinetics of response in lymphoid tissues to antiretroviral therapy of HIV-1 infection, *Science,* 276, 960–996, 1997.

53. Derdeyn, C.A., Kilby, J.M., Diego Miralles, G., Li, L-F., Sfakianos, G., Saag, M.S., Hockett, R.D., and Bucy, R.P., Evaluation of distinct blood lymphocyte populations in human immunodeficiency virus type 1 infected subjects in the absence or presence of effective therapy, *J. Infect. Dis.,* 180, 1851–1862, 1999.

54. Harrigan, P.R., Whaley, M., and Montaner, J.S.G., Rate of HIV-1 RNA rebound upon stopping antiretroviral therapy, *AIDS,* 13, F59–F62, 1999.

55. Montaner, J.S.G., Harris, M., Mo, T., and Harrigan, R.P., Rebound of plasma HIV viral load following prolonged suppression with combination therapy, *AIDS,* 12, 1398–1399, 1998.

56. Lisziewicz, J., Rosenberg, E., Lieberman, J., Jessen, H., Lopalco, L., Siliciano, R., Walker, B., and Lori, F., Control of HIV despite the discontinuation of antiretroviral therapy, *N. Engl. J. Med.,* 340, 1683–1684, 1999.

57. Chun, T.-W., Davey, R.T., Ostrowsi, M.Jr., Justement, J.S., Engel, D., Mullins, J.I., and Fauci, A.S., Relationship between pre-existing viral reservoirs and the re-emergence of plasma viremia after discontinuation of highly active anti-retroviral therapy, *Nat. Med.,* 6, 757–761, 2000.

58. Zhang, L., Chung, C., Hu, B-S., He, T., Guo, Y., Kim, A.J., Skulsky, E., Jin, X., Hurley, A., Ramratnam, B., Markowitz, M., and Ho, D.D., Genetic characterization of rebounding HIV-1 after cessation of highly active anti-retroviral therapy, *J. Clin. Invest.,* 106, 839–845, 2000.

59. Zhang, H., Dornadula, G., Wu, Y., Havlir, D., Richman, D.D., and Pomerantz, R.J., Kinetic analysis of intravirion reverse transcription in the blood plasma of human immunodeficiency virus type 1-infected individuals: direct assessment of resistance to reverse transcriptase inhibitors *in vivo*, *J. Virol.,* 70, 628–634, 1996.

60. Dornadula, G., Zhang, H., VanUitert, B., Stern, J., Livornese, L., Ingerman, M.J., Witek, J. Jr., Kedanis, R.J., Natkin, J., Desimone, J., and Pomerantz, R.J., Residual HIV-1 RNA in the blood plasma of patients on suppressive highly active anti-retroviral therapy (HAART), *JAMA,* 282, 1627–1632, 1999.

61. Finzi, D., Blankson, J., Siliciano, J.S., Margolick, J.B., Chadwick, K., Person, T., Smith, K., Lisziewicz, J., Lori, F., Flexner, C., Quinn, T.C., Chaisson, R.E., Rosenberg, E., Walker, B., Gange, S., Gallant, J., and Siliciano, R.F., Latent infection of CD4+ T-cells provides a mechanism for lifelong persistence of HIV-1 even in patients on effective combination therapy, *Nat. Med.,* 5, 512–517, 1999.

62. Ramratnam, B., Mittler, J.E., Zhang, L., Boden, D., Hurley, A., Fang, F., Macken, C.A., Perelson, A.S., Markowitz, M., and Ho, D.D., The decay of the latent reservoir of replication-competent HIV-1 is inversely correlated with the extent of residual viral replication during prolonged anti-retroviral therapy, *Nat. Med.,* 6, 82–85, 1999.

63. Havlir, D.V., Bassett, R., Levitan, D., Gilbert, P., Tebas, P., Collier, A.C., Hirsch, M.S., Ignacio, C., Condra, J., Gunthard, H.F., Richman, D.D., and Wong, J.K., Prevalence and predictive value of intermittent viremia with combination HIV therapy, *JAMA,* 286, 171–179, 2001.

64. Sklar, P.A., Ward, D.J., Baker, R.K., Wood, K.C., Gafoor, Z., Alzola, C.F., Moorman, A.C., and Holmberg, S.D., HIV Outpatient Study (HOPS) Investigators. Prevalence and clinical correlates of HIV viremia ("blips") in patients with previous suppression below the limits of quantification, *AIDS,* 16, 2035–2041, 2002.

65. Easterbrook, P.J., Ives, N., Waters, A., Mullen, J., O'Shea, S., Peters, B., and Gazzard, B.G., The natural history and clinical significance of intermittent viraemia in patients with initial viral suppression to <400 copies/ml, *AIDS,* 16, 1521–1527, 2002.

66. Pomerantz, R.J., Kuritzkes, D.R., de la Monte, S.M., and Hirsch, M.S., Infection of the retina by human immunodeficiency virus type 1, *N. Engl. J. Med.,* 317, 1643–1647, 1987.

67. Huisman, M.T., Smit, J.W., and Schinkel, A.H., Significance of P-glycoprotein for the pharmacology and clinical use of HIV protease inhibitors, *AIDS,* 14, 237–242, 2000.

68. Pomerantz, R.J., Residual HIV-1 infection during antiretroviral therapy: the challenge of viral persistence, *AIDS,* 15, 1201–1211, 2001.
69. Gupta, P., Mellors, J., Kingsley, L., Riddler, S., Singh, M.K., Schreiber, S., Cronin, M., Rinaldo, C.R., High viral load in semen of human immunodeficiency virus type 1-infected men at all stages of disease and its reduction by therapy with protease and nonnucleoside reverse transcriptase inhibitors, *J. Virol.,* 71, 6271–6275, 1997.
70. Cu-Uvin, S., Caliendo, A.M., Reinert, S., Chang, A., Juliano-Remollino, C., Flanigan, T.P., Mayer, K.H., and Carpenter, C.C.J., Effect of highly active antiretroviral therapy on cervicovaginal HIV-1 RNA, *AIDS,* 14, 415–421, 2000.
71. Coombs, R.W., Reichelderfer, P.S., and Landay, A.L., Recent observations on HIV type-1 infection in the genital tract of men and women, *AIDS,* 17, 455–480, 2003.
72. Bagasra, O., Farzadegan, H., Seshamma, T., Oakes, J.W., Saah, A., and Pomerantz, R.J., Detection of HIV-1 proviral DNA in sperm from HIV-1-infected men, *AIDS,* 8, 1669–1674, 1994.
73. Quayle, A.J., Xu, C., Mayer, K.H., and Anderson, D.J., T lymphocytes and macrophages, but not motile spermatozoa, are a significant source of human immunodeficiency virus in semen, *J. Infect. Dis.,* 176, 960–968, 1997.
74. Zhu, T., Wang, N., Carr, A., Nam, D.S., Moor-Jankowski, R., Cooper, D.A., and Ho, D.D., Genetic characterization of human immunodeficiency virus type 1 in blood and genital secretions: evidence for viral compartmentalization and selection during sexual transmission, *J. Virol.,* 70, 3098–3107, 1996.
75. Liuzzi, G., Chirianni, A., Clementi, M., Bagnarelli, P., Valenza, A., Cataldo, P.T., and Piazza, M., Analysis of HIV-1 load in blood, semen and saliva: evidence for different viral compartments in a cross-sectional and longitudinal study, *AIDS,* 10, F51–F56, 1996.
76. Paranjpe, S., Craigo, J., Patterson, B., Ding, M., Barroso, P., Harrison, L., Montelaro, R., and Gupta, P., Subcompartmentalization of HIV-1 quasispecies between seminal cells and seminal plasma indicates their origin in distinct genital tissues, *AIDS Res. and Human Retroviruses,* 17, 1271–1280, 2002.
77. Kiessling, A.K., Zheng, G., and Eyre, R.C., Semen producing organs are an isolated reservoir of HIV-1 which may play a significant role in the development of drug resistant strains, *J. Hum. Virol.,* 2, 193, 1992.
78. Curran, R., and Ball, J.K., Concordance between semen-derived HIV-1 proviral DNA and viral RNA hypervariable region 3 (V3) envelope sequences in cases where semen populations are distinct from those present in blood, *J. Med. Virol.,* 67, 9–19, 2002.
79. Pomerantz, R.J., de la Monte, S.M., Donegan, S.P., Rota, T.R., Vogt, M.W., Craven, D.E., and Hirsch, M.S., Human immunodeficiency virus (HIV) infection of the uterine cervix, *Ann. Intern. Med.,* 108, 321–327, 1988.
80. Shaheen, F., Sison, A.V., McIntosh, L., Mukhtar, M., and Pomerantz, R.J., Analysis of HIV-1 in the cervicovaginal secretions and blood of pregnant and nonpregnant women, *J. Hum. Virol.,* 2, 154–166, 1999.
81. Wu, Z., Chen, Z., and Phillips, D.M., Human genital epithelial cells capture cell-free human immunodeficiency virus type 1 and transmit the virus to CD4+ cells: implications for mechanisms of sexual transmission, *J. Infect. Dis.,* 188, 1473–1482, 2003.
82. Asin, S.N., Wildt-Perinic, D., Mason, S.I., Howell, A.L., Wira, C.R., and Fanger, M.W., Human immunodeficiency virus type 1 infection of human uterine epithelial cells: viral shedding and cell contact-mediated infectivity, *J. Infect. Dis.,* 187, 1522–1533, 2003.
83. Fiore, J.R, Suligoi, B., Saracino, A., Di Stefano, M., Bugarini, R., Lepera, A., Favia, A., Monno, L., Angarano, G., and Pastore, G., Correlates of HIV-1 shedding in cervicovaginal secretions and effects of antiretroviral therapies, *AIDS,* 17, 2169–2176, 2003.
84. Kovacs, A., Wasserman, S.S., Burns, D., Wright, D.J., Cohn, J., Landay, A., Weber, K., Cohen, M., Levine, A., Minkoff, H., Miotti, P., Palefsky, J., Young, M., and Reichelderfer, P., DATRI Study Group, WIHS Study Group, Determinants of HIV-1 shedding in the genital tract of women, *Lancet,* 358, 1593–1601, 2001.
85. Kemal, K.S., Foley, B., Burger, H., Anastos, K., Minkoff, H., Kitchen, C., Philpott, S.M., Gao, W., Robison, E., Holman, S., Dehner, C., Beck, S., Meyer, W.A., III, Landay A., Kovacs, A., Bremer, J., and Weiser, B., HIV-1 in genital tract and plasma of women: compartmentalization of viral sequences, coreceptor usage, and glycosylation, *Proc. Natl. Acad. Sci. U.S.A.,* 100, 12972–12977, 2003.

86. De Pasquale, M.P., Leigh Brown, A.J., Uvin, S.C., Allega-Ingersoll, J., Caliendo, A.M., Sutton, L., Donahue, S., and D'Aquila, R.T., Differences in HIV-1 pol sequences from female genital tract and blood during antiretroviral therapy, *J. Acquir. Immune Defic. Syndr.,* 34, 37–44, 2003.

87. Chun, T.-W., Engel, D., Mizell, S.B., Hallahan, C.W., Fischette, M., Park, S., Davey, R.T., Dybul, M. Jr., Kovacs, J.A., Metcalf, J.A., Mican, J.M., Berrey, M.M., Corey, L., Lane, H.C., and Fauci, A.S., Effect of interleukin-2 on the pool of latently infected resting CD4+ T cells in HIV-1-infected patients receiving highly active anti-retroviral therapy, *Nat. Med.,* 5, 651–655, 1999.

88. Prins, J.M., Jurrianns, S., van Praag, R.M.E., Blaak, H., R.V.R., Schellekens, P.T.A., Berge, I.J.M.T., Yong, S-L., Fox, C.H., Roos, M.T.L., Goudsmit, F.D.W.J., Schuitemaker, H., and Lange, J.M.A., Immunoactivation with anti-CD3 and recombinant human IL-2 and HIV-1-infected patients on potent antiretroviral therapy, *AIDS,* 13, 2405–2410, 1999.

89. Chun, T.-W., Davey, R.T., Engel, D. Jr., Lane, H.C., and Fauci, A.S., Re-emergence of HIV after stopping therapy, *Nature,* 401, 874–875, 1999.

90. Davey, R.T., Bhat, N., Jr., Yoder, C., Chun, T-W., Metcalf, J. A., Dewar, R., Natarajan, V., Lempicki, R.A., Adelsberger, J.W., Miller, K.D., Kovacs, J.A., Polis, M.A., Walker, R.E., Falloon, J., Masur, H., Gee, D., Baseler, M., Dimitrov, D.S., Fauci, A.A., and Lane, H.C., HIV-1 and T cell dynamics after interruption of highly active antiretroviral therapy (HAART) in patients with a history of sustained viral suppression, *Proc. Natl. Acad. Sci. U.S.A.,* 96, 15109–15114, 1999.

91. Vella, A.T., Dow, S., Potter, T.A., Kappler, J., and Marrack, P., Cytokine-induced survival of activated T cells *in vitro* and *in vivo*, *Proc. Natl. Acad. Sci. U.S.A.,* 95, 3810–3815, 1998.

92. Vella, A., Teague, T.K., Ihle, J., Kappler, J., and Marrack, P. Interleukin 4 (IL-4) or IL-7 prevents the death of resting T cells: stat6 is probably not required for the effect of IL-4, *J. Exp. Med.,* 186, 325–330, 1997.

93. Boise, L.H., Minn, A.J., June, C.H., Lindsten, T., and Thompson, C.B. Growth factors can enhance lymphocyte survival without committing the cell to undergo cell division, *Proc. Natl. Acad. Sci. U.S.A.,* 92, 5491–5495, 1995.

94. Komschlies, K.L., Gregorio, T.A., Gruys, M.E., Back, T.C., Faltynek, C.R., and Wiltrout, R.H., Administration of recombinant human IL-7 to mice alters the composition of B-lineage cells and T cell subsets, enhances T cell function, and induces regression of established metastases, *J. Immunol.,* 152, 5776–5784, 1994.

95. Grzegorzewski, K.J., Komschlies, K.L., Jacobsen, S.E., Ruscetti, F.W., Keller, J.R., and Wiltrout, R.H., Mobilization of long-term reconstituting hematopoietic stem cells in mice by recombinant human interleukin 7, *J. Exp. Med.,* 181, 369–374, 1995.

96. Carini, C., and Essex, M., Interleukin 2-independent interleukin 7 activity enhances cytotoxic immune response of HIV-1-infected individuals, *AIDS Res. Hum. Retroviruses,* 10, 121–130, 1994.

97. Bui, T., Dykers, T., Hu, S.L., Faltynek, C.R., and Ho, R.J. Effect of MTP-PE liposomes and interleukin-7 on induction of antibody and cell-mediated immune responses to a recombinant HIV-envelope protein, *J. Acquir. Immune. Defic. Syndr.,* 7, 799–806, 1994.

98. Napolitano, L.A., Grant, R.M., Deeks, S.G., Schmidt, D., De Rosa, S.C., Herzenberg, L.A., Herndier, B.G., Andersson, J., and McCune, J.M., Increased production of IL-7 accompanies HIV-1-mediated T-cell depletion: implications for T-cell homeostasis, *Nat. Med.,* 7, 73–79, 2001.

99. Smithgall, M.D., Wong, J.G., Critchett, K.E., and Haffar, O.K., IL-7 up-regulates HIV-1 replication in naturally infected peripheral blood mononuclear cells, *J. Immunol.,* 156, 2324–2330, 1996.

100. Chene, L., Nugeyre, M.T., Guillemard, E., Moulian, N., Barre-Sinoussi, F., and Israel, N., Thymocyte-thymic epithelial cell interaction leads to high-level replication of human immunodeficiency virus exclusively in mature CD4(+) CD8(−) CD3(+) thymocytes: a critical role for tumor necrosis factor and interleukin-7, *J. Virol.,* 73, 7533–7542, 1999.

101. Moran, P.A., Diegel, M.L., Sias, J.C., Ledbetter, J.A., and Zarling, J.M., Regulation of HIV production by blood mononuclear cells from HIV-infected donors: I. Lack of correlation between HIV-1 production and T cell activation, *AIDS Res. Hum. Retroviruses,* 9, 455–464, 1993.

102. Scripture-Adams, D.D., Brooks, D.G., Korin, Y.D., and Zack, J.A., Interleukin-7 induces expression of latent human immunodeficiency virus type 1 with minimal effects on T-cell phenotype, *J. Virol.,* 76, 13077–13082, 2002.

103. Soares, M.V., Borthwick, N.J., Maini, M.K., Janossy, G., Salmon, M., and Akbar, A.N., IL-7-dependent extrathymic expansion of CD45RA+ T cells enables preservation of a naive repertoire, *J. Immunol.,* 161, 5909–5917, 1998.

104. Unutmaz, D., KewalRamani, V.N., Marmon, S., and Littman, D.R., Cytokine signals are sufficient for HIV-1 infection of resting human T lymphocytes, *J. Exp. Med.,* 189, 1735–1746, 1999.

105. Havlir, D.V., Gilbert, P.B., Bennett, K., Collier, A.C., Hirsch, M.S., Tebas, P., Adams, E.M., Wheat, L.J., Goodwin, D., Schnittman, S., Holohan, M.K., and Richman, D.D., Effects of treatment intensification with hydroxyurea in HIV-infected patients with virologic suppression, *AIDS,* 15, 1379–1388, 2001.

106. Lori, F., Malykh, A., Cara, A., Sun, D., Weinstein, J.N., Lisziewicz, J., and Gallo, R.C., Hydroxyurea as an inhibitor of human immunodeficiency virus type I replication, *Science,* 266, 801–804, 1994.

107. Vila, J., Nugier, F., Bargues, G., Vallet, T., Peyramond, D., Hamedi-Sangsari, F., and Seigneurin, J.-M., Absence of viral rebound after treatment of HIV-infected patients with didanosine and hydroxy-carbamide, *Lancet,* 2, 88–89, 1997.

108. Zala, C., Salomon, H., Ochoa, C., Kijak, G., Federico, A., Perez, H., Montaner, J.S., and Cahn, P., Higher rate of toxicity with no increased efficacy when hydroxyurea is added to a regimen of stavudine plus didanosine and nevirapine in primary HIV infection, *J. Acquir. Immune. Defic. Syndr.,* 29, 368–373, 2002.

109. DeBoer, R.J., Target cell availability and the successful suppression of HIV by hydroxyurea and didanosine, *AIDS,* 12, 1567–1570, 1998.

110. Smith, C., Lilly, S., and Miralles, G.D., Treatment of HIV infection with cytoreductive agents, *AIDS Res. Human Retroviruses,* 14, 1305–1313, 1998.

111. Berger, E.A., Moss, B., and Pastan, I., Reconsidering targeted toxins to eliminate HIV infection: you gotta have HAART, *Proc. Natl. Acad. Sci. U.S.A.,* 95, 11511–11513, 1998.

112. Deeks, S.G., Barbour, J.D., Martin, J.N., Swanson, M.S., and Grant, R.M., Sustained CD4+ T-cell response after virologic failure to protease inhibitor- based regimens in patients with human immunodeficiency virus infection, *J. Infect. Dis.,* 181, 946–953, 2000.

113. McCune, J.M., Loftus, R., Schmidt, D.K., Carroll, P., Webster, D., Swor-Yim, L.B., Francis, I.R., Gross, B.H., and Grant, R.M., High prevalence of thymic tissue in adults with human immunodeficiency virus 1 infection, *J. Clin. Invest.,* 101, 2301–2308, 1998.

114. Chapius, A.G., Rizzzardi, G.P., D'Agostino, C., Attinger, A., Knabenhans, C., Fleury, S., Acha-Orbea, H., and Pantaleo, G., Effects of mycophenolic acid on human immunodeficiency virus infection *in vitro* and *in vivo, Nat. Med.,* 6, 762–768, 2000.

115. Douek, D.C., McFarland, R.D., Keiser, P.H., Gage, E.A., Massey, J.M., Haynes, B.F., Polis, M.A., Haase, A.T., Feinberg, M.B., Sullivan, J.L., Jamieson, B.D., Zack, J.A., Picker, L.J., and Koup, R.A., Changes in thymic function with age and during the treatment of HIV infection, *Nature,* 396, 690–695, 1998.

116. Ortiz, G.M., Nixon, D.F., Trkola, A., Binley, J., Jin, X., Bonhoeffer, S., Kuebler, P.J., Donahoe, S.M., Demoitie, M-A., Kakimoto, W.M., Ketas, T., Clas, B., Heymann, J.J., Zhang, L., Cao, Y., Hurley, A., Moore, J.P., Ho, D.D., and Markowitz, M., HIV-1-specific immune responses in subjects who temporarily contain virus replication after discontinuation of highly active antiretroviral therapy, *J. Clin. Invest.,* 104, R13–R18, 1999.

117. Kalams, S.A., Goulder, P.J., Shea, A.K., Jones, N.G., Trocha, A.K., Ogg, G.S., and Walker, B.D., Levels of human immunodeficiency virus type 1 specific cytotoxic T-lymphocyte effector and memory responses decline after suppression of viremia with highly active antiretroviral therapy, *J. Virol.,* 73, 6721–6728, 1999.

118. Rosenberg, E.S., Billingsley, J.M., Calieno, A.M., Boswell, S.L., Sax, P.E., Kalams, S.A., and Walker, B.D., Vigorous HIV-1-specific CD4+ T-cell responses associated with control of viremia, *Science,* 278, 11447–11450, 1997.

119. Al-Haarthi, L., Siegel, J., Spritzler, J., Pottage, J., Agnoli, M., and Landay, A., Maximum suppression of HIV replication leads to the restoration of HIV-specific responses in early HIV disease, *AIDS,* 14, 761–770, 2000.

120. Deeks, S.G., Wrin, T., Liegler, T., Hoh, R., Hayden, M., Barbour, J.D., Hellmann, N.S., Petropoulos, C.J., McCune, J.M., Hellerstein, M.K., and Grant, R.M., Virologic and immunologic consequences of discontinuing combination antiretroviral-drug therapy in HIV-infected patients with detectable viremia, *N. Engl. J. Med.,* 344, 472–480, 2001.

121. Garcia, F., Plana, M., Ortiz, G.M., Bonhoeffer, S., Soriano, A., Vidal, C., Cruceta, A., Arnedo, M., Gil, C., Pantaleo, G., Pumarola, T., Gallart, T., Nixon, D.F., Miro, J.M., and Gatell, J.M., The virological and immunological consequences of structured treatment interruptions in chronic HIV-1 infection, *AIDS,* 15, F29–F40, 2001.

122. Ioannidis, J.P., Havlir, D.V., Tebas, P., Hirsch, M.S., Collier, A.C., and Richman, D.D., Dynamics of HIV-1 viral load rebound among patients with previous suppression of viral replication, *AIDS,* 14, 1481–1488, 2000.

123. Carcelain, G., Tubiana, R., Samri, A., Calvesz, V., Delaugerre, C., Augt, H., Katlama, C., and Autran, B., Transient mobilization of human immunodeficiency virus (HIV)-specific CD4 T-helper cells fails to control virus rebounds during intermittent antiretroviral therapy in chronic HIV type 1 infection, *J. Virol.,* 75, 234–241, 2001.

124. Dybul, M., Nies-Kraske, E., Daucher, M., Hertogs, K., Hallahan, C.W., Csako, G., Yoder, C., Ehler, L., Sklar, P.A., Belson, M., Hidalgo, B., Metcalf, J.A., Davey, R.T., Rock Kress, D.M., Powers, A., and Fauci, A.S., Long-cycle structured intermittent vs. continuous highly active antiretroviral therapy for the treatment of chronic infection with human immunodeficiency virus: effects on drug toxicity and on immunologic and virologic parameters, *J. Infect. Dis.,* 188, 388–396, 2003.

125. Ortiz, G.M., Wellons, M., Brancato, J., Vo, H.T., Zinn, R.L., Clarkson, D.E., Van Loon, K., Bonhoeffer, S., Miralles, G.D., Montefiori, D., Bartlett, J.A., and Nixon, D.F., Structured antiretroviral treatment interruptions in chronically HIV-1-infected subjects, *Proc. Natl. Acad. Sci. U.S.A.,* 98, 13288–13293, 2001.

126. Markowitz, M., Vesanen, M., Tenner-Racz, K., Cao, Y., Binley, J.M., Talal, A., Hurley, A., Ji, X., Chaudhry, R., Yaman, M., Frankel, S., Heath-Chiozzi, M., Leonard, J.M., Moore, J.P., Racz, P., Nixon, D.F., and Ho, D.D., The effect of commencing combination antiretroviral therapy soon after human immunodeficiency virus type 1 infection on viral replication and antiviral immune responses, *J. Infect. Dis.,* 179, 525–537, 1999.

127. Rosenberg, E.S., Altfeld, M., Poon, S.H., Phillips, M.N., Wilkes, B.M., Eldridge, R.L., Robbins, G.K., D'Aquila, R.T., Goulder, P.J.R., and Walker, B.D., Immune control of HIV-1 after early treatment of acute infection, *Nature,* 407, 523–526, 2000.

128. Hel, Z., Venzon, D., Poudyal, M., Tsai, W-P., Giuliani, L., Woodward, R., Chougnet, C., Shearer, G., Altman, J.D., Watkins, D., Bischofberger, N., Abimiku, A., Markham, P., Tartaglia, J., and Franchini, G., Viremia control following antiretroviral treatment and therapeutic immunization during primary SIV251 infection of macaques, *Nat. Med.,* 6, 1140–1146, 2000.

27 Therapeutic Approaches to Modulation of Cell Death (non-HIV)

Dara Ditsworth
Rebecca L. Elstrom
Craig B. Thompson

CONTENTS

INTRODUCTION

Cell death contributes to the pathogenesis of a wide variety of diseases. Three distinct forms of cell death are recognized to occur in mammalian cells. Necrosis was traditionally considered an unregulated, rapidly induced form of pathological cell death. Often resulting from bioenergetic failure of the cell, necrosis is manifested by cell swelling, rupture of the plasma membrane, and disintegration and release of cellular contents, leading to a marked inflammatory response. In contrast, the genetically programmed apoptotic form of cell death requires maintenance of adenosine triphosphate (ATP) levels for execution and is characterized by cell shrinkage, chromatin condensation and internucleosomal DNA fragmentation, and lack of inflammation. Autophagy is another form of cell death that is increasingly recognized as having an important role during development and tumor suppression.

Most, if not all, cancer cells contain aberrations in the genetic pathway of programmed cell death in order to enable deregulated survival. An altered ratio of antiapoptotic to proapoptotic factors can occur due to overexpression of antiapoptotic proteins (via chromosomal translocations or amplifications), loss of proapoptotic proteins (via deletions or loss of function point mutations), or excessive survival signaling (via activating mutations that render survival kinases constitutively active).

This chapter will highlight a few important examples of these aberrations and discuss current therapeutic strategies targeted at modulating cell death summarized in Tables 27.1 and 27.2. Our intent is to provide a broad overview of the approaches being taken to therapeutically regulate cell survival. For a more detailed discussion, readers are referred to specific review articles as indicated.

TABLE 27.1
Strategies to Induce Cell Death

Category	Target/Strategy	Drug	Examples of Relevant Diseases	Trial Status	Ref.
Death receptors	TRAIL receptor	TRAIL	Multiple malignancies	Preclinical (in vivo)	11–17
IAPs	XIAP antisense		Ovarian cancer, lung cancer	Preclinical (in vivo)	23–25
	Survivin antisense	ISIS 23722	Colon, brain, lung, and skin cancers	Phase I	26
	Small molecule inhibitors of XIAP		Multiple malignancies	Preclinical (in vivo)	27–28
	Smac peptides		Glioblastoma, non-small cell lung cancer	Preclinical (in vivo)	29, 30

TABLE 27.1
Strategies to Induce Cell Death

Category	Target/Strategy	Drug	Examples of Relevant Diseases	Trial Status	Ref.
Caspases	Caspase activator		Multiple malignancies	Preclinical (in vitro)	39–40
Bcl-2 family	Bcl-2 antisense	Genasense	Non-Hodgkin's lymphoma, leukemia, melanoma, and small cell lung carcinoma	Phase III, Phase I with chemotherapy	52–55
	Small molecule inhibitor of Bcl-xL BH3 binding	BH3I-1, BH3I-2		Preclinical (in vivo)	58
	Small molecule inhibitor of Bcl-2 BH3 binding	Antimycin A		Preclinical (in vivo)	59
	Natural inhibitor of Bcl-xL BH3 binding	Chelerythrine		Preclinical (in vitro)	61
p53	p53 gene replacement	INGN201, SCH58500, Gendicine	Head and neck cancer, ovarian cancer	Clinical trials, approved (China)	83–86
	Oncolytic virus	ONYX-015	Head and neck cancer, oral dysplasia	Phase II, Phase I	88–89
	p53 C-terminus peptide	TAT-p53C	Peritoneal lymphoma	Preclinical (in vivo)	95
	Rescue of mutant p53 conformation/function	CP-31398		Preclinical (in vivo)	90–92
	Restoration of mutant p53 function	Prima 1		Preclinical (in vivo)	78, 93
	Disruption of p53-MDM2 interaction	Nutlins		Preclinical (in vivo)	104
	Disruption of MDM2	MDM2 antisense	Colon cancer, prostate cancer, and lymphoid cancers	Clinical trials	101–103
Survival kinase pathways	H-ras antisense	ISIS 2503	Advanced carcinoma	Phase I	113
	HER2 monoclonal antibody	Herceptin	Breast cancer	Approved	114, 115
	Inhibitor of Bcr-Abl, c-KIT, PDGFR	Imatinib mesylate (Gleevec)	CML, gastrointestinal stromal tumors	Approved	116–118
	EGFR inhibitor	Iressa	Non-small cell lung carcinoma	Approved	119

TABLE 27.1 (Continued)
Strategies to Induce Cell Death

Category	Target/Strategy	Drug	Examples of Relevant Diseases	Trial Status	Ref.
	mTOR inhibitors	RAD001, CC1779	Breast, pancreatic, brain, and renal cancers	Phase II, Phase III	123
	BRAF inhibitor	BAY-43-9006	Melanoma	Phase III	127
	RAF antisense	ISIS 5132	NSCLC, prostate cancer	Phase II	128
	IKK inhibitor	Various	Hematologic malignancies	Preclinical	135
	E1A expression to inhibit multiple kinases	Tg-DCC-E1A	Breast, ovarian, head, and neck cancers	Phase I, Phase II	145
Proteasome	26S proteasome inhibitor	PS-341	Hematologic malignancies	Clinical trials	159
Gene silencing	DNA methylation inhibitor	Decitabine, 5-azacytidine	Hematologic malignancies	Phase II, Phase III	164, 165
	HDAC inhibitor	Depsipeptide	Hematologic malignancies		166, 167
ROS	Photodynamic therapy	Porphyrin, chlorin	Solid tumors	Phase II	204

CELL DEATH BY MURDER

DEATH RECEPTOR PATHWAYS

In some cases, apoptosis is initiated through ligation of specific cell-surface receptors. These death receptors belong to the tumor necrosis factor receptor (TNFR) family, and the best studied members include Fas and TNFR1. Death receptors exist as trimers at the cell surface and are activated upon binding of their respective ligands. Upon ligation, the intracellular domains of these receptors recruit a death-inducing signaling complex (DISC), which initiates a caspase cascade by activating caspase-8 or caspase-10. Caspase-8 cleavage of Bid further amplifies the death signal via cross talk with the mitochondrial pathway, as discussed in earlier chapters.

Ligand binding to TNFR can also activate adapter proteins that recruit components of the mitogen-activated protein kinase (MAPK) cascade. This leads to activation of Jun-*N*-terminal kinase (JNK) and c-jun, which participate in receptor-mediated apoptosis by controversial mechanisms.[1] Other adapter proteins recruit signaling components of the NF-κB pathway, which prevents apoptosis by inducing expression of survival molecules and proinflammatory cytokines. TNFR ligation primarily stimulates inflammation but can induce apoptosis when NF-κB signaling is blocked.

Another ligand–receptor pair that is increasingly understood to play a role in immune regulation and cancer is the TNF-related apoptosis-inducing ligand (TRAIL) and its receptors. TRAIL-induced signaling pathways are similar to those activated by TNFR1. However, TRAIL can also bind to decoy receptors that lack intracellular signaling capability.[2] The decoy receptors sequester TRAIL from competent signaling receptors and, therefore, block apoptosis. Expression of decoy receptors provides one mechanism by which normal cells escape TRAIL-induced death. On the other hand,

TABLE 27.2
Strategies to Prevent Cell Death

Category	Target/Strategy	Drug	Examples of Relevant Diseases	Trial Status	Ref.
IAPs	IAP gene therapy	Ad-XIAP, Ad-NAIP	Cerebral ischemia	Preclinical (in vivo)	32, 33
Caspases	Caspase inhibitors	IDN 6556, z-VAD	Myocardial infarction, sepsis, liver disease	Phase I	34–37
Bcl-2 family	Bcl-xL protein expression	TAT-Bcl-XL	Cerebral ischemia	Preclinical (in vivo)	68
	Calcineurin inhibitor	FK506, cyclosporine A	Traumatic brain injury, ischemia, neurodegeneration	Preclinical (in vivo)	69, 71–73
	Inhibitors of Bax	Propranolol, dibucaine	Ischemia	Preclinical (in vitro)	74
Survival kinase pathways	Neurotrophins conjugated to BBB-delivery vehicle	Bio-bFGF/ OX26-SA	Ischemia, neurodegenerative diseases	Preclinical (in vivo)	150, 151
	BBB-permeable neurotrophin mimetics		Ischemia, neurodegenerative diseases	Preclinical (in vivo)	152
	Recombinant EPO	rhEPO	Stroke, cardiac ischemia	Phase I, Phase II	155
Ca2+ homeostasis	NMDAR antagonist	Memantine	Alzheimer's disease	Approved	174, 175
ROS	Antioxidant mimetic	EUK-8, EUK-134	Ischemia, neurodegenerative diseases	Preclinical (in vivo)	198–202
	Antioxidant chemical	N-acetyl-cysteine, tocopherol	Neurodegenerative diseases	Preclinical (in vivo)	188
	Inhibitors of ceramide synthesis	ISP-1	Neurodegenerative diseases	Preclinical (in vitro)	188
PARP/necrosis	PARP inhibitors	3-AB, DHIQ, DPQ, GPI-6150	Ischemia, septic shock	Preclinical (in vivo)	208

tumor cells exhibit lower expression of decoy receptors due to promoter methylation.[3] This may be one explanation for increased sensitivity of tumor cells to TRAIL-induced apoptosis.

Proposed physiological functions for endogenous TRAIL involve immune surveillance against oncogene-transformed and virus-infected cells. A role in immune surveillance against tumors is supported by the increased incidence of metastases in mice with mutagen-induced tumors when endogenous TRAIL is eliminated by gene knockout[4] or neutralizing antibody.[5] TRAIL was also proposed to play a role in the interferon (IFN)-dependent host defense against viral infection. IFNs prevent viral replication by inducing apoptosis in infected cells, and they regulate TRAIL expression.[6] For example, in response to IFN, cytomegalovirus-infected human fibroblasts become more sensitive to TRAIL, whereas neighboring uninfected cells increase expression of the TRAIL ligand and downregulate death receptors. *In vivo* studies have shown that TRAIL expression by natural killer cells is critical for limiting viral replication and, furthermore, that TRAIL induces apoptosis of infected cells in a mouse model of adenoviral hepatitis.[7] Some viruses directly target the TRAIL pathway to overcome immune surveillance and protect infected cells from apoptotic death, as adenoviruses were shown to induce receptor internalization.[8]

Therapeutic Approaches

Modulation of the death receptor pathway occurs at several levels. Decoy receptors compete with death receptors for ligand binding and were described for both TRAIL and Fas.[9] Downstream of receptor ligation, cellular-FLICE inhibitory protein (c-FLIP) competes with caspase-8 and -10 for DISC binding, and inhibitors of apoptosis proteins (IAPs) block effector caspase activation. Each of these steps can be deregulated in tumor cells to promote survival, and each can be pharmacologically targeted to induce cell death.

Whereas attempts at developing death-receptor-targeted chemotherapeutics, such as anti-Fas antibodies or recombinant TNF ligand, were hindered by toxicity,[10] strategies using TRAIL hold more promise. Multiple studies in mice and nonhuman primates showed minimal toxicity to most normal cell types when treated with TRAIL,[11,12] although some concern was raised regarding the toxicity to normal human hepatocytes.[13] Preclinical studies of TRAIL characterized its use as a single agent[14] and showed significant synergistic effects with traditional chemotherapeutics, such as gemcitabine[15] or CPT-11,[16] in both cell lines and in rodent xenograft models. Combination therapy of TRAIL with the proteasome inhibitor PS-341 further enhances apoptosis by blocking NF-κB signaling.[17]

INHIBITOR OF APOPTOSIS PROTEINS (IAPs)

IAP family proteins were originally discovered in baculovirus-infected cells and are characterized by one or more 70-80 amino acid baculoviral IAP repeat (BIR) domains.[18] Subsequent studies identified cellular homologues[19,20] and showed that IAPs inhibit caspase activation via two mechanisms. IAPs can block self-cleavage of initiator caspases by direct binding of the BIR domain, and they can also target activated effector caspases for proteasomal degradation with the E3 ubiquitin ligase activity of their RING domains. IAPs are inhibited by the proapoptotic molecules Smac/DIABLO and Omi/HtrA2, which are released from the mitochondria upon the loss of mitochondrial membrane integrity during apoptosis.

Inhibition of IAPs in Cancer Therapy

IAP family members are highly expressed in many types of cancer.[21] Beyond overexpression correlations, direct genetic evidence for an oncogenic role of IAP members is found in mucosa-associated lymphoid tissue (MALT) B cell lymphomas, in which translocation events result in fusion of the BIR domains of cIAP-2 to the MALT protein.[22] In many cancers, downregulation of the overexpressed IAP protein alone might be sufficient to induce apoptosis or at least sensitize cells to other chemotherapeutics. For example, antisense oligonucleotides targeting X-linked IAP (XIAP) were shown to induce cell death in chemotherapy-resistant ovarian cancer cell lines[23,24] and to cooperate with chemotherapy in animal studies.[25] Antisense against the IAP survivin (ISIS 23722) is currently in Phase I studies.[26]

More recent developments include the identification of small molecule antagonists of IAPs, which target either caspase binding or ubiquitin ligase activity. Compounds targeting XIAP were shown to sensitize multiple cancer cell lines to TRAIL-induced apoptosis and to be effective as single agents in tumor xenograft models, although their precise mechanism of action remains unclear.[27] One concern with pharmacologic strategies that specifically target ligase activity is that IAPs might still be able to bind caspases and prevent their activation. Therefore, ligase inhibitors would work better in combination with specific inhibitors of IAP–caspase binding[28] or with agents that induce Smac release. Along these lines, injection of Smac peptides in combination with TRAIL ligand was shown to induce complete tumor regression in human glioma cells in nude mice.[29,30] Although IAP family proteins clearly represent an important target for modulation of cell death, current therapeutic options are still in the experimental stages of development. Strategies to inhibit specific IAPs with small molecules will rapidly advance with more detailed structural studies.

Activation of IAPs to Prevent Neuronal Loss

In contrast to cancer, overexpression of IAP family members can be a protective therapeutic approach to prevent apoptosis in neurodegenerative diseases. Loss of function studies in animal models provided the first evidence that neuronal apoptosis inhibitory protein (NAIP) expression is critical to prevent cell loss.[31] Attempts to therapeutically increase IAP expression showed that injection of adenoviral vectors expressing NAIP or XIAP into the rat hippocampus provide protective effects in a global ischemia model.[32] Adenoviral vectors encoding IAP family members were also shown to suppress apoptosis in a sciatic axotomy model.[33]

CASPASES

Development of therapeutics that modulate apoptosis downstream of mitochondrial permeability primarily focused on the identification of small molecule inhibitors of caspases. Caspase activity is fundamental to execution of both intrinsic and extrinsic apoptotic pathways. Broad-spectrum caspase inhibitors proved to be protective in animal models of ischemia-reperfusion and excitotoxicity[34,35] and reached clinical trials for treatment of diseases such as acute myocardial infarction, sepsis, and liver disease.[36,37] However, to date, caspase inhibitors are unsuccessful at maintaining cell survival in more chronic diseases associated with cell death, such as neurodegenerative diseases or chronic heart failure.

The therapeutic benefit of global caspase inhibition also depends on the specific mechanism contributing to a given disease and can actually exacerbate cell death in disorders involving caspase-independent mechanisms. For example, caspase inhibition was shown to increase toxicity of TNF-α-induced shock *in vivo* by enhancing oxidative stress and mitochondrial damage.[38] In this situation, caspase activity might actually serve a protective function by cleaving phospholipase A_2 and preventing further generation of reactive oxygen species.

In contrast, small molecules that induce apoptosis downstream of mitochondrial permeability were identified through cell-free assays for caspase activation. This strategy identified chemicals that specifically induce oligomerization of apoptotic protease-activating factor 1 (Apaf-1) into the mature apoptosome, causing activation of caspase-9 and -3 without the need for cytochrome *c* release.[39,40] Compounds identified in these screens showed cytotoxic activity against multiple cancer cell lines without affecting normal cells. However, further studies in animal models are needed to determine the efficacy of this strategy for future drug development.

Furthermore, controversy remains over the importance of caspase activation following mitochondrial disruption in determining cell death vs. orchestrating cellular disposal.

CELL DEATH BY SUICIDE

The importance of mitochondria in initiating apoptosis was realized in 1996, when Liu and colleagues demonstrated that cytochrome *c*, a component of the electron transport chain, could initiate apoptosis when released into the cytosol.[41] Upon release from mitochondria, cytochrome *c* forms a complex with Apaf-1 and ATP to activate caspase-9, initiating the intrinsic caspase cascade. Subsequent studies showed that mitochondrial dysfunction and loss of membrane integrity cause release of multiple proteins, which contribute to the cell death program. For example, Smac/DIABLO and Omi/HtrA2 inhibit IAP family proteins to reinforce caspase-9 activation. EndoG and AIF are proposed to be components of caspase-independent pathways leading to nuclear fragmentation.

The mitochondrial pathway of apoptosis is initiated by internal cellular stress, such as DNA damage, cytokine deprivation, hypoxia, and other stimuli. Controversy exists over the mechanism by which mitochondria lose integrity and release their contents into the cytosol. Some hypotheses suggest that mitochondrial dysfunction leads to matrix swelling and eventual rupture of the outer membrane. Others propose the formation of specific pores large enough to release intermembrane components. Under either model, the importance of Bcl-2 family members in regulating mitochondrial membrane integrity is paramount. Another input for regulation occurs at the transcriptional level, by p53, and at the posttranslational level, by survival kinase signaling pathways.

Bcl-2 Family

The first molecularly characterized example of an apoptotic genetic defect in cancer was the t(14:18) chromosomal translocation found in follicular lymphoma patients that juxtaposes the Bcl-2 gene with the immunoglobulin enhancer.[42] Later work revealed that the Bcl-2 gene product promotes oncogenesis by a novel mechanism: rather than inducing cell proliferation, Bcl-2 inhibits the normal programmed cell death of B cells.[43,44] The mechanism by which antiapoptotic proteins Bcl-2 and Bcl-xL promote survival is not completely clear, but one hypothesis involves suppression of the activation of proapoptotic members Bax and Bak.[45] Bax and Bak can initiate apoptosis when they oligomerize at the mitochondria,[46] disrupting membrane integrity to release cytochrome c, or at the endoplasmic reticulum,[47,48] causing disruption of Ca^{2+} homeostasis. Alternatively, Bcl-2 and Bcl-xL were proposed to preserve mitochondrial membrane integrity by promoting the exchange of metabolic substrates across the mitochondrial membrane and allowing maintenance of respiration.[49]

Whereas these main apoptotic effectors contain multiple Bcl-2 homology (BH) domains, other family members share only the BH3 region. Subsequent investigations defined distinct roles for individual proapoptotic BH3-only proteins,[50] with functions that can be divided into two groups based on binding specificity. Bid-like domains exert proapoptotic functions by activating Bax and Bak, and Bad-like domains sensitize cells to apoptosis by occupying the binding pocket of antiapoptotic members Bcl-2 or Bcl-xL.[51] Therefore, the ratio of proapoptotic to antiapoptotic Bcl-2 family proteins determines the status of mitochondrial permeability and cell survival (Figure 27.1).

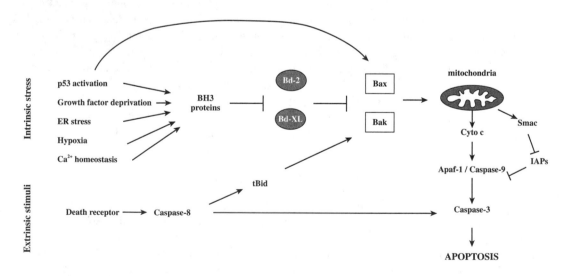

FIGURE 27.1 Regulation of apoptosis by the Bcl-2 family proteins. After cell autonomous stress, the balance between BH3-only and Bcl-2 family proteins regulates mitochondrial membrane permeability and apoptosis. In response to external death receptor signaling, cleavage of Bid amplifies the mitochondrial death pathway.

Induction of Cell Death with Bcl-2 Antisense or BH3-Mimetic Strategies

Bcl-2 overexpression can cause resistance of cells to various apoptotic stimuli, including chemotherapy, hypoxia, radiation, Ca^{2+} overload, ceramide, and growth factor deprivation. One approach to modulating the ratio of antiapoptotic to proapoptotic proteins is to target the expression of antiapoptotic molecules with antisense strategies. For example, antisense oligonucleotides that reduce Bcl-2 expression (such as Genasense) effectively sensitize cancer cells to chemotherapy and are being used in clinical trials for hematologic malignancies[52,53] and several solid tumors.[54,55]

As a promising alternative strategy to antisense approaches, structural studies of BH3 dimerization domains[56] led to the discovery of small molecule inhibitors that prevent protein–protein interactions of Bcl-2 family members. Initial studies used cell-permeable peptide inhibitors to bind Bcl-2 and sensitize cancer cells to apoptosis.[57] The first small molecule inhibitors, BH3I-1 and BH3I-2, were discovered by screening a chemical library for compounds that disrupted the interaction between Bcl-xL and the Bak BH3 peptide.[58] Another study serendipitously found that Antimycin A competes with Bak for interaction with Bcl-xL or Bcl-2.[59] Several other studies utilized synthetic BH3-mimetic peptides[60] or naturally derived chemicals[61] to induce mitochondrial membrane permeabilization by blocking Bcl-2 or Bcl-xL function.

Prevention of Cell Death by Maintenance of Mitochondrial Membrane Integrity

In contrast to cancer, diseases characterized by excessive cell death would be improved by therapies that enhance the activity of survival proteins such as Bcl-2 and Bcl-xL. Bcl-2 overexpression was shown to prevent cell death in multiple animal models of neurodegeneration, ischemic injury, and septic shock.[62–64] Besides gene therapy, an alternative strategy to increase protein expression can be achieved by using the HIV Tat protein conjugation system for intracellular delivery of peptides.[65,66] Although the mechanism of Tat transduction across the membrane is not entirely clear, it is thought to involve lipid-raft-dependent macropinocytosis.[67] Tat-mediated protein transduction of Bcl-xL showed protective effects against cell death in a rodent model of global ischemia.[68]

As an alternative to altering expression of antiapoptotic proteins, mitochondrial membrane integrity can also be preserved with drugs that block their activity either directly or through upstream signaling pathways. For example, immunosuppressants such as FK506 and cyclosporine A are emerging as clinically useful therapeutic agents for neurodegenerative diseases. Because their immunophilin receptors are expressed 10- to 100-fold higher in central and peripheral nervous system tissue than in immune tissue, they can be protective in neurons at low doses without affecting the immune system. The neuroprotective effect of FK506 was first reported in ischemic brain injury[69] and later found to function by blocking calcineurin. Although both FK506 and cyclosporine A can block calcineurin,[70] cyclosporine A can also inhibit the mitochondrial matrix prolyl-isomerase cyclophilin D and prevent activation of the mitochondrial permeability transition pore. Because FK506 lacks the ability to bind cyclophilin D, cyclosporine A provides more neuroprotective effects in animal models of traumatic brain injury,[71] ischemia,[72] and neurodegeneration.[73] Other drugs might preserve mitochondrial membrane integrity by directly disrupting Bax or Bak function. For example, dibucaine and propranolol were proposed to block Bax-induced changes in lipid structure.[74] Further investigation into the mechanism of outer-membrane permeabilization induced by BH3 death domain protein–Bax interaction will likely lead to the development of more specific antiapoptotic therapeutics relevant for neuroprotection.

P53 STATUS

The tumor suppressor p53 is another major point of control in the initiation of apoptosis, and loss of p53 function in tumor cells contributes to resistance to chemotherapy and radiation. p53 stability and subcellular localization are regulated by multiple posttranslational modifications and protein–protein interactions in response to stresses including DNA damage, activation of oncogenic signaling, and hypoxia.[75] p53 stabilization leads to cell cycle arrest or apoptosis and exerts these effects in both transcription-dependent and transcription-independent ways. p53 directly activates multiple pathways through transcription of apoptotic genes, including some death receptors, cytosolic proteins Bid and Apaf-1, and mitochondrial effectors Bax and Bak. p53 further promotes apoptosis through transcriptional repression of antiapoptotic genes, such as Bcl-2 and survivin.[76] Early studies also suggested a transcription-independent role for p53.[77] Recent reports found that p53 localizes directly at the mitochondria, though further research is required to understand the significance of this finding.[78–81]

Induction of Cell Death by Restoration of p53 Function

Restoration of disrupted apoptotic pathways can be accomplished by multiple methods. If a gene product is lost or mutated, as for example p53, gene therapy may provide direct restoration of expression of the missing wild-type protein. The most common approach taken for delivery has been use of an adenoviral vector.[82] Intratumoral injections have shown some success in inducing tumor regression. Several approaches are currently in clinical trials, including a construct from Introgen Therapeutics (INGN201) for head and neck cancer[83] and a construct from Schering-Plough (SCH58500) for advanced ovarian cancer.[84,85] Another construct, called Gendicine, was recently approved for commercial use in China and shows significant tumor regression in combination with radiotherapy for head and neck squamous cell carcinoma.[86]

Another genetic approach to target cancer cells lacking functional p53 is the use of oncolytic viruses. This strategy takes advantage of the fact that adenoviruses must inactivate p53 in order to replicate in cells. Wild-type adenovirus accomplishes this inactivation through the viral protein E1Bp55, which binds and inactivates p53. A virus lacking E1Bp55, therefore, selectively replicates in and causes lysis of cancer cells that lack p53.[87] This approach is being used clinically with the drug ONYX-015, which showed success in combination with chemotherapy in phase II trials when injected into squamous cell carcinomas of the head and neck[88] and in phase I studies of oral dysplasia.[89] Although these approaches seem promising for topical application or intratumoral delivery of solid tumors, systemic gene therapy poses additional challenges, because feasibility is limited by efficiency of transgene expression. Furthermore, the overall safety of adenoviral vectors remains an issue.

In addition to gene deletion, a common mechanism by which p53 function is lost is through development of loss-of-function point mutations. To target mutant p53 in cancer, one strategy was to use small molecules that restore p53 function. CP-31398 is a drug that was found through a chemical screen of compounds that stabilize the active conformation of the DNA binding domain of p53, rescuing the conformation and function of mutant p53 in tumor cells.[90] Subsequent studies on the mechanism by which this drug causes apoptosis yielded conflicting results, with some finding restoration of p53 function to be responsible for cell death and others citing nonspecific toxicity.[91,92] Prima1 is another small molecule compound that restores the DNA binding ability of mutant p53,[93] and follow-up studies on its specificity are forthcoming. Recent reports suggested that small molecule drugs might also activate transcription-independent functions of p53 in inducing apoptosis.[78]

As an alternative to gene therapy and small molecule drugs, a more controversial approach to restoring p53 function involves systemic administration of a transducible p53-activating peptide. A peptide derived from the C-terminal end of p53 was recognized as an activator of endogenous wild-type p53 and some mutants.[94] Because this peptide functions by restoring DNA-binding ability, it can induce apoptosis in tumor cells containing wild-type or DNA-binding mutants of p53 but not in cells deficient for p53 or containing structural mutants. A recent study showed that delivery of this peptide into cells by Tat-mediated transduction can selectively activate p53 and induce apoptosis in cancer cells but not normal cells.[95] Importantly, this approach showed *in vivo* efficacy in animal models of human terminal malignancy, such as peritoneal carcinomatosis and peritoneal lymphoma. Whereas current clinical use of macromolecular agents is limited to extracellular factors, the evidence from this study suggests that delivery of intracellular targets by Tat-mediated transduction might be another viable therapeutic option.

Induction of Cell Death by Activating Wild-Type p53

Recent attempts at targeting wild-type p53 to induce cell death showed that small molecule drugs can be designed to specifically disrupt the interaction between p53 and its negative regulator MDM2. MDM2 functions as an E3 ubiquitin ligase to promote p53 nuclear export and degradation and specifically binds to a 15-amino-acid segment of the p53 transactivation domain to block transcriptional activity.[96] In turn, the tumor suppressor ARF negatively regulates MDM2 by sequestering it in the nucleolus. Abnormalities of these upstream regulators tend to occur exclusively of each other[97] and of p53 mutations, demonstrating that the p53/MDM2/ARF pathway can be disrupted at multiple points to functionally inactivate p53 activity. Blockade of the MDM2-p53 interaction might be a particularly useful therapy in tumors where MDM2 is overexpressed, such as in some non-small cell lung carcinomas, breast cancers, and brain tumors[98,99] or where ARF is deleted or mutated, such as in melanomas and colon cancer.[100]

Studies investigating inhibition of MDM2 established that disrupting its interaction with p53 can lead to a p53 response and tumor growth inhibition. However, inhibition of MDM2 expression with antisense oligonucleotides suppressed growth of both wild-type and mutant-p53-containing cancer cells in animal studies[101,102] and in clinical trials.[103] The possibility that small molecules might effectively target this protein interaction arose when structural studies revealed that only three amino acid side chains from p53 were responsible for its interaction with MDM2.[96] In order to identify compounds that could specifically disrupt this interaction and elicit p53-dependent apoptosis, Vassilev et al.[104] screened a library of synthetic chemicals for the ability to displace recombinant p53 from MDM2. They identified a series of compounds, termed nutlins, that specifically block this interaction as verified by crystal structure analysis. These compounds induced cell cycle arrest and apoptosis in human cancer cell lines with wild-type p53 but not mutant p53 and effectively suppressed growth of human tumor xenografts in nude mice.

Wild-type p53 stabilization might also be achieved by blocking its interaction with the negative regulator, Parc.[105] Parc was recently identified as a cytoplasmic anchor for p53, which controls p53 subcellular localization and function. Many neuroblastoma cell lines that exhibit a primarily cytosolic localization of wild-type p53 were found to overexpress Parc. Knockdown of endogenous Parc expression in these cells by RNA interference caused apoptosis by inducing redistribution of p53 to the nucleus and sensitized neuroblastoma cells to genotoxic stress. The feasibility of targeting this interaction with small molecule inhibitors depends on specificity and awaits further structural studies.

SURVIVAL KINASES

Another mechanism by which cancer cells evade apoptosis is through constitutive activation of cell survival signaling cascades, including the PI3K/Akt, MAPK, and NF-κB pathways (Figure 27.2).

PI3K/Akt/mTOR

The phosphatidylinositol-3-kinase (PI3K) pathway plays a major role in signaling for growth factors such as interleukin-2, interleukin-3, platelet-derived growth factor (PDGF), and insulin-like growth factor (IGF). Upon growth factor binding, receptor tyrosine kinases (RTKs) dimerize and autophosphorylate tyrosine residues in their cytoplasmic tails. This signal can activate PI3K directly or can recruit adapter proteins that activate the Ras small GTPase to activate PI3K. Once activated, PI3K phosphorylates phosphatidylinositide-4,5-bisphosphate (PIP2) to phosphatidylinositide-3,4,5-triphosphate (PIP3), which acts as a second messenger to activate downstream kinases, such as the

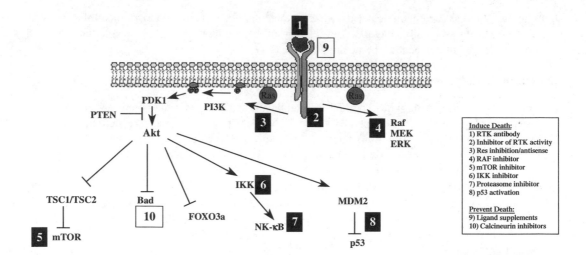

FIGURE 27.2 Therapeutic modulation of survival kinase signaling pathways. Strategies to induce cell death (black boxes 1–8) and to prevent cell death (white boxes 9–10) are highlighted. Survival factors or transforming events activate PI3K signaling, which generates phosphoinositides that recruit Akt to the plasma membrane. After activation, Akt phosphorylates numerous effector molecules that promote cell survival. Current therapeutics target both upstream and downstream components of this pathway to modulate cell survival.

serine/threonine kinase Akt. Akt promotes cell survival by multiple mechanisms, including phosphorylation and inactivation of the proapoptotic BH3-only protein Bad,[106,107] inhibition of Forkhead transcription factors that regulate Bim and Fas expression,[108] and inhibition of the tuberous sclerosis complex proteins (TSC1/TSC2) that block mTOR activation.[109] In addition, Akt promotes glucose uptake and glycolysis upon growth factor stimulation, leading to sustained growth and proliferation. Akt's role in maintaining substrate availability for mitochondrial respiration was proposed to protect mitochondrial membrane integrity and prevent apoptosis.[110,111]

Constitutive Akt signaling can occur through amplification of one of the three Akt genes or through loss of its negative regulator, the phosphatase and tensin homologous on chromosome 10 (PTEN). PTEN is a dual specificity protein and lipid phosphatase that degrades PIP3 into PIP2, terminating the signal of PI3K and preventing Akt activation.[112] Second only to p53 mutations, loss of PTEN is one of the most frequent aberrations seen in cancer and is found in the majority of glioblastomas and prostate cancers.

PI3K/Akt/mTOR Targeted Therapeutics

Deregulation of protein kinase cascades may occur by amplification or overexpression of RTKs (i.e., HER2/Neu in breast cancers) or by activating point mutations in cytoplasmic signaling kinases (i.e., BRAF in melanoma). RTKs were targeted with monoclonal antibodies that either directly block ligand binding or prevent receptor dimerization. Another approach to kinase inhibition involves small molecule antagonists that directly target the kinase domain, acting as either ATP competitors or substrate mimetics. Alternatively, a more indirect strategy was to target molecular chaperones, such as Hsp90, in order to reduce kinase stability and promote degradation. Antisense oligonucleotides are being investigated for inactivation of H-ras, a critical signaling molecule between RTK and PI3K/Akt activation.[113]

An early example of a specific RTK inhibitor to become FDA approved was Herceptin, a monoclonal antibody against the HER2 receptor tyrosine kinase for the treatment of breast cancer.[114,115] Imatinib mesylate (Gleevec), a small molecule kinase inhibitor that targets Bcr-Abl,[116] c-Kit, and

PDGFR, was approved for the treatment of chronic myelogenous leukemia in 2001[117] and also showed efficacy in gastrointestinal stromal tumors.[118] Another drug currently in use is Iressa, a small molecule inhibitor of the epidermal growth factor (EGFR), for treatment of non-small cell lung carcinoma.[119]

RTK inhibitors likely achieve some of their antitumor effects by targeting upstream signaling of the PI3K pathway. However, their success is hampered in tumors that harbor activating mutations in downstream signaling components, such as Akt or PTEN.[120] Direct inhibitors of PI3K, such as LY294002 and wortmannin, were extensively used in research studies but lack specificity among various isoforms of PI3K. Pending development of structure-based specific inhibitors, alternative strategies to directly inhibit PI3K include peptido-mimetics of the p85 SH2 binding site for phosphotyrosine[121] or inhibitors that disrupt the Ras-PI3K interaction.[122] Alternatively, restoration of PTEN function by gene therapy might provide growth-suppressive effects in appropriate tumors.

Specific targeting of downstream effectors of Akt responsible for cell growth and survival might allow tumor targeting while sparing other functions of Akt not involved in tumorigenesis, such as insulin signaling in metabolism. Of these, the best-studied therapeutic target is mTOR (mammalian target of rapamycin).[123] mTOR signaling controls cell growth by regulating translation initiation. In response to nutrients and signal transduction, mTOR phosphorylates 4EBP1, activating cap-dependent translation, and S6K, inducing ribosome biogenesis. Its direct inhibitor rapamycin was one of the first natural antiproliferative agents to be discovered. Clinical studies are underway for the related compounds RAD001 and CC1779, with the latter in phase III trials for treatment of breast, pancreatic, brain, and renal cancers.

In parallel to the Akt/mTOR signaling pathway, the Pim-2 kinase was shown to promote survival signaling by cytokines.[124] Although its importance in human cancer is still under investigation, Pim-2 might represent an important target for tumors that show resistance to rapamycin treatment.

MAPK Signaling and BRAF

Multiple survival kinases were demonstrated to be activated by single-point mutations to render a signaling cascade active. One of the most exciting targets recently identified is BRAF, which was found to be mutated in 67% of melanomas.[125] BRAF is an effector of the MAPK cascade, which involves sequential activation of Ras/Raf/MEK/ERK to ultimately alter gene transcription in favor of proliferation and tumor invasion when corresponding growth suppressors (such as p53 and Arf) are absent. The most common activating mutation found in BRAF was a valine-to-glutamate substitution at residue 599. Melanoma cells harboring this point mutation are dependent on its activity for survival, as suppression of mutant BRAF by RNA interference induces apoptosis.[126] A small molecule inhibitor against BRAF is currently in clinical trials for treatment of patients with melanoma and other malignancies (Bay 43-9006, in Phase III).[127] One possible hurdle for this drug will be specificity, as it inhibits both BRAF and CRAF. Another approach currently in clinical trials is the antisense oligonucleotide ISIS 5132.[128]

NF-κB Pathway

The NF-κB pathway plays a pivotal role in cell survival, proliferation, and inflammation by activating transcription of antiapoptotic genes and proinflammatory cytokines.[129] Constitutive activation of NF-κB contributes to chemotherapy resistance by increasing expression of multidrug resistance P-glycoproteins. Constitutive activation of the NF-κB pathway can occur in response to oncogenic stimuli, such as Ras-mediated transformation.[130] Chromosomal aberrations affecting NF-κB genes are frequent in hematologic malignancies, such as amplification of c-Rel in B cell non-Hodgkin's lymphoma.[131] In chronic myeloid leukemia, the Bcr-Abl fusion protein activates NF-κB by promoting nuclear translocation.[132] Activation of IKK by the Tax oncoprotein is important for virally induced transformation of CD4 T cells by HTLV-1.[133] Disruption of NF-κB signaling is an important therapeutic strategy in multiple contexts, including viral transformation.

Under basal conditions, NF-κB is maintained in an inactive state and sequestered in the cytosol by binding the inhibitor of NF-κB, IκB. Activating stimuli release NF-κB from inhibition through the phosphorylation of IκB by IκB kinase (IKK). Phosphorylated IκB is then targeted for ubiquitination and degradation by the 26S proteasome, which unmasks the nuclear localization signal of NF-κB to allow nuclear translocation and transcriptional activation. An essential activator of this pathway is the IKK enzyme complex that contains catalytic subunits IKKα and IKKβ and the regulatory subunit IKKγ. Knockout studies showed that IKKβ is responsible for IκB degradation in response to environmental stimuli[134] and, therefore, represents a potential specific drug target.

NF-κB-Targeted Therapeutics

Large-scale screens by pharmaceutical companies identified several small molecule inhibitors of IKK.[135] Although these drugs were shown to induce apoptosis in preclinical studies, many are hampered by a lack of specificity.[136] Apart from specific chemical inhibitors, many natural products were found to inhibit the NF-κB pathway. Examples include curcumin, the antioxidant N-acetyl-cysteine, and nonsteroidal anti-inflammatory drugs such as aspirin.[137] One mechanism by which such inhibitors may function is by targeting oxidation-sensitive cysteine residues. For example, the cyclopentenone prostaglandin 15d-PGJ$_2$ inhibits the NF-κB pathway by direct covalent modification of a cysteine residue in the kinase activation loop of IKKβ[138] or in the DNA-binding subunits of NF-κB, p50, and RelA.[139] Treatment with 15d-PGJ$_2$ was shown to inhibit inflammation in several animal models, such as autoimmune encephalitis[140] and adjuvant-induced arthritis.[141]

Adenoviral Gene Therapy to Inhibit Multiple Kinases

Another strategy for targeting survival kinase cascades is gene therapy with the E1A tumor suppressor. The E1A gene was shown to exert antitumor effects through multiple mechanisms, including suppression of RTKs such as Her2/Neu and inhibition of Akt/NFκB signaling. The E1A gene was also proposed to cooperate with chemotherapeutic drugs to induce cell death in otherwise-resistant cell lines, for example, sensitizing hepatocellular carcinoma cells to gemcitabine.[142] The mechanism behind this function may involve inhibition of p21 to induce cell death rather than cell cycle arrest following DNA damage.[143]

TgDCC-E1A is a lipid-based delivery system developed by Targeted Genetics for treatment of solid tumors overexpressing Her2/Neu.[144] Preclinical studies of TgDCC-E1A in mouse models of breast and ovarian cancer demonstrated reduced Her2/Neu expression and tumor formation and increased survival. This therapy showed modest tumor regression in Phase I and II clinical trials in patients with breast and ovarian cancer and with head and neck cancer.[145]

Activation of Survival Kinase Cascades by Ligand Supplementation

Activation of survival kinase signaling is a therapeutic goal for treatment of neurodegenerative diseases. There is strong rationale for therapeutic use of neurotrophins in multiple neurological disorders, such as nerve growth factor (NGF) in the treatment of Alzheimer's disease and neuropathies of the peripheral nervous system or IGF and ciliary neurotrophic factor in motor neuron atrophy.[146] Although the protective effects of growth factor supplementation were proven in various animal models, clinical trials with neurotrophins were disappointing, largely due to poor pharmacokinetics and other problems associated with the use of large polypeptides as drugs. Other methods for delivery of intact neurotrophins were attempted, including gene therapy[147] and microencapsulation of genetically engineered cells that secrete neurotrophins.[148] However, these strategies are still limited by the ability to cross the blood–brain barrier (BBB) and require surgical intervention.[149] Alternatively, the improved delivery of neurotrophins by conjugation to a BBB-delivery vector showed protection in animal models of ischemia.[150,151] BBB-permeable small molecule mimetics of neurotrophins are also under development.[152]

Erythropoietin (EPO) is another cytokine that showed protective effects in animal models of cell injury, including cerebral and cardiac ischemia.[153,154] Although mostly known for its roles in regulating hematopoiesis through activation of PI3K/Akt, MAPK, and JAK/STAT signaling pathways, EPO is increasingly being recognized as an important endogenous protective factor produced in the brain following oxidative stresses. Unlike neurotrophins, recombinant human EPO (rhEPO) can cross the BBB and showed significant improvement in early clinical trials with stroke patients when administered intravenously.[155] Because prolonged exposure to rhEPO might have deleterious effects on red blood cell mass, other nonerythropoietic variants were proposed to selectively provide neuroprotection.[156]

TARGETING PROTEIN DEGRADATION

The 26S proteasome was implicated in numerous cellular pathways and may be particularly important for the survival of cancer cells.[157] Although proteasome inhibition likely impacts many targets, NF-κB activation is tightly regulated by IκB degradation and seems to be highly sensitive to this strategy. The proteasomal inhibitor PS-341 was shown to effectively inhibit NF-κB activity and tumor cell growth in the HTLV-1 Tax transgenic tumor model *in vitro*, although the correlation *in vivo* is less consistent.[158] PS-341 is currently being used in clinical trials and has shown success in the treatment of hematologic malignancies.[159]

In addition to its role in the NF-κB pathway, the proteasome controls degradation of the TSC1 and TSC2 gene products after Akt signaling to allow mTOR activation. p53 degradation in MDM2 overexpressing cells also depends on the proteasome, as does caspase degradation after IAP binding, among other targets. Therefore, although proteasome inhibition is a nonspecific strategy, it seems to selectively induce death in cancer cells.

GENERAL INHIBITORS OF GENE SILENCING

Epigenetic gene silencing is emerging as an important mechanism by which tumors suppress proapoptotic pathways. Apoptotic elements that are silenced during tumorigenesis include Apaf-1 in melanomas[160] and ARF in multiple types of cancer.[161] Another example is promoter methylation of caspase-8 in some childhood neuroblastomas.[162] In this situation, reexpression of caspase-8 by gene transfer was shown to sensitize cells to death receptor signaling.[163] Alternatively, treatment with DNA methylation inhibitors such as 5-azacytidine[164,165] or histone deacetylase inhibitors such as depsipeptide[166,167] may rescue expression. As with proteasome inhibition, although strategies to reverse gene silencing are nonspecific, they seem to preferentially impact the survival of cancer cells.

ER STRESS AND CA²⁺ HOMEOSTASIS

In addition to the classic intrinsic and extrinsic pathways for apoptosis, the endoplasmic reticulum (ER) is increasingly being recognized as an important organelle in initiating and propagating apoptotic signals.[168] Members of Bcl-2 and BH3-only family members are found directly localized to the ER.[48] *In vivo*, ER perturbation occurs in response to viral stress or the accumulation of unfolded proteins. The unfolded protein response is a highly conserved signal transduction pathway that halts protein synthesis and upregulates chaperone proteins and other factors in the secretory pathway in order to allow the ER to regain folding capacity. However, when ER stress exceeds the ability of this program to restore ER function, apoptosis is initiated. Experimentally, excessive ER stress can be induced with pharmacologic agents that prevent N-linked glycosylation, block ER to Golgi transport, impair disulfide bond formation, or disrupt ER Ca^{2+} stores. After such stimuli, the ER-localized caspase-12 is activated[169] and can directly activate caspase-9, independent of Apaf-1 and mitochondrial membrane permeability. The relevance of caspase-12 signaling in humans remains controversial due to reports of truncation polymorphism in humans.[170]

ER stress leads to release of Ca^{2+} stores, primarily through the 1,4,5-triphosphate receptor (IP_3R) or Ryanodine receptor (RyR) families. This is an important step in amplifying the ER apoptosis signal through the mitochondrial pathway, as disruption of calcium homeostasis can lead to activation of Ca^{2+}-sensitive calpains, calcineurin and Bad activation, and upregulation of factors that induce cytochrome *c* release, such as the nuclear orphan receptor Nur77/TR3.[171] Furthermore, excessive intracellular calcium can directly cause Ca^{2+} influx to the mitochondria and subsequent membrane swelling and dysfunction.

ER Stress in Neurodegenerative Disorders

Many neurodegenerative diseases are characterized by misfolded protein aggregates, such as huntingtin in Huntington's disease, amyloid- in Alzheimer's disease, and α-synuclein in Parkinson's disease. Therapeutic strategies to target this initial stress include peptido-mimetics that block aggregate seeding and fibril formation.[172] Although subsequent steps in ER stress signaling can be targeted *in vitro*, for example, by using Ca^{2+} chelators, specific drugs targeting this pathway that are applicable to human use are still under development.

Another source of disruption of Ca^{2+} homeostasis occurs from excessive Ca^{2+} influx at the plasma membrane. In some neurodegenerative diseases, overactivation of glutamate receptors leads to sustained elevations in calcium, activation of calcineurin and progressive mitochondrial dysfunction, and, ultimately, neuronal degeneration.[173] Memantine is an *N*-methyl-D-aspartate (NMDA) receptor antagonist that prevents this glutamate influx. It was recently approved for treatment of patients with moderate to severe Alzheimer's disease and showed efficacy as a single agent or in combination with cholinesterase inhibitors.[174,175]

OTHER FORMS OF CELL DEATH: NECROSIS AND AUTOPHAGY

Although knowledge of alterations in the genetically programmed apoptotic cell death pathway is crucial for understanding cell survival in cancer and cell death in degenerative disease, many pathological forms of cell death involve nonapoptotic mechanisms. Furthermore, there are many currently used drugs that modulate cell death via nonapoptotic programs.

NECROSIS

Necrosis is morphologically characterized by a loss of plasma membrane integrity and leakage of cellular contents into the intercellular space, which stimulates an inflammatory response. This form of cell death can be initiated by both external stimuli and internal cellular damage. Although the mechanisms behind death-receptor-induced necrosis are not completely understood,[176–180] an important mediator common to both signaling sources appears to be the accumulation of reactive oxygen species (ROS).

Reactive Oxygen Species

Oxidative stress results from an imbalance between production of ROS (such as superoxide anion, hydrogen peroxide, and the hydroxyl radical) and reduction of ROS by endogenous antioxidant enzymes (such as superoxide dismutase, glutathione peroxidase, catalase, and thioredoxin). When the antioxidant capacity of the cell is exceeded, ROS accumulation damages DNA, lipids, and protein. Free radicals induce DNA single-strand breaks and can initiate the DNA damage response to induce cell death or repair, depending on the extent of damage. Lipid peroxidation products include F_2-isoprostanes, free-radical catalyzed isomers of prostaglandins. Lipid peroxidation can have profound effects on structural and metabolic properties of cell membranes and ultimately might lead to disruption of Ca^{2+} homeostasis and subsequent apoptotic cell death. Protein oxidation by free radicals can cause dysfunction due to cross-linking between oxidized cysteine residues or, when combined with nitric oxide, can lead to tyrosine nitration or S-nitrosylation to affect protein function.[181] Cell death induced by ROS was shown to occur through both mitochondria-dependent and mitochondria-independent pathways.

ROS in Alzheimer's Disease

Alzheimer's disease (AD) is characterized by progressive accumulation of the amyloid-β peptide and subsequent neuronal degeneration in regions of the brain affecting learning and memory. In addition to the presence of abundant senile plaques and neurofibrillary tangles caused by protein aggregation seen in the AD brain, factors thought to contribute to disease pathogenesis include increased oxidative stress, inflammation, and alterations in lipid metabolism. Animal models and human studies have suggested that oxidative damage is an early functional event in AD pathogenesis,[182,183] although the initiating events that cause increased ROS in AD remain controversial. Some studies have suggested a role for enzymes involved in arachidonic acid metabolism, such as 12/15-lipoxygenase,[184] and found increased expression and activity of this enzyme in affected areas of AD brains when compared to unaffected regions.[185,186] Regulation of expression and activity of this enzyme is controlled by inflammatory cytokines,[187] which are known to mediate inflammation in AD.

Downstream of ROS production, multiple studies have shown that increased oxidative stress and ceramide synthesis can lead to accumulation of cholesterol in cells, and this process was observed in brains of AD patients compared with age-matched controls.[188] Furthermore, a role for increased cholesterol levels in AD is supported by the increased risk of AD in people with an apolipoprotein E4 allele[189] and decreased risk in people prescribed cholesterol-lowering drugs.[190] The cellular toxicity of high levels of cholesterol was shown by the ability of inhibitors of cholesterol-metabolizing enzymes to induce apoptosis[191] and by the ability of statins to protect neurons against oxidative injury.[192] Cell culture experiments showed that antioxidants (i.e., tocopherol) or inhibitors of ceramide synthesis (ISP-1) can protect neurons from amyloid-β-induced cell death *in vitro*.[188]

ROS in Ischemia-Reperfusion Injury

Ischemic-reperfusion injury occurs after restoration of oxygen and metabolic substrates to energetically deprived tissue. Abundant evidence has shown that onset of cell death occurs during reperfusion, not ischemia,[193] and possible mechanisms behind cell death include increased ROS,[194] imbalance of calcium homeostasis,[195] chemotactic cytokines,[196] and complement.[197] For example, in myocardial injury following reperfusion, sources of ROS generation include xanthine oxidase in endothelial cells, NADPH oxidase in inflammatory cells, and the mitochondrial electron transport chain in myocytes. Another source of damage to cells during reperfusion injury is the influx of Ca^{2+} through activated glutamate receptors and through voltage-sensitive calcium channels. One consequence of this intracellular rise in Ca^{2+} is activation of nitric oxide synthase (NOS) enzymes to increase levels of NO. Combination of NO with superoxide generates the highly reactive species peroxynitrite, which can directly cause DNA breaks and damage to proteins and lipids.

Protection against Reactive Oxygen Species

Protection against ROS can be achieved by enhancing the reducing capacity of the cell. Small molecules were designed that have the catalytic activity of both superoxide dismutase and catalase and, therefore, remove both superoxide anions and hydrogen peroxide. These agents, EUK-8 and EUK-134, showed protective effects in rodent models of ischemia-reperfusion of the brain,[198] kidney,[199] and liver,[200] in a murine model of amyotrophic lateral sclerosis[201] and in a rat model of endotoxic shock.[202] Additional antioxidant chemicals that can prevent cell death induced by ROS include N-acetyl-cysteine and tocopherol, among others.

Cancer Therapeutics That Induce ROS and Cell Death

In cancer treatment, many currently used therapeutics function by promoting oxidative stress. For example, photodynamic therapy (PDT) uses a photosensitizing compound, such as porphyrin or chlorin, that selectively accumulates in target cancer cells and induces cell death only locally after exposure to laser light with a specific wavelength.[203] Absorption of light by the photosensitizing drug excites it to an extremely unstable state, which generates reactive free radical intermediates upon interaction with endogenous oxygen. Although approved for use in some malignancies, other applications of PDT are currently in Phase II trials for the treatment of solid tumors.[204] Other chemopreventive agents were shown to increase ROS generation *in vitro*, including tamoxifen, deguelin, and the cyclooxygenase inhibitors celecoxib and indomethacin.[205]

Modulation of PARP Activity in Necrotic Cell Death

The fact that DNA-alkylating agents are effective at killing cancer cells that have defective apoptotic machinery suggests that they can initiate another form of cell death. For example, follicular lymphoma cells are sensitive to DNA-alkylating agents, despite overexpression of Bcl-2.[206] Many studies have shown that DNA damage can induce a form of cell death that depends on the activity of poly(ADP)-ribose polymerase (PARP). PARP is a nuclear enzyme that senses DNA strand breaks and catalyzes poly(ADP)-ribosylation of a variety of proteins, using NAD^+ as a substrate. When activated by DNA damage, PARP activity contributes to the loosening of chromatin structure and recruitment of DNA repair enzymes. However, after excessive DNA damage, prolonged PARP activity consumes cellular NAD^+ and leads to a necrotic cell death due to ATP depletion. It has been shown that alkylating agents selectively kill proliferating cells and not quiescent cells, due to their differential dependence on cytosolic NAD^+ to generate ATP.[207] During apoptosis, PARP is one of the first targets cleaved by caspase-3 and is a key determinant in the decision to die by necrosis or apoptosis. Therefore, PARP activation may represent a therapeutic target relevant for the treatment of cancer.

On the other hand, PARP inhibition shows protective effects in disorders characterized by necrotic cell death. Many PARP inhibitors were developed based on nicotinamide analogues and function by blocking catalytic activity of all PARP family members. These inhibitors have been protective in animal models of septic shock, cardiac ischemia, traumatic brain injury, and NMDA-mediated excitotoxicity, among others.[208]

AUTOPHAGY

Autophagy is increasingly being recognized as an important alternative mechanism for cell death during development and in response to nutrient deprivation.[209] Autophagy controls protein and organelle turnover through rearrangement of subcellular membranes to deliver cargo to the lysosome or autophagic vacuole for degradation. This process is critical for removal of organelles damaged by oxidative stress and likely plays an important role in limiting cellular exposure to free radicals that cause genotoxic stress.[210] Under nutrient-rich conditions, autophagy is inhibited by mTOR signaling. Upon starvation, the loss of mTOR signaling allows dephosphorylation and activation of proteins involved in autophagy initiation and formation of the double-membrane autophagosome. Rapamycin treatment stimulates autophagy, even in nutrient-rich conditions, by inhibiting mTOR. Vinblastine is another agent used experimentally to induce autophagy, whereas inhibitors of autophagy include chloroquine and 3-methyladenine.

Proteins that are involved in the formation of the autophagosome, such as Beclin 1, were identified as tumor suppressors. Beclin 1 is the human homologue of yeast autophagy gene Apg6 and was found to interact with Bcl-2.[211] Cells deficient in Beclin 1 can undergo apoptosis but not autophagy. The fact that Beclin 1 is a dose-dependent tumor suppressor found to be monoallelically deleted in breast and ovarian cancers suggests that the process of autophagy is important in tumor suppression.[212] Transfection of Beclin 1 into breast cancer cell lines expressing low levels of it reduced tumorigenicity in nude mice. Therefore, Beclin 1 represents a potential therapeutic target to modulate autophagic activity.

The link between autophagy and cancer is further supported by the involvement of known tumor suppressors in autophagic pathways, including PTEN and DAP-kinase. During cancer progression, activation of the Akt pathway, mTOR signaling, and loss of PTEN may promote transformation by simultaneously blocking apoptosis and autophagy.[213] Furthermore, an increasing number of studies suggest that the mechanism of action of some anticancer agents involves induction of autophagic cell death.

Potential cross talk between the pathways of apoptosis and autophagy has been reported, as inhibition of caspase-8 activity induces autophagic cell death.[214] This is relevant to the host response to viral pathogens, which have caspase inhibitors. Thus, the induction of autophagic death following caspase inhibition might provide a security mechanism that ensures destruction of an infected cell via nonapoptotic cell death.

Paradoxically, although autophagy is often considered an alternative form of cell death, in some situations, increased autophagic activity can promote cell survival. For example, in diseases characterized by the accumulation of misfolded proteins, autophagy may be an important mechanism for aggregate clearance and cell survival. This model was recently supported in a transgenic fly model of Huntington's disease, in which rapamycin treatment increased autophagy, improved clearance of huntingtin aggregates, and enhanced cell survival.[215]

SUMMARY

Recent advances in our understanding of the mechanisms behind programmed cell death have greatly informed the design of targeted therapeutic strategies. Future research into other forms of cell death will be equally important in order to assemble a full complement of agents that induce or prevent cell death. As these agents reach clinical trials, it is becoming increasingly clear that proper diagnosis of disease based on a molecular characterization will significantly contribute to therapeutic efficacy and treatment success.

REFERENCES

1. Varfolomeev, E.E. and Ashkenazi, A., Tumor necrosis factor: an apoptosis JuNKie? *Cell,* 116 (4), 491–497, 2004.
2. Pan, G., Ni, J., Wei, Y.F., Yu, G., Gentz, R., and Dixit, V.M., An antagonist decoy receptor and a death domain-containing receptor for TRAIL, *Science,* 277 (5327), 815–818, 1997.
3. van Noesel, M.M., van Bezouw, S., Salomons, G.S., Voute, P.A., Pieters, R., Baylin, S.B., Herman, J.G., and Versteeg, R., Tumor-specific down-regulation of the tumor necrosis factor-related apoptosis-inducing ligand decoy receptors DcR1 and DcR2 is associated with dense promoter hypermethylation, *Cancer Res,* 62 (7), 2157–2161, 2002.
4. Cretney, E., Takeda, K., Yagita, H., Glaccum, M., Peschon, J.J., and Smyth, M.J., Increased susceptibility to tumor initiation and metastasis in TNF-related apoptosis-inducing ligand-deficient mice, *J. Immunol.,* 168 (3), 1356–1361, 2002.
5. Takeda, K., Hayakawa, Y., Smyth, M.J., Kayagaki, N., Yamaguchi, N., Kakuta, S., Iwakura, Y., Yagita, H., and Okumura, K., Involvement of tumor necrosis factor-related apoptosis-inducing ligand in surveillance of tumor metastasis by liver natural killer cells, *Nat. Med.,* 7 (1), 94–100, 2001.
6. Almasan, A. and Ashkenazi, A., Apo2L/TRAIL: apoptosis signaling, biology, and potential for cancer therapy, *Cytokine Growth Factor Rev.,* 14 (3–4), 337–348, 2003.

7. Mundt, B., Kuhnel, F., Zender, L., Paul, Y., Tillmann, H., Trautwein, C., Manns, M.P., and Kubicka, S., Involvement of TRAIL and its receptors in viral hepatitis, *FASEB J.,* 17 (1), 94–96, 2003.

8. Tollefson, A.E., Toth, K., Doronin, K., Kuppuswamy, M., Doronina, O.A., Lichtenstein, D.L., Hermiston, T.W., Smith, C.A., and Wold, W.S., Inhibition of TRAIL-induced apoptosis and forced internalization of TRAIL receptor 1 by adenovirus proteins, *J. Virol.,* 75 (19), 8875–8887, 2001.

9. Jenkins, M., Keir, M., and McCune, J.M., A membrane-bound Fas decoy receptor expressed by human thymocytes, *J. Biol. Chem.,* 275 (11), 7988–7993, 2000.

10. Ashkenazi, A., Targeting death and decoy receptors of the tumour-necrosis factor superfamily, *Nat. Rev. Cancer,* 2 (6), 420–430, 2002.

11. Ashkenazi, A., Pai, R.C., Fong, S., Leung, S., Lawrence, D.A., Marsters, S.A., Blackie, C., Chang, L., McMurtrey, A.E., Hebert, A., DeForge, L., Koumenis, I.L., Lewis, D., Harris, L., Bussiere, J., Koeppen, H., Shahrokh, Z., and Schwall, R.H., Safety and antitumor activity of recombinant soluble Apo2 ligand, *J. Clin. Invest.,* 104 (2), 155–162, 1999.

12. Walczak, H., Miller, R.E., Ariail, K., Gliniak, B., Griffith, T.S., Kubin, M., Chin, W., Jones, J., Woodward, A., Le, T., Smith, C., Smolak, P., Goodwin, R.G., Rauch, C.T., Schuh, J.C., and Lynch, D.H., Tumoricidal activity of tumor necrosis factor-related apoptosis-inducing ligand *in vivo, Nat. Med.,* 5 (2), 157–163, 1999.

13. Jo, M., Kim, T.H., Seol, D.W., Esplen, J.E., Dorko, K., Billiar, T.R., and Strom, S.C., Apoptosis induced in normal human hepatocytes by tumor necrosis factor-related apoptosis-inducing ligand, *Nat. Med.,* 6 (5), 564–567, 2000.

14. Kelley, S.K., Harris, L.A., Xie, D., Deforge, L., Totpal, K., Bussiere, J., and Fox, J.A., Preclinical studies to predict the disposition of Apo2L/tumor necrosis factor-related apoptosis-inducing ligand in humans: characterization of *in vivo* efficacy, pharmacokinetics, and safety, *J. Pharmacol. Exp. Ther.,* 299 (1), 31–38, 2001.

15. Xu, Z.W., Kleeff, J., Friess, H., Buchler, M.W., and Solioz, M., Synergistic cytotoxic effect of TRAIL and gemcitabine in pancreatic cancer cells, *Anticancer Res.,* 23 (1A), 251–258, 2003.

16. Ray, S. and Almasan, A., Apoptosis induction in prostate cancer cells and xenografts by combined treatment with Apo2 ligand/tumor necrosis factor-related apoptosis-inducing ligand and CPT-11, *Cancer Res.,* 63 (15), 4713–4723, 2003.

17. Zhang, H.G., Wang, J., Yang, X., Hsu, H.C., and Mountz, J.D., Regulation of apoptosis proteins in cancer cells by ubiquitin, *Oncogene,* 23 (11), 2009–2015, 2004.

18. Clem, R.J. and Miller, L.K., Control of programmed cell death by the baculovirus genes p35 and iap, *Mol. Cell Biol.,* 14 (8), 5212–5222, 1994.

19. Uren, A.G., Pakusch, M., Hawkins, C.J., Puls, K.L., and Vaux, D.L., Cloning and expression of apoptosis inhibitory protein homologs that function to inhibit apoptosis and/or bind tumor necrosis factor receptor-associated factors, *Proc. Natl. Acad. Sci. U.S.A.,* 93 (10), 4974–4978, 1996.

20. Rothe, M., Pan, M.G., Henzel, W.J., Ayres, T.M., and Goeddel, D.V., The TNFR2-TRAF signaling complex contains two novel proteins related to baculoviral inhibitor of apoptosis proteins, *Cell,* 83 (7), 1243–1252, 1995.

21. Liston, P., Fong, W.G., and Korneluk, R.G., The inhibitors of apoptosis: there is more to life than Bcl2, *Oncogene,* 22 (53), 8568–8580, 2003.

22. Baens, M., Maes, B., Steyls, A., Geboes, K., Marynen, P., and De Wolf-Peeters, C., The product of the t(11;18), an API2-MLT fusion, marks nearly half of gastric MALT type lymphomas without large cell proliferation, *Am. J. Pathol.,* 156 (4), 1433–1439, 2000.

23. Sasaki, H., Sheng, Y., Kotsuji, F., and Tsang, B.K., Down-regulation of X-linked inhibitor of apoptosis protein induces apoptosis in chemoresistant human ovarian cancer cells, *Cancer Res.,* 60 (20), 5659–5666, 2000.

24. Li, J., Feng, Q., Kim, J.M., Schneiderman, D., Liston, P., Li, M., Vanderhyden, B., Faught, W., Fung, M.F., Senterman, M., Korneluk, R.G., and Tsang, B.K., Human ovarian cancer and cisplatin resistance: possible role of inhibitor of apoptosis proteins, *Endocrinology,* 142 (1), 370–380, 2001.

25. Hu, Y., Cherton-Horvat, G., Dragowska, V., Baird, S., Korneluk, R.G., Durkin, J.P., Mayer, L.D., and LaCasse, E.C., Antisense oligonucleotides targeting XIAP induce apoptosis and enhance chemotherapeutic activity against human lung cancer cells *in vitro* and *in vivo, Clin. Cancer Res.,* 9 (7), 2826–2836, 2003.

26. Holmlund, J.T., Applying antisense technology: Affinitak and other antisense oligonucleotides in clinical development, *Ann. N.Y. Acad. Sci.,* 1002, 244–251, 2003.

27. Schimmer, A.D., Welsh, K., Pinilla, C., Wang, Z., Krajewska, M., Bonneau, M.J., Pedersen, I.M., Kitada, S., Scott, F.L., Bailly-Maitre, B., Glinsky, G., Scudiero, D., Sausville, E., Salvesen, G., Nefzi, A., Ostresh, J.M., Houghten, R.A., and Reed, J.C., Small-molecule antagonists of apoptosis suppressor XIAP exhibit broad antitumor activity, *Cancer Cell.,* 5 (1), 25–35, 2004.

28. Wu, T.Y., Wagner, K.W., Bursulaya, B., Schultz, P.G., and Deveraux, Q.L., Development and characterization of nonpeptidic small molecule inhibitors of the XIAP/caspase-3 interaction, *Chem. Biol.,* 10 (8), 759–767, 2003.

29. Fulda, S., Wick, W., Weller, M., and Debatin, K.M., Smac agonists sensitize for Apo2L/TRAIL- or anticancer drug-induced apoptosis and induce regression of malignant glioma *in vivo, Nat. Med.,* 8 (8), 808–815, 2002.

30. Yang, L., Mashima, T., Sato, S., Mochizuki, M., Sakamoto, H., Yamori, T., Oh-Hara, T., and Tsuruo, T., Predominant suppression of apoptosome by inhibitor of apoptosis protein in non-small cell lung cancer H460 cells: therapeutic effect of a novel polyarginine-conjugated Smac peptide, *Cancer Res.,* 63 (4), 831–837, 2003.

31. Liston, P., Roy, N., Tamai, K., Lefebvre, C., Baird, S., Cherton-Horvat, G., Farahani, R., McLean, M., Ikeda, J.E., MacKenzie, A., and Korneluk, R.G., Suppression of apoptosis in mammalian cells by NAIP and a related family of IAP genes, *Nature,* 379 (6563), 349–353, 1996.

32. Xu, D.G., Crocker, S.J., Doucet, J.P., St.-Jean, M., Tamai, K., Hakim, A.M., Ikeda, J.E., Liston, P., Thompson, C.S., Korneluk, R.G., MacKenzie, A., and Robertson, G.S., Elevation of neuronal expression of NAIP reduces ischemic damage in the rat hippocampus, *Nat. Med.,* 3 (9), 997–1004, 1997.

33. Perrelet, D., Ferri, A., Liston, P., Muzzin, P., Korneluk, R.G., and Kato, A.C., IAPs are essential for GDNF-mediated neuroprotective effects in injured motor neurons *in vivo, Nat. Cell Biol.,* 4 (2), 175–179, 2002.

34. Iwata, A., Harlan, J.M., Vedder, N.B., and Winn, R.K., The caspase inhibitor z-VAD is more effective than CD18 adhesion blockade in reducing muscle ischemia-reperfusion injury: implication for clinical trials, *Blood,* 100 (6), 2077–2080, 2002.

35. Hara, H., Friedlander, R.M., Gagliardini, V., Ayata, C., Fink, K., Huang, Z., Shimizu-Sasamata, M., Yuan, J., and Moskowitz, M.A., Inhibition of interleukin 1 converting enzyme family proteases reduces ischemic and excitotoxic neuronal damage, *Proc. Natl. Acad. Sci. U.S.A.,* 94 (5), 2007–2012, 1997.

36. Natori, S., Higuchi, H., Contreras, P., and Gores, G.J., The caspase inhibitor IDN-6556 prevents caspase activation and apoptosis in sinusoidal endothelial cells during liver preservation injury, *Liver Transpl.,* 9 (3), 278–284, 2003.

37. Valentino, K.L., Gutierrez, M., Sanchez, R., Winship, M.J., and Shapiro, D.A., First clinical trial of a novel caspase inhibitor: anti-apoptotic caspase inhibitor, IDN-6556, improves liver enzymes, *Int. J. Clin. Pharmacol. Ther.,* 41 (10), 441–449, 2003.

38. Cauwels, A., Janssen, B., Waeytens, A., Cuvelier, C., and Brouckaert, P., Caspase inhibition causes hyperacute tumor necrosis factor-induced shock via oxidative stress and phospholipase A2, *Nat. Immunol.,* 4 (4), 387–393, 2003.

39. Nguyen, J.T. and Wells, J.A., Direct activation of the apoptosis machinery as a mechanism to target cancer cells, *Proc. Natl. Acad. Sci. U.S.A.,* 100 (13), 7533–7538, 2003.

40. Jiang, X., Kim, H.E., Shu, H., Zhao, Y., Zhang, H., Kofron, J., Donnelly, J., Burns, D., Ng, S.C., Rosenberg, S., and Wang, X., Distinctive roles of PHAP proteins and prothymosin- in a death regulatory pathway, *Science,* 299 (5604), 223–226, 2003.

41. Liu, X., Kim, C.N., Yang, J., Jemmerson, R., and Wang, X., Induction of apoptotic program in cell-free extracts: requirement for dATP and cytochrome c, *Cell,* 86 (1), 147–157, 1996.

42. Tsujimoto, Y., Finger, L.R., Yunis, J., Nowell, P.C., and Croce, C.M., Cloning of the chromosome breakpoint of neoplastic B cells with the t(14;18) chromosome translocation, *Science,* 226 (4678), 1097–1099, 1984.

43. Vaux, D.L., Cory, S., and Adams, J.M., Bcl-2 gene promotes haemopoietic cell survival and cooperates with c-myc to immortalize pre-B cells, *Nature,* 335 (6189), 440–442, 1988.

44. McDonnell, T.J., Deane, N., Platt, F.M., Nunez, G., Jaeger, U., McKearn, J.P., and Korsmeyer, S.J., Bcl-2-immunoglobulin transgenic mice demonstrate extended B cell survival and follicular lymphoproliferation, *Cell,* 57 (1), 79–88, 1989.

45. Cory, S. and Adams, J.M., The Bcl2 family: regulators of the cellular life-or-death switch, *Nat. Rev. Cancer,* 2 (9), 647–656, 2002.

46. Wei, M.C., Zong, W.X., Cheng, E.H., Lindsten, T., Panoutsakopoulou, V., Ross, A.J., Roth, K.A., MacGregor, G.R., Thompson, C.B., and Korsmeyer, S.J., Proapoptotic BAX and BAK: a requisite gateway to mitochondrial dysfunction and death, *Science,* 292 (5517), 727–730, 2001.

47. Scorrano, L., Oakes, S.A., Opferman, J.T., Cheng, E.H., Sorcinelli, M.D., Pozzan, T., and Korsmeyer, S.J., BAX and BAK regulation of endoplasmic reticulum Ca^{2+}: a control point for apoptosis, *Science,* 300 (5616), 135–139, 2003.

48. Zong, W.X., Li, C., Hatzivassiliou, G., Lindsten, T., Yu, Q.C., Yuan, J., and Thompson, C.B., Bax and Bak can localize to the endoplasmic reticulum to initiate apoptosis, *J. Cell Biol.,* 162 (1), 59–69, 2003.

49. Vander Heiden, M.G., Chandel, N.S., Schumacker, P.T., and Thompson, C.B., Bcl-xL prevents cell death following growth factor withdrawal by facilitating mitochondrial ATP/ADP exchange, *Mol. Cell,* 3 (2), 159–167, 1999.

50. Bouillet, P. and Strasser, A., BH3-only proteins — evolutionarily conserved proapoptotic Bcl-2 family members essential for initiating programmed cell death, *J. Cell Sci.,* 115 (Pt. 8), 1567–1574, 2002.

51. Letai, A., Bassik, M.C., Walensky, L.D., Sorcinelli, M.D., Weiler, S., and Korsmeyer, S.J., Distinct BH3 domains either sensitize or activate mitochondrial apoptosis, serving as prototype cancer therapeutics, *Cancer Cell,* 2 (3), 183–192, 2002.

52. Webb, A., Cunningham, D., Cotter, F., Clarke, P.A., di Stefano, F., Ross, P., Corbo, M., and Dziewanowska, Z., BCL-2 antisense therapy in patients with non-Hodgkin lymphoma, *Lancet,* 349 (9059), 1137–1141, 1997.

53. Marcucci, G., Byrd, J.C., Dai, G., Klisovic, M.I., Kourlas, P.J., Young, D.C., Cataland, S.R., Fisher, D.B., Lucas, D., Chan, K.K., Porcu, P., Lin, Z.P., Farag, S.F., Frankel, S.R., Zwiebel, J.A., Kraut, E.H., Balcerzak, S.P., Bloomfield, C.D., Grever, M.R., and Caligiuri, M.A., Phase 1 and pharmacodynamic studies of G3139, a Bcl-2 antisense oligonucleotide, in combination with chemotherapy in refractory or relapsed acute leukemia, *Blood,* 101 (2), 425–432, 2003.

54. Jansen, B., Wacheck, V., Heere-Ress, E., Schlagbauer-Wadl, H., Hoeller, C., Lucas, T., Hoermann, M., Hollenstein, U., Wolff, K., and Pehamberger, H., Chemosensitisation of malignant melanoma by BCL2 antisense therapy, *Lancet,* 356 (9243), 1728–1733, 2000.

55. Rudin, C.M., Kozloff, M., Hoffman, P.C., Edelman, M.J., Karnauskas, R., Tomek, R., Szeto, L., and Vokes, E.E., Phase I study of G3139, a Bcl-2 antisense oligonucleotide, combined with carboplatin and etoposide in patients with small-cell lung cancer, *J. Clin. Oncol.,* 22 (6), 1110–1117, 2004.

56. Sattler, M., Liang, H., Nettesheim, D., Meadows, R.P., Harlan, J.E., Eberstadt, M., Yoon, H.S., Shuker, S.B., Chang, B.S., Minn, A.J., Thompson, C.B., and Fesik, S.W., Structure of Bcl-xL-Bak peptide complex: recognition between regulators of apoptosis, *Science,* 275 (5302), 983–986, 1997.

57. Wang, J.L., Zhang, Z.J., Choksi, S., Shan, S., Lu, Z., Croce, C.M., Alnemri, E.S., Korngold, R., and Huang, Z., Cell permeable Bcl-2 binding peptides: a chemical approach to apoptosis induction in tumor cells, *Cancer Res.,* 60 (6), 1498–1502, 2000.

58. Degterev, A., Lugovskoy, A., Cardone, M., Mulley, B., Wagner, G., Mitchison, T., and Yuan, J., Identification of small-molecule inhibitors of interaction between the BH3 domain and Bcl-xL, *Nat. Cell Biol.,* 3 (2), 173–182, 2001.

59. Tzung, S.P., Kim, K.M., Basanez, G., Giedt, C.D., Simon, J., Zimmerberg, J., Zhang, K.Y., and Hockenbery, D.M., Antimycin A mimics a cell-death-inducing Bcl-2 homology domain 3, *Nat. Cell Biol.,* 3 (2), 183–191, 2001.

60. Polster, B.M., Kinnally, K.W., and Fiskum, G., BH3 death domain peptide induces cell type-selective mitochondrial outer membrane permeability, *J. Biol. Chem.,* 276 (41), 37887–37894, 2001.

61. Chan, S.L., Lee, M.C., Tan, K.O., Yang, L.K., Lee, A.S., Flotow, H., Fu, N.Y., Butler, M.S., Soejarto, D.D., Buss, A.D., and Yu, V.C., Identification of chelerythrine as an inhibitor of BclXL function, *J. Biol. Chem.,* 278 (23), 20453–20456, 2003.

62. Dubois-Dauphin, M., Frankowski, H., Tsujimoto, Y., Huarte, J., and Martinou, J.C., Neonatal motoneurons overexpressing the Bcl-2 protooncogene in transgenic mice are protected from axotomy-induced cell death, *Proc. Natl. Acad. Sci. U.S.A.,* 91 (8), 3309–3313, 1994.

63. Linnik, M.D., Zahos, P., Geschwind, M.D., and Federoff, H.J., Expression of Bcl-2 from a defective herpes simplex virus-1 vector limits neuronal death in focal cerebral ischemia, *Stroke,* 26 (9), 1670–1674, discussion 1675, 1995.

64. Kostic, V., Jackson-Lewis, V., de Bilbao, F., Dubois-Dauphin, M., and Przedborski, S., Bcl-2: prolonging life in a transgenic mouse model of familial amyotrophic lateral sclerosis, *Science,* 277 (5325), 559–562, 1997.

65. Fawell, S., Seery, J., Daikh, Y., Moore, C., Chen, L.L., Pepinsky, B., and Barsoum, J., Tat-mediated delivery of heterologous proteins into cells, *Proc. Natl. Acad. Sci. U.S.A.,* 91 (2), 664–668, 1994.

66. Schwarze, S.R., Ho, A., Vocero-Akbani, A., and Dowdy, S.F., *In vivo* protein transduction: delivery of a biologically active protein into the mouse, *Science,* 285 (5433), 1569–1572, 1999.

67. Wadia, J.S., Stan, R.V., and Dowdy, S.F., Transducible TAT-HA fusogenic peptide enhances escape of TAT-fusion proteins after lipid raft macropinocytosis, *Nat. Med.,* 10 (3), 310–315, 2004.

68. Asoh, S., Ohsawa, I., Mori, T., Katsura, K., Hiraide, T., Katayama, Y., Kimura, M., Ozaki, D., Yamagata, K., and Ohta, S., Protection against ischemic brain injury by protein therapeutics, *Proc. Natl. Acad. Sci. U.S.A.,* 99 (26), 17107–17112, 2002.

69. Sharkey, J. and Butcher, S.P., Immunophilins mediate the neuroprotective effects of FK506 in focal cerebral ischaemia, *Nature,* 371 (6495), 336–339, 1994.

70. Liu, J., Farmer, J.D., Jr., Lane, W.S., Friedman, J., Weissman, I., and Schreiber, S.L., Calcineurin is a common target of cyclophilin-cyclosporin A and FKBP-FK506 complexes, *Cell,* 66 (4), 807–815, 1991.

71. Okonkwo, D.O. and Povlishock, J.T., An intrathecal bolus of cyclosporin A before injury preserves mitochondrial integrity and attenuates axonal disruption in traumatic brain injury, *J. Cereb. Blood Flow Metab.,* 19 (4), 443–451, 1999.

72. Uchino, H., Minamikawa-Tachino, R., Kristian, T., Perkins, G., Narazaki, M., Siesjo, B.K., and Shibasaki, F., Differential neuroprotection by cyclosporin A and FK506 following ischemia corresponds with differing abilities to inhibit calcineurin and the mitochondrial permeability transition, *Neurobiol. Dis.,* 10 (3), 219–233, 2002.

73. Keep, M., Elmer, E., Fong, K.S., and Csiszar, K., Intrathecal cyclosporin prolongs survival of late-stage ALS mice, *Brain Res.,* 894 (2), 327–331, 2001.

74. Polster, B.M., Basanez, G., Young, M., Suzuki, M., and Fiskum, G., Inhibition of Bax-induced cytochrome c release from neural cell and brain mitochondria by dibucaine and propranolol, *J. Neurosci.,* 23 (7), 2735–2743, 2003.

75. El-Deiry, W.S., The role of p53 in chemosensitivity and radiosensitivity, *Oncogene,* 22 (47), 7486–7495, 2003.

76. Hoffman, W.H., Biade, S., Zilfou, J.T., Chen, J., and Murphy, M., Transcriptional repression of the anti-apoptotic survivin gene by wild type p53, *J. Biol. Chem.,* 277 (5), 3247–3257, 2002.

77. Caelles, C., Helmberg, A., and Karin, M., p53-dependent apoptosis in the absence of transcriptional activation of p53-target genes, *Nature,* 370 (6486), 220–223, 1994.

78. Chipuk, J.E., Kuwana, T., Bouchier-Hayes, L., Droin, N.M., Newmeyer, D.D., Schuler, M., and Green, D.R., Direct activation of Bax by p53 mediates mitochondrial membrane permeabilization and apoptosis, *Science,* 303 (5660), 1010–1014, 2004.

79. Dumont, P., Leu, J.I., Della Pietra, A.C., III, George, D.L., and Murphy, M., The codon 72 polymorphic variants of p53 have markedly different apoptotic potential, *Nat. Genet.,* 33 (3), 357–365, 2003.

80. Mihara, M., Erster, S., Zaika, A., Petrenko, O., Chittenden, T., Pancoska, P., and Moll, U.M., p53 has a direct apoptogenic role at the mitochondria, *Mol. Cell,* 11 (3), 577–590, 2003.

81. Leu, J.I., Dumont, P., Hafey, M., Murphy, M.E., and George, D.L., Mitochondrial p53 activates Bak and causes disruption of a Bak-Mcl1 complex, *Nat. Cell Biol.,* 6 (5), 443–450, 2004.

82. Nishizaki, M., Meyn, R.E., Levy, L.B., Atkinson, E.N., White, R.A., Roth, J.A., and Ji, L., Synergistic inhibition of human lung cancer cell growth by adenovirus-mediated wild-type p53 gene transfer in combination with docetaxel and radiation therapeutics *in vitro* and *in vivo, Clin. Cancer Res.,* 7 (9), 2887–2897, 2001.

83. Merritt, J.A., Roth, J.A., and Logothetis, C.J., Clinical evaluation of adenoviral-mediated p53 gene transfer: review of INGN 201 studies, *Semin. Oncol.,* 28 (5, Suppl. 16), 105–114, 2001.

84. Buller, R.E., Shahin, M.S., Horowitz, J.A., Runnebaum, I.B., Mahavni, V., Petrauskas, S., Kreienberg, R., Karlan, B., Slamon, D., and Pegram, M., Long term follow-up of patients with recurrent ovarian cancer after Ad p53 gene replacement with SCH 58500, *Cancer Gene Ther.,* 9 (7), 567–572, 2002.

85. Zhang, J.Y., Apoptosis-based anticancer drugs, *Nat. Rev. Drug Discov.,* 1 (2), 101–102, 2002.

86. Pearson, S., Jia, H., and Kandachi, K., China approves first gene therapy, *Nat. Biotechnol.,* 22 (1), 3–4, 2004.

87. Dobbelstein, M., Replicating adenoviruses in cancer therapy, *Curr. Top. Microbiol. Immunol.,* 273, 291–334, 2004.

88. Khuri, F.R., Nemunaitis, J., Ganly, I., Arseneau, J., Tannock, I.F., Romel, L., Gore, M., Ironside, J., MacDougall, R.H., Heise, C., Randlev, B., Gillenwater, A.M., Bruso, P., Kaye, S.B., Hong, W.K., and Kirn, D.H., A controlled trial of intratumoral ONYX-015, a selectively-replicating adenovirus, in combination with cisplatin and 5-fluorouracil in patients with recurrent head and neck cancer, *Nat. Med.,* 6 (8), 879–885, 2000.

89. Rudin, C.M., Cohen, E.E., Papadimitrakopoulou, V.A., Silverman, S., Jr., Recant, W., El-Naggar, A.K., Stenson, K., Lippman, S.M., Hong, W.K., and Vokes, E.E., An attenuated adenovirus, ONYX-015, as mouthwash therapy for premalignant oral dysplasia, *J. Clin. Oncol.,* 21 (24), 4546–4552, 2003.

90. Foster, B.A., Coffey, H.A., Morin, M.J., and Rastinejad, F., Pharmacological rescue of mutant p53 conformation and function, *Science,* 286 (5449), 2507–2510, 1999.

91. Rippin, T.M., Bykov, V.J., Freund, S.M., Selivanova, G., Wiman, K.G., and Fersht, A.R., Character-ization of the p53-rescue drug CP-31398 *in vitro* and in living cells, *Oncogene,* 21 (14), 2119–2129, 2002.

92. Takimoto, R., Wang, W., Dicker, D.T., Rastinejad, F., Lyssikatos, J., and el-Deiry, W.S., The mutant p53-conformation modifying drug, CP-31398, can induce apoptosis of human cancer cells and can stabilize wild-type p53 protein, *Cancer Biol. Ther.,* 1 (1), 47–55, 2002.

93. Bykov, V.J., Issaeva, N., Shilov, A., Hultcrantz, M., Pugacheva, E., Chumakov, P., Bergman, J., Wiman, K.G., and Selivanova, G., Restoration of the tumor suppressor function to mutant p53 by a low-molecular-weight compound, *Nat. Med.,* 8 (3), 282–288, 2002.

94. Hupp, T.R., Sparks, A., and Lane, D.P., Small peptides activate the latent sequence-specific DNA binding function of p53, *Cell,* 83 (2), 237–245, 1995.

95. Snyder, E.L., Meade, B.R., Saenz, C.C., and Dowdy, S.F., Treatment of terminal peritoneal carcino-matosis by a transducible p53-activating peptide, *PLoS Biol.,* 2 (2), E36, 186–193, 2004.

96. Chene, P., Inhibiting the p53-MDM2 interaction: an important target for cancer therapy, *Nat. Rev. Cancer,* 3 (2), 102–109, 2003.

97. Eymin, B., Gazzeri, S., Brambilla, C., and Brambilla, E., Mdm2 overexpression and p14(ARF) inactivation are two mutually exclusive events in primary human lung tumors, *Oncogene,* 21 (17), 2750–2761, 2002.

98. Bueso-Ramos, C.E., Manshouri, T., Haidar, M.A., Yang, Y., McCown, P., Ordonez, N., Glassman, A., Sneige, N., and Albitar, M., Abnormal expression of MDM-2 in breast carcinomas, *Breast Cancer Res. Treat.,* 37 (2), 179–188, 1996.

99. Momand, J., Jung, D., Wilczynski, S., and Niland, J., The MDM2 gene amplification database, *Nucleic Acids Res.,* 26 (15), 3453–3459, 1998.

100. Burri, N., Shaw, P., Bouzourene, H., Sordat, I., Sordat, B., Gillet, M., Schorderet, D., Bosman, F.T., and Chaubert, P., Methylation silencing and mutations of the p14ARF and p16INK4a genes in colon cancer, *Lab. Invest.,* 81 (2), 217–229, 2001.

101. Tortora, G., Caputo, R., Damiano, V., Bianco, R., Chen, J., Agrawal, S., Bianco, A.R., and Ciardiello, F., A novel MDM2 anti-sense oligonucleotide has anti-tumor activity and potentiates cytotoxic drugs acting by different mechanisms in human colon cancer, *Int. J. Cancer,* 88 (5), 804–809, 2000.

102. Zhang, R. and Wang, H., Antisense oligonucleotide inhibitors of MDM2 oncogene expression, *Meth-ods Mol. Med.,* 85, 205–222, 2003.

103. Capoulade, C., Mir, L.M., Carlier, K., Lecluse, Y., Tetaud, C., Mishal, Z., and Wiels, J., Apoptosis of tumoral and nontumoral lymphoid cells is induced by both mdm2 and p53 antisense oligodeoxynu-cleotides, *Blood,* 97 (4), 1043–1049, 2001.

104. Vassilev, L.T., Vu, B.T., Graves, B., Carvajal, D., Podlaski, F., Filipovic, Z., Kong, N., Kammlott, U., Lukacs, C., Klein, C., Fotouhi, N., and Liu, E.A., *In vivo* activation of the p53 pathway by small-molecule antagonists of MDM2, *Science,* 303 (5659), 844–848, 2004.

105. Nikolaev, A.Y., Li, M., Puskas, N., Qin, J., and Gu, W., Parc: a cytoplasmic anchor for p53, *Cell,* 112 (1), 29–40, 2003.

106. Datta, S.R., Dudek, H., Tao, X., Masters, S., Fu, H., Gotoh, Y., and Greenberg, M.E., Akt phosphory-lation of BAD couples survival signals to the cell-intrinsic death machinery, *Cell,* 91 (2), 231–241, 1997.

107. Dudek, H., Datta, S.R., Franke, T.F., Birnbaum, M.J., Yao, R., Cooper, G.M., Segal, R.A., Kaplan, D.R., and Greenberg, M.E., Regulation of neuronal survival by the serine-threonine protein kinase Akt, *Science,* 275 (5300), 661–665, 1997.

108. Burgering, B.M. and Medema, R.H., Decisions on life and death: FOXO Forkhead transcription factors are in command when PKB/Akt is off duty, *J. Leukoc. Biol.,* 73 (6), 689–701, 2003.

109. Manning, B.D. and Cantley, L.C., United at last: the tuberous sclerosis complex gene products connect the phosphoinositide 3-kinase/Akt pathway to mammalian target of rapamycin (mTOR) signalling, *Biochem. Soc. Trans.,* 31 (Pt. 3), 573–578, 2003.

110. Plas, D.R., Talapatra, S., Edinger, A.L., Rathmell, J.C., and Thompson, C.B., Akt and Bcl-xL promote growth factor-independent survival through distinct effects on mitochondrial physiology, *J. Biol. Chem.,* 276 (15), 12041–12048, 2001.

111. Vander Heiden, M.G., Plas, D.R., Rathmell, J.C., Fox, C.J., Harris, M.H., and Thompson, C.B., Growth factors can influence cell growth and survival through effects on glucose metabolism, *Mol. Cell Biol.,* 21 (17), 5899–5912, 2001.

112. Stambolic, V., Suzuki, A., de la Pompa, J.L., Brothers, G.M., Mirtsos, C., Sasaki, T., Ruland, J., Penninger, J.M., Siderovski, D.P., and Mak, T.W., Negative regulation of PKB/Akt-dependent cell survival by the tumor suppressor PTEN, *Cell,* 95 (1), 29–39, 1998.

113. Cunningham, C.C., Holmlund, J.T., Geary, R.S., Kwoh, T.J., Dorr, A., Johnston, J.F., Monia, B., and Nemunaitis, J., A phase I trial of H-ras antisense oligonucleotide ISIS 2503 administered as a continuous intravenous infusion in patients with advanced carcinoma, *Cancer,* 92 (5), 1265–1271, 2001.

114. Baselga, J., Tripathy, D., Mendelsohn, J., Baughman, S., Benz, C.C., Dantis, L., Sklarin, N.T., Seidman, A.D., Hudis, C.A., Moore, J., Rosen, P.P., Twaddell, T., Henderson, I.C., and Norton, L., Phase II study of weekly intravenous recombinant humanized anti-p185HER2 monoclonal antibody in patients with HER2/neu-overexpressing metastatic breast cancer, *J. Clin. Oncol.,* 14 (3), 737–744, 1996.

115. Slamon, D.J., Leyland-Jones, B., Shak, S., Fuchs, H., Paton, V., Bajamonde, A., Fleming, T., Eiermann, W., Wolter, J., Pegram, M., Baselga, J., and Norton, L., Use of chemotherapy plus a monoclonal antibody against HER2 for metastatic breast cancer that overexpresses HER2, *N. Engl. J. Med.,* 344 (11), 783–792, 2001.

116. Druker, B.J., Tamura, S., Buchdunger, E., Ohno, S., Segal, G.M., Fanning, S., Zimmermann, J., and Lydon, N.B., Effects of a selective inhibitor of the Abl tyrosine kinase on the growth of Bcr-Abl positive cells, *Nat. Med.,* 2 (5), 561–566, 1996.

117. Deininger, M.W. and Druker, B.J., Specific targeted therapy of chronic myelogenous leukemia with imatinib, *Pharmacol. Rev.,* 55 (3), 401–423, 2003.

118. Demetri, G.D., von Mehren, M., Blanke, C.D., Van den Abbeele, A.D., Eisenberg, B., Roberts, P.J., Heinrich, M.C., Tuveson, D.A., Singer, S., Janicek, M., Fletcher, J.A., Silverman, S.G., Silberman, S.L., Capdeville, R., Kiese, B., Peng, B., Dimitrijevic, S., Druker, B.J., Corless, C., Fletcher, C.D., and Joensuu, H., Efficacy and safety of imatinib mesylate in advanced gastrointestinal stromal tumors, *N. Engl. J. Med.,* 347 (7), 472–480, 2002.

119. Lynch, T.J., Bell, D.W., Sordella, R., Gurubhagavatula, S., Okimoto, R.A., Brannigan, B.W., Harris, P.L., Haserlat, S.M., Supko, J.G., Haluska, F.G., Louis, D.N., Christiani, D.C., Settleman, J., and Haber, D.A., Activating mutations in the epidermal growth factor receptor underlying responsiveness of non-small-cell lung cancer to gefitinib, *N. Engl. J. Med.,* 350 (21), 2129–2139, 2004.

120. Bianco, R., Shin, I., Ritter, C.A., Yakes, F.M., Basso, A., Rosen, N., Tsurutani, J., Dennis, P.A., Mills, G.B., and Arteaga, C.L., Loss of PTEN/MMAC1/TEP in EGF receptor-expressing tumor cells counteracts the antitumor action of EGFR tyrosine kinase inhibitors, *Oncogene,* 22 (18), 2812–2822, 2003.

121. Eaton, S.R., Cody, W.L., Doherty, A.M., Holland, D.R., Panek, R.L., Lu, G.H., Dahring, T.K., and Rose, D.R., Design of peptidomimetics that inhibit the association of phosphatidylinositol 3-kinase with platelet-derived growth factor- receptor and possess cellular activity, *J. Med. Chem.,* 41 (22), 4329–4342, 1998.

122. Djordjevic, S. and Driscoll, P.C., Structural insight into substrate specificity and regulatory mechanisms of phosphoinositide 3-kinases, *Trends Biochem. Sci.,* 27 (8), 426–432, 2002.

123. Bjornsti, M.A. and Houghton, P.J., The TOR pathway: a target for cancer therapy, *Nat. Rev. Cancer,* 4 (5), 335–348, 2004.

124. Fox, C.J., Hammerman, P.S., Cinalli, R.M., Master, S.R., Chodosh, L.A., and Thompson, C.B., The serine/threonine kinase Pim-2 is a transcriptionally regulated apoptotic inhibitor, *Genes Dev.,* 17 (15), 1841–1854, 2003.

125. Davies, H., Bignell, G.R., Cox, C., Stephens, P., Edkins, S., Clegg, S., Teague, J., Woffendin, H., Garnett, M.J., Bottomley, W., Davis, N., Dicks, E., Ewing, R., Floyd, Y., Gray, K., Hall, S., Hawes, R., Hughes, J., Kosmidou, V., Menzies, A., Mould, C., Parker, A., Stevens, C., Watt, S., Hooper, S.,

Wilson, R., Jayatilake, H., Gusterson, B.A., Cooper, C., Shipley, J., Hargrave, D., Pritchard-Jones, K., Maitland, N., Chenevix-Trench, G., Riggins, G.J., Bigner, D.D., Palmieri, G., Cossu, A., Flanagan, A., Nicholson, A., Ho, J.W., Leung, S.Y., Yuen, S.T., Weber, B.L., Seigler, H.F., Darrow, T.L., Paterson, H., Marais, R., Marshall, C.J., Wooster, R., Stratton, M.R., and Futreal, P.A., Mutations of the BRAF gene in human cancer, *Nature,* 417 (6892), 949–954, 2002.

126. Hingorani, S.R., Jacobetz, M.A., Robertson, G.P., Herlyn, M., and Tuveson, D.A., Suppression of BRAF(V599E) in human melanoma abrogates transformation, *Cancer Res.,* 63 (17), 5198–5202, 2003.

127. Lyons, J.F., Wilhelm, S., Hibner, B., and Bollag, G., Discovery of a novel Raf kinase inhibitor, *Endocr. Relat. Cancer,* 8 (3), 219–225, 2001.

128. Bollag, G., Freeman, S., Lyons, J.F., and Post, L.E., Raf pathway inhibitors in oncology, *Curr. Opin. Investig. Drugs,* 4 (12), 1436–1441, 2003.

129. Ghosh, S., May, M.J., and Kopp, E.B., NF-B and Rel proteins: evolutionarily conserved mediators of immune responses, *Annu. Rev. Immunol.,* 16(1), 225–260, 1998.

130. Finco, T.S., Westwick, J.K., Norris, J.L., Beg, A.A., Der, C.J., and Baldwin, A.S., Jr., Oncogenic Ha-Ras-induced signaling activates NF-B transcriptional activity, which is required for cellular transformation, *J. Biol. Chem.,* 272 (39), 24113–24116, 1997.

131. Karin, M., Cao, Y., Greten, F.R., and Li, Z.W., NF-B in cancer: from innocent bystander to major culprit, *Nat. Rev. Cancer,* 2 (4), 301–310, 2002.

132. Reuther, J.Y., Reuther, G.W., Cortez, D., Pendergast, A.M., and Baldwin, A.S., Jr., A requirement for NF-B activation in Bcr-Abl-mediated transformation, *Genes Dev.,* 12 (7), 968–981, 1998.

133. Chu, Z.L., DiDonato, J.A., Hawiger, J., and Ballard, D.W., The tax oncoprotein of human T-cell leukemia virus type 1 associates with and persistently activates IκB kinases containing IKKα and IKKβ, *J. Biol. Chem.,* 273 (26), 15891–15894, 1998.

134. Karin, M. and Ben-Neriah, Y., Phosphorylation meets ubiquitination: the control of NF-κB activity, *Annu. Rev. Immunol.,* 18(1), 621–663, 2000.

135. Karin, M., Yamamoto, Y., and Wang, Q.M., The IKK NF-κB system: a treasure trove for drug development, *Nat. Rev. Drug Discov.,* 3 (1), 17–26, 2004.

136. Haefner, B., NF-κB: arresting a major culprit in cancer, *Drug Discov. Today,* 7 (12), 653–663, 2002.

137. Epinat, J.C. and Gilmore, T.D., Diverse agents act at multiple levels to inhibit the Rel/NF-B signal transduction pathway, *Oncogene,* 18 (49), 6896–6909, 1999.

138. Castrillo, A., Diaz-Guerra, M.J., Hortelano, S., Martin-Sanz, P., and Bosca, L., Inhibition of IB kinase and IB phosphorylation by 15-deoxy-Delta(12,14)-prostaglandin J(2) in activated murine macrophages, *Mol. Cell Biol.,* 20 (5), 1692–1698, 2000.

139. Cernuda-Morollon, E., Pineda-Molina, E., Canada, F.J., and Perez-Sala, D., 15-Deoxy-12,14-prostaglandin J2 inhibition of NF-B-DNA binding through covalent modification of the p50 subunit, *J. Biol. Chem.,* 276 (38), 35530–35536, 2001.

140. Diab, A., Deng, C., Smith, J.D., Hussain, R.Z., Phanavanh, B., Lovett-Racke, A.E., Drew, P.D., and Racke, M.K., Peroxisome proliferator-activated receptor-gamma agonist 15-deoxy-(12,14)-prostaglandin J(2) ameliorates experimental autoimmune encephalomyelitis, *J. Immunol.,* 168 (5), 2508–2515, 2002.

141. Kawahito, Y., Kondo, M., Tsubouchi, Y., Hashiramoto, A., Bishop-Bailey, D., Inoue, K., Kohno, M., Yamada, R., Hla, T., and Sano, H., 15-deoxy-(12,14)-PGJ(2) induces synoviocyte apoptosis and suppresses adjuvant-induced arthritis in rats, *J. Clin. Invest.,* 106 (2), 189–197, 2000.

142. Lee, W.P., Tai, D.I., Tsai, S.L., Yeh, C.T., Chao, Y., Lee, S.D., and Hung, M.C., Adenovirus type 5 E1A sensitizes hepatocellular carcinoma cells to gemcitabine, *Cancer Res.,* 63 (19), 6229–6236, 2003.

143. Chattopadhyay, D., Ghosh, M.K., Mal, A., and Harter, M.L., Inactivation of p21 by E1A leads to the induction of apoptosis in DNA-damaged cells, *J. Virol.,* 75 (20), 9844–9856, 2001.

144. Wagner, J.A., Technology evaluation: tgDCC-E1A, targeted genetics/MD Anderson, *Curr. Opin. Mol. Ther.,* 1 (2), 266–270, 1999.

145. Villaret, D., Glisson, B., Kenady, D., Hanna, E., Carey, M., Gleich, L., Yoo, G.H., Futran, N., Hung, M.C., Anklesaria, P., and Heald, A.E., A multicenter phase II study of tgDCC-E1A for the intratumoral treatment of patients with recurrent head and neck squamous cell carcinoma, *Head Neck,* 24 (7), 661–669, 2002.

146. Hefti, F., Neurotrophic factor therapy for nervous system degenerative diseases, *J. Neurobiol.,* 25 (11), 1418–1435, 1994.

147. Haase, G., Kennel, P., Pettmann, B., Vigne, E., Akli, S., Revah, F., Schmalbruch, H., and Kahn, A., Gene therapy of murine motor neuron disease using adenoviral vectors for neurotrophic factors, *Nat. Med.,* 3 (4), 429–436, 1997.

148. Aebischer, P., Schluep, M., Deglon, N., Joseph, J.M., Hirt, L., Heyd, B., Goddard, M., Hammang, J.P., Zurn, A.D., Kato, A.C., Regli, F., and Baetge, E.E., Intrathecal delivery of CNTF using encapsulated genetically modified xenogeneic cells in amyotrophic lateral sclerosis patients, *Nat. Med.,* 2 (6), 696–699, 1996.

149. Pan, W., Banks, W.A., and Kastin, A.J., Permeability of the blood–brain barrier to neurotrophins, *Brain Res.,* 788 (1–2), 87–94, 1998.

150. Wu, D. and Pardridge, W.M., Neuroprotection with noninvasive neurotrophin delivery to the brain, *Proc. Natl. Acad. Sci. U.S.A.,* 96 (1), 254–259, 1999.

151. Song, B.W., Vinters, H.V., Wu, D., and Pardridge, W.M., Enhanced neuroprotective effects of basic fibroblast growth factor in regional brain ischemia after conjugation to a blood–brain barrier delivery vector, *J. Pharmacol. Exp. Ther.,* 301 (2), 605–610, 2002.

152. Pollack, S.J. and Harper, S.J., Small molecule Trk receptor agonists and other neurotrophic factor mimetics, *Curr. Drug Targets CNS Neurol. Disord.,* 1 (1), 59–80, 2002.

153. Digicaylioglu, M. and Lipton, S.A., Erythropoietin-mediated neuroprotection involves cross-talk between Jak2 and NF-B signalling cascades, *Nature,* 412 (6847), 641–647, 2001.

154. Parsa, C.J., Matsumoto, A., Kim, J., Riel, R.U., Pascal, L.S., Walton, G.B., Thompson, R.B., Petrofski, J.A., Annex, B.H., Stamler, J.S., and Koch, W.J., A novel protective effect of erythropoietin in the infarcted heart, *J. Clin. Invest.,* 112 (7), 999–1007, 2003.

155. Ehrenreich, H., Hasselblatt, M., Dembowski, C., Cepek, L., Lewczuk, P., Stiefel, M., Rustenbeck, H.H., Breiter, N., Jacob, S., Knerlich, F., Bohn, M., Poser, W., Ruther, E., Kochen, M., Gefeller, O., Gleiter, C., Wessel, T.C., De Ryck, M., Itri, L., Prange, H., Cerami, A., Brines, M., and Siren, A.L., Erythropoietin therapy for acute stroke is both safe and beneficial, *Mol. Med.,* 8 (8), 495–505, 2002.

156. Erbayraktar, S., Grasso, G., Sfacteria, A., Xie, Q.W., Coleman, T., Kreilgaard, M., Torup, L., Sager, T., Erbayraktar, Z., Gokmen, N., Yilmaz, O., Ghezzi, P., Villa, P., Fratelli, M., Casagrande, S., Leist, M., Helboe, L., Gerwein, J., Christensen, S., Geist, M.A., Pedersen, L.O., Cerami-Hand, C., Wuerth, J.P., Cerami, A., and Brines, M., Asialoerythropoietin is a nonerythropoietic cytokine with broad neuroprotective activity *in vivo, Proc. Natl. Acad. Sci. U.S.A.,* 100 (11), 6741–6746, 2003.

157. Adams, J., The development of proteasome inhibitors as anticancer drugs, *Cancer Cell,* 5 (5), 417–421, 2004.

158. Mitra-Kaushik, S., Harding, J.C., Hess, J., and Ratner, L., Effects of the proteasome inhibitor PS-341 on tumor growth in HTLV-1 tax transgenic mice and tax tumor transplants, *Blood,* 104(3), 802–809, 2004.

159. Orlowski, R.Z., Stinchcombe, T.E., Mitchell, B.S., Shea, T.C., Baldwin, A.S., Stahl, S., Adams, J., Esseltine, D.L., Elliott, P.J., Pien, C.S., Guerciolini, R., Anderson, J.K., Depcik-Smith, N.D., Bhagat, R., Lehman, M.J., Novick, S.C., O'Connor, O.A., and Soignet, S.L., Phase I trial of the proteasome inhibitor PS-341 in patients with refractory hematologic malignancies, *J. Clin. Oncol.,* 20 (22), 4420–4427, 2002.

160. Soengas, M.S., Capodieci, P., Polsky, D., Mora, J., Esteller, M., Opitz-Araya, X., McCombie, R., Herman, J.G., Gerald, W.L., Lazebnik, Y.A., Cordon-Cardo, C., and Lowe, S.W., Inactivation of the apoptosis effector Apaf-1 in malignant melanoma, *Nature,* 409 (6817), 207–211, 2001.

161. Esteller, M., Cordon-Cardo, C., Corn, P.G., Meltzer, S.J., Pohar, K.S., Watkins, D.N., Capella, G., Peinado, M.A., Matias-Guiu, X., Prat, J., Baylin, S.B., and Herman, J.G., p14ARF silencing by promoter hypermethylation mediates abnormal intracellular localization of MDM2, *Cancer Res.,* 61 (7), 2816–2821, 2001.

162. Teitz, T., Wei, T., Valentine, M.B., Vanin, E.F., Grenet, J., Valentine, V.A., Behm, F.G., Look, A.T., Lahti, J.M., and Kidd, V.J., Caspase 8 is deleted or silenced preferentially in childhood neuroblastomas with amplification of MYCN, *Nat. Med.,* 6 (5), 529–535, 2000.

163. Fulda, S., Kufer, M.U., Meyer, E., van Valen, F., Dockhorn-Dworniczak, B., and Debatin, K.M., Sensitization for death receptor- or drug-induced apoptosis by re-expression of caspase-8 through demethylation or gene transfer, *Oncogene,* 20 (41), 5865–5877, 2001.

164. Flynn, J., Fang, J.Y., Mikovits, J.A., and Reich, N.O., A potent cell-active allosteric inhibitor of murine DNA cytosine C5 methyltransferase, *J. Biol. Chem.*, 278 (10), 8238–8243, 2003.

165. Lyons, J., Bayar, E., Fine, G., McCullar, M., Rolens, R., Rubinfeld, J., and Rosenfeld, C., Decitabine: development of a DNA methyltransferase inhibitor for hematological malignancies, *Curr. Opin. Investig. Drugs*, 4 (12), 1442–1450, 2003.

166. Piekarz, R.L., Robey, R., Sandor, V., Bakke, S., Wilson, W.H., Dahmoush, L., Kingma, D.M., Turner, M.L., Altemus, R., and Bates, S.E., Inhibitor of histone deacetylation, depsipeptide (FR901228), in the treatment of peripheral and cutaneous T-cell lymphoma: a case report, *Blood*, 98 (9), 2865–2868, 2001.

167. Richon, V.M. and O'Brien, J.P., Histone deacetylase inhibitors: a new class of potential therapeutic agents for cancer treatment, *Clin. Cancer Res.*, 8 (3), 662–664, 2002.

168. Breckenridge, D.G., Germain, M., Mathai, J.P., Nguyen, M., and Shore, G.C., Regulation of apoptosis by endoplasmic reticulum pathways, *Oncogene*, 22 (53), 8608–8618, 2003.

169. Nakagawa, T., Zhu, H., Morishima, N., Li, E., Xu, J., Yankner, B.A., and Yuan, J., Caspase-12 mediates endoplasmic-reticulum-specific apoptosis and cytotoxicity by amyloid-, *Nature*, 403 (6765), 98–103, 2000.

170. Saleh, M., Vaillancourt, J.P., Graham, R.K., Huyck, M., Srinivasula, S.M., Alnemri, E.S., Steinberg, M.H., Nolan, V., Baldwin, C.T., Hotchkiss, R.S., Buchman, T.G., Zehnbauer, B.A., Hayden, M.R., Farrer, L.A., Roy, S., and Nicholson, D.W., Differential modulation of endotoxin responsiveness by human caspase-12 polymorphisms, *Nature*, 429 (6987), 75–79, 2004.

171. Li, H., Kolluri, S.K., Gu, J., Dawson, M.I., Cao, X., Hobbs, P.D., Lin, B., Chen, G., Lu, J., Lin, F., Xie, Z., Fontana, J.A., Reed, J.C., and Zhang, X., Cytochrome *c* release and apoptosis induced by mitochondrial targeting of nuclear orphan receptor TR3, *Science*, 289 (5482), 1159–1164, 2000.

172. Cohen, F.E. and Kelly, J.W., Therapeutic approaches to protein-misfolding diseases, *Nature*, 426 (6968), 905–909, 2003.

173. Ankarcrona, M., Dypbukt, J.M., Orrenius, S., and Nicotera, P., Calcineurin and mitochondrial function in glutamate-induced neuronal cell death, *FEBS Lett.*, 394 (3), 321–324, 1996.

174. Koch, H.J., Szecsey, A., and Haen, E., NMDA-antagonism (memantine): an alternative pharmacological therapeutic principle in Alzheimer's and vascular dementia, *Curr. Pharm. Des.*, 10 (3), 253–259, 2004.

175. Tariot, P.N., Farlow, M.R., Grossberg, G.T., Graham, S.M., McDonald, S., and Gergel, I., Memantine treatment in patients with moderate to severe Alzheimer disease already receiving donepezil: a randomized controlled trial, *JAMA*, 291 (3), 317–324, 2004.

176. Chandel, N.S., Schumacker, P.T., and Arch, R.H., Reactive oxygen species are downstream products of TRAF-mediated signal transduction, *J. Biol. Chem.*, 276 (46), 42728–42736, 2001.

177. Sakon, S., Xue, X., Takekawa, M., Sasazuki, T., Okazaki, T., Kojima, Y., Piao, J.H., Yagita, H., Okumura, K., Doi, T., and Nakano, H., NF-B inhibits TNF-induced accumulation of ROS that mediate prolonged MAPK activation and necrotic cell death, *EMBO J.*, 22 (15), 3898–3909, 2003.

178. Holler, N., Zaru, R., Micheau, O., Thome, M., Attinger, A., Valitutti, S., Bodmer, J.L., Schneider, P., Seed, B., and Tschopp, J., Fas triggers an alternative, caspase-8-independent cell death pathway using the kinase RIP as effector molecule, *Nat. Immunol.*, 1 (6), 489–495, 2000.

179. Chan, F.K., Shisler, J., Bixby, J.G., Felices, M., Zheng, L., Appel, M., Orenstein, J., Moss, B., and Lenardo, M.J., A role for tumor necrosis factor receptor-2 and receptor-interacting protein in programmed necrosis and antiviral responses, *J. Biol. Chem.*, 278 (51), 51613–51621, 2003.

180. Lin, Y., Choksi, S., Shen, H.M., Yang, Q.F., Hur, G.M., Kim, Y.S., Tran, J.H., Nedospasov, S.A., and Liu, Z.G., Tumor necrosis factor-induced nonapoptotic cell death requires receptor-interacting protein-mediated cellular reactive oxygen species accumulation, *J. Biol. Chem.*, 279 (11), 10822–10828, 2004.

181. Radi, R., Nitric oxide, oxidants, and protein tyrosine nitration, *Proc. Natl. Acad. Sci. U.S.A.*, 101 (12), 4003–4008, 2004.

182. Pratico, D., Uryu, K., Leight, S., Trojanoswki, J.Q., and Lee, V.M., Increased lipid peroxidation precedes amyloid plaque formation in an animal model of Alzheimer amyloidosis, *J. Neurosci.*, 21 (12), 4183–4187, 2001.

183. Nunomura, A., Perry, G., Aliev, G., Hirai, K., Takeda, A., Balraj, E.K., Jones, P.K., Ghanbari, H., Wataya, T., Shimohama, S., Chiba, S., Atwood, C.S., Petersen, R.B., and Smith, M.A., Oxidative damage is the earliest event in Alzheimer disease, *J. Neuropathol. Exp. Neurol.*, 60 (8), 759–767, 2001.

184. Funk, C.D. and Cyrus, T., 12/15-lipoxygenase, oxidative modification of LDL and atherogenesis, *Trends Cardiovasc. Med.,* 11 (3–4), 116–124, 2001.

185. Pratico, D., Zhukareva, V., Yao, Y., Uryu, K., Funk, C.D., Lawson, J.A., Trojanowski, J.Q., and Lee, V.M., 12/15-lipoxygenase is increased in Alzheimer's disease: possible involvement in brain oxidative stress, *Am. J. Pathol.,* 164 (5), 1655–1662, 2004.

186. Li, Y., Maher, P., and Schubert, D., A role for 12-lipoxygenase in nerve cell death caused by glutathione depletion, *Neuron,* 19 (2), 453–463, 1997.

187. Huang, J.T., Welch, J.S., Ricote, M., Binder, C.J., Willson, T.M., Kelly, C., Witztum, J.L., Funk, C.D., Conrad, D., and Glass, C.K., Interleukin-4-dependent production of PPAR- ligands in macrophages by 12/15-lipoxygenase, *Nature,* 400 (6742), 378–382, 1999.

188. Cutler, R.G., Kelly, J., Storie, K., Pedersen, W.A., Tammara, A., Hatanpaa, K., Troncoso, J.C., and Mattson, M.P., Involvement of oxidative stress-induced abnormalities in ceramide and cholesterol metabolism in brain aging and Alzheimer's disease, *Proc. Natl. Acad. Sci. U.S.A.,* 101 (7), 2070–2075, 2004.

189. Wolozin, B., Kellman, W., Ruosseau, P., Celesia, G.G., and Siegel, G., Decreased prevalence of Alzheimer disease associated with 3-hydroxy-3-methyglutaryl coenzyme A reductase inhibitors, *Arch. Neurol.,* 57 (10), 1439–1443, 2000.

190. Fassbender, K., Simons, M., Bergmann, C., Stroick, M., Lutjohann, D., Keller, P., Runz, H., Kuhl, S., Bertsch, T., von Bergmann, K., Hennerici, M., Beyreuther, K., and Hartmann, T., Simvastatin strongly reduces levels of Alzheimer's disease -amyloid peptides Abeta 42 and Abeta 40 *in vitro* and *in vivo, Proc. Natl. Acad. Sci. U.S.A.,* 98 (10), 5856–5861, 2001.

191. Kellner-Weibel, G., Geng, Y.J., and Rothblat, G.H., Cytotoxic cholesterol is generated by the hydrolysis of cytoplasmic cholesteryl ester and transported to the plasma membrane, *Atherosclerosis,* 146 (2), 309–319, 1999.

192. Buxbaum, J.D., Geoghagen, N.S., and Friedhoff, L.T., Cholesterol depletion with physiological concentrations of a statin decreases the formation of the Alzheimer amyloid Abeta peptide, *J. Alzheimers Dis.,* 3 (2), 221–229, 2001.

193. Zhao, Z.Q., Oxidative stress-elicited myocardial apoptosis during reperfusion, *Curr. Opin. Pharmacol.,* 4 (2), 159–165, 2004.

194. Li, C. and Jackson, R.M., Reactive species mechanisms of cellular hypoxia-reoxygenation injury, *Am. J. Physiol. Cell Physiol.,* 282 (2), C227–C241, 2002.

195. Deshpande, J.K., Siesjo, B.K., and Wieloch, T., Calcium accumulation and neuronal damage in the rat hippocampus following cerebral ischemia, *J. Cereb. Blood Flow Metab.,* 7 (1), 89–95, 1987.

196. Verma, S., Fedak, P.W., Weisel, R.D., Butany, J., Rao, V., Maitland, A., Li, R.K., Dhillon, B., and Yau, T.M., Fundamentals of reperfusion injury for the clinical cardiologist, *Circulation,* 105 (20), 2332–2336, 2002.

197. De Vries, B., Matthijsen, R.A., Wolfs, T.G., Van Bijnen, A.A., Heeringa, P., and Buurman, W.A., Inhibition of complement factor C5 protects against renal ischemia-reperfusion injury: inhibition of late apoptosis and inflammation, *Transplantation,* 75 (3), 375–382, 2003.

198. Baker, K., Marcus, C.B., Huffman, K., Kruk, H., Malfroy, B., and Doctrow, S.R., Synthetic combined superoxide dismutase/catalase mimetics are protective as a delayed treatment in a rat stroke model: a key role for reactive oxygen species in ischemic brain injury, *J. Pharmacol. Exp. Ther.,* 284 (1), 215–221, 1998.

199. Gianello, P., Saliez, A., Bufkens, X., Pettinger, R., Misseleyn, D., Hori, S., and Malfroy, B., EUK-134, a synthetic superoxide dismutase and catalase mimetic, protects rat kidneys from ischemia-reperfusion-induced damage, *Transplantation,* 62 (11), 1664–1666, 1996.

200. Pucheu, S., Boucher, F., Sulpice, T., Tresallet, N., Bonhomme, Y., Malfroy, B., and de Leiris, J., EUK-8 a synthetic catalytic scavenger of reactive oxygen species protects isolated iron-overloaded rat heart from functional and structural damage induced by ischemia/reperfusion, *Cardiovasc. Drugs Ther.,* 10 (3), 331–339, 1996.

201. Jung, C., Rong, Y., Doctrow, S., Baudry, M., Malfroy, B., and Xu, Z., Synthetic superoxide dismutase/catalase mimetics reduce oxidative stress and prolong survival in a mouse amyotrophic lateral sclerosis model, *Neurosci. Lett.,* 304 (3), 157–160, 2001.

202. McDonald, M.C., d'Emmanuele di Villa Bianca, R., Wayman, N.S., Pinto, A., Sharpe, M.A., Cuzzocrea, S., Chatterjee, P.K., and Thiemermann, C., A superoxide dismutase mimetic with catalase activity (EUK-8) reduces the organ injury in endotoxic shock, *Eur. J. Pharmacol.,* 466 (1–2), 181–189, 2003.

203. Luksiene, Z., Photodynamic therapy: mechanism of action and ways to improve the efficiency of treatment, *Medicina (Kaunas),* 39 (12), 1137–1150, 2003.

204. Lustig, R.A., Vogl, T.J., Fromm, D., Cuenca, R., Alex Hsi, R., D'Cruz, A.K., Krajina, Z., Turic, M., Singhal, A., and Chen, J.C., A multicenter phase I safety study of intratumoral photoactivation of talaporfin sodium in patients with refractory solid tumors, *Cancer,* 98 (8), 1767–1771, 2003.

205. Sun, S.Y., Hail, N., Jr., and Lotan, R., Apoptosis as a novel target for cancer chemoprevention, *J. Natl. Cancer Inst.,* 96 (9), 662–672, 2004.

206. Lister, T.A., The management of follicular lymphoma, *Ann. Oncol.,* 2 (Suppl. 2), 131–135, 1991.

207. Zong, W.X., Ditsworth, D., Bauer, D.E., Wang, Z.Q., and Thompson, C.B., Alkylating DNA damage stimulates a regulated form of necrotic cell death, *Genes Dev.,* 18 (11), 1272–1282, 2004.

208. Greenberg, J.H., LaPlaca, M.C., and McIntosh, T.K., PARP inhibitors as neuroprotective agents for brain injury, in *PARP as a Therapeutic Target,* Zhang, J., Ed., CRC Press, Boca Raton, FL, 2002, pp. 83–115.

209. Klionsky, D.J. and Emr, S.D., Autophagy as a regulated pathway of cellular degradation, *Science,* 290 (5497), 1717–1721, 2000.

210. Edinger, A.L. and Thompson, C.B., Defective autophagy leads to cancer, *Cancer Cell.,* 4 (6), 422–424, 2003.

211. Liang, X.H., Kleeman, L.K., Jiang, H.H., Gordon, G., Goldman, J.E., Berry, G., Herman, B., and Levine, B., Protection against fatal Sindbis virus encephalitis by beclin, a novel Bcl-2-interacting protein, *J. Virol.,* 72 (11), 8586–8596, 1998.

212. Liang, X.H., Jackson, S., Seaman, M., Brown, K., Kempkes, B., Hibshoosh, H., and Levine, B., Induction of autophagy and inhibition of tumorigenesis by beclin 1, *Nature,* 402 (6762), 672–676, 1999.

213. Gozuacik, D. and Kimchi, A., Autophagy as a cell death and tumor suppressor mechanism, *Oncogene,* 23 (16), 2891–2906, 2004.

214. Yu, L., Alva, A., Su, H., Dutt, P., Freundt, E., Welsh, S., Baehrecke, E.H., and Lenardo, M.J., Regulation of an ATG7-beclin 1 program of autophagic cell death by caspase-8, *Science,* 304 (5676), 1500–1502, 2004.

215. Ravikumar, B., Vacher, C., Berger, Z., Davies, J.E., Luo, S., Oroz, L.G., Scaravilli, F., Easton, D.F., Duden, R., O'Kane, C.J., and Rubinsztein, D.C., Inhibition of mTOR induces autophagy and reduces toxicity of polyglutamine expansions in fly and mouse models of Huntington disease, *Nat. Genet.,* 36 (6), 585–595, 2004.

28 Immunotherapy of HIV Disease

Franco Lori
Laurene M. Kelly
Andrea Cossarizza
Julianna Lisziewicz

CONTENTS

INTRODUCTION

Apoptosis has been described as a homeostatic mechanism to limit the expression of activated lymphocytes and may contribute to the elimination of T cells after viral disease resolution.[1] However, it has become evident that accelerated apoptosis, which follows immune system overactivation, plays a prominent role in human immunodeficiency virus (HIV) infection. The concept that chronic immune system hyperactivation represents a fundamental mechanism in acquired immunodeficiency syndrome (AIDS) pathogenesis is supported by a large body of data. The first report indicating such a relationship was published as early as 1988, with the 6-year longitudinal study of the Edinburgh Hemophilia/HIV cohort. The investigators reported a correlation between disease progression and plasma concentrations of interleukin-2 receptor (IL-2R; a marker of lymphocyte activation) and later with the number and percentage of circulating death receptor (DR+)-positive activated T cells.[2,3] It was subsequently observed that inappropriate induction of cell death during continued, chronic activation of the immune system eventually deletes reactive T cells during pathogenic HIV/simian immunodeficiency virus (SIV) infection.[4,5]

The concept was refined by demonstrating that cell death occurs mainly in bystander cells, which are not infected by the virus,[6] and increased susceptibility to apoptosis correlated with disease progression.[7] The authors concluded that "chronic activation of the immune system is the primary mechanism for the cell deletion process."[7] Furthermore, it was demonstrated that apoptosis is massive during primary infection and involves mitochondria functionality.[8,9] It was then elegantly shown that T cell activation is more closely associated with disease progression than plasma virus burden during HIV infection.[10] In the case of HIV, according to the "autoimmunity" hyperactivation model, the immune system is eventually exhausted.[11] For example, during late-stage AIDS, dramatic increases in virus titers are often observed followed by overwhelming infection, which may activate the majority of CD8 effector cells, depleting this HIV-specific subset.[11] The fact that CD4 T cell depletion correlates more closely with levels of immune activation than viral load was confirmed during HIV-1 infection and extended to HIV-2 infection.[12] Interestingly, patients with similar CD4 depletion had dramatically different viral loads but strikingly similar levels of immune activation, regardless of whether they were infected by HIV-1 or HIV-2. Another study, the "Amsterdam Cohort,"[13] focused on the level of CD4 and CD8 T cell activation at different time points after infection before and after documented seroconversion. This research team demonstrated "that low pre-conversion numbers of CD4 T cells and increased levels of immune activation were associated with an increased risk to develop AIDS after seroconversion."[13]

In a compelling manuscript,[14] researchers analyzed the importance of immune activation for total body CD4 count in 43 rhesus macaques. It was shown that despite reduced CD4 counts in the blood and reduced CD4:CD8 in the blood and lymph nodes, the total body CD4 counts and the number of proliferating CD4 significantly increased throughout the entire phase of asymptomatic infection. This was due mainly to CD4 increases in the secondary lymphoid organs. Total body CD8 counts and the number of proliferating CD8 increased even more significantly, again due mainly to increases in the secondary lymphoid organs, thus lowering the CD4:CD8 (a reduced CD4:CD8 was already observed in human lymph nodes[15,16]; however, this was incorrectly interpreted by the field as the result of a selective CD4 decrease). The context changed during AIDS, due to reduced proliferation and complete exhaustion of thymic function. The authors concluded that a drastically increased turnover in the early stages of HIV infection, driven by a generalized immune activation (with an important bystander component) rather than a homeostatic response to CD4 T cell destruction, is followed by exhaustion of the regenerative capacity of the immune system. In other words, the antigen drives the immune system, which, in turn, activates homeostatic mechanisms to control its expansion, and the whole immune machinery drives at maximum speed until exhaustion.

These data helped substantially in interpreting HIV pathogenesis and may have relevant clinical implications. In this chapter, we will discuss chronic immune activation; the exaggerated, precocious apoptosis of T cell subsets; CD95-induced apoptosis and pathogenic implications; and alternative therapy options within the context of highly active antiretroviral therapy (HAART) that include cytokines and cytostatic drugs (i.e., hydroxyurea, mycophenolic acid, cyclosporin).

STATE-OF-THE-ART HAART TREATMENT

Standard of care anti-HIV treatment dictates the combined use of three or more of the 21 approved antiretroviral agents. These derive from four different classes of drugs: nucleoside reverse transcriptase inhibitors (NRTIs), nonnucleoside reverse transcriptase inhibitors (NNRTIs), protease inhibitors (PIs), and entry inhibitors (Figure 28.1). Treatment goals are durable viral suppression, restoration of immunologic function, reduction in HIV pathogenesis, and improvement in the quality of life, with minimal drug-related toxicities. For the most part, clinicians rate the success of these combinations by looking at viral suppression and recovery of CD4 T cells in the absence of toxicity and viral escape. However, if patients stop all therapy, viral load rebounds and the incomplete state of immune reconstitution recovered during HAART quickly erodes.[17] Furthermore, it is generally accepted that although the advent of HAART greatly reduced HIV morbidity and mortality, these

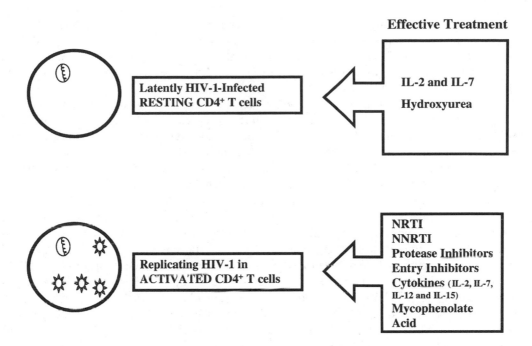

FIGURE 28.1 HIV-1 treatment. Targeting replicating HIV-1 in activated lymphocytes may be accomplished by the combination of the four approved classes of antivirals (NRTI, NNRTI, PI, and entry inhibitors) and likely by the addition of cytokines and virostatics. The ultimate challenge is to eliminate reservoir virus in latently infected resting cells (CD4 and macrophage). Some cytokines may "tickle out" the virus, and hydroxyurea may have indirect or direct antiviral effects in resting lymphocytes.

drug combinations are not free of serious drawbacks (i.e., drug-related toxicities, emergence of drug-resistant strains, difficulty of adhering to regimens, and extremely high costs, that make them inaccessible in developing nations). Thus, we need to devise maintenance therapies that are easier to tolerate, contain viral expansion (to what minimal level may need to be revised), and fully restore the immune system. Moreover, we need to investigate and address what is happening in virally suppressed patients: why is immune reconstitution incomplete, and how can we remedy this situation?

Many studies confirm the observations mentioned in the introduction and pose additional relationships that correlate HIV infection with induced apoptosis and immune dysfunction, including the following: T cell population subsets and T cell receptor repertoire are skewed during HIV-1 infection, and HAART does not always restore preinfection homeostasis; it is probable that different mechanisms induce apoptosis in CD4 and CD8; HAART reduces apoptosis, but the mechanisms responsible remain elusive; and immune dysfunction may be indirectly linked to apoptosis.

CD4 AND CD8 T LYMPHOCYTES DURING HIV-INDUCED CHRONIC IMMUNE ACTIVATION

T cell proliferation occurs after cells receive appropriate antigenic and co-stimulatory signals; the strength and duration of antigen presentation are related to their commitment to clonal expansion and determine the response type (Th-1 or Th-2).[18] The CD8 subset has a much lower activation threshold and produces a larger burst size than CD4. Moreover, chronically stimulated T cells in the absence of adequate co-stimulation (i.e., cytokine levels) lead to the clonal expansion of anergic CD4.[19] Finally, differences in the kinetics of CD4/CD8 death were observed in a recent study of

10 patients who were followed during a 12-week structured treatment interruption (STI).[20] After a brief period of proliferation, spontaneous apoptosis of CD4 T cells seemed to be biphasic and upregulated earlier than CD8 apoptosis, and neither was related to viral load. Despite a slight decrease in CD8 numbers, there was an increase in the CD8+CD38+ subpopulation, with CD38+ expression reflecting activation and corresponding with a poor prognosis.[20]

Lack of complete immune recovery may, in part, be related to HIV-stimulated apoptosis or activation-induced cell death in T lymphocytes, which is reflected by changes in the distribution equilibrium of specific lymphocyte subsets.[7] Gougeon et al., as early as 1996, demonstrated that lymphocyte populations from both healthy uninfected and HIV-1-infected patients undergo apoptosis *in vitro*. However, the demographics of apoptotic CD4 from HIV-infected patients drastically change with the stage of disease.[7] For example, in uninfected controls, CD4 make up 25% of the apoptotic population that decreases to 18% in asymptomatic patients and 4% during AIDS, whereas the CD8 subset represents a less variable but very large portion ranging from 37 to 45%. However, these percentages do not reflect the overall decrease in total CD4 numbers, which plummet dramatically. It was also found that in the chronic phase of HIV infection, 50 to 60% of the apoptotic cells exhibited an activated phenotype (positive human leukocyte antigen-death receptor [HLA-DR+], CD38+, and CD95+), and the CD45R0+ subset seemed to be more susceptible to apoptosis in HIV-positive persons. A significant correlation was found between the intensity of anti-CD3-induced apoptosis in both CD4 and CD8 T cells from HIV-infected persons and their *in vivo* expression of CD45RO and HLA-DR molecules. A significant correlation was found between the intensity of spontaneous or anti-CD3-induced apoptosis in total lymphocytes and disease progression. The fact that an increased susceptibility to apoptosis of peripheral T cells from HIV-infected persons correlates with disease progression strongly supports the hypothesis that the chronic activation of the immune system occurring throughout HIV infection is the primary mechanism responsible for this cell deletion process.[7]

The timing of treatment initiation also seems to influence the level of immune restoration. Hess et al. compared the maturation phenotype of immunodominant HIV-1-specific CD8 T cells in patients undergoing structured treatment interruptions during the acute phase of HIV-1 infection with untreated HIV-1 subjects for a 4-year period.[21] Using the CD45RA (isoform of the CD45 molecule, which is characteristic of virgin/unprimed T lymphocytes) and CCR7 (a receptor for lymph tissue homing) markers to follow lineage differentiation, the authors found that patients exhibiting maximum viral suppression during treatment interruption expanded fully differentiated effector CD8 T cells for various HLA class I tetramer accessible epitopes, whereas those exhibiting a lack of immune control had a paucity of fully differentiated CD8 T cells.[21] In addition, Cossarizza et al. studied two treatment-naive populations (acutely infected and chronically infected treatment-naive populations). Of these two, only the acutely treated were able to restore T cell repertoire in both CD4 and CD8 subsets, whereas the latter group restored perturbations only in the CD8 subset repertoire.[22] These results imply that HIV-1 generates apoptosis in CD4/CD8 subsets through different mechanisms. It is possible that the CD4 class may be subject to a global activation leading to generalized cell death whereas specific CD8 subsets are activated, expand, and progress to apoptosis, thus altering the composition of the total CD8 pool. The latter observation can be confirmed only by looking at the particular CD8 subclasses: naive, HIV-specific effector, central memory, and effector memory, which can be easily performed by using CD45RA and CCR7 markers and fluorescence-activated cell sorter (FACS) analysis.

HIV-1-INDUCED APOPTOSIS

Apoptosis is considered a main defense mechanism of the organism and generally falls under one of two major headings: intrinsic (mitochondrial) or extrinsic (receptor mediated); both signaling pathways converge at the level of specific proteases.[23–25] In general, lymphocyte apoptotic propensity is determined largely by the expression of certain apoptotic proteins that include, but are not limited to, the following: CD95, CD178 (note that CD95 ligand has been renamed CD178), DR-4 and DR-5, and tumor necrosis factor receptor type-1 (TNF-R1), which may be regulated by antiapoptotic proteins such as Bcl-2 or Bcl-xL. T cell activation results in the increased expression of CD178, which may bind to CD95, expressed on target cells. Activated cells expressing CD178 may kill target cells[26,27] or, in an autocrine loop, activate their own CD95, thus committing suicide.[23] The DR-4 and DR-5 bind to and are activated by TNF-related apoptosis-inducing ligand (TRAIL), whereas TNF-R, when activated, can signal for proliferation as well as apoptosis. All of these cellular receptors contain a cytosolic death domain that, when stimulated, may activate the caspase cascade.

The majority of reports investigating HAART and HIV-induced apoptosis involve studies with the CD95/CD178 ligand mechanism. However, conflicting results on CD95-induced apoptosis have been reported in the literature, probably due to differences in study designs. In an early study, spontaneous apoptosis (SA) and CD95 expression in unstimulated peripheral blood mononuclear cells (PBMCs) were correlated with an advanced stage of HIV-1.[28] The majority of apoptotic cells from negative controls and patients with CD4 counts above 100/μL did not express detectable amounts of CD95, whereas patients with lower CD4 T cell counts (in an advanced disease stage) tended to have an elevated number of apoptotic cells, most of which were CD95+. In a second prospective study of pretreated patients undergoing a change in therapy after virological failure, responders (defined as CD4 cell recovery >100 cells/μL) were compared with nonresponders. After 3 months of therapy, there was a significant and durable decrease in spontaneous apoptosis in the CD4 population in the responders, whereas CD95-induced apoptosis significantly declined through the first month but quickly returned to study-baseline values by month 3 in nonresponders.[29]

The existence of a so-called fratricide mechanism further complicates our understanding of this process. Badley et al. originally demonstrated *in vitro* that HIV-infected macrophages can provide a source of CD178,[30] and they later proposed that CD95-induced apoptosis is dependent on two factors: CD95 expression on the target cell and CD178 susceptibility.[31] In one study, the investigators studied CD95-mediated apoptosis in PBMCs and CD178 expression in tonsillar tissue at various time points before, during, and after HAART. Therapy reduced apoptosis in both lymphoid tissue and peripheral blood, and CD95 susceptibility in CD4 lymphocytes was reduced, whereas CD178 expression remained unchanged during HAART. Differences in susceptibility to apoptosis were also observed for infected macrophages and infected T lymphocytes. Pinti et al. used the established cell lines A301 and U937 and their respective HIV-infected clones ACH-2 and U1 to investigate the effects of apoptotic stimuli on survival and expression of CD95. The pre-monocytic cell line U1 was less susceptible to apoptosis with a diminished level of CD95 expression as compared with its noninfected parental cell line, whereas both infected and noninfected lymphocyte cultures were susceptible to apoptosis.[32] From these data, it is tempting to speculate that HIV-1 facilitates its own survival by downregulating the expression of CD95 in monocytes and macrophages, which are known reservoirs for latent virus. However, these studies need to be confirmed with primary cell cultures.

ANTIRETROVIRAL DRUGS AND APOPTOSIS

Because of the early implementation of NRTIs in the treatment of HIV-1 infection, the addition of protease inhibitors (PIs) and NNRTIs has greatly improved clinical outcomes. Other than improving the resistance profile of these multidrug combinations, it has also been proposed that some possess immunomodulatory and antiapoptotic properties (for a review, see Phenix et al.[33]). There are, however, differences in the mechanisms of action of individual drugs from the PI class. Independent of their role in HIV suppression, it has been demonstrated that both ritonavir and saquinavir modulate susceptibility to apoptosis *in vitro* and *in vivo*,[34–36] presumably by effecting the cleavage of key caspase proteins. Chavan et al. clearly demonstrated in one *in vivo* study that nelfinavir (Nfv) contributes to the rescue of peripheral T lymphocytes from spontaneous apoptosis.[37] Interestingly, in another study, Nfv was found to inhibit apoptosis in Jurkat T cells at minimal doses after only 1 h, which indicates that it acts through a mechanism independent of transcription or protein synthesis (including pro- or antiapoptotic proteins).[38] Additional experiments revealed that Nfv inhibits apoptosis by preventing permeability transition pore complex (PTPC) opening, which maintains the mitochondrial transmembrane potential and prevents the release of apoptogenic factors.[38] Indinavir (IDV), on the other hand, does not appear to have any direct effect on apoptosis but may prolong cell survival by inhibiting entry into the cell cycle.[39]

A decrease in mitochondrial membrane potential can occur in apoptotic cells, including activated T lymphocytes, even if this is not the rule for cells undergoing cell death.[40] HIV PI, independent from any viral infection, can hinder lymphocyte apoptosis by influencing mitochondrial homeostasis. An interesting study was undertaken in both resting and activated human lymphocytes exposed to different apoptotic stimuli.[41] T cell activation was found to be accompanied by a significant increase in mitochondrial membrane potential, or hyperpolarization, which was undetectable in resting cells. PIs were able to block activation-associated mitochondria hyperpolarization, which, in turn, was paralleled by an impairment of cell cycle progression. Remarkably, PIs also prevented zidovudine-mediated mitochondrial toxicity. Cells from drug-naive HIV-positive patients behaved identically to activated T cells, displaying hyperpolarized mitochondria, whereas lymphocytes from HIV patients under HAART, which included HIV PIs, seemed to react as resting cells, indicating that the hyperpolarization state of mitochondria represents a prerequisite for the sensitization of lymphocytes to activation-induced cell death (AICD).

LYMPHOCYTE DYSFUNCTION

It was observed that CD4 dysfunction, which occurs early after infection with HIV-1 (even before any decline in absolute numbers of CD4 T cells), involves an early loss of HIV-1-specific responses in terms of both proliferation and IL-2 production.[42-44] The inhibition of IL-2 secretion was linked to the induction of clonal anergy in T cells, possibly due to the development of a proximal signal transduction defect in the IL-2 gene.[45] It is also well known that IL-2 is the main survival factor for T lymphocytes, and any impairment of its production or utilization can result in cell death. It was observed that IL-2, in a dose-dependent manner, prevents T cells from entering apoptosis induced by -irradiation, mitomycin C, or dexamethasone.[46] Results from another study suggest that the balance between cytokine (IL-2) and serum-induced Bcl-2 expression and cytokine-induced Bax expression may determine whether a cell undergoes cytokine-induced apoptosis.[47]

Even in healthy uninfected individuals, regulation of CD8 and CD4 T cell responses seems fundamentally different. CD8 T cell memory is stable, whereas specific CD4 T cell memory gradually declines.[48] In uninfected subjects, however, this CD4 T decay is not associated with loss of immunologic function but, rather, is linked to the reduced expression of Bcl-2 and Bcl-xL.[48] Conditions precipitating CD4 T cell loss (i.e., HIV) might compromise immunity despite maintenance of specific CD8 memory. As Pitcher et al.[17] demonstrates, the "overall frequencies of functional HIV-1 specific CD4 memory cells are considerably diminished in a cohort of long term HAART-treated subjects, indicating that prolonged viral suppression is associated with a decline in functional HIV-1-specific CD4 T cell memory."[17] Interestingly, one study also indicated that CD4 cell rescue (increase in CD4 lymphocyte numbers) is dependent on the CD8 subset rather than directly related to inhibition of apoptosis in the same CD4 population.[49] Hansjee et al.[50] noted a strong inverse relationship between SA rates in PBMCs and CD8 with the recovery of CD4, indicating that there is a very strong interdependent balance between these two lymphocyte classes and lymphocyte dysfunction is somehow linked to apoptosis.

From these studies, it may be inferred that viral replication leads to activation and apoptosis of both CD4 and CD8 T lymphocytes, albeit by different mechanisms, and the magnitude of the CD8 cytotoxic T lymphocyte (CTL) response may be diminished due to a lack of CD4 "help." Controlling the virus with HAART therapy reduces viral load but does not completely restore the preinfection equilibrium of T cell subpopulations, and it does not seem to restore the functionality of the T cells, which persist, even though certain therapies (i.e., Nfv) protect the organism from apoptosis. Moreover, integrated deoxyribonucleic acid (DNA) provirus in latently infected resting memory CD4 T cells are transcriptionally inactive; thus, they escape both immune recognition and the effects of antiviral therapy,[51] underscoring the need for alternative therapeutic strategies that may restore immune function and limit immune pathology associated with this exaggerated immune activation.

ALTERNATIVE HIV-1 THERAPEUTIC STRATEGIES

CYTOKINES

Cytokines are soluble, low-molecular-weight, secreted proteins that regulate the amplitude and duration of the inflammatory immune response. They are produced in a transient manner, tightly regulated by the presence of foreign material, and most act locally. In general, they are pleiotropic, with multiple effects on growth and differentiation of a variety of cell types. There also is considerable redundancy among different types.[52] Mossman and Coffman first reported that cloned murine T lymphocytes (CD4+) could be divided into functional subsets on the basis of the immune-regulatory cytokines they produced: Th-1 producing interferon (IFN)-γ,[53] Th-2 producing IL-4, and Th-0 producing both.[53,54] An inverse relationship was found between cell-mediated immunity (CMI) (mainly sustained by Th-1 cytokines) and humoral antibody (Ab) production (mainly sustained by Th-2 cytokines), with the exception of IgG2a, which is enhanced by Th-1 cytokines.[55] This antagonistic relationship between CMI and Ab production may be attributed to the production and cross-regulation of Th-1 and Th-2 cytokines. The nomenclature has since evolved to Th-1-like and Th-2-like to reflect the functional plasticity of the two groups and indicate that other cytokines contribute to the regulation of the immune system.[56,57] Subsequent studies demonstrated that, in addition to CD4+ T cells, many different cell types produce cytokines, namely, CD8+ T cells, monocytes/macrophages, natural killer (NK) cells, B cells, mast cells, eosinophils, and antigen-presenting cells (APCs). Some researchers also believe that a switch from a Th-1- to Th-2-like immune response may contribute to clinical progression to AIDS.[58,59] It was also proposed that Th-1-like cytokines offer some protection from apoptotic cell death, whereas Th-2-like cytokines promote apoptosis of CD4+ cells.[60]

IL-2 and IL-15

Although the mechanism by which IL-2 treatment increases CD4 T cells is not fully understood, it was shown that this increase occurs by the peripheral expansion of existing naive CD4 T cell populations rather than by increased thymic output.[61] The intermittent administration of IL-2 to patients receiving continuous HAART was shown to substantially decrease replication-competent HIV in resting CD4 T cells.[61] Based on the observation that lymphocytes from HIV-infected individuals express proteins characteristic of G_1/S to G_2/M, despite a DNA content and metabolic profile characteristic of the G_0 cell-cycle phase, Paiardini and Silvestri et al. proposed that this cell-cycle perturbation may lower the threshold for activation-induced apoptosis.[62] They demonstrated that the addition of exogenous IL-2 to HIV-infected lymphocytes induces a profound normalizing effect on the expression of cell-cycle-dependent proteins, thus reducing progression to apoptosis. Data from the same study also indicate that although normal quantities of IL-2 production are observed during HIV infection, its functionality is impaired two- to threefold, and the addition of exogenous IL-2 makes this defect normal.[62] This drug, however, should be used with caution in patients with advanced disease, because National Institute of Allergy and Infectious Diseases (NIAID) studies indicate that IL-2 can accelerate disease progression in patients with CD4 counts below 200 cells/$\mu\mu^3$ by substantially increasing HIV levels.[63]

Both IL-15 and -2 drive T cell proliferation *in vitro* but are produced by different cell types (IL-2 is produced by activated T cells, and IL-15 is produced by many cell types, including antigen-presenting cells, keratinocytes, and epithelial cells). Proliferation experiments for CD4 and CD8 showed that these subsets exhibit specific sensitivities at different stages of differentiation to IL-2 and IL-15, respectively.[64] IL-15 promotes the proliferation and long-term survival of CD8 memory but not CD4 memory T cells in an antigen-independent fashion.

Both IL-2 and -15 upregulate intracellular levels of Bcl-2 and induce the expression of the HIV co-receptor CCR5 (the latter is an important determinant of the level of HIV infection).[65] Thus, it is likely that both have discordant (the upregulation of Bcl-2 is a positive feature whereas an increase in the expression of CCR5 may benefit the virus) effects on HIV-infected individuals.[66,67] In one study by Zaunders et al.,[64] the addition of IL-2 or IL-15 significantly inhibited caspase activation, increased Bcl-2 expression, and rapidly initiated proliferation of CD8 T cells expressing CCR5 *in vitro*. Although the authors indicate that inhibition of apoptosis proceeded via the Bcl-2/ cytokine-regulated mitochondrial pathway, they also noted a substantial unexplained upregulation of CD95 expression.[64] Another serious consideration with IL-15 administration was raised after one study demonstrated that the expansion of memory cells to concurrent infection in *Mycobacterium bovis*-IL-15-infected mice coincided with the attrition of existing memory.[68] This could obviously have deleterious consequences for the already immune-compromised HIV-1 patient.

The production of different cytokines is crucial from a functional point of view and can represent a useful marker to analyze the composition of the pool of virus-specific CD4 T cells. On the basis of the ability to secrete IL-2 and IFN-γ, three functionally distinct populations of CD4 T cells can be identified as follows: IL-2-secreting cells, IL-2/IFN-γ-secreting cells, and IFN-γ-secreting cells.[69] This pivotal study showed that cytomegalovirus (CMV)-specific CD4 T cells were almost equally distributed within the three functionally distinct cell populations. However, a skewing toward IFN-γ-secreting cells (70% of HIV-1-specific CD4 T cells) was observed in subjects with progressive HIV infection, and IL-2-secreting and IL-2/IFN-γ-secreting cells were almost absent. The frequencies of IL-2-secreting and IL-2/IFN-γ-secreting HIV-1-specific CD4 T cells were negatively correlated with the levels of viremia. Prolonged antiretroviral therapy was able to correct the skewed representation of different populations of HIV-1-specific CD4 T cells but was associated with only a partial recovery of IL-2-secreting cells. These results underline the importance and role of a decreased capacity to produce IL-2.

IL-12

IL-12 links the innate immune response and adaptive immunity by inducing the early production of IFN-γ by NK cells and later by T cells and is also required for the optimal differentiation of cytotoxic T lymphocytes.[70] For these reasons, this cytokine is being used as an adjuvant in several preclinical vaccine trials (mostly cancer studies).[71,72] Because vaccine-elicited responses should include long-lived memory cells that can rapidly expand in number following reexposure to antigen, one study was designed to analyze the kinetics, magnitude, and durability of plasmid IL-12-augmented DNA vaccine–elicited responses in mice.[73] Mice were immunized with a gp120 vaccine with or without plasmid IL-12 on days 4, 7, 10, and 14. The authors reported a notable increase in CTL responses in the IL-12 group only at day 10. However, there was no expansion in CTL when boosted with gp120 alone at day 130. Therefore, it is likely that the strong response at day 10 resulted from the expansion of effector memory (instead of central memory) and highlights the importance of timing with cytokine adjuvants.

IL-7

IL-7 is another immune-modulating HIV therapeutic candidate. This interleukin provides survival signals to early thymocyte progenitors and protects mature thymocytes from dexamethasone-induced apoptosis,[74] which was linked to the upregulation of Bcl-2 in peripheral T cells.[75] Unlike IL-2 or phytohaemagglutinin (PHA) treatment, IL-7 can induce the *in vitro* expansion of naive T cells and maintains their naive phenotype.[76] This mechanism was confirmed in the analysis of a small number of chronically infected severely immune-impaired patients who initiated HAART and were capable of reconstituting their pool of CD4 T cells, even in the apparent absence of thymic regeneration.[77] Expanded CD4 populations displayed a high percentage of phenotypically naive cells with a very low expression of CD95[+], CD45R0[+], and CD38[+]. Plasma levels of IL-7 were significantly high, and most CD4/CD8 cells expressed significant levels of CD-127 (IL-7 receptor) characteristic of healthy donors and had low CD95 levels. There are, however, conflicting reports regarding susceptibility to HIV after IL-7 treatment. The results of one *in vitro* study indicated that IL-7 promotes both T-tropic and M-tropic HIV replication in naive T cells.[76] On the contrary, Napolitano et al. demonstrated in the severe combined immunodeficient with co-engrafted human fetal liver and thymus (SCID-hu Thy/ Liv) mouse model that IL-7 does not increase HIV replication.

VIROSTATICS

A new concept of antiviral/cytostatic ("virostatics") drugs was proposed within the context of HAART to restrict virus target populations (CD4[+] T lymphocytes), target viral reservoirs, and possibly restore immune functions, by reducing excess immune activation. These virostatics include drugs such as hydroxyurea, mycophenolic acid, and cyclosporin. They use multiple novel mechanisms of action to impede HIV by targeting host cellular proteins that are not susceptible to mutation. Therefore, their resistance profile seems to be favorable.

Reducing the hyperactivation of the immune system that accompanies chronic HIV infection might be advantageous. Giorgi et al.[10] convincingly showed that immunologic (T cell activation markers) and not virological factors may be more accurate correlates of disease progression. More recently, Silvestri et al. observed minimal immune activation (CD8) and no CD4 T cell depletion in the state of chronically elevated viremia in clinically healthy SIV-infected sooty mangabey monkeys.[78] Based on these observations, the authors propose that AIDS progression may involve more than direct virus killing of CD4 T cells. They maintain that excessive turnover rates for both pools of CD4 and CD8 effector cells is most likely the result of AICD[6,7,79–83] and constitutes a drain on the naive T cell pool; a reduction in the establishment of long-lived memory cells, which protect the host from secondary infection; and the exaggerated production of effector-generated inflammatory cytokines, which exacerbate immunopathology of HIV and increase viral target populations (CD4 T cells).[78]

The absence of CD4 T cell depletion despite chronically high levels of virus replication observed in the sooty mangabey model supports the concept that alternative indicators of disease progression may exist. There is substantial evidence that HIV-1 replication requires factors that are made available when host CD4 T cells are activated.[84,85] Consistent with this data, it was noted that HIV-1 patients who receive vaccinations against influenza, hepatitis B, or tetanus toxoid (thus, in a state of immune activation) generally experience increases in viral load.[84,85] The predator–prey model also substantiates this correlation between the number of available CD4 target cells and virus expansion. Mathematical models based on the predator–prey equilibrium predict that a sustained expansion of the CD4 T lymphocytes subset—"the prey"—increases the probability that HIV—"the predator"—will spread.[86] It was proposed that the cytostatic properties of virostatics may limit the resources made available to the virus, thus keeping their numbers in check.[87] Therefore, the quantification and qualitative analysis of CD4 and CD8 T cell subsets and their degree of activation or stimulation may be of clinical interest in predicting progression to AIDS.

Although virostatics should function through a common cytostatic mechanism, each needs to be closely evaluated for ulterior immune-modulating properties. For example, both hydroxyurea and mycophenolate acid reduce the nucleotide pool, but each drug acts at a different point in purine synthesis sequence and, thus, there are substantial differences between the two drug mechanisms. Mycophenolic acid (MPA) acts early in the synthesis of purines and has an effect on both transcription and DNA synthesis, whereas hydroxyurea, which interferes with ribonucleotide reductase, primarily effects DNA synthesis (Figure 28.2). Data from recent *in vitro* and *in vivo* studies

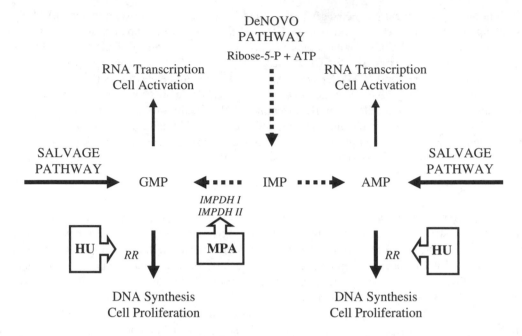

FIGURE 28.2 General synthesis scheme for purines. Two pathways exist for purine synthesis: *de novo* (dashed lines), mainly used by activated lymphocytes, and salvage (continuous lines), mainly used by resting lymphocytes. Hydroxyurea (HU) inhibits ribonucleotide reductase, thus limiting the production of dNTP synthesis and DNA elongation. Mycophenolic acid (MPA) blocks both RNA transcription and DNA synthesis by targeting the different isoforms of IMPDH (type I and type II). Type I is expressed in resting lymphocytes, whereas type II is primarily expressed in activated lymphocytes and is five times more sensitive to MPA. (See also Konno, Y. et al., *J. Biol. Chem.*, 266, 506, 1991; Nagai, M., Natsumeda, Y., and Weber, G., *Cancer Res.*, 52, 258, 1992; Allison, A.C. and Eugui, E.M., *Immunopharmacology*, 47, 85, 2000.) For these reasons, MPA is expected to be a more potent inhibitor of dNTP synthesis than hydroxyurea in activated lymphocytes, whereas hydroxyurea should be more potent than MPA in quiescent lymphocytes. (RR: ribonucleotide reductase; IMPDH: inosine monophosphate dehydrogenase; IMP: inosine monophosphate; and GMP and AMP are nucleoside precursors.)

clearly indicate that these drugs also effect other changes at the molecular level in various cell types. This is not surprising, because purines not only serve as precursors for DNA/ ribonucleic acid (RNA) synthesis, but they also play an important role in lipid synthesis and protein glycosylation and serve as a source of energy for many other metabolic processes.

Hydroxyurea

Hydroxyurea (hydroxycarbamide) was used in hematology to treat chronic leukemia for more than 40 years and more recently was approved for sickle-cell anemia therapy.[88,89] Hydroxyurea uses multiple mechanisms of action to impede the replication of HIV. First, hydroxyurea inhibits the cellular enzyme ribonucleotide reductase, thus blocking the transformation of ribonucleotides into deoxyribonucleotides, depleting the intracellular deoxynucleotide triphosphate (dNTP) pool, and arresting the cell cycle in the G1/S phase,[88,89] which directly halts HIV DNA synthesis. Second, synergistic anti-HIV activity was demonstrated when hydroxyurea was combined with NRTIs.[90–93] This augmentation in antiretroviral activity may be explained by a favorable change in the proportion of activated NRTI (ddNTP) to dNTP and an enhancement of NRTI phosphorylation.[94] Because NRTIs compete with cellular dNTPs for incorporation into the growing HIV DNA, using hydroxyurea to decrease the concentration of cellular dNTP creates a competitive advantage for the NRTIs.[95] Consequently, a higher proportion of the NRTI eventually becomes incorporated by reverse transcriptase into viral DNA and thus blocks DNA synthesis. Third, because cell division must progress to the G1b phase of the cell cycle for completion of HIV reverse transcription,[96] the virus will only replicate in activated T lymphocytes.[97] Thus, when T lymphocytes are treated with hydroxyurea, they remain quiescent and become refractory to productive HIV infection.

Finally, two other interesting, indirect, immune-mediated antiviral characteristics of hydroxyurea were discovered. Heredia et al. found that the arrest of T lymphocytes in the G1 phase of the cell cycle, in response to hydroxyurea, results in the accumulation of -chemokines exerting antiviral activity. These -chemokines, RANTES, MIP-1 and MIP-1 are the natural ligands of the HIV-1 co-receptor CCR5 and compete with R5 HIV-1 virus (CCR5 using strains) for receptor binding.[98] In addition, hydroxyurea-containing antiretroviral regimens were shown to restore functional aspects of the immune response impaired by HIV infection, such as reverting the defect of CD3- (the signaling chain of the T cell receptor complex) expression, restoring responses to allogeneic and influenza virus antigens,[99] and inducing vigorous HIV-specific T-helper responses.[100,101]

This drug was extensively evaluated in all three stages of infection (in more than 500 patients) in conjunction with a wide variety of NRTIs. Hydroxyurea was most often paired with didanosine (ddI), a nucleoside analogue, because it augments the activity of this NRTI. Moreover, even though this combination does not prevent the emergence of mutant viral strains, it renders these strains sensitive to ddI.[92,102] However, due to the potential for hydroxyurea to increase the toxicity of certain NRTI analogues, a dosing study of the drug was warranted. RIGHT 702 was a randomized, controlled, hydroxyurea dose-regimen study of 115 HIV-infected patients, to compare safety and efficacy of hydroxyurea administered at three daily doses (600, 800 to 900, or 1200 mg/day) and at three dosing intervals—qd, bid, or three times daily (tid)—in combination with ddI and stavudine (d4T). The primary objectives of this study were to determine the safety and tolerability of hydroxyurea administered at different daily doses as well as compare antiretroviral activity at these same doses in combination with ddI and d4T. This study provided the most convincing data supporting the clinical value of an appropriately dosed hydroxyurea/ddI regimen. Surprisingly, the lowest 600-mg dose in the RIGHT 702 study proved the most efficacious in suppressing viremia and was associated with fewer adverse events and the highest CD4 count increase.

Our laboratory recently showed in an *in vitro* analysis of T lymphocyte proliferation that hydroxyurea has no relevant effect on the expression of activation markers. Moreover, this drug was found to increase the number of apoptotic cells as visualized with Annexin V staining (see MPA section) only at concentrations above 10 μL.[103] Considering that *in vivo* pharmacokinetic analyses indicate that a 1000-mg total daily dose corresponds with an *in vitro* dosage of 135 μL, and the drug demonstrates linear pharmacokinetics (absorption and elimination),[104] a lower 600-mg total daily dose should only minimally induce apoptosis. This, of course, needs to be confirmed with new *in vivo* pharmacokinetic studies at this concentration.

Mycophenolate Acid

MPA is another interesting candidate for the treatment of HIV/AIDS. Mycophenolate mofetil (CellCept™), the prodrug of MPA, is licensed for the prevention of renal transplant rejection and was extensively investigated in different models of allogeneic transplantation, autoimmune skin disorders, and rheumatoid arthritis.[105–107] Its principal mechanism of action is the inhibition of inosine monophosphate dehydrogenase (IMPDH), which is the rate-limiting enzyme in the *de novo* synthesis of guanosine nucleotides. Because MPA selectively inhibits synthesis of guanosine nucleotides, it potentiates the antiviral activity of guanosine inhibitors such as abacavir. There also exist different isoforms of IMPDH (type I and type II): type I is expressed in resting lymphocytes, whereas type II is primarily expressed in activated lymphocytes and is five times more sensitive to MPA.[108–110] Based on this information, MPA should limit cellular proliferation as well as the expansion of viruses in activated T lymphocytes but likely has no effect on latently infected resting cells (Figure 28.1).

One common method for investigating apoptosis is to look at the phospholipid phosphati-dylserine, which is translocated to the outer plasma membrane during the early stages of apoptosis. This molecule will bind the marker Annexin V, which may be visualized and quantified by FACS analysis. Chapuis and Pantaleo et al., using this system, found that MPA suppresses proliferation in the presence of IL-2 by inducing apotosis in activated, but not resting, T cells.[111]

Cyclosporin

Cyclosporin is a known immunosuppressant that is used extensively to prevent organ transplantation rejection and was proposed to have a possible role in HIV treatment. Cyclosporin forms a complex with cyclophilin A, which inhibits the calcium and calmoduline-dependent protein phosphatase calcineurin, thus ultimately blocking the activation of genes for IL-2, -4, and the -2 receptor in T cells.[112] Rizzardi et al. maintain "that shutdown of T cell activation (with a drug such as cyclosporin) in the early phases of primary HIV-1 infection can have long-term beneficial effects and establish a more favorable immunologic set-point."[112,113] Regarding apoptosis, one recent study with anti-CD3/anti-CD28-stimulated T lymphocytes from healthy donors demonstrated with Annexin V staining that cyclosporin inhibits AICD.[114] This drug also directly acts against HIV by preventing proper HIV virion mutation. Finally, it was postulated that cyclosporin synergizes with protease inhibitors by augmenting their absorption. However, due to this drug's intrinsic immuno-suppressive effects (blocking the activation of genes for IL-2, the -2 receptor, and -4), it should be scrutinized to test whether it could become a candidate for HIV treatment.

CONCLUSION

Although HAART has clearly brought about an enormous improvement in the survival rates of HIV-1-infected patients, it is clear that it does not completely eradicate the virus or completely restore immune responses. Moreover, in the near future, we will also likely be faced with an exponential increase in the number of HIV-1-infected patients who will have exhausted their treatment options due to the emergence of antiviral-resistant strains. Therefore, we need to revise our approach to this disease. We need to go beyond the state of the art in HIV treatment and implement changes in current strategies. HIV interventions may require a bimodal approach, targeting the virus and limiting host-related hyperimmune activation, which acerbates immune pathology. Agents acting on different apoptotic pathways are clearly of clinical interest, and with a better understanding of HIV-induced apoptosis and how the immune system may be reconstituted, we may devise new therapeutic regimens that include cytokines or virostatics.

REFERENCES

1. Akbar, A.N., Borthwick, N., Salmon, M., Gombert, W., Bofill, M., Shamsadeen, N., Pilling, D., Pett, S., Grundy, J.E., and Janossy, G., The significance of low Bcl-2 expression by CD45RO T cells in normal individuals and patients with acute viral infections. The role of apoptosis in T cell memory, *J. Exp. Med.,* 178, 427, 1993.
2. Steel, C.M., Ludlam, C.A., Beatson, D., Peutherer, J.F., Cuthbert, R.J., Simmonds, P., Morrison, H., and Jones, M., HLA haplotype A1 B8 DR3 as a risk factor for HIV-related disease, *Lancet,* 1, 1185, 1988.
3. Simmonds, P., Beatson, D., Cuthbert, R.J., Watson, H., Reynolds, B., Peutherer, J.F., Parry, J.V., Ludlam, C.A., and Steel, C.M., Determinants of HIV disease progression: six-year longitudinal study in the Edinburgh haemophilia/HIV cohort, *Lancet,* 338, 1159, 1991.
4. Ameisen, J.C. and Capron, A., Cell dysfunction and depletion in AIDS: the programmed cell death hypothesis, *Immunol. Today,* 12, 102, 1991.

5. Meyaard, L., Otto, S.A., Jonker, R.R., Mijnster, M.J., Keet, R.P., and Miedema, F., Programmed death of T cells in HIV-1 infection, *Science*, 257, 217, 1992.

6. Finkel, T.H., Tudor-Williams, G., Banda, N.K., Cotton, M.F., Curiel, T., Monks, C., Baba, T.W., Ruprecht, R.M., and Kupfer, A., Apoptosis occurs predominantly in bystander cells and not in productively infected cells of HIV- and SIV-infected lymph nodes, *Nat. Med.*, 1, 129, 1995.

7. Gougeon, M.L., Lecoeur, H., Dulioust, A., Enouf, M.G., Crouvoiser, M., Goujard, C., Debord, T., and Montagnier, L., Programmed cell death in peripheral lymphocytes from HIV-infected persons: increased susceptibility to apoptosis of CD4 and CD8 T cells correlates with lymphocyte activation and with disease progression, *J. Immunol.*, 156, 3509, 1996.

8. Cossarizza, A., T-cell repertoire and HIV infection: facts and perspectives, *AIDS*, 11, 1075, 1997.

9. Cossarizza, A., Stent, G., Mussini, C., Paganelli, R., Borghi, V., Nuzzo, C., Pinti, M., Pedrazzi, J., Benatti, F., Esposito, R., Rosok, B., Nagata, S., Vella, S., Franceschi, C., and De Rienzo, B., Deregulation of the CD95/CD95L system in lymphocytes from patients with primary acute HIV infection, *AIDS*, 14, 345, 2000.

10. Giorgi, J.V., Hultin, L.E., McKeating, J.A., Johnson, T.D., Owens, B., Jacobson, L.P., Shih, R., Lewis, J., Wiley, D.J., Phair, J.P., Wolinsky, S.M., and Detels, R., Shorter survival in advanced human immunodeficiency virus type 1 infection is more closely associated with T lymphocyte activation than with plasma virus burden or virus chemokine coreceptor usage, *J. Infect. Dis.*, 179, 859, 1999.

11. Zinkernagel, R.M. and Hengartner, H., T-cell-mediated immunopathology vs. direct cytolysis by virus: implications for HIV and AIDS, *Immunol. Today*, 15, 262, 1994.

12. Sousa, A.E., Carneiro, J., Meier-Schellersheim, M., Grossman, Z., and Victorino, R.M., CD4 T cell depletion is linked directly to immune activation in the pathogenesis of HIV-1 and HIV-2 but only indirectly to the viral load, *J. Immunol.*, 169, 3400, 2002.

13. Hazenberg, M.D., Otto, S.A., van Benthem, B.H., Roos, M.T., Coutinho, R.A., Lange, J.M., Hamann, D., Prins, M., and Miedema, F., Persistent immune activation in HIV-1 infection is associated with progression to AIDS, *AIDS*, 17, 1881, 2003.

14. Sopper, S., Nierwetberg, D., Halbach, A., Sauer, U., Scheller, C., Stahl-Hennig, C., Matz-Rensing, K., Schafer, F., Schneider, T., ter Meulen, V., and Muller, J.G., Impact of simian immunodeficiency virus (SIV) infection on lymphocyte numbers and T-cell turnover in different organs of rhesus monkeys, *Blood*, 101, 1213, 2003.

15. Cajigas, A., Suhrland, M., Harris, C., Chu, F., McGowan, J., Golodner, M., Seymour, A.W., and Lyman, W.D., Correlation of the ratio of CD4+/CD8+ cells in lymph node fine needle aspiration biopsies with HIV clinical status. A preliminary study, *Acta Cytol.*, 41, 1762, 1997.

16. Tedla, N., Dwyer, J., Truskett, P., Taub, D., Wakefield, D., and Lloyd, A., Phenotypic and functional characterization of lymphocytes derived from normal and HIV-1-infected human lymph nodes, *Clin. Exp. Immunol.*, 117, 92, 1999.

17. Pitcher, C.J., Quittner, C., Peterson, D.M., Connors, M., Koup, R.A., Maino, V.C., and Picker, L.J., HIV-1-specific CD4+ T cells are detectable in most individuals with active HIV-1 infection, but decline with prolonged viral suppression, *Nat. Med.*, 5, 518, 1999.

18. Kaech, S.M., Wherry, E.J., and Ahmed, R., Effector and memory T-cell differentiation: implications for vaccine development, *Nat. Rev. Immunol.*, 2, 251, 2002.

19. Tanchot, C., Barber, D.L., Chiodetti, L., and Schwartz, R.H., Adaptive tolerance of CD4+ T cells *in vivo*: multiple thresholds in response to a constant level of antigen presentation, *J. Immunol.*, 167, 2030, 2001.

20. Roger, P.M., Durant, J., Ticchioni, M., Halfon, P., Breittmayer, J.P., Brignone, C., Chaillou, S., Dunais, B., Dellamonica, P., and Bernard, A., Apoptosis and proliferation kinetics of T cells in patients having experienced antiretroviral treatment interruptions, *J. Antimicrob. Chemother.*, 52, 269, 2003.

21. Hess, C., Altfeld, M., Thomas, S.Y., Addo, M.M., Rosenberg, E.S., Allen, T.M., Draenert, R., Eldrige, R.L., van Lunzen, J., Stellbrink, H.J., Walker, B.D., and Luster, A.D., HIV-1 specific CD8+ T cells with an effector phenotype and control of viral replication, *Lancet*, 363, 863, 2004.

22. Cossarizza, A., Poccia, F., Agrati, C., D'Offizi, G., Bugarini, R., Pinti, M., Borghi, V., Mussini, C., Esposito, R., Ippolito, G., and Narciso, P., Highly active antiretroviral therapy restores CD4+ V T-cell repertoire in patients with primary acute HIV infection but not in treatment-naive HIV+ patients with severe chronic infection, *J. Acquir. Immune Defic. Syndr.*, 35, 213, 2004.

23. Lawen, A., Apoptosis—an introduction, *Bioessays*, 25, 888, 2003.

24. Reed, J.C., Mechanisms of apoptosis, *Am. J. Pathol.*, 157, 1415, 2000.

25. Rossi, D. and Gaidano, G., Messengers of cell death: apoptotic signaling in health and disease, *Haematologica,* 88, 212, 2003.

26. Yonehara, S., Ishii, A., and Yonehara, M., A cell-killing monoclonal antibody (anti-Fas) to a cell surface antigen co-downregulated with the receptor of tumor necrosis factor, *J. Exp. Med.,* 169, 1747, 1989.

27. Trauth, B.C., Klas, C., Peters, A.M., Matzku, S., Moller, P., Falk, W., Debatin, K.M., and Krammer, P.H., Monoclonal antibody-mediated tumor regression by induction of apoptosis, *Science,* 245, 301, 1989.

28. Patki, A.H., Georges, D.L., and Lederman, M.M., CD4+-T-cell counts, spontaneous apoptosis, and Fas expression in peripheral blood mononuclear cells obtained from human immunodeficiency virus type 1-infected subjects, *Clin. Diagn. Lab. Immunol.,* 4, 736, 1997.

29. Roger, P.M., Breittmayer, J.P., Durant, J., Sanderson, F., Ceppi, C., Brignone, C., Cua, E., Clevenbergh, P., Fuzibet, J.G., Pesce, A., Bernard, A., and Dellamonica, P., Early CD4(+) T cell recovery in human immunodeficiency virus-infected patients receiving effective therapy is related to a down-regulation of apoptosis and not to proliferation, *J. Infect. Dis.,* 185, 463, 2002.

30. Badley, A.D., McElhinny, J.A., Leibson, P.J., Lynch, D.H., Alderson, M.R., and Paya, C.V., Upregulation of Fas ligand expression by human immunodeficiency virus in human macrophages mediates apoptosis of uninfected T lymphocytes, *J. Virol.,* 70, 199, 1996.

31. Badley, A.D., Dockrell, D.H., Algeciras, A., Ziesmer, S., Landay, A., Lederman, M.M., Connick, E., Kessler, H., Kuritzkes, D., Lynch, D.H., Roche, P., Yagita, H., and Paya, C.V., *In vivo* analysis of Fas/FasL interactions in HIV-infected patients, *J. Clin. Invest.,* 102, 79, 1998.

32. Pinti, M., Biswas, P., Troiano, L., Nasi, M., Ferraresi, R., Mussini, C., Vecchiet, J., Esposito, R., Paganelli, R., and Cossarizza, A., Different sensitivity to apoptosis in cells of monocytic or lymphocytic origin chronically infected with human immunodeficiency virus type-1, *Exp. Biol. Med.,* 228, 1346, 2003.

33. Phenix, B.N., Cooper, C., Owen, C., and Badley, A.D., Modulation of apoptosis by HIV protease inhibitors, *Apoptosis,* 7, 295, 2002.

34. Weichold, F.F., Bryant, J.L., Pati, S., Barabitskaya, O., Gallo, R.C., and Reitz, M.S., Jr., HIV-1 protease inhibitor ritonavir modulates susceptibility to apoptosis of uninfected T cells, *J. Hum. Virol.,* 2, 261, 1999.

35. Sloand, E.M., Kumar, P.N., Kim, S., Chaudhuri, A., Weichold, F.F., and Young, N.S., Human immunodeficiency virus type 1 protease inhibitor modulates activation of peripheral blood CD4(+) T cells and decreases their susceptibility to apoptosis *in vitro* and *in vivo, Blood,* 94, 1021, 1999.

36. Sloand, E.M., Maciejewski, J., Kumar, P., Kim, S., Chaudhuri, A., and Young, N., Protease inhibitors stimulate hematopoiesis and decrease apoptosis and ICE expression in CD34(+) cells, *Blood,* 96, 2735, 2000.

37. Chavan, S.J., Tamma, S.L., Kaplan, M., Gersten, M., and Pahwa, S.G., Reduction in T cell apoptosis in patients with HIV disease following antiretroviral therapy, *Clin. Immunol.,* 93, 24, 1999.

38. Phenix, B.N., Lum, J.J., Nie, Z., Sanchez-Dardon, J., and Badley, A.D., Antiapoptotic mechanism of HIV protease inhibitors: preventing mitochondrial transmembrane potential loss, *Blood,* 98, 1078, 2001.

39. Chavan, S., Kodoth, S., Pahwa, R., and Pahwa, S., The HIV protease inhibitor Indinavir inhibits cell-cycle progression *in vitro* in lymphocytes of HIV-infected and uninfected individuals, *Blood,* 98, 383, 2001.

40. Cossarizza, A., Kalashnikova, G., Grassilli, E., Chiappelli, F., Salvioli, S., Capri, M., Barbieri, D., Troiano, L., Monti, D., and Franceschi, C., Mitochondrial modifications during rat thymocyte apoptosis: a study at the single cell level, *Exp. Cell Res.,* 214, 323, 1994.

41. Matarrese, P., Gambardella, L., Cassone, A., Vella, S., Cauda, R., and Malorni, W., Mitochondrial membrane hyperpolarization hijacks activated T lymphocytes toward the apoptotic-prone phenotype: homeostatic mechanisms of HIV protease inhibitors, *J. Immunol.,* 170, 6006, 2003.

42. Lane, H.C., Depper, J.M., Greene, W.C., Whalen, G., Waldmann, T.A., and Fauci, A.S., Qualitative analysis of immune function in patients with the acquired immunodeficiency syndrome. Evidence for a selective defect in soluble antigen recognition, *N. Engl. J. Med.,* 313, 79, 1985.

43. Wahren, B., Morfeldt-Mansson, L., Biberfeld, G., Moberg, L., Sonnerborg, A., Ljungman, P., Werner, A., Kurth, R., Gallo, R., and Bolognesi, D., Characteristics of the specific cell-mediated immune response in human immunodeficiency virus infection, *J. Virol.,* 61, 2017, 1987.

44. Clerici, M., Stocks, N.I., Zajac, R.A., Boswell, R.N., Lucey, D.R., Via, C.S., and Shearer, G.M., Detection of three distinct patterns of T helper cell dysfunction in asymptomatic, human immunodeficiency virus-seropositive patients. Independence of CD4+ cell numbers and clinical staging, *J. Clin. Invest.,* 84, 1892, 1989.

45. Telander, D.G., Malvey, E.N., and Mueller, D.L., Evidence for repression of IL-2 gene activation in anergic T cells, *J. Immunol.,* 162, 1460, 1999.

46. Mor, F. and Cohen, I.R., IL-2 rescues antigen-specific T cells from radiation or dexamethasone-induced apoptosis. Correlation with induction of Bcl-2, *J. Immunol.,* 156, 515, 1996.

47. Shi, Y., Wang, R., Sharma, A., Gao, C., Collins, M., Penn, L., and Mills, G.B., Dissociation of cytokine signals for proliferation and apoptosis, *J. Immunol.,* 159, 5318, 1997.

48. Homann, D., Teyton, L., and Oldstone, M.B., Differential regulation of antiviral T-cell immunity results in stable CD8+ but declining CD4+ T-cell memory, *Nat. Med.,* 7, 913, 2001.

49. Grelli, S., D'Ettore, G., Lauria, F., Montella, F., Di Traglia, L., D'Agostini, C., Lichtner, M., Vullo, V., Favalli, C., Vella, S., Macchi, B., and Mastino, A., CD4+ lymphocyte increases in HIV patients during potent antiretroviral therapy are dependent on inhibition of CD8+ cell apoptosis, *Ann. N.Y. Acad. Sci.,* 1010, 560, 2003.

50. Hansjee, N.E.A., Persistent apoptosis in HIV-1 infected individuals receiving potent antiretroviral therapy is associated with poor recovery of CD4 lymphocyes, *J. Acquir. Immune Defic. Syndr.,* 36, 671, 2004.

51. Lum, J.J., Pilon, A.A., Sanchez-Dardon, J., Phenix, B.N., Kim, J.E., Mihowich, J., Jamison, K., Hawley-Foss, N., Lynch, D.H., and Badley, A.D., Induction of cell death in human immunodeficiency virus-infected macrophages and resting memory CD4 T cells by TRAIL/Apo2l, *J. Virol.,* 75, 1128, 2001.

52. Roitt, I.M., *Essential Immunology,* 7th ed., Blackwell Scientific, Oxford, 1991, p. 141.

53. Mosmann, T.R., Cherwinski, H., Bond, M.W., Giedlin, M.A., and Coffman, R.L., Two types of murine helper T cell clone. I. Definition according to profiles of lymphokine activities and secreted proteins, *J. Immunol.,* 136, 2348, 1986.

54. Paul, W.E. and Seder, R.A., Lymphocyte responses and cytokines, *Cell,* 76, 241, 1994.

55. Parish, C.R., The relationship between humoral and cell-mediated immunity, *Transplant Rev.,* 13, 35, 1972.

56. Clerici, M., Hakim, F.T., Venzon, D.J., Blatt, S., Hendrix, C.W., Wynn, T.A., and Shearer, G.M., Changes in interleukin-2 and interleukin-4 production in asymptomatic, human immunodeficiency virus-seropositive individuals, *J. Clin. Invest.,* 91, 759, 1993.

57. Bloom, B.R., Salgame, P., and Diamond, B., Revisiting and revising suppressor T cells, *Immunol. Today,* 13, 131, 1992.

58. Clerici, M. and Shearer, G.M., The Th1-Th2 hypothesis of HIV infection: new insights, *Immunol. Today,* 15, 575, 1994.

59. Clerici, M. and Shearer, G.M., A TH1-->TH2 switch is a critical step in the etiology of HIV infection, *Immunol. Today,* 14, 107, 1993.

60. Clerici, M., Sarin, A., Coffman, R.L., Wynn, T.A., Blatt, S.P., Hendrix, C.W., Wolf, S.F., Shearer, G.M., and Henkart, P.A., Type 1/type 2 cytokine modulation of T-cell programmed cell death as a model for human immunodeficiency virus pathogenesis, *Proc. Natl. Acad. Sci. U.S.A.,* 91, 11811, 1994.

61. Chun, T.W., Engel, D., Mizell, S.B., Hallahan, C.W., Fischette, M., Park, S., Davey, R.T., Jr., Dybul, M., Kovacs, J.A., Metcalf, J.A., Mican, J.M., Berrey, M.M., Corey, L., Lane, H.C., and Fauci, A.S., Effect of interleukin-2 on the pool of latently infected, resting CD4+ T cells in HIV-1-infected patients receiving highly active anti-retroviral therapy, *Nat. Med.,* 5, 651, 1999.

62. Paiardini, M., Galati, D., Cervasi, B., Cannavo, G., Galluzzi, L., Montroni, M., Guetard, D., Magnani, M., Piedimonte, G., and Silvestri, G., Exogenous interleukin-2 administration corrects the cell cycle perturbation of lymphocytes from human immunodeficiency virus-infected individuals, *J. Virol.,* 75, 10843, 2001.

63. NIH press release, NIAID News: IL-2 Increases CD4+ Cells, March 1995.

64. Zaunders, J.J., Moutouh-de Parseval, L., Kitada, S., Reed, J.C., Rought, S., Genini, D., Leoni, L., Kelleher, A., Cooper, D.A., Smith, D.E., Grey, P., Estaquier, J., Little, S., Richman, D.D., and Corbeil, J., Polyclonal proliferation and apoptosis of CCR5+ T lymphocytes during primary human immunodeficiency virus type 1 infection: regulation by interleukin (IL)-2, IL-15, and Bcl-2, *J. Infect. Dis.,* 187, 1735, 2003.

65. Reynes, J., Portales, P., Segondy, M., Baillat, V., Andre, P., Reant, B., Avinens, O., Couderc, G., Benkirane, M., Clot, J., Eliaou, J.F., and Corbeau, P., CD4+ T cell surface CCR5 density as a determining factor of virus load in persons infected with human immunodeficiency virus type 1, *J. Infect. Dis.,* 181, 927, 2000.

66. Unutmaz, D., KewalRamani, V.N., Marmon, S., and Littman, D.R., Cytokine signals are sufficient for HIV-1 infection of resting human T lymphocytes, *J. Exp. Med.,* 189, 1735, 1999.

67. Perera, L.P., Goldman, C.K., and Waldmann, T.A., IL-15 induces the expression of chemokines and their receptors in T lymphocytes, *J. Immunol.,* 162, 2606, 1999.

68. Chapdelaine, Y., Smith, D.K., Pedras-Vasconcelos, J.A., Krishnan, L., and Sad, S., Increased CD8+ T cell memory to concurrent infection at the expense of increased erosion of pre-existing memory: the paradoxical role of IL-15, *J. Immunol.,* 171, 5454, 2003.

69. Harari, A., Petitpierre, S., Vallelian, F., and Pantaleo, G., Skewed representation of functionally distinct populations of virus-specific CD4 T cells in HIV-1-infected subjects with progressive disease: changes after antiretroviral therapy, *Blood,* 103, 966, 2004.

70. Trinchieri, G., Interleukin-12: a cytokine produced by antigen-presenting cells with immunoregulatory functions in the generation of T-helper cells type 1 and cytotoxic lymphocytes, *Blood,* 84, 4008, 1994.

71. Spadaro, M., Lanzardo, S., Curcio, C., Forni, G., and Cavallo, F., Immunological inhibition of carcinogenesis, *Cancer Immunol. Immunother.,* 53, 204, 2004.

72. Portielje, J.E., Gratama, J.W., van Ojik, H.H., Stoter, G., and Kruit, W.H., IL-12: a promising adjuvant for cancer vaccination, *Cancer Immunol. Immunother.,* 52, 133, 2003.

73. Seaman, M.S., Peyerl, F.W., Jackson, S.S., Lifton, M.A., Gorgone, D.A., Schmitz, J.E., and Letvin, N.L., Subsets of memory cytotoxic T lymphocytes elicited by vaccination influence the efficiency of secondary expansion *in vivo, J. Virol.,* 78, 206, 2004.

74. Napolitano, L.A., Stoddart, C.A., Hanley, M.B., Wieder, E., and McCune, J.M., Effects of IL-7 on early human thymocyte progenitor cells *in vitro* and in SCID-hu Thy/Liv mice, *J. Immunol.,* 171, 645, 2003.

75. Pallard, C., Stegmann, A.P., van Kleffens, T., Smart, F., Venkitaraman, A., and Spits, H., Distinct roles of the phosphatidylinositol 3-kinase and STAT5 pathways in IL-7-mediated development of human thymocyte precursors, *Immunity,* 10, 525, 1999.

76. Steffens, C.M., Managlia, E.Z., Landay, A., and Al-Harthi, L., Interleukin-7-treated naive T cells can be productively infected by T-cell-adapted and primary isolates of human immunodeficiency virus 1, *Blood,* 99, 3310, 2002.

77. Mussini, C.., Pinti, M., Borghi, V., Nasi, M., Amorico, G., Monterastelli, E., Moretti, L., Troiano, L., Esposito, R., and Cossarizza, A, Features of "CD4-exploders," HIV-positive patients with an optimal immune reconstitution after potent antiretroviral therapy, *AIDS,* 16, 1609, 2002.

78. Silvestri, G. and Feinberg, M.B., Turnover of lymphocytes and conceptual paradigms in HIV infection, *J. Clin. Invest.,* 112, 821, 2003.

79. Roederer, M., Getting to the HAART of T cell dynamics, *Nat. Med.,* 4, 145, 1998.

80. Haase, A.T., Population biology of HIV-1 infection: viral and CD4+ T cell demographics and dynamics in lymphatic tissues, *Annu. Rev. Immunol.,* 17, 625, 1999.

81. McCune, J.M., The dynamics of CD4+ T-cell depletion in HIV disease, *Nature,* 410, 6831, 2001.

82. Grossman, Z., Meier-Schellersheim, M., Sousa, A.E., Victorino, R.M., and Paul, W.E., CD4+ T-cell depletion in HIV infection: are we closer to understanding the cause? *Nat. Med.,* 8, 319, 2002.

83. Douek, D.C., McFarland, R.D., Keiser, P.H., Gage, E.A., Massey, J.M., Haynes, B.F., Polis, M.A., Haase, A.T., Feinberg, M.B., Sullivan, J.L., Jamieson, B.D., Zack, J.A., Picker, L.J., and Koup, R.A., Changes in thymic function with age and during the treatment of HIV infection, *Nature,* 396, 690, 1998.

84. Zack, J.A., Arrigo, S.J., Weitsman, S.R., Go, A.S., Haislip, A., and Chen, I.S., HIV-1 entry into quiescent primary lymphocytes: molecular analysis reveals a labile, latent viral structure, *Cell,* 61, 213, 1990.

85. Stevenson, M., Stanwick, T.L., Dempsey, M.P., and Lamonica, C.A., HIV-1 replication is controlled at the level of T cell activation and proviral integration, *EMBO J.,* 9, 1551, 1990.

86. Wein, L.M., D'Amato, R.M., and Perelson, A.S., Mathematical analysis of antiretroviral therapy aimed at HIV-1 eradication or maintenance of low viral loads, *J. Theor. Biol.,* 192, 81, 1998.

87. Lori, F., Hydroxyurea and HIV: 5 years later—from antiviral to immune-modulating effects, *AIDS,* 13, 1433, 1999.

88. Donehower, R.C., An overview of the clinical experience with hydroxyurea, *Semin. Oncol.,* 19, 11, 1992.

89. Charache, S., Effect of hydroxyurea on the frequency of painful crises in sickle cell anemia. Investigators of the Multicenter Study of Hydroxyurea in Sickle Cell Anemia, *N. Engl. J. Med.,* 332, 1317, 1995.

90. Lori, F., Malykh, A., Cara, A., Sun, D., Weinstein, J.N., Lisziewicz, J., and Gallo, R.C., Hydroxyurea as an inhibitor of human immunodeficiency virus-type 1 replication, *Science,* 266, 801, 1994.

91. Malley, S.D., Grange, J.M., Hamedi-Sangsari, F., and Vila, J.R., Suppression of HIV production in resting lymphocytes by combining didanosine and hydroxamate compounds, *Lancet,* 343, 1292, 1994.

92. Palmer, S., Shafer, R.W., and Merigan, T.C, Hydroxyurea enhances the activities of didanosine, 9-[2-(phosphonylmethoxy)ethyl]adenine, and 9-[2- (phosphonylmethoxy)propyl]adenine against drug-susceptible and drug-resistant human immunodeficiency virus isolates, *Antimicrob. Agents Chemother.,* 43, 2046, 1999.

93. Lori, F. and Lisziewicz, J., Hydroxyurea: mechanisms of HIV-1 inhibition, *Antiviral Ther.,* 3, 25, 1998.

94. Gao, W.Y., Johns, D.G., Chokekuchai, S., and Mitsuya, H., Disparate actions of hydroxyurea in potentiation of purine and pyrimidine 2,3-dideoxynucleoside activities against replication of human immunodeficiency, *Proc. Natl. Acad. Sci. U.S.A.,* 92, 8333, 1995.

95. Gao, W.Y., Cara, A., Gallo, R.C., and Lori, F., Low levels of deoxynucleotides in peripheral blood lymphocytes: a strategy to inhibit human immunodeficiency virus type 1 replication, *Proc. Natl. Acad. Sci. U.S.A.,* 90, 8925, 1993.

96. Korin, Y.D. and Zack, J.A., Progression to the G1b phase of the cell cycle is required for completion of human immunodeficiency virus type 1 reverse transcription in T cells, *J. Virol.,* 72, 3161, 1998.

97. Zagury, D., Bernard, J., Leonard, R., Cheynier, R., Feldman, M., Sarin, P.S., and Gallo, R.C., Long-term cultures of HTLV-III-infected T cells: a model of cytopathology of T-cell depletion in AIDS, *Science,* 231, 4740, 1986.

98. Heredia, A., Davis, C., Amoroso, A., Dominique, J.K., Le, N., Klingebiel, E., Reardon, E., Zella, D., and Redfield, R.R., Induction of G1 cycle arrest in T lymphocytes results in increased extracellular levels of -chemokines: a strategy to inhibit R5 HIV-1, *Proc. Natl. Acad. Sci. U.S.A.,* 100, 4179, 2003.

99. Lori, F., Jessen, H., Lieberman, J., Clerici, M., Tinelli, C., and Lisziewicz, J., Immune restoration by combination of a cytostatic drug (hydroxyurea) and anti-HIV drugs (didanosine and indinavir), *AIDS Res. Hum. Retroviruses,* 15, 619, 1999.

100. Lori, F., Jessen, H., Lieberman, J., Finzi, D., Rosenberg, E., Tinelli, C., Walker, B., Siliciano, R.F., and Lisziewicz, J., Treatment of human immunodeficiency virus infection with hydroxyurea, didanosine, and a protease inhibitor before seroconversion is associated with normalized immune parameters and limited viral reservoir, *J. Infect. Dis.,* 180, 1827, 1999.

101. Lori, F., Pollard, R.B., Whitman, L., Bakare, N., Blick, G., Shalit, P., Foli, A., Peterson, D., Tennenberg, A., Schrader, S., Rashbaum, B., Farthing, C., Herman, D., Norris, D., Greiger, P., Frank, I., Groff, A., Lova, L., Asmuth, D., and Lisziewicz, J., A Low Hydroxyurea Dose (600 mg Daily) Is Found Optimal in a Randomized, Controlled Trial (RIGHT 702), The XIV International AIDS Conference, Barcelona, July 7–12, 2002, Abstract #TuPeB446.

102. De Antoni, A., Mutations in the pol gene of human immunodeficiency virus type 1 in infected patients receiving didanosine and hydroxyurea combination therapy, *J. Infect. Dis.,* 176, 899, 1997.

103. Gwilt, P.R. and Tracewell, W.G., Pharmacokinetics and pharmacodynamics of hydroxyurea, *Clin. Pharmacokinet.,* 34, 347, 1998.

104. Colic, M., Stojic-Vukanic, Z., Pavlovic, B., Jandric, D., and Stefanoska, I, Mycophenolate mofetil inhibits differentiation, maturation and allostimulatory function of human monocyte-derived dendritic cells, *Clin. Exp. Immunol.,* 134, 63, 2003.

105. Margolis, D., Heredia, A., Gaywee, J., Oldach, D., Drusano, G., and Redfield, R., Abacavir and mycophenolic acid, an inhibitor of inosine monophosphate dehydrogenase, have profound and synergistic anti-HIV activity, *J. Acquir. Immune Defic. Syndr.,* 21, 362, 1999.

106. Lui, S.L., Ramassar, V., Urmson, J., and Halloran, P.F, Mycophenolate mofetil reduces production of interferon-dependent major histocompatibility complex induction during allograft rejection, probably by limiting clonal expansion, *Transpl. Immunol.,* 6, 23, 1998.

107. Konno, Y., Natsumeda, Y., Nagai, M., Yamaji, Y., Ohno, S., Suzuki, K., and Weber, G., Expression of human IMP dehydrogenase types I and II in *Escherichia coli* and distribution in human normal lymphocytes and leukemic cell lines, *J. Biol. Chem.,* 266, 506, 1991.

108. Nagai, M., Natsumeda, Y., and Weber, G., Proliferation-linked regulation of type II IMP dehydrogenase gene in human normal lymphocytes and HL-60 leukemic cells, *Cancer Res.,* 52, 258, 1992.

109. Allison, A.C. and Eugui, E.M., Mycophenolate mofetil and its mechanisms of action, *Immunophar-macology,* 47, 85, 2000.

110. Chapuis, A.G., Paolo Rizzardi, G., D'Agostino, C., Attinger, A., Knabenhans, C., Fleury, S., Acha-Orbea, H., and Pantaleo, G., Effects of mycophenolic acid on human immunodeficiency virus infection *in vitro* and *in vivo, Nat. Med.,* 6, 762, 2000.

111. Ravot, E., Lisziewicz, J., and Lori, F., New uses for old drugs in HIV infection: the role of hydroxyurea, cyclosporin and thalidomide, *Drugs,* 58, 953, 1999.

112. Rizzardi, G.P., Harari, A., Capiluppi, B., Tambussi, G., Ellefsen, K., Ciuffreda, D., Champagne, P., Bart, P.A., Chave, J.P., Lazzarin, A., and Pantaleo, G., Treatment of primary HIV-1 infection with cyclosporin A coupled with highly active antiretroviral therapy, *J. Clin. Invest.,* 109, 681, 2002.

113. Takahashi, K., Reynolds, M., Ogawa, N., Longo, D.L., and Burdick, J, Augmentation of T-cell apoptosis by immunosuppressive agents, *Clin. Transplant,* 18, 72, 2004.

Index